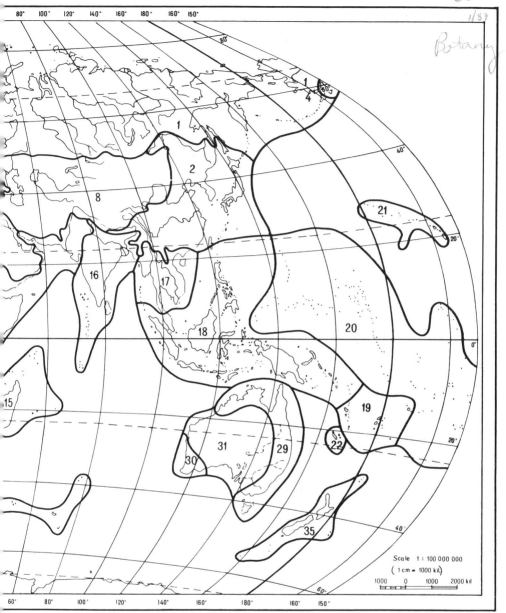

Floristic Regions of the World

FLORISTIC REGIONS OF THE WORLD

Armen Takhtajan

Translated by Theodore J. Crovello with the assistance and
collaboration of the author and under the editorship
of Arthur Cronquist

UNIVERSITY OF CALIFORNIA PRESS
Berkeley   Los Angeles   London

University of California Press
Berkeley and Los Angeles, California

University of California Press, Ltd.
London, England

Copyright © 1986 by The Regents of the University of California

Library of Congress Cataloging in Publication Data

Takhtajan, A. L. (Armen Leonovich)
  Floristic regions of the world.

  Translation of: Floristicheskie oblasti Zemli.
  Bibliography.
  Includes index.
  1. Phytogeography.   I. Crovello, Theodore J.   II. Cronquist, Arthur.
III. Title.
QK101.T313   1986   582.09   85-8731
ISBN 0-520-04027-9 (alk. paper)

Printed in the United States of America

1 2 3 4 5 6 7 8 9

To Alice

CONTENTS

Editor's Preface	xv
Author's Preface to the English Translation	xvii
PREFACE	xxi
INTRODUCTION	1
I. HOLARCTIC KINGDOM (HOLARCTIS)	9
A. BOREAL SUBKINGDOM	10
1. Circumboreal Region	10
1. Arctic Province	12
2. Atlantic-European Province	16
3. Central European Province	17
4. Illyrian or Balkan Province	21
5. Euxine Province	22
6. Caucasian Province	25
7. Eastern European Province	27
8. Northern European Province	29
9. Western Siberian Province	29
10. Altai-Sayan Province	30
11. Middle Siberian Province	31
12. Transbaikalian Province	32

13. Northeastern Siberian Province		33
14. Okhotsk-Kamchatka Province		33
15. Canadian Province		34
2. Eastern Asiatic Region		39
1. Manchurian Province		48
2. Sakhalin-Hokkaido Province		52
3. Japanese-Korean Province		53
4. Volcano-Bonin Province		56
5. Ryukyu or Tokara-Okinawa Province		58
6. Taiwanian Province		59
7. Northern Chinese Province		62
8. Central Chinese Province		63
9. Southeastern Chinese Province		65
10. Sikang-Yünnan Province		65
11. Northern Burmese Province		68
12. Eastern Himalayan Province		70
13. Khasi-Manipur Province		73
3. North American Atlantic Region		78
1. Appalachian Province		84
2. Atlantic and Gulf Coastal Plain Province		91
3. North American Prairies Province		95
4. Rocky Mountain Region		99
1. Vancouverian Province		101
2. Rocky Mountain Province		105
B. TETHYAN (ANCIENT MEDITERRANEAN) SUBKINGDOM		110
5. Macaronesian Region		112
1. Azorean Province		114
2. Madeiran Province		115
3. Canarian Province		115
4. Cape Verde Province		118
6. Mediterranean Region		119
1. Southern Moroccan Province		124
2. Southwestern Mediterranean Province		124
3. South Mediterranean Province		125

		4. Iberian Province	126
		5. Balearic Province	127
		6. Liguro-Tyrrhenian Province	128
		7. Adriatic Province	130
		8. East Mediterranean Province	131
		9. Crimean-Novorossiysk Province	131
	7. Saharo-Arabian Region		132
		1. Saharan Province	134
		2. Egyptian-Arabian Province	134
	8. Irano-Turanian Region		135
	8A. Western Asiatic Subregion		142
		1. Mesopotamian Province	142
		2. Central Anatolian Province	143
		3. Armeno-Iranian Province	144
		4. Hyrcanian Province	150
		5. Turanian or Aralo-Caspian Province	152
		6. Turkestanian Province	152
		7. Northern Baluchistanian Province	153
		8. Western Himalayan Province	154
	8B. Central Asiatic Subregion		155
		1. Central Tien Shan Province	156
		2. Dzungaro-Tien Shan Province	156
		3. Mongolian Province	156
		4. Tibetan Province	157
C. MADREAN SUBKINGDOM			158
	9. Madrean Region		159
		1. Great Basin Province	165
		2. Californian Province	171
		3. Sonoran Province	178
		4. Province of the Mexican Highlands	183
II. PALEOTROPICAL KINGDOM (PALEOTROPIS)			197
A. AFRICAN SUBKINGDOM			197
	10. Guineo-Congolian Region		198
		1. Upper Guinea Province	199

      2. Nigerian-Cameroonian Province   199
      3. Congolian Province   199
11. Uzambara-Zululand Region   200
      1. Zanzibar-Inhambane Province   201
      2. Tongoland-Pondoland Province   201
12. Sudano-Zambezian Region   202
  12A. Zambezian Subregion   202
      1. Zambezian Province   204
  12B. Sahelo-Sudanian Subregion   204
      1. Sahelian Province   204
      2. Sudanian Province   205
  12C. Eritreo-Arabian Subregion   205
      1. Somalo-Ethiopian Province   206
      2. South Arabian Province   208
      3. Socotran Province   209
  12D. Omano-Sindian Subregion   209
      1. Province of Oman   209
      2. South Iranian Province   210
      3. Sindian Province   211
13. Karoo-Namib Region   211
      1. Namib Province   213
      2. Namaland Province   213
      3. Western Cape Province   213
      4. Karoo Province   214
14. St. Helena and Ascension Region   214

B. MADAGASCAN SUBKINGDOM   216

15. Madagascan Region   222
      1. Eastern Madagascan Province   223
      2. Western Madagascan Province   223
      3. Southern and Southwestern Madagascan Province   223
      4. Comoro Province   224
      5. Mascarene Province   224
      6. Seychelles Province   225

C. INDOMALESIAN SUBKINGDOM — 226

16. Indian Region — 227
- 1. Sri Lanka Province — 229
- 2. Malabar Province — 230
- 3. Deccan Province — 232
- 4. Upper Gangetic Plain Province — 233
- 5. Bengal Province — 233

17. Indochinese Region — 234
- 1. South Burmese Province — 236
- 2. Andamanese Province — 237
- 3. South Chinese Province — 237
- 4. Thailandian Province — 237
- 5. North Indochinese Province — 238
- 6. Annamese Province — 238
- 7. South Indochinese Province — 238

18. Malesian Region — 238
18A. Malesian Subregion — 241
- 1. Malay Province — 241
- 2. Kalimantan (Bornean) Province — 241
- 3. Philippinean Province — 241
- 4. Sumatran Province — 242
- 5. South Malesian Province — 242

18B. Papuan Subregion — 242
- 1. Celebesian (Sulawesian) Province — 243
- 2. Moluccan Province — 243
- 3. Papuan Province — 243
- 4. Bismarckian Province — 243

19. Fijian Region — 243
- 1. New Hebridean Province — 244
- 2. Fijian Province — 244

D. POLYNESIAN SUBKINGDOM — 245

20. Polynesian Region — 245
- 1. Micronesian Province — 246
- 2. Polynesian Province — 246

21. Hawaiian Region	246
1. Hawaiian Province	248
E. NEOCALEDONIAN SUBKINGDOM	248
22. Neocaledonian Region	248
1. Neocaledonian Province	253
III. NEOTROPICAL KINGDOM (NEOTROPIS)	254
23. Caribbean Region	255
1. Central American Province	256
2. West Indian Province	256
3. Galapageian Province	256
24. Region of the Guayana Highlands	257
1. Guayana Province	258
25. Amazonian Region	258
1. Amazonian Province	259
2. Llanos Province	259
26. Brazilian Region	259
1. Caatinga Province	259
2. Province of Uplands of Central Brazil	260
3. Chacoan Province	260
4. Atlantic Province	260
5. Paraná Province	260
27. Andean Region	261
1. Northern Andean Province	262
2. Central Andean Province	262
IV. CAPE KINGDOM (CAPENSIS)	263
28. Cape Region	263
1. Cape Province	267
V. AUSTRALIAN KINGDOM (AUSTRALIS)	268
29. Northeast Australian Region	270
1. North Australian Province	270
2. Queensland Province	270
3. Southeast Australian Province	271
4. Tasmanian Province	272

30. Southwest Australian Region	273
1. Southwest Australian Province	274
31. Central Australian or Eremaean Region	274
1. Eremaean Province	275
VI. HOLANTARCTIC KINGDOM (HOLANTARCTIS)	276
32. Fernándezian Region	280
1. Fernándezian Province	282
33. Chile-Patagonian Region	283
1. Northern Chilean Province	286
2. Middle Chilean Province	286
3. Pampean Province	288
4. Patagonian Province	288
5. Magellanian Province	289
34. Region of the South Subantarctic Islands	291
1. Tristan-Goughian Province	292
2. Kerguelenian Province	293
35. Neozeylandic Region	293
1. Lord Howean Province	297
2. Norfolkian Province	298
3. Kermedecian Province	299
4. Northern Neozeylandic Province	299
5. Central Neozeylandic Province	300
6. Southern Neozeylandic Province	301
7. Chathamian Province	301
8. Province of New Zealand Subantarctic Islands	301
APPENDIX: List of Families of Living Vascular Plants	305
BIBLIOGRAPHIES	357
General	357
Holarctic Kingdom	361
Paleotropical Kingdom	377
Neotropical Kingdom	390
Cape Kingdom	392
Australian Kingdom	394
Holantarctic Kingdom	396
INDEX	401

EDITOR'S PREFACE

This book began as a translation of Academician Takhtajan's Russian-language book, *The Floristic Regions of the World,* published by Soviet Sciences Press in 1978. Like other translations of his works, it has been substantially rewritten and transformed by the author during the process, becoming a new book based on the old. At his request, I expanded the section on temperate North America, retaining his emphasis on endemic taxa, but with more attention to the dominant plant communities and the probable evolutionary history of the flora. Thus this area, which may be expected to provide the most readers of the present volume, also has the most detailed treatment.

It will not escape attention that, even aside from North America, some regions are treated more fully than others. Obviously the author can write with more assurance about the areas which he is personally familiar. The published record for other areas varies in breadth and depth, and these differences are reflected in the treatment here presented.

In 1983 the author asked me to become the English-language editor for the new book. I reviewed the entire text. Because of my long and cordial association with Academician Takhtajan, I felt free to diverge in places from Dr. Crovello's literal translation to a more idiomatic version, retaining the thrust but not always the phraseology of the original. Inevitably the manuscript acquired some of the elements of my own style of writing. In the summer of 1984 I spent a week with the author in Leningrad, going over the revised version page by page, to ensure that his views were adequately presented.

The massive index has been prepared by my wife, Mabel A. Cronquist, to whom we all owe our thanks.

Students of the general system of classification of plants will find the appendix as interesting as the text itself. Here we have the latest version of Academician Takhtajan's scheme for the families and higher taxa of vascular plants. A number of the families are more narrowly limited than in the past, and several new families are described. The author continues to be one of the most important contributors to the ongoing major reorganization of the general system.

2 January 1985   Arthur Cronquist

AUTHOR'S PREFACE TO THE ENGLISH TRANSLATION

More than six years have elapsed since publication of the Russian original of this book. It is therefore quite natural that I could not leave the work without any changes and additions. In fact, I have extensively revised much of the book and have introduced many changes and additions into the English manuscript; some parts are even written anew. Furthermore, inasmuch as the English translation is being published in the United States, I asked my friend and colleague Arthur Cronquist to revise and considerably enlarge the North American parts. He has written the description of the floristic provinces (including the Canadian Province, the North American Atlantic Region, the Rocky Mountain Region, and the Madrean Region) almost anew, and has also prepared the floristic map of the area. In addition, Professor Cronquist edited the final manuscript of the English translation and supplied it with his preface. I wish to express my sincere gratitude to him for his assistance and cooperation, and to Mabel A. Cronquist for the preparation of the index.

I have found it useful to supply the book with an appendix—a list of families of living vascular plants, with an indication of the number (frequently only approximate) of genera and species of each family and its geographical distribution. The list of families is a revised version of my system of vascular plants, which differs in some respects from the one published in Russian in the last three volumes of *The Life of Plants* (*Zhizn Rasteniy*, vols. 4–6, 1978–1982) and in my article on the classification of flowering plants published in *Botanical Review* (1980, vol. 46, no. 3). A consideration of new information and new thoughts accumulated during the last six years compelled me to make some realignments in the macrosystem of vascular plants. In this new version of my system I have favored smaller and better outlined orders and families. It is my considered opinion that heterogenous orders and families are less convenient both for phylogenetic and phytogeographic purposes.

In conclusion, I wish to thank many of my colleagues who have helped me in one way or another during the preparation of the English version of my book. I should like to acknowledge particularly helpful advice and suggestions from the following people: Vandica Avetissian (Erevan), Victor Botschantzev (Leningrad), Eleonora Gabrielian (Erevan), and Rudolph Kamelin (Leningrad), all of whom looked through the list of Irano-Turanian endemic genera, while Avetissian and Botschantzev also examined the list of Saharo-Arabian endemic genera. Bassett Maguire (New York) and Ghillean Prance (New York) read through the Neotropical part of the English manuscript; M. S. Mani (Agra, India) looked through the section on the Indian Region; Ledyard Stebbins (Davis) read the whole of the Russian original; C. K. Rao (Bangalore, India) looked through the list of endemic genera of the Indian Region; Jules Vidal (Paris) read through the section on the Indochinese Region; the late Per Wendelbo (Göteborg) reviewed the Armeno-Iranian and Hyrcanian provinces and offered many valuable comments; and Gennadiy Yakovlev looked through lists of endemic fabaceous genera in various parts of the book. Of course, the conclusions presented are my own, and do not always reflect the advice received.

I am particularly indebted to my wife, Alice, who has facilitated my work in the preparation of the English version, and to whom I dedicate this book.

I would like to thank my assistant Tatiana Wielgorskaya, who has not only skillfully typed the corrected pages of the English manuscript but also maintained and typed the bibliography.

For the excellent map drawings I am indebted to Galina Katenina of the Laboratory of Botanical Cartography of the Komarov Botanical Institute in Leningrad.

Last but not least, I wish to express my sincere gratitude to Professor Theodore Crovello, who kindly undertook the onerous task of translation. I heartily thank him also for his patience and constant cooperation.

July 1984   Armen Takhtajan

PREFACE

Biogeography can play a basic role in the solution of one of today's most important challenges—protection of the plant world. Along with ecological geography, floristic geography (and the corresponding faunistic geography) is of the utmost importance, especially that branch which I call phytochorionomy.

As is true of every specific ecosystem, the preservation of the entire global ecosystem (ecosphere) is practically impossible without preservation of the taxonomic diversity of its components. The current, large-scale impoverishment of taxonomic components and the structural simplification of many ecosystems is depriving them of their optimal stages of function and stability. The disappearance not only of numerous populations but also of many species and even genera (the danger is increasing even for several small families) means irreversible erosion of germ plasm and the loss of genetic material of potentially very great value for humanity. The fate of the flora of those regions of the world where a very large number of endemic taxa of various ranks are concentrated is especially distressing.

As is well known, the choice of territory to be protected can be made by means of very diverse criteria, ranging from the extremely practical to the purely aesthetic. But no matter what the path leading to the choice of a protected territory, it is necessary everywhere, in any part of the world and with any environment, to protect from the outset the flora of those territories which represent the richest storehouse of the world's unique genetic material—namely, the endemic forms of life.

Concern about the genetic resources of the world cannot be limited by some kind of regional framework, even one as broad as the borders of an entire continent. As in every global problem, protection must be based on a planned and carefully implemented global strategy. To accomplish this, it is important to know where and what to protect. To do this, we must have before us an adequately clear map of the geography of these genetic materials, including a map of the phytochoria. Thus one of the most classical botanical disciplines, floristic geography, has now acquired new importance. Because it accumulates information on the geographic distribution of genetic material, it becomes the theoretical basis for its protection. It is exactly this circumstance that served as my principal motivation in writing this book, which I now offer to the attention of the reader.

This book has its history. The first version of my developing floristic system was published in English as a supplement to the book, *Flowering Plants, Origin and Dispersal* (Takhtajan 1969). The second, considerably more detailed version, which in contrast to the first, gives an index of floristic provinces, appeared as a separate chapter in my book, *Origin and Dispersal of Flowering Plants* (Takhtajan 1970, in Russian). Finally, I wrote a chapter on floristic geography (Takhtajan 1974, in Russian) especially for the multivolume, semipopular Russian work, *Life of Plants*. The present book is the result of still further development of my floristic system. The map of floristic kingdoms and regions has also undergone considerable change. The first version was published in 1969 and the most recent in 1974.

Readers undoubtedly will notice some unevenness in the treatment of different phytochoria, especially at the provincial level. This is due to both objective and subjective reasons. To me, it is of the greatest importance to characterize in detail those territories that are richer in endemic taxa, and thus that are more important from the point of view of protecting genetic resources. Unfortunately, we often failed to achieve this goal, especially in tropical countries, where floristic, phytochorionomic studies are still very poorly developed. Of course this unevenness of exposition reflects also the author's personal interests.

In the designation of phytochoria, excluding kingdoms, references are given to that publication in which a given region, subregion, and province was first established, and also to those works of authors who accepted them (frequently with somewhat different boundaries). This is followed by the most important synonyms of these phytochoria. I try as much as possible to avoid mention of the names of units of purely geobotanical subdivisions, especially those with borders that do not agree with the boundaries of my floristic choria.

For convenience of use, the literature cited is arranged according to geographical principles (in this case, chorionomically). However, works of a general nature are given only once, at the beginning of the list. As a rule, the bibliographies include works to which reference is made in the text, as well as recent publications that seem useful for further elaboration of phytochorionomy. I did not include in the bibliography the numerous works devoted to phytochorionomy of limited territories, such as subprovinces, because I only touch upon subprovinces in some cases, and do not consider districts at all. Also omitted are numerous monographs of genera and

PREFACE

"floras" of specific territories of the world, without examination of which (just as without examination of herbarium material) it would be impossible for an author to create a list of endemic taxa or of individual phytochoria. Inclusion of all such works in the bibliography would have duplicated reference books already in existence.

It is a pleasure to note my debt to those sympathetic to this work. I acknowledge the attention I received from my many colleagues and friends in various countries of the world. Without it, I truly would not have been able to carry out this extremely laborious task. Whatever its shortcomings might be (and they are truly great), they would be even greater had I worked without such support.

One of my greatest difficulties was the absence in our library of many important works produced by smaller publishers around the world. Numerous colleagues kindly sent me copies of many of these works, which allowed me to use the literature more fully. In particular, I must mention Peter S. Ashton, Aberdeen (now Cambridge, Mass.); M. M. J. van Balgooy, Leiden; Nancy T. Burbidge, Canberra; Francisco Blasco, Pondicherry, India; J. Patrick M. Brenan, London; Henri Gaussen, Toulouse; W. Greuter, Genève (now Berlin); Roland Good, Hull, England; Peter H. Davis, Edinburgh; Valerio Giacomini, Roma; Lawrence A. S. Johnson, Sydney; Michael Zohary, Jerusalem; Angel L. Cabrera, La Plata; Pierre Quezel, Marseille; Hsuan Keng, Singapore; Boris Kitanov, Sofia; Arthur Cronquist, New York; Leon Croizat, Caracas; G. Ll. Lukas, Kew; Bassett Maguire, New York; V. M. Meher-Homji, Pondicherry; Guido Moggi, Firenze; David M. Moore, Reading; Jack Major, University of California, Davis; Andre M. Aubreville, Paris; Robert Ornduff, University of California, Berkeley; Duncan M. Porter, St. Louis (now Blacksburg, Va.); Vishwambhar Puri, Meerut; Peter H. Raven, St. Louis; Salvador Rivas-Martinez, Madrid; G. Ledyard Stebbins, University of California, Davis; William T. Stearn, London; Arne K. Strid, Copenhagen; Robert F. Thorne, Claremont, California; Takasi Tuyama, Tokyo; K. Thothathri, Calcutta; Frank White, Oxford; S. Max Walters, Cambridge; F. Raymond Fosberg, Washington, D.C.; Joao do Amaral Franco, Lisbon; Alfred Hansen, Copenhagen; Hiroshi Hara, Tokyo; Olov Hedberg, Uppsala; F. N. Hepper, Kew; and Maurice Schmid, Paris.

In addition I am sincerely grateful to Dr. B. Maguire, with whom I had the pleasure to discuss several problems of neotropical phytogeography in London in May 1976. The map in the book by A. Cabrera and A. Willink, *Biogeografia de America Latina* (1973), was very useful and is cited in the text. I also am grateful to B. Kitanov (Sofia), with whom in Leningrad in August 1977 I had the pleasure of discussing several questions about the phytogeography of the Balkan Peninsula, especially Bulgaria.

I consider it a great pleasure to offer my sincere thanks to my colleagues at the Komarov Botanical Institute of the USSR Academy of Sciences,

V. P. Bochantsev, V. R. Grubov, and R. I. Kamelin, for reviewing lists of endemic genera of the Irano-Turanian Region and for valuable comments. In addition, V. P. Bochantsev read through and modified somewhat a list of endemic genera of the Saharo-Arabian Region. In map design, especially the map of the floristic regions of the world, G. D. Katenina gave me invaluable help.

If this book contributes to the further development of the floristic system and to its use for the sound protection of the plant world, the author will consider his labor justified.

Leningrad
August 1977

A. Takhtajan

INTRODUCTION

The comparative study of floras of different countries necessitates division of the world into natural floristic units—namely, the creation of a floristic system. One of the first attempts, which was extraordinarily successful for its time, was by the Danish botanist J. F. Schouw. In his book, *Grundzüge Einer Allgemeinen Pflanzengeographie,* published in 1823 (a German translation of the Danish original published in 1822), Schouw divided the flora of the world into 25 kingdoms, several of which were further subdivided into provinces—terms which are widely accepted in biogeography to this day. In his divisions, Schouw (like subsequent authors) considered the degree of endemism of taxa of various categories: for the recognition of this or that territory as a special floristic kingdom, at least half of the species and one quarter of the genera had to be endemics. In addition, Schouw believed that a floristic kingdom must also contain endemic families, or at least families having their diversity in its territory. Contemporary procedures of floristic classification are in many ways close to the principles Schouw formulated years ago.

A floristic system represents a hierarchial classification of coordinated natural floristic areas, or choria[1] of various ranks. As in any other classification, it serves for the storage and retrieval of information. Thus delimitation of choria and the establishment of their ranks must be such that classification may most effectively fulfill these functions. The number of choria must not be too large. Their optimal number, especially of choria of the highest categories, must be such as to provide for the easiest retrieval of information. To do this, the floristic system must be well composed and easily comprehensible.

Over the one hundred fifty years since the publication of Schouw's book, floristic geography has attained very great success. One of its essential achievements has been the development of a well-thought-out, logically faultless system of chorionomic categories, along with several principles of their delimitation. The highest chorionomic category is the *kingdom* or the *realm* (German, Florenreich; French, empire). The rank of kingdom is used for choria that are characterized by endemic taxa of the highest categories and by the maximum observable diversity and distinctiveness of their flora. Kingdoms are characterized by endemic families, subfamilies, and tribes, and by very high generic and species endemism. The number of kingdoms customary in plant geography is not great. Thus Engler (1882, 1899, 1903, 1912, 1924) and then Turrill (1959), Tolmatchev (1974), and a series of other authors adopted only four kingdoms of the terrestrial flora: Northern, or Boreal (termed Holarctic in Tolmatchev 1974); Paleotropic; Central and South American, or Neotropical; and Australian, or Southern. Engler recognized a fifth kingdom, the Oceanic—that is, the kingdom of the shallow water flora of the world's oceans. Diels (1908, 1918) and, along with him, the majority of contemporary authors (including Hayek 1926; Rikli 1934; Good 1947, 1974; Schmithüsen 1961; Mattick 1964; Neil 1969; Walter and Straka 1970; Takhtajan 1970, 1974; and Ehrendorfer 1971) subdivided the flora of the terrestrial world into six kingdoms: Holarctic, Paleotropical, Neotropical, Cape, Australian, and Antarctic (termed Holantarctic in Takhtajan). Several of these authors (e.g., Mattick 1964; Walter and Straka 1970), along with Engler, recognized a separate Oceanic kingdom. Thus at the present time division of the terrestrial flora into six kingdoms may be considered widely accepted. This division adequately reflects the broadest types of flora and therefore is quite strongly established in phytogeography. However, the boundaries of floristic kingdoms are continually being specified and sometimes are subjected to major changes. Many questions arise, especially in connection with the southern boundaries of the Holarctic Kingdom and the northern boundaries of the Holantarctic.

A kingdom is divided into *regions* (Russian, oblast; German, Gebiet, or Region; French, région), which are established on the basis of high amounts of species and generic endemism, and sometimes even of endemic taxa of higher rank (up to families and orders). In addition, as Tolmatchev (1974: 236) remarked, floristic regions are characterized by a definite series of families that occupy a predominant position in the region, the quantitative correlations between these dominant families being relatively stable.

Engler (1912, 1924) subdivided the flora of the world into 29 regions. Tolmatchev (1974) accepted 30 regions, while the author of the present book recognizes 35. Good (1947, 1974) and Schmithüsen (1961) used 37, while Mattick (1964) increased their number to 43. Thus the number of regions in different floristic systems, beginning with Engler's, has ranged from 29 and 43, with the mean being 36. This number is apparently close to optimal.

Regions may in turn be subdivided into *provinces* (English, province or domain; German, Provinz; French, domaine). For provinces, generic endemism is less characteristic than for regions. If endemic genera exist, they are usually monotypic or oligotypic. Species endemism is characteristic, but it occurs at a lower frequency than in regions. It is also important to note that, in addition to endemism, provinces are characterized by statistically distinct assemblages of species. The number of provinces surpasses the number of regions by several times. In the final version of Engler's floristic system there are 102 provinces, and in Good's system there are 120. In my system the flora of the world is divided into 153 provinces (including regions with one province). It appears to me that this last number approaches a maximum, beyond which further division threatens to reduce the value of this chorionomic category.

The lowest chorionomic category of the floristic system is the *district* (German, Bezirk; French, district), characterized mainly by subspecific endemism. Species endemism is weakly expressed or is absent. In several cases the district may even possess endemic, monotypic genera, but one should attribute the generic endemism not so much to the district itself as to the province in which it lies.

Along with the basic chorionomic categories just mentioned, intermediate categories are widely used: subkingdom, subregion, subprovince, and sometimes even subdistricts. The actual practice of classifying phytochoria shows that these categories are totally sufficient for adequate expression of the hierarchy of choria, and therefore the introduction of additional categories is superfluous.[2]

In phytochorionomy, usually we intuitively follow the rule that floristic differences between choria of any category (e.g., provinces) are inversely proportional to their size. In other words, the smaller the territory constituting our provinces (or choria of any other category), the stronger must be the expression of its floristic distinctness, and vice versa.[3] Thus such provinces as the Eastern European, Northern European, and Western Siberian are not separated by any appreciative floristic peculiarities, and endemism in those provinces is poorly expressed (especially in the last two). But since these provinces occupy a vast territory, their separation is considered justified. Along these lines, we cannot recognize the division of the Caucasus into 19 provinces (9 in the northern Caucasus and 10 in the Transcaucasus), as Kuznetsov (1909) did many years ago. The establishment of the majority of these provinces on the basis of floristic relations is completely unfounded, and is at the present time of historical interest only. Grossheim and Sosnovsky (1928) divided the Caucasian Isthmus into 10 provinces, thus reducing by almost one-half the number of provinces created by Kuznetsov. Later, Grossheim (1948) modified the number of these provinces to 9 (see Holarctic bibliography). The author of this book recognizes only 7 (Takhtajan 1974).

Unfortunately, in many narrow regional chorionomic works in which floristic division is carried out in isolation from a global, floristic system, we note a tendency to exaggerate the rank of intraprovincial differences, and to erect subprovinces and even districts as independent provinces. But the floristic differences between different parts of a province may be fully expressed by differences in their respective number of subprovinces and districts. Any floristic division of a specific country must harmonize with a global floristic system. In contrast to this harmony, chorionomic inflation threatens to make the floristic system very cumbersome, difficult to visualize, and unfit for use. This will serve only to discredit the very idea of phytochorionomy in the eyes of botanists of other specialties.

Floristic relations are usually very complex, and the boundaries between territories with floras of different types are often not well defined. In many cases choria merge from one into another so gradually that between them lies a more or less broad (sometimes very broad) transitional zone. Therefore any floristic division is to a large degree arbitrary, and is characterized to some degree by simplification of the true floristic relationships. Based on this, several authors reject the usefulness of floristic maps, while others designate regions of mutual penetration of different floras with strokes or other signs. To us, it seems that the drawing of borders of choria on a map, although only conditional, is useful both for scientific and educational purposes.

Floristic lines separating adjacent choria on a floristic map correspond more or less closely to the pronounced coincidence of the distributions of taxa of various ranks (cf. Szafer 1952). The largest overlap of borders of distribution is observed when the distribution of plants is combined with zonation, especially vertical zonation. These boundaries are the most difficult, and the broadest belt of overlap occurs in those cases where the boundaries cross flat land in a latitudinal direction. The plains of eastern Europe, western Siberia, and Kazakhstan are striking examples of cases where more or less definite floristic lines are drawn only with great difficulty. Therefore these lines are somewhat conditional.

Statistical analysis of the flora has great importance in phytochorionomy. Different methods of mathematical analysis are being used more and more, sometimes with the aid of computers,[4] but in this area, quantitative methods are often overrated. As Schmithüsen (1961) remarked most aptly, "qualitative differences in the composition of systematic units appear considerably more essential than quantitative differences." The taxonomic and biogeographic study of the composition of systematic units appears to be much more important. Hence it is most valuable to carry out monographic studies on the systematics and geography of specific taxa. Without adequate numbers of such studies, statistical analyses can not give useful results. Chorionomy must, above all, be based on systematics.

A basic principle of biotic classification (as in floristics, so also in faunistics) is the fact that it must be based on taxonomic and geographic studies of systematic units. Speaking about principles of zoogeographic classification, Geptner (1936:429, in Russian) wrote that "neither the outline of the land and the purely geographic subdivision of it, nor the landscapes and soil or botanical regions, and so on may serve as its basis." This applies equally well to floristic classification. Furthermore, this principle has been understood for a long time (e.g., Engler 1899) and has been convincingly substantiated by Tolmatchev (1974). Neither geomorphology and climate nor soil and fauna nor even vegetation can serve as the sole basis for floristic division or classification.

It is well known that similar types of vegetation may be formed on totally different floristic bases. In addition, within the limits of a given floristic province—and even more so within a region—very different plant formations are encountered. Therefore, one cannot help but agree that Tolmatchev (1974:230, in Russian) was quite right in opposing "the at first glance tempting tendency to unite geobotanical and floristic classification of the earth's surface in a unified botanico-geographical classification, which takes into account to equal degree both the features of the vegetation and the composition and genesis of the floras of the world." It is important to emphasize that Tolmatchev did not deny a definite relationship between the development of flora and of vegetation. He fully admitted the importance to floristic division of data about vegetative cover, especially in the delimitation of floristic units (choria) of the lowest ranks. More emphatic is the point of view of Schmithüsen (1961), according to whom floristic division must not be influenced by vegetative formation, "but arises only from distribution of taxonomic units."

Unquestionably, distributions of taxa form the basis of floristic divisions, but one should also consider plant formations. Many characteristic plant communities appear as the best indicators of floristic regions and provinces. For example, the peculiar sclerophyllous formation, well known under the term maquis, consists of typical Mediterranean elements. Thus it is completely natural that the appearance of this formation-type immediately provides evidence that the given territory belongs to the Mediterranean Floristic Region. In a similar way its ecological analogue, the California chaparral, appears as one of the characteristic indicators of the California Floristic Province. The broad sclerophyll forest is also such an indicator, though its floristic composition differs strongly from that of chaparral. The more diversified the vegetation of a given phytochorion, the greater the number of such indicator plant formations. Every such formation appears as a distinctive floristic complex that is characteristic of the flora of a given phytochorion. To summarize, in phytochorionomy we should always consider the plant formations of a given area. Acquaintance with a vegetation map, and especially direct observation in nature, provide much valuable information required for the delimitation of boundaries of phytochoria. Naturally, the ideal material for phytochorionomic studies would be accurate maps of the ranges of all species.

To this day, a generalized system of choria still has not been created that is suitable for plants, let alone for all organisms. Several biogeographers (predominantly zoogeographers) do not even believe that such a universal chorionomic system is generally possible. Since the highest taxa of the living world have unequal geological ages, have had different centers of geographic radiation, and consequently have migrated in different directions, doubt has arisen as to whether the paleogeographic changes of the world can be properly reflected by a uniform interpretation of the current distribution of these diverse taxa. Indeed, the current distribution of reptiles and ferns, for example, is in many ways different from the distribution of the geologically considerably younger mammals and flowering plants. From such observations, some conclude that it is necessary to create a separate system of choria for every large systematic group. But it is difficult to agree with such an extreme conclusion. As the zoogeographer Darlington (1957:422) pointed out, "the system of faunal regions, then, represents the average, gross pattern of distribution of many different animals with more or less different distributions." Darlington emphasized that this average type of distribution is completely real, and that it indicates both how animals are adapted to the environment in their distribution and how they are so strongly influenced in their distribution by climate and barriers, so that "a natural system of faunal regions would, I think, be worth making for this reason alone" (Darlington 1957). He also points out that a system of faunistic regions is itself a kind of standard geographic reference system, which could help us measure, describe, and compare the distributions of different animals: "Deviations from the standard are expected and informative. If, for example, a particular group of animals reaches the Australian region, but shows less than standard differentiation there, this suggests recent dispersal and power of crossing water gaps." All these observations are equally applicable to the system of floristic regions and of phytochoria of all other ranks. Thus one of the basic (but far from unique) values of the system of choria is that it appears as its own type of standard or model (paradigm) in the study of the geographic distribution of organisms.

A system of choria as a paradigm for the study of the geographic distribution of very different organisms will be more efficient as the taxonomic and biogeographic material on which it is based becomes broader. In the ideal situation, such a system must be based on the synthesis of data about the distribution of all living organisms, excluding cosmopolitan species. But this is currently impractical, for the simple reason that the systematics and geographic distribution of many organisms (e.g., many invertebrates, fungi, and even groups of higher plants, including the flowering plants) are still very inadequately known. Therefore the faunistic system is based mainly on the geographic distribution of freshwater fishes, amphibians, reptiles, birds, and especially mammals, while the floristic system is based on the distribution of higher plants (mainly flowering plants). Nevertheless, neither system of choria appears to provide an adequate standard for studying the geography of other groups of terrestrial organisms.

As mentioned in the preface, phytochorionomy has great value in the development of a general geographic strategy for protecting the genetic resources of the plant world. Thus the floristic system is not only a paradigm for the chorological study of taxa of various categories but also one of the most important scientific bases for the protection of the world's flora. It should also be completely obvious that both the development of a system of phytochoria and the protection of the flora require detailed study of the chorology of the maximum possible number of taxa. For this, not only the study and mapping of areas of taxa of different ranks but also the phylogenetic approach to their chorology—which Engler (1899) termed "phylogenetische Pflanzengeographie"—is important. Of course, the phylogenetic (or even more broadly, the evolutionary) approach assumes the existence of detailed monographic studies of different taxa. Furthermore, evolutionary and floristic protection problems require the study of the structure of species and of their population diversity over their entire area of distribution. The remarkable work of N. I. Vavilov, in the study of geographic centers of origin of cultivated plants, convincingly showed the theoretical and practical importance of the use of his "differential botanico-geographic method," the essence of which includes "clarification of the distribution of hereditary variations of forms of a given species throughout regions and countries," and "the establishment of geographic centers of concentration of basic diversity" (Vavilov 1935:8). In the final analysis, this is the direction of study which phytochorionomy must take, as must all strategies for the protection of the flora in all of the richness of its gene pool.

NOTES

1. The term *chorion* (Greek chorion, pl. choria) was coined by Turrill (1958). Along with this general biogeographic term, Turrill introduced subdivisions of it—*phytochorion* and *zoochorion*. I now define chorionomy to include knowledge about choria, as well as the methods and principles of their study and classification. Botanical chorionomy may be referred to as *phytochorionomy*. Phytochorionomy is an important part of floristic geography.

2. In the literature, especially the French (e.g., Braun-Blanquet 1923*b*, in the Holarctic bibliography; Gaussen 1954; Dupont 1962), the term *sector* is used, corresponding to subprovince.

3. An analogous "inverse ratio" rule is recommended in systematics (e.g., Mayr 1969:92).

4. The interesting work of Peters (1971) is devoted to the use of computers. The works of Clayton and Hepper (1974) and Clayton and Panigrahi (1974) are successful examples of their use.

I
HOLARCTIC KINGDOM
(HOLARCTIS)

The Holarctic Kingdom[1] is the largest of all floristic kingdoms and occupies more than half of the entire terrestrial world. It embraces all of Europe, extratropical northern Africa, all of extratropical Asia, and almost all of North America. Despite the vastness of its territory, the flora of each region of this kingdom is closely related to the flora of the others, and shares much in common with them.

Nearly 60 endemic or nearly endemic families of vascular plants are restricted to the Holarctic flora. These include Ginkgoaceae, Cephalotaxaceae, Sciadopityaceae, Calycanthaceae s. str., Sargentodoxaceae, Nandinaceae, Glaucidiaceae, Hydrastidaceae, Circaeasteraceae, Hypecoaceae, Paeoniaceae, Trochodendraceae, Tetracentraceae, Cercidiphyllaceae, Eupteleaceae, Platanaceae (1 species in North Vietnam), Simmondsiaceae, Eucommiaceae, Rhoipteleaceae (reaching North Vietnam), Stachyuraceae, Parnassiaceae, Diapensiaceae, Dipentodontaceae, Fouquieriaceae, Cannabaceae, Crossosomataceae, Tetradiclidaceae, Peganaceae, Leitneriaceae, Bretschneideraceae, Biebersteiniaceae, Limnanthaceae, Kirengeshomaceae, Pterostemonaceae, Davidiaceae, Aucubaceae, Toricelliaceae, Viburnaceae, Adoxaceae, Morinaceae, Theligonaceae, Trapellaceae, Helwingiaceae, Hippuridaceae, Cynomoriaceae, Butomaceae s. str., Scheuchzeriaceae, Liliaceae s. str., Ruscaceae, Hemerocallidaceae, Hesperocallidaceae, Funkiaceae, Ixioliriaceae, Aphyllanthaceae, and Trilliaceae. None of these families is large, the overwhelming majority of them consisting of one, often monotypic genus.

In the Holarctic flora, the following families are richly represented: Magnoliaceae, Lauraceae, Ranunculaceae, Berberidaceae, Caryophyllaceae, Chenopodiaceae, Polygonaceae, Plumbaginaceae, Hamamelidaceae, Fagaceae, Betulaceae, Juglandaceae, Theaceae, Salicaceae, Brassicaceae, Ericaceae, Primulaceae, Malvaceae, Euphorbiaceae, Thymelaeaceae, Rosaceae, Fabaceae, Cornaceae, Araliaceae, Apiaceae, Rhamnaceae, Gentianaceae, Boraginaceae, Scrophulariaceae, Lamiaceae, Campanulaceae, Asteraceae, Iridaceae, Hyacinthaceae, Alliaceae, Asphodelaceae, Orchidaceae, Juncaceae, Cyperaceae, and Poaceae.

Among the conifers, there are many species of the Pinaceae and the Cupressaceae. The ferns are represented mainly by the Pteridaceae, Polypodiaceae, Aspleniaceae, and Aspidiaceae. The majority of these families contain many endemic Holarctic genera and a multitude of endemic species.

The Holarctic Kingdom is subdivided into three subkingdoms: the Boreal; the Tethyan or Ancient Mediterranean; and the Madrean or Sonoran (Takhtajan 1969, 1970, 1974).

A. BOREAL SUBKINGDOM

The Boreal Subkingdom is the most extensive of all the Holarctic subkingdoms. It has the richest flora and is somewhat higher in endemic families and genera than are the two other subkingdoms. A considerable number of ancient and primitive families and genera characterizes several regions of the Boreal Subkingdom. The Boreal Subkingdom consists of four regions: the Circumboreal (European-Canadian) Region, the Eastern Asiatic Region, the North American Atlantic Region, and the Rocky Mountain Region.

1. Circumboreal Region

Takhtajan 1974; Région eurosibérienne-boréoaméricaine, Braun-Blanquet 1919, 1923b; Région holarctique au sens strict, Chevalier et Emberger 1937; Boreal Region, Popov 1940, 1949.[2]

This is the largest floristic region of the world, a considerable part of which is located in the territory of the Soviet Union. It includes Europe (with the exception of parts that are assigned to the Mediterranean Region); northern Anatolia; the Caucasus (with the exception of the arid parts and Talish); the Urals; Siberia (with the exception of the southeastern part along the course of the Amur River); Kamchatka; northern Sakhalin; the northern Kurile Islands to the north of the island of Iturup; the Aleutian Islands; most of Alaska; and a large part of Canada.

In the majority of floristic systems, the Eurasian and American parts of the

Circumboreal Region are placed in different regions. Only a few authors recognize just one region.

No endemic families are found in the flora of the Circumboreal Region, and the number of endemic genera is comparatively small. But there are many genera that are found only in it and in the Eastern Asiatic Region as well. The endemic and subendemic genera include the following:[3]

Ranunculaceae: *Miyakea* (1, Sakhalin; very close to *Pulsatilla*)
Portulacaceae: *Claytoniella* (2, northeastern Asia; northwestern America)
Caryophyllaceae: *Petrocoptis* (6, Pyrénées)
Brassicaceae: *Borodinia* (1, eastern Siberia; close to *Christolea*), *Gorodkovia* (1, northeastern Siberia), *Microstigma* (1, Altai, very close to *Matthiola*), *Pachyphragma* (1, western Caucasus, northeastern Anatolia), *Pseudovesicaria* (1, Caucasus), *Redowskia* (1, Yakutia), *Rhizobotrya* (1, southern Alps), *Schivereckia* (4–6, eastern and southeastern Europe, northern Anatolia)
Primulaceae: *Soldanella* (11, Europe), *Sredinskya* (1, Caucasus; close to *Primula*)
Malvaceae: *Kitaibela* (1, Yugoslavia)
Apiaceae: *Agasyllis* (1, Caucasus), *Astrantia* (10), *Chymsidia* (1), *Dethawia* (1, Pyrénées), *Endressia* (2, Pyrénées, northern Spain), *Hacquetia* (1, eastern Alps to northern Carpathians and southern Poland), *Hladnikia* (1, western Slovenia), *Symphyoloma* (1, Caucasus), *Thorella* (1, France, western Portugal)
Valerianaceae: *Pseudobetckea* (1, Caucasus)
Boraginaceae: *Halacsya* (1, western part of the Balkan Peninsula), *Megacaryon* (1, northern and northeastern Anatolia), *Pulmonaria* (14, Europe), *Trachystemon* (1, eastern Bulgaria, European part of Turkey, northern Anatolia, western Caucasus), *Trigonocaryum* (1, Caucasus)
Scrophulariaceae: *Erinus* (1, Pyrénées, Alps)
Gesneriaceae: *Haberlea* (1, Bulgaria, northeastern Greece), *Jankaea* (1, central Greece), *Ramonda* (3, Pyrénées, northeastern Spain, the Balkan Peninsula)
Campanulaceae: *Astrocodon* (1, western shores of the Okhotsk Sea), *Physoplexis* (1, southern Alps; close to *Phyteuma*)
Asteraceae: *Berardia* (1, southwestern Alps), *Cladochaeta* (2, Caucasus; close to *Helichrysum*), *Telekia* (2, central Europe to the Caucasus), *Tridactylina* (1, southern coast of Lake Baikal)
Alismataceae: *Luronium* (1, Europe)
Hydrocharitaceae: *Stratiotes* (1, Europe)
Anthericaceae: *Paradisea* (2, mountains of central and southern Europe)
Orchidaceae: *Chamorchis* (1, Europe)
Poaceae: *Dupontia* (2, the Arctic)

The majority of genera endemic to the Circumboreal Region are concentrated in the Pyrénées, the Alps, and the Caucasus. The Pyrénées, the Alps, the Carpathians, the Caucasus, the mountains of Siberia, and the Alaska-Yukon area are richest in endemic species.

The most characteristic coniferous species are species of *Pinus, Picea, Abies,* and *Larix*. In Canada, we also find *Tsuga* and *Thuja*. Among the numerous broad-leaved species are common species of *Quercus, Fagus, Betula, Alnus, Acer, Carpinus, Populus, Salix, Fraxinus, Ulmus, Tilia, Juglans, Celtis, Ostrya,* and *Cornus*. Other woody plants include species of *Prunus, Crataegus, Pyrus, Malus, Sorbus, Spiraea, Staphylea, Rhododendron, Lonicera, Viburnum, Sambucus, Rhamnus,* and *Vaccinium*. Along with the broad-leaved and coniferous forests are widely distributed meadows, which change into steppes in the southern areas of the European part of the Soviet Union and in Siberia. Northern areas are characterized by vast bogs and tundra. In the mountains, forest vegetation forms belts, or zones, with the upper parts usually consisting of conifer forests. Above the forest belts is a zone of high montane vegetation (subalpine and alpine), very rich in endemic taxa.

The vast Circumboreal Floristic Region is broken up into a series of more or less clearly defined provinces:

1. **Arctic Province** (Krasnov 1899; Engler 1903, 1924; Popov 1949; Takhtajan 1974; Fedorov 1979). In many ways this phytochorion appears to be one of the most disputable. There is disagreement not only about its range and boundaries but also about its very existence. The majority of authors (Engler 1882, 1924; Drude 1890; Wulff 1944; Good 1947, 1974; Turrill 1959; Mattick [in Engler's *Syllabus*] 1964; Walter and Straka 1970; Tolmatchev 1974; Yurtsev et al. 1978; and many others) not only do not doubt its reality but also consider it to be a separate floristic region (the Arctic or Arctic-Subarctic). Diels (1908), however, does not accept this phytochorion at all, and considers the arctic and subarctic flora simply as zonal occurrences in the boundaries of two regions of the Holarctic Kingdom—namely, the Eurasian (Eurasiaticum) and the North American (Septamericanum). Nor is the Arctic flora treated as an independent phytochorion in the floristic system of Chevalier and Emberger (1937). Vasiliev (1956) gives special attention to the question of the phytochorionomic status of the Arctic flora. In his botanico-geographic division of eastern Siberia, Vasiliev does not assign the Arctic Province a place, and describes the arctic tundra as a zone, but not as a phytochorion. Fedorov (1979) also considers that "the Arctic Province is to a greater degree a zonal geobotanical concept rather than a floristic one."

I cannot agree completely with Vasiliev and Fedorov. Though in relation to the Subarctic and even the Low Arctic flora, they are correct to a considerable degree, the same cannot be said about the circumpolar flora of the

HOLARCTIC KINGDOM

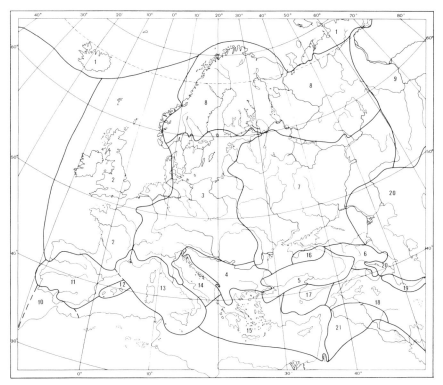

Map 2. FLORISTIC PROVINCES OF EUROPE, ASIA MINOR, AND THE CAUCASUS

1, Arctic Province. 2, Atlantic-European Province. 3, Central European Province. 4, Illyrian or Balkan Province. 5, Euxine Province. 6, Caucasian Province. 7, Eastern European Province. 8, Northern European Province. 9, Western Siberian Province. 10, Southwestern Mediterranean Province. 11, Iberian Province. 12, Balearic Province. 13, Liguro-Tyrrhenian Province. 14, Adriatic Province. 15, East Mediterranean Province. 16, Crimean-Novorossiysk Province. 17, Central Anatolian Province. 18, Armeno-Iranian Province. 19, Hyrcanian Province. 20, Turanian or Aralo-Caspian Province. 21, Mesopotamian Province.

High Arctic, which obviously constitutes a rather distinct phytochorion (Yurtsev et al. 1978). The floristic unity of the circumpolar territory of the Arctic is shown in the predominance of circumpolar species in the composition of its flora. As regards the Subarctic and "lower" Arctic, there are rather clearly expressed close ties with the more southerly disposed phytochoria. This is seen very clearly in northeastern Asia (to the east of the Taimyr Peninsula) and in Alaska, especially in the Beringian regions. As Young (1978) pointed out, the Beringian Arctic occurs in close contact with a series of potentially important migration routes that permit possible penetration into it of preadapted alpine species from the mountain regions of the temperate zones of the Eastern and Western Hemispheres.

In connection with these great latitudinal differences in composition of the arctic flora, a question arises about the boundary of the Arctic phytochorion, irrespective of its phytochorionomic rank. Good (1947, 1974), Mattick (1964), Tolmatchev (1974), and others include in the Arctic region all of Greenland; Iceland; the northern treeless region of Norway; the northern treeless region of the European part of the Soviet Union and of Siberia; the northern treeless region of North America; and all islands to the north of the Arctic Circle. In the Bering Sea, the southern boundary of the arctic flora extends to the island of Saint Matthew or farther south (right up to the Pribilof Islands, or even to and including the Aleutian Islands and Kodiak Island, as shown on Mattick's map of floristic kingdoms and regions).

Such a broad understanding of the Arctic phytochorion is not acceptable to everyone. Thus Böcher (1978) considers that large parts of southern, southwestern, and the southern part of eastern Greenland do not have an arctic character, but a boreal-oceanic one, and should be brought together with Iceland, the Faroe Islands, and with similar territories of western Europe. In a joint work, Yurtsev, Tolmatchev, and Rebristaya (1978) also accept a narrower limit for the Arctic region than appears in the earlier works of Tolmatchev, excluding the following: the Pribilof Islands; the region of large trailing shrubs and dwarf trees ("stlaniks") in the Anadyr and the Penzhina basins; the treeless northern montane regions of Fennoscandia (including the Murmansk shores); Iceland; the Faroe Islands; and south, southwest, and southeast Greenland. As a basis for such a boundary, they considered the exceptionally large role both of the boreal—especially the oceanic-boreal—species, and of the oceanic-subarctic, alpine, and arctic-alpine species in the composition of the flora of these areas.

With such boundaries, the Arctic phytochorion ("region," according to these authors) evidently appears more natural, more integral. However, difficulties arise here in connection with the boreal-oceanic territories, especially with Greenland and Iceland. Both Yurtsev et al. and Böcher realized this. As Böcher said, it raises the "big question" as how to delimit the true arctic territories from the boreal-oceanic ones. Unfortunately, we do not receive a completely definite and clear reply to this question; such an answer is probably not possible. If we separate the boreal-oceanic flora from the Arctic flora—especially that of Greenland and Iceland—a most difficult phytochorionomic problem arises. Actually, it would be very difficult to incorporate these territories (except the Faroe Islands)[4] into any of their neighboring provinces, and there is insufficient basis for separating them into independent provinces of their own. Thus it seems to me that it is better to keep these territories within the Arctic Province as separate, transitional subprovinces. A similar situation may be observed in many other phytochoria, so that the existence of transitional territories is the rule. Finally, the location of the borders of the Arctic flora on the northern shores of Fennoscandia and northeastern Asia can hardly be disputed.

The flora of the Arctic Province is very poor, and consists of no more than 1,000 species of vascular plants (see Yurtsev et al. 1978). Floristically, the richest areas are the Chukot Autonomous Okrug, where about 900 species and subspecies grow (Yurtsev et al. 1978), and Greenland, where about 500 species grow (Böcher 1978). The Poaceae, Cyperaceae, Asteraceae, Brassicaceae, and Caryophyllaceae play the largest role in the flora of the Arctic Province. The province contains two monotypic, endemic genera: the circumboreal genus *Dupontia (D. fischeri)*, and the Canadian genus *Parrya* s.str. *(P. arctica)*. Endemism at the species level is rather high. The number of endemic and subendemic species is more than 120, and along with endemic subspecies this figure probably approaches 150. The number of strictly endemic species probably is less than 100, but their calculation becomes difficult due both to the less than completely clear taxonomic status of many endemic forms, and to the vagueness and controversy of the boundaries of the arctic flora in northeastern Asia and Alaska. Endemic and subendemic species include the following:

Ranunculus chamissonis, R. punctatus, R. sabinii, Papaver augustifolium, P. anjuicum, P. dahlianum, P. gorodkovii, P. keelei, P. pulvinatum, P. ushakovii, P. walpolei, Stellaria crassipes, Cerastium regelii, Silene soczaviana, S. triflora, Arabidopsis tschuktschorum, Braya intermedia, B. linearis, B. purpurascens, B. thorildwulfii, Smelowskia jurtzevii, S. media, S. porsildii, Cardamine digitata, C. hyperborea, C. sphenophylla, Draba bellii, D. chamissonis, D. glacialis, D. gredinii, D. groenlandica, D. kjellmanii, D. macrocarpa, D. oblongata, D. pohlei, D. subcapitata, D. taimyrensis, Cochlearia groenlandica, Salix arctophila, S. uva-ursi, Primula beringensis, Androsace andersonii, A. triflora, Saxifraga nathorstii, Potentilla anachoretica, P. beringensis, P. pulchella, P. rubella, P. rubricaulis, P. vahliana, Oxytropis arctica, O. arctobia, O. bellii, O. hyperborea, O. terrae-novae, Astragalus gorodkovii, A. richardsonii, Gentiana arctica, G. detonsa, Mertensia drummondii, Lagotis hultenii Castilleja arctica, C. septentrionalis, C. vorkutensis, Pedicularis amoena, P. dasyantha, P. hirsuta, Campanula tschuktschorum, Aster pygmaeus, Erigeron muirii, Antennaria affinis, A. canescens, A. glabrata, A. hansii, A. intermedia, Artemisia comata, A. flava, A. hyperborea, A. richardsonii, A. senjavinensis, Saussurea tschuktschorum, Taraxacum amphiphron, T. hyparcticum, T. pumilum, T. tundricola etc., *Hieracium* spp., the endemic Greenland "iris" *Sisyrinchium groenlandicum, Luzula beringensis, L. tundricola, Eriophorum triste, Kobresia hyperborea, Carex ursina* and other species of the genus, *Calamagrostis chordorrhiza, C. hyperborea, C. poluninii, Elymus virescens, Poa abbreviata, P. hartzii, P. sublanata, P. trautvetteri, P. vrangelica, Puccinellia andersonii, P. angustata, P. brugemanii, P. byrrangensis, P. gorodkovii, P. groen-*

landica, P. lenensis, P. poacea, P. porsildii, P. rosenkrantzii, P. svalbardensis, P. tenella, P. vaginata, and *P. vahliana.*

The Arctic flora is an integral part of the circumboreal flora, being its northernmost and most depauperate variant. The presence and predominance in the High Arctic of the circumpolar element provides some basis for the recognition of a separate Arctic Province, but the floristic uniqueness of the Arctic flora is evidently inadequate for the elevation of this province to the rank of a distinct floristic region.

Several attempts at floristic division of the Arctic have been made. Engler (1924) recognized 6 subprovinces in the boundaries of his Arctic Province. Polunin (1951, 1959) separated the Arctic flora into ten "sectors." In the most recent scheme of the classification of the Arctic, Yurtsev et al. (1978) divided it into 6 provinces and 19 subprovinces. These "provinces" should more properly be described as subprovinces, and several of them probably as districts.

A basic and most characteristic plant formation of the Arctic Province is the tundra, which to the north gives way to polar deserts. In the south, especially in those areas with the greatest snowfall, the tundra is characterized by shrubby communities. Shrubs are represented mainly by arctic species of *Salix* and *Betula.* Also very characteristic are shrubs of the Ericaceae (especially species of *Vaccinium* and *Arctostaphylos*) and herbaceous plants of the Cyperaceae (species of *Eriophorum* and *Carex*). In more northerly regions, the vegetation consists almost exclusively of one-layered communities. Species of *Dryas* and several prostrate willows (*Salix polaris* etc.) are especially characteristic, as are perennial herbs of the Poaceae, Brassicaceae, and Caryophyllaceae. Also very characteristic are species of *Saxifraga, Pedicularis,* and *Potentilla,* and *Oxyria digyna.* Annual herbs are very rare and frequently are completely absent.

2. Atlantic-European Province (Engler 1882, 1903, 1924; Braun-Blanquet 1923*b*, 1928, 1964; Schmid 1944; Gaussen 1954, 1965; Dupont 1962; Jäger 1968; Roisin 1969; Zohary 1973; Atlantische Florenregion, Walter and Straka 1970).[5] This province extends from the northwest and northern parts of the Iberian Peninsula (including the Pyrénées) to the southwestern and western shores of Norway (north approximately to the islands of Hitra and Froya). It includes a large part of France (the Armorican Massif, the Aquitanian and Parisian Basins, the Central Massif); the lower Rhine; a large part of the north German Lowlands; the Jutland Peninsula;[6] all of Great Britain and Ireland, and their adjoining islands; and the Faroe Islands.[7] Its eastern boundary conforms approximately to the western boundary of *Ilex aquifolium* in western Europe. In northeastern Europe, a series of Atlantic species reaches Poland (Czeczott 1926) and even the Gulf of Finland. Atlantic species are distributed in the Parisian Basin as far as Paris (several even to Champagne) and occupy the low parts of the Central Massif.

There are four endemic genera in the flora of this province: *Petrocoptis, Dethawia, Endressia,* and *Thorella.* The genus *Daboecia* (2) appears as a common Azores-Atlantico-European endemic. There are scores of endemic species and subspecies, of which we list the following:

Ranunculus tripartitus, Meconopsis cambrica, Corydalis claviculata, Fumaria occidentalis, F. purpurea, Spergularia rupicola, Dianthus gallicus, Rumex rupestris, Limonium recurvum, Viola lactea, V. hispida, Arabis brownii, Rhynchosinapis monensis, R. wrightii, Salix hibernica, Erica mackaiana, E. vagans, Daboecia cantabrica, Primula scotica, Androsace pyrenaica, A. ciliata, Soldanella villosa, Saxifraga spathularis, S. praetermissa, S. canaliculata, S. hariotii, S. pubescens, S. hartii, S. conifera, Potentilla montana, Alchemilla fulgens, A. minima, Cytisus commutatus, C. cantabricus, Genista hystrix, G. berberidea, Ulex gallii, U. micranthus, Astragalus baionensis, Vicia argentea, Geranium endressii, Erodium manescavi, Eryngium viviparum, Seseli nanum, S. cantabricum, Angelica heterocarpa, A. razulii, A. pachycarpa, Peucedanum gallicum, Centaurium scilloides, C. chloodes, Gentianella anglica, Buglossoides gastonii, Onosma bubanii, Pulmonaria affinis, Myosotis alpina, Omphalodes nitida, O. littoralis, Valeriana pyrenaica, Linaria thymifolia, L. faucicola, L. arenaria, Euphrasia spp., *Bartsia spicata, Senecio cambrensis, Crepis lampsanoides, Baldellia alpestris, Narthecium ossifragum, Lilium pyrenaicum, Scilla verna, Endymion non-scriptus, Allium palentinum, A. pyrenaicum, Narcissus cyclamineus, N. bicolor, Iris latifolia, Epipactis dunensis, Carex trinervis, C. durieui, Bromus interruptus,* and *Calamagrostis scotica.*

A series of arctic species (glacial relicts) grows along with warmloving species in the northern part of the province (especially on the Faroe Islands). The forest consists mainly of deciduous trees—namely, oak and beech. In western and central France (Normandy), *Abies alba* also appears, and *Castanea sativa* is frequently seen. The forest is characterized by the presence in the understory of evergreen plants (*Ilex aquifolium, Ruscus aculeatus*) and usually of *Hedera helix* as well. Along with the forest, one finds areas of shrublands and heathlands.

3. Central European Province (Braun-Blanquet 1923*b*, 1928; Kulczynski 1924; Popov 1949; Sjörs 1956; Zohary 1973).[8] This province extends from the eastern edge of the Central Massif; the western edge of the Lorraine Plateau and the Massif of Ardennes; the eastern shores of the Jutland Peninsula; the southwestern and southeastern shores of Norway; and the southern part of Sweden (where to the west the border extends somewhat south of 60° north latitude, and to the east it turns somewhat to the north). It extends north of the Åland Islands to the shores of the Gulf of Finland, the western shores of the Karelian Isthmus, the environs of Leningrad, and the entire southern coast of the gulf. The entire western shore of Estonia is also included (Laasimer 1959). Further on, the border of the province moves in a southerly direction to the west of Riga; cuts through Latvia; passes somewhat south of Vilnius; turns from the east toward the Belovezh Forest; goes west of Lvov; passes to the south of the Dniester River, and along it through the environs of Chernovtsy; and proceeds along the Prut River to the lower Danube Lowlands (Popov 1949:87). To the south, the Central European Province includes the Alps, the northern Apennines (including Abruzzi), the Transylvanian Alps, and the Carpathians.

The typical central European flora usually is not found further east than the Belovezh Forest and the eastern foothills of the Carpathians, but individual elements reach the Valdai Hills and even to Sverdlovsk (Gorchakovsky 1968, 1969). These elements fade out rather gradually, so that it is very difficult to determine the eastern boundary of the province, which corresponds approximately to the common areas of distribution of such plants as *Taxus baccata* (one of its large populations being around the city of Kolomyya in the southwestern Ukraine), *Abies alba* (approaching the Belovezh Forest foothills), *Fagus sylvatica, Quercus petraea, Hedera helix, Astrantia major, Isopyrum thalictroides, Lysimachia nemorum, Geranium phaeum, Salvia glutinosa, Scopolia carniolica, Allium vineale, Hordelymus europaeus,* and *Carex distans.*

Species endemism is highest in the eastern Alps, where it reaches 18%, compared to about 13% in the western Alps (Favarger 1972). Species endemism is about 12% in the Carpathians (Pawlowski 1970; Favarger 1972). *Rhizobotrya, Hacquetia, Hladnikia,* and *Berardia* are the monotypic, endemic genera of the province. Sets of endemic sections, subsections, and species also exist (cf. Pawlowski 1970).

Characteristic Central European Province species and subspecies (several of which penetrate into neighboring provinces) include the following:

Isoëtes malinverniana (northwestern Italy), *Asplenium seelosii* subsp. *seelosii* (central and eastern Alps), *Pinus cembra* (the Alps and Carpathians), *Callianthemum anemonoides* (northeastern Alps), *C. kerneranum* (southern Alps), *Trollius europaeus* subsp. *transsilvanicus* (Carpathians); *Aconitum firmum* (central Europe east to Rumania), *A. paniculatum* (central Europe, reaching central Italy and Rumania), *A. variegatum* (mountains of central Europe, reaching central Italy, Bulgaria, and the western Ukraine), *Delphinium dubium* (southwestern Alps), *D. oxysepalum* (western Carpathians), *Hepatica transsilvanica* (central Rumania), *Ranunculus carpaticus* (eastern Carpathians), *Aquilegia alpina* (Alps, northern Apennines), *A. einseleana* (Alps), *A. thalictrifolia* (northern Italy), *Papaver burseri* (northern Alps, northern Carpathians), *P. corona-sancti-stephani* (eastern and southern Carpathians), *P. sendtneri* (central and eastern Alps), *Arenaria huteri* (northeastern Italy), *Moehrin-*

gia dielsiana (northern Italy), *M. glaucovirens* (southern Alps), *M.* **markgrafii** (northern Italy), a series of species of *Minuartia, Pseudostellaria europaea* (southeastern Austria, northern Yugoslavia, northwestern Italy), a series of species of *Cerastium, Lychnis flos-jovis* (Alps), *L. nivalis* (eastern Carpathians), *Silene dinarica* (southern Carpathians), *S. zawadzkii* (eastern Carpathians), *Saponaria pumilio* (eastern Alps, southeastern Carpathians), *Dianthus nitidus* (western Carpathians), a series of species of *Viola, Murbeckiella zanonii* (northern Apennines), *Braya alpina* (eastern Alps), species of *Erysimum, Hesperis nivea* (Carpathians), *H. oblongifolia* (eastern Carpathians), *Cardaminopsis neglecta* (Carpathians), *Arabis* spp., *Alyssum wulgenianum* (southeastern Alps), *A. alpestre* (western and central Alps), a whole set of species of *Draba, Cochlearia tatrae* (Tatra Mountains), *Rhizobotrya alpina* (southern Alps), *Thlaspi* spp., a series of species of *Salix, Rhododendron hirsutum* (central and eastern Alps, mountains of northwestern Yugoslavia), *Rhodothamnus chaemaecistus* (eastern Alps), *Primula spectabilis* (southern Alps) and a series of other species of the genus, *Androsace* spp., *Soldanella austriaca* (Austrian Alps), *S. carpatica* (western Carpathians), *Euphorbia valliniana* (southwestern Alps), *Daphne arbuscula* (eastern Czechoslovakia), *Prunus bigantina* (southwestern Alps), *Cytisus emeriflorus* (southern Alps), *Astragalus leontinus* (Alps and mountains of northwestern Yugoslavia), *A. roemeri* (eastern Carpathians), *Medicago pironae* (northeastern Italy), *Trifolium saxatile* (Alps), *Hedysarum boutignyanum* (southwestern Alps), *Oenothera silesiaca* (central Europe), *Epilobium fleischeri* (Alps), *Astrantia bavarica* (eastern Alps), *A. carniolica* (southeastern Alps), *Eryngium spinalba* (southwestern Alps), *Chaerophyllum elegans* (Alps), *C. villarsii* (central Europe), *Seseli leucospermum* (western Hungary), *Athamanta cortiana* (northern Italy), *Hladnikia pastinacifolia* (northwestern Yugoslavia, mountains to the north of Ajdovščina), *Bupleurum petraeum* (southern and eastern Alps), *Heracleum austriacum* (eastern Alps, one locality in Switzerland), *H. carpaticum* (eastern Carpathians), *H. minimum* (mountains of southeastern France), *H. sphondylium* subsp. *transsilvanicum* (eastern and southern Carpathians), *Laserpitium nitidum* (northern Italy), *L. peucedanoides* (southeastern Alps, mountains of northwestern Yugoslavia), *Syringa josikaea* (mountains of Transylvania and the Ukrainian Carpathians), *Thesium rostratum, T. kerneranum* (southern Carpathians), *Valeriana celtica* (Alps), *V. elongata* (eastern Alps), *Knautia* spp., *Scabiosa vestina* (southern Alps, northern Apennines), *Gentiana* spp., *Gentianella* spp., *Onosma* spp., *Moltkia suffruticosa* (mountains of northern Italy), *Pulmonaria* spp., *Symphytum cordatum* (Carpathians, central Rumania, western Ukraine), *Myosotis rehsteineri* (Alps), *Linaria loeselii* (southern shores of the Baltic Sea), *Veronica allionii* (southwestern Alps), *Melampyrum subalpinum* (eastern Alps), *Paederota bonarota* (eastern Alps), *P. lutea* (eastern Alps), *Euphrasia* spp., *Pedicularis* spp., *Rhinanthus* spp., *Orobanche lucorum* (Alps), *Campanula* spp. including *C. alpestris* (southwestern Alps), *C. car-*

patica (Carpathians) and *C. zoysii* (southeastern Alps), *Phyteuma* spp., *Adenostyles leucophylla* (Alps), *Doronicum* spp., *Senecio cordatus* (Alps, Apennines), *S. incanus* (Alps, northern Apennines, Carpathians), *S. persoonii* (western Alps), *Erigeron nanus* (Carpathians), *Achillea atrata* (Alps), *Leucanthemum discoideum* (western Alps), *Artemisia glacialis* (southwestern Alps), *A. oelandica* (Aland Islands), *A. umbelliformis* (Alps, northern Apennines), *Telekia speciosissima* (northern Italy), *Saussurea porcii* (eastern Carpathians), *S. pygmaea* (eastern Alps, eastern Carpathians), *Carduus* spp., *Cirsium spinosissimum* (Alps, northern and central Apennines), an entire set of species of *Centaurea*, *Berardia subacaulis* (southwestern Alps), *Leontodon schischkinii* (Carpathians), *Crepis rhaetica* (Alps), *Hieracium glaciale* (Alps), *Andryala levitomentosa* (eastern Carpathians), *Fritillaria burnatii* (Alps), *F. tubiformis* (southwestern Alps), *Allium* spp., *Carex* spp., *Festuca* spp., *Poa granitica* (Carpathians), *Sesleria ovata* (Alps), *Trisetum fuscum* (Carpathians), *Koeleria* spp., and *Helictotrichon* spp.

The flora of the Central European Province has much in common with the flora of the Atlantic-European Province, especially with that of the Pyrénées (cf. Favarger 1972; Küpfer 1974, where all the important literature on this question is found).[9] Common to both provinces are a series of endemic genera, many common endemic species, and a considerable number of vicariant taxa. The Pyrénées are also characterized by considerable endemism (generic but especially specific)—about 14% (Favarger 1972). Based on this, Engler (1882, 1924) created an independent province, "Provinze der Pyrenäen," which many subsequent authors also recognized. In contrast, Gaussen (1954, 1965) united all the central European mountains (including the Pyrénées) into a separate province ("domaine") called the "Hautes montagnes de l'Europe," which thus became geographically discontinuous and was totally unacceptable for purely methodological reasons.

In my opinion, Braun-Blanquet (1923b, 1928, 1964) arrived at the best solution to this problem. He included the Pyrénées in the Atlantic-European Province because the flora of the Pyrénées is even closer to that of this province than to the flora of the Alps. Formation of the montane floras of Europe was basically completed before the Pleistocene (Kulczynski 1924), but during the Pleistocene glaciation further changes occurred, due mainly to migrations (this interpretation has been confirmed in part by karyosystematic investigations of many taxa of the Pyrénées and the Alps; see Favarger 1972; Küpfer 1974). However, this intensive exchange did not predominantly change the Atlantic character of the flora of the Pyrénées. But even if the flora of the Pyrénées is considered to be predominantly central European, in this case it is more expedient to describe it as an enclave of a central European flora in the Atlantic-European Province, since the Pyrénées are separated from the Alps by a considerable area, which is occupied by the Mediterranean flora.

Braun-Blanquet (1928, 1964) subdivided the Central European Province into several subprovinces ("sectors"): Alpine, Pannonian, Carpathian, and Baltic (occupying the entire northern part of the province). To these may be added the Apennine Subprovince, a territory that Braun-Blanquet included in his "Alpen-Sektor." Many authors describe these subprovinces as independent provinces.[10]

4. **Illyrian or Balkan Province** (Adamovič 1909; Braun-Blanquet 1928, 1964; Gajewski 1937; Giacomini [in Giacomini and Fenaroli] 1958; Meusel et al. 1965; Takhtajan 1974; Balkan Province, Kitanov 1976). I accept this province, which follows closely the boundaries established by Braun-Blanquet. It includes a large part of Yugoslavia (with the exclusion of almost all of the Adriatic Coast), the northwest (the Karawanken and the Julian Alps), and northern regions, where the northern boundary runs mostly along the valley of the Sava River and continues along the southern Yugoslavian segment of the Danube River; the northeastern and southeastern regions of Albania; part of the lower Danube lowlands; almost all of Bulgaria (with the exclusion of the northeastern steppe regions and a little territory belonging to the Euxine Province); part of European Turkey (excluding regions with Mediterranean or Euxine floras); and a considerable part of northern Greece (including the Pindus Mountains).

The flora of the Illyrian Province contains four endemic genera (*Petteria, Halacsya, Haberlea, Jankaea*) and is distinguished by high species endemism (especially in northeastern Albania, northern Macedonia, Serbia, and the Rhodope Mountains). Many of these endemics appear to be ancient Tertiary relicts, having remote geographic affinities with eastern Asiatic and Himalayan species (Turrill 1929, 1958; Stojanoff 1930; Stefanov 1943). Among the most interesting are:

Picea omorika (central Yugoslavia), *Pinus peuce* (mountains of the Balkan Peninsula from 41° to 43° north latitude; a species close to the Himalayan *P. griffithii*), *Aguilegia aurea* and several other species of this genus, *Pulsatilla halleri* subsp. *rhodopaea*, species of *Dianthus, Silene* spp., *Rheum rhaponticum* (southwestern Bulgaria), *Rumex balcanicus, Limonium asterotrichum* (Bulgaria), *Viola* spp., *Arabis ferdinandi-corburgii* (Bulgaria), *Alyssum doerfleri* (Macedonia), *Thlaspi bellidifolium, Malcolmia illyrica, Primula deorum, P. frondosa, Euphorbia gregersenii, E. velenovskyi*, species of *Sedum* and *Sempervivum, Saxifraga ferdinandi-coburgii, S. stribrnyi, Geum bulgaricum, Sorbus baldaccii, Petteria ramentacea* (Yugoslavia, northern Albania), *Astragalus* spp. (especially *A. physocalyx*), *Oxytropis prenja, Lathyrus pancicii, Acer heldreichii, Seseli* spp., *Eryngium palmatum, E. serbicum, Pimpinella serbica, Bupleurum karglii, Forsythia europaea* (northern Albania and adjoining parts of Yugoslavia), *Scabiosa rhodopensis* (the Rhodope Mountains), *Asperula capitata, Galium stojanovii* (southwestern Bulgaria), *Onosma thracica* (Bulgaria), *Molt-*

kia doerfleri (mountains of northeastern Albania), species of *Alkanna*, *Halacsya sendtneri* (central Yugoslavia, northern Albania), *Solenanthus scardicus* (eastern Albania and western Macedonia), species of *Verbascum*, *Scrophularia aestivalis*, *Wulfenia baldaccii* (northern Albania), *Lathraea rhodopea*, *Veronica rhodopaea* (mountains of southern Bulgaria), *V. saturejoides* (the Rhodope Mountains, northeastern Greece), *Haberlea rhodopensis* (Bulgaria and northeastern Greece), *Ramonda nathaliae*, *R. serbica*, *Jankaea heldreichii* (Olympus, a mountain range in Thessaly, Greece), *Marrubium friwaldskyanum*, *Sideritis scardica*, *Stachys iva*, *S. milanii*, *Satureja rumelica* (Bulgaria), *Micromeria frivaldszkyana*, species of *Campanula* and *Edraianthus*, *Jasione bulgarica*, *Anthemis* spp., *Amphoricarpos neumayeri*, *Jurinea tzarferdinandii* (Bulgaria), a whole set of species of *Centaurea*, *Cicerbita pancicii*, *Crepis albanica*, *C. macedonica*, *C. schachtii*, *Narthecium scardicum* (Albania, northern Macedonia, Herzegovina), *Colchicum callicymbium*, *Merendera rhodopaea*, *Lilium jankae*, *L. rhodopaeum*, *Fritillaria drenovskii*, *Tulipa pavlovii*, *T. urumoffii*, *Allium melanantherum*, *Dioscorea balcanica* (northern Albania, Herzegovina), species of *Crocus* and *Iris*, *Bromus moesiacus*, and *Poa pirinica*.

Many relict species occur in the Ilyrian Province, including *Ostrya carpinifolia*, *Juglans regia*, *Quercus* spp., *Diospyros lotus*, *Buxus sempervirens*, *Staphylea pinnata*, *Vitis vinifera*, *Ilex aquifolium*, *Comandra elegans*, *Syringa vulgaris*, and so on.

5. **Euxine Province** (Gajewski 1937, pro parte; Maleev 1938, 1941; Davis 1965, 1971; Meusel et al. 1965; Takhtajan 1974). This province embraces the western Caucasus (Adzharia, Abkhazia, the Kutaisi area in western Georgia, and the coastal parts of the Krasnodar Krai, where the boundary then passes somewhat further north to Tuapse); northern Anatolia (where the boundary continues along the ranges dividing watersheds, delimiting the coastal belt, and includes a large part of Lazistan;[11] Paphlagonia and Bithynia, and also the northeastern part of Myzia);[12] the Black Sea shores of European Turkey (where the boundary continues along the peninsula of Pasaeli and farther along the Istranca Range); and the southern part of the Black Sea shores of Bulgaria (where the boundary to the northeast approaches the Gulf of Burgas).[13] Several elements, such as *Trachystemon orientale*, extend somewhat farther north.

It was pointed out long ago that the flora of the southern part of eastern Bulgaria and the European part of Turkey is much more similar to that of Lazistan and Colchis (western Georgia) than to that of the nearby parts of the Balkan Peninsula (see literature in Maleev 1938). On this basis, Gajewski (1937) created a separate phytochorion, the Euxine Province, uniting all of these so-called Colchian floras. But Gajewski included in his Euxine Province not only the flora of the Colchian type but also the flora of the Crimean-

Novorossiysk Province, and even that of the Hyrcanian Province (which I include in the Irano-Turanian Region). The Euxine Province *sensu* Maleev (1938), who created it independently of Gajewski, was more natural, in that he did not include the Hyrcanian flora. However, Maleev included in his Euxine Province the territory of the Crimean-Novorossiysk Province, which I prefer to put in the Mediterranean Region.

Two monotypic endemic genera are found in the flora of the Euxine Province: *Chymsidia and Megacaryon*. We should also mention the subendemic monotypic genus *Pachyphragma* (Brassicaceae). In addition, *Agasyllis* and *Sredinskya* are endemic to the Euxine and Caucasian provinces. There are many endemic species (over 200) among the large number of ancient relict species (Czeczott 1932, 1938–1939; Grossheim 1936; Maleev 1941; Kolakovsky 1962). Among the more interesting endemic or subendemic taxa of the Euxine flora we can mention:

Asplenium hermannii-christii (Adzharia), *A. woronowii* (Colchis), *Dryopteris liliana* (Lazistan, Adzharia), *Abies nordmanniana* subsp. *bornmuelleriana* (northwestern and northern Anatolia), *A. nordmanniana* subsp. *nordmanniana* (Lazistan, Colchis; subendemic), *Picea orientalis* (Lazistan, Colchis; subendemic), *Aristolochia pontica* (Colchis), *Helleborus abchasicus* (Colchis), *Aquilegia colchica* (Colchis), *Epimedium pubigerum* (southeastern Bulgaria, northern Anatolia; encountered also in southwestern Bulgaria), *E. pinnatum* subsp. *colchicum* (Lazistan, Colchis), *Corydalis vittae* (Abkhazia), *Buxus colchica* (Colchis), *Quercus pontica* (Lazistan, Colchis), *Q. hartwissiana* (extreme southeast Bulgaria, the eastern European part of Turkey, northern Anatolia, Colchis, Belaya River Basin in the northern Caucasus), *Betula medwedewii* (Lazistan in the region of Artvin, Adzdharia, Guria), *B. megrelica* (Mengrelia), *Corylus colchica* (northern Colchis), *C. pontica* (Lazistan, Colchis), *Minuartia abchasica* (Abkhazia), *M. rhodocalyx* (northern Colchis), *Silene boissieri* (Colchis), *S. physocalyx* (Adzharia), *Gypsophila steupii* (northern Colchis), *Dianthus imereticus* (Colchis), *Paeonia macrophylla* (Adzharia), *P. wittmanniana* (Abkhazia), *Hypericum bithynicum* (northern Anatolia, Lazistan, Colchis; subendemic), *H. xylosteifolium* (northern Anatolia, Lazistan, Colchis), *Viola orthoceras* (Colchis), *Pachyphragma macrophyllum* (Lazistan, Colchis), *Draba subsecunda* (Colchis), *Arabis colchica* (Abkhazia), *A. sakhokiana* (northern Colchis), *A. nordmanniana* (Colchis), *Erysimum contractum* (Lazistan, Adzharia), *Rhododendron smirnowii* (Lazistan, Adzharia), *R. ungernii* (Lazistan, Adzharia), *Epigaea gaultherioides* (Lazistan, southern Adzharia), *Cyclamen colchicum* (Abkhazia), *C. parviflorum* (Lazistan), *Primula abchasica* (Abkhazia), *P. longipes* (Lazistan), *P. megaseifolia* (Lazistan, Adzharia), *Hibiscus ponticus* (Colchis), *Alcea abchasica* (Abkhazia), *Andrachne colchica* (from the southernmost part of the Krasnodar Krai to Kutaisi), *Euphorbia pontica* (Adzharia), *Saxifraga abchasica* (Abkhazia), *Sorbus colchica* (Colchis), *Pyrus balansae* (Colchis), *Astragalus aszharicus* (Lazistan, Adzharia), *A. bachmarensis*

(Lazistan, Adzharia), *Vicia freyniana* (northern Anatolia), *Trifolium euxinum* (northern Anatolia), *T. polyphyllulm* (Lazistan, Colchis), *T. sintenisii* (Lazistan), *Staphylea colchica* (Colchis; subendemic), *Trapa colchica* (Colchis), *Hedera colchica* (northern Anatolia, Lazistan, Colchis; subendemic), *Astrantia pontica* (northern Colchis), *Chaerophyllum schmalhausenii* (Abkhazia), *Scaligeria lazica* (Lazistan), *Pimpinella idae* (Abkhazia), *Chymsidia agasylloides* (northern Colchis), *Seseli foliosum* (Lazistan, Adzharia), *S. rupicola* (southern part of the Krasnodar Krai, northern Abkhazia), *Bupleurum rischavii* (northern Colchis), *Ligusticum caucasicum* (Abkhazia), *L. physospermifolium* (Colchis), *Pastinaca aurantiaca* (northern Colchis), *Heracleum aconitifolium* (northern Colchis), *H. calcareum* (northern Colchis), *H. mantegazzianum* (Colchis; subendemic), *H. paphlagonicum* (northern Anatolia), *H. scabrum* (northern Colchis), *Ilex colchica* (northern Anatolia, Lazistan, Colchis), *Rhamnus imeretinus* (Lazistan, Colchis; subendemic), *Osmanthus decorus* (Lazistan, Adzharia), *Scabiosa olgae* (from Tuapse to Bzyb), *Cephalaria calcarea* (Colchis), *Vinca major* subsp. *hirsuta* (northern Anatolia, Lazistan, Colchis), *Gentiana kolakovskyi* (northern Colchis), *G. paradoxa* (Abkhazia), *Omphalodes cappadocica* (Lazistan, Colchis), *O. caucasica* (northern Colchis), *O. kuznetzovii* (Abkhazia), *Symphytum sylvaticum* (Lazistan), *Trachystemon orientalis* (the Black Sea shores of Bulgaria and European Turkey, northern Anatolia, Lazistan, Colchis and bordering regions of western Transcaucasia), *Scrophularia sosnovskyi* (Lazistan, Adzharia), *Rhamphicarpa medwedewii* (Lazistan, Colchis), *Veronica monticola* (Lazistan, Colchis), *V. turrilliana* (Istranca Mountain Range), *Melampyrum alboffianum* (Adzharia), *Pedicularis olympica* (northwestern Anatolia), *Teucrium multinodum* (Abkhazia), *T. trapezunticum* (Lazistan, Adzharia), *Satureja bzybica* (Abkhazia), *Stachys macrophylla* (Colchis), *S. trapezuntica* (Lazistan, Colchis), *Salvia forskaohlei* (southeastern part of the Balkan Peninsula, northern Anatolia), *Scutellaria helenae* (northern Colchis), *S. pontica* (Lazistan, Adzharia), *Betonica abchasica* (Abkhazia), *Campanula calcarea* (Abkhazia), *C. dzaaku* (Abkhazia), *C. dzyschrica* (Abkhazia), *C. jadvigae* (Abkhazia), *C. kluchorica* (Abkhazia), *C. kolakovskyi* (Abkhazia), *C. longistyla* (Colchis), *C. makaschwilii* (Adzharia), *C. mirabilis* (Abkhazia), *C. panjutinii* (Abkhazia), *C. paradoxa* (Abkhazia), *C. schistosa* (Abkhazia), *C. sphaerocarpa* (Abkhazia), *C. symphytifolia* (Abkhazia), *Symphyandra lazica* (Lazistan), *Inula magnifica* (Lazistan, Colchis), *Doronicum balansae* (Lazistan), *D. macrolepis* (Lazistan), *D. tobeyi* (Lazistan), *Senecio lazicus* (Lazistan), *S. trapezunticus* (Lazistan), *Anthemis zyghia* (Colchis), *Cirsium adjaricum* (Lazistan, Adzharia), *Centaurea abchasica* (Abkhazia), *C. albovii* (Adzharia), *C. appendicigera* (Lazistan), *C. barbeyi* (Abkhazia), *C. helenioides* (Lazistan), *C. kolakowskyi* (Abkhazia), *Amphoricarpos elegans* (Colchis), *Narthecium balansae* (Colchis), *Colchicum liparochiadys* (Colchis), *Scilla monanthus* (Lazistan, Colchis), *S. winogradowii* (Colchis), *Muscari alpanicum* (Colchis), *M. colchicum* (Adzharia), *M. dolichanthum* (Abkhazia), *M. pendulum* (Colchis),

Lilium szovitsianum subsp. *ponticum* (Lazistan, Adzharia), *Polygonatum obtusifolium* (Colchis), *Ruscus colchicus* (Istranca Range, northern Anatolia, Colchis), *Allium candolleanum* (Abkhazia), *A. graciliensis* (Colchis), *Crocus autranii* (Abkhazia), *C. scharojanii* (Colchis), *C. vallicola* (Lazistan, Colchis), *Iris lazica* (Lazistan, Adzharia), *Galanthus krasnovii* (Lazistan, Colchis), *Dioscorea caucasica* (Abkhazia), *Orchis viridifusca* (Colchis), *Pycreus colchicus* (Colchis), *Rhynchospora caucasica* (Adzharia), *Carex latifrons* (Colchis), *C. mingrelica* (Colchis), and *C. pontica* (Colchis).

Many endemic species, such as *Gentiana paradoxa*, *Campanula mirabilis*, and *C. paradoxa*, are restricted to limestone rocks. Several Colchian elements are found further to the east, such as *Abies nordmanniana* in the Kabardino-Balkarian ASSR.

Predominant plant formations in the Euxine Province are forests—mainly oak, beech (*Fagus orientalis*), and beech-chestnut forests, which higher in the mountains are replaced by forests of fir. The evergreen elements (*Rhododendron ponticum, Laurocerasus officinalis*) are characteristic, as are such interesting deciduous shrubs as *Vaccinium arctostaphylos*, which is very close to *V. padifolium* (found on the oceanic island of Madeira). Among the lianas, the most characteristic are *Hedera colchica, H. helix*, and *Smilax excelsa*. High mountain meadows are found above the forests. Lowland mixed and alder forests and swamps are developed along seashores. The Euxine flora is best preserved in Adzharia and Lazistan (the largest concentration of forest relicts), and is most impoverished on the Balkan Peninsula (Stefanov 1924; Maleev 1938).

6. Caucasian Province (Provinz des Caucasus und Elbrus, Engler 1882; Provinz des Kaukasus, Engler 1899, 1903, 1924; the Caucasian Province, Grossheim 1948;[14] Takhtajan [in Takhtajan and Fedorov] 1972; Takhtajan 1974). This province embraces the mountain systems of the Greater and Lesser Caucasus, but without its western part (which is included in the Euxine Province), and without the arid southwestern, southern, and southeastern parts (which are in the Armeno-Iranian Province). Here also the southern premontane part of the Kuban Lowlands enters to the south from the Kuban River, a large part of the Stavropol Hills, the Tersko-Sunzhenskian Hills and all of Dagestan, and also central Transcaucasia. The Rioni Lowlands, the Apsheron Peninsula, the Kura and Lenkoran Lowlands, the Lake Sevan Basin, the Araks River Valley, and the Talish Mountains are not included in the territory of the Caucasian Province.

The flora of the Caucasian Province, especially of its western part, has much in common with the flora of the Euxine Province, especially Colchis and Lazistan. It has an entire series of common endemics, including two endemic genera (*Agasyllis* and *Sredinskya*), as well as species like *Rhododendron caucasicum, Vaccinium arctostaphylos, Daphne pontica*, and *Paris incompleta*. In its high mountain flora there is much in common with the floras of the Pyrénées, the Apennines, the Alps, the Carpathians, and the mountains of the Balkan Peninsula (the Illyrian Province). In Dagestan and in several parts of the Lesser Caucasus (especially in the Seven Lake Basin, in Zangesur and Karabakh), we find a very strongly expressed influence of the Armeno-Iranian and (in part) the Turanian floras.

In the flora of the Caucasian Province, the most numerous families are the Asteraceae, followed in order by the Poaceae, Fabaceae, Rosaceae, Caryophyllaceae, Brassicaceae, Apiaceae, and the Lamiaceae. There are five endemic genera: *Pseudovesicaria, Symphyoloma, Pseudobetckea, Trigonocaryum,* and *Cladochaeta.* Species endemism is rather high, especially in the high mountains of the central part of the Greater Caucasus (Balkaria, Digoria) and in Dagestan (Grossheim 1936). Some of the endemics are:

Pinus brutia subsp. *eldarica, Anemone kuznetzowii, Gymnospermium smirnovii, Papaver bracteatum, P. oreophilum, Corydalis smirnowii, C. emanuelii, Betula raddeana, Cerastium kasbek, C. argenteum, C. multiflorum, Minuartia inamoena, Silene akinfievii, S. lacera, Gypsophila acutifolia, Dianthus fragrans, Sobolewskia caucasica, Paeonia mlokosewitschii, Draba bryoides, D. elisabethae, D. supranivalis, D. mollissima, D. ossetica, Primula bayernii, P. darialica, P. juliae, P. renifolia, Androsace lehmanniana, Stellera caucasica, Sedum stevenianum, Saxifraga subverticillata, Pyrus zangezura, Oxytropis owerinii, Astragalus* spp., *Geranium renardii, Gentiana grossheimii, G. lagodechiana, G. marcowiczii, Veronica caucasica, Campanula andina, C. ardonensis, C. dolomitica, C. kryophila, C. ossetica, Edrajanthus owerinianus, Anthemis saguramica, Tanacetum akinfievii, Centaurea amblyolepis, Lilium monadelphum, Ornithogalum arcuatum, O. magnum, Colchicum lactum, Asphodeline tenuior, Gagea helenae, Allium grande, Galanthus caucasicus, G. lagodechianus, G. latifolius, Crocus adamii, Calamagrostis caucasica,* and many others.

The vegetative cover of the Caucasian Province is very mixed. Of the forest formations, the riparian and marsh forests occupy the lowest altitudes, which are followed by montane broad-leaved forests that are more characteristic of the province. The forests of beech and rock oak (*Quercus petraea*) are especially characteristic. The upper montane belt and, in part, the middle belt are occupied by conifer forests in the western half of the Greater Caucasus. A low montane belt of forest steppes is usually found in the central and eastern parts of the Greater Caucasus. Here oak stands alternate with meadow steppes. It is assumed that in earlier times oak forests occupied much wider areas. Beech forests (*Fagus orientalis*) occupy considerable areas. Above the beech forests lies a belt of subalpine parklike forests with *Acer trautvetteri* and *Quercus macranthera* forming the basis. Here also is a well-developed "krivolesie"—a forest of more or less crooked birch, usually occupying the rockiest and steepest slopes. Even higher are found the alpine meadows.

Steppe vegetation is developed mainly along the lower ranges in Eastern Ciscaucasia and in the Armenian Highlands.

7. **Eastern European Province** (Lipmaa 1935; Stoyanov 1950; Takhtajan 1974).[15] In the west, the Eastern European Province includes the eastern part of the Baltic Republics, almost all of Belorussia (except some of its western parts), most of the Ukraine (except its southwestern part and southern Crimea), and the lower Danube Lowlands (including northeastern and eastern parts of Romania and the northeastern most part of Bulgaria). In the north, it is bordered by the zone of conifer forests of the Northern European Province (see the next section); in the south, it stretches to the northern shores of the Black Sea (with the exclusion of the southern shore of the Crimea), the northern and eastern shores of the Sea of Azov, the valley of the Kuban River, the Stravropolsk Hills, Volgograd and Uralsk. In the east, its boundary agrees approximately with the eastern boundary of oak forests that cross the mountain range in the region of the southern Urals. On the whole, the northern, eastern, and southeastern boundaries of the province correspond roughly to the distribution of *Quercus robur, Acer platanoides, Corylus avellana, Euonymus verrucosus, Asarum europaeum, Astragalus arenarius, Vicia cassubica, Lathyrus sylvestris, Campanula persicifolia, Convallaria majalis,* and *Carex montana.*

In spite of the vastness of its territory, only one monotypic endemic genus, *Cymbochasma,* is found here, and it is only weakly separated from *Cymbaria.* In the southern steppe region, there are endemic or almost endemic sections of the genus *Centaurea.* There are a number of endemic species and subspecies. They include:

Pinus sylvestris subsp. *cretacea, Anemone uralensis, Delphinium uralense, Thalictrum uncinatum, Papaver maeoticum, Corydalis paczoskii, Cerastium uralense, Minuartia helmii, M. krascheninnikovii, Silene hellmannii, S. cretacea, S. baschkirorum, Gypsophila belorossica, G. uralensis, Dianthus eugeniae, D. krylovianus, D. carbonatus, D. marschallii, D. humilis, D. uralensis, Sisymbrium volgense, Syrenia taljevii, Schivereckia podolica, S. berteroides, S. monticola, S. kusnezovii, Crambe aspera, Androsace koso-poljanskii, Alchemilla nemoralis, Potentilla eversmanniana, Pyrus rossica, Crataegus ucrainica, Cotoneaster alaunicus, Chamaecytisus paczoskii, C. skrobiszewskii, C. blockianus, C. podolicus, Calophaca wolgarica, Astragalus clerceanus, A. karelinianus, A. pubiflorus, A. tanaiticus, A. henningii, A. pallescens, Oxytropis hippolyti, O. gmelinii, Hedysarum cretaceum, H. grandiflorum, H. ucrainicum, Lathyrus litvinovii, Erodium beketowii, Libanotis sibirica, Knautia tatarica, Cephalaria litvinovii, Onosma guberlinensis, Eritrichium uralense, Scrophularia cretacea, Linaria biebersteinii, L. macroura, Cymbochasma borysthenicum (= Cymbaria borysthenica), Hyssopus cretaceus, Thymus* spp., *Achillea glaberrima, Cicerbita uralen-*

sis, *Centaurea pseudoleucolepis, C. talievii, Senecio igoschinae, Elymus uralensis* subsp. *uralensis,* and *Zingeria biebersteiniana.*

In the Pleistocene, the flora of the Russian Plain underwent considerable changes due to glaciation. The most ancient flora was preserved only in uplands not exposed to glaciation. Such relict areas include the Podolsk Hills; the Donets Range; the granite outcrops in the region of Mariupol-Berdyansk; the right bank of the Dnieper; several elevated areas in the Kursk, Voronezh, Kharkov, and the Tula regions with outcrops of chalk and limestone; uplands in the Volga River Basin (in particular, the Zhiguli Hills and the Penza area); and the central and southern Urals (Litvinov 1891, 1927; Korzhinsky 1899; Paczoski 1910; Lavrenko 1930, 1938; Kozo-Poljansky 1931; Sprygin 1936, 1941; Krasheninnikov 1937, 1939; Kleopov 1941; Igoshina 1943; Wulff 1944; Grosset 1967; Gorchakovsky 1963, 1968, 1969; Golitsin and Doronin 1970; Doronin 1973). In postglacial times, the Russian Plain was inhabited both by plants of these relict regions and by those of relict centers of the Central European and Caucasian provinces.

Today, the plant cover of the Eastern European Province consists mainly of broad-leaved and mixed conifer–broad-leaved forests, and to the south of meadows, forest steppes, and steppes. The mixed conifer–broad-leaved forests occupy broad areas in the western part of the province (from southwest of the Leningrad region in the north to Brest and Bryansk in the south), and only a narrow belt extending to the east. Here spruce forests predominate, but with a considerable admixture of broad-leaved species. Beginning with the Karelian Isthmus (north of the Otradnoe railroad station), *Quercus robur* is found. The northern boundary of this species extends even further, reaching Lake Lodoga, on the islands of which it grows north to Priozersk. Along the eastern shores of Lake Ladoga, the northern boundary of *Quercus robur* intersects the environs of Tikhvin continuing across the Vologda region (around Cherepovets) to the Kirovsk region, from which it approaches the Ural foothills in the Perm and Sverdlovsk regions. At the southern terminus of the Urals, the oak crosses low mountains and continues a comparatively long distance outside the eastern boundary of the Eastern European Province. In the oak forests one encounters *Fraxinus excelsior,* maples, pears, elms, *Tilia cordata, Alnus glutinosa,* and others. Among the maples the most widely distributed is *Acer platanoides,* which follows the oaks and enters from the southern Urals to the Zilair Plateau (Koz'yakov 1962). To the north, it crosses the eastern slope of the Urals close to Lake Ufa. *Acer campestre* also occurs as a component of oak forests, except that its northern border passes considerably more to the south, although it does not reach the shores of the Volga. *Acer tataricum* is similar in distribution to *Acer campestre.* The eastern border of *Acer tataricum* reaches Chkalov. *Pyrus communis* extends to the east to the Volga.

In the northern part of their distributions, all of these species represent an admixture to the spruce forests and form part of the composition of mixed forests. More to the south, they form a subzone of broad-leaved forests, in the western part of which both *Quercus robur* and *Quercus petraea* participate. In all the subzones, pine (*Pinus sylvestris*) forests occupy considerable areas in various places. In the Volynskoe Polesie (the Ukraine), and also in the Mozyr region of Belorussia, we find *Rhododendron luteum*, a Tertiary relict the basic distribution of which is now found in western Transcaucasia and northern Anatolia.

In certain places in the southern part of the Russian Plain, forest steppes occupy great areas. *Quercus robur* usually appears as the predominant woody representative of the forest members found in the forest steppe of the Russian Plain, but in the southeastern part, birch becomes dominant in certain places. The zone of steppe vegetation stretches more to the south.

8. Northern European Province (Nordeuropaische Provinz, Engler 1882; Braun-Blanquet 1964; Provinz Subarktisches Europa, Engler 1899, 1903, 1924).[16] This province stretches from Norway to the northern Urals.

There are not many endemic species or even subspecies in this province, and the flora here is very young. Of the endemic taxa we can mention *Cerastium gorodkovianum* (northern Urals), *Corispermum algidum*, *Helianthemum arcticum*, *Cotoneaster cinnabarinus*, *Anthylis kuzeneviae*, *Castilleja schrenkii*, *Arnica alpina*, *Gagea samojedorum* (northern Urals), and *Carex scandinavica*.

The coniferous forest is the most characteristic for this province, with the main forest formation species being *Picea abies* with two of its subspecies— subspecies *abies* and subspecies *obovata*. The latter is basically a Siberian race, more commonly found in the eastern part of the province. *Pinus sylvestris* can often be added to the spruce. This pine frequently forms pure stands in sandy places, sometimes occupying large areas. Along with the spruce and pine we often find *Betula pendula* and *Populus tremula*. To the east, *Abies sibirica* (east of the Vaga River) and *Larix sibirica* (to the east of Lake Onega) also occur. Several Siberian elements accompany them, such as *Clematis alpina* subsp. *sibirica*, *Actaea erythrocarpa*, and *Paeonia anomala*.

9. Western Siberian Province (Krylov 1919; Schischkin 1947; Shumilova 1962). To the east of the Northern European and Eastern European provinces stretches the broad Western Siberian Province, extending to the east up to the Yenisey.[17] In the north it borders the Arctic Province, and in the south the semideserts of Kazakhstan. In the northeast its border approaches Tomsk and passes somewhat east of Novosibirsk, climbing upward along the valley of the Ob River close to Biysk, and then turning to the southeast and further to the south, extending at 50° north latitude to the west of Ust-Kamenogorsk.

The largest families of the Western Siberian flora are Asteraceae, Fabaceae, Poaceae, Cyperaceae, Rosaceae, Caryophyllaceae, Scrophulariaceae, Ranunculaceae, and Lamiaceae. Despite the vastness of the territory, species endemism is not great, and generic endemism is completely absent. For this province, the absence of *Quercus, Fraxinus, Acer, Corylus,* and many other broad-leaved taxa is characteristic. Only *Tilia cordata* is met in the forests of the Tobolsk region, at Narym, and at the east, it approaches Krasnoyarsk. Treelike species of alder give way to the shrubby alder, *Alnus viridis* subsp. *fruticosa.*

The largest part of the province consists of dense coniferous taiga. The usual forest-forming species is *Picea abies* subsp. *obovata,* along with which *Abies sibirica* and *Pinus sibirica* grow. *Larix sibirica* and pines also play a large role in the composition of the western Siberian forest. Of the broad-leaved species in these forests, *Betula pendula, B. pubescens,* and *Populus tremula* are characteristic. In the north the taiga passes into the woodland and the forest tundra, and in the south into the forest steppe and steppes (Shumilova 1962).

10. Altai-Sayan Province (Kuznetzov 1912; Krylov 1919; Pavlov 1929; Reverdatto 1931; Kuminova 1960, 1969, 1973; Shumilova 1962; Peshkova 1972; Takhtajan 1974).[18] This province includes the Salair Ridge and the Kuznetsk Alatau; the Altai Mountains, including part of northwestern Mongolia; the Khangai Plateau; montane Shoria; the Minusinsk Basin; western and eastern Sayan; the Khamar-Daban Range; mountainous slopes surrounding the southern part of Baikal south of the mouth of the Barguzin River; and the area between the River Selenga and Chikoi.

Two monotypic endemic genera are found in the flora of the province—*Microstigma* (Central Altai) and *Tridactylina* (the southern shores of Baikal)—and more than 120 endemic species are present, especially in the genera *Oxytropis* and *Astragalus.* The following are among the interesting endemics:

Callianthemum sajanense, Eranthis sibirica, Aquilegia borodinii, Delphinium mirabile, D. sajanense, D. inconspicuum, Aconitum altaicum, A. krylovii, Anemone baicalensis, Ranunculus sajanensis, Gymnospermium altaicum subsp. *altaicum, Betula kellerana, Stellaria martjanovii, S. imbricata; S. irrigua, Silene turgida, Rheum altaicum, Viola incisa, Megadenia bardunovii, Aphragmus involucratus, Parrya grandiflora, Erysimum inense, Microstigma deflexum, Eutrema parviflorum, Salix nasarovii, S. sajanensis, Euphorbia alpina, E. altaica, E. tshuiensis, Ribes graveolens, Rhodiola algida, Sedum populifolium, Chrysosplenium alberti, C. baicalense, C. filipes, Cotoneaster lucidus, Caragana altaica, Astragalus olchonensis, Oxytropis jurtzevii, O. kusnetzovii, O. sajanensis, Hedysarum zundukii, Vicia lilacina; Lathyrus frolovii, L. krylovii, Linum violascens,*

Bupleurum martjanovii, Polemonium pulchellum, Brunnera sibirica, Scrophularia altaica, Veronica sajanensis, Pedicularis brachystachys, Schizonepeta annua, Dracocephalum fragile subsp. *fragile, Valeriana petrophila, Tridactylina kirilowii, Tanacetum lanuginosum, Brachanthemum baranovii, Echinops humilis, Saussurea frolovii, S. sajanensis, S. squarrosa, Allium pumilum, Carex tatjanae; Koeleria atroviolacea, K. geniculata, Poa altaica, P. ircutica, P. sajanensis,* and *Agropyron sajanense.*

The vegetative cover is characterized by a distinctly expressed vertical zonation. Dominant formations are the larch forests of *Larix sibirica* and the fir forests of *Abies sibirica,* which are replaced high in the mountains by alpine meadows. Fir forests are distributed in middle mountain belts of the Altai, the Kuznetsk Alatau, Salair, and also in the Khamar-Daban and in a less typical form in western Sayan. In the herbaceous cover of the fir forests, side by side with common taiga species, several species usually appear which are characteristic of the broad-leaved forests of the European part of the Soviet Union, or of the Far East. They include *Dryopteris filix-mas, Asarum europaeum, Osmorhiza aristata* (the major distribution of which is in the Far East), *Sanicula europaea, Epilobium montanum, Circaea lutetiana, Asperula odorata, Campanula trachelium,* and *Festuca gigantea.* The dark conifer forests of the Kuznetsk Alatau are the richest in such relicts (Krapivkina 1973). *Populus tremula* is often found with the firs. *Pinus sibirica* is also often present, as sometimes are spruce and birch. In several places of the Kuznetsk Alatau (at the source of the Tomsk River), *Tilia sibirica* becomes part of the dark conifer forests, forming almost pure stands in a small area (the Kuzodeev "Islands of Linden"). For low mountain and foothill areas, a parklike open forest of *Larix sibirica* with a rich herbaceous cover is characteristic. In the understory one encounters *Rhododendron dahuricum* along with other shrubs. Forests of *Pinus sylvestris* are found rather often is some places. In the premontane and midmontane basins are characteristic islands of steppes, which become larger to the south. In the steppes, besides grasses and various forbs, shrubs in many cases play a significant role. These include *Caragana arborescens* and *Prunus tenella. Leontopodium ochroleucum* grows in the mountains.

11. Middle Siberian Province (Krylov 1919; Vasiliev 1956; Shumilova 1962; Peshkova 1972).[19] This province includes a large part of northern Siberia from the lower and middle course of the Yenisey River, the area between the upper course of the Angara and Lena Rivers, and the mouth of the Barguzin River, up to the Verkhoyansk Range and the eastern shores of the Aldan River.

In western Siberia woody species of the Altai center of postglacial dispersal predominate (e.g., *Albies sibirica, Pinus sibirica,* and *Larix sibirica,* but in the Middle Siberian Province there is a noticeable occurrence of woody species of the Transbaikal and partly of the Manchurian centers, among which *Larix gmelinii* (= *L. dahurica*) represents the basic forest-forming type. From the point of view of floristic relations, the borders of the province are not very precise. Especially for its western part, it is characteristic that many eastern boundaries of western species occur in its territory (Popov 1957). For example, the eastern boundary of *Corydalis halleri* is at the Yenisey, at the Birius is *Lathyrus vernus*, at the Angara is *Silene otites*, and so on.

An endemic monotypic genus is found in the flora of the Middle Siberian Province (*Redowskia*), as are a relatively small number of endemic species, concentrated mainly in Yakutia. Of greatest interest are *Ceratoides lenensis, Polygonum amgense, Redowskia sophiifolia, Androsace gorodkovii, Potentilla tollii, Eritrichium czekanowskii,* and *Adenophora jakutica*.

As Shumilova (1962) remarked, Middle Siberia is characterized by an almost longitudinal extension of horizontal zones to the north of the sixtieth parallel, from the dark conifer taiga close or near the Yenisey to the larch forests farther east and the pretundra woodlands at the Yenisey-Lena watershed. To the south it also has the most complete development of a zone of pine forests in Eurasia.

12. Transbaikalian Province (Shumilova 1962; Peshkova 1972; Takhtajan 1974).[20] This province occupies territory extending to the east and southeast from the Patomsk and Northbaikalian Plateaus, and includes the northern part of the Prebaikal area, a large part of Transbaikalia, as well as part of northeastern Mongolia.

In the flora of the province, only one monotypic genus, *Borodinia*, which is close to *Parrya*, is subendemic. It is found in the southern part of the Baikal Range, in the southern part of the Barguzin Range, in the Kodar Range, and also in the Ayano-Maisk region of the Khabarovsk Krai. In spite of the vastness of its territory, there are few endemic species and subspecies. These include:

Aconitum montibaicalensis, Draba baicalensis, Salix berberifolia subsp. *fimbriata, Saxifraga algisii, Potentilla adenotricha, Astragalus trigonocarpus, Oxytropis heterotricha, O. kodarensis, O. komarovii, O. oxyphylloides, O. adenophylla, Dracocephalum fragile* subsp. *crenatum, Mertensia serrulata, Saussurea poljakovii, Carex malyschevii,* and *Calamagrostis kalarica*.

The predominant vegetation of the Transbaikalian Province is montane larch forest. Almost everywhere the larch is *Larix gmelinii*, with the exception of the southwestern part, where *L. sibirica* grows. The Transbaikalian Province, especially its southern part, is characterized by islands of meadow-steppe, steppe, and forest-steppe vegetation, which occur in the intermontane depressions. The most northern are the Barguzinsk steppes. Intermingled with the steppes (in which different species of grass predominate, including *Leymus chinensis*), one encounters a kind of steppe in which dicotyledonous herbs predominate. Here *Hemerocallis minor* is often encountered; sometimes it even predominates. The Eastern Asiatic flora shows a noticeable influence on the flora of meadow-steppes and forest-steppes, and this influence grows as one moves east.

13. **Northeastern Siberian Province** (Vasiliev 1956; Takhtajan 1974).[21] The Northeastern Siberian Province occupies a broad expanse of land from the western and southwestern foothills of the Verkhoyansk Range to the Okhotsk and Bering seas (with the exclusion of Kamchatka and areas related to the Arctic Province).

The monotypic genus *Gorodkovia* is endemic to the province, and there are many endemic species and subspecies (Yurtsev 1974), including the following:

Papaver rivale, Corydalis gorodkovii, Cardamine conferta, Arabis turczaninovii, Salix jurtsevii, S. khokhrjakovii, Androsace gorodkovii, Ribes kolymense, Saxifraga anadyrensis, S. multiflora, S. redowskii, Potentilla anadyrensis, and *P. tollii.* On the shores of the Okhotsk Sea (the Pyagin Peninsula and to the south), we find a considerable number of Okhotsk-Kamchatka elements.

For the most part, the territory of this province is dominated by open and not very tall larch forests (reaching 10–15 m). The highest elevation of these forests is usually 600–700 m above sea level, but this varies. In the northern forest regions they only occur up to 200–400 m, but in the less cold southern regions they occasionally grow at elevations of 1100–1400 m. In the understory of larch forests the following are often found: low-growing birches, *(Betula rotundifolia, B. middendorfii,* and *B. fruticosa), Salix* spp., *Alnus fruticosa, Juniperus communis* subsp. *nana, Rhododendron parvifolium,* and *R. adamsii.* The valley forests differ from these open forests of rocky, mountainous slopes. The valley forests consist primarily of *Populus suaveolens* and *Chosenia arbutifolia.* Above the forest zone there are thick growths of cedar, prostrate dwarf stone pine *(Pinus pumila),* or shrublike alders. These gradually give way to mountain tundra, in which sedge-grass alpine meadows form scattered communities.

14. **Okhotsk-Kamchatka Province** (Schischkin 1947; Shumilova 1962).[22] This province includes Kamchatka (without its forest tundra regions); Karagin Island; the western shores of the Okhotsk Sea (Okhotia) and adjacent islands; the western shores of the Tatar Straits to the south, approximately to the Bay of Soviet Gavan; the lower Amur River; northern Sakhalin (south approximately to 51° 30′ north latitude); the Kurile Islands northward from the so-called line of Miyabe;[23] the Commander Islands; and the western and middle Aleutian Islands, up to approximately 165° west longitude to the east.

An isolated population of *Abies nephrolepis* is preserved in a small grove of about 20 hectares in Kamchatka, but the species is more widely distributed in the continental part of the province, where to the north it occurs up to the southern shores of the Okhotsk Sea, and to the west up to the basin of the Zeya River. It also grows in Primorski Krai, in northeastern China, and in the southern part of the Korean Peninsula. Other woody species found in the province are *Picea jezoensis, Clematis (Atragene) ochotensis, Betula ermannii, B. middendorfii, Alnus crispa* subsp. *sinuata, A. hirsuta, Populus tremula, P. suaveolens, Chosenia arbutifolia,* and *Salix* spp. In the flora of the lower Amur we find many Manchurian elements (representatives of the Eastern Asiatic flora; see the section on that region). To the north, the flora of Okhotia has many elements in common with the flora of Kamchatka and especially that of the Commander Islands, which are rich with Beringian–North-American elements.

This province contains one endemic genus (*Miyakea*) and a number of endemic species, including:

Aconitum ochotense (Okhotia), *Miyakea integrifolia* (Sakhalin), *Pulsatilla magadanensis* (Okhotia), *Corydalis gorodkovii* (Okhotia), *Cerastium aleuticum* (Aleutian Islands), *Silene ajanensis* (Okhotia, Sakhalin), *Alnus crispa* subsp. *sinuata* (Kamchatka, Okhotia, northern Kurile Islands), *Salix berberifolia* subsp. *kamtschatica, Cardamine pedata* (Okhotia), *Smelowskia tilingii* (Okhotia), *Sedum stephanii* (Okhotia), *S. purpurascens, Oxytropis ajanensis* (Okhotia), *O. erecta* (Kamchatka), *O. kamtschatica* (Kamchatka, northern Kurile Islands), *O. tilingii* (Okhotia), *O. trautvetteri* (Okhotia), *Conioselinum victoris* (northern Okhotia), *Gentiana sugawarae* (Sakhalin), *Valeriana ajanensis* (Okhotia), *Arnica sachalinensis* (Okhotia, Sakhalin), *Senecio* spp., *Saussurea* spp., *Cirsium kamtschaticum, Achillea alpina* subsp. *kamtschatica, Artemisia* spp., *Leontopodium kamtchaticum, Taraxacum* spp., *Eleocharis globularis* (Kamchatka, Commander Islands, lower Amur Basin), *Carex flavocuspis* (Kamchatka, Kurile Islands, northern Sakhalin), *C. pyrophila* (Kamchatka, Commander Islands, northern Kurile Islands), *C. ramenskii* (Kamchatka, Okhotia, Sakhalin), *Fimbristylis ochotensis* (Kamchatka), and *Poa shumushuensis* (Kamchatka and northern Kurile Islands).

15. Canadian Province (Engler 1899, 1903, 1924; Dice 1943; Alekhin 1944; Good 1947, 1974; Gleason and Cronquist 1964; Takhtajan 1974).[24] The Canadian Province forms a broad band across Canada and Alaska, south of the Arctic Province and north of the Rocky Mountain and North American Atlantic regions. It also includes a considerable part of Maine, a smaller part of northern Minnesota, and minimal northern parts of Vermont and Michigan.

Aside from apomictic microspecies in such genera as *Antennaria* and *Taraxacum*, there do not appear to be more than about two dozen endemic species of vascular plants in the Canadian Province. Most of these are in the

Alaska-Yukon area, where mountains rise well above the lowlands, providing a variety of habitats not well represented in the central part of the province. Among such western endemic species are *Claytonia bostockii, Silene williamsii, Smelowskia pyriformis, Douglasia gormanii, Astragalus williamsii, Podistera yukonensis, Phacelia mollis, Cryptantha shacklettiana, Castilleja annua, Synthyris borealis, Aster yukonensis, Haplopappus macleanii,* and *Agropyron yukonense. Stellaria alaskana, Castilleja yukonis, Campanula aurita,* and a few other species might be added to the list if small marginal overlaps into adjacent provinces are ignored. There are also a few other described species of doubtful taxonomic status, which will be endemics if their specific rank is confirmed.

At the eastern end of the Canadian Province, in southeastern Canada and adjacent Maine, there are also a few endemic or very nearly endemic species. These include *Amelanchier fernaldii, Scutellaria churchilleana* (a vegetatively reproducing hybrid between the more widespread species *S. galericulata* and *S. lateriflora*), *Pedicularis furbishiae* (a rare species of New Brunswick and northern Maine, whose status as an endangered species played an important role in preventing the construction of a large dam), *Aster crenifolius, Hieracium robinsonii,* and *Solidago calcicola.*

Although the western end and, to a lesser extent, the eastern end of the Canadian Province are mountainous, the broad central heartland is a land of low relief, with many lakes and slow streams, and extensive boggy areas covered with *Sphagnum.* The number of kinds of plants that are adapted to such conditions is not large, and in North America only the Arctic Province has fewer species. A considerable number of the species (although not the trees) are circumboreal. Some others also occur in the Arctic Province, and a great many extend southward at increasing elevations into the Rocky Mountain and/or Appalachian provinces. Thus we may say that the Canadian Province is an area of widespread species and little endemism. There are no endemic genera and only a handful of endemic species.

The Canadian Province is dominated by coniferous forest, similar in aspect to the coniferous forest in the last seven provinces of the Circumboreal Region just discussed. The trees are mostly rather small—seldom over 20 m tall or with a trunk more than 6 decimeters thick—but they often form very dense cover.

The most characteristic tree of the Canadian Province is *Picea glauca* (white spruce). East of the cordillera, it is commonly associated with *Abies balsamea* (balsam fir), and one thinks immediately of spruce-fir forest when the Canadian Province is mentioned. *Larix laricina* and *Picea mariana* are also common in the Canadian Province, especially in the wetter sites. Drier places often support a forest consisting mainly of *Pinus banksiana,* which—like its cordilleran relative, *P. contorta*—is frequently a fire-tree. Like *Abies balsamea, Pinus banksiana* occurs wholly to the east of the western cordillera.

Three species of broad-leaved trees also occur in some abundance in the

Canadian Province. These are *Betula papyrifera, Populus tremuloides,* and *Populus balsamifera.* They are especially common in burned or otherwise distributed areas, but they often occur in moister sites than *Pinus banksiana.* In the absence of disturbance, all of these species tend to be replaced eventually by spruce and fir.

Picea glauca is closely related to *P. abies*—a larger species of Europe and Siberia—and to *P. engelmannii* of the Rocky Mountain Province, but it has smaller cones than either of these species. *Abies balsamea* is closely related to *A. sibirica* of Eurasia and to *A. lasiocarpa* of the Rocky Mountains. *Larix laricina* is closely related to *L. gmelinii* of Eurasia. *Betula papyrifera* is closely related to *B. pubescens* of Eurasia, and *Populus tremuloides* is closely related to the Eurasian *P. tremula* and to the Appalachian *P. grandidentata.*

In the denser forests of the Canadian Province there is very little undergrowth except for mosses and mycorhizal herbs. Where the forest is more open, or around the margins of ponds and streams, there are a number of shrubs. Many of these are circumboreal, or are closely related to Eurasian species. Among the common and widespread circumboreal shrubby species (under a broad specific concept) are *Juniperus communis, Myrica gale, Arctostaphylos uva-ursi, Cassandra (Chamaedaphne) calyculata, Chimaphila umbellata, Vaccinium oxycoccos, V. uliginosum, V. vitis-idaea, Salix glauca, Ribes triste* (to eastern Asia only), and *Potentilla fruticosa.* Some other shrubs of the Canadian Province are so closely related to Eurasian species that some authors prefer to treat them at an infraspecific rank. Among these are *Alnus crispa* (close to *A. incana), Betula glandulosa* (close to *B. rotundifolia), Salix planifolia* (close to *S. phylicifolia), Cornus sericea* (close to *C. alba), Lonicera villosa* (close to *L. caerulea),* and *Sambucus pubens* (close to *S. racemosa).* Some other species, such as *Kalmia polifolia, Ribes glandulosum, Ribes lacustre,* and *Viburnum edule,* are more distinctly American.

The herbaceous flora of the Canadian Province likewise consists of a mixture of circumboreal species, species closely allied to Eurasian species, and wholly American species. Like the woody species, the characteristic widespread herbs generally extend into one or more of the adjoining American provinces in suitable habitats. Among the more or less circumboreal species (again with the benefit of a broad definition in some cases) are *Coptis trifolia, Stellaria longipes, Arenaria lateriflora, Moneses uniflora,* several species of *Pyrola, Rubus chamaemorus, Cornus suecica, Oxalis acetosella* (absent from the western half of the province; the American plants are often treated as a distinct species *O. montana), Linnaea borealis, Calypso bulbosa, Corallorhiza trifida, Listera cordata,* and a number of species of *Potamogeton, Juncus, Luzula, Carex, Eriophorum, Scirpus,* and *Calamagrostis.*

Some of the more distinctly American elements in the herbaceous flora are *Nuphar variegatum, Anemone canadensis, Cornus canadensis, Pedicularis groenlandica, Solidago multiradiata, Maianthemum canadense, Smilacina trifolia,*

Cypripedium acaule, Habenaria obtusata, Listera borealis, and species of a number of genera of grasses and sedges.

Most of the Canadian Province was repeatedly glaciated during the Pleistocene. Only a portion of Alaska and the Yukon appears to have escaped glaciation and provided a refugium for plants capable of withstanding a frigid climate. Otherwise, the flora—along with that of the Arctic Province to the north—was wiped out, and the area had to be revegetated from the south. There has obviously been a strong post-Pleistocene interchange between the American and Eurasian portions of the Circumboreal Region, and doubtless there was similar interchange during the interglacial periods. The bulk of the present flora of northern Eurasia, however, as well as that of northern North America, must have immigrated from the south along the retreating ice-front.

It was at one time thought that portions of eastern Quebec, notably the higher parts of the Gaspé Peninsula, also escaped glaciation. Fernald (1925) based his well-known nunatak theory on that supposition, which he used to explain the occurrence in the Gaspé region of a considerable number of basically cordilleran species. Fernald also considered that many of the species of the Gaspé Peninsula were local endemics, with their nearest relatives in the Rocky Mountains.

Time has not been kind to the nunatak theory. It now appears that the whole of the Gaspé Peninsula, as well as the rest of Quebec, was fully glaciated, and that no refugium existed there. Scarcely any of the supposed local endemics are now generally accepted as such. Either they are disjunct local populations of predominantly cordilleran species (as for example *Agoseris aurantiaca,* including *A. gaspensis*); or they have a somewhat wider—even if fairly restricted—range (as for example *Amelanchier fernaldii,* including *A. gaspensis*); or they do not appear distinctly different from more widespread species on more thorough study.

There is no denying that a number of cordilleran species occur also on the Gaspé Peninsula, or in northern Michigan, or on the north side of Lake Superior in northeastern Minnesota and adjacent Ontario, or in two or all three of these areas. *Arnica cordifolia,* for example, occurs in northern Michigan, and *Adenocaulon bicolor* occurs in northeastern Minnesota as well as in northern Michigan. Neither of these basically cordilleran species reaches Quebec. *Polystichum scopulinum* is likewise cordilleran but also occurs on Mt. Albert on the Gaspé Peninsula.

Arnica lonchophylla may provide a particularly instructive example. It occurs in four disjunct areas: (1) the Rocky Mountains, and more especially the intermontane valleys of Alberta and adjacent British Columbia, and far northward in the interior lowlands of western Canada; (2) northeastern Minnesota and adjacent Ontario, and northeast to the vicinity of James Bay; (3) Newfoundland to the Gaspé Peninsula of Quebec; and (4) the Black Hills of South Dakota and the east slope of the Big Horn Mountains in Wyoming.

Some authors have seen as many as seven species here, but only the Black Hills–Big Horn Mountains population now appears sufficiently distinctive to warrant infraspecific segregation.

It is now generally believed that the conditions on the moraines along the retreating ice-front during the wane of the Wisconsin glaciation provided an opportunity for rapid eastward migration of a number of cordilleran species. Further amelioration of the climate permitted Appalachian and Prairie species to outcompete these cordilleran travelers, except in sites particularly favorable to the latter. The greatest concentration of such sites is in the mountainous Gaspé Peninsula, but there are some others, especially near the shores of Lake Superior.

It is doubtless significant that cordilleran species isolated on the Gaspé Peninsula occur mainly on limited outcrops of basic rock (serpentine, dolomite, limestone), and not on the surrounding granitic rocks, although they are not necessarily so limited in the West (see Wynne-Edwards 1937).

There is a broad transition zone between the Canadian Province and the Arctic Province. As one goes northward in this "land of little sticks," the trees are progressively smaller and more confined to protected sites. Locally the border may be abrupt, but generally there is no sharp arctic tree line comparable to the timberline of mountains farther south.

Many circumboreal species characteristic of the Arctic Province extend south at increasing elevations or in exposed habitats to the southern limits of the Canadian Province; some of them even extend into the Rocky Mountain or the Appalachian Province, or both. Among the more conspicuous of such tundra species in the Canadian Province are *Silene acaulis, Harrimanella hypnoides, Loiseleuria procumbens, Rhododendron lapponicum, Diapensia lapponica, Saxifraga oppositifolia, Epilobium latifolium,* and *Campanula uniflora.*

On its southern margin the Canadian Province adjoins four other provinces (in two different regions). None of the boundaries is sharp, and the delimitation of the Canadian Province from the Rocky Mountain and Appalachian provinces is particularly vexing.

The northern part of the Rocky Mountain Province resembles a southern extension of the Canadian Province. Spruce and fir are still the most common trees, although the species are different. Farther south the Rocky Mountain Province is much more complex, containing many elements that have little or nothing to do with the boreal forest. Southward the spruce-fir forests are progressively confined to the upper elevations. As one climbs a mountain anywhere in the Rockies, the uppermost zone of forest, just below timberline, is likely to be dominated by species of spruce and fir closely related to the *Picea glauca* and *Abies balsamea* of the north. The representation of boreal species in the understory of these Rocky Mountain spruce-fir forests diminishes progressively toward the south.

A similar but more attenuated arm of the boreal forest extends southward at increasing elevations in the Appalachian Mountain area as far south as the Great Smoky Mountains of North Carolina and Tennessee. Here, as in the western mountains, the dominant genera are still *Picea* and *Abies,* but again, the species are different (this Appalachian extension of the boreal forest is discussed further under the Appalachian Province of the North American Atlantic Region).

2. Eastern Asiatic Region

Diels 1901, 1908, 1918; Wulff 1944; Turrill 1959; Schmithüsen 1961; Mattick 1964; Fedorov (in Grubov and Fedorov) 1964; Takhtajan 1970, 1974; Tolmatchev 1974.[25]

This region includes the eastern Himalayas (extending eastward from approximately 83° east longitude); several parts of the northeastern border regions of India; mountainous northern Burma; montane northern Tonkin; a considerable part of continental China and the Island of Taiwan; the Korean Peninsula; the Ryukyu Islands; the Islands of Kyushu, Shikoku, Honshu, Hokkaido, Bonin, and Volcano; the southern Islands of the Kurile series, to the south from the so-called line of Miyabe; the south and central parts of Sakhalin Island to the south from 51°30' north latitude; Primorski Krai and a considerable part of the basin of the Amur River; the southeastern part of the Transbaikalia; and pieces in the northeast and eastern extremes of Mongolia.

The flora of the Eastern Asiatic Region is extremely rich and distinctive, including more than 20 endemic families and over 300 endemic genera, not to mention the huge number of endemic species. The endemic families are Ginkgoaceae, Sciadopityaceae, Cephalotaxaceae, Nandinaceae, Glaucidiaceae, Circaeasteraceae, Trochodendraceae, Tetracentraceae, Cercidiphyllaceae, Eupteleaceae, Eucommiaceae, Rhoipteleaceae, Stachyuraceae, Dipentodontaceae, Kirengeshomaceae, Bretschneideraceae, Davidiaceae, Aucubaceae, Toricelliaceae, Helwingiaceae, Trapellaceae, and Funkiaceae. Among the endemic or almost endemic genera are the following:

Marattiaceae: *Archangiopteris* (10, Yunnan, Tonkin, Taiwan)
Pteridaceae: *Pleurosoriopsis* (1, Primorski Krai, Japan, Korea, northern China), *Sinopteris* (2, continental China)
Polypodiaceae: *Drymotaenium* (2, Japan, continental China, Taiwan), *Saxiglossum* (1, continental China)
Aspleniaceae: *Ceterachopsis* (2, eastern Himalayas to western China), *Cheilanthopsis* (1, northeastern Himalayas, western China), *Lithostegia* (1, eastern Himalayas to southwestern China), *Phanerophlebiopsis* (4, central China; very close to *Polystichum*)

Ginkgoaceae: *Ginkgo* (1, eastern China)
Cephalotaxaceae: *Cephalotaxus* (6–7, from Assam to Vietnam, Taiwan, Korea and Japan)
Taxaceae: *Pseudotaxus* (1, eastern China)
Pinaceae: *Cathaya* (1, southern and western China), *Pseudolarix* (1, eastern China)
Sciadopityaceae: *Sciadopitys* (1, Japan)
Taxodiaceae: *Cryptomeria* (1, Japan, continental China), *Cunninghamia* (2–3, western and southern China, Taiwan), *Metasequoia* (1, central China), *Taiwania* (2, northern Burma, Yünnan, Taiwan)
Cupressaceae: *Microbiota* (1, southern part of Sikhote-Alin), *Thujopsis* (1, Japan)
Magnoliaceae: *Alcimandra* (1, eastern Himalayas, Khasi Hills and Naga Hills to North Vietnam), *Tsoongiodendron* (1, southern China: Kwangsi, Kiangsi, Fukien, Hainan, Tonkin)
Calycanthaceae: *Chimonanthus* (3, continental China)
Saururaceae: *Gymnotheca* (2, continental China: Szechuan)
Aristolochiaceae: *Saruma* (1, northwestern and southwestern China)
Lardizabalaceae: *Akebia* (4–5, continental China, Taiwan, Korea, Japan), *Decaisnea* (2, eastern Himalayas, Tibet, continental China), *Holboellia* (8–10, Himalayas, continental China to Indochina), *Sinofranchetia* (1, west and central China), *Stauntonia* (15, Assam, Burma, continental China, Taiwan, Ryukyu Islands, Japan, Korea, 1 species on Hainan)
Sargentodoxaceae: *Sargentodoxa* (1, continental China, northern Laos, Tonkin)
Menispermaceae: *Sinomenium* (1, continental China, Korea, Japan, Ryukyu Islands)
Berberidaceae: *Ranzania* (1, Japan)
Nandinaceae: *Nandina* (1, continental China, Japan)
Circaeasteraceae: *Kingdonia* (1, China), *Circaeaster* (1, from the Himalayas across southeastern Tibet to western China)
Ranunculaceae: *Anemonopsis* (1, Japan), *Asteropyrum* (2, continental China), *Beesia* (3, northern Burma, continental China), *Calathodes* (3, eastern Himalayas, continental China, Taiwan), *Megaleranthis* (1, South Korea), *Paroxygraphis* (1, eastern Nepal to Bhutan), *Schlagintweitiella* (2, Tibet, western China), *Semiaquilegia* (7), *Souliea* (1, eastern Himalayas, northern Burma, continental China), *Urophysa* (2, continental China)
Glaucidiaceae: *Glaucidium* (1, Japan)
Papaveraceae: *Eomecon* (1, continental China), *Hylomecon* (3, continental China, Japan, Korea; very close to *Chelidonium*), *Macleaya* (2, continental China, Taiwan, Japan), *Pteridophyllum* (1, Japan)
Polygonaceae: *Pteroxygonum* (1, continental China)
Trochodendraceae: *Trochodendron* (1, Japan, South Korea, Ryukyu Islands, Taiwan)

Tetracentraceae: *Tetracentron* (1, from eastern Nepal to northern Burma, southwestern and central China)
Cercidiphyllaceae: *Cercidiphyllum* (2, continental China, Japan and Kunashir Island)
Eupteleaceae: *Euptelea* (2, eastern Himalayas, Assam, Mishmi Hills, central and southwestern China, Japan)
Eucommiaceae: *Eucommia* (1, continental China: provinces situated along the middle course of the Yangtze River)
Hamamelidaceae: *Corylopsis* (7–10, eastern Himalayas, Khasi Hills, Manipur, eastern Asia), *Disanthus* (1, central China, Japan), *Eustigma* (3–4, southern China, Taiwan, northern Indochina), *Fortunearia* (1, central and eastern China), *Loropetalum* (3–4, eastern Himalayas, Khasi Hills, southern China, Japan), *Sinowilsonia* (1, central China), *Tetrathyrium* (1, Hong Kong)
Rhoipteleaceae: *Rhoiptelea* (1, southwestern China, Tonkin)
Juglandaceae: *Cyclocarya* (1, continental China; very close to *Pterocarya* and probably congeneric), *Platycarya* (1, Japan, continental China and Taiwan)
Stachyuraceae: *Stachyurus* (16, eastern Himalayas, Assam, northern Burma, continental China, Taiwan, Japan, northern Indochina)
Ochnaceae: *Sinia* (1, southeastern China)
Actinidiaceae: *Clematoclethra* (11–12, western, southwestern, and central China)
Ericaceae: *Diplarche* (2, eastern Himalayas and southwestern China), *Enkianthus* (10, from eastern Himalayas to Japan, 1 species on Hainan), *Monotropastrum* (4, eastern Himalayas, Khasi Hills, southeastern Tibet, northern Burma, continental China, Taiwan, Okinawa, Korean Peninsula, Japan, Sakhalin, Primorski Krai), *Tripetaleia* (2, Sakhalin, Kurile Islands, Japan), *Tsusiophyllum* (1, Japan)
Diapensiaceae: *Berneuxia* (2, eastern Tibet, northern Burma, southwestern China)
Styracaceae: *Alniphyllum* (3, southwestern and southern China, Taiwan, northern Indochina), *Huodendron* (6, northeastern India [NEFA], southeastern Tibet, southern China, Burma, northern Indochina), *Melliodendron* (3, southern and southwestern China), *Pterostyrax* (4, northern Burma, continental China, Japan), *Rehderodendron* (6–8, southern and western China, northern Indochina), *Sinojackia* (3, continental China)
Primulaceae: *Bryocarpum* (1, eastern Himalayas, Assam, southern Tibet), *Omphalogramma* (15, eastern Himalayas and Assam to western China and northern Burma), *Pomatosace* (1, northwestern China), *Stimpsonia* (1, Japan, Okinawa, central and eastern China, Taiwan)
Flacourtiaceae: *Carrierea* (3, western and southern China, Tonkin), *Idesia* (1, Japan, Korea, Ryukyu Islands, central and western China, Taiwan), *Poliothyrsis* (1, continental China)

Dipentodontaceae: *Dipentodon* (1, eastern Himalayas, northeastern India, Upper Burma, southeastern Tibet, southwestern China to Kweichow and Kwangsi Chuang)

Cucurbitaceae: *Biswarea* (1, eastern Himalayas, Nepal, Sikkim), *Bolbostemma* (1-2, continental China), *Edgaria* (1, Himalayas: Gorhwal to Bhutan), *Neoluffa* (1, eastern Himalayas), *Herpetospermum* (1, Himalayas, southern Tibet, Assam, southern China), *Schizopepon* (4, Primorski Krai, the basin of Amur River, Sakhalin; Japan, Korea, continental China, eastern Himalayas, Assam, northern Burma)

Brassicaceae: *Berteroella* (1, Japan, Korea, northern China), *Coelonema* (1, southwestern China), *Chrysobraya* (1, from eastern Nepal to Bhutan; close to *Braya*), *Dipoma* (2, southeastern Tibet, southwestern China), *Hemilophia* (2, southwestern China), *Lepidostemon* (1, eastern Himalayas), *Lignariella* (1, eastern Himalayas, Tibet), *Loxostemon* (5, eastern Himalayas to western China), *Neomartinella* (1, continental China), *Parryodes* (1, southern Tibet, eastern Himalayas), *Platycraspedum* (1, eastern Tibet), *Solms-Laubachia* (9, eastern Himalayas, Tibet, southwestern China), *Synstemon* (1, northwestern China), *Staintoniella* (2, Nepal, Tibet, southwestern China)

Sterculiaceae: *Corchoropsis* (3, continental China, Korean Peninsula, Japan), *Craigia* (2, southwestern China, Tonkin; sometimes included in the Tiliaceae), *Reevesia* (15, Himalayas to Taiwan)

Ulmaceae: *Hemiptelea* (1, northern China, Korea), *Pteroceltis* (1, northern China, Mongolia)

Urticaceae: *Nanocnide* (3), *Smithiella* (1, eastern Himalayas)

Euphoribiaceae: *Speranskia* (3, continental China)

Thymelaeaceae: *Edgeworthia* (2-3, from Nepal to Bhutan, Khasi Hills, northern Burma, western China), *Pentathymelaea* (1, eastern Tibet)

Saxifragaceae: *Astilboides* (1, northern China, northern Korea), *Mukdenia* (2, northern China, Korean Peninsula), *Oresitrophe* (1, northern China), *Peltoboykinia* (2, Japan; close to *Boykinia*), *Rodgersia* (5-6), *Tanakea* (2, continental China, Japan)

Rosaceae: *Chaenomeles* (5, continental China, Japan), *Dichotomanthes* (1, southwestern China), *Kerria* (1, continental China, Japan), *Maddenia* (4, eastern Himalayas, Mishmi Hills, southern Tibet, continental China), *Pentactina* (1, Korea), *Plagiospermum* (2, Primorskī Krai; north and northwestern China, Mongolia; very close to *Prinsepia*), *Prinsepia* (3-4, Himalayas and Khasi Hills to northern China and Taiwan), *Rhaphiolepis* (14, continental China, Taiwan, Ryukyu Islands, Korea, Japan, Bonin Islands), *Rhodotypos* (1, continental China, Korea, Japan), *Spenceria* (2, western China), *Stephanandra* (4, continental China, Taiwan, Korea, Japan)

Lythraceae: *Orias* (1, western China)
Melastomataceae: *Barthea* (2, southern China, Taiwan), *Cyphotheca* (1, southwestern China), *Plagiopetalum* (4, continental China), *Styrophyton* (1, southwestern China)
Fabaceae: *Cochlianthus* (2, central Nepal to southwestern China), *Craspedolobium* (1, southwestern China), *Lysidice* (1, southern China, Taiwan, Tonkin), *Maackia* (10, Primorski Krai, the basin of Amur River, Shikatan, Japan, continental China, Taiwan, Korea, Ryukyu Islands), *Neodielsia* (1, continental China), *Piptanthus* (9, from Simla to Bhutan, Assam, Burma, western and southwestern China), *Salweenia* (1, southeastern Tibet)
Rutaceae: *Boninia* (2, Bonin Islands), *Orixa* (1, Japan), *Phellodendron* (10, Primorski Krai, the basin of the Amur River, southern Sakhalin, southern Kuriles, Japan, Korea, continental China, Taiwan), *Poncirus* (1, central and northern China, Korea), *Psilopeganum* (1, central China), *Skimmia* (8–9, Himalayas in the west to Afghanistan, Khasi Hills, continental China, Taiwan, Ryukyu Islands, Japan, southern Sakhalin, southern Kurile Islands, 1 species on Luzon)
Podoaceae: *Dobinea* (2, Nepal to Bhutan, Assam, northern Burma, southern China)
Staphyleaceae: *Euscaphis* (4, Japan, Korea, continental China, Taiwan, Hainan, Ryukyu Islands), *Tapiscia* (2, southwestern and central China)
Sapindaceae: *Delavaya* (1, southwestern China), *Eurycorymbus* (1, southern China, Taiwan), *Handeliodendron* (1, continental China), *Xanthoceras* (1, northern China)
Aceraceae: *Dipteronia* (2, central and southern China)
Bretschneideraceae: *Bretschneidera* (1, continental China)
Linaceae: *Anisadenia* (2, Kumaun to Bhutan and Khasi Hills to central China), *Tirpitzia* (2, southwestern China, Tonkin; close to *Reinwardtia*)
Icacinaceae: *Hosiea* (2, western and central China, Japan)
Celastraceae: *Monimopetalum* (1, continental China), *Sinomerrillia* (1, southwestern China), *Tripterygium* (3, continental China, Taiwan, Japan)
Rhamnaceae: *Hovenia* (probably monotypic; continental China, Korea, Japan)
Escalloniaceae: *Pottingeria* (1, Naga Hills, Upper Burma, northwestern Thailand; stands apart in the family and in several characteristics calls to mind the Celastraceae)
Hydrangeaceae: *Cardiandra* (4, central China, Taiwan, Japan), *Deinanthe* (2, central China, Japan), *Pileostegia* (2, eastern Himalayas and Khasi Hills to continental China, Taiwan and Ryukyu Islands), *Platycrater* (2, continental China, Japan), *Schizophragma* (3, continental China, Taiwan, Korea, Japan)

Kirengeshomaceae: *Kirengeshoma* (1, Japan, Korea and continental China)
Davidiaceae: *Davidia* (1, Tibet, western and central China)
Nyssaceae: *Camptotheca* (1, Tibet, continental China)
Aucubaceae: *Aucuba* (5, eastern Himalayas, Assam, Manipur, northern Burma, continental China, Ryukyu Islands, Taiwan, Korea, Japan)
Toricelliaceae: *Toricellia* (3, from western Nepal to Bhutan, northern Burma and continental China)
Caprifoliaceae: *Dipelta* (4, western and central China), *Heptacodium* (2, central and eastern China), *Kolkwitzia* (1, central China), *Leycesteria* (6, from Kashmir to Bhutan, Assam, northern Burma, southeastern Tibet, western China), *Silvianthus* (2, from Assam to southwestern China and Burma), *Weigela* (14, Primorski Krai, the basin of Amur River, Okhotia, Sakhalin, Kurile Islands, Japan, Korea, continental China; very close to *Diervilla*)
Adoxaceae: *Sinadoxa* (1, China), *Tetradoxa* (1, China)
Helwingiaceae: *Helwingia* (4, from central Nepal to Bhutan, Khasi Hills, Manipur, northern Burma, continental China, Taiwan, northern Tonkin, Okinawa, Japan)
Araliaceae: *Diplopanax* (1, continental China), *Evodiopanax* (2, continental China, Japan), *Fatsia* (2, Japan, Taiwan), *Gamblea* (2, from eastern Nepal to Bhutan, southeastern Tibet, Burma), *Kalopanax* (2, Primorski Krai, Sakhalin, Kurile Islands, Japan, Korea, continental China, Ryukyu Islands), *Merrilliopanax* (1–2, from eastern Nepal to Bhutan, Assam, southwestern China), *Sinopanax* (1, Taiwan), *Tetrapanax* (1, western and southern China, Taiwan), *Woodburnia* (1, northern Burma)
Apiaceae: *Acronema* (15, from western Nepal to Bhutan, Assam, northern Burma, western China), *Apodicarpum* (1, Japan), *Carlesia* (1, continental China), *Chaerophyllopsis* (1, western China), *Chamaele* (1, Japan), *Chamaesium* (7, eastern Himalayas, Tibet, western China), *Changium* (1, eastern China; close to *Conopodium*), *Cortiella* (2, central Nepal to Bhutan, Tibet), *Dickinsia* (1, southwestern China), *Halosciastrum* (1, Primorski Krai, Korea, and possibly northeastern China), *Haploseseli* (1, western China), *Haplosphaera* (1, southwestern China), *Harrysmithia* (1, southwestern China), *Macrochlaena* (central China), *Melanosciadum* (1, western China), *Nothosmyrnium* (1, continental China), *Notopterygium* (4, continental China), *Oreochorte* (1, continental China; close to *Anthriscus*), *Pternopetalum* (26, eastern Himalayas, Assam, Tibet, continental China, Tonkin, Korea, Japan), *Pterygopleurum* (1, Korea, Japan), *Sinocarum* (8, continental China), *Sinodielsia* (1, southwestern China), *Sinolimprichtia* (1, eastern Tibet)
Rubiaceae: *Damnacanthus* (5, Assam, continental China, Taiwan, Thailand, Ryukyu Islands, Korea, Japan), *Dunnia* (1, southeastern China), *Luculia* (5, from western Nepal to Assam, Tibet, northern Burma and south-

western China, Tonkin), *Pseudopyxis* (2, Japan), *Trailliaedoxa* (1, southwestern China)
Apocynaceae: *Chunechites* (1, southeastern China), *Formosia* (1, Taiwan), *Neohenrya* (2, continental China)
Asclepiadaceae: *Aphanostelma* (1, continental China), *Biondia* (6, continental China), *Metaplexis* (6, Primorski Krai, the basin of Amur River, southern Kuriles, Japan, Korea, continental China), *Treutlera* (1, eastern Nepal to Bhutan, Khasi Hills)
Gentianaceae: *Latouchea* (1, eastern China), *Megacodon* (1, eastern Nepal to Assam, southern Tibet, southwestern China), *Parajaeschkea* (1, eastern Himalayas), *Veratrilla* (2, eastern Himalayas, Assam, western China)
Oleaceae: *Abeliophyllum* (1, Korea; close to *Fontanesia*)
Boraginaceae: *Ancistrocarya* (1, Japan), *Antiotrema* (1, southwestern China), *Brachybotrys* (1, Primorski Krai, northeastern China), *Chionocharis* (1, from eastern Nepal to Bhutan, southern Tibet, western China), *Henryettana* (1, southwestern China), *Pedinogyne* (1, eastern Himalayas), *Schistocaryum* (1, southwestern China), *Sinojohnstonia* (2, western China), *Thyrocarpus* (3, continental China)
Solanaceae: *Physaliastrum* (6, continental China, Korea, Japan)
Scrophulariaceae: *Calorhabdos* (4–5, from eastern Nepal to Taiwan), *Centrantheropsis* (2, continental China), *Deinostema* (2, northern China, Korea, Japan; close to *Gratiola*), *Monochasma* (4, continental China, Japan), *Oreosolen* (3, from western Nepal to Bhutan, Tibet), *Phacelanthus* (2, Primorski Krai, continental China, Korea, Japan), *Phtheirospermum* (7, Himalayas, continental China, Taiwan, Korea, Japan), *Pterygiella* (3, continental China), *Rehmannia* (7–8, continental China, Korea; several characteristics call to mind members of the Gesneriaceae, so placement in the Scrophulariaceae is somewhat questionable), *Scrofella* (1, northwestern China), *Tienmuia* (1, eastern China), *Triaenophora* (2, continental China)
Bignoniaceae: *Paulownia* (8–12, continental China, Taiwan; often referred to the Scrophulariaceae, from which it differs in several characters of the gynoecium), *Shiuyinghua* (1, central China; close to *Paulownia*)
Trapellaceae: *Trapella* (1, Primorski Krai, the basin of Amur River, Japan, Korea, continental China)
Gesneriaceae: *Ancylostemon* (5, Tibet, western China), *Briggsia* (about 23, eastern Himalayas, northern Burma, southern China), *Corallodiscus* (18, from Kumaun to Bhutan, Khasi Hills, Tibet, southwestern China, northern Indochina), *Isometrum* (2, eastern Tibet, continental China), *Loxostigma* (3, from eastern Nepal to the Khasi Hills, Manipur, northern Burma, western China, northern Indochina), *Opithandra* (6, continental China, Japan), *Oreocharis* (20, continental China, Japan), *Petrocodon* (1,

continental China), *Rhabdothamnopsis* (1, continental China), *Tengia* (1, southwestern China), *Titanotrichum* (1, southern China, Taiwan, southern Ryukyus), *Tremacron* (3, southwestern China), *Whytockia* (2, southwestern China, Taiwan)

Acanthaceae: *Codonacanthus* (2, Khasi Hills, southern China, Taiwan, Ryukyu Islands, Japan)

Verbenaceae: *Caryopteris* (15, western Himalayas to Japan and Taiwan)

Lamiaceae: *Bostrychanthera* (1, continental China), *Cardioteucris* (1, southwestern China), *Chelonopsis* (15, eastern Tibet, continental China, Taiwan, Japan), *Hanceola* (3, continental China), *Heterolamium* (1, continental China), *Holocheila* (1, southwestern China), *Keiskea* (5, continental China, Taiwan, Japan), *Leucosceptrum* (5, from Kumaun to Bhutan, northern Burma, continental China, Taiwan, Japan), *Loxocalyx* (2, continental China), *Ombrocharis* (1, continental China), *Paralamium* (1, southwestern China), *Schnabelia* (1, southwestern China), *Stiptanthus* (1, eastern Himalayas, Assam)

Campanulaceae: *Cyananthus* (30, from Garhwal to Bhutan, Assam, southern Tibet, northern Burma, western China), *Hanabusaya* (2, Korea), *Leptocodon* (1, from eastern Nepal to NEFA, northern Burma, western China), *Platycodon* (1, Primorski Krai, the basin of Amur River, continental China, Korea, Japan), *Popoviocodonia* (1, southeastern Khabarovski Krai, Primorski Krai, Sakhalin)

Asteraceae: *Atractylodes* (about 7, Primorski Krai, the basin of Amur River, continental China, Korea, Japan), *Callistephus* (1, Primorski Krai, continental China), *Cavea* (1, eastern Himalayas), *Cremanthodium* (55, Kashmir to Bhutan, Assam, Tibet, China), *Dendrocacalia* (1, Bonin Islands), *Diplazoptilon* (1, southwestern China), *Dubyaea* (10, Kumaun, Garhwal to Bhutan, Assam, northern Burma, southeastern Tibet, western China), *Faberia* (5, southwestern China), *Formania* (1, southwestern China), *Heteroplexis* (1, eastern China), *Miricacalia* (1, Japan), *Myripnois* (3, northern China), *Nannoglottis* (8, from western Nepal to Bhutan, Tibet, western China), *Nouelia* (1, southwestern China), *Sheareria* (1, eastern China), *Symphyllocarpus* (1, the basin of Amur River, northern China), *Syneilesis* (6, Primorski Krai, the basin of Amur River, Japan, Korea, northern and eastern China, Taiwan; close to *Cacalia*), *Takeikadzuchia* (1, continental China), *Vladimiria* (12, southwestern China), *Xanthopappus* (2, continental China)

Melanthiaceae: *Chionographis* (4, Japan, Cheju Island, continental China), *Heloniopsis* (2, Sakhalin, Japan, Korea, Ryukyu, Taiwan), *Japonolirion* (1, Japan), *Ypsilandra* (5, from central Nepal to Bhutan, Tibet, northern Burma, western China)

Uvulariaceae: *Tricyrtis* (10 or more, from central Nepal to Bhutan, Khasi Hills, northern Burma, continental China, Taiwan, Japan)

Iridaceae: *Belamcanda* (2, Primorski Krai, continental China, Japan)
Liliaceae s. str.: *Cardiocrinum* (3–4, Himalayas from Garhwal to NEFA, Khasi Hills, Manipur, northern Burma, continental China, Japan, Sakhalin, southern Kuriles)
Alliaceae: *Milula* (1, central Nepal, Tibet)
Funkiaceae: *Hosta* (about 40 or more, Primorski Krai, the basin of Amur River, Sakhalin; Kurile Islands, Japan, Korea, continental China)
Amaryllidaceae: *Lycoris* (10, eastern Himalayas, northern Burma, continental China, Korea, Japan, Ryukyu Islands, Taiwan)
Asphodelaceae: *Diuranthera* (2, western China)
Anthericaceae: *Alectorurus* (1, Japan), *Anemarrhena* (1, eastern Mongolia, northern and northeastern China, southern Korea), *Terauchia* (1, northern Korea)
Convallariaceae: *Aspidistra* (11, eastern Himalayas, Assam, northern Tonkin, southwestern and southern continental China, Hainan, Taiwan), *Campylandra* (9, from central Nepal to Bhutan, Mishmi, Naga Hills, Khasi Hills, western China, northern Tonkin), *Disporopsis* (6, continental China, Taiwan, northern Thailand), *Liriope* (8, Japan, Korea, continental China, Tonkin, Ryukyu Islands, Taiwan), *Reineckea* (1, continental China, Japan), *Rohdea* (3, continental China, Japan), *Speirantha* (1, eastern China), *Theropogon* (1, Himalayas, Khasi Hills, southwestern China)
Trilliaceae: *Kinugasa* (1, Japan)
Orchidaceae: *Aceratorchis* (2, Tibet, continental China), *Bletilla* (9, continental China, Taiwan, Ryukyu Islands, Japan, Korea), *Bulleyia* (1, eastern Himalayas to southwestern China), *Changnienia* (1, continental China), *Cremastra* (7, Sakhalin, southern Kurile, Japan, Himalayas), *Dactylostalix* (1, southern Kurile Islands, Japan), *Didiciea* (2, eastern Himalayas, Japan), *Diplolabellum* (1, Korea), *Eleorchis* (1, southern Kurile Islands, Japan), *Ephippianthus* (2, Khabarovski Krai, Primorski Krai, Sakhalin, Kurile Islands, Japan, Korea), *Hancockia* (1, continental China), *Ischnogyne* (1, continental China), *Neofinetia* (1, continental China, Japan, Ryukyu Islands), *Smithorchis* (1, continental China), *Symphyosepalum* (1, continental China), *Tangtsinia* (1, southwestern China), *Tsaiorchis* (1, continental China), *Vexillabium* (4, Korea, Japan), *Yoania* (3, eastern Himalayas, Naga Hills, Manipur, Japan, Taiwan)
Commelinaceae: *Streptolirion* (1, eastern Himalayas, Khasi Hills and Patkoi Hills to northern Indochina, Korea)
Poaceae: *Anisachne* (1, southwestern China), *Brachystachyum* (1, southeastern China), *Brylkinia* (1, Sakhalin, Kurile Islands, Korea, Japan, northeastern China), *Chikusichloa* (2, Japan, Ryukyu Islands, southwestern China), *Cyathopus* (1, eastern Himalayas), *Eccoilopus* (4, Himalayas, Khasi Hills, Naga Hills, Taiwan, Japan), *Fargesia* (2, continental China), *Hakonechloa* (1, Japan; close to *Phragmites*), *Oreocalamus* (2, western China), *Phaeno-*

sperma (1, Japan, Korea, continental China, Taiwan), *Phyllostachys* (about 50, Korea to southern China and Indochina), *Sasa* (about 50, Sakhalin, Kurile Islands, Japan, eastern China), *Shibataea* (3, eastern China, Japan, Okinawa), *Sinobambusa* (8, Upper Assam, Naga Hills, continental China, Okinawa), *Sinochasea* (1, continental China)
Arecaceae: *Satakentia* (1, southern Ryukyu Islands), *Trachycarpus* (6, Himalayas, Khasi Hills, Naga Hills, Manipur, northern Burma, continental China, southern Japan)
Araceae: *Pinellia* (7, continental China, Korea, Japan, Ryukyu Islands)

Thus the Eastern Asiatic flora is characterized by an exceptionally large number of endemic genera, which are in large part monotypic or oligotypic. These genera belong to very different families, both archaic and advanced. Yet many of these endemic genera belong to the archaic subclasses Magnoliidae, Ranunculidae, and the Hamamelididae. This underscores the antiquity of the flora, which contains not only many ancient relict endemics of various taxonomic ranks, but also a series of endemic families and even suborders.

The flora of the Eastern Asiatic Region holds special interest for the study of the history of temperate floras of the Northern Hemisphere, as well as for the solution of many problems of evolution and systematics of the flowering plants. The Eastern Asiatic Region appears simultaneously as one of the major centers of development of higher plants (especially the gymnosperms and the angiosperms), and as one of the centers of the preservation of ancient forms (gigantic refuges of "living fossils"). Here one finds many archaic and intermediate forms that are of great importance in the elucidation of phylogenetic problems. Many genera, and even families, that are well differentiated in the flora of the Circumboreal Region are here connected by intermediate taxa, and often merge.

1. Manchurian Province (Krasnov 1899; Krylov 1919; Good 1947, 1974; Schischkin 1947; Fedorov [in Grubov and Fedorov] 1964; Tamura 1966; Takhtajan 1970, 1974).[26] This province includes the southeastern part of the Khabarovski Krai, including the middle course of the Amur River; the Primorski Krai, in the USSR (the system of mountain ranges of Sikhote-Alin, and the region of Lake Khanka situated west of it, which extends in the south up to the Amur Gulf); the northern part of the Korean Peninsula; the basin of the middle and upper course of the Yalu River in the northeastern part of China; the southeastern part of the Onon-Argun steppes in the Transbaikalia; and localities in northeastern Mongolia (in the basins of the Uldza and Onon Rivers) and in the extreme east of Mongolia, in the foothills of Khingan Mountains. In China, the western boundary skirts around from

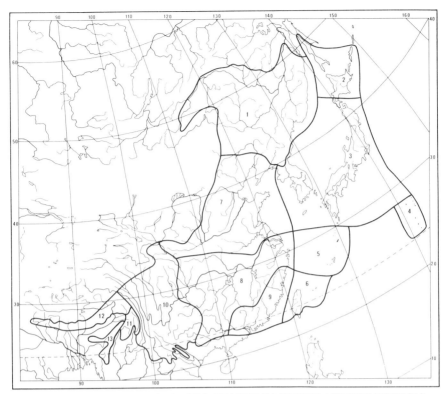

Map 3. FLORISTIC PROVINCES OF THE EASTERN ASIATIC REGION

1, Manchurian Province. 2, Sakhalin-Hokkaido Province. 3, Japanese-Korean Province. 4, Volcano-Bonin Province. 5, Ryukyu or Tokara-Okinawa Province. 6, Taiwanian Province. 7, North Chinese Province. 8, Central Chinese Province. 9, Southeastern Chinese Province. 10, Sikang-Yünnan Province. 11, Northern Burmese Province. 12, Eastern Himalayan Province. 13, Khasi-Manipur Province.

the west of the Great Khingan Mountains, proceeds further to the east from Shenyang (formerly Mukden), continues in an arch in a southeastern direction, intersects the Yalu River between 40° and 41° north latitude, and forms a tongue extending into the northern part of the Korean Peninsula to approximately the latitude of Pyongyang. From here the border turns to the north, approaching 40° latitude and passing along the Puch'ong Range, reaching the Sea of Japan somewhat north of the Cape of Orandan. In the Soviet Union, the border of the Manchurian Province occurs approximately along the line of the interlacing boundaries of such Eastern Asiatic (including Manchurian) species as *Pinus koraiensis, Schisandra chinensis, Corylus heterophylla, Juglans mandshurica, Quercus mongolica, Ulmus macrocarpa, Actinidia kolomikta, Tilia mandshurica, Phellodendron amurense, Acanthopanax senticosus, Vitis amurensis,* and *Fraxinus mandshurica.*

The rich woody flora of the Manchurian Province consists almost entirely of deciduous forms. Endemism in this province is rather high, particularly along the middle Amur, in the spurs of the Bureyan Uplands and the Little Khingan Mountains, and also along the shores of the northern part of the Sea of Japan (from Olga to Adimi). There are several endemic genera: *Microbiota* (the mountains of Sikhote-Alin), *Mukdenia, Astilboides, Halosciastrum, Brachybotrys, Omphalothrix, Hanabusaya* (Korean Peninsula), *Terauchia* (Korean Peninsula), *Anemarrhena,* and a very large number of endemic or subendemic species (and subspecies), of which I mention:

Abies holophylla, Larix olgensis, Picea obovata subsp. *koraiensis, P. manshurica, Thuja koraiensis, Microbiota decussata, Aristolochia manshuriensis, Ceratophyllum manshuricum, Trollius macropetalus, Eranthis stellata, Semiaquilegia manshurica, Cimicifuga heracleifolia, Delphinium maackianum, Aconitum alboviolaceum, A. arcuatum, A. crassifolium, A. desoulavyi, A. jaluense, A. kirinense, A. monanthum, A. paishanense, A. raddeanum, A. sichotense, Clematis koreana, C. mandschurica, C. serratifolia, Ranunculus amurensis, R. hainganensis, Epimedium macrosepalum, Gymnospermium microrrhinchum, Jeffersonia dubia, Adlumia asiatica, Corydalis buschii, C. watanabei, Dianthus morii, Silene capitata, S. koreana, S. olgae, Stellaria neo-palustris, Rheum coreanum* (Korea), *Girardinia cuspidata, Betula costata, Paeonia lactiflora, Viola amurica, V. diamantica, V. interposita, V. kusnezowiana, V. savatieri, V. yamatsutai, Megadenia speluncarum, Salix kangensis, S. maximowiczii, Rhododendron schlippenbachii, Primula saxatilis, Tilia amurensis, T. mandshurica, T. semicostata* (northern Korea), *Euphorbia hakutosanensis, E. komaroviana, E. lucorum, E. mandshurica, E. savaryi, Daphne koreana, Mukdenia acanthifolia, M. rossii, Astilbe koreana, Astilboides tabularis, Bergenia pacifica, Saxifraga manshuriensis, S. sichotensis, Sedum viviparum, Ribes komarovii, R. maximoviczianum, R. ussuriensis, Exochorda giraldii, E. serratifolia, Physocarpus amurensis, P. ribesifolius, Neillia uekii, Spiraea pubescens, S. ussuriensis, Aruncus parvulus, Sanguisorba magnifica, Fragaria concolor, Potentilla (Pentaphylloides) mandshurica, Pyrus corymbifera, P. ussuriensis* subsp. *ussuriensis, Sorbus manshurensis, S. schneideriana, Crataegeus komarovii, Prunus mandschurica, P. nakaii, Prinsepia sinensis, Keyserlingia koreensis* (Korea), *Caragana fruticosa, C. ussuriensis, Astragalus satoi, A. setsureianus, Oxytropis arnertii, Hedysarum ussuriense, Phellodendron amurense, Acer barbinerve, A. mandshuricum, A. pseudosieboldianum, A. tegmentosum, A. triflorum, Linum amurense, Geranium hattai, G. koreanum, G. maximowiczii, Deutzia amurensis, D. glabrata, Acanthopanax sessiliflorus, Oplopanax elatus, Aralia continentalis, Panax ginseng, Bupleurum euphorbioides, Halosciastrum melanotilingia, Angelica gigas, A. anomala, Peucedanum elegans, P. paishanense, Heracleum lanatum* subsp. *voroschilovii, Euonymus pauciflora,*

Rhamnus diamantiaca, *R. ussuriensis*, *Vitis amurensis*, *Galium manshuricum*, *G. platygalium*, *Rubia sylvatica*, *Vincetoxicum volubile*, *Ophelia tscherskyi*, *Gentiana komarovii*, *G. manshurica*, *Forsythia mandschurica*, *Syringa patula*, *S. reticulata* subsp. *amurensis*, *S. wolfii*, *Lonicera maximowiczii*, *Abelia coreana*, *Triosteum sinuatum*, *Valeriana subpennatifolia*, *Scabiosa lachnophylla*, *Polemonium racemosum*, *Brachybotrys paridiformis*, *Eritrichium mandshuricum*, *Bothriospermum decumbens*, *Minulus stolonifer*, *Scutellaria moniliorrhiza*, *Nepeta koreana*, *N. manchuriensis*, *Thymus przewalskii*, *Elsholtzia serotina*, *Plectranthus excisus*, *P. serra*, *Scrophularia amgunensis*, *S. mandshurica*, *Omphalothrix longipes*, *Pedicularis mandshurica*, *Campanula chinganensis*, *Adenophora crispata*, *A. gmelinii*, *A. grandiflora*, *Hanabusaya asiatica* (northern and central Korea), *Symphyllocarpus exilis*, *Leontopodium palibinianum*, *Chrysanthemum* spp., species of *Artemisia*, *Cacalia*, *Senecio*, *Ligularia* (including *L. splendens*), *Saussurea*, and *Cirsium*, *Serratula komarovii*, *Echinops dissectus*, *E. manshuricus*, *Taraxacum* spp., *Prenanthes blinii*, *Hieracium coreanum*, *Potamogeton mandschuriensis*, *Veratrum maackii*, *V. dahuricum*, *V. dolichopetalum*, *Fritillaria maximowiczii*, *F. ussuriensis*, *Gagea hiensis*, *G. nakaiana*, *G. pauciflora*, *Lilium cernuum*, *Anemarrhena asphodeloides*, *Terauchia anemarrhenefolia*, *Polygonatum stenophyllum*, *Disporum ovale*, *Hosta ensata*, *Iris koreana*, *I. mandshurica*, *I. minutiaurea*, *I. typhifolia*, *Hemerocallis coreana*, *Paris manshurica*, *Carex* spp., *Eriocaulon chinorossicum*, *E. ussuriense*, *Hystrix coreana*, *Helictotrichon trisetoides*, *Calamagrostis distantiflora*, *C. hsinganensis*, *C. mongolicola*, *Neomolinia mandshurica*, and *Arisaema manshuricum*.

The most characteristic plant formations of the Manchurian Province are the broad-leaved and the mixed forests, by which the vegetation of this province is sharply distinguished from that of the neighboring regions of Siberia. Of the conifers, the most common are *Larix gmelinii*, *Picea jezoensis* and *P. obovata* subsp. *koraiensis*, *Abies nephrolepis*, and *A. holophylla*. The broad-leaved forests are represented essentially by oak forests (*Quercus mongolica*) and, in valleys of mountain rivers, by elm-ash forests. Of the broad-leaved types found in the forests, besides the Mongolian oak (which despite its name is not found in Mongolia), we also find *Fraxinus mandshurica*, *Carpinus cordata*, *Juglans mandshurica*, *Pyrus ussuriensis*, several maple species (especially *Acer mono*, *A. ginnala*, *A. ukurunduense*, *A. pseudosieboldianum*, *A. tegmentosum*, and *A. mandshuricum*), *Philadelphus tenuifolius* and representatives of the Araliaceae (such as *Oplopanax elatus*, *Acanthopanax senticosus*, *Kalopanax septemlobus*, and *Aralia elata*), *Maackia amurensis*, *Deutzia amurensis*, and so on.

Along with the forest vegetation, the forest steppe, various types of steppe, and meadow steppes play an important role in the vegetative cover of the Manchurian Province. Steppes are characteristic of the southeastern parts of Transbaikalia (i.e., the Onon and Tsugol Steppes, the basins of the Turga, Borzya, and Uruliunguy Rivers, and from the upper Borzya to the village of

Argunsk), the northeastern tip of Mongolia (the Uldz District), and considerable expanses in northeastern China. Two small, isolated steppe areas are found in the limits of the Soviet Far East in the Prekhankian Valley (the basin of Lake Khanka and the middle part of the basin of the Razdolni River) and in the Zeya and Bureya valleys. Particularly widespread are steppes with either the presence or the predominance of *Filifolium sibiricum* (= *Artemisia sibirica*). Among the characteristic species are:

Clematis hexapetala, Arenaria juncea, Polygonum alopecuroides, Paeonia lactiflora, Euphorbia komaroviana, Lespedeza juncea, L. davurica, Astragalus dahuricus, Ledebouriella divaricata, Veronica linariifolia, V. dahurica, Scutellaria baicalensis, Hemerocallis minor, Allium senescens, Iris dichotoma, Spodiopogon sibiricus, Arundinella hirta, and *Cleistogenes kitagawae.*

In Dahuria,[27] besides the *Filifolium* steppes we find the characteristic forest of *Larix dahurica*, and occasionally meet *Betula platyphylla* forests and pine forests. Everywhere, especially along steep stony slopes, one finds steppe shrubs and trees: *Ulmus macrocarpa, Prunus sibirica,* species of *Spiraea* and of *Rosa, Ribes diacantha,* and others. In the extreme southeast, from the Upper Borzya River to the village of Afgunsc, the birch forests and the peony steppes play an increasing role in the vegetation cover. In the forest steppes one finds the more or less common trees and shrubs characteristic of the surrounding forests. These include *Quercus mongolica, Corylus heterophylla, Betula dahurica, Prunus glandulosa, Securinega suffruticosa, Lespedeza bicolor, Euonymus maackii, Rhamnus ussuriensis, Lonicera chrysantha,* and so on.

In the western part of the province, in the forest steppe flora and especially in the flora of steppes, strong influence is evident of the neighboring Circumboreal Region and of the Mongolian Province (in the Irano-Turanian Region). But to the east, the flora of the Manchurian Province becomes more and more typically eastern Asiatic. Some botanists have separated the forest steppe and steppe segments of the Manchurian Province into an independent Dahurian or Dahurian-Manchurian Province, but this hardly seems advisable.

2. Sakhalin-Hokkaido Province (Tolmatchev 1959; Fedorov [in Grubov and Fedorov] 1964; Takhtajan 1974).[28] This province includes a large part of Sakhalin Island (to the south from 51°30' north latitude);[29] a large part of the island of Hokkaido to the northeast, approximately from the Kuromatsunai Depression; and the southern islands of the Kurile series— namely, Kunashir, Shikotan, and Iturup. Of the endemic and subendemic species, we list the following:

Abies sachalinensis (Sakhalin, north to Okha, all of Hokkaido, and the southern Kurile Islands), *Trollius pulcher* (Sakhalin, Rishiri Island, Hokkaido), *Callianthemum sachalinense* (Sakhalin), *C. mitabeanum* (Hokkaido), *Aconitum sachalinense* (Sakhalin, Hokkaido, southern Kurile Islands), *A. yezoense* (Hokkaido), *A. kurilense* (southern Kurile Islands), *A. neosacha-*

linense (southern Sakhalin), *Pulsatilla sachalinensis* (Sakhalin), *Ranunculus yesoensis* (Hokkaido, southern Kurile Islands), *Thalictrum integrilobum* (Hokkaido), *Papaver fauriei* (Rishiri Island), *Corydalis curvicalcarata* (Hokkaido), *Rumex regelii* (Sakhalin), *Betula tatewakiana* (Hokkaido), *Salix pauciflora* (Hokkaido), *S. yezoalpina* (Hokkaido), *S. paludicola* (Hokkaido), *Cardamine schinziana* (Hokkaido), *C. yezoensis* (Hokkaido, Sakhalin, Iturup), *Primula hidakana* (Hokkaido), *Fragaria yezoensis* (Hokkaido, southern Kurile Islands), *Potentilla miyabei* (Hokkaido, Kurile Islands), *Astragalus yamamotoi* (Hokkaido), *A. japonicus* (Hokkaido, southern Kurile Islands), *A. sachalinensis* (Sakhalin), *Oxytropis kudoana* (Hokkaido), *O. yezoensis* (Hokkaido), *O. megalantha* (Hokkaido, Sakhalin, Kurile Islands), *O. hidakamontana* (Hokkaido), *O. rishiriensis* (Hokkaido), *O. schokanbetsuensis* (Hokkaido), *Gentiana sugawarae* (Sakhalin), *G. yuparensis* (Hokkaido), *Leontopodium kurilense* (Shikotan and Iturup Island), *Saussurea chionophylla* (Hokkaido), *S. fauriei* (Hokkaido, southern Kurile Islands), *Artemisia limosa* (Sakhalin), *Stenanthella sachalinense* (Sakhalin), *Gagea vaginata* (Hokkaido, Shikotan), *Carex flavicuspis, C. pyrophila, C. ramenskii, Fimbristylis ochotensis,* and *Sasa rivularis* (Sakhalin, Iturup Island, Hokkaido).

The majority of endemics of the Sakhalin-Hokkaido Province have a limited, often narrowly local distribution. Most of these endemics play only an unimportant role in the vegetation cover. The flora of the Sakhalin-Hokkaido Province is closest to the flora of the Japanese-Korean Province (especially to the northern parts of the island of Honshu), but it also shows much in common with the flora of the Manchurian Province (particularly with the flora of the mountains of northern Sikhote-Alin), and also with the Okhotsk-Kamchatka flora.

3. Japanese-Korean Province (Good 1947, 1974;[30] Takhtajan 1970, 1974). In this province I include the southwestern part of the island of Hokkaido to the southeast from the Kuromatsunai Depression;[31] the island of Honshu; the island of Kyushu; the island of Tanega; the island of Yaku; the Goto-retto islands; the Tsushima islands; and a large part of the Korean Peninsula.

This province is remarkably rich in endemics. The endemic families Sciadopityaceae and Glaucidiaceae are found here, as well as the following endemic genera: *Thujopsis, Ranzania, Anemonopsis, Pteridophyllum, Tsusiophyllum, Peltoboykinia, Pterygopleurum, Abeliophyllum, Pseudopyxis, Ancistrocarya, Vexillabium, Hakonechloa, Japonolirion, Alectorurus,* and *Kinugasa*. Among the many endemic and subendemic species are:

Isoëtes japonica (Japan, Korea), *Ophioglossum kawamurae, Botrychium nipponicum, Cheilanthes krameri, Pleopeltis onoei* (Japan, Korea), *Polypodium someyae, Grammitis sakaguchiana, Microlepia yakusimensis, Trichomanes titibuensis,* species of *Athyrium, Polystichum, Arachnioides, Dryopteris, Thelypteris,* and

Blechnum, Torreya nucifera, Abies homolepis, A. firma, A. koreana (south of Korea and Cheju Island), *A. mariesii, A. veitchii, Pseudotsuga japonica, Tsuga sieboldii, T. diversifolia, Picea bicolor, P. polita, P. maximowiczii, P. koyamae, Larix leptolepis, Pinus parviflora, P. densiflora* (Japan, Korea), *P. thunbergii* (Japan, south of Korea), *Thuja standishii, Thujopsis dolabrata, Chamaecyparis obtusa* subsp. *obtusa, C. pisifera, Magnolia sieboldii* (Japan, Korea), *M. salicifolia, M. kobus* (Japan, Korea), *M. stellata, Lindera sericea, L. praecox, Asarum* spp., *Aristolochia onoei, Nuphar* spp., *Schisandra repanda* (Japan, Korea), *Trollius hondoensis, Aconitum* spp., *Anemonopsis macrophylla, Coptis* spp., *Cimicifuga,* spp., *Eranthis pinnatifida, Thalictrum* spp., *Anemone* spp., *Pulsatilla nipponica,* species of *Clematis* and *Ranunculus, Glaucidium palmatum* (on the island of Hokkaido, it extends beyond the border of the province), species of *Berberis* and *Epimedium, Ranzania japonica,* and *Achlys japonica* (a second species of this genus is found in Pacific North America), *Hylomecon japonicum, Corydalis* spp., *Pteridophyllum racemosum, Cercidiphyllum magnificum, Euptelea polyandra, Hamamelis japonica, Corylopsis* spp., *Celtis jessoensis* (Japan, Korea), *Morus tiliifolia* (Japan, Korea), *Broussonetia kaempferi, Buckleya lanceolata, Laportea macrostachya,* species of *Pilea, Elatostema, Pellionia,* and *Boemeria, Fagus crenata, F. japonica, Quercus* spp., *Castanea crenata, Castanopsis cuspidata* (Japan, Korea), Carpinus spp., *Betula* spp., *Pterocarya rhoifolia, Juglans ailanthifolia, Minuartia hondoensis, Pseudostellaria palibiniana* (Japan, Korea), *Cerastium schizopetalum,* species of *Stellaria, Dianthus, Lychnis,* and *Silene, Chenopodium koraiense* (Japan, Korea), species of *Suaeda, Rumex madaio,* species of *Polygonum, Camellia rusticana,* species of *Stewartia,* including *S. monadelpha* (Japan, Cheju Island to the south of Korea), species of *Hypericum* and *Viola, Trichosanthes multiloba, Actinostemma lobatum, Melothria japonica* (Japan, Cheju Island to the south of Korea), species of *Cardamine, Eutrema bracteata, Rorippa nikoensis,* species of *Draba, Arabis,* and *Salix, Actinidia hypoleuca, Clethra barbinervis* (Japan, Cheju Island), species of *Menziesia* and *Rhododendron,* including *R. tschonoskii* (Japan, Korea) and *R. weyrichii* (Japan, Cheju Island), *Phyllodoce nipponica, Epigaea asiatica,* species of *Enkianthus, Gaultheria japonica,* species of *Vaccinium,* including *V. japonicum* (Japan, south of Korea), *Shortia uniflora, Styrax shiraiana, Pterostyrax hispida,* species of *Symplocos,* including *S. coreana* (Japan, Cheju Island) and *S. tanakana* (south of Korea), species of *Lysimachia,* including *L. leucantha* (Japan, Korea) and *L. acroadenia* (Japan, Cheju Island), species of *Primula* and *Tilia, Buxus microphylla* subsp. *microphylla,* species of *Euphorbia, Daphne,* and *Wikstroemia,* including *W. trichotoma* (Japan, Korea), *Itea japonica,* species of *Ribes, Schizophragma hydrangeoides* (Japan, Korea), *Deinanthe bifida, Platycrater arguta, Cardiandra alternifolia,* species of *Hydrangea, Philadelphus satsumi,* species of *Deutzia, Kirengeshoma koreana* (Korea), *Pittosporum illicioides,* species of *Sedum, Rodgersia podophylla* (Japan, Korea), species of *Astilbe, Tanakaea radicans,* species of *Saxifraga, Pelto-*

boykinia tellimoides, P. watanabei, Boykinia lycoctonifolia, species of *Chrysosplenium* and *Mitella, Parnassia alpicola, Stephanandra incisa* (Japan northward to the province of Khidaka on Hokkaido, Korea), *S. tanakae,* species of *Spiraea, Fragaria nipponica* (Japan, Cheju Island), *Potentilla* spp., species of *Rubus, Filipendula, Sanguisorba, Rosa,* and *Prunus, Photinia glabra, Chaenomeles japonica, Malus* spp., *Amelanchier asiatica* (Japan, Korea), species of *Sorbus, Cladrastis sikokiana, Euchresta japonica, Lespedeza maximowiczii* (Tsushima Island, Korea), *L. homoloba, Rhynchosia acuminatifolia, Dumasia truncata, Wisteria floribunda, W. japonica* (Japan, Korea), species of *Astragalus, Oxytropis japonica,* species of *Hydrobryum* and *Cladopus, Lagerstroemia fauriei, Rotala elatinomorpha, Trapa incisa,* species of *Ludwigia, Haloragis walkeri, Myriophyllum oguraense, Zanthoxylum fauriei* (Japan, south of Korea), *Coriaria japonica,* species of *Acer,* including *A. palmatum* (Japan, Korea), *Aesculus turbinata,* species of *Meliosma* and *Geranium,* including *G. tripartitum* (Japan, Cheju Island), *Polygala reinii, Aralia glabra, Panax japonicus,* species of *Acanthopanax, Evodiopanax innovans, Oplopanax japonicus, Hydrocotyle ramiflora* (Japan, Korea), *H. yabei, Bupleurum nipponicum* (in the north of Japan to the mountains of the province of Khidaka), *Sanicula kaiensis, S. tuberculata* (Japan, Korea), *Pternopetalum tanakae* (Japan, south of Korea), *Pimpinella calycina, P. nikoensis, Apium ikenoi, Pterygopleurum neurophyllum* (Japan, Korea), *Ligusticum tsusimense* (Tsu Island), *Dystaenia ibukiensis* (a second species of this genus on Korea), species of *Angelica, Heracleum moellendorffii,* species of *Ilex, Tripterygium doianum, Celastrus stephanotiifolius* (Japan, south of Korea), species of *Euonymus* and *Berchemia,* species of *Rhamnus, Vitis,* and *Ligustrum,* including *L. salicinum* (Japan, south of Korea), *Forsythia koreana* and *F. ovata* (Korea), *F. japonica* (Japan, Korea), species of *Fraxinus,* including *F. sieboldiana* (Japan, Korea), *Buckleya lanceolata, Loranthus tanakae* (Japan, south of Korea), *Taxillus kaempferi,* species of *Elaeagnus* and *Viburnum,* including *V. carlesii* (Japan, Korea), species of *Abelia, Weigela* and *Lonicera,* including *L. vidalii* (Japan, south of Korea), *Patrinia triloba, Valeriana flaccidissima, Scabiosa japonica, Gardneria insularis* (southern Korea; reported to occur also in Kyushu), *Apocynum basikurumon, Trachelospermum asiaticum* (Japan, Korea), species of *Cynanchum, Tylophora, Gentiana,* and *Swertia, Uncaria rhynchophylla, Mussaenda shikokiana, Lasianthus satsumensis, Damnacanthus macrophyllus, Leptodermis pulchella, Pseudopyxis depressa, P. heterophylla, Mitchella undulata* (Japan, south of Korea), *Rubia hexaphylla* (Japan, Korea), species of *Galium, Asperula trifida, Theligonum japonicum,* species of *Polemonium* and *Omphalodes, Ancistrocarya japonica,* species of *Trigonotis, Scopolia japonica, Physalis chamaesarachoides, Physaliastrum savatieri, Buddleja japonica,* species of *Scrophularia,* including *S. kakudensis* (Japan, Korea), *Mazus miquelii, Deinostema adenocaulum* (Japan, Cheju Island), species of *Veronica,* including *V. kiusiana* (Japan, Korea), *Monochasma japonicum,* species of *Euphrasia* and *Pedicularis, Lathraea japonica, Opithandra primuloides,* species of *Utricularia,* including *U. pilosa* (Japan, south of Korea), *Strobilanthes oligantha, Plantago hakusanensis,* species of *Callicarpa,* including *C. mollis* (Japan, Korea), species of *Ajuga, Teucrium japonicum* (Japan, Korea), species of *Scutellaria, Meehania montis-koyae, Nepeta subsessilis, Prunella prunelliformis, Chelonopsis longipes, Lamium ambiguum, L. humile,* species of *Salvia, Mosla japonica* (Japan, Korea), *Clinopodium macranthum, C. micranthum, Mentha japonica,* species of *Leucoseptrum, Dysophylla yatabeana* (Japan, Korea), *Keiskea japonica,* species of *Plectranthus,* including *P. japonicus* (Japan, Korea), species of *Adenophora,* including *A. tashiroi* (Japan, Cheju), *Campanula hondoensis,* species of *Leontopodium* and *Carpesium,* including *C. rosulatum* (Japan, Cheju), species of *Pertya, Ainsliaea,* and *Eupatorium,* species of *Erigeron* and *Aster, Myriactis japonensis, Arnica mallotopus,* species of *Ligularia, Miricacalia makineana,* species of *Cacalia, Senecio, Chrysanthemum, Artemisia, Cirsium, Saussurea, Synurus,* and *Taraxacum, Hieracium japonicum, H. krameri, Prenanthes acerifolia, P. tanakae,* species of *Ixeris, Youngia yoshinoi, Blyxa leiosperma, Zostera caulescens* (Japan, Korea), *Potamogeton fryeri* (Japan, Korea), species of *Najas* and *Sciaphila, Chionographis japonica* (Japan, Korea), *C. koidzumiana,* species of *Tofieldia, Veratrum, Tricyrtis, Hosta,* and *Hemerocallis, Anthericum yedoense,* species of *Allium, Nothoscordum inutile,* species of *Lilium, Erythronium japonicum* (Japan, Korea), *Tulipa latifolia, Fritillaria amabilis, F. japonica, Asparagus kiusianus,* species of *Polygonatum, Smilacina,* and *Disporum, Kinugasa japonica,* species of *Ophiopogon, Lycoris squamigera,* species of *Croomia* and *Dioscorea, Iris gracilipes, Glaziocharis abei, Zingiber mioga, Apostasia nipponica, Cypripedium debile,* species of *Orchis* and *Platanthera, Tulotis iinumae,* species of *Amitostigma* and *Habenaria, Androcorys japonensis, Listera makineana,* species of *Lecanorchis, Nervilia nipponica, Stigmatodactylus sikokianus,* species of *Gastrodia, Goodyera,* and *Hetaeria, Cheirostylis okabeana,* species of *Vexillabium, Odontochilus hatusimanus, Tropidia nipponica, Oberonia japonica, Tipularia japonica, Didiciea japonica* (a second species of this genus is in Sikkim), *Ephippianthus sawadanus,* species of *Liparis* and *Calanthe, Dactylostalix ringens,* species of *Bulbophyllum* and *Cymbidium, Luisia teres, Sarcanthus scolopendrifolius* (Japan, south of Korea), *Taeniophyllum aphyllum, Sarcochilus japonicus,* species of *Juncus, Cyperus, Eleocharis, Carex, Eriocaulon, Semiarundinaria, Arundinaria,* and *Sasa, Agrostis nipponensis,* species of *Calamagrostis, Helictotrichon hideoi, Hystrix japonica, Brachyelytrum japonicum* (Japan, Cheju Island), *Festuca takedana, Puccinellia nipponica* (Japan, Korea), species of *Poa, Neomolinia japonica* (Japan, Cheju), *Hakonechloa macra,* species of *Miscanthus, Trachycarpus fortunei, Amorphophallus kiusianus,* and species of *Arisaema.*

4. Volcano-Bonin Province (Engler 1912, 1924;[32] Takhtajan 1970; the Ogasawara [Bonin] Floral Region, of Hara 1959; the Bonin or Ogasawara

Region, Maekawa 1974). This province consists of two isolated groups of Pacific Ocean Islands—the Bonin Islands and the Volcano Islands. They are located 840 km southeast of the Island of Honshu. Both groups of islands are of volcanic origin, but although the Bonin Island group is of Eocene age, the Volcano Islands presumably arose in the Pleistocene (see Tuyama 1953). Although there is a considerable admixture of tropical elements in the flora of the province, both on the Bonin (Wilson 1919; Nakai 1930; Hosokawa 1934; Tuyama 1953; Balgooy 1960, 1971) and on the Volcano Islands (Tuyama 1953; Tuyama [in Tuyama and Asami] 1970; Hara 1959), they are characterized by the predominance of endemics of Eastern Asiatic stock: "They obviously form part of the East Asiatic Region" (Balgooy 1971:103).

There are two endemic monotypic genera—*Boninia* and *Dendrocacalia* (close to *Cacalia*). About 80% of the flowering plant species in this province are endemic (Tuyama 1953). The following list of endemic species is based mainly on work of Tuyama (in Tuyama and Assami 1970) and Kobayashi (1978):[33]

Angiopteris boninensis, Marattia boninensis, M. tuyamae, Actinostachys (Schizaea) boninensis, Trichomanes auto-obtusum, T. boninense, T. bonincola, Adiantum ogasawarense, Lindsaea repanda, Cyathea mertensiana, C. ogurae, Bolbitis boninensis, Ctenitis microlepigera, Diplazium bonincola, D. longicarpum, D. subtripinnatum, Dryopteris insularis, Lomariopsis boninensis, Thelypteris boninensis, T. ogasawarensis, Asplenium ikenoi, A. micantifrons, Pleopeltis boniniensis, Loxogramme boninensis, L. toyoshimae, Microsorium masaskei, M. subnormale, Vittaria bonincola, V. ogasawarensis, Pteris boninensis, Cinnamomum pseudopedunculatum, Machilus boninensis, M. kobu, M. pseudokobu, Neolitsea boninensis, Peperomia boninsimensis, Piper postelsianum, Clematis boninensis, Distylium lepidotum, Celtis boninensis, Ficus boninsimae, F. iidaiana, F. mishimurae, Morus boninensis, Boehmeria boninensis, Procris boninensis, Portulaca boninensis, Eurya boninensis, Schima mertensiana, Stachyurus macrocarpus, Trichosanthes boninensis, Rhododendron boninense, Vaccinium boninense, Symplocos boninense, S. kawakamii, S. pergracilis, Pouteria boninensis, Myrsine maximowiczii, M. okabeana, Lysimachia rubida, Elaeocarpus pachycarpus, E. photiniifolius, Hibiscus glaber, Phyllanthus boninsimae, Claoxylon centenarium, Drypetes integerrima, Wikstroemia pseudoretusa, Pittosporum beecheyi, P. boninense, P. chichijimense, P. parvifolium, Sedum boninense, Osteomeles boninensis, O. lanata, Rhaphiolepis integerrima, Rubus nakaii, Erythrina boninensis, Metrosideros boninensis, Syzygium cleyeraefolium, Melastoma tetramerum, Boninia glabra, B. grisea, Evodia kawagaiana, E. inermis, E. mishimurae, Fagara boninsimae, Sapindus boninensis, Fatsia oligocarpella, Peucedanum boninense, Ilex beecheyi, I. matanoana, I. mertensii, I. percoriacea, Euonymus boninensis, Ligustrum micranthum, Santalum boninense, Elaeagnus rotundata, Viburnum boninsimense, Geniostoma glabrum, Excavatia hexandra, Ochrosia nakaiana, Trachelospermum foetidum,

Gardenia boninensis, Hedyotis grayi, H. leptopetala, H. mexicana, H. pachyphylla, Morinda boninensis, Psychotria boninensis, P. homalosperma, Tarenna subsessilis, Lycium griseolum, Lycianthes boninensis, Capsicum boninense, Myoporum boninense, Orobanche boninsimae, Callicarpa glabra, C. nishimurae, C. subpubescens, Ajuga boninsimae, Scutellaria longituba, Lobelia boninensis, Dendrocacalia crepidifolia, Cirsium boninense, C. toyoshimae, Ixeris ameristophylla, I. grandicolla, I. linguaefolia, I. longirostrata, Sciaphila boninensis, S. okabeana, Alpinia bilamellata, A. boninsimensis, Calanthe hattorii, Cirrhopetalum boninense, Corymborchis subdensa, Eulophia toyoshimae, Gastrodia boninensis, Goodyera augustini, G. boninensis, Liparis hostaefolia, Luisia boninensis, L. brachycarpa, Malaxis (Microstylis) boninensis, Platanthera boninensis, Zeuxine boninensis, Carex augustinii, C. hattoriana, C. toyoshimae, Rhynchospora boninensis, Aristida boninensis, Digitaria platycarpha, Ischaemum ischaemoides, Miscanthus boninensis, Paspalidium tuyamae, Clinostigma savoryana, Livistona chinensis var. *boninensis, Freycinetia boninensis, Pandanus boninensis,* and *P. farusei.*

Despite a series of endemics with tropical relationships (e.g., *Procris boninensis, Claoxylon centenarium, Melastoma tetramerum, Excavatia hexandra, Ochrosia nakaiana, Hedyotis leptopetala, Morinda boninensis, Tarenna subsessilis, Clinostigma savoryana, Freycinetia boninensis, Pandanus boninensis,* and several orchids), by far the majority of endemic species are related to Eastern Asiatic taxa, principally Japanese and Chinese. As far as the nonendemic taxa are concerned, it is as Tuyama (1953:212) said: they almost all belong to the subtropical Asiatic element, and only few of them are related to pantropical or Polynesian elements. Hara (1959) came to a similar conclusion. The most tropical character is found in the littoral vegetation, which consists of *Scaevola frutescens, Vitex rotundifolia, Ipomoea pes-caprae, Hibiscus tiliaceus, Messerschmidia argentea, Caesalpinia bonduc, Sporobolus virginicus, Thuarea involuta, Hernandia peltata, Calophyllum inophyllum,* and *Terminalia catappa* (Hara 1959; Tuyama [in Tuyama and Asami] 1970). All of these species are widely distributed in the warmer areas of the Pacific; their seeds were carried by ocean currents, and possibly also by the wind and birds (Hara 1959).

As noted already by Warburg (1891), the flora of the Bonin and Volcano Islands has rather more in common with the floras of the Ryukyu Islands, Taiwan, and southern China than with the flora of the closer lying Mariana Islands. Subsequent studies (Nakai 1930; Hara 1959; Tuyama [in Tuyama and Assami] 1970; Balgooy 1971) fully support Warburg's opinion.

5. Ryukyu or Tokara-Okinawa Province (Engler 1912, 1924; Takhtajan 1970, 1974).[34] In the north, this province is separated from the Volcano-Bonin Province by the Strait of Tokara, and its southern boundary passes between the island groups of Okinawa and Sakishima.

The flora of this province like that of the Taiwanian Province (see next section) is to some extent transitional between the Holarctic and paleotropic floras, and is rather strongly permeated by tropical elements, the number of which gradually increases southward. *Kandelia candel*, a typical mangrove plant, even reaches the shores of the southwestern end of Kyushu Island, and two others (*Barringtonia racemosa* and *Bruguiera gymnorrhiza*) extend north to the Amami Islands. In the Ryukyu Province, especially on Okinawa, there is also a series of purely tropical taxa, pointed out particularly in the works of Hara (1959) and of Maekawa (1974). According to Hara, the appearance in the Ryukyus of such ancient Japanese elements as *Aucuba* and *Heloniopsis* shows that they were broadly distributed in the Middle Tertiary period, when Japan and the Ryukyus constituted the edge of the Asiatic continent. As for the tropical elements in the flora of the Ryukyus, they have a recent character and are represented almost exclusively by nonendemic taxa.

Endemic species and subspecies of the Ryukyus belong predominantly to Holarctic genera. These include *Corydalis tashiroi*, *Schima wallichii* subsp. *liukiuensis*, *Adinandra ryukyuensis*, *Viola amamiana*, *V. maculicola*, *V. utchinensis*, *Vaccinium amamianum*, *Deutzia amanoi*, *Cardiandra amamioshimensis*, *Hydrangea scandens* subsp. *liukiuensis*, *Zanthoxylum amamiense*, *Acer itoanum*, *Ilex dimorphophylla*, *I. poneantha*, *Rhamnus kanagusuki*, *Viburnum tashiroi*, and *Aster miyagii*. But there are also endemics of tropical origin.

6. **Taiwanian Province** (Fedorov [in Grubov and Fedorov] 1964; Takhtajan 1970, 1974).[35] This province includes Taiwan, except for its southern tip (the Hengchun Peninsula); the Sakishima Islands; and all other neighboring islands, excluding Lanyu and Lutao.[36] The Taiwanian Province is usually included in the Paleotropical Kingdom (Engler 1882; Diels 1908, 1918; Hayek 1926; Good 1947, 1974; Mattick 1964; Fedorov [in Grubov and Fedorov] 1964; and many others); only rarely is it related to the Holarctic (Gaussen 1954; Meusel et al. 1965; Takhtajan 1970, 1974; Tolmatchev 1974).

Throughout Taiwan (including the tropical peninsula of Hengchun), the paleotropic element constitutes only about 6%, and the pantropical about 1.5% of the woody flora (Li 1963:15). The pantropical element is represented by weedy or strand plants, which were dispersed by ocean currents. Paleotropic species are usually encountered in valleys and in lowlands (especially in the secondary forests), and in all probability are relatively recent arrivals (Li 1963). The Eastern Asiatic elements clearly predominate in the flora of the Taiwanian Province, especially in the primary vegetation. According to Li (1963), about 18% of the woody flora of all of Taiwan consists of species widely distributed in eastern Asia, including the Holarctic parts of China, Korea, Japan, and adjoining territories. In addition, no less than 18%

of the species (both subtropical and temperate) constituting the woody flora of Taiwan are found in southern and eastern China as well, growing at lower and middle altitudes. Finally, 1.5% of the woody flora consists of species found also in central and western China; about 2.5% are also found in Japan (some also in Korea); and about 2.2% also occur on the Ryukyu Islands (Li 1963). All of this indicates the predominance of the Eastern Asiatic element over the tropical ones. But especially important is the composition of endemics, which among the woody flora of Taiwan number not less than 35%. According to Li (1963:20), "in general, species endemism is more strongly indicated in temperate genera than tropical ones." Suffice it to say that among the endemics of the woody flora of Taiwan are species of the genera *Cephalotaxus, Amentotaxus, Abies, Picea, Pseudotsuga, Cunninghamia, Taiwania, Chamaecyparis, Sassafras, Schisandra, Akebia, Stauntonia, Corylopsis, Distylium, Ulmus, Fagus, Quercus, Carpinus, Camellia, Stachyurus, Salix, Actinidia, Rhododendron, Styrax, Reevesia, Daphne, Stellera, Ribes, Hydrangea, Cardiandra, Deutzia, Spiraea, Prunus, Prinsepia, Rubus, Rosa, Cotoneaster, Pyracantha, Stranvaesia, Sorbus, Eriobotrya, Pyrus, Phellodendron, Acer, Koelreuteria, Fatsia, Hedera, Jasminum, Abelia, Lonicera, Sambucus*, and others. Therefore, despite the presence of tropical enclaves and a series of endemic species of tropical genera, the flora of the Taiwanian Province is basically Eastern Asiatic.

There are three monotypic, endemic genera in the flora of the province: *Sinopanax, Formosia*, and *Satakentia* (Sakishima Islands). There is also a multitude of endemic species, among which we mention:

Isoëtes taiwanensis, Archangiopteris itoi, Vittaria mediosora, species of *Pteris, Crypsinus*, and *Polypodium, Loxogramme biformis*, species of *Microlepia, Hymenophyllum alishanense, Trichomanes palmifolium*, species of *Asplenium, Athyrium, Diplazium, Ctenitis, Polystichum*, and *Dryopteris, Cephalotaxus wilsoniana, Amentotaxus formosana, Abies kawakamii, Picea morrisonicola, Pinus morrisonicola, P. taiwanensis, Pseudotsuga wilsoniana, Cunninghamia konishii, Taiwania cryptomerioides, Chamaecyparis formosensis, Calocedrus formosana, Magnolia kachirachirai, Lindera akoensis, Nothaphoebe konishii, Phoebe formosana, Sassafras randaiense*, species of *Cinnamomum, Litsea, Neolitsea, Persea*, and *Piper, Peperomia makaharai*, species of *Aristolochia* and *Asarum, Nuphar shimadai, Illicium arborescens, Schisandra arisanensis, Akebia longeracemosa*, species of *Stauntonia, Cissampelos (Paracyclea) ochiaiana, Pericampylus trinervatus, Isopyrum arisanense, Aconitum bartletii*, species of *Ranunculus, Clematis, Thalictrum*, and *Berberis, Mahonia oiwakensis*, species of *Corydalis, Corylopsis stenopetala, Distylium gracile, Sycopsis formosana, Celtis nervosa, Ulmus uyematsui*, species of *Ficus, Boehmeria taiwaniana, Fagus hayatae*, species of *Lithocarpus, Castanopsis, Quercus, Alnus*, and *Carpinus, Cerastium* spp., *Dianthus pygmaeus*, species of *Silene* and *Stellaria*, *Pyrenaria shinkoensis*, species of *Adinandra*, *Camellia*, and *Eurya*, *Hypericum formosanum* and several other species of this genus, *Viola formosana*, *V. nagasawai*, *Thladiantha punctata*, species of *Begonia*, *Capparis formosana*, species of *Arabis*, *Barbarea taiwaniana*, species of *Cardamine*, *Draba sekiyana*, *Cochlearia formosana*, species of *Salix*, *Actinidia* spp., *Pieris taiwanensis*, species of *Rhododendron* and *Vaccinium*, *Styrax matsumuraei*, species of *Symplocos*, *Ardisia cornudentata*, *Embelia lenticellata*, *Lysimachia fragrans*, *Grewia rhombifolia*, *Reevesia formosana*, *Hibiscus taiwanensis*, *Pachysandra axillaris* var. *tricarpa*, species of *Acalypha* and *Breynia*, *Drypetes karapinensis*, *Phyllanthus takaoensis*, *Euphorbia formosana*, *E. tashiroi*, *Daphne arisanensis*, *Wikstroemia mononectaria*, *Stellera formosana*, species of *Chrysosplenium*, *Mitella formosana*, *Itea parviflora*, *Ribes formosanum*, *Deutzia taiwanensis*, *Cardiandra formosana*, *Hydrangea longifolia*, *Pittosporum daphniphylloides*, *Kalanchoë gracilis*, *Astilbe macroflora*, species of *Spiraea* and *Prunus*, *Prinsepia scandens*, species of *Rubus*, *Rosa*, and *Cotoneaster*, *Pyracantha koidzumii*, *Stranvaesia niitakayamensis*, *Sorbus randaiensis*, *Photinia chingshuiensis*, *P. lasiopetala*, *P. lucida*, *Pyrus kawakamii*, *Ormosia formosana*, *Milletia taiwaniana*, *Lespedeza pubescens*, species of *Syzygium*, *Otanthera scaberrima*, species of *Bredia*, *Barthea formosana*, *Pachycentria formosana*, species of *Zanthoxylum*, *Evodia merrillii*, *Murraya euchrestifolia*, *Turpinia formosana*, species of *Acer*, *Koelreuteria henryi*, *Sabia transarisanensis*, *Meliosma callicarpaefolia*, *Aralia taiwaniana*, *Schefflera taiwaniana*, *Fatsia polycarpa*, *Hedera formosana*, *Dendropanax pellucidopunctatus*, *Pentapanax castanopsisicola*, *Sinopanax formosanus*, *Bupleurum kaoi*, *Sanicula petagnioides*, *Peucedanum formosanum*, *Oreomyrrhis involucrata*, *Angelica hirsutiflora*, *A. morii*, *A. morrisonicola*, species of *Ilex*, *Perrottetia arisanensis*, species of *Euonymus*, *Sageretia randaiensis*, species of *Rhamnus*, *Tetrastigma umbellatum*, *Jasminum hemsleyi*, *Osmanthus lanceolatus*, species of *Ligustrum*, *Taxillus matsudai*, species of *Scurula*, *Viscum alniformosanae*, *V. multinerve*, *Balanophora spicata*, species of *Elaeagnus*, *Helicia formosana*, species of *Viburnum*, *Sambucus formosana*, *Lonicera oiwakensis*, *L. kawakamii*, *Gardneria shimadai*, *Formosia benthamiana*, *Heterostemma brownii*, *Gentiana arisanensis* and several other species of this genus, species of *Rubia*, *Mussaenda taiwaniana*, species of *Lasianthus*, *Galium echinocarpum*, *G. formosense*, *Buddleja formosana*, *Veronica morrisonicola*, *Strobilanthes formosanus*, *Dicliptera longiflora*, species of *Callicarpa*, *Plectranthus lasiocarpus*, *Ajuga dictyocarpa*, *Clinopodium laxiflorum*, *Codonopsis kawakamii*, *Adenophora morrisonensis*, species of *Aster*, *Cacalia nokoensis*, species of *Chrysanthemum* and *Cirsium*, *Ligularia kojimae*, *Petasites formosanus*, *Prenanthes formosana*, species of *Saussurea* and *Senecio*, *Lilium formosanum*, *Heloniopsis arisanensis*, *Smilacina formosana*, *Tricyrtis formosana*, species of *Smilax*, *Heterosmilax seisuiensis*, numerous representatives of the Orchidaceae, species of *Carex*, *Isachne debilis*,

species of *Panicum, Bambusa,* and *Phyllostachys, Arundinaria usawai, Chikusichloa brachyanthera* (Iriomote Island), and *Satakentia riukiuensis* (Iriomote and Ishigaki).
Despite the considerable role of the tropical element, the Holarctic element dominates among the endemics of the Taiwanian flora.

7. **Northern Chinese Province** (Engler 1924;[37] Handel-Mazzetti 1926–27, 1931; Li 1944; Good 1947, 1974; Fedorov [in Grubov and Fedorov] 1964; Tamura 1966; Takhtajan 1970, 1974; North China region, Wu Cheng-yih 1979). The boundary of this province agrees approximately with the temperate zone of summergreen forests (aestisilvae) in the geobotanical classification of K'ien Ch'ung-shu, Wu Cheng-i, and Ch'en Ch'ang-tao (1957), and also with the boundaries designated by Wu Cheng-yih (1979) for his North China region.
The Northern Chinese Province extends borders beyond the Mongolian and Manchurian provinces, and includes the southeastern part of the Loess Plateau, the Shansi (Shanxi) Plateau, the North China Plain, the Jehol Mountains (North Hopeh Mountain Area), the Liaotung (Liaodong) Peninsula, and the Shantung (Shandong) Peninsula. In its floristic composition this province has much in common both with the Manchurian and Japanese-Korean provinces and with the Central Chinese Province. The predominant vegetation in the Loess Plateau is the forest steppe. The strong influence of the flora of the Mongolian Province is expressed. There is only one endemic genus (*Oresitrophe*), but a considerable number of endemic and subendemic species, of which I mention only the following:
Gymnopteris borealisinensis, Woodsia rosthorniana, Picea mastersii, P. meyeri, Phoebe sheareri, Aconitum pekinense, Aquilegia yabeana, Thalictrum przewalskii, Epimedium pubescens, Meconopsis racemosa, Macleaya microcarpa, Corydalis spp., *Hypecoum chinense, H. erectum, Pteroceltis tatarinowii, Carpinus chowii, Ostrya liana, Ulmus davidiana, Stellaria davidii, Silene tatarinowii, Corispermum stenolepis, Limonium franchetii, Viola rhodosepala, Actinostemma paniculatum, Cheiranthus aurantiacus, Erysimum stigmatosum, Orychophragmus (Brassica) violaceus, Tamarix juniperina, Populus hopeinensis, Andrachne chinensis, Euphorbia croizatii, Oresitrophe rupifraga, Sedum dumulosum, S. stellariifolium, S. tatarinowii, Spiraea dasyantha, S. nishimurae, Crataegus kansuensis, Pyrus betulaefolia, P. bretschneideri, P. corymbifera, Potentilla ancistrifolia, Prunus davidiana, Caragana litwinowii, C. zahlbruckneri, Guldenstaedtia maritima, Astragalus* spp., *Biebersteinia heterostemon, Eriocycla (Pimpinella) albescens, Pimpinella brachystyla, Euonymus kiautschovicus, Fraxinus bungeana, Syringa microphylla, S. pekinensis, S. pubescens, Cynanchum bungei, C. chinense, Vincetoxicum mukdenense, V. versicolor, Abelia biflora, Ligustrum suave, Lonicera*

kungeana, Sambucus peninsularis, Patrinia heterophylla, Galium bungei, G. pauciflorum, Leptodermis oblonga, Bothriospermum chinense, Eritrichium borealisinense, Trigonotis amblyosepala, Elsholtzia integrifolia, E. stauntonii, Nepeta pseudokoreana, Schizonepeta tenuifolia, Salvia umbratica, Pedicularis tatarinowii, Rehmannia glutinosa, Adenophora pinifolia, Serratula cupiliformis, S. ortholepis, Chrysanthemum jucundum, C. namikawanum, Artemisia spp., *Senecio* spp., *Ligularia sinica, Myripnois dioica, Saussurea* spp., *Veratrum mandshuricum, Tricyrtis puberula, Polygonatum kansuense, P. platyphyllum, Asparagus longiflorus, Allium* spp., *Iris kobayashii, Smilax pekingensis,* and so on.

8. Central Chinese Province (Diels 1901; Good 1947, 1974; Fedorov [in Grubov and Fedorov] 1964; Meusel et al. 1965; Takhtajan 1970, 1974; East China Region, Central China Region, and Yunnan, Guizhon, Guanxi Region, Wu Cheng-yih 1979). The territory of this province includes the Tapa Shan Mountains, the Szechwan Basin, a part of the Kweichow Hills (Kweichow Plateau), and the southern part of the Great Chinese Plain eastward up to the East China Sea.[38] I draw its western boundary approximately according to Handel-Mazzetti (1926–27; see his map), with some corrections based on Wu Cheng-yih 1979.

The flora of the Central Chinese Province is extraordinarily rich in endemic taxa, among which are two monotypic families—the Ginkgoaceae[39] and the Eucommiaceae (Kiangsi, Shensi, Shanse, Hupeh, Anhwei, and Chekiang). Endemic genera include *Pseudotaxus, Pseudolarix, Metasequoia, Sinofranchetia, Eomecon, Fortunearia, Sinowilsonia, Eucommia, Poliothyrsis, Psilopeganum, Changium, Heptacodium,* and *Kolkwitzia.* With the exception of *Heptacodium* (two species), all are monotypic. The number of endemic species is very great, and includes the following:

Ginkgo biloba, Pseudotaxus chienii, Torreya grandis, Abies fargesii, A. chensiensis, Keteleeria fortunei, Pseudolarix amabilis, Metasequoia glyptostroboides, Magnolia sprengeri, M. zenii, M. cylindrica, Liriodendron chinense, species of *Machilus, Litsea,* and *Lindera, Sassafras tzumu, Gymnotheca chinensis, Asarum* spp., *Aristolochia heterophylla, Illicium henryi, Kadsura longipedunculata, Schisandra sphenanthera, Clematis apiifolia, Holboellia fargesii, Sinofranchetia chinensis,* species of *Berberis* and *Corydalis, Corylopsis veitchiana, C. platypetala, Fortunearia sinensis, Sinowilsonia henryi, Hamamelis mollis, Eucommia ulmoides, Celtis biondii, C. labilis, C. julianae, Zelkova schneideriana, Fagus engleriana, Quercus* spp., *Carpinus fargesii, C. oblongifolia, Pterocarya hupehensis, P. paliurus, Juglans cathayensis, Stewartia sinensis, Poliothyrsis sinensis, Stachyurus retusus, S. szechuanensis, Thladiantha nudiflora, Begonia henryi, Salix heterochroma, Actinidia melanandra, Clethra monostachya, Rhododendron* spp., *Vaccinium,* spp. *Styrax veitchiorum, S. dasyantha, Sinojackia xylocarpa, S. reh-*

deriana, Diospyros armata, species of *Myrsine* and *Ardisia, Primula faberi, Lysimachia* spp., *Tilia oliveri, T. tuan, T. henryana,* species of *Euphorbia* and *Wikstroemia, Itea ilicifolia, Ribes henryi, Deutzia discolor, D. globosa, D. mollis, D. hypoglauca, Hydrangea strigosa, H. sargentiana, Spiraea* spp., *Neillia sinensis,* species of *Prunus, Rubus, Rosa, Cotoneaster,* and *Crataegus, Osteomeles subrotunda, Sorbus koehneana, S. dunnii, S. folgneri, S. xanthoneura, S. zahlbruckneri, Photinia parvifolia, P. amphidoxa, Eriobotrya japonica, Pyrus* spp., *Bauhinia hupeana, Cercis chinensis, C. racemosa, Gleditsia macracantha, G. sinensis, Gymnocladus chinensis, Caesalpinia szechuensis, Maackia chinensis, Cladrastis wilsonii, Psilopeganum sinense, Evodia hupehensis, Phellodendron chinense, Staphylea holocarpa, Dipteronia sinensis, Acer fulvescens, A. amplum, A. sinense, A. wilsonii, A. oliverianum, A. flabellatum, A. robustum, A. davidii, A. grosseri, A. maximowiczii, A. franchetii, A. henryi, Meliosma flexuosa, M. veitchiorum, M. beaniana, Aesculus wilsonii, Polygala mariesii, P. wattersii, Nyssa sinensis, Cornus* spp., *Acanthopanax simonii, A. henryi, Changium smyrnioides, Ilex cornuta, I. szechwanensis, Euonymus* spp., *Celastrus hypoleuca, C. loeseneri, Paliurus hemsleyanus, Rhamnella franguloides, Rhamnus utilis, Vitis wilsonae, Ampelopsis* spp., *Parthenocissus laetivirens, P. henryana, Fraxinus* spp., *Syringa reflexa, Ligustrum* spp., *Jasminum giraldii, Balanophora* spp., a whole set of species of *Viburnum, Symphoricarpos sinensis, Dipelta floribunda, Abelia uniflora, A. chinensis, Heptacodium jasminoides, H. miconioides, Kolkwitzia amabilis, Lonicera* spp., *Patrinia angustifolia, Sindechites henryi, Holostemma sinense, Cynanchum linearifolium, C. stenophyllum,* species of *Gentiana* and *Swertia, Myrioneuron faberi, Diplospora fruticosa, Nertera sinensis, Omphalodes cordata, Trigonotis mollis, Solanum pittosporifolium, Scopolia sinensis, Buddleja lindleyana, B. albiflora, Scrophularia henryi, S. ningpoensis, Mazus* spp., *Rehmannia rupestris,* species of *Calorhabdos* and *Pedicularis, Trapella sinensis, Lysionotus ophiorrhizoides, Phylloboea sinensis, Oreocharis auricola, Didissandra* spp., *Chirita fauriei, Hemiboea henryi, Strobilanthes* spp., *Justicia latiflora, Callicarpa bodinieri, Premna ligustroides, Orthosiphon debilis, Plectranthus* spp., *Salvia maximowicziana, Dracocephalum faberi, D. henryi, Scutellaria obtusifolia, S. sessilifolia, Stachys adulterina, Phlomis albiflora, P. gracilis, Loxocalyx urticifolius, Hanceola sinensis, Leucosceptrum sinense,* species of *Teucrium, Adenophora,* and *Aster, Leontopodium sinense, Carpesium minus, Senecio* spp., *Saussurea lamprocarpa, S. microcephala, Ainsliaea glabra, A. ramosa, Crepis* spp., *Faberia sinensis, Lactuca elata, L. triflora, Prenanthes faberi, Cardiocrinum cathayanum, Polygonatum* spp., *Smilacina* spp., *Chlorophytum chinense, Allium* spp., *Lilium* spp., *Disporum uniforum, Daiswa* spp., *Aletris* spp., *Ophiopogon clavatus, Smilax* spp., *Stemona erecta, Dioscorea zingiberensis, Iris henryi, Cypripedium* spp., *Myrmechis chinensis, Goodyera henryi, Herminium souliei, Habenaria* spp., *Diplomeris chinensis, Luzula chinensis, Juncus* spp., *Eriocaulon faberi, Stipa henryi, Oryzopsis obtusa, Deyeuxia* spp. (a genus close to *Calamagrostis*), *Poa prolixior,*

Arundinaria spp., *Phyllostachys* spp., *Dendrocalamus affinis, Pinellia cordata, P. integrifolia,* and *Arisaema* spp.

9. **Southeastern Chinese Province** (Fukien Subprovince of the Himalayan-Chinese Province, Fedorov [in Grubov and Fedorov] 1964 p.p.; South Region, Wu Cheng-yih 1979). This province includes the easternmost part of Kwangsi; Kwantung, except its southern tropical parts, but including Hong Kong; the southernmost part of Hunan; the southernmost part of Kiangsi; almost all of Fukien, with adjacent islands; and the southern part of Chekiang, with adjacent islands.

The flora of this province is very rich and contains many characteristic Chinese taxa, though it also contains a considerable number of species of tropical, especially Indomalesian, genera. There are a number of endemic genera, including *Tetrathyrium, Dunnia,* and *Latouchea,* and numerous endemic species. Characteristic plants include representatives of the Taxodiaceae (including *Glyptostrobus*), Magnoliaceae (including *Tsoongiodendron*), Annonaceae, Lauraceae, Moraceae, Fagaceae, Theaceae (including species of *Tutcheria*), Clusiaceae, Euphorbiaceae, Fabaceae, Rutaceae, Anacardiaceae, Meliaceae, and Sapindaceae.

Mixed evergreen and deciduous forests, as well as forests of *Keteleeria, Fokienia,* and *Podocarpus nagi* are characteristic for this province. The most important deciduous oaks are *Quercus acutissima* and *Q. variabilis.* In the mountains, especially in the high mountains, there are conifer forests consisting of *Tsuga longibracteata* (Nanling Range) and *Pinus kwangtungensis.*

10. **Sikang-Yünnan Province.**[40] This province embraces a large part of western Szechwan (west from the Szechwan Depression), including the Szechwan Alps, the Yünnan Plateau, and the bordering regions of the mountains of northeastern Burma, northern Laos, and northwestern Tonkin.

The flora of this province is very rich and diverse. It includes one endemic, monotypic family (Rhoipteleaceae) and a series of endemic genera, including *Rhoiptelea, Coelonema, Berneuxia, Craigia, Delavaya, Tirpitzia, Dickinsia, Haplosphaera, Harrysmithia, Sinodielsia, Sinolimprichtia, Chaerophyllopsis, Sinomerrillia, Antiotrema, Henryettana, Schistocaryum, Tengia, Cardioteucris, Holocheila, Paracheila, Paralamium, Diplazoptilon, Faberia, Formania, Nouelia,* and *Vladimiria.* Of the numerous endemic species, I mention the following:

Species of *Angiopteris, Archangiopteris, Adiantum, Polypodium, Polystichum,* and of many other fern genera,[41] *Amentotaxus yunnanensis, Torreya yunnanensis, Abies* spp., *Tsuga yunnanensis, Picea* spp., *Pinus* spp., *Juniperus changii, Manglietia chingii, M. forrestii, M. grandis, M. megaphylla, M. rufibarbata, M. tenuipes, M. wangii, M. yunnanensis, Magnolia delavayi, M. henryi, M. shangpaensis, M. wilsonii, M. sinensis, M. dawsoniana, Michelia wilsonii, M.

chapaensis, M. yunnanensis and several other species of the family, *Knema yunnanensis, Chimonanthus yunnanensis, Beilschmiedia yunnanensis, Cryptocarya yunnanensis, Lindera tonkinensis, Machilus yunnanensis, Gymnotheca involucrata, Illicium* spp., *Schisandra* spp., *Kadsura* spp., *Holboellia grandiflora, H. marmorata, Trollius yunnanensis,* species of *Delphinium, Aconitum, Ranunculus, Thalictrum, Anemone, Clematis, Berberis, Mahonia, Meconopsis,* and *Corydalis, Corylopsis yunnanensis* and other species of this genus, *Distylium pingpiensis, Sycopsis dunnii, Celtis* spp., *Morus yunnanensis, Castanopsis* spp., *Lithocarpus* spp., *Quercus* spp., *Betula delavayi, Alnus cremastogyne, Carpinus austro-yunnanensis, Corylus yunnanensis, Rhoiptelea chiliantha, Arenaria yunnanensis, Platycarya tonkinensis, Carya tonkinensis, Lychnis yunnanensis, Silene yunnanensis, Stellaria delavayi, S. yunnanensis, Polygonum* spp., *Paeonia czechuanica, P. delavayi, P. lutea, P. potaninii, P. yunnanensis, P. mairei, P. yui,* species of *Camellia* and *Stewartia, Pyrenaria yunnanensis, Eurya* spp., *Hypericum yunnanensis, Viola* spp., *Passiflora henryi,* species of *Thladiantha, Hemsleya,* and *Begonia, Capparis yunnanensis, Cleome yunnanensis, Sisymbrium yunnanensis, Cardamine yunnanensis, Camelina yunnanensis, Eutrema yunnanensis, Loxostemon delavayi, Thlaspi yunnanensis, Draba* spp., *Erysimum yunnanense, Populus yunnanensis, P. szechuanuca, P. bonatii, Salix* spp., *Actinidia* spp., *Clethra delavayi,* numerous species of *Rhododendron, Enkianthus pauciflorus, Cassiope* spp., *Craibiodendron henryi, C. yunnanense, Lyonia* spp., *Pieris forrestii, Leucothoe tonkinensis, Gaultheria* spp., *Agapetes* spp., *Vaccinium* spp., *Berneuxia thibetica, B. yunnanensis, Diapensia* spp., *Shortia sinensis, S. thibetica, Styrax* spp., *Alniphyllum eberhardtii* (distribution not completely clear), *Huodendron* spp., *Rehderodendron fengii, R. tsiangii, Melliodendron xylocarpum,* species of *Pterostyrax, Symplocos, Diospyros,* and *Ardisia,* numerous *Primula* spp., *Lysimachia* spp., *Androsace* spp., species of *Omphalogramma, Grewia, Daphne, Wikstroemia, Deutzia, Hydrangea, Philadelphus, Schizophragma,* and *Spiraea, Sibiraea tomentosa, Neillia* spp., *Maddenia yunnanensis, Prunus* spp., species of *Rubus, Potentilla,* and *Rosa, Dichotomanthes tristaniicarpa, Cotoneaster* spp., *Pyracantha angustifolia, Osteomeles schwerinae, Stranvaesia scandens, Sorbus* spp., species of *Photinia, Pyrus,* and *Bauhinia, Cercis yunnanensis, Gleditsia delavayi, Ormosia yunnanensis, Cladrastis delavayi, Piptanthus tomentosus,* species of *Crotalaria, Millettia, Dalbergia,* and *Indigofera, Caragana franchetiana, Nephelotrophe polystichoides, Astragalus* spp., *Oxytropis yunnanensis, Glycyrrhiza yunnanensis, Pueraria* spp., *Erythrina arborescens, Vicia* spp., species of *Trifolium, Desmodium, Lespedeza,* and *Campylotropsis, Oxyspora yunnanensis, Rhus delavayi, Zanthoxylum yunnanense, Evodia* spp., *Clausena yunnanensis, Ailanthus vilmoriniana, Acer* spp., *Delavaya yunnanensis, Koelreuteria bipinnata, Meliosma* spp., *Tirpitzia sinensis, Geranium* spp., *Impatiens* spp., *Polygala* spp., *Cornus* spp., *Toricellia angulata, Aralia* spp., *Acanthopanax wilsonii, Pentapanax yunnanensis,* species of *Angelica, Hera-*

cleum, Ligusticum, Pleurospermum, and Pternopetalum, Sinodielsia yunnanensis, Sinolimprichtia alpina, Sinocarum spp., Physospermopsis forrestii, Notopterygium forrestii, Harrysmithia dissecta, Haplosphaera phaea, Dickinsia hydrocotyloides, Chaerophyllopsis huai, Acronema spp., Pterocyclus rivularum, Seseli yunnanense, Trachydium spp., Trachyspermum scaberulum, Vicatia conifolia, Ilex yunnanensis and other species of this genus, Euonymus spp., Celastrus spp., Gymnosporia berberoides, Monocelastrus monosperma, Sinomerrillia bracteata, Ziziphus spp., Berchemia yunnanensis, Rhamnella forrestii, R. longifolia, Sageretia spp., species of Rhamnus, Vitis, Tetrastigma, and Fraxinus, Syringa yunnanensis, S. tomentella, S. potaninii, Ligustrum spp., Siphonosmanthus delavayi, Osmanthus spp., Jasminum spp., Linociera henryi, Olea spp., Elaeagnus spp., Helicia spp., Heliciopsis henryi, Sambucus wightiana, Viburnum spp., Dipelta yunnanensis, Abelia spp., Lonicera spp., Valeriana spp., Morina delavayi, Biondia yunnanensis, Cynanchum spp., Hoya yunnanensis, Tylophora yunnanensis, Cotylanthera yunnanensis, numerous Gentiana spp., Gentianopsis spp., Swertia spp., Veratrilla baillonii, Galium spp., Hymenopogon oligocarpus, Ixora henryi, I. yunnanensis, species of Lasianthus, Leptodermis, Luculia, Mussaenda, Oldenlandia, Psychotria, Randia, and Rubia, Trailliaedoxa gracilis, Ipomoea yunnanensis, Actinocarya tibetica, Antiotrema dunnianum, Ehretia spp., Microula spp., Onosma spp., Thyrocarpus sampsonii, Trigonotis spp., Anisodus luridus, Cyphotheca betacea, Buddleja spp., Calorhabdos sutchuensis, Lagotis yunnanensis, Lancea tibetica, Linaria yunnanensis, Mazus spp., Mimulus szechuanensis, Pedicularis spp., Pterygiella spp., Scrophularia spp., Veronica spp., Catalpa fargesii, C. tibetica, Incarvillea spp., Paulownia duclouxii, P. fortunei, Radermachera yunnanensis, Ancylotemon spp., Briggsia spp., Chirita spp., Corallodiscus spp., Didymocarpus yunnanensis, Lysionotus spp., Oreocharis spp., Petrocosmea spp., Phylloboea henryi, Tremacron forrestii, T. rubrum, Whytockia chiritaeflora, Gleadovia yunnanensis, Orobanche yunnanensis, Cystacanthus yunnanensis, Justicia spp., Strobilanthes spp., Caryopteris forrestii, Clerodendrum spp., Gmelina delavayana, Premna spp., Vitex yunnanensis, Chelonopsis spp., species of Colquhounia, Dracocephalum, Elsholtzia, and Gomphostemma, Hanceola sinensis, Leucosceptrum plectranthoideum, Loxocalyx urticifolius, Microtoena spp., Orthodon spp., Paralamium gracile, Paraphlomis hispida, P. robusta, species of Phlomis, Plectranthus, Salvia, Scutellaria, Teucrium, Adenophora, Campanula, Codonopsis, and Cyananthus, Heterocodon brevipes, Pentaphragma sinense, species of Lobelia, Ainsliaea, Anaphalis, Artemisia, Aster, Blumea, Cacalia, Cirsium, and Cremanthodium, Crepis bodinieri, Dubyaea spp., Inula spp., Ixeris stebbinsiana, Faberia ceterach, F. lancifolia, Formania mekongensis, species of Gerbera, Lactuca, Leontopodium, and Ligularia, Mazettia salweenensis, Myriactis spp., Nannoglottis yunnanensis, Nouelia insignis, numerous species of Saussurea, Serratula forrestii, species of Soroseris, Tanacetum, and Veronica, Vladimiria salwinensis, Wardaster lanuginosus, Youngia spp., Ottelia yunnanensis, species of *Aletris, Chlorophytum,* and *Disporopsis, Diuranthera minor, Fritillaria delavayi, Hemerocallis forrestii,* species of *Lilium, Nomocharis, Ophiopogon, Polygonatum,* and *Smilacina, Tofieldia yunnanensis, Veratrum yunnanensis, Ypsilandra yunnanensis,* species of *Daiswa, Allium, Smilax,* and *Heterosmilax, Stemona vagula,* species of *Dioscorea, Iris, Hedychium,* and *Roscoea,* many taxa of the Orchidaceae, *Juncus* spp., numerous species of *Carex,* and species of *Cyperus, Eleocharis, Aneilema,* and *Arundinella.*

11. Northern Burmese Province (Meusel et al. 1965).[42] This province embraces Upper Burma, or the "Burmese Oberland," as Ward 1946a called it. Inaccessible and sparsely populated northern Burma, with its steep mountain slopes and deep ravines, is among the least studied lands of Asia.[43] Therefore it is impossible to provide any exact description of the boundaries of the Northern Burmese Province. Based mainly on the work of Ward (especially Ward 1944), it is possible to designate the boundaries as follows, but in a very approximate way. Beginning in the south, apparently somewhat south of the city of Myitkyina (i.e., at about 25° north latitude), the Northern Burmese Province occupies the upper part of the Irrawaddy River Basin within the national boundaries of Burma; only in the middle course of the Taron River (the upper course of the Nmai River) does it enter extreme northwestern Yünnan. In the west, the province occupies the eastern slopes of the Patkai Hills (Patkai Bum), thus also including the massif of Mount Saramati (3,824 m). However, a large part of the Chindwin River Basin probably should be excluded because, according to Ward (1944), the flora here is more nearly Assamese than Burmese, which is especially clear in relation with the lower Chindwin. Thus the territory of the flora of Upper Burma is found between 25° and 29°30' north latitude and between 94°30' and 98°40' east longitude.

Upper Burma is greatly dissected by the sources of the Irrawaddy River, which makes it difficult of access. It represents the only remnant of a once widespread, undulating plateau, including the Yünnan Plateau to the east and the Tibetan Plateau in the north. During the Pleistocene, its surface was smoothed out by ice, then subsequently eroded by waterflow and so strongly dissected by rivers that its resemblance to a plateau has been lost (Ward 1944:555). A large amount of geomorphological data indicates that the territory of Upper Burma southward to 25°30', at least along its eastern side, was subjected to Pleistocene glaciation. The Pleistocene glaciation created the present set of rivers (Ward 1944:557).

Situated between the eastern Himalayas and southwestern China, Upper Burma is under the influence of various floristic centers. Over long periods of time it was the crossroad of large-scale floristic migrations, especially during the alternation of glacial and interglacial epochs. This left an imprint on its flora, which to a considerable degree exhibits a transitional character from the eastern Himalayan and Assamese to the Yünnanian, and from the eastern Asiatic to the Indomalesian (see the section on the Indomalesian Subkingdom, in the Paleotropic Kingdom). As Ward (1944) remarked, the flora of Upper Burma has basically an Eastern Asiatic ("Sino-Himalayan") character, which is especially well expressed in the temperate zone. Although in the southern part of the province the Indomalesian flora plays a considerable role, it is characterized by a considerable mixture of eastern Asiatic elements.[44] The Eastern Asiatic character of the flora of the Northern Burmese Province is underscored by the appearance of such plants as:

Podocarpus macrophyllus, Taxus wallichiana, Abies spp., *Larix potaninii, Tsuga yunnanensis, Pinus armandii, Taiwania flousiana, Manglietia insignis, Magnolia campbellii, M. globosa, M. griffithii, M. rostrata, Alcimandra cathcartii, Michelia doltsapa, M. floribunda, Machilus odoratissima, Lindera cercidifolia, L. vernayana, Aristolochia griffithii, Illicium simonsii, Schisandra neglecta, S. rubriflora, Kadsura interior, Decaisnea fargesii, Mahonia lomariifolia, Berberis* spp., *Coptis* spp., *Isopyrum adiantifolium, Aquilegia* spp., *Caltha* spp., *Trollius micranthus, T. pumilus, T. yunnanensis,* species of *Delphinium, Aconitum,* and *Thalictrum, Anemone vitifolia, Clematis buchaniana, C. fasciculo.ris, C. nepalensis* etc., *Ranunculus* spp., *Souliea vaginata,* species of *Berberis* and *Meconopsis, Corydalis leptocarpa, Dicentra paucinervia, Tetracentron sinense, Corylopsis manipurensis, Ulmus lanceaefolia,* a series of Himalayan-Chinese species of *Quercus* and *Castanopsis, Betula* spp., *Carpinus viminea, Corylus ferox, Engelhardtia spicata, Rheum* spp., *Camellia kissi, Gordonia axillaris, Schima argentea, S. khasiana, Eurya* spp., *Hypericum elodeoides, H. hookerianum, Stachyurus himalaicus, Populus ciliata, Salix lindleyana, Actinidia* spp., *Saurauia napaulensis, Clethra delavayi, Rhododendron* (many species, including *R. delavayi, R. simsii, R. sulfureum,* and *R. yunnanense), Enkianthus* spp., *Lyonia ovalifolia, Pieris formosa,* species of *Gaultheria, Vaccinium, Agapetes,* and *Pyrola, Monotropastrum humile, Diapensia himalaica, Berneuxia thibetica, Styrax serrulatus, Myrsine semiserrata,* species of *Lysimachia, Primula,* and *Androsace, Sarcococca saligna, Edgeworthia gardneri, Daphne* spp., *Ribes glaciale, Deutzia compacta, D. glomeriflora, D. purpurascens, D. wardiana, Hydrangea* spp., *Schizophragma* spp., *Sedum multicaule, Astilbe* spp., *Rodgersia* spp., *Bergenia purpurascens, Saxifraga* spp., *Chrysosplenium forrestii, C. lanuginosum, C. nepalense, Parnassia* spp., *Spiraea arcuata, S. canescens, Neillia thyrsiflora, Prunus* spp., *Prinsepia utilis, Rubus* spp., *Potentilla peduncularis, Rosa* spp., *Sorbus wardii, Cotoneaster distichus, C. horizontalis, C. rubens* and other species of the genus, *Photinia integrifolia, Docynia indica, Sophora praseri, Skimmia laureola, S. melanocarpa, Dobinea vulgaris, Boenninghausenia albiflora, Turpinia nepalensis, Acer campbellii, A. hookeri, A laevigatum, A. sikkimense, A. tetramerum* and other species of this genus, *Impatiens* spp., *Aesculus assamica, Epilobium sikkimense* subsp.

ludlowianum, Cornus capitata, Aucuba himalaica, Helwingia himalaica, Hedera nepalensis, Panax pseudoginseng, Ilex spp., including *I. corallina, Celastrus hookeri, Euonymus* spp., *Tripterygium wilfordii, Fraxinus floribunda, Syringa* spp., *Osmanthus fragrans, O. suavis, Dipentodon sinicus, Leycesteria* spp., *Viburnum coriaceum*, species of *Lonicera, Valeriana, Gentiana, Swertia*, and *Myosotis, Buddleja myriantha, Pedicularis* spp., *Luculia intermedia, Wightia speciosissima, Briggsia kurzii, Lysionotus serrata, Callicarpa rubella, Ajuga lobata, Notochaete hamosa, Orthosiphon incurvus, Teucrium quadrifarium, Plectranthus* spp., *Pogostemon brachystachys, Dracocephalum* spp., *Phlomis breviflora, Lobelia pyramidalis, Peracarpa carnosa, Streptopus simplex, Smilacina fusca, S. oleracea, Lilium bakerianum, L. giganteum*, species of *Nomocharis, Ophiopogon, Polygonatum, Paris, Allium, Smilax, Dioscorea, Iris, Luzula, Eleocharis, Cyperus*, and *Carex, Streptolirion volubile, Dendrocalamus hamiltonii*, and *Arisaema* spp.

The flora of the Northern Burmese Province apparently has several endemic, monotypic genera (including *Woodburnia*) and a considerable number of endemic species, constituting probably no less than 25% of the entire vascular plant flora. Of the endemic species we mention only the following: *Juniperus coxii, Magnolia nitida, Litsea brachypoda, L. cuttingiana, Lindera wardii, Illicium burmanicum, I. wardii, Mahonia aristata, Berberis burmanica, B. coxii, B. hypokerina, B. rufescens, B. venusta, Meconopsis violacea, Dactylicapnos grandifolia, Camelia irrawadiensis, C. wardii, Gordonia axillaris, Eurya urophylla, E. wardii, Stachyurus cordatula, Begonia hymenophylloides, Saurauia subspinosa*, no less than 25 species of *Rhododendron*, including *R. dendricola, R. imperator, R. insculptrum, R. magnificum, R. myrtilloides*, and *R. taggianum*, several species of *Agapetes, Diplycosia alboglauca, D. pauciseta, Symplocos araioura, Maesa marianae, Primula burmanica, P. densa, P. distyophylla* and other species of this genus, *Wickstroemia floribunda, Saxifraga anisophylla, S. calopetala, S. heteroclada, S. virgularis, Pygeum cordatum, Rubus chaetocalyx, R. wardii, Sorbus apicidem, S. detergibilis, S. paucinervia, Photinia myriantha, Eriobotrya platyphylla, E. wardii, Syzygium stenurum, Epilobium kermodei, Acer chionophyllum, A. chloranthus, A. pinnatinervum, Woodburnia penduliflora, Dendropanax burmanicus, Gamblea longipes, Brassaiopsis trilobata, Ilex cyrtura, Euonymus griffithii, E. kachinensis, Jasminum farreri, Leycesteria insignis, Viburnum cuttingianum, Gentiana gradata, Ophiorrhiza lignosa, Brachycome wardii, Ixora kingdon-wardii, Lasianthus wardii, Aeschynanthus wardii, Strobilanthes stramineus, Vernonia adenophylla, Aster helenae, Senecio pentanthus, Prenanthes volubilis, Lactuca gracilipetiolata, Peliosanthes longibracteata, Allium acidoides, Stemona wardii, Paphiopedilum wardii*, and *Coelogyne ecarinata*.

Exclaves of the Northern Burmese flora are found beyond the limits of the province, especially on the highest crests of the Shan Uplands.

12. Eastern Himalayan Province (Hooker [1904] 1907; Stearn 1960;

Meusel et al. 1965; Takhtajan 1970, 1974).[45] This province embraces a large part of eastern Nepal, in the west approximately up to the valley of the Kali River Valley—that is, around 83° east longitude (Stearn 1960; Raven 1962; Gabrielyan 1978), with the exception of areas of tropical vegetation usually found lower than 1,000 meters above sea level. It also includes Darjeeling and Sikkim; Bhutan; a large part of the Assam Himalayas; some parts of extreme south and southeastern Tibet (several of the more humid mountains chains open to monsoon winds, the extreme eastern part of the Tsangpo River Basin to the east of 92° east longitude, the warm and humid valleys of the Trisuli, Torsa, and other rivers, and a large part of Mönyul, with the exception of the tropical zone). In the east the territory of the province includes Mt. Namcha-Barva (7,756 m above sea level) and approaches the valley of the Dihang River (a continuation of the Tsangpo River).[46] While the western boundary seems rather well defined, the eastern boundary is rather indefinite. Here we observe a gradual transition from the eastern Himalayan flora to the Chinese. A number of Sikang-Szechwan and montane Yünnan species penetrate into the Tsangpo and the Mönyul valleys. Likewise, a considerable number of eastern Himalayan plants are also encountered in southwestern China.

Geologically, the Himalayas are younger than the mountains of western and southwestern China and of the Shillong Plateau, and the flora of the eastern Himalayan Province is also younger. After the glacial period, the eastern Himalayas were colonized basically from the east and southeast, but the flora also contains a considerable number of endemic species and even a series of endemic genera. This indicates that the glaciation was not absolutely catastrophic, and that many taxa may have developed indigenously since as far back as the Pliocene.

The flora of the Eastern Himalayan Province contains a number of endemic and subendemic genera, including *Paroxygraphis, Smithiella, Neoluffa, Edgaria, Biswarea, Bryocarpum,* and *Milula.* Of the numerous endemic and subendemic species I mention:

Lycopodium subuliferum, Pyrrosia boothii, Abies densa, Pinus bhutanica, Larix griffithiana, Cupressus corneyana, Lindera heterophylla (Nepal to Bhutan, Assam), *L. venosa* (Nepal to Bhutan), *Machilus edulis* (Sikkim), *M. gammieana* (Sikkim), *Aristolochia griffithii* (Nepal to NEFA, southern Tibet), *A. nakaoi* (eastern Nepal to Bhutan), *Asarum himalaicum* (Nepal to NEFA), species of *Delphinium, Aconitum, Thalictrum, Anemone, Clematis,* and *Ranunculus, Paroxygraphis sikkimensis* (Nepal to Bhutan), *Mahonia griffithii* (Bhutan), *M. hicksii* (eastern Bhutan), *M. monyulensis* (southeastern Tibet), *M. sikkimensis* (Sikkim), several species of *Berberis,* species of *Meconopsis* and *Corydalis, Corylopsis himalayana* (Bhutan), species of *Stellaria* and *Arenaria, Rheum nobile* (Nepal to Bhutan), *Ceratostigma griffithii* (Bhutan), *Symplocos pyrifolia*

(Nepal to NEFA), *Hypericum tenuicaule* (Nepal, Sikkim), *Homalium napaulense* (Nepal), *Viola* spp., *Salix bhutanensis* (Nepal to Bhutan), *S. plectilis* (eastern Nepal), *Begonia cathcartii* (Nepal to Bhutan), *B. gemmipara* (Nepal, Sikkim), *B. griffithii* (Nepal to Bhutan), *Eutrema himalaicum* (Sikkim, Bhutan, southern Tibet), *Microsisymbrium axillare* (Nepal to Bhutan), *Cardamine* spp., *Draba bhutanica* (Bhutan), *Actinidia strigosa* (Nepal, Sikkim), *Saurauia fasciculata* (Nepal, Sikkim), many species of *Rhododendron*, *Agapetes* spp., (including *A. bhutanica*), *Pyrola sikkimensis*, *Androsace* spp., a series of species of *Primula*, *Bryocarpum himalaicum*, *Omphalogramma elwesiana* (Nepal to Assam, southeastern Tibet), *Baliospermum nepalense* (eastern Nepal), *Euphorbia himalayensis* (Nepal to Bhutan, Tibet), *E. sikkimensis* (Nepal to Bhutan, Tibet), a series of species of *Rhodiola* and *Saxifraga*, *Rodgersia nepalensis* (eastern Nepal, southern Tibet), *Maddenia himalaica* (Nepalto Bhutan, southern Tibet), a series of species of *Rubus*, *Potentilla bhutanica* (Bhutan), *P. bryoides* (Bhutan), *Geum sikkimense* (Nepal to Bhutan), *Cotoneaster taylorii*, *Sorbus griffithii*, *S. hedlundii* (eastern Nepal to Bhutan), *S. kurzii* (eastern Nepal, Sikkim), *Eriobotrya hookeriana* (eastern Nepal to Bhutan), *Sophora bhutanica* (Bhutan), *Astragalus* spp., *Acer hookeri* (from Nepal to NEFA), *A. stachyophyllum* (Nepal to Bhutan, southern Tibet), an entire series of species of *Impatiens*, *Brassaiopsis alpina* (eastern Nepal, Sikkim), *Heracleum wallichii* (Nepal to Bhutan), *Euonymus macrocarpus* (Sikkim), *E. tibeticus* (Sikkim, southeastern Tibet), *Berchemia flavescens*, *Ampelocissus nervosa* (Nepal, Sikkim), *Thesium emodi* (Nepal, Bhutan), *Elaeagnus caudata* (Nepal to Bhutan), *Lonicera myrtilloides* (Nepal, Sikkim), *Dipsacus atratus* (Nepal to Bhutan), *Ichnocarpus himalaicus* (Nepal to Bhutan), *Ceropegia bhutanica* (Bhutan), *Hoya bhutanica* (Bhutan), *H. polyneura* (Sikkim, Bhutan), *H. serpens* (Nepal, Sikkim), a series of species of *Gentiana* and *Swertia*, *Luculia grandifolia* (Bhutan), *Ophiorrhiza prostrata*, *O. steintonii* (Nepal, Sikkim), *Actinocarya bhutanica* (Bhutan), *Ehretia wallichiana*, *Microula bhutanica* (Bhutan), *Onosma bhutanica* (Bhutan), *Buddleja bhutanica* (Bhutan), *B. colvilei* (Nepal to NEFA), *B. tibetica* (Bhutan and southern Tibet), *Euphrasia bhutanica* (Bhutan), *E. simplex* (Nepal, Bhutan), *Lagotis clarkei* (Nepal to Bhutan), *Lindenbergia bhutanica* (Bhutan), numerous *Pedicularis* spp., *Scrophularia sikkimensis* (Sikkim, Bhutan), *Veronica deltigera* (Nepal, Sikkim), *V. lanuginosa* (Nepal to Bhutan, southeastern Tibet), *V. robusta* (Nepal to Bhutan), *Chirita primulacea* (Nepal, Sikkim), *Corallodiscus bhutanicus* (Bhutan), *Didymocarpus* spp., *Acanthus carduaceus* (Sikkim, Bhutan), *Pteracanthus lachenensis* (Nepal, Sikkim), *Strobilanthes laevigatus* (Sikkim, Bhutan), *S. thomsonii*, *Callicarpa lobata* (Nepal, Sikkim), *Pogostemon tuberculosus* (Nepal to Bhutan), *Salvia sikkimensis* (Nepal to Bhutan, Tibet), *Codonopsis subsimplex*, *C. thalictrifolia*, *Cyananthus pedunculatus* (Nepal, Sikkim, southern Tibet), *Lobelia nubigena* (Bhutan), *Cacalia mortonii*, *Cicerbita macrantha* (Nepal to

Bhutan), *Cirsium eriophoroides* (Sikkim, Bhutan), *Conyza angustifolia* (Nepal to Bhutan), *Ligularia pachycarpa* (Sikkim), *Saussurea conica* (Sikkim, Bhutan), *Senecio tetranthus* (Nepal to Bhutan), *Tofieldia himalaica* (Nepal to Bhutan), *Aletris gracilis, Iris clarkei* (Nepal to Bhutan), *Polygonatum brevistylum* (Nepal, Sikkim), *Lilium sherriffiae* (Nepal to Bhutan), *Notholirion macrophyllum* (Nepal to Bhutan, southeastern Tibet), *Allium phariense* (southeastern Tibet), *A. rhabdotum* (Bhutan), species of *Bulbophyllum* and *Calanthe, Chiloschista usneoides* (Nepal, Bhutan), *Coelogyne occultata* (Sikkim, Bhutan), *C. longipes* (Nepal, Sikkim), *Cymbidium hookerianum* (Nepal to Bhutan, southeastern Tibet), *Diplomeris hirsuta* (Nepal to Bhutan), *Gastrochilus affinis* (Nepal, Sikkim), *G. dasypogon* (Nepal, Sikkim), *G. distichus* (Nepal to Bhutan), *Goodyera hemsleyana* (Nepal, Sikkim), *G. vittata* (Nepal to Bhutan, Tibet), *Habenaria juncea* (Sikkim), *Liparis perpusilla* (Nepal, Sikkim), *L. pygmaea* (Nepal, Sikkim, southeastern Tibet), *Nervilia macroglossa* (Sikkim, Bhutan), *N. scottii* (Nepal, Sikkim), *Pholidota recurva* (Nepal, Sikkim), *Platanthera exelliana* (Nepal, Sikkim, southeastern Tibet), *P. juncea* (Nepal, Sikkim), *P. leptocaulon* (Nepal to Bhutan, southeastern Tibet), *P. sikkimensis* (Nepal, Sikkim), *Spathoglottis ixioides* (Nepal to Bhutan), *Sunipia paleacea* (Nepal to Bhutan), *Tipularia josephii* (Nepal to Bhutan, southwestern Tibet), species of *Juncus, Carex, Kobresia,* and *Eriocaulon, Agrostis* spp., *Bromus himalaicus* (Nepal to Bhutan), *Anthoxanthum hookeri* (Nepal, Sikkim), *A. sikkimense* (Nepal, Sikkim), *Festuca* spp., and *Arisaema* spp.

The flora of the Eastern Himalayan Province appears to be one of the youngest in the Eastern Asiatic Region. It is considerably younger than the flora of the Khasi Hills or of southwestern China, and was formed as a result of dispersal of plants to the west from the more ancient mountain ranges of China, Burma, and Assam (Diels 1913; Wulff 1944; Fedorov 1957). It has much in common with the floras of the Sikang-Yünnan and Northern Burmese provinces and also with the flora of the Khasi-Manipur Province (especially that of Nagaland).

13. Khasi-Manipur Province.[47] Insufficient and, in places, very superficial studies of the eastern and especially northeastern areas of India prevent me from accurately indicating the boundaries of this floristic province (which I have, nevertheless, established). It undoubtedly includes a large part of the Shillong Plateau—namely, the Khasi Hills, the Jaintia Hills, and Manipur, and apparently also the Naga Hills,[48] the Patkai Hills, and possibly the Mikir Hills. In the east, the boundary probably passes through the entire eastern slope of the Naga Hills (within the border of Burma). The Shillong Plateau joins with the Manipur Mountains and the Naga Hills to constitute part of the system of the Assam-Burmese Mountains, through the Barail Range, which is not very high and is covered with predominantly tropical vegetation.

In the south, the Manipur Mountains extend to the Mizo Hills and to the Chin Hills. Tropical vegetation predominates up to approximately 900 m above sea level, then changes into subtropical. According to Rao (1974), the temperate zone begins approximately at 1,300 m.

The Khasi-Manipur Province is usually placed in the Paleotropical Kingdom (e.g., Good 1974), but Clayton and Panigrahi (1974) with good reason attribute their "Naga-Khasia Endemic Centre" to the Holarctic Kingdom and include it in the "Himalayan Region." Thus they assign it to the western part of the Eastern Asiatic Region.

The nucleus of the flora of the Khasi-Manipur Province consists of eastern Asiatic elements. Such plants include:

Polypodium lachnopus (Khasi Hills), *Arthromeris himalayensis* (Khasi Hills), *A. wardii* (Naga Hills), *Cephalotaxus griffithii* (Khasi Hills, Naga Hills; found also in Upper Assam), *Taxus wallichiana* (Khasi Hills, Naga Hills), *Alcimandra cathcartii* (Khasi Hills, Naga Hills, Manipur), *Manglietia caveana* (Khasi Hills), *M. insignis* (Khasi Hills, Naga Hills, Manipur), *Magnolia campbellii* (Manipur), *Michelia doltsopa* (Khasi Hills, Manipur), *M. manipurensis* (Khasi Hills, Manipur), *M. punduana* (Khasi Hills), *M. velutina* (Khasi Hills, Naga Hills, Manipur), *Actinodaphne reticulata* (Khasi Hills), *A. sikkimensis* (Manipur), *Lindera latifolia* (Khasi Hills), *L. nacusua* (Manipur), *L. pulcherrima* (Khasi Hills), *Litsea cubeba* (Khasi Hills), *L. elongata* (Khasi Hills), *L. kingii* (Khasi Hills), *L. oblonga* (Manipur), *L. sericea* (Khasi Hills, Manipur), *Neolitsea lanuginosa* (Khasi Hills), *N. umbrosa* (Khasi Hills, Manipur), *Piper nepalense* (Khasi Hills, Manipur), *Peperomia heyneana* (Khasi Hills, Manipur), *Mitrastemon yamamotoi* (Khasi Hills, Naga Hills), *Illicium griffithii* (Khasi Hills, Naga Hills, Manipur), *I. simonsii* (Naga Hills), *Schisandra neglecta* (Khasi and Jaintia Hills), *S. propinqua* (Khasi Hills), *Kadsura roxburghiana* (Khasi Hills, Manipur), *Stauntonia brunoniana* (Khasi Hills, Manipur), *Holboellia latifolia* (Khasi Hills, Jaintia Hills, Manipur), *Stephania elegans* (Khasi Hills, Naga Hills, Manipur), *S. glandulifera* (Khasi Hills), *Trollius pumilus* (Khasi Hills), *Delphinium altissimum* (Khasi Hills, Manipur), *D. stapeliosmum* (Khasi Hills), *Thalictrum foliolosum* (Khasi Hills, Manipur), *T. punduanum* (Khasi Hills), *Anemone elongata* (Khasi Hills), *A. rivularis* (Khasi Hills), species of *Clematis*, *Ranunculus diffusus* (Khasi Hills, Manipur), *Mahonia sikkimensis* (Khasi Hills, Naga Hills), *Berberis griffithiana* (Khasi Hills), *B. sublevis* (Khasi Hills, Manipur), *B. wallichiana* (Khasi Hills), *Dicentra paucinervia* (Khasi Hills, Naga Hills), *D. roylei* (Khasi Hills, Jaintia Hills), *D. scandens* (Khasi Hills, Manipur) and several other species of this genus, *Corydalis chaerophylla* (Naga Hills, Manipur), *C. longipes* (Khasi Hills, Manipur), *C. himalayana*(Khasi Hills), *Corylopsis himalayana* (Khasi Hills), *Loropetalum chinense* (Khasi Hills), *Sycopsis griffithiana* (Khasi Hills, Jaintia Hills), about 10 species of *Quercus,* including *Q. dealbata* (Khasi Hills, Naga Hills), *Q. fenestrata* (Khasi Hills), *Q. griffithii* (Khasi Hills) and *Q. lineata* (Khasi

Hills), *Alnus nepalensis* (Khasi Hills, Naga Hills, Manipur), *Betula alnoides* (Khasi Hills, Manipur), *Carpinus viminea* (Khasi Hills), *Myrica farguhariana* (*M. Sapida*) (Khasi Hills, Naga Hills, Manipur), *Juglans regia, Stilbanthus scandens* (Naga Hills), *Polygonum* spp., *Camellia caduca* (Khasi Hills), *C. kissi* (Khasi Hills, Naga Hills, Manipur), *Eurya cerasifolia* (Manipur), *Schima wallichii* (Khasi Hills, Manipur), *Cleyera japonica* (Khasi Hills), *Hypericum elodeoides* (Khasi Hills), *H. hookerianum* (Khasi Hills, Manipur), *H. sampsonii* (Khasi Hills, Manipur, Chin Hills in Burma), *Stachyurus himalaicus* (Naga Hills, Manipur), *Viola sikkimensis* (Khasi Hills, Nagaland), *V. thomsonii* (Manipur), *Salix eriophylla* (Khasi Hills), *Saurauia napaulensis* (Khasi Hills, Manipur), *Rhododendron arboreum* (Khasi Hills, Naga Hills, Manipur), *R. lindleyi* (Manipur) and several other species of this genus, *Gaultheria griffithiana* (Khasi Hills). *G. nummularoides* (Khasi Hills), *Pieris formosa* (Khasi Hills, Manipur), species of *Vaccinium, Pyrola decorata* (Manipur), *Monotropastrum humile* (Khasi Hills, Manipur), *Styrax hookeri* (Manipur), *Symplocos* spp., *Maesa* spp., *Ardisia macrocarpa* (Khasi Hills), *Myrsine semiserrata* (Khasi Hills, Manipur), *Primula listeri* (Manipur), species of *Lysimachia, Sarcococca hookeriana* (Khasi Hills), *Daphniphyllum himalayense* (Khasi Hills), *Edgeworthia gardneri* (*E. tomentosa*) (Manipur), *Daphne papyracea* (Khasi Hills, Manipur), *Itea macrophylla* (Khasi Hills, Manipur), *I. chinensis* (Khasi Hills, Jaintia Hills, Manipur), *Ribes glaciale* (Naga Hills, Manipur), *Dichroa febrifuga* (Khasi Hills, Manipur), *Hydrangea heteromalla, H. macrophylla* (Manipur), *Pittosporum napaulense* (Khasi Hills, Manipur), *Astilbe rivularis* (Khasi Hills, Manipur), *A. rubra* (Khasi Hills), *Saxifraga brachypoda* (Naga Hills), *Chrysosplenium lanuginosum* (Manipur), *C. nepalense* (Manipur), *Spiraea micrantha* (Khasi Hills, Manipur), *Neillia thyrsiflora* (Khasi Hills, Manipur), species of *Prunus, Prinsepia utilis* (Khasi Hills), species of *Rubus, Potentilla,* and *Rosa, Stranvaesia nussia* (Khasi Hills, Jaintia Hills), *Eriobotrya angustissima* (Garo Hills, Khasi Hills, Jaintia Hills), *Docynia indica* (Khasi Hills, Manipur), *Pyrus pashia* (Khasi Hills, Jaintia Hills, Manipur) and other species of *Pyrus, Hydrobryum griffithii* (Khasi Hills, Manipur), *Sophora benthamii* (Khasi Hills, Manipur), *Syzygium* spp., *Circaea alpina* subsp. *imaicola* (Khasi Hills), *Epilobium brevifolium* subsp. *trichoneurum, E. royleanum* (Khasi Hills), *E. wallichianum* (Khasi Hills, Nagaland), *Rhus chinensis* (Khasi Hills, Naga Hills), *Zanthoxylum* spp., *Evodia fraxinifolia* (Khasi Hills, Jaintia Hills), *Skimmia laureola* (Khasi Hills, Jaintia Hills), *Citrus latipes* (Khasi Hills), *Coriaria nepalensis* (Manipur), *Turpinia nepalensis* (Khasi Hills), *Acer laevigatum* (Khasi Hills, Manipur), *A. oblongum* (Khasi Hills, Jaintia Hills, Manipur), *Anisadenia pubescens* (Khasi Hills), *A. saxatilis* (Khasi Hills, Manipur), *Reinwardtia indica* (Manipur), *Oxalis acetosella* subsp. *griffithii* (Khasi Hills, Manipur), *Impatiens* spp., *Cornus capitata* (Khasi Hills, Jaintia Hills, Naga Hills), *C. controversa* (Khasi Hills, Manipur), *C. macrophylla* (Khasi Hills, Manipur), *C. oblonga* (Khasi Hills, Manipur), *Aucuba himalaica* (Manipur), *Helwingia himalaica* (Khasi Hills, Naga Hills, Manipur), *H. lanceolata* (Naga Hills, Manipur), *Tupidanthus calyptratus* (Khasi Hills), *Pentapanax racemosus* (Khasi Hills), *Acanthopanax aculeatus* (Khasi Hills), *Panax pseudoginseng* (Khasi Hills, Manipur), *Hedera nepalensis* (Khasi Hills), *Brassaiopsis* spp., *Macropanax undulatus* (Khasi Hills), *Pimpinella diversifolia* (Khasi Hills, Manipur), *P. sikkimensis* (Manipur), *Peucedanum ramosissimum* (Khasi Hills, Jaintia Hills), *Heracleum obtusifolium* (Khasi Hills, Jaintia Hills), species of *Ilex* and *Euonymus*, *Celastrus hookeri* (Khasi Hills, Manipur), *Ampelocissus sikkimensis* (Khasi Hills), *Parthenocissus semicordata* (Khasi Hills, Naga Hills), *Tetrastigma rumicispermum* (Khasi Hills), *Cayratia japonica* (Khasi Hills, Manipur), *C. thomsonii* (Khasi Hills), *Fraxinus floribunda* (Khasi Hills, Naga Hills), *Ligustrum confusum* (Khasi Hills, Manipur), *Osmanthus suavis* (Manipur), *Jasminum amplexicaule* (Khasi Hills), *J. dispermum* (Khasi Hills, Manipur), *J. lanceolarium* (Khasi Hills, Jaintia Hills, Naga Hills), *J. nepalense* (Khasi Hills), *Loranthus odoratus* (Manipur), *Leycesteria formosa* (Khasi Hills), *Lonicera glabrata* (Manipur), *L. ligustrina* (Khasi Hills), *L. macrantha* (Khasi Hills), *Viburnum cylindricum* (Khasi Hills, Manipur), *V. foetidum* (Khasi Hills, Manipur), *Sambucus adnata* (Khasi Hills), *Periploca calophylla* (Khasi Hills), *Gentiana speciosa* (Khasi Hills, Manipur), *Swertia chirayita* (Khasi Hills), *S. macrosperma* (Khasi Hills), *S. nervosa* (Khasi Hills, Naga Hills), *Ophiorrhiza fasciculata* (Khasi Hills), *O. treutleri* (Khasi Hills), *Luculia pineana* (Khasi Hills), *Galium elegans* (Khasi Hills, Manipur), *Rubia manjith* (Khasi Hills), *R. sikkimensis* (Naga Hills, Manipur), *R. wallichiana* (Khasi Hills), *Buddleja macrostachya* (Khasi Hills), *Centranthera grandiflora* (Khasi Hills), *Hemiphragma heterophyllum* (Khasi Hills), *Wightia speciosissima* (Khasi Hills, Manipur), *Aeschynanthus bracteatus* (Khasi Hills), *Chirita pumila* (Khasi Hills, Manipur), *Corallodiscus lanuginosus* (Khasi Hills), *Loxostigma griffithii* (Khasi Hills, Manipur), *Lysionotus serrata* (Khasi Hills, Manipur), *Teucrium quadrifarium* (Khasi Hills, Manipur), *Ajuga lobata* (Khasi Hills, Manipur), *A. macrosperma* (Khasi Hills, Manipur), *Scutellaria discolor* (Khasi Hills, Manipur), *Colquhounia vestita* (Khasi Hills, Manipur), *Craniotome furcata* (Khasi Hills), *Elsholtzia ciliata* (Khasi Hills), *Pogostemon amaranthoides* (Khasi Hills), *Orthosiphon incurvus* (Khasi Hills, Naga Hills), *Codonopsis viridis* (Khasi Hills), *Lobelia pyramidalis* (Khasi Hills, Naga Hills, Manipur), *Peracarpa carnosa* (Khasi Hills, Manipur), *Senecio chrysanthemoides* (Khasi Hills, Manipur), *Inula nervosa* (Khasi Hills), *Gerbera piloselloides* (Khasi Hills, Manipur), *Ophiopogon intermedius* (Khasi Hills, Manipur), *O. leptophyllus* (Khasi Hills), *Theropogon pallidus* (Khasi Hills), *Polygonatum cathcartii* (Khasi Hills, Manipur Nagaland), *P. cirrifolium* (Manipur), *P. punctatum* (Manipur), *Smilacina fusca* (Khasi Hills, Manipur), *Cardiocrinum giganteum* (Khasi Hills, Manipur), *Dioscorea kumaonensis* (Khasi Hills, Manipur), *Iris nepalensis* (Khasi Hills), *I. wattii* (Manipur), *Coelogyne* spp., species of *Juncus* and *Luzula*, *Carex* spp., *Streptolirion cordifolium* (Khasi Hills, Manipur), *Commelina sik-*

kimensis (Khasi Hills), species of *Eriocaulon, Sinobambusa elegans* (Naga Hills), *Trachycarpus martianus* (Khasi Hills, Manipur), *Gonatanthus ornatus* (Khasi Hills), *G. pumilus* (Khasi Hills, Manipur), and *Arisaema* spp.

This extensive but far from complete list of eastern Asiatic components of the flora of the Khasi-Manipur Province testifies to its strong ties with the floras of the eastern Himalayas, Upper Burma, and China.[49] The flora of the Khasi Hills, the one most richly saturated by eastern Asiatic elements, is itself one of the most important centers of survival of the Tertiary flora of eastern Asia. As Bor (1942a:194) wrote,

> It must be remembered that the Khasi Hills are more ancient geologically than either the Naga Hills or the Himalaya. In fact this plateau has stood firm during all the upheavals and earth movements which took place after the close of the Cretaceous Period. It has seen the retreat of the Himalayan flora to the southeast during the glacial epoch and welcomed it back on its return. The plateau has altered very little in all these thousands of years.

But it seems to me that this does not allow us to conclude that the Khasi Hills were only a refuge for the retreating Himalayan flora. The geological antiquity of the Khasi Hills, as well as the taxonomic composition of the flora and its richness and peculiarity, indicate that long before glacial times, these mountains already constituted an important part of the area where the nucleus of the eastern Asiatic flora originated.

According to the somewhat obsolete data of Deb (1958), the vascular plant flora of the entire state of Manipur contains 2,191 species, of which only 27 species are endemic (i.e., 1.3% of the entire flora). The flora of the Shillong Plateau and of the Khasi Hills in particular is considerably richer, and its endemism is much higher. But this is still very approximate since exact statistical calculations cannot yet be made. Even less well known is the flora of Nagaland. Of the endemic species in the Khasi-Manipur Province I mention the following:

Cephalotaxus mannii (Khasi Hills, Naga Hills), *Mahonia magnifica* (Manipur), *M. manipurensis* (Manipur), *M. pycnophylla* (Khasi Hills), *M. simonsii* (Khasi Hills), *Berberis feddei* (Manipur), *B. khasiana* (Khasi Hills), *B. manipurana* (Manipur), *B. wardii* (Naga Hills), *Distylium indicum* (Khasi Hills), *Corylopsis manipurensis* (Manipur), *Boehmeria hamiltoniana* (Khasi Hills), *B. macrophylla* (Khasi Hills), *B. polystachya* (Khasi Hills), *B. sidaefolia* (Khasi Hills), *Polygonum paleaceum, P. rude, Schima khasiana* (Khasi Hills), *Daphne shillong* (Khasi Hills), *Rhododendron elliottii* (Manipur), *R. johnstoneanum* (Manipur), *R. manipurense* (Manipur), *R. wattii* (Manipur), *Ardisia khasiana* (Khasi Hills), *A. polycephala* (Khasi Hills, Naga Hills, Manipur), *A. quinquangularis* (Khasi Hills), *A. rhynchophylla* (Khasi Hills), *A. virens, Phyllanthus griffithii* (Khasi Hills, Manipur), *Kalanchoë rosea* (Manipur), *Rubus assamensis* (Khasi Hills), *R. opulifolium* (Khasi Hills, Manipur), *Potentilla manipurensis*

(Manipur), *Cotoneaster simonsii* (Khasi Hills), *Sorbus khasiana* (Khasi Hills), *Justicia khasiana* (Khasi Hills, Manipur), a whole set of species of *Impatiens, Carum khasianum* (Khasi Hills), *Pimpinella flaccida* (Manipur), *Euonymus attenuatus* (Khasi Hills), *Ligustrum myrsinites* (Khasi Hills, Manipur), *Jasminum dumicola* (Naga Hills, Manipur), *Dipsacus asper* (Khasi Hills, Manipur), *Gentiana campanulacea* (Khasi Hills, Manipur), *Lacaitaea khasiana* (Khasi Hills, Manipur), *Scutellaria khasiana* (Khasi Hills), *Aeschynanthus superba* (Khasi Hills, Manipur), *Strobilanthes acrocephalus* (Khasi Hills, Naga Hills, Manipur), *S. maculatus* (Khasi Hills, Manipur), *Callicarpa psilocalyx* (Khasi Hills, Manipur), *Veronica cylindriceps* (Manipur), *Aster wattii* (Manipur), *Senecio filifolius* (Khasi Hills, Manipur), *S. nagensium* (Naga Hills, Manipur), *S. rhabdos* (Naga Hills, Manipur), *Ainsliea angustifolia* (Khasi Hills, Manipur), *Lilium mackliniae* (Manipur), *Smilax myrtillus* (Khasi Hills, Naga Hills, Manipur), *Iris bakeri* (Manipur), *I. wattii* (Manipur), *Hedychium* spp., an entire series of members of the Orchidaceae, *Carex manipurensis* (Manipur), *Hierochloë clarkei* (Khasi Hills, Naga Hills, Manipur), *Brachiaria villosa* (Khasi Hills, Nagaland, Manipur), *Pogonatherum rufobarbatum* (Khasi Hills, Manipur), and *Cymbopogon khasianus* (Khasi Hills, Naga Hills, Manipur).

The endemic species are related mainly to Holarctic taxa. However, among the endemics of the Khasi-Manipur Province there is also a series of representatives of typical paleotropic genera, of which one of the most remarkable is *Nepenthes khasiana* (Khasi Hills, Jaintia Hills), which is the northernmost location in the distribution of the genus. On the whole, the flora of the province is of mixed Indomalesian–Eastern Asiatic character, but above 900 m it is predominantly Holarctic. I had the opportunity to visit this area in the spring of 1966, during joint trips with Dr. A. S. Rao to the Shillong Plateau.

Isolated, sometimes rather large exclaves of the Eastern Asiatic flora are found in the southern regions of Assam and Burma, reaching the Mizo Hills in Assam, the southern Chin Hills, and even the Arakan Yoma. An exceptionally interesting exclave of the Holarctic flora was found on Mt. Victoria (3,170 m above sea level) in the northern Arakan Yoma. Having studied the flora of Mt. Victoria, Ward (1959) concluded that it was of a relict character, the remainder of a larger flora that migrated in the Pleistocene to the south and then was cut off as a result of the fractionation of the plateau into individual blocks. There is a considerable basis to assume that the other exclaves of the Holarctic flora are also of Pleistocene age.

3. North American Atlantic Region

Gebiet des Atlantischen Nordamerika, Engler 1882, 1899, 1902, 1903, 1924; Hayek 1926; Nördliches atlantisches Nordamerika, Mattick 1964;

Atlantic North American Region, Good 1947, 1974; Appalachian Region, Tolmatchev 1974.

The North American Atlantic Region stretches across North America from the Atlantic Ocean to the Rocky Mountains, and from the Gulf of Mexico to southern Canada. On the north it adjoins the Canadian Province of the Circumboreal Region. It consists of three provinces—the Appalachian Province (the largest and most characteristic of the three); the Atlantic and Gulf Coastal Plain Province; and the North American Prairies Province. There is a broad transition zone between the Appalachian Province and the Canadian Province of the Circumboreal Region, especially from the Atlantic coast to Lake Huron, and different phytogeographers have drawn different boundaries in this area (see further comments under the Appalachian Province).

The flora of the North American Atlantic Region is fairly rich and is characterized by high endemism. There is only one endemic family (Leitneriaceae, in the Atlantic and Gulf Coastal Plain Province), but there are almost 100 endemic or nearly endemic genera of vascular plants. Some of these occur in all three provinces, some in only the Appalachian Province or only the Coastal Plain Province, and some in both of these latter provinces collectively. Some have very restricted geographic ranges, occurring in only a small part of one province. Only the monotypic genus *Daucosma* is endemic to the North American Prairies Province. The following are among the endemic or nearly endemic genera in the region:

Annonaceae: *Asimina* (3)
Aristolochiaceae: *Hexastylis* (6)
Menispermaceae: *Calycocarpum* (1)
Ranunculaceae: *Anemonella* (1), *Xanthorhiza* (1)
Caryophyllaceae: *Stipulicida* (1)
Polygonaceae: *Brunnichia* (1), *Polygonella* (9)
Hamamelidaceae: *Fothergilla* (4)
Myricaceae: *Comptonia* (1)
Theaceae: *Franklinia* (1; known today only in cultivation)
Ericaceae: *Elliottia* (1), *Leiophyllum* (1), *Monotropsis* (1), *Oxydendrum* (1), *Zenobia* (1)
Cyrillaceae: *Cliftonia* (1)
Empetraceae: *Ceratiola* (1)
Diapensiaceae: *Galax* (1), *Pyxidanthera* (2)
Brassicaceae: *Leavenworthia* (7), *Warea* (4)
Malvaceae: *Napaea* (1)
Cistaceae: *Hudsonia* (3)
Ulmaceae: *Planera* (1)
Moraceae: *Maclura* (1)

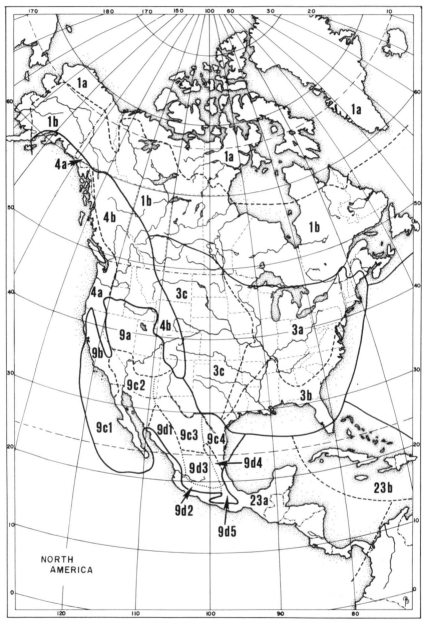

Map 4. FLORISTIC REGIONS AND PROVINCES OF NORTH AMERICA

LEGEND FOR MAP 4

1. Circumboreal Region.
   1a. Arctic Province.
   1b. Canadian Province.
3. North American Atlantic Region.
   3a. Appalachian Province.
   3b. Atlantic and Gulf Coastal Plain Province.
   3c. North American Prairies Province.
4. Rocky Mountain Region.
   4a. Vancouverian Province.
   4b. Rocky Mountain Province.
9. Madrean Region.
   9a. Great Basin Province.
   9b. Californian Province.
   9c. Sonoran Province.
      9c1. Baja Californian Subprovince.
      9c2. Sonoran Subprovince.
      9c3. Chihuahuan Subprovince.
      9c4. Tamaulipan Subprovince.
   9d. Mexican Highlands Province.
      9d1. Sierra Madre Occidental Subprovince.
      9d2. Trans-Mexican Volcanic Belt Subprovince.
      9d3. Mexican Altiplano Subprovince.
      9d4. Sierra Madre Oriental Subprovince.
      9d5. Sierra Madre del Sur Subprovince.
23. Caribbean Region.
    23a. Central American Province.
    23b. West Indian Province.

Euphorbiaceae: *Crotonopsis* (2)
Crassulaceae: *Diamorpha* (1)
Droseraceae: *Dionaea* (1)
Rosaceae: *Gillenia* (2), *Neviusia* (1)
Lythraceae: *Decodon* (1)
Onagraceae: *Stenosiphon* (1)
Fabaceae: *Strophostyles* (3)
Leitneriaceae: *Leitneria* (1)
Caprifoliaceae: *Diervilla* (2)
Apiaceae: *Cynosciadium* (1), *Daucosma* (1), *Erigenia* (1), *Limnosciadium* (2), *Polytaenia* (2), *Taenidia* (2), *Thaspium* (4), *Trepocarpus* (1)
Santalaceae: *Nestronia* (1)
Rubiaceae: *Pinckneya* (1)
Apocynaceae: *Thyrsanthella* (1)
Gentianaceae: *Bartonia* (3), *Obolaria* (1)
Convolvulaceae: *Stylisma* (6)
Scrophulariaceae: *Dasistoma* (1), *Leucospora* (1), *Schwalbea* (1), *Tomanthera* (2)
Acanthaceae: *Yeatesia* (1)
Verbenaceae: *Stylodon* (1)
Lamiaceae: *Blephilia* (2), *Collinsonia* (5), *Conradina* (4), *Macbridea* (2), *Rhododon* (1), *Synandra* (1)
Asteraceae: *Balduina* (2), *Bigelowia* (2), *Brintonia* (1), *Carphephorus* (7), *Chrysogonum* (1), *Chrysoma* (1), *Dracopis* (1), *Echinacea* (4), *Garberia* (1), *Hartwrightia* (1), *Jamesianthus* (1), *Krigia* (7), *Liatris* (30 +, also extending shortly into the Rocky Mountain and Sonoran provinces), *Marshallia* (8), *Phoebanthus* (2), *Sclerolepis* (1), *Silphium* (15), *Stokesia* (1)
Melanthiaceae: *Chamaelirium* (1), *Helonias* (1; very closely allied to the eastern Asiatic genus *Heloniopsis*), *Pleea* (1), *Stenanthium* (1), *Uvularia* (4)
Liliaceae: *Medeola* (1)
Haemodoraceae: *Lophiola* (2)
Orchidaceae: *Cleistes* (1), *Isotria* (2)
Cyperaceae: *Cymophyllus* (1)
Poaceae: *Anthaenantia* (2), *Hydrochloa* (1), *Triplasis* (2)
Araceae: *Orontium* (1), *Peltandra* (3)

In addition, a number of essentially North American Atlantic genera (e.g., *Sanguinaria, Sarracenia, Nemopanthus*, and *Chelone*) extend for a greater or lesser distance into the southern part of the Canadian Province of the Circumboreal Region. The almost endemic genus *Aureolaria* (Scrophulariaceae) has one species in Mexico, in addition to about 10 in the North American

Atlantic Region, and one species of the almost endemic genus *Rhexia* (Melastomataceae, 13 spp.) occurs in Cuba as well as in Florida.

In spite of the large number of endemic genera, there is a remarkable similarity between the flora of eastern Atlantic North America (especially the southern Appalachian Mountains) and eastern Asia.[50] This similarity has attracted the attention of botanists for more than two-hundred years (see Graham 1972). The relationship is shown by a rather large number of identical or closely allied genera, and by closely related vicariant species. Among the more famous examples are *Liriodendron tulipifera,* widespread in the eastern United States, and *L. chinense,* known from a much more limited area in China; and *Shortia galacifolia,* restricted to a local area in the Blue Ridge Mountains of North and South Carolina, *S. uniflora* of Japan, and *S. sinensis,* found very locally in southern Yünnan, China. There are many other examples, in a wide range of families.

Gray (1846, 1859) and Hooker (1879) invoked historical causes to explain the similarity between the two floras. They postulated the past existence of a close botanical relationship between Asia and North America at a time when the climate at higher latitudes was considerably warmer than it is now, so that the northern parts of both continents supported a temperate flora. Here we have the seed of the Arcto-Tertiary geoflora concept of Chaney (1940), subsequently expounded by Chaney and by Axelrod in numerous papers. Although the geoflora concept may have been overworked (as has been pointed out by several recent authors), it does appear to provide the right explanation for the eastern American–eastern Asian similarities.

The floristic exchange between Asia and America was doubtless in both directions, but migration from Asia to North America seems to have been more important than the reverse. Taken as a whole, the flora of eastern Asia is considerably richer and contains more archaic taxa than that of eastern North America.

The North American Atlantic Region embraces a considerable diversity in the dominant aspect of the vegetation, from the grasslands of the North American Prairies Province to the deciduous forests of the Appalachian Province and the pine or live-oak–magnolia forests of the Coastal Plain Province. Aside from the endemic genera that occur in two or all three of the provinces, an important unifying feature of the region is the abundance and variety of species of *Aster* and *Solidago.* Although both of these genera extend far beyond the North American Atlantic Region, they reach their best development here, and from midsummer into fall they are ubiquitous and conspicuous. Several other common genera of Asteraceae—notably *Helianthus, Rudbeckia,* and *Silphium* (all in the tribe Heliantheae)—also center in the North American Atlantic Region, and contribute to a late-season dominance of Asteraceae in the aspect of the herbaceous cover.

1. **Appalachian Province** (Engler 1882; Provinz des sommergrünen Mississippi- und Alleghany-Waldes mit den Alleghanies, southern part of the Seenprovinz, Engler 1902, 1903, 1924; Eastern Deciduous Forest Province, Gleason and Cronquist 1964). This province includes a large part of the eastern United States (excluding the Atlantic and Gulf Coastal Plain) and a small part of southern Canada (southeastern Ontario and the southernmost part of Quebec).[51] In the south the Appalachian Province extends to central Georgia, central Alabama, much of Arkansas, and a part of eastern Texas, while on the west it reaches Minnesota,[52] eastern Iowa, the Ozark Plateau, and the Quachita Mts.

The northern part of the Appalachian Province (and of the North American Prairies Province as well) was repeatedly glaciated during the Pleistocene epoch. The Missouri and Ohio rivers mark the approximate southern limit of the glaciation. Farther east, the terminal moraine runs the length of Long Island, off the coast of New York. To the south, the coastal plain has been repeatedly inundated, in whole or in considerable part, not only during the Pleistocene interglacials, but also at intervals during the Tertiary period. The Mississippi Embayment, an extension of the coastal plain, reaches north to southern Illinois.

The combination of glaciation on the north and flooding on the south has repeatedly pinched off the Ozarkian area (mainly southern Missouri and northern or northwestern Arkansas) from the main body of the Appalachian Province. On the west, the forests of the Ozarkian area are limited by increasing aridity, as the Appalachian Province gives way to the North American Prairies Province. Thus there has tended to be a Southern Appalachian center, and a smaller, drier, less diversified Ozarkian center in which plants of the Appalachian Province have survived the vicissitudes of climate, and within which new taxa have evolved during the course of geologic time. During the interglacial periods some of these species have extended their ranges northward into the areas freed from ice-cover. Others have not been able to expand their ranges, either because of specific habitat requirements, or because of depletion of genotypes during times of stress (these being perhaps different aspects of the same problem), or for other reasons as yet undetermined. Vicariant populations migrating northward from these two refugia may or may not differ significantly, and if they do so differ, they may or may not retain their differences when their ranges come into contact and they have a chance to hybridize.

It is often possible to recognize a species-pair—one member in or originating from each centre—even though their ranges may now overlap. Among such species-pairs are the following, with the Appalachian species listed first in each case: *Hamamelis virginiana* and *H. vernalis; Castanea pumila* and *C. ozarkensis; Carya glabra* and *C. texana; Hypericum punctatum* and *H. pseudomaculatum; Ribes rotundifolium* and *R. missouriense; Asclepias perennis* and *A. texana; Aster sagittifolius* and *A. drummondii; Echinacea laevigata* and *E. purpurea; Helenium brevifolium* and *H. campestre; Helianthus angustifolius* and *H. salicifolius; Helianthus atrorubens* and *H. silphioides; Marshallia ramosa* and *M. caespitosa; Parthenium auriculatum* and *P. hispidum; Prenanthes roanensis* and *P. aspera; Prenanthes serpentaria* and *P. barbata*. Sometimes the differentiation between the Appalachian and Ozarkian taxa has reached (or maintained) only an intraspecific level.

The flora of the Appalachian Province has a great many endemic and nearly endemic species. A small sample follows:

Abies fraseri, Tsuga canadensis, T. caroliniana, Pinus pungens, P. rigida, P. virginiana, Magnolia acuminata, M. fraseri, Asimina triloba, Asarum canadense, Hexastylis shuttleworthii, Aristolochia macrophylla, Menispermum canadense, Aconitum noveboracense, A. reclinatum, A. uncinatum, Actaea alba, Cimicifuga americana, C. racemosa, C. rubifolia, Clematis spp., *Hepatica acutiloba, H. americana, Delphinium exaltatum, D. tricorne, Hydrastis canadensis, Ranunculus* spp., *Thalictrum* spp., *Trautvetteria carolinensis, Trollius laxus, Berberis canadensis, Caulophyllum thalictroides, Diphylleia cymosa, Jeffersonia diphylla, Adlumia fungosa, Corydalis flavula, Dicentra canadensis, D. cucullaria, D. eximia, Stylophorum diphyllum, Fothergilla major, Hamamelis vernalis, Celtis occidentalis, Ulmus serotina, U. thomasii, Maclura pomifera, Laportea canadensis, Castanea dentata, C. ozarkensis,* many spp. of *Quercus* and several spp. of *Betula, Comptonia peregrina,* many species of *Carya, Juglans cinerea, Claytonia caroliniana, C. virginica,* a series of species of *Polygonum, Stewartia ovata, Hypericum* spp., *Triadenum virginicum, Hybanthus concolor, Populus grandidentata, Salix* spp., *Clethra acuminata, Chimaphila maculata, Eubotrys recurva, Gaylussacia brachycera, Leiophyllum buxifolium, Leucothoe fontanesianum, Menziesia pilosa, Monotropsis odorata, Pieris floribunda, Rhododendron arborescens, R. calendulaceum, R. catawbiense, R. cumberlandense, R. maximum, R. periclymenoides, R. prinophyllum, Vaccinium erythrocarpum, V. pallidum, Galax aphylla* (also found in spots in the Atlantic and Gulf Coastal Plain Province), *Pyxidanthera barbulata, Shortia galacifolia, Halesia tetraptera, Lysimachia* spp., *Tilia americana, T. heterophylla, Napaea dioica, Croton alabamensis, Crotonopsis elliptica, Euphorbia* spp., *Sedum pusillum, Ribes* spp., *Philadelphus* spp., *Astilbe biternata, Boykinia aconitifolia, Heuchera* spp., *Saxifraga* spp., *Tiarella cordifolia, Parnassia grandifolia, Amelanchier* spp., *Aruncus dioicus, Crataegus* spp., *Geum* spp., *Gillenia stipulata, G. trifoliata, Neviusia alabamensis, Potentilla* spp., *Prunus* spp., *Rosa* spp., *Spiraea virginiana, Waldsteinia fragarioides, Gymnocladus dioicus, Cladrastis kentuckea, Robinia hispida, R. pseudoacacia, R. viscosa, Thermopis mollis, Trifolium reflexum, T. virginicum, Staphylea trifolia, Acer nigrum, A. pensylvanicum, A. saccharum, Aesculus glabra, A. neglecta, Rhus typhina, Linum virginianum, Panax quinquefolium, P. trifolium, Angelica triquinata, Erigenia bulbosa, Taenidia integerrima, Ilex montana, Euonymus obovatus, Pachystima canbyi, Buckleya distichophylla, Pyrularia pubera, Galium* spp., *Gentiana* spp., *Forestiera acuminata, Fraxinus quadrangulata, Diervilla rivularis,* D. *sessilifolia, Viburnum alnifolium, Hydrophyllum appendiculatum, Phacelia fimbriata, Blephilia ciliata, Conradina verticillata, Meehania cordata, Pycnanthemum montanum* and other spp., *Scutellaria* spp., *Stachys* spp., *Synandra hispidula, Aureolaria grandiflora, A. laevigata, A. patula, Collinsia verna, Dasistoma macrophylla, Leucospora multifida, Pedicularis furbishiae, Penstemon* spp., numerous species of *Aster* and *Solidago, Eupatorium* spp., *Helenium virginicum, Marshallia grandiflora, Senecio schweinitzianus, Veronica glauca, Clintonia umbellata, Disporum lanuginosum,* D. *maculatum, Lilium grayi, Polygonatum pubescens, Veratrum parviflorum, V. woodii, Allium tricoccum, Trillium* spp., *Iris* spp., and numerous species of *Carex.*

Many species are endemic to the Appalachian Province and the Atlantic and Gulf Coastal Plain Province collectively. Among these are *Liriodendron tulipifera, Carpinus caroliniana, Fagus grandifolia, Castanea pumila, Betula nigra,* several species of *Quercus* and *Carya, Vaccinium arboreum, Bumelia lycioides, Acer barbatum, A. leucoderme, Toxicodendron vernix, Nyssa sylvatica, Cornus florida,* C. *stricta, Aralia spinosa, Chionanthus virginiana, Senecio anonymus,* and *Orontium aquaticum.*

The dominant type of climax community in the Appalachian Province is deciduous forest. This forest attains its most favorable development in the moist valleys and slopes of the southern Appalachian Mountains, especially in the Cumberland Mountains of eastern Kentucky and the Great Smoky Mountains (part of the Blue Ridge Geological Province) of Tennessee and North Carolina. More than two dozen kinds of trees are abundant and self-perpetuating under natural conditions in this mixed mesophytic forest. Among there are species of *Acer, Aesculus, Betula, Carya, Fagus, Liriodendron, Quercus,* and *Tilia.*

The number of common dominant trees in relatively undisturbed forest thins out in all directions from the southern Appalachian center. To the north, *Fagus grandifolia* (American beech) and *Acer saccharum* (sugar maple) are increasingly abundant on the better soils, with species of *Quercus* and *Carya* (hickory) often occupying the drier or more exposed sites. To the far northwest, in Wisconsin and Minnesota, the beech-maple community gives way to the maple-basswood community as the beech drops out and is replaced by *Tilia americana.* The oak-hickory community, confined to the poorer sites in the beech-maple and maple-basswood regions, is the most abundant type in the southern part of the deciduous forest (outside of the mountain region). Toward the western edge of the oak-hickory forest, as one approaches the grassland, the trees are smaller and often more widely spaced, and especially toward the southwest there is some open woodland vegetation, with the trees widely scattered among the grasses. Farther north, the forest-prairie transition zone generally shows a mosaic of the two vegetation types, rather than a blend.

East of the mixed mesophytic forest, there was once a long fringe of oak-chestnut (*Quercus-Castanea*) forest, extending from Georgia to southern New England. The destruction of the chestnut, since 1900, by the chestnut blight has changed most of this to an oak or oak-hickory community.

In much of New England, *Tsuga canadensis* grows intermingled with broad-leaved deciduous trees, especially beech and sugar maple. This beech-maple-hemlock or hemlock-hardwood community extends west to the Great Lakes region, just south of the coniferous forest (Canadian Province); it extends south in parts of the Appalachian Mountains to Georgia.

In pioneer days there were large stands of *Pinus strobus* (white pine) in the same general area as the beech-maple-hemlock community. The pine forests developed mainly after fire or some other disturbance, and each stand tended to be replaced eventually by hemlock-hardwood forest. Most of the once extensive stands of white pine have long since fallen to the lumberman's axe. White pine also occurs in the more southern part of the Canadian Province in southeastern Canada and adjacent United States, as a fire-tree that eventually gives way to spruce and fir. The less abundant *Pinus resinosa* has a range similar to that of *P. strobus.*

The difficulty of drawing a clear border between the Appalachian Province and the Canadian Province is exacerbated by the distributional patterns of several common trees, in addition to the two pines just noted. *Betula papyrifera* is surely a Canadian element, being common almost throughout that province, but it is also common in New England and in much of New York and Pennsylvania. *Populus tremuloides* has a similar pattern of distribution in the north and east, but also extends well to the south at upper altitudes in the western cordillera. No matter how abundant they may be locally, all of these trees must be ignored in any effort to delimit the Appalachian and Canadian provinces.

Picea rubens occurs mainly in the southeastern part of the Canadian Province and in adjacent parts of the Appalachian Province. It is frequently associated with *Abies balsamea,* and sometimes also with *Picea glauca.* Thus it appears to belong with the Canadian Province element, but it also extends southward at increasing elevations in the mountains. In the White Mountains of New Hampshire it is the principal spruce of the spruce-fir community. The White Mountains are completely surrounded at the base by deciduous forest and are here assigned to the Appalachian Province, but they lie only a few kilometers to the south of the more nearly continuous boreal forest of the Canadian Province. Farther south, at high elevations in the Southern Appalachian Mountains (especially the Blue Ridge), *Picea rubens* associates with *Abies fraseri* to form enclaves of spruce-fir forest that look very much like the spruce-fir forests of the Canadian Province. *Abies fraseri* is scarcely more than a southern variety of *A. balsamea.*

Some of the understory plants in the Southern Appalachian spruce-fir forests—such as *Ribes glandulosum, Oxalis acetosella,* and *Maianthemum canadense*—are the same as the species in the spruce-fir forests of the Canadian Province, although the *Maianthemum,* to be sure, also extends well into the Appalachian Province in other communities, especially toward the north. Some species commonly found here—such as *Viburnum alnifolium, Aster acuminatus,* and *Clintonia borealis*—have their principal range in the northern part of the Appalachian Province and the southeastern part of the Canadian Province, growing in both coniferous and deciduous forest. Other members of these *Abies fraseri–Picea rubens* forests, such as *Rhododendron vaseyi, Prenanthes roanensis,* and *Solidago lancifolia,* are the same as or closely related to species of the deciduous forest. *Vaccinium erythrocarpum,* which has its nearest relatives in China and Japan, reaches its best development here, but also extends down into the upper part of the deciduous forest. *Cacalia rugelia,* endemic to openings in the *Abies fraseri–Picea rubens* community, finds its nearest relatives in Mexico.

The Southern Appalachian Mountains form a geologically complex, uplifted region extending from southern Pennsylvania southwest to northern Georgia and northeastern Alabama. For some purposes it is convenient to think of these mountains as a single region. Many species of plants that have a wider range in the northeastern part of the Appalachian Floristic Province extend southward at increasing elevations in the Southern Appalachian Mountains, without being confined to any one geologic segment of them. *Betula lutea, Acer pensylvanicum,* and *Sorbus americana* are among the woody plants with such a distribution. Within the single genus *Aster, A. acuminatus, A. lowrieanus, A. macrophyllus* and *A. puniceus* show such a pattern, and there are many others in other genera and families. Other species are more or less widespread in these mountains, but do not extend much (if at all) beyond them. Within the family Asteraceae, one may cite *Aster surculosus, Echinacea laevigata, Helianthus laevigatus, Liatris turgida, Marshallia grandiflora, Parthenium auriculatum,* and *Solidago curtisii* as examples.

For other purposes it is useful to recognize that the Southern Appalachians consist of three long, narrow, parallel geologic provinces: the Blue Ridge Province on the eastern side; the Folded Appalachians, or Ridge and Valley Province, in the middle; and the Appalachian (or Cumberland) Plateau Province toward the west. Each of these geologic provinces harbors a set of endemic species, as well as many that are more widespread.

The Blue Ridge Geological Province consists largely of pre-Cambrian granitic and metamorphic rocks. The Blue Ridge itself is continuous from southern Pennsylvania to northern Georgia, reaching its highest elevations in western North Carolina. Mount Mitchell, at a height of about 2,000 m, is the highest point in eastern North America. Toward the north, the province consists solely of the Blue Ridge itself and is only a few kilometers wide. Farther south, in western North Carolina and adjacent Tennessee, the province widens somewhat, and some other ranges to the west of the Blue Ridge—notably the Great Smoky Mountains—reach comparable heights. The only tree that is essentially confined to the Blue Ridge Province is *Tsuga caroliniana*, the Carolina hemlock. Among the many species of herbaceous plants endemic (or almost endemic) to the Blue Ridge Geological Province, we here list only those from the single family Asteraceae:—namely, *Aster curtisii, Cacalia rugelia, Coreopsis latifolia, Helianthus glaucophyllus, Krigia montana, Prenanthes roanensis, Senecio millefolium, Solidago glomerata,* and *S. spithamea*. *Shortia* (Diapensiaceae), with one species on the Blue Ridge, one in Japan, and one in China, has already been mentioned.

The spruce-fir forests of the Southern Appalachians are largely confined to the highest parts of the Blue Ridge Geological Province, in eastern Tennessee, western North Carolina, and adjacent southwestern Virginia. Some other summits in the same area—for the most part not quite so high as those clothed with spruce-fir forest—are bare of trees and are locally known as balds.

The Folded Appalachians Geological Province has an anticline-syncline structure and consists largely of Paleozoic sedimentary rocks. The most distinctive habitat in the Folded Appalachians is the shale-barrens, which occur in scattered small patches in western Virginia, eastern West Virginia, western Maryland, and southwestern Pennsylvania. A hard, noncalcareous shale of mainly Upper Devonian age (especially the Brallier shale) provides the specific substrate. The forest trees and many of the understory plants on the shale-barrens are mostly the same as those on other sites, but the trees are relatively small and more widely spaced. Often the pines are more abundant than in non-shale-barren habitats. The habitat is barren only by comparison with the more densely wooded sites nearby; a person acquainted with the arid, interior western United States would not think of it as barren.

A considerable number of species and varieties of herbaceous plants are endemic (or nearly so) to the shale-barrens. Among these are *Clematis albicoma, C. coactilis, C. viticaulis, Eriogonum allenii, Trifolium virginicum, Oenothera argillicola, Taenidia montana, Phlox buckleyi, Senecio antennariifolius, Antennaria virginica,* and *Solidago arguta* var. *harrisii*. Some other species, such as *Helianthus laevigatus* and *Hieracium trailii*, are most commonly found on shale-barrens but occasionally grow in other habitats. Most of the shale-barren endemics are related to more widespread species of the Appalachian Floristic Province. Others, such as *Trifolium virginicum* and *Phlox buckleyi*, have no obvious close relatives. A few, such as *Eriogonum allenii* and *Senecio antennariifolius*, relate to western American species. The progenitors of these latter may be presumed to have reached the drier sites in the Appalachian country during the postglacial hypsithermal period, and to have differentiated as distinct species in the shale-barren habitat during the past several thousand years. One of the noteworthy features of the shale-barrens is the seemingly helter-skelter pattern of distribution of their endemics on the separate outcrops. Any individual species occurs on some shale-barrens but not others; conversely, every shale-barren bears some of the characteristic endemics but not others. The pattern presumably reflects the vicissitudes of chance dispersal from one shale-barren to another.

The Appalachian (or Cumberland) Plateau Geological Province lies just to the west of the Folded Appalachians, from which it differs topographically in consisting of a series of uplifted plateaus, with flat-lying Paleozoic sedimentary rocks. A considerable number of species are endemic or almost endemic to this area. Among these are *Thalictrum mirabile, Silene rotundifolia, Coreopsis pulchra, Silphium brachiatum, S. mohrii, Arenaria cumberlandensis, Eupatorium luciae-brauniae,* and *Solidago albopilosa*. The last three of these are limited to the "rock-houses" under great, overhanging sandstone cliffs.

The granite flatrocks of the Piedmont Geological Province provide another notable special habitat in the Appalachian Floristic Province. The flatrocks are scattered from North Carolina to Alabama, but are most numerous in Georgia. Collectively they may cover about 35 km². Except for Stone Mountain—a gigantic boulder some 235 m high in north central Georgia—they are mostly fairly flat, as the name implies. The granitic substrate is exposed, with thin patches of soil and with shallow pools of water that dry up in the summer. A dozen or more species, including two monotypic genera (*Diamorpha* and *Amphianthus*), are nearly or quite confined to the flatrocks. Among these endemic or nearly endemic species are *Isoetes melanospora, I. tegetiformans, Quercus georgiana, Portulaca smallii, Hypericum splendens, Sedum pusillum, Aster avitus, Viguiera porteri, Juncus georgianus,* and *Rhynchospora saxicola*. Most of these have fairly close relatives in more ordinary habitats in the Appalachian Floristic Province, but *Amphianthus* is taxonomically rather isolated in the Scrophulariaceae, and *Viguera porteri* is an eastern outlier of a genus otherwise distributed from the western United States and Mexico to South America. *Diamorpha* may be distantly related to *Sedum pusillum*, but the latter species has no evident close relatives in its genus. Several species that are more widespread in the Appalachian Floristic Province have varieties that are mainly or wholly restricted to the flatrocks. Among these are *Phacelia dubia* var. *georgiana* and *Coreopsis grandiflora* var. *saxicola*. Some other species that are especially abundant and well developed on the flatrocks are also more or less widespread in dry habitats elsewhere in the Appalachian Province. Among these are *Talinum teretifolium, Arenaria brevifolia, Crotonopsis elliptica,* and *Lotus helleri*.

The cedar glades of central Tennessee and adjacent Kentucky and Alabama form another special community in the Appalachian Floristic Province. They have a characteristic aspect, with scattered trees of *Juniperus virginiana* (a widespread species) in a grassy, herbaceous cover. A number of species are endemic to the cedar glades, and others reach their best development there, although they have a wider distribution. Among the more or less strictly endemic species are *Talinum calcaricum, Leavenworthia stylosa, Lesquerella perforata, L. stonensis, Astragalus tennesseensis, Dalea gattingeri,* and *Pediomelum subacaule.* Some taxa that occur mainly in the North American Prairies Province have eastern outliers in the cedar glades. Among these is *Echinacea pallida* var. *angustifolia;* the cedar glade population has been thought by some enthusiasts to constitute a distinct species, *E. tennesseensis.* Presumably such plants reached the present cedar glades during the postglacial hypsithermal period, when many prairie species extended their range eastward.

2. **Atlantic and Gulf Coastal Plain Province** (Immergrüne Provinz der südatlantischen Staaten, Engler 1899, 1902, 1903, 1924; Südliches atlantisches Nordamerika, Mattick 1964; Coastal Plain Province, Gleason and Cronquist 1964). This province occupies the geologic coastal plain of the Atlantic and Gulf coast states of the United States, from New Jersey to Florida and west to southeastern Texas. Like the geologic coastal plain, it also extends up the valley of the Mississippi River (geologically, the Mississippi Embayment) to the southern tip of Illinois. The tropical southern tip of Florida, although it is geologically coastal plain, is floristically better associated with the West Indian Province of the Caribbean Region, as shown on map 4. To the north, a progressively attenuated element of the coastal plain flora is found on southern Long Island (south of the glacial moraine), on Cape Cod in Massachusetts and even on the southern tip of Nova Scotia. Some coastal plain elements are found in sandy soil around the southern end of Lake Michigan and the western end of Lake Erie. For example, *Euthamia remota* of this inland area (belonging to the Appalachian Province) is distinguishable only with difficulty from the coastal plain species *E. tenuifolia.* Some other species once thought to be confined to the coastal plain are now known to occur locally in the southern Appalachian Mountains as well, and this latter area may indeed be their original home.

The Coastal Plain Province has in general the sharpest boundaries of any of the North American floristic provinces, because a clear geologic line coincides with a difference in vegetation that is partly governed by soil type. It would be much more difficult to draw the boundaries solely on the basis of the vegetation, both because of the progressive depauperization of the coastal plain flora toward the north and because many of the characteristic species of the Appalachian Province occur also on the Coastal Plain.

Much of the coastal plain is characterized by the presence of extensive pine forests. To the north, as in New Jersey, *Pinus rigida* is the dominant species. *Pinus rigida* also extends well north into New England, and south in the mountain country inland, but in these parts of its range it is seldom dominant. Farther south, *Pinus palustris, P. elliottii, P. taeda,* and *P. echinata* are common. These latter two species also extend well into the piedmont (Appalachian Province), but the bulk of the coastal plain flora does not go with them. One of the most typical coastal plain communities is an open stand of *Pinus palustris*, with a thick carpet of *Aristida stricta* under the trees. The continued dominance of pine on much of the coastal plain depends on the repeated fires to which the region has been subjected for thousands (perhaps millions) of years. When protected from fire, the pine forests of the coastal plain tend to give way to hardwood forests. In upland, fairly well-drained sites, species of oak and hickory (notably *Quercus alba, Q. nigra, Q. falcata, Carya tomentosa,* and *C. glabra*) are common constituents of such hardwood forests. Many of these species, including all of those just listed, are also common in parts of the Appalachian Province, and these upland hardwood forests on the coastal plain are floristically not very different from some of the hardwood forests in the southern part of the Appalachian Province. Dry sandhills on the inner coastal plain often support a thicket of various species of small oaks. Some of these stump-sprout after fire and grow intermingled with pines.

In fairly moist, lowland but not swampy places on the coastal plain, the characteristic hardwoods that tend to replace the pines when protected from fire are beech (*Fagus grandifolia*), sweet bay (*Magnolia virginiana*, an evergreen species) and several species of live (i.e., evergreen) oak, notably *Quercus virginiana* and *Q. laurifolia*. Beech is also widespread in the Appalachian Province, but the other species are largely or wholly restricted to the coastal plain. These several species dominate many small areas that have escaped fire in the recent past.

On the outer coastal plain the water table is seldom very far below the surface, and many areas are periodically or permanently flooded. Marshes with permanently standing water are usually dominated by coarse grasses and rushes. Places where water stands for most but not all of the year commonly support forests of *Taxodium distichum*. In areas flooded for a shorter period, broad-leaved trees such as *Nyssa aquatica* and *Fraxinus caroliniana* are mixed with the bald cypress, or replace it entirely.

Some coastal plain species with relatives that are mainly tropical are limited northward by low temperatures. Three conspicuous species of this group are *Sabal minor, S. palmetto,* and *Tillandsia usneoides* (Spanish moss). The two palmettos are endemic to the southern part of the Coastal Plain Province. *Sabal minor* is the more widespread of the two, and extends north along the coast occasionally to northern North Carolina, whereas *S. palmetto* barely reaches southernmost North Carolina. *Tillandsia usneoides* is widespread in the American tropics and extends up the coastal plain to Virginia.

Much of the flora of the Coastal Plain Province is considerably younger than that of the Appalachian Province, because of the history of inundation on the coastal plain. Even so, it contains one endemic, monotypic family, the Leitneriaceae, and a series of endemic genera. In comparison with the

Appalachian Province, the number of species with eastern Asiatic affinities is not very great, but there are some. *Stewartia malacodendron* and *Croomia pauciflora* may be cited as examples, but the former species has outlying stations in the mountains, in addition to its major area on the coastal plain.

The Atlantic and Gulf Coastal Plain Province has hundreds of endemic and nearly endemic species—almost 100 in the Asteraceae alone, including 10 species of *Aster*, 7 of *Liatris*, all 7 species of *Carphephorus*, 5 species of *Solidago*, and 5 of *Eupatorium*. Among the endemic and nearly endemic trees and shrubs are *Taxus floridana, Torreya taxifolia, Pinus clausa, P. elliottii, P. glabra, P. palustris, P. serotina, Taxodium distichum, Juniperus silicicola, Magnolia grandiflora, M. pyramidata, Persea borbonia, P. palustris, Illicium floridanum, I. parviflorum, Planera aquatica, Quercus arkansana, Q. chapmanii, Q. laevis, Q. laurifolia, Q. virginiana, Leitneria floridana* (constituting the only endemic family in the entire Appalachian Region), *Carya aquatica, C. floridana, Castanea alnifolia, Myrica inodora, Franklinia alatamaha, Gordonia lasianthus, Salix floridana, Lyonia ferruginea, Bumelia tenax, Halesia diptera, H. parviflora, Prunus caroliniana, Gleditsia aquatica, Ilex amelanchier, I. coriacea, I. myrtifolia, Nyssa ogeche, N. sylvatica, Fraxinus caroliniana, Viburnum obovatum,* and *Serenoa repens.* Some of the other interesting endemic and nearly endemic species are *Schisandra coccinea, Sarracenia* spp., *Dionaea muscipula, Asclepias* spp., *Conradina* spp., *Pinckneya pubens, Yucca gloriosa, Iris hexagona, I. tripetala, Xyris* spp., and *Lachnocaulon* spp.

The flora of the coastal plain was largely destroyed by inundation in some of the interglacial periods of the Pleistocene, and at intervals before that during the Tertiary. It has subsequently been revegetated by colonization from its own inner margins, from the Southern Appalachian and Ozarkian refugia, to a lesser extent from the North American Prairies Province, and to a considerable degree from the West Indies and tropical parts of Mexico. Some of the coastal plain plants with tropical affinities are spp. of *Chaptalia, Vernonia, Cyperus, Rhynchospora, Scleria, Panicum, Paspalum, Sabal, Serenoa,* and *Tillandsia usneoides.*

Sarracenia provides an example of an Appalachian origin for a seemingly coastal plain genus. Four of the 8 species of *Sarracenia* appear to be restricted entirely to the coastal plain—namely, *S. alata, S. leucophylla, S. minor,* and *S. psittacina. Sarracenia purpurea* is found on much of the coastal plain, but on the north it extends far into glaciated territory in the northeastern United States and southeastern Canada. This northward extension is obviously post-Pleistocene and implies nothing about the origin of the genus or species. More importantly, this species also occurs rarely on the piedmont and in the Blue Ridge Mountains of North and South Carolina and Georgia. *Sarracenia flava,* although found mainly on the coastal plain, also occurs at scattered stations on the piedmont in North and South Carolina. *Sarracenia rubra* is found mainly on the coastal plain, but also occurs rarely in the Blue Ridge

Mountains of North and South Carolina. Finally, *S. oreophila* is primarily a montane species, found mainly toward the southern end of the Appalachian Plateau in northeastern Alabama. Although the other species grow mainly in bogs and similarly wet places, *S. oreophila* grows in dense woods and along streambanks, in a habitat that might be thought to be the evolutionary precursor of the more specialized habitat of the other species. Thus it seems likely that the diversification of the genus occurred in the southern Appalachian region, and that the invasion of the coastal plain took place some time later. The taxonomic connection of *Sarracenia* to *Darlingtonia* and *Heliamphora* of course suggests a long history prior to the Appalachian radiation. The family is surely of Gondwanaland rather than Laurasian origin.

Alnus maritima—a species of disjunct, bicentric distribution—exemplifies some of the problems that beset a consideration of the origin of the coastal plain flora. This species occurs in a small area in Delaware and Maryland, on the Delmarva Peninsula, where it appears to be an endemic coastal plain element. It also occurs, however, in an even smaller area in southern Oklahoma. The Oklahoma station lies in the North American Prairies Province, but in a broader sense the area might be taken as a western extension of the Ozark refugium. Presumably *Alnus maritima* had a wider distribution at one time, and presumably its Oklahoma range is older than its Delmarva range. It might be supposed that the Delmarva occurrence reflects migration from Oklahoma, but the supposition is not necessarily correct. Although the Delmarva Peninsula was flooded during some of the interglacial periods, might not the species have migrated landward as the sea rose, and seaward again as the water dropped? Perhaps the recent fossil record will eventually provide an answer, but at present we simply do not know. If the species had not survived in Oklahoma, its history would be even more of a mystery.

Leitneria floridana provides an even more puzzling case. This is a taxonomically isolated species, forming a monotypic family. It is entirely restricted to the coastal plain, where it occurs at scattered stations in Florida, Georgia, Missouri (on the Mississippi Embayment), Arkansas, and Texas. In Missouri and Arkansas, its range abuts the Ozarkian refugium but does not transgress significantly on it. It is a bit difficult, though perhaps not impossible, to visualize the recent entry of *Leitneria* into the coastal plain from an Ozarkian or other refugium from which it has subsequently disappeared.

There is a local center of endemism along the lower Apalachicola River in western Florida and adjacent Georgia, well removed from any other floristic province. *Torreya taxifolia* grows along the bluffs above the river, in sites that must have been repeatedly inundated during the geologic past. In addition to the Florida species, there is one species in California, and four in eastern Asia. *Taxus floridana* is another species in the same habitat that presumably relates to the Arcto-Tertiary flora. The family Asteraceae has several species nearly or quite limited to the Apalachicola center, although they occur on the lowlands rather than on the bluffs. These are *Aster chapmanii, A. eryngiifolius, A. spinulosus, Cacalia sulcata, Chrysopsis flexuosa, C. oligantha, Liatris provincialis, Rudbeckia graminifolia, Phoebanthus tenuifolius,* and *Verbesina chapmanii.* There are other such endemics in other families.

A number of species are nearly or quite limited to peninsular Florida, and would be very amenable to the concept of an Orange Island refugium. In the single family Asteraceae these include *Aster reticulatus, Berlandiera subacaulis, Cacalia floridana, Carphephorus carnosus, C. corymbosus, Chrysopsis latisquamea, Hartwrightia floridana* (a monotypic genus), *Hieracium megacephalum, Liatris ohlingerae* (a taxonomically very distinctive species), *Phoebanthus grandiflorus, Pluchea longifolia, Polypteris feayi,* and *P. integrifolia.*

Some species have a continuous distribution that links the Orange Island and Apalachicola centers, without being widespread elsewhere. Among these (again choosing examples from the Asteraceae) are *Balduina angustifolia, Chrysopsis scabrella, C. subulata, Liatris chapmanii,* and *Lygodesmia aphylla.* The genus *Phoebanthus* has only two species, one in central peninsular Florida, the other in the Apalachicola area.

3. North American Prairies Province (Prärien-Provinz, Engler 1899, 1902, 1903, 1924; Amerikanische Prairienprovinz, Hayek 1926; North American steppe (prairie) subregion, Alekhin 1944; Grassland Province, Gleason and Cronquist 1964). This province includes the largely treeless prairies and plains east of the Rocky Mountains, bounded on the north by the coniferous forests of Alberta, Saskatchewan, and Manitoba, on the east by the deciduous forest (Appalachian Province), and on the south and southwest by the deserts of southern and western Texas and southern New Mexico (Sonoran Province).

The Colorado Front Range and the mountains of northern New Mexico form an abrupt western boundary for part of the North American Prairies Province, but farther north, in Wyoming and Montana, the Great Plains and their grassland flora are continuous with broad intermontane valleys. It is convenient to draw the boundaries to exclude all of the major mountains (except the Black Hills) from the Prairies Province, but a considerable segment of the flora must then be admitted to penetrate well into the Rocky Mountain Floristic Province.

Along the northern boundary of the province there is a strip of aspen-grass woodland, in which *Populus tremuloides* occupies the north slopes and other relatively moist habitats, with grassland in between. It has no special flora of its own, however, and from the standpoint of floristic provinces it is no more than a transition zone between the Prairies Province and the Canadian Province.

The eastern boundary of the North American Prairies Province is, if anything, more vague than the western one. In pioneer days the central and northern part of the eastern boundary was a broad belt in which patches of prairie and forest formed a great mosaic. The forest extended into the grassland along all the streams, and the grassland extended into the forest on the uplands. The individual boundaries between forest and grassland were usually sharp, but the boundary between the two provinces as a whole was not.

Grasses appear to have been an important part of the flora of the territory of the present North American Prairies Province, at least since the middle of the Miocene epoch. The mosaic of forests (in moister sites) and grassland (in drier sites) of the middle Miocene gave way to extensive grasslands during the Miocene-Pliocene transition of about five to seven million years ago. Following the climatic and vegetational fluctuations of the Pleistocene, the present grassland region took shape only after the most recent glaciation (Axelrod 1985).

The flora of the North American Province is not large by comparison with the adjoining Appalachian and Rocky Mountain provinces, but a great many species have their principal range in the grassland and reach their best development there. Presumably, many of these species evolved in the general area where they grow today. At the same time, the vast majority of the characteristic prairie species extend also into one or more of the adjoining provinces. Plants typical of the more eastern part of the province commonly occur at drier sites well into the Appalachian Province, occupying habitats not suitable for trees. Species of the higher, drier western part of the province commonly extend into the drier, lowland habitats in the Rocky Mountain Province, or southward into the more favored sites in the desert. Doubtless some of these species migrated into the grassland from their original home in an adjoining province, but the proportion of immigrant versus autochthonous species is uncertain. The virtual absence of endemic genera suggests that a large number of the species are immigrants.

Aside from the Edwards Plateau in Texas (discussed later in this section), the number of truly endemic and very nearly endemic species in the North American Prairies Province probably does not exceed 50. Among these are *Aschisma kansanum, Eriogonum correllii, E. vischeri, Lesquerella angustifolia, L. arenosa, L. auriculata, L. calcicola, L. engelmannii, L. ovalifolia, L. recurvata, L. sessilis, Amorpha nana, Astragalus barrii, A. gracilis, A. hyalinus, A. pectinatus, A. plattensis, Lespedeza leptostachya, Psoralea cuspidata, P. hypogaea, P. linearifolia, Oenothera fremontii, Daucosma laciniatum, Eryngium leavenworthii, Eurytaenia texana, Phlox andicola, P. oklahomensis, Tomanthera densiflora, Valeriana texana, Haplopappus engelmannii,* and *Carex hookerana.*

Although grasses dominate the Prairies Province and give character to the landscape, they are by no means the only plants. Various sorts of perennial and even annual herbs, many with showy flowers, occur in considerable abundance. Members of the Asteraceae are prominent throughout the season, from *Senecio* and *Gaillardia* in the spring to *Helianthus, Rudbeckia, Silphium* and *Ratibida* in midsummer and *Aster* and *Solidago* in late summer and fall. Members of the Fabaceae are also especially abundant in many places.

The North American Prairies Province is roughly divisible into three north-south strips—the tall-grass prairie to the east, the short-grass prairie to the west, and the mixed-grass prairie in between. The zonation reflects increasing aridity from east to west. The boundaries between these three types of prairie are irregular and unstable. In drier, more exposed sites, the short-grass vegetation extends into the mixed-grass type, and the mixed-grass into the tall-grass. In the swales, with a better moisture supply, the situation is reversed. A series of dry years shoves all the boundaries eastward, and wet years push them west again.

Before the advent of the plowman, the grasses of the tall-grass prairie formed a dense cover commonly 1–2 or even 3 m tall, and at least in the more favored sites, a tough, thick sod was formed. *Andropogon gerardii, A. scoparius,* and *Sorghastrum nutans* were among the most common of the many dominant species, but there was a great deal of local and regional variation. The deep, dark, fertile soil of the tall-grass prairie region is as good as any soil in the world, and the original vegetation has now been almost entirely replaced by farms and cities. The prairie survives only in small fragments.

The high plains east of the Rocky Mountains, where the buffalo once roamed, are short-grass country. *Buchloe dactyloides* and *Bouteloua gracilis* are the most common species in relatively undisturbed sites. These and other common grasses in the region are generally less than 3 dm tall, although taller species occur in the more favorable sites, Rainfall in the short-grass region is relatively scanty and irregular—commonly only 25–40 cm annually—and the later part of the summer is often very dry.

The mixed-grass prairie is composed chiefly of elements from the tall-grass and short-grass prairies that flank it. The climate is favorable enough to permit many of the tall-grass species to survive (although they do not reach full size, seldom much over 1 m), but not favorable enough to permit them to crowd out the shorter, more drought-resistant species from farther west.

Fire has for thousands (and probably millions) of years been an important natural factor affecting the grassland community. All of the successful species are adapted to survive fire, and it appears that the tall-grass prairie cannot be maintained without it. The tall-grass prairie—or at least the more eastern part of it—tends to be replaced by forest when protected from fire over a period of some decades. Although the boundary between the deciduous forest (Appalachian Province) and the prairie is in a general sense set by water relations, in a more specific sense it is largely governed by fire. Fires also occur in the forest, but not so often as in the prairie. In areas moist enough so that fire is infrequent, the forest has a better chance to regenerate.

Several thousand years ago, during what is called the postglacial hypsithermal (or xerothermic) period, the climate in north temperate regions was somewhat warmer and/or drier than now. During that time the forest-prairie boundary was farther east in North America than now, and a broad prairie peninsula extended across Illinois and Indiana and into Ohio. Subsequently the forest reoccupied much of the land, so that when the first white men came to the scene the prairie peninsula was broken up, toward the west, by patches of timber, and in its easternmost portion it was represented only by detached fragments of prairie on the driest sites. The former extent of the prairie peninsula is shown indirectly by the large patches of black prairie soil, which does not develop under trees, that were occupied by ordinary deciduous forest before they were more recently put to the plow. The climatic basis of the prairie peninsula is shown even on modern meteorogical maps: the more or less north-south lines of similar precipitation/evaporation ratios bulge eastward across Illinois and Indiana and into Ohio. In the Asteraceae, at least one species (*Chrysopsis camporum*) and one variety (*Hymenoxys acaulis* var. *glabra*) with western affinities occur primarily in vestigial bits of prairie in the prairie peninsula.

The prairies are not completely treeless. A thin border of *Populus deltoides* var. *occidentalis* (Great Plains cottonwood)—and sometimes other species— forms a gallery forest along most of the larger streams. A dwarf phase of the bur oak (*Quercus macrocarpa* var. *depressa*) occurs wherever the moisture conditions are a bit more favorable than usual on the northern half of the plains. Usually hardly more than a shrub, it sometimes becomes a small tree several meters tall. Other small trees occur here and there in favorable sites.

Several groups of wooded hills or low mountains rise out of the plains in the northern half of the North American Prairies Province. Among the most noteworthy of these are the Black Hills of western South Dakota, the Turtle Mountains of northern North Dakota and adjacent Manitoba, and the Cypress Hills of southern Saskatchewan. The Black Hills are the largest and have attracted the most attention.

The Black Hills form a domelike outlier of the western American cordillera. They reach an elevation of nearly 2,200 m—not high enough to have a timberline. The lower slopes are largely dominated by *Pinus ponderosa*, but at upper altitudes this gives way to *Picea glauca*. Aside from widely distributed species, the flora of the Black Hills is largely a mixture of prairie, cordilleran, eastern American, and boreal elements. Among the trees, *Pinus ponderosa*, *P. contorta*, *P. flexilis*, and *Juniperus scopulorum* are cordilleran species; *Picea glauca*, *Betula papyrifera*, and *Populus balsamifera* are boreal; and *Ostrya virginiana* is eastern. *Populus tremuloides*, also common in the Black Hills, is a basically boreal species that extends well south into the western cordillera and occurs at diverse outlying stations in the northern half of the prairies.

The herbaceous flora of the Black Hills is likewise a mixture of western, northern, and eastern as well as prairie elements. *Aconitum columbianum*, *Lithophragma bulbifera*, *Microsteris gracilis*, *Balsamorhiza sagittata*, and

HOLARCTIC KINGDOM 99

Calochortus nuttallii are some of the many western herbaceous species that reach the Black Hills. *Saxifraga cernua, Astragalus alpinus, Circaea alpina, Adoxa moschatellina, Aster sibiricus,* and *Maianthemum canadense* are some of the northern ones; *Isopyrum biternatum, Thalictrum dioicum, Sanguinaria canadensis, Silene nivea, Lathyrus venosus, Aralia nudicaulis, Asclepias ovalifolia,* and *Allium canadense* are some of the eastern ones.

The Edwards Plateau in south central Texas forms a district at once distinctive and transitional to the Sonoran Province. Much of it is obviously prairie, and it bears a full complement of prairie grasses, including species of *Andropogon, Bouteloua, Buchloe,* and other genera. On the other hand, it shares a number of species with the Tamaulipan and Chihuahuan deserts of the Sonoran Province, notable among them *Prosopis glandulosa*. It takes much of its ecological character from the rugged terrain, with many large canyons. The canyon sides are commonly clothed with a dense scrub-oak–juniper thicket. Limestone and granite underlie different parts of the area. There are many special habitats, and more than 20 endemic species. Among these are *Anemone edwardsensis, Quercus texana, Streptanthus bracteatus, Styrax platanifolia, S. tenax, Lythrum ovatum, Galium correllii, Matelea edwardsensis, Forestiera reticulata, Physostegia correllii, Buddleja racemosa, Campanula reverchonii, Erigeron mimegletes, Verbesina lindheimeri, Dasylirion heteracanthium, Yucca rupicola,* and *Tridens buckleyanus*.

4. Rocky Mountain Region

Rocky Mountain Region, Gray and Hooker 1880; Harshberger 1911; Takhtajan 1969, 1974; Nordliches pazifisches Nordamerica, Mattick 1964; Cordilleran Forest Province, Gleason and Cronquist 1964.

The Rocky Mountain Region includes the Rocky Mountain and Cascade-Sierran mountain systems of the western United States and southwestern Canada, together with the coast ranges, the Olympic Mountains of Washington, and a narrow strip of southern Alaska as far west as Kodiak Island. It consists of two major parts—the Vancouverian Province on the west, and the Rocky Mountain Province farther inland. Each of these provinces may be further subdivided. The region is bordered by the North American Prairies Province (North American Atlantic Region) on the east, by the Canadian Province (Circumboreal Region) on the north and northeast, and by provinces of the Madrean Region on the south.

The flora of the Rocky Mountain Region includes one endemic family of liverworts (the Gyrothyraceae, native to the northern coast ranges of California, in the Vancouverian Province), but no endemic families of vascular plants. There are about 40 endemic or nearly endemic genera of vascular plants, including the following:

Taxodiaceae: *Sequoia* (1, along the Pacific coast from the southwestern corner of Oregon to Monterey County, California, thus extending shortly into the Californian Province of the Madrean Region), *Sequoiadendron* (1, western slopes of the Sierra Nevada in California)
Berberidaceae: *Vancouveria* (3, Vancouverian Province, south into Monterey County, California)
Sarraceniaceae: *Darlingtonia* (1, Vancouverian, near the coast in southwestern Oregon and northwestern California)
Ericaceae: *Allotropa* (1, Vancouverian), *Cladothamnus* (1, Vancouverian), *Hemitomes* (1, Vancouverian), *Kalmiopsis* (1, Vancouverian), *Pityopus* (1, Vancouverian), *Pleuricospora* (1, Vancouverian), *Sarcodes* (1, Vancouverian, south to the San Jacinto Mountains in the Californian Province)
Brassicaceae: *Anelsonia* (1, both provinces)
Saxifragaceae: *Bolandra* (2, both provinces), *Conimitella* (1, Rocky Mountain), *Elmera* (1, Vancouverian), *Peltiphyllum* (1, Vancouverian), *Suksdorfia* (2, both provinces), *Tellima* (1, both provinces), *Tolmiea* (1, Vancouverian)
Rosaceae: *Kelseya* (1, Rocky Mountain); *Luetkea* (1, both provinces, extending also into the mountains of Alaska and the Yukon); *Oemlera* (1, Vancouverain)
Hydrangeaceae: *Whipplea* (1, Vancouverian)
Apiaceae: *Harbouria* (1, Rocky Mountain), *Neoparrya* (1, Rocky Mountain)
Hydrophyllaceae: *Draperia* (1, Vancouverian), *Romanzoffia* (4, Vancouverian)
Boraginaceae: *Dasynotus* (1, Rocky Mountain)
Scrophulariaceae: *Chionophila* (1, Rocky Mountain), *Nothochelone* (1, Vancouverian), *Tonella* (2, both provinces)
Campanulaceae: *Howellia* (1, both provinces)
Asteraceae: *Adenocaulon* (1, both provinces; also disjunct about Lake Superior), *Crocidium* (1, both provinces), *Luina* (4, both provinces), *Orochaenactis* (1, Sierra Nevada, in the Vancouverian Province), *Raillardella* (5, Vancouverian; one species also south to the San Bernardino Mountains of southern California)
Uvulariaceae: *Scoliopus* (2, Vancouverian)
Poaceae: *Scribneria* (1, Vancouverian, and south to San Luis Obispo County, California)

Thus on the one hand the majority of the endemic genera pertain to the Vancouverian Province, a smaller number occur in both provinces, and only a few are restricted to the Rocky Mountain Province. On the other hand, both provinces have large numbers of endemic species, and many other species are endemic to the two provinces collectively.

In spite of the very many endemic species and a considerable number of small endemic genera, the Rocky Mountain Region is not the principal home or major center of diversity of many large genera. The largest genus is *Carex*, with perhaps 200 species, but *Carex* is nearly cosmopolitan and has more than a thousand species in all. The large genera *Delphinium, Ranunculus, Silene, Eriogonum, Salix, Draba, Ribes, Saxifraga, Potentilla, Astragalus, Lupinus,* and *Penstemon* are well represented in the region, but they all have major centers elsewhere. *Lomatium, Castilleja, Arnica,* and *Erigeron* are the principal large genera that have their greatest concentration of species in the Rocky Mountain Region, but they are also very well represented in the Great Basin, and might perhaps better be thought of as cordilleran.

The dominant vegetative cover of most of the Rocky Mountain Region is coniferous forest, containing especially species of *Abies, Picea, Pinus, Pseudotsuga, Thuja,* and *Tsuga,* and sometimes *Chamaecyparis, Cupressus, Juniperus, Larix, Libocedrus, Sequoia, Sequoiadendron,* and *Taxus*. No other floristic region of the New World has such a great diversity of conifers. Deciduous trees form a much smaller part of the community, although groves of *Populus tremuloides* are abundant in the Rocky Mountain Province. The highest parts of the mountains in both provinces form an alpine tundra, with a progressive diminution of arctic species toward the south and a progressive increase in species related to those of lower altitudes. Although conifers dominate the region, the only individual species that are common and widespread throughout most of both provinces are *Pinus contorta* and *Pseudotsuga menziesii*. Both are taxonomically complex species—the pine with four geographic varieties, *P. menziesii* with two.

1. Vancouverian Province (Vancouverian Province, Stebbins 1978, personal communication; Sitka, British Columbia, Washington, and Oregon, Good 1947, 1974; Sitka Province, Takhtajan 1970; Sitkan Province, Dice 1943; Oregonian Province, Dice 1943, p.p.; Munz 1959, p.p.; Sitka-Oregonian Province, Takhtajan 1978).

The Vancouverian Province extends from Alaska to California, spanning some 25 degrees of latitude, but it is not at any place more than about 350 km wide. In Alaska it forms a narrow strip near the coast from Kodiak Island eastward and southward through the panhandle. It continues southward through coastal British Columbia, and in Washington and Oregon it extends far enough inland to include the Cascade Mountains as well as the Coast Ranges and the Olympic Mountains. In California it goes south in the Coast Ranges nearly to San Francisco, and to the southern tip of the Sierra Nevada, thus bordering the northern part of the Californian Province on both the east and the west. The Warner Mountains in northeastern California and adjacent Oregon contain a mixture of Vancouverian and Great Basin species, and are here arbitrarily assigned to the Vancouverian Province.

Although the conifers set the character of the forests, broad-leaved trees are not wholly lacking from the Vancouverian Province. *Alnus rubra, Populus trichocarpa, Cornus nuttallii, Acer macrophyllum,* and *Fraxinus latifolia* occur in favorable sites throughout much of the province. *Acer circinatum* is common along streams, but it is scarcely a tree. Several species that are more abundant in the Californian Province extend well into the southern part of the Vancouverian Province. Notable among these are *Quercus garryana* and *Q. kellogii,* which even become locally dominant at the southern end of the Willamette Valley. Two evergreen (broad sclerophyll) angiosperm trees may also be noted as Californian elements. *Arbutus menziesii* (madrone) extends up the Puget Trough to southernmost British Columbia, and *Umbellularia californica* (California laurel) reaches the Coast Range in southwestern Oregon.

Some of the common shrubs in the Vancouverian Province are evergreen. *Berberis aquifolium, Gaultheria shallon,* and *Rhododendron macrophyllum* are notable examples. *Vaccinium parvifolium* is nearly evergreen. *Polystichum munitum* (Christmas fern) is an abundant evergreen herb. Other common shrubs are deciduous—namely, *Alnus sinuata, Amelanchier florida, Holodiscus discolor, Rubus spectabilis,* and *Oplopanax horridum.*

Meadows and slopes near and above timberline in the Vancouverian Province support a wealth of showy flowers in many genera, just as in the Rocky Mountain and Great Basin provinces. Natural openings in the forest, especially southward, also have a variety of flowering shrubs and herbs. In the dense forests, in contrast, such species are relatively few. *Trillium ovatum* is abundant and showy in the forests of Oregon and Washington, but it barely reaches British Columbia. *Achlys triphylla,* with inconspicuous flowers, is another characteristic forest species. Several of the endemic genera of Ericaceae (*Allotropa, Hemitomes, Pityopus, Pleuricospora,* and *Sarcodes*) are mycorhizal herbs without chlorophyll, which thrive in the humus-rich soil and do not require sun. The more widespread species *Hypopytis monotropa, Monotropa uniflora,* and *Pterospora andromedea* are taxonomically allied to these endemics and ecologically similar.

The western slopes of the Olympic Mountains in Washington support a temperate rain forest. Epiphytic bryophytes coat the larger branches of the trees, and hanging strands of *Usnea* (old man's beard lichen) are sometimes a meter long. The dominant species are *Picea sitchensis, Thuja plicata,* and *Tsuga heterophylla.* These are all large trees.

The spruce-cedar-hemlock forest continues northward near the coast in progressively less luxuriant form. *Thuja* drops out at the south end of the Alaska Panhandle, but *Tsuga heterophylla* continues to the Kenai Peninsula, and *Picea sitchensis* to Kodiak Island, the last outpost of the Vancouverian Province. *Thuja mertensiana,* a subalpine species in the more southerly portion of its range, grows at progressively lower elevations northward, and in coastal Alaska it mingles with *T. heterophylla* and *Picea sitchensis. Acer mac-*

rophyllum and *Cornus nuttallii*, which grow with the spruce-cedar-hemlock community on the Olympic Peninsula, as well as with other conifers elsewhere in the province, drop out in British Columbia, but *Alnus rubra* continues throughout the Alaska Panhandle.

In Oregon and Washington (aside from the west side of the Olympic Peninsula), summer drought, increasing southward, is a significant factor in determining the composition of the forest. One of the effects is to increase the frequency of fire. Here we find the great forests of *Pseudotsuga menziesii* (Douglas fir), a very large tree with thick, fire-resistant bark. In the absence of fire, Douglas fir tends to be replaced eventually by *Tsuga heterophylla* and *Thuja plicata*.

The east side of the Cascade Mountains is not so moist as the west side, and the forest is correspondingly more open. Here *Pinus ponderosa* becomes an important or dominant tree. This species is also common in the Rocky Mountain and Great Basin provinces and in the more southerly parts of the Vancouverian Province.

In southwestern Oregon and adjacent California, the Puget Trough is interrupted by a series of low mountains connecting the Cascade Range to the Coast Range. This area, called the Klamath region, is topographically and geologically diversified and harbors a relatively large number of local endemic species. Many of the endemics are associated with the serpentine or peridotite rock that is interspersed, often in large blocks, with granitoid substrates. Among the Klamath endemics are *Kalmiopsis leachiana*, three species of *Iris* (*I. bracteata*, *I. innominata*, and *I. thompsonii*), and more than a dozen species of Asteraceae, in addition to species of various genera in other families. The Asteraceae include *Arnica cernua*, *A. spathulata*, *Aster brickellioides*, *A. paludicola*, *A. siskiyouensis*, *Antennaria suffrutescens*, *Brickellia greenei*, *Erigeron cervinus*, *E. delicatus*, *E. flexuosus*, *Hieracium bolanderi*, *H. greenei*, and *Microseris howellii*. *Picea breweri* (Brewer spruce) is also nearly confined to the Klamath region. The Klamath region shows a considerable infiltration of species from the Californian Province, and some phytogeographers assign it to that province. In either province it is transitional to the other, but the ultrabasic rocks give it a special quality of its own.

There are also a number of other local centers of endemism in the Vancouverian Province, including the Olympic Mountains, the Wenatchee Mountains (an eastern spur of the Cascade Mountains in Washington, with considerable outcrops of serpentine), the Columbia River Gorge, and the southern Cascade Mountains, especially about Mt. Shasta and Mt. Eddy. The Sierra Nevada requires a more extended discussion.

The Sierra Nevada, in California and adjacent Nevada, forms a very distinctive element within the Vancouverian Province, topographically but not geologically continuous with the southern end of the Cascade Mountains. The western foothills, sometimes up to more than 1,000 m above sea level, belong with the Californian Province, but on the east side the Sierran community commonly extends all the way to the base of the mountains.

Typically the lowest forest zone in the Sierra Nevada is dominated by *Pinus ponderosa*. A little higher up, at about 1,500 m, ponderosa pine gives way to *Pinus jeffreyi*, *P. lambertiana*, *Abies concolor*, and *A. magnifica*. *Pinus contorta* is an important fire-tree at upper middle elevations. Although the zonation is far from exact, these several species in turn give way at higher elevations to *Pinus monticola*, *Tsuga mertensiana*, *Pinus albicaulis*, and *P. flexilis* (mainly on the east side of the mountains), which here become the timberline trees. None of these species is restricted to the Sierra Nevada, although *Abies magnifica* and *Pinus lambertiana* reach their best development here.

Sequoiadendron giganteum (California big tree or giant sequoia) occurs in scattered groves at middle or lower middle elevations (1,400 to 2,500 m) on the west slope of the Sierra Nevada, often with Jeffrey pine or red fir. It is certainly the most massive living thing, reaching a trunk diameter of 12 m and a height of 80 or even 100 m. It also reaches a great age—about thirty-five hundred years. This figure is known to be exceeded only by *Pinus longaeva*, a much smaller tree of high elevations in the mountains of the Great Basin (ca. 5,000 years).

Several monotypic genera and a hundred or more species are nearly or quite restricted to the Sierra Nevada. The Asteraceae alone include more than a dozen endemic or nearly endemic species—namely, *Aster peirsonii*, *Chaenactis alpigena*, *C. nevadensis*, *Erigeron petiolaris*, *E. miser*, *Eriophyllum nubigenum*, *Haplopappus eximius*, *H. peirsonii*, *H. whitneyi*, *Hulsea brevifolia*, *Madia yosemitana*, *Orochaenactis thysanocarpha*, *Phalacroseris bolanderi*, *Senecio clarkianus*, and *S. pattersonensis*.

There are of course also many widespread species in the Sierra Nevada. Some of the trees, such as *Pinus albicaulis*, *P. contorta*, *P. flexilis*, and *P. ponderosa*, are in many of the mountains of the western United States and adjacent Canada. *Pinus monticola* extends northward to southern British Columbia and reaches its best development in northern Idaho and adjacent parts of the Rocky Mountain Province. Some of the herbaceous species tend to link the Sierra Nevada with one or more of the following areas: the southern Cascade Mountains; the east (relatively dry) side of the Cascade Mountains in Oregon and Washington; the White Mountains of California (Great Basin Province, but separated from the Sierra Nevada only by a deep narrow trench); the Toiyabe and Toquima mountains (central Nevada, in the Great Basin Province); Steens Mountain (southeastern Oregon, in the Great Basin Province); the Wallowa Mountains (northeastern Oregon, Rocky Mountain Province); and the mountains of central Idaho. Other herbs, such as *Ranunculus eschscholtzii* and *Juncus mertensianus*, occur in suitable habitats nearly throughout the western cordillera. Others, such as *Pedicularis groenlandica*, *Senecio pauciflorus*, and *Spiranthes romanzoffianum*, are widespread in northern North America as well as extending south in the mountains. And a considerable number—such as *Cryptogramma crispa, Phleum alpinum, Oxyria digyna, Potentilla fruticosa, P. palustris*, and *Menyanthes trifoliata*—are more or less circumboreal at high latitudes.

The northern coast ranges of California form an attenuated southern arm of the Vancouverian Province. On the seaward side they are bathed in fog during the summer as well as the winter, although the actual precipitation is not very great. Here is the home of *Sequoia sempervirens*, the coast redwood, arguably the tallest tree in the world, although the trunk does not reach the diameter of *Sequoiadendron. Pseudotsuga menziesii* intermingles with the *Sequoia* or replaces it locally. *Polystichum munitum, Rhododendron macrophyllum, Gaultheria shallon, Vaccinium ovatum, Oxalis oregona, Vancouveria parviflora*, and *Whipplea modesta* are prominent understory plants. All of these species are also common farther north in Oregon and Washington. In contrast, *Umbellularia californica, Myrica californica*, and *Lithocarpus densiflora* are evergreen angiosperms (broad sclerophylls) that reflect the influence of the Californian Province.

Inland from the fog belt, the North Coast Ranges of California show an increasing influence of the Californian Province, with many broad sclerophyll shrubs and trees growing intermingled with or locally replacing Douglas fir, sugar pine, and big-leaf maple. Among these California elements are *Umbellularia californica, Lithocarpus densiflora, Quercus chrysolepis, Q. kelloggii, Castanopsis chrysophylla*, and *Arbutus menziesii*.

The Coast Ranges are not very high, and do not reach timberline. At upper elevations the forest in the North Coast Ranges of California rather resembles that of similar elevations in the Sierra Nevada, with *Abies magnifica, Pinus jeffreyi*, and *P. monticola* as prominent species.

Many basically Vancouverian species extend eastward through northern Washington and adjacent British Columbia into northern Idaho and northwestern Montana, or they have a disjunct occurrence in the more inland area. This segment of the Rocky Mountain Province is relatively moist, and at lower elevations not very cold, so that Vancouverian species can survive there. Among the trees with this distribution pattern are *Thuja plicata, Abies grandis*, and *Tsuga heterophylla. Pinus monticola* is similar, but may be even more at home in northern Idaho than farther west. Some shrubs and herbs that show a similar pattern are *Polystichum munitum, Berberis aquifolium, Salix sitchensis, Leptarrhena pyrolifolia, Mitella breweri, M. caulescens, Tellima grandiflorum, Ribes howellii, R. laxiflorum, Philadelphus lewisii, Holodiscus discolor, Rubus ursinus, Adenocaulon bicolor*, and *Maianthemum dilatatum*.

2. Rocky Mountain Province (Provinz der Rocky Mountains, Engler 1882, p.p., 1902, p.p., 1903 p.p., 1924 p.p.; Good 1947, 1974; Takhtajan 1970).

This province includes the Rocky Mountains and associated ranges from northernmost British Columbia to central Oregon (the Blue and Wallowa mountains), northeastern Utah (Uinta Mountains), and north central New Mexico (the south end of the southern Rocky Mountains).

The flora of the portion of the Rocky Mountain Province that lies in Canada is neither very large nor very rich in endemic species. Nearly all of this more northern portion of the province was heavily glaciated, and the flora has had to be reconstituted during the past ten thousand years. The dominant trees are mostly *Picea engelmannii* and *Abies lasiocarpa*, with increasing amounts of *Pseudotsuga menziesii* var. *glauca* southward. *Pinus contorta* is an important fire-tree. *Populus tremuloides*, *P. trichocarpa* (especially southwestward) and *Betula papyrifera* (especially northward) are also locally abundant. All of these trees, except *Betula papyrifera* and *Populus trichocarpa*, are also dominant species at increasing elevations southward, not only in the Rocky Mountain Province but also in the Great Basin Province. Many of the understory shrubs and herbs in the northern part of the Rocky Mountain Province are species that are widespread in the Canadian Province or in Eurasia as well.

Many or most of the alpine and subalpine species of the northern part of the Rocky Mountain Province are arctic (often circumpolar or boreal) species that also occur in the western, mountainous part of the Canadian Province as well as in the northern tundra. Among these are *Arenaria rossii* (North American), *Cerastium beeringianum*, *Silene acaulis*, *Oxyria digyna*, *Draba crassifolia*, *Cassiope tetragona*, *Saxifraga aizoides*, *S. bronchialis*, *S. caespitosa*, *S. cernua*, *S. flagellaris*, *S. nivalis*, *S. oppositifolia*, *S. tricuspidata* (North American), *Dryas integrifolia*, *D. octopetala*, *Potentilla nivea*, *Sibbaldia procumbens*, *Gentiana glauca*, *Pedicularis capitata*, *P. lanata*, *Campanula uniflora*, *Aster alpinus*, *Crepis nana*, *Erigeron humilis*, *Senecio resedifolius*, *Poa arctica*, *Carex capitata*, and *Kobresia simpliciuscula*. Some of these—such as *Silene acaulis*, *Oxyria digyna*, *Draba crassifolia*, *Saxifraga flagellaris*, and *Poa arctica*—extend south at increasing elevations nearly to the southern limits of the province. Others drop out progressively to the south, being replaced by species of more limited distribution. Among the arctic and alpine tundra species that probably do not extend south of the Canadian border are *Saxifraga aizoides*, *S. nivalis*, *S. tricuspidata*, *Pedicularis capitata*, and *P. lanata*. Some other species, such as *Cassiope tetragona* and *Gentiana glauca*, reach only to northwestern Montana.

In contrast, species such as *Draba argyraea*, *D. sphaerocarpa*, *Conimitella williamsii*, *Kelseya uniflora*, *Phacelia lyallii*, *Castilleja nivea*, *Chionophila tweedyi*, *Pedicularis pulchella*, *Penstemon montanus*, *Synthyris canbyi*, *Chaenactis evermannii*, *Cirsium tweedyi*, *Erigeron evermannii*, *E. flabellifolius*, and *E. rydbergii* are restricted to high elevations in the mountains of Montana, Idaho, and northern Wyoming, or to some part of that area. Still other endemic alpine species occur farther south.

The Rocky Mountain Province fades into the Canadian Province at the north and northeast. The transition is marked by the disappearance of *Picea engelmannii, Abies lasiocarpa,* and *Pinus contorta* as forest dominants, and their replacement by *Picea glauca, P. mariana,* and *Larix laricina. Abies balsamea,* a close relative of *A. lasiocarpa,* is common in the eastern and central parts of the Canadian Province, but hardly reaches the western cordillera. Likewise *Pinus banksiana,* of the Canadian Province, which is taxonomically and ecologically similar to *P. contorta,* scarcely extends into the cordillera. Not surprisingly, these specific replacements do not all occur at the same place. *Picea glauca* and *P. mariana* replace *P. engelmannii* in central British Columbia, but *Abies lasiocarpa* and *Pinus contorta* extend northward to the central Yukon.

A number of characteristic Rocky Mountain (or more broadly cordilleran) species extend northward in the Canadian Province well into the Yukon, or even into eastern Alaska. Among these are *Sedum lanceolatum, Parnassia fimbriata, Potentilla diversifolia, Geranium richardsonii, Penstemon procerus, Valeriana sitchensis, Agoseris aurantiaca, Arnica cordifolia, A. latifolia, Senecio triangularis,* and *Carex phaeocephala.*

The portion of the Rocky Mountain Province that lies in the United States is ecologically more diversified than the portion in Canada and supports a larger number of species. Glaciation was local rather than continental here, and this history is reflected in the flora. Each of the separate mountain ranges differs floristically to some extent from the others, and is likely to harbor one or more endemic (not necessarily alpine) species.

One of the outstanding features of the Rocky Mountain Province is the vertical zonation of the plant communities. The portion that lies in Canada is highly mountainous, with narrow valleys. This area is nearly all forested at lower elevations, but the higher parts of the mountains reach well above timberline and support an arctic-alpine tundra. Farther south, in the United States, there are fairly large unforested valleys and plains between the forested mountains. There may be two or even three timbered zones at increasing elevations, and the mountaintops are alpine.

South of northern Washington, northern Idaho, and northwestern Montana, the forests in the Rocky Mountain Province mostly do not reach the valley floors. Often, one may stand in a valley and see a band of timber across the face of the mountains. Below, it is too dry for forest; above, too cold. The lowest zone of forest often consists of an open stand of *Pinus ponderosa.* This gives way upward to *Pseudotsuga menziesii* var. *glauca,* which sometimes extends to timberline but usually does not. The uppermost timbered zone is typically dominated by *Abies lasiocarpa* and *Picea engelmannii. Pinus contorta* and *Populus tremuloides* are also common, especially at middle levels (the pine mainly as a fire-tree). Toward the south *Picea pungens* is a frequent component of the forests, usually at a slightly lower level than *P. engelmannii*. Often there are scattered junipers (mainly *Juniperus scopulorum*) in the foothills, and these small trees may extend well up into the drier and more open parts of the forest. In Colorado and New Mexico there is often a band of oak brush (*Quercus gambelii*) below the forest proper. This species is more characteristic of the eastern part of the Great Basin Province than of the Rocky Mountain Province.

On the east, in the United States and in southern Alberta, the Rocky Mountain Province adjoins the North American Prairies Province. In Montana and Wyoming the prairie flora is continuous with the broad intermontane valleys of the Rocky Mountain Province. Farther to the north or south, in Alberta, Colorado, and New Mexico, the mountain front is more continuous and the Colorado Front Range, in particular, rises abruptly from the plains. The presence of a number of Rocky Mountain species in the Black Hills of South Dakota is noted in the discussion of the North American Prairies Province.

West of its southern tip in New Mexico, the Rocky Mountain Province adjoins the Great Basin Province (in the Madrean Region) on the south. Many Great Basin elements extend for some distance into the Rocky Mountain Province. *Artemisia tridentata*, a characteristic Great Basin species, also occurs in suitable habitats throughout much of the part of the Rocky Mountain Province that lies in the United States, and it even extends into the Okanogan Valley in southern British Columbia. It is convenient to draw the southern boundary of the Rocky Mountain Province to include the Wallowa, Blue, and associated mountains of central and east central Oregon. Many Great Basin species reach the southern base of these low mountains but do not extend significantly into them.

In spite of their limited flora, the sagebrush lands of central Washington have several endemic species. Among these are *Astragalus leibergii*, *A. speirocarpus*, *Lomatium tuberosum*, *Erigeron basalticus*, and *E. piperianus*.

The John Day Valley in north central Oregon also has several endemic species, notably on the barren clay slopes derived from volcanic tuff. Among these are *Castilleja xanthotricha*, *Chaenactis nevii*, and *Allium pleianthum*.

A considerable part of southeastern Washington and a bit of adjacent Idaho are covered by a deep, fine-grained, fertile loess with a dunelike topography. This palouse country, as it is called, was once a lush grassland, with *Agropyron spicatum*, *Festuca idahoensis*, and *Poa sandbergii* as dominant species. It is now largely devoted to the cultivation of wheat. Despite the relatively recent (certainly postglacial) origin of the habitat, several species are nearly or quite confined to the palouse. Among these are *Astragalus* arrectus, Polemonium pectinatum, Aster jessicae, Cirsium brevifolium, and Haplopappus liatriformis.

The Snake River Canyon forms a deep, narrow, dry trench between the mountains of northeastern Oregon and those of west central Idaho. Here again there are a number of local species, some of which also extend shortly southward into the Great Basin Province along the Oregon-Idaho border. Among these endemics are *Astragalus arthurii, A. cusickii, A. vallaris, Lomatium rollinsii, L. serpentinum, Phlox colubrina, P. viscida, Nemophila kirtleyi, Hackelia hispida* (also, in another variety, disjunct in the Grand Coulee of central Washington), *Penstemon elegantulus, P. triphyllus, Erigeron disparipilus, Haplopappus radiatus,* and *Camassia cusickii.*

A broad swath of high, dry plains and mesas across southern Wyoming separates the northern Rocky Mountains (central and northern Wyoming and northward) from the southern Rocky Mountains (southeastern Wyoming and southward). This visually barren area, dominated by sagebrush, is host to a number of local species, and the western end of it shares some species with the Uinta Basin in Utah (Great Basin Province). In relatively dry areas such as this, the nature of the rock plays an even more important role in governing the habitat and flora than it does in moister regions. The frequent seleniferous outcrops are of particular significance here, as they are in the Colorado Plateau segment of the Great Basin Province in Utah. Among the species nearly or quite confined to southern Wyoming are *Lesquerella fremontii, L. macrocarpa, Physaria condensata, Astragalus drabelliformis, A. nelsonianus, A. proimanthus, A. simplicifolius, Penstemon acaulis, Artemisia porteri, Haplopappus contractus, Parthenium alpinum,* and *Tanacetum simplex.* Some species that in the field appear to be characteristic of special habitats in southern Wyoming actually have a wider range and occur in similar habitats in eastern Utah and western Colorado. *Machaeranthera glabriuscula* is such a species.

The southern Rocky Mountains extend from southeastern Wyoming to north central New Mexico and form a distinctive district within the Rocky Mountain Province. Narrow at the north and south ends, they fan out more broadly in Colorado, and parallel ranges are separated by open valleys locally known as parks. The mountains rise well above timberline over large areas.

Upward of 50 species are nearly or quite confined to the southern Rocky Mountains, and many more center there but also extend into one or more other mountain ranges or groups of ranges to the north, west, or south. Among the endemic or nearly endemic species are *Aquilegia saximontana, Delphinium ramosum, Draba exunguiculata, D. grayana, Physaria bella, Primula angustifolia, Heuchera bracteata, H. hallii, Astragalus anisus, A. molybdenum, A. parryi, Aletes humilis, Angelica ampla, A. grayi, Harbouria trachyphylla, Cryptantha virgata, C. weberi, Penstemon alpinus, P. hallii, P. har-*

bourii, P. secundiflorus, P. virens, Castilleja puberula, Chionophila jamesii, Artemisia pattersonii, Aster porteri, Cirsium parryi, Erigeron melanocephalus, E. pinnatisectus, E. vetensis, Haplopappus pygmaeus, Helianthus pumilus, Senecio fendleri, S. taraxacoides, Townsendia eximia, T. fendleri, T. glabella, T. rothrockii, Carex oreocharis, and *Veratrum tenuipetalum.*

The southern Rocky Mountains are the southern outpost for many arctic or boreal (often circumpolar) species. Some of these are well scattered in the mountains from Alaska to Colorado and even New Mexico, but others are more or less strongly disjunct from their more northern occurrence. The more rare and highly disjunct species mostly do not occur at the very highest elevations or on the most exposed sites. Instead they are likely to be in moist meadows or other protected sites, sometimes well below timberline. Some of the species in Colorado—which are not otherwise known to occur within the contiguous United States, or which approach no closer than northwestern Montana—are *Koenigia islandica, Viola biflora, Braya humilis, Primula egaliksensis* (North America), *Saxifraga foliolosa, S. hirculus, S. rivularis, Parnassia kotzebuei* (North America), *Phippsia algida*, and *Juncus biglumis*. Except as indicated, these are all circumboreal species.

The Uinta Mountains, which run east-west in northeastern Utah, rise high above timberline. At the west end they are nearly contiguous with the Wasatch Mountains (Great Basin Province), but geologically and floristically they belong with the Rocky Mountains. In the density of their forest cover, especially the large stands of *Pinus contorta*, they are reminiscent of the northern Rocky Mountains. At upper altitudes, especially near or above timberline, there are a considerable number of arctic and boreal species, such as *Isoetes lacustris, Silene acaulis, Gentiana algida, Carex bipartita, C. paupercula, C. rupestris, Eriophorum polystachion, E. scheuchzeri*, and *Kobresia myosuroides*. All of these also occur in the mountains of Colorado, but some of them are missing from other high mountains in Wyoming, Montana, and Idaho. Surprisingly, there are but few endemics, of which only *Parrya rydbergii* and *Penstemon uintahensis* have come to our attention.

B. TETHYAN (ANCIENT MEDITERRANEAN) SUBKINGDOM

Takhtajan 1969, 1970, 1974.

The Tethyan Subkingdom[53] stretches from Macaronesia in the west through the entire Mediterranean and through western and central Asia to the Gobi in the east. Despite the fact that the plant world of this vast territory is extraordinarily varied and consists of different ecological types, the history of the development of the Tethyan flora is characterized by several common traits. In his work on the geographic distribution of the genus *Rhus*, Engler (1881) introduced the concept of the "flora of the Ancient Mediterranean," meaning that formed along the seashore of the Tethys. Popov (1927, 1929) developed the idea of a Tethyan flora in more detail, but unfortunately he expanded the borders of this flora all the way to California and Mexico, which can hardly be considered correct.

The Tethyan flora developed primarily due to migration. Almost the entire territory of the receding Tethys Sea was an arena of migration and mixing of floras of completely different origins. The Tethyan flora developed at the juncture of boreal and tropical floras, but the overwhelming majority of the Tethyan plants have boreal, and in part eastern Asiatic, origins. However, various modified immigrants from the tropical flora play an evident role in this subkingdom, especially in the Macaronesian and Saharo-Arabian regions. Such tropical elements include most notably: representatives of the Lauraceae; the palms—*Phoenix canariensis* in Macaronesia, and *P. theophrastii* on Crete Island; *Chamaerops humilis* in the western Mediterranean; and *Nannorhops*, distributed from Arabia to northwestern India. They also include *Cytinus hypocistis* and *Pilostyles haussknechtii* of the Rafflesiales, representative of the families Capparaceae, Loranthaceae, Santalaceae and Zygophyllaceae, the genus *Cynomorium*, the Macaronesian endemic *Sideroxylon marmulano* and the Moroccan endemic *Argania spinosa*—both of the Sapotaceae, and others. The Tethyan flora exhibits a well-developed relationship with the African, particularly with the Cape, flora. Many xerophytic and especially geophytic elements of the Tethyan flora have close relatives in South Africa. Thus the conifer *Tetraclinis articulata*, growing in northwestern Africa, southeastern Spain, and Malta is close to the genus *Widdringtonia* (tropical and South Africa) and also to the genus *Callitris* (Australia and New Caledonia). Among the dicotyledons, the African relationship is shown best in the genera *Limonium, Dianthus, Silene, Linum, Pelargonium, Erica, Argyrolobium, Convolvulus, Stachys, Salvia, Scabiosa, Cephalaria, Filago, Micropus, Ifloga, Helichrysum, Echinops,* and others; and among the monocots; in *Androcymbium, Colchicum, Scilla, Urginea, Dipcadi, Ornithogalum, Gladiolus, Iris, Gynandriris,* and others. Engler (1879) explained these relationships by a common origin from the original tropical ancestors; however, it is much more likely that a floristic exchange took place for a longer time between the Tethys area and South Africa. This exchange occurred via the mountains of East Africa, Ethiopia, and farther through the Drakensberg Mountains, which formed a bridge joining the Tethyan flora with the South African. In the Pliocene, mountain ranges of northeastern Africa extended to Asia Minor and the Balkan Peninsula, which explains the dominance among the Mediterranean elements of Africa of eastern Mediterranean taxa over those of the western Mediterranean. Perhaps this floristic exchange was reciprocal, although we are far from being able to identify the direction of migrations in all cases.

The Tethyan Subkingdom is divided into four regions: the Macaronesian; the Mediterranean; the Saharo-Arabian, and the Irano-Turanian.

5. Macaronesian Region

Makaronesisches Uebergangsgebiet, Engler 1882, 1899, 1903, 1924; Graebner 1910; Hayek 1926; Chevalier (in Chevalier and Cuenot) 1932; Good 1947, 1974; Mattick 1964; Takhtajan 1969, 1970, 1974; Ehrendorfer 1971; Bramwell 1972, 1976; Knapp 1973; Zohary 1973; Humphries 1979; Sunding 1979.

The Macaronesian Region includes the Azores Islands, the Madeira Islands, the Salvage Islands, the Canary Islands, and the Cape Verde Islands. An exclave of the Macaronesian flora occurs on the African mainland in southern Morocco and Spanish West Africa (Sunding 1979). According to Sunding (1979:14), a great number of plant and animal taxa within this mainland area shows greater affinity to taxa in the Macaronesian islands than to taxa on the mainland outside the "Macaronesian Enclave."

More than half of the flora of the region consists of Mediterranean species. There are comparatively few endemic genera (see Bramwell 1976 and Humphries 1979), and more than half of them inhabit the Canary Islands. Endemic genera include the following:

Caryophyllaceae: *Dicheranthus* (1, Canary Islands)
Theaceae: *Visnea* (1, Madeira, Canary Islands)
Brassicaceae: *Parolinia* (3, Canary Islands), *Sinapidendron* (about 9, Madeira, Canary Islands, Cape Verde Islands)
Myrsinaceae: *Pleiomeris* (1, Madeira, Canary Islands)
Urticaceae: *Gesnouinia* (1, Canary Islands)
Crassulaceae: *Aichryson* (14), *Greenovia* (4, Canary Islands)[54]
Rosaceae: *Bencomia* (5, Madeira, Canary Islands; very close to *Sanguisorba*), *Dendriopoterium* (2, Canary Islands; very close to *Sanguisorba*), *Chamaemeles* (1, Madeira), *Marcetella* (2, Madeira, Canary Islands; very close to *Sanguisorba*)
Fabaceae: *Spartocytisus* (2, Canary Islands, very close to *Cytisus*)
Cneoraceae: *Neochamaelea* (1, Canary Islands)
Apiaceae: *Melanoselinum* (2, close to *Thapsia*), *Tinguarra* (1, Canary Islands), *Todaroa* (2, Canary Islands), *Tornabenea* (1, Cape Verde Islands).[55]
Oleaceae: *Picconia* (2, Azores, Madeira, Canary Islands)
Santalaceae: *Kunkeliella* (2, Canary Islands)
Gentianaceae: *Ixanthus* (1, Canary Islands)
Rubiaceae: *Phyllis* (2, Madeira, Canary Islands), *Plocama* (1, Canary Islands)

Scrophulariaceae: *Isoplexis* (4, Madeira, Canary Islands; close to *Digitalis*).
Lamiaceae: *Cedronella* (1, Azores, Madeira, Canary Islands)
Campanulaceae: *Azorina* (1, Azores), *Musschia* (2, Madeira)
Asteraceae: *Allagopappus* (2, Canary Islands), *Argyranthemum* (about 22, Canary Islands; close to the Cape shrubs of *Chrysanthemum*), *Gonospermum* (2, Canary Islands), *Heywoodiella* (1, Canary Islands), *Lactucosonchus* (1, Canary Islands; close to *Sonchus*), *Schizogyne* (2, Canary Islands), *Sventenia* (1, Canary Islands; related to *Sonchus*), *Vieraea* (1, Canary Islands)
Ruscaceae: *Semele* (5–6, Madeira and Canary Islands)
Poaceae: *Monachyron* (1, Cape Verde Islands).

The flora of the Macaronesian Region also contains a number of endemic sections, some of which are often elevated to the rank of genus. Endemic sections are found in the genera *Crambe* (section Dendrocrambe), *Descurainia* (section Sisymbriodendron), *Aeonium*, *Teucrium* (section Teucropsis), *Sideritis* (section Leucophae), *Convolvulus* (sections Frutescentes and Floridi), *Echium* (sections Simplicia and Gigantea), *Sonchus* (section Atalanthus), *Senecio*, and others.

The number of native species of the Macaronesian Region is relatively small (approximately 3,200 species of flowering plants; Humphries 1979). The Canary Islands contain the richest flora, numbering about 1,860 species of vascular plants (Sunding 1979), followed by Madeira with 1,141 species (Sunding 1979) and the Azores with 843 species (Sunding 1979). The Salvage Islands have only 87 (Sunding 1979) or 91 species (Humphries 1979). The percentage of endemic species is very high (about 680 species of flowering plants; Humphries 1979). Among the endemic taxa are a considerable number of ancient relict endemics, with their greatest affinities amongst the Tertiary flora (Engler 1879; Wulff 1944; Takhtajan 1969; Bramwell 1972, 1976; Sunding 1972, 1979). It is characteristic of these relict endemics that they are either in general systematically rather isolated or, if they have some close relatives, these usually grow in more or less remote geographical regions. Some of the most interesting examples of disjunct taxa are:

Pinus canariensis (Macaronesia)—*P. roxburghii* (Himalayas)
Apollonias barbusana (Macaronesia)—*A. arnottii* (southern India)
Persea indica (Macaronesia)—species of *Persea* in South, Southeast, and East Asia, and also in America
Visnea (Macaronesia)—*Eurya* (East Asia, Indian, and Malesia)
Clethra arborea (Macaronesia)—related species in America
Heberdenia excelsa (Macaronesia)—*H. penduliflora* (Mexico)
Drusa (Macaronesia)—*Bowlesia* and *Homalocarpus* (America)
Ilex canariensis (Macaronesia)—related species in America

Picconia (Macaronesia)—*Notelaea* (eastern Australia)
Phyllis (Macaronesia)—*Galopina* (South Africa)
Bystropogon subgenus Bystropogon (Macaronesia)—*Bystropogon* subgenus Minthostachys (mountains of South America, mainly the Andes)
Canarina canariensis (Macaronesia)—*C. abyssinica* and *C. eminii* (East Africa)
Dracaena draco (Macaronesia)—*D. cinnabari* (Socotra)

These disjunctions show how diverse the geographic relationships of elements of the ancient nucleus of the flora of Macaronesia are. Of great particular interest are the American relations, which may be explained in some measure by continental drift (Bramwell 1976).

Among the endemic species of the Macaronesian Region are a large number of secondary shrubs and treelike forms, which are usually represented outside Macaronesia by herbaceous species.

In Macaronesia, the most characteristic vegetative cover is the evergreen laurel forest (laurisilvae), which among all contemporary plant formations is closest to the late Miocene and Pliocene evergreen forests of Europe and the Caucasus (Wulff 1944; Meusel et al. 1965, Ciferri 1962; Takhtajan 1969; Bramwell 1974).

1. Azorean Province (Engler 1882, 1899, 1903, 1924). This province includes the Azores. There is only one endemic genus, *Azorina*, but there are about 40 endemic species, constituting about 5% of the native flora of vascular plants (Humphries 1979; Sunding 1979).

Until the colonization of the islands by Europeans, they were covered by evergreen forests dominated by the Macaronesian endemic *Persea indica* and by the Macaronesian subendemic *Laurus azorica* (encountered also in Morocco; see Barbero et al. 1980). Currently, only small fragments of these laurel forests are preserved. In terms of floristic relations, the most interesting components of the forest and shrub communities are the Azores endemics *Juniperus brevifolia, Erica scoparia* subsp. *azorica, Hypericum foliosum, Sanicula azorica, Vaccinium cylindraceum* (close to the Caucasus bilberry *V. arctostaphylos*), *Daboecia azorica, Picconia azorica, Rubus hochstetterorum,* and *Prunus lusitanica* subsp. *azorica.* Then there are the characteristic Macaronesian endemics, *Ilex perado* and *Frangula azorica*, as well as *Myrica faya*, which in addition to Macaronesia is also found in central and southern Portugal, where it presumably grows wild. The presence in the Azores flora of a representative of the Culcitaceae is very interesting. It is the Macaronesian-Iberian endemic *Culcita macrocarpa*, which in addition to the Azores is also found in the Canary Islands, Madeira, northern Portugal, and southern Spain. Also of interest is the appearance of the evergreen shrub, *Myrsine africana* (outside of the Azores, its distribution includes central and southern

Africa, and from Arabia to Taiwan). In contrast with the remaining provinces of Macaronesia, the xerophytic flora of African origin is absent in the Azores, but one finds a marked number of species in common with the flora of the Iberian Peninsula.

2. Madeiran Province (Engler 1882, 1899, 1903, 1924). This province includes the island of Madeira; Porto Santo Island; and two groups of uninhabited, rocky islands, the Desertas and Selvagens (the Salvage Islands).

The flora of the Madeiran Province numbers about 1,158 species of vascular plants (Sunding 1979), among which are two endemic genera (*Chamaemeles* and *Musschia*) and over 120 endemic species (Hansen 1969; Sunding 1979). But according to Humphries (1979), it has less than 100 endemic species, including the following:

Polystichum drepanum, P. falcinellum, P. maderense, P. webbianum, Berberis maderensis, Cerastium vagans, Beta patula, Armeria maderensis, Viola paradoxa, Crambe fruticosa, Erysimum arbuscula, E. tenuifolium, Matthiola maderensis, Sinapidendron angustifolium, S. frutescens, S. rupestre, Erica cinerea, Vaccinium maderense, Euphorbia piscatoria, Aeonium glandulosum, A. glutinosum, Aichryson divaricatum, A. dumosum, A. villosum, Sedum brissemoretii, S. farinosum, S. fusiforme, S. nudum, Saxifraga maderensis, S. portosanctana, Chamaemeles coriacea, Rubus spp., *Sanguisorba maderensis, Sorbus maderensis, Anthyllis lemanniana, Cytisus maderensis, C. paivae, C. tener, Lotus loweanus, L. macranthus, Ononis costae, Vicia atlantica, V. capreolata, V. pectinata, Bunium brevifolium, Melanoselinum edulis, Oenanthe pteridifolia, Peucedanum lowei, Maytenus dryandri* (= *Catha dryandri*), *Jasminum azoricum, J. odoratissimum, Galium productum, Convolvulus massonii, Echium nervosum, Solanum trisectum, Isoplexis sceptrum, Odontites holliana, Scrophularia hirta, S. pallescens, S. racemosa, Sibthorpia peregrina, Plantago leiopetala, P. malato-belizii, P. subspathulata, Bystropogon maderensis, B. piperitus, Teucrium abutiloides, T. betonicum, Musschia aurea, M. wollastonii, Andryala crithmifolia, Artemisia argentea, Calendula maderensis, Carduus squarrosus, Centaurea massoniana, Chrysanthemum barretii, C. dissectum, C. haematomma, C. mandonianum, C. pinnatifidum, Cirsium latifolium, Crepis divaricata, C. noronhaea, Helichrysum devium, H. melanopthalmum, H. monizii, H. obconicum, Lactuca patersonii, Senecio maderensis, Sonchus pinnatus, S. squarrosus, S. ustulatus, Tolpis macrorhiza, Scilla maderensis, Semele madeirensis, S. menezesii, S. pterygophora, S. tristonis, Ruscus streptophyllus, Carex lowei, C. malato-belizii, Dactylorhiza foliosa, Goodyera macrophylla, Orchis scopulorum, Deschampsia argentea, Festuca albida, F. donax, Helictotrichon sulcatum, Lolium lowei,* and *Phalaris maderensis.*

3. Canarian Province (Engler 1882, 1899, 1903, 1924). This province includes the Canary Islands. There are about 20 genera endemic to the

Canary Islands. In addition, there is a series of endemic sections and some 470 (Bramwell 1972, 1976) or 460 (Humphries 1979) endemic species. The most ancient elements of the Canary Islands (including the most ancient endemics) are found in the evergreen laurel forests. Until colonization of the islands by Europeans, these forests were much more widely distributed than at present. Unfortunately, almost nothing of this great expanse of laurel forests has been preserved. Different stages of degradation are observable, especially where the moistening influence of fog is weaker. The following Macaronesian endemics contribute to the composition of these forests: *Laurus azorica, Persea indica, Ocotea foetens,* and *Apollonias barbusana,* together with *Ilex canariensis* (Canary-Madeira endemic), *Arbutus canariensis* (Canary endemic), *Rhamnus glandulosa* (Canary-Madeira endemic), *Viburnum rigidum* (Canary endemic), the Macaronesian-Mediterranean species *Erica arborea, Visnea maconera* (Canary-Madeira endemic), *Myrica faya* (Macaronesian endemic), *Pleiomeris canariensis* (Canary endemic), and *Salix pedicellata* subsp. *canariensis* (Canary-Madeira endemic). In addition, there is the interesting epiphytic fern *Davallia canariensis* (Canary Islands, Madeira, Portugal, and western Spain), and the characteristic lianas *Hedera canariensis* (Macaronesian endemic) and *Semele androgyna* (Canary-Madeira endemic).

A belt of pine forest is commonly associated with the laurel forest, at higher elevations and in drier places. The endemic Canary pine, *Pinus canariensis,* is very similar to the Himalayan pine, *P. roxburghii.* It is accompanied by *Erica arborea,* endemic species of *Cistus,* the endemic *Chamaecytisus proliferus,* and *Adenocarpus viscosus.* The endemic *Juniperus cedrus* grows at the upper boundaries of the pine forests. It is very close to *J. oxycedrus.*

A rather sharp separation exists between the forest vegetation and the formation of montane, shrubby xerophytes ("mountain desert") developing above the cloud layer. In the eastern Canary Islands, the tree vegetation is almost absent and only formations of the semidesert and desert types are developed. Here, scattered among rocky soils, are the very characteristic huge shrubs of the endemic *Spartocytisus supranubius.* Among the endemic plants encountered in this formation, the following merit special attention: *Echium* spp., *Descurainia bourgaeana,* and *Cheiranthus scoparius.* Over 70% of the species of this formation are endemic.

At heights of more than 2,600 m above sea level, one finds the unique xerophytic and cold-resistant vegetation represented by such endemic species as *Viola cheiranthifolia.* In the lower parts of the island of Tenerife and also in comparable conditions on other islands, a xerophytic flora flourishes in which Saharo-Arabian elements are encountered. Of an entirely African character is the formation of succulents, which is developed in the lower parts of slopes of Tenerife and several other islands. It includes the very characteristic *Euphorbia canariensis,* as well as the endemics *E. regis-jubae* and *Senecio kleinia.* Another less xerophytic, succulent formation is found in more widely

distributed, moister, rocky sites. Here grow endemic representatives of the genera *Aeonium, Sonchus,* and others. One of the most characteristic Macaronesian endemics, *Dracaena draco,* grows in this belt (Canary Islands, Madeira Island, Cape Verde Islands). The endemic *Phoenix canariensis* also grows here. In addition to the endemics just mentioned we cite the following:
Aspidium canariense, Ranunculus cortusifolius, Parietaria filamentosa, Gesnouinia arborea, Forsskaolea angustifolia, Minuartia spp., *Cerastium sventenii, Buffonia teneriffae, Silene berthelotiana, S. nocteolens* and other species of this genus, *Polycarpaea* spp., *Paronychia canariensis, Dicheranthus plocamoides, Herniaria canariensis, Beta webbiana, Rumex lunaria, Limonium arborescens* and a series of other species of this genus, *Hypericum canariense, H. coadunatum, H. reflexum, Viola palmensis, Cistus* spp., *Helianthemum* spp., *Bryonia verrucosa, Crambe* spp. (including *C. arborea* and *C. gigantea*), species of *Descurainia* and *Parolinia, Matthiola bolleana, Erucastrum canariense, Reseda crystallina, R. scoparia, Lavatera phoenicea,* about 10 species of *Euphorbia* (including several shrubs), *Umbilicus heylandianus,* the majority of species of the Canary-Moroccan genus *Monanthes,* the majority of the species of *Aichryson,* many species of *Aeonium,* 4 species of the endemic *Greenovia,* several species of *Bencomia, Anagyris latifolia,* species of *Adenocarpus, Cytisus, Dorycnium, Lotus,* and *Vicia, Ruta oreojasme, R. pinnata, Neochamaelea pulverulenta* (= *Cneorum pulverulentum*), *Zygophyllum fontanesii, Drusa glandulosa, Bupleurum salicifolium, Ruthea herbanica, Astydamia latifolia, Ferula linkii, F. lancerottensis, Seseli webbii, Tinguarra cervariifolia, T. montana, Todaroa aurea, Pimpinella* spp., *Cryptotaenia elegans, Ilex platyphylla* (very close to *I. perado*), *Maytenus canariensis, Rhamnus crenulata, R. integrifolia, Kunkeliella canariensis, K. psilotoclada, Sambucus palmensis, Pterocephalus* spp., *Ceropegia* spp., *Ixanthus viscosus, Plocama pendula, Phyllis viscoa, Convolvulus canariensis* and other species of this genus, *Messerschmidia fruticosa,* about 20 species of *Echium, Solanum nava, S. vespertilio, Scrophularia* spp., *Campylanthus salsoloides, Sutera* (*Lyperia*) *canariensis, Ixoplexis canariensis, Kickxia* spp., *Globularia* spp., *Plantago arborescens, P. webbii, Lavandula canariensis, L. minutolii, Bystropogon* spp., *Thymus origanoides,* over 10 species of *Micromeria, Salvia canariensis, S. broussonetii, Nepeta teydea,* over 10 species of *Sideritis, Canarina canariensis, Laurentia canariensis, Phagnalon purpurascens, P. umbelliforme, Gnaphalium webbii, Helichrysum gossypium, H. monogynum, Allagopappus* spp., *Vieraea laevigata, Pulicaria burchardii, P. canariensis, Schizogyne* spp., *Asteriscus* spp., *Gonospermum* spp., over 10 species of *Argyranthemum* and *Chrysanthemum, Artemisia canariensis,* over 10 species of *Senecio, Carlina canariensis, C. xeranthemoides, Onopordum nogalasii, Carduus* spp., *Centaurea* spp., *Rhaponticum canariensis, Andryala* spp., *Hypochoeris oligocephala, Tolpis* spp., *Crepis canariensis, Lactucosonchus webbii, Sventenia bupleuroides, Erigeron cabrerae, Prenanthes pendula,* over 20 species of *Sonchus* (including *S. arboreus* and *S. canariensis*), *Lactuca* spp., *Reichardia* spp., *Pancratium canariensis,*

Androcymbium psammophilum, Scilla haemorrhoidalis, Asparagus arborescens and several other species of this genus, *Habenaria tridactylites, Orchis canariensis, Luzula canariensis, Carex canariensis, C. perraudieriana, Agrostis canariensis, Phalaris canariensis, Festuca bornmuelleri, Avena canariensis, A. occidentalis, Dactylis smithii, Brachypodium arbuscula, Melica canariensis, M. teneriffae,* and *Dracunculus canariensis.*

Among the Canary Islands endemics, species with Mediterranean relationships predominate. According to Sunding (1970), in the flora as a whole, the Mediterranean element constitutes 81% (1,282 species), and of the endemics it constitutes 68% (or 401 species). In second place is the East African element (6% of the total composition of the flora and 16% of the endemic composition). Following these in total composition of the flora are species of a "Saharo-Sindian" or, more specifically, a Saharo-Arabian element (68 species, constituting 4% of the total flora). In third place in terms of numbers of endemics is the South African element, with 48 species, or 16% (Sunding 1970).

4. Cape Verde Province (Engler 1882, 1899, 1903, 1924). This province encompasses the 14 islands of Cape Verde. The climate of these islands is considerably drier than that of the other islands of Macaronesia. Nevertheless, according to testimony of the first Europeans who visited these islands, they were "well-forested." Today the forest vegetation is absent. On the dunes of the seashores one finds *Suaeda vermiculata* and *Sporobolus robustus.* Behind the dunes grow *Tamarix gallica, Euphorbia tuckeyana, Zygophyllum fontanesii,* and *Launaea spinosa.* Grasslands occur in the interior of the eastern islands. *Dracaena draco* grows in the mountains. In the mountains the flora reflects a predominantly Macaronesian character, but along the shores the African element (desert and savannah) predominates.

According to Sunding (1973, 1979), in its total composition, the vascular flora of the Cape Verde Islands shows a marked difference from that of the northern islands of Macaronesia, and its floristic spectrum is quite different. As much as 54% of the species belong to a tropical flora element, with a significant part being neotropical. In addition, several of the endemics have tropical affinities (Sunding 1979:34).

The number of indigenous species of vascular plants is about 650 (Sunding 1979). Two endemic genera (*Tornabenea* and *Monachyron*) occur, as well as 92 endemic species (Sunding, 1979). Among these are *Papaver gorgoneum, Paronychia illecebroides, Polycarpaea gayi, Limonium brunneri, L. braunii, Helianthemum gorgoneum, Sinapidendron glaucum, S. hirtum, Matthiola capoverdeana, Sideroxylon marmulana, Euphorbia tuckeyana, Umbilicus schmidtii, Aeonium gorgoneum, Lotus purpureus, L. jacobaeus, Melanoselinum insulare, Sarcostemma daltonii, Echium stenosiphon, E. lindbergii, E. glabrescens, Celsia insularis, C. cystolithica, Campylanthus glaber, C. benthamii, Linaria brunneri,*

HOLARCTIC KINGDOM

Lythanthus amygdalifolius, Lavandula rotundifolia, Campanula jacobaea, C. bravensis, Nidorella varia, Artemisia gorgonea, Odontospermum daltonii, O. vogelii, Launaea picridioides, L. melanostigma, and *Aristida paradoxa.*

6. Mediterranean Region

Candolle (A. P. de) 1808; Candolle (Alph. de) 1855; Boissier, 1867; Grisebach 1872, 1884; Hayek 1926; Eig, 1931; Good 1947, 1974; Braun-Blanquet 1923b, 1937; Mattick 1964; Davis 1965; Polunin and Huxley 1965; Takhtajan 1969, 1970, 1974; Ehrendorfer 1971; Zohary 1973; Hedge 1976; Quézel 1978.

The northern and southern boundaries of the Mediterranean Region are on the whole relatively clear, and disagreement among authors appears only in relation to comparatively small areas—mainly the Iberian and Balkan peninsulas. In contrast, the establishment of both the western and especially the eastern boundaries is very difficult. Therefore the boundaries of this phytochorion are not generally agreed upon, although the majority of authors subscribe to the classical understanding of the Mediterranean Region, which goes back to Alphonse de Candolle and Edmond Boissier.

The chorionomic boundary of the Mediterranean Region on the whole coincides rather well with the boundary of the typical Mediterranean forests of *Quercus ilex* and with the basic (primary) area of olive cultivation. It corresponds rather well with the general distribution of such plants as *Asplenium petrarchae, Pinus halepensis, P. brutia* and *P. pinea, Aristolochia sempervirens, Quercus coccifera, Arbutus andrachne* and *A. unedo, Erica multiflora, Styrax officinalis, Coris monspeliensis, Euphorbia dendroides, Poterium spinosum, Cercis siliquastrum, Anagyris foetida, Calycotome infesta, Genista cinerea, Anthyllis hermanniae, Phillyrea angustifolia, P. latifolia, Cynomorium coccineum, Globularia alypum, Rosmarinus officinalis, Prasium majus, Thymus capitatus, Putoria calabrica, Ampelodesmos tenax, Chamaerops humilis,* and *Arisarum vulgare.* This list of Mediterranean species could be multiplied many times.

Comparative study of the areas of these species leads one to include in the Mediterranean Region a large part of the Iberian Peninsula (up to the mountain barrier of the Pyrénées and the Cantabrian Mountains to the north), the maritime part of southeastern France, the Apennines, the Balkan Peninsula, the Mediterranean Islands, Morocco, northern Algeria, Tunisia, northwestern Tripolitania, Cyrenaica, the shores of Levant (the greater part of Palestine and Lebanon), western Syria, and western Anatolia. Small and improverished disjunct exclaves of the Mediterranean flora occur in the southern mountainous parts of the Crimea and along the Black Sea shores of the Caucasus, especially in northern and northeastern Anatolia. According to

Alphonse de Candolle (1855:1305), the Mediterranean Region extends from Spain to Syria and from Morocco to the Black Sea, which agrees completely with contemporary ideas. The famous author of the "Flora Orientalis," Boissier (1867:VI), defined the winding, eastern boundaries of the region very exactly. He wrote:

> En Orient la région Mediterranéene comprend le littoral et la zone inférieure de la Grèce et de la Turquie d'Europe, les îles de la Méditerranée, la côte méridionale de la Crimée, les côtes occidentales et méridionales de l'Anatolie, enfin celles de la Syrie et de la Palestine.

This wonderfully exact delimitation of the eastern boundaries of the Mediterranean Region testifies to the deep understanding that Boissier had of the flora. Much the same understanding of the Mediterranean Region was shown by Grisebach (1872), who indicated its boundary on a map. However, he enlarged the boundaries of the region somewhat, especially in the east.

Engler (1882, 1924) and many of his successors moved the eastern boundaries of the Mediterranean Region considerably, extending them to Afghanistan and Baluchistan. Rikli (1934, 1943), author of a three-volume monograph about the plant world of the Mediterranean, also interpreted this region rather broadly. He included in it not only the entire Iberian Peninsula and all of the sub-Mediterranean parts of southwestern and southern France, of northern Italy, and of the Balkan Peninsula but also northern Libya, northern Egypt, the entire Sinai Peninsula, all of Anatolia, all of the Caucasus and Ciscaucasia, the southern shores of the Caspian Sea, all of the Crimea, and the Black Sea steppes. Some authors, including Gaussen (1954) and Tolmatchev (1974), included Macaronesia in the Mediterranean Region. In general, they had a very broad concept of the boundaries of this phytochorion (especially Gaussen).[56] However, with such an interpretation of the Mediterranean Region, its boundaries became extremely blurred. Therefore I consider it better to adhere to the boundaries previously outlined by Alphonse de Candolle and Boissier, and specified by numerous contemporary authors as well.

The flora of the Mediterranean Region contains only one endemic family, the Aphyllanthaceae (which many authors include in the Liliaceae), but the number of endemic genera probably reaches 150. These endemic or almost endemic Mediterranean genera include the following:

Cupressaceae: *Tetraclinis* (1, mainly in the Haut Atlas range in Morocco, but small populations also are encountered near Cartagena in southeastern Spain, from Morocco to Cyrenaica, and on Malta)
Fumariaceae: *Ceratocapnos* (2, southwestern Spain, northwestern Africa,

Syria, Palestine), *Rupicapnos* (about 30, southwestern Spain, northwestern Africa), *Sarcocapnos* (4, western Mediterranean)
Caryophyllaceae: *Bolanthus* (8, Greece to Palestine), *Ortegia* (1, Portugal, Spain, Italy), *Thurya* (1, southeastern Anatolia)
Chenopodiaceae: *Oreobliton* (1, Algeria), *Traganopsis* (1, Morocco)
Polygonaceae: *Bucephalophora* (1; close to *Rumex*)
Plumbaginaceae: *Limoniastrum* (10)
Brassicaceae: *Bivonaea* (1, western Mediterranean east to Sicily; close to *Ionopsidium*), *Boleum* (1, eastern Spain), *Ceratocnemum* (1, Morocco), *Cordylocarpus* (1, Morocco, Algeria), *Crambella* (1, northeastern Morocco), *Degenia* (1, northwestern Yugoslavia), *Euzomodendron* (1, Province of Almeria in southern Spain), *Fezia* (1, Morocco), *Guiraoa* (1, southeastern Spain), *Hemicrambe* (1, northern Morocco), *Hutera* (2, southeastern Spain), *Ionopsidium* (5, Portugal, Spain, central and southeastern Italy, Sicily, northwestern Africa), *Kremeriella* (1, northeastern Morocco, northwestern Algeria), *Lycocarpus* (1, southeastern Spain), *Morisia* (1, Corsica, Sardinia), *Ochthodium* (1, eastern Mediterranean), *Otocarpus* (1, Algeria), *Psychine* (1, Algeria, Tunisia, Morocco), *Raffenaldia* (*Cossonia*) (2, Morocco, Algeria), *Rytidocarpus* (1, Morocco), *Sisymbrella* (5, western and central Mediterranean, north to western France), *Succowia* (1, Balearic Islands, Corsica, Spain, Italy, Sardinia, Sicily, Morocco, Algeria), *Trachystoma* (3, Morocco)
Resedaceae: *Sesamoides* (1–2, western Mediterranean, north as far as central France)
Sapotaceae: *Argania* (1, Morocco)
Malvaceae: *Stegia* (1, close to *Lavatera*)
Cistaceae: *Halimium* (14)
Urticaceae: *Soleirolia* (1, Balearic Islands, Corsica, Tuscan Archipelago, Sardinia)
Rosaceae: *Sarcopoterium* (1, eastern Mediterranean, west to Tunisia and Sardinia; close to *Poterium*)
Fabaceae: *Argyrocytisus* (1, Morocco), *Benedictella* (1, Morocco), *Calicotome* (6–7), *Chronanthus* (2, Spain, Balearic Islands, northwestern Africa, western Anatolia), *Cytisopsis* (1, southwestern and southern Anatolia, western Syria), *Gonocytisus* (1, eastern Mediterranean), *Hammatolobium* (2, northwestern Africa, southern Anatolia), *Hesperolaburnum* (1, Morocco), *Lyauteya* (1, northwestern Africa), *Lygaeum* (1), *Passaea* (1), *Stauracanthus* (2, Iberian Peninsula, western Morocco)
Droseraceae: *Drosophyllum* (1, Portugal, southern Spain, Morocco; taxonomically a very isolated genus)
Apiaceae: *Ammiopsis* (2, northwestern Africa), *Ammoides* (2), *Astoma* (1, eastern Mediterranean), *Balansaea* (2, Spain, Morocco; close to

Chaerophyllum), *Bonannia* (1, Calabria, Sicily, and locally also in southern Greece and on the Aegean Islands), *Chaetosciadium* (1, eastern Mediterranean), *Chrysophae* (2, eastern Mediterranean), *Elaeoselinum* (10), *Hellenocarum* (2, central Mediterranean), *Keracia* (1, western Mediterranean), *Kunamannia* (1), *Magydaris* (2, Portugal, central and southern Spain, Balearic Islands, southern Italy, Sardinia, Sicily), *Meopsis* (2, western Mediterranean; very close to *Daucus*), *Microsciadium* (1, western Anatolia and neighboring islands), *Naufraga* (1, Balearic Islands), *Olymposciadium* (1, northwestern Anatolia), *Pachyctenium* (1, Cyrenaica), *Petagnia* (1, Sicily), *Physocaulis* (1), *Portenschlagiella* (1, southern Italy, western Yugoslavia, northwestern Albania), *Rouya* (1, Corsica, Sardinia, northern Africa), *Synelcosciadium* (1, Syria) *Thapsia* (6)
Valerianaceae: *Fedia* (1)
Rubiaceae: *Putoria* (3), *Warburgina* (1, Syria, Palestine)
Boraginaceae: *Echiochilopsis* (northwestern Africa), *Elizaldia* (5, western Mediterranean), *Procopiania* (3, southern Greece, Crete, Aegean Islands)
Scrophulariaceae: *Anarrhinum* (12), *Lafuentia* (2, southern Spain, Morocco), *Macrosyringion* (2)
Lamiaceae: *Amaracus* (15, eastern Mediterranean), *Cleonia* (2, western Mediterranean), *Dorystoechas* (1, southwestern Anatolia), *Pitardia* (2, northwestern Africa), *Prasium* (1), *Preslia* (1, western Mediterranean), *Rosmarinus* (3), *Saccocalyx* (1, northwestern Africa; close to *Satureja*)
Campanulaceae: *Feeria* (1, Morocco; close to *Trachelium*), *Petromarula* (1, Crete)
Asteraceae: *Aaronsohnia* (1, Palestine), *Ammanthus* (5), *Andryala* (25), *Anvilleina* (1, Morocco), *Bellium* (6), *Carduncellus* (20), *Catananche* (5), *Cladanthus* (4, southern Spain, northwestern Africa), *Cyanopsis* (1, southern Spain), *Daveaua* (1, northwestern Africa, southern Portugal), *Evacidium* (1, northwestern Africa, northern Sicily), *Fontquera* (1, Morocco), *Glossopappus* (1, northwestern Africa, southern Portugal, southwestern Spain), *Heteranthemis* (1, southern Portugal, southwestern Spain, northwestern Africa), *Hispidella* (1, central Spain, northern Portugal), *Hymenonema* (2, eastern Mediterranean), *Hymenostemma* (1, southwestern Spain), *Hyoseris* (3), *Lepidophorum* (1, Portugal, southwestern and northwestern Spain), *Leuzea* (4), *Lonas* (1, northwestern Africa, Sicily, southeastern Italy), *Lyrolepis* (2, Crete, Aegean islands), *Mantisalca* (5), *Mecomischus* (1, northern Africa), *Nananthea* (2, Corsica, Sardinia, Algeria), *Ormenis* (10), *Otospermum* (1, central and southern Portugal, southwestern Spain), *Palaeocyanus* (1, Malta, Gozo), *Perralderia* (4, northwestern Africa), *Phalacrocarpum* (2, Iberian Peninsula), *Plagius* (1, Corsica, Sardinia), *Prolongoa* (1, central and southern Spain), *Rothmaleria* (1, southern Spain), *Santolina* (10, western Mediterranean), *Staehelina*

(5), *Tourneuxia* (1, Algeria), *Wagenitzia* (1, Crete)
Iridaceae: *Hermodactylus* (1, Algeria-Cyrenaica, southeastern France, northern Italy, Corsica, Sicily, western Anatolia, Lebanon, Palestine; close to *Iris*),[57] *Siphonostylis* (3).
Hyacinthaceae: *Strangweia* (1, southern and western Greece)
Amaryllidaceae: *Hannonia* (1, southwestern Morocco), *Lapiedra* (1, southern and southeastern Spain and Morocco)
Aphyllanthaceae: *Aphyllanthes* (1, from Portugal to Italy, northern Africa)
Orchidaceae: *Barlia* (1)
Poaceae: *Ampelodesmos* (1), *Chaetopogon* (2), *Colobanthium* (1), *Lamarckia* (1), *Libyella* (1, Cyrenaica), *Lycochloa* (1, Syria), *Lygeum* (1), *Narduroides* (1), *Parvotrisetum* (1, northern Italy to European Turkey), *Triplachne* (1, Iberian Peninsula, islands of Mediterranean Region), *Vulpiella* (1), *Wangenheimia* (1, Iberian Peninsula, northern Africa)
Arecaceae: *Chamaerops* (1, southern Portugal, southern Spain, Balearic Islands, Sardinia, southwestern and southern Italy, Sicily, Malta, northern Africa; also southern France between Nice and Menton where, however, it may be introduced)
Araceae: *Ambrosina* (1, central and southern Italy and islands of Mediterranean Region), *Biarum* (15), *Dracunculus* (2)

One notices the absence of representatives of archaic subclasses of flowering plants in the composition of the endemic genera. The majority of endemic genera belong to such advanced families as the Brassicaceae, Fabaceae, Apiaceae, Asteraceae, and Poaceae.

Species endemism of the Mediterranean flora reaches 50%. Endemism here is of the dispersed type—that is, the endemics are broadly scattered among many genera.

In comparison with the flora of the Macaronesian Region, the Mediterranean flora contains fewer ancient, Tertiary relicts and considerably greater numbers of younger, progressive endemics, the origin of which is related to increasing adaptation to aridity. As in the flora of Macaronesia, the influence of the African flora is clearly evident here, especially in the xerophytic formations and among the bulbous plants, where the relationship with South Africa is obvious. However, there are fewer representatives of purely tropical families and they are preserved mainly in the western Mediterranean, especially in Morocco. Relationships with the eastern Asiatic flora (dating from Tertiary times) are quite obvious. The circumboreal flora also exerts an extremely great influence on the Mediterranean flora. Many of these circumboreal elements penetrated into the Mediterranean, especially its mountain regions, during the glaciation of northern Eurasia. In drier and warmer interglacial periods, many of the Mediterranean elements penetrated to the north, where some of them are now preserved as relict species. Examples include areas in central France, in various regions of Switzerland, and even in southern Ireland, where *Arbutus unedo* still grows.

The natural vegetation of the Mediterranean is preserved in a relatively undisturbed state only in a few mountain regions. In the plains and foothills it has been greatly altered or more frequently almost completely annihilated and replaced by cultivated plants. Most characteristic of the natural vegetation of the Mediterranean are conifers and evergreen sclerophyll forests of low trees with short, thick trunks, as well as various formations of shrubs, shrublets, and subshrubs. The character of the vegetation changes noticeably from west to east and from north to south. In both directions the vegetation becomes more xerophytic

1. **Southern Moroccan Province** (Engler 1908, 1912, 1924).[58] According to Engler (1908), this province lies approximately between 29° and 32°31' north latitude. From the shores of the Atlantic Ocean it extends to the western foot of the Haut Atlas and Anti Atlas Mountains.

Mediterranean elements predominate in the Southern Moroccan Province. These include such common Mediterranean (or predominantly Mediterranean) plants as *Juniperus oxycedrus, J. phoenicea, Tetraclinis articulata, Ephedra altissima, E. cossoniana, Aristolochia baetica, Clematis cirrhosa, Quercus ilex, Hypericum aegyptiacum, Helianthemum canariense, Argania spinosa* (the major part of its distribution is in this province), *Andrachne maroccana, Euphorbia terracina, Thymelaea antatlantica, Anagyris foetida, Ceratonia siliqua, Cytisus albidus, C. mollis, Genista demnutensis, G. ferox, G. webbii, Laburnum platycarpum, Pistacia lentiscus, Rhus albida, R. oxyacantha, R. pentaphylla, Polygala balansae, Bunium mauritianum, Bupleurum dumosum, Thapsia decussata, Rhamnus alaternus, R. lycioides, R. oleoides, Ziziphus lotus, Jasminum fruticans, Olea europaea* subsp. *oleaster, Phillyrea media, Osyris quadrialata, Lonicera biflora, Nerium oleander, Rubia peregrina, Lycium intricatum, Withania frutescens, Linaria ventricosa, Ballota hirsuta, Lavandula dentata, L. maroccana, L. multifida, Prasium majus, Salvia interrupta, Teucrium fruticans, Thymus satureoides, Artemisia arborescens, Launea spinosa, Phagnalon calycinum, Scorzonera undulata, Sonchus tenerrimus, Dipcadi serotinum, Asparagus acutifolius, A. albus, A. aphyllus, A. stipularis, Smilax aspera, Stipa tenacissima,* and *Chamaerops humilis.*

The flora of the Southern Moroccan Province has been subjected to considerable influence from the Saharo-Arabian flora, along with the floras of Macaronesia and of tropical Africa. Interestingly, *Acacia gummifera* grows here, and also such species of *Euphorbia* as *E. resinifera, E. officinarum, E. echinus,* and *E. beaumierana.*

2. **Southwestern Mediterranean Province** (Engler 1899, 1903, 1924; Takhtajan 1970, 1974). This province occupies the southern pre-Atlan-

tic part of the Iberian Peninsula, the shores of the Straits of Gibraltar, and a large part of Morocco, including the Haut Atlas and the Anti Atlas Mountains.

Its endemic genera include *Rubicapnos, Ceratocnemum* (Morocco), *Crambella* (northeastern Morocco), *Euzomodendron* (the province of Almeria in southern Spain), *Hemicrambe* (northern Morocco), *Rytidocarpus* (Morocco), *Trachystoma* (Morocco), *Argania, Benedictella* (Morocco), *Stauracanthus, Drosophyllum, Lafuentia* (2, southern Spain and Morocco), *Fontquera* (Morocco), and *Homalachne*. The only European site of *Psilotum nudum* is found along the shores of Gibraltar. Here also grow the Macaronesian species *Culcita macrocarpa, Davallia canariensis* (encountered also in northern Morocco), and *Diplazium caudatum*. Morocco has about 15 species in common with Macaronesia. *Abies pinsapo* is characteristic for the province. One of its subspecies (*A. pinsapo* subsp. *pinsapo*) grows close to Ronda in southwestern Spain, and another (*A. pinsapo* subsp. *maroccana*) grows in Morocco.

The natural vegetation of this province has been affected over the centuries by an extremely strong influence in the form of human economic activity, so that forests such as oak and pine are preserved only in small, isolated tracts. Usually, such forests became woodlands or were replaced by maquis and other derived formations. In the most humid regions of Morocco and Algeria, one finds forests of *Cedrus atlantica*, which are also characteristic for Algeria. Stands of *Tetraclinis articulata* occupy comparatively large areas in the drought-prone parts of eastern Rif. *Argania spinosa* is encountered in the northern part of Morocco. Finally, we mention the various types of xerophytic grasslands and shrub communities; in which the most characteristic plants include *Stipa tenacissima, Artemisia herba-alba, Aristida pungens, Lygeum spartum,* and *Atriplex halimus*.

3. **South Mediterranean Province** (Engler 1899, 1903, 1924; Takhtajan 1974). This province includes northern Algeria; a greater part of Tunisia in the south approximately to the latitude of the city of Sfax (Houérou 1959); northwestern Tripolitania (Djebel-Nefoussa and Djebel-Garion); and northern Cyrenaica south to approximately 32° north latitude.[59]

There are several endemic genera, including *Oreobliton, Kremeriella, Otocarpus,* and *Lyauteya*. Almost all are concentrated in Algeria, the flora of which is rather richer than that of the remaining parts of the province. According to the Quézel's (1964b) data, Algeria's vascular plant flora consists of 2,840 species, among which 247 are endemic. Toward the east, the number of endemic species falls sharply. Among the Algerian and Algerian-Tunisian endemic and semi-endemic species, we mention the following:

Abies numidica, Ephedra altissima, Epimedium perralderianum, Papaver malviflorum, Rupicapnos muricaria, R. numidicus, Quercus afares, Herniaria mauritania, Spergularia spp., *Buffonia chevalieri,* about two dozen species of *Silene, Oreobliton thesioides, Limonium* spp., *Hypericum afrum, Crambe kralikii,*

Kremeriella cordylocarpus, Otocarpus virgatus, Iberis peyerimhoffii, Lepidium rigidum, Alyssum macrocalyx, Brassica spp., *Malcolmia arenaria, Lavatera stenopetala, Euphorbia reboudiana, E. hieroglyphica, Sedum multiceps, Genista* spp., *Crotalaria vialattei, Adenocarpus* spp., *Ononis* spp., *Lyauteya ahmedi, Lotus* spp., *Coronilla atlantica, Hedysarum* spp., *Astragalus* spp., *Epilobium numidicum, Zygophyllum cornutum, Erodium* spp., *Bupleurum plantagineum, Hohenackeria polyodon, Pimpinella battandieri, Ammoides atlantica, Ammiopsis aristidis, Carum* spp., *Bunium* spp., *Lonicera kabylica, Fedia sulcata,* species of *Valerianella, Scabiosa,* and *Galium, Convolvulus durandoi, Echium suffruticosum, Solenanthus tubiflorus, Rindera gymnandra, Digitalis atlantica, Linaria* spp., *Scrophularia tenuipes, Pedicularis numidica, Odontites* spp., *Cistanche mauritanica, Orobanche* spp., *Plantago tunetana, Teucrium* spp., *Rosmarinus tournefortii, Salvia* spp., *Saccocalyx satureoides, Sideritis maura, Marrubium alyssoides, Thymus* spp., *Calamintha* spp., *Phlomis cabbaleroi, P. bovei, Stachys* spp., *Nepeta algeriensis, Origanum* spp., *Laurentia bicolor, Specularia julianii, Wahlenbergia bernardii, Campanula* spp., *Filago* spp., *Pulicaria* spp., *Senecio gallerandianus, Calendula monardii, Mecomischus pedunculatus, Anthemis chrysantha, Anacyclus linearilobus, Kremeria grandis, K. multicaulis, Carlina atlantica, Carduus balansae, Cirsium kirbense, Galactites mutabilis, Onopordum algeriense, Centaurea* spp., *Mantisalca delestrei, Carthamus strictus, Carduncellus* spp., *Hypochoeris* spp., *Leontodon djurdjurae, Andryala* spp., *Bellevalia pomelii, Romulea* spp., *Gagea mauritanica, Allium* spp., *Trisetaria nitida, Avena* spp., *Koeleria balansae, Cynosurus* spp., *Festuca algeriensis,* and *Agropyropsis lolium.*

4. Iberian Province (Engler 1882, 1903, 1924; Takhtajan 1970, 1974).[60] A large part of the Iberian Peninsula is included in this province, with the exception of the territory assigned to the Atlantic European Province and the Southwestern Mediterranean Province.[61]

There are several endemic genera, including *Boleum, Guiraoa, Hutera, Lycocarpus,* and *Hispidella.* There are very many endemic species. In the coastal belt of Portugal, especially in the area of the Serra de Sintra ridge, small areas of forests are preserved. They have a series of elements in common with the flora of the Macaronesian Region. In particular, we mention *Davallia canariensis* and *Myrica faya.*

In the past, forests occupied considerable areas in the Iberian Peninsula, especially in Portugal and in Estremadura (Spain). Within historical time, but especially beginning with the sixteenth century, these forests were intensively logged, and at present are represented only by scattered tracts of evergreen oaks and pines. One encounters *Pinus pinaster, P. nigra* subsp. *salzmannii, P. sylvestris, P. halepensis,* and *P. pinea.* Near Cartagena one encounters isolated stands of *Tetraclinis articulata.* Among the evergreen oaks,

Quercus coccifera is encountered most often. Also encountered are *Q. suber, Q. ilex* and *Q. rotundifolia* (very frequent), while among the deciduous oaks there are *Q. pyrenaica, Q. faginea* and *Q. canariensis*. On dry and rocky slopes from the shores of the sea up to considerable altitudes (occurring in the mountains and on the plateau of Castilla la Nueva and Castilla la Vieja), various formations of medium-height evergreen forests and evergreen shrubs, shrublets, and subshrubs, as well as steppes are widely distributed.

    5. **Balearic Province.**[62] This province includes the Balearic Islands (consisting of two large islands, Mallorca and Menorca), the small Cabrera Island, and the Ibiza (Iviza) or Pithyusan Islands (located to the southwest of Mallorca). Despite the fact that the Balearic Islands are not far from the eastern shores of the Iberian Peninsula and that the natural conditions are very similar to those of Valencia and Catalonia (the Spanish Levant), their flora is not completely Iberian. Floristically, the Balearic Islands occupy an intermediate position between the Iberian (and Ibero-Maghreb) flora and the flora of the Liguro-Tyrrhenian Province (see next section), and are linked especially with the flora of Corsica and Sardinia (including many schizoendemics; see Cardona and Contandriopoulos 1979). Whereas the flora of the Ibiza Islands has much in common with the flora of the Iberian Peninsula (Font Quer 1927; Bolós 1958), the flora of Mallorca and Menorca has many elements in common with the flora of the islands of the Tyrhennian Sea (Chodat 1924; Bolós 1958). Thus it is difficult to assign the flora of the Balearic Islands to either of these two provinces. The best solution appears to be recognition of the Balearic Islands as a separate, transitional province.

    Of those species in common with the flora of Spain and of northern Morocco, the following are of special interest: *Helianthemum caput-felis, H. origanifolium, Satureja barceloi,* and *Lonicera pyrenaica* (Bolós 1958). Among these species in common with the Spanish Levant, the most remarkable is *Rhamnus ludovici-salvatoris* (on the Iberian Peninsula; found only near Valencia). The distribution of several species encompasses not only the Balearic Islands but also Spain and the islands of the Liguro-Tyrrhenian Province. The best-known example is *Buxus balearica* (found on the Balearic Islands, in several localities in southern and eastern Spain, and on Sardinia). Of the extensive number of Balearic–Liguro-Tyrrhenian taxa, we mention *Helleborus trifolius, Delphinium pictum, Arenaria balearica, Cymbalaria aequitriloba, Lavatera triloba* subsp. *pallescens* (Menorca and San Pietro Island close to Sardinia), *Sesleria insularis* subsp. *insularis, Micromeria filiformis, Cephalaria squamiflora* subsp. *balearica, Bellium bellidioides,* and *Arum pictum*. In addition, the genera *Soleirolia* and *Helicodiceros* are endemics common to the Balearic and Luguro-Tyrrhenian provinces.

    The flora of the Balearic Islands contains only one endemic genus, *Nau-*

fraga,[63] and about 50 endemic species and subspecies, two species of which apparently have already disappeared (Lucas and Walters 1976). Among the endemic species are the following:

Aristolochia bianorii, Helleborus trifolius subsp. *lividus, Ranunculus weyleri, Limonium biflorum, L. caprariense, L. majoricum, Paeonia cambessedesii, Hypericum balearicum, Viola jaubertiana, Brassica balearica, Primula vulgaris* subsp. *balearica, Lysimachia minoricensis* (now extinct), *Euphorbia maresii, Daphne rodriguezii, Thymelaea myrtifolia* (including *T. velutina?*), *Astragalus balearicus, Genista lucida, G. dorycnifolia, G. acanthoclada* subsp. *balearica, Vicia bifoliolata* (probably extinct), *Lotus tetraphyllus, Anthyllis fulgurans, Hippocrepis balearica, Erodium reichardii, Naufraga balearica, Pastinaca lucida, Pimpinella bicknellii* (= *Spiroceratium bicknellii*), *Bupleurum barceloi, Lonicera pyrenaica* subsp. *majoricensis, Galium balearicum, G. crespianum, Cephalaria balearica, Phlomis italica, Scutellaria balearica, Teucrium asiaticum, T. cossonii, T. subspinosum, Thymus richardii* subsp. *ebusitanus, Digitalis dubia, Sibthorpia africana, Globularia combessedesii, Plantago coronopus* subsp. *purpurascens, Centaurea balearica, Crepis triasii, Helichrysum ambiguum, Launaea cervicornis, Aetheorhiza bulbosa* subsp. *willkommii, Senecio rodriguezii, Crocus cambessedesii, Allium grossii,* and *Carex rorulenta.*

6. **Liguro-Tyrrhenian Province** (Engler 1882, 1899, 1903, 1924; Giacomini [in Giacomini and Fenaroli] 1958; Takhtajan, 1970, 1974). This province includes Roussillon (where the Iberian flora still exerts a strongly pronounced influence); the lower Corbières, the Lower Languedoc, the lower Rhone (north to Montélimar), the maritime Province, the French shores of the Ligurian Sea (Côte d'Azur), Liguria, the entire Italian seacoast of the Ligurian and Tyrrhenian seas, all of the peninsula of Calabria, the Tuscan Archipelago, Corsica, Sardinia, Sicily, Malta, the Lipari Islands (Isole Eolie), the Egadi Islands, the Isole Pelagie (Lampedusa and Linosa), and Pantelleria Island.[64]

In the Liguro-Tyrrhenian Province there are only four endemic genera: *Morisia, Petagnia, Melitella,* and *Palaeocyanus.* There are, however, a great many endemic species, the overwhelming majority of which are concentrated on the islands of Corsica, Sardinia, and Sicily. The endemic species include the following:

Abies nebrodensis (Sicily), *Aristolochia sicula* (Sicily), *Quercus congesta* (southern France, Sardinia, Sicily, Lipari, southern Calabria), *Q. sicula* (Sicily), *Urtica rupestris* (Sicily), *Helleborus trifolius* subsp. *corsicus* (Corsica, Sardinia), *Aquilegia bernardii* (Corsica, Sardinia), *A. litardierei* (Corsica), *Thalictrum calabricum, Ranunculus marschlinsii* (Corsica), *R. revelierei* (southern France, Corsica, Sardinia) and several other species of this genus, *Arenaria cinerea* (southeastern France), *A. provincialis* (southeastern France between Marseilles and Toulon), *Cerastium soleirolii* (Corsica), *Sagina pilifera* (Corsica, Sardinia), *Silene* spp., *Polygonum scoparium* (Corsica, Sardinia), several species of *Armeria* and *Limonium*, *Viola aethnensis* (Sicily), *V. nebrodensis* (Sicily), *Barbarea rupicola* (Corsica, Sardinia), *Alyssum robertianum* (Corsica, Sardinia), *A. nebrodense* (Sicily), *Ptilotrichum halimifolium* (southeastern France, northwestern Italy), *P. macrocarpum* (southern France), *Thlaspi brevistylum* (Corsica, Sardinia), *Iberis semperflorens* (western shores of Italy and Sicily), *I. stricta* (southeastern France, Liguria), *Biscutella rotgesii* (Corsica), *Brassica* spp., including *B. insularis* (Corsica, Sardinia), *Erucastrum virgatum* (southern Italy, Sicily), *Morisia monanthos* (Corsica, Sardinia), *Primula palinuri* (southwestern Italy), *Malva corsica* (Corsica, Sardinia), *Mercurialis corsica* (Corsica), *Euphorbia ceratocarpa* (southern Italy, Sicily), *Ribes sardoum* (Sardinia), *Potentilla crassinervis* (Corsica, Sardinia), *Cytisus aeolicus* (Lipari Islands), *Genista* spp., *Astragalus heutii* (Sicily), *A. maritimus* (Sardinia), *A. sirinicus* subsp. *genargenteus* (Corsica, Sardinia), *A. verrucosus* (Sardinia), *Lathyrus odoratus* (southern Italy, Sicily), *Trifolium bivonae* (Sicily), *T. brutium* (southern Italy), *Ruta corsica* (Corsica, Sardinia), *Erodium gussonii* (southern Italy), *E. corsicum* (Corsica, Sardinia), *E. rodiei* (southeastern France), *Polygala preslii* (Sicily), *P. sardoa* (Sardinia), *P. apiculata* (southern Italy), *Petagnia saniculifolia* (Sicily), *Oenanthe lisae* (Sardinia), *Bupleurum elatum* (Sicily), *Apium crassipes* (Corsica, Sardinia, Sicily, southern Italy), *Lereschia thomasii* (Calabria), *Ammi crinitum* (southern Italy, Sicily), *Ligusticum corsicum* (Corsica), *Peucedanum paniculatum* (Corsica, Sardinia), *Pastinaca latifolia* (Corsica), *Rhamnus persicifolia* (Sardinia), *Thesium italicum* (Corsica, Sardinia), *Scabiosa parviflora* (Sicily), *Centranthus trinervis, Asperula* spp., including *A. crassifolia* (Sardinia, vicinity of Naples), *Galium* spp., *Buglossoides calabra* (southern Italy), *B. minima* (southwestern Italy, Sicily, Sardinia), *Lithospermum minimum, Onosma lucana* (southern Italy), *Symphytum gussonei* (Sicily), *Anchusa crispa* (Corsica, Sardinia), *Borago pygmaea* (Corsica, Sardinia, Capraia), *Myosotis ruscinonensis* (southern France), *M. soleirolii* (Corsica), *M. corsicana* (Corsica), *Verbascum siculum* (southwestern Italy, Sicily), *Anarrhinum corsicum* (Corsica), *Antirrhinum siculum* (Sicily, Malta, and probably southwestern Italy), *Linaria capraria* (Tuscan Archipelago), *Cymbalaria hepaticifolia* (Corsica), *C. muelleri* (Sardinia), *Odontites bocconei* (Sicily), *O. corsica* (Corsica, Sardinia), *Globularia neapolitana* (vicinity of Naples), *Orobanche chironii* (Sicily), *Pinguicula corsica* (Corsica), *Ajuga acaulis, Phlomis ferruginea* (southern Italy), *Lamium corsicum* (Corsica, Sardinia), *Stachys corsica* (Corsica, Sardinia), *S. glutinosa* (Corsica, Sardinia, Capraia), *Nepeta agrestis* (Corsica), *N. foliosa* (Sardinia), *Acinos corsicus* (Corsica), *Micromeria filiformis* subsp. *cordata, Thymus herba-barona* (Corsica, Sardinia), *Mentha requienii* (Corsica, Sardinia, Montecristo), *Campanula forsythii* (Sardinia), *Phyteuma serratum* (Corsica), *Doronicum corsicum* (Corsica), *Bellis bernardii* (Corsica), *Bellium crassifolium* (Sardinia), *Anthemis* spp., *Leucanthemum corsicum* (Corsica), *Plagius flosculosus* (Corsica, Sardinia), *Nananthea perpusilla* (Corsica, Sardinia and nearby islands), *Evax rotundata* (Corsica, Sardinia), *Helichrysum frigidum* (Corsica, Sardinia), *H. montelinasum* (Sardinia), *H. saxatile* (Sardinia, Pantelleria), *Phagnalon metlesicsii, Buphthalmum inuloides* (Sardinia and nearby islands), *Carlina macrocephala, Carduus fasciculiflorus* (Corsica, Sardinia Montecristo), *Lamyropsis microcephala* (Sardinia), *Centaurea* spp., *Palaeocyanus crassifolia* (Malta, Gozo), *Hyoseris taurina* (Sardinia), *Hypochoeris robertia, Leontodon siculus* (southwestern Italy, Sicily), *Lactuca longidentata* (Sardinia), *Colchicum corsicum* (Corsica), *Narthecium reverchonii* (Corsica), *Muscari gussonei* (Sicily, Calabria), *Allium parciflorum* (Corsica, Sardinia), *Crocus corsicus* (Corsica), *C. etruscus* (Tuscan Archipelago and adjoining mainland), *C. longiflorus* (southwestern Italy, Sicily, Malta), *C. minimus* (Corsica, Sardinia, Capraia), *Romulea ligustica* (coasts of northwestern Italy, Corsica, Sardinia), *R. melitensis* (Malta, Gozo), *R. requienii* (Corsica, Sardinia, westcentral Italy), *R. revelierei* (Corsica, Capraia, islands of northern Sardinia), *Leucojum longifolium* (Corsica), *L. roseum* (Corsica, Sardinia), *Pancratium illyricum* (Corsica, Sardinia, Capraia), *Serapias nurrica* (Sardinia), *Juncus requienii* (Corsica), *Festuca morisiana* (Sardinia), *F. sardoa* (Corsica, Sardinia), *Avena saxatilis* (islands near Sicily), *Trisetum gracile* (Corsica, Sardinia), and *Holcus notarisii* (near Genova).

Engler (1912, 1934) divided this province into seven subprovinces: (1) Provencal; (2) Ligurian, (3) northern Tyrrhenian, the western part of central Italy and adjacent small islands; (4) southern Tyrrhenian, including Naples and Calabria, (5) Corsican, (6) Sardinian, and (7) Sicilian-Maltese. Giacomini (in Giacomini and Fenaroli 1958)—who considered only the eastern part of the province (without southern France)—recognized the following subprovinces (referring to them as "distretti"): (1) Tyrrhenian, corresponding approximately with the Ligurian, northern Tyrrhenian, and southern Tyrrhenian subprovinces of Engler; (2) Sardinian-Corsican, including Sardinia, Corsica, Elba, the remaining islands of the Tuscan Archipelago (except Gorgona Island, which belongs to the Ligurian Subprovince), and two small peninsulas located opposite the Tuscan Archipelago; and (3) Sicilian, including Sicily, Malta and Gozo, the Isole Pelagie, and Pantelleria. This smaller number of subprovinces seems more acceptable to me.

7. Adriatic Province (Giacomini [in Giacomini e Fenaroli], 1958; Horvatič 1967, Takhtajan 1974).[65] This province includes the Adriatic coast of Italy (northward approximately to the Ravenna area), and in the south the entire Salentine Peninsula, the Istra Peninsula (and northward, including the environs of Trieste), almost all of the maritime region of Yugoslavia, and shores of Albania (south approximately to Vlorë).

The Adriatic Province has only two endemic genera (*Degenia* and *Portenschlagiella*). Species endemism is comparatively low, but includes the following: *Phyllitis hybrida* (islands of northwestern Yugoslavia), *Aristolochia croatica* (Pag Island), *Silene reichenbachii* (western Yugoslavia), *Dianthus ciliatus, Limonium anfractum* (southern Yogoslavia, Albania), *L. japygicum* (southeastern Italy), *Crambe maritima, C. croatica* (Velebit), *Degenia velebitica* (Velebit), *Alyssum leucadeum, Euphorbia triflora* (Velebit), *Portenschlagiella ramosissima* (southern Italy, western Yugoslavia, northeastern Albania), *Asperula staliana* (islands along the northwestern coast of Yugoslavia), *A. garganica* (southeastern Italy), *Cerinthe glabra* subsp. *smithiae, Linaria microsepala* (Dalmatia), *Ornithogalum visianicum* (Palagruza Island), *Allium horvatii* (Kvarner Archipelago), and *Stipa mayeri* (southwestern Yugoslavia).

8. **East Mediterranean Province** (Engler 1882, p.p.; Hayek 1926, p.p.: Davis 1965, p.p.).[66] Hayek, and especially Engler, interpreted this province very broadly, and Davis extended it from the eastern half of Italy to Lebanon. In contrast with Davis, I exclude the shores of the Adriatic Sea from the East Mediterranean Province. In my opinion, this province includes part of southern Albania, a large part of peninsular Greece (including all of Peloponnesus), the Ionian Islands, all islands of the Aegean Sea, Crete, Karpathos, Cyprus, the Gallipoli Peninsula, a narrow belt along the northern shores of the Sea of Marmara, the Aegean and Mediterranean shores of Asia Minor, the shores of Syria, Lebanon, and a large part of Palestine.[67]

As shown repeatedly in the literature (e.g., Maleev 1938; Davis 1965), the East Mediterranean flora forms a series of enclaves along the Black Sea shores of Anatolia—enclaves in which one encounters such species as *Pinus pinea, Laurus nobilis, Quercus ilex, Cistus creticus, C. salviifolius, Erica arborea, Arbutus andrachne, Spartium junceum,* and *Myrtus communis*. Several of these (e.g., *Pinus pinea, Cistus creticus, C. salviifolius,* and *Arbutus andrachne*) reach the ravines of the middle course of the Coruh and its tributaries. The eastern and parts of the northern and southern boundaries of the East Mediterranean Province coincide rather well with the distribution of such species as *Quercus coccifera, Arbutus unedo, Styrax officinalis, Pistacia lentiscus, Salvia triloba,* and especially *Poterium spinosum*.

There are over 20 endemic genera (*Bolanthus, Thurya, Didesmus, Ochthodium, Cytisopsis, Gonocytisus, Astoma, Chaetosciadium, Microsciadium, Synelcosciadium, Warburgina, Procopiania, Dorystoechas, Petromarula, Trochocodon, Aaronsohnia, Hymenonema, Lyrolepis, Lycochloa,* et al.), and a great many endemic species, including *Phoenix theophrastii* (Crete). The largest concentration of endemics is found in Greece and on the islands of the southern part of the Aegean Sea, especially on Crete (Rechinger 1949–50).

9. **Crimean-Novorossiysk Province** (Kuznetsov 1901, 1909; Takhtajan 1974, Fedorov 1979).[68] This province includes southern Crimea (a narrow maritime belt of the southern slope of the first ridges of the Crimean Moun-

tains from Sevastopol to Feodosiya, bordered in the north by the Yaila), and the northern part of the western Transcaucasia, from Anapa to the vicinity of Tuapse in the south.[69]
With its relatively small area and its rather impoverished Mediterranean flora (especially in the Caucasian parts), this area would hardly merit recognition as a separate province were it not for its considerable remoteness from the major area of the Mediterranean flora. Among the Mediterranean plants extending into the Crimean-Novorossiysk Province (mainly in southern Crimea), we mention: *Pinus brutia*,[70] *Cistus incanus* (*C. tauricus*), *Capparis spinosa, Brassica cretica, Arbutus andrachne, Euphorbia rigida, Trifolium grandiflorum* (= *T. speciosum*), *Crithmum maritimum, Lonicera etrusca* (the Novorossiysk-Anapa area), *Vitex agnuscastus, Asphodeline lutea*, and *Ruscus hypoglossum*. Species endemism is rather high, although for the most part the endemics are confined to the Crimea. There are considerably fewer endemics in the Caucasian part of the province. There is also a number of endemics common to the Crimean and Transcaucasian parts of the province: *Thlaspi macranthum, Hesperis steveniana, Crambe koktebelica, Hedysarum candidum, Medicago rupestris, Asperula taurica, Onosma polyphyllum, O. rigidum, Nonnea taurica, Centaurea declinata, Crocus tauricus, Himantoglossum caprinum*, and others.

7. Saharo-Arabian Region

Saharo-Arabian Region, Zohary 1963, 1973; Takhtajan 1974; Quézel 1978.

This region[71] includes the entire extratropical part of the Sahara, from the Atlantic shores to Egypt (along the shores of the Gulf of Sidra and eastern Cyrenaica right up to the Mediterranean Sea); the Sinai Peninsula; all of the vast, extratropical part of the Arabian Peninsula; part of southern Palestine; part of Jordan; the southern part of the Syrian Desert (to the south from the Rutba-Habbaniya Road and lower Mesopotamia, approximately between Baquba and Basra); and the southern desert to the south from the railroad line of Samawa-Ur.[72] In the Sahara, as in the Arabian Peninsula, the southern boundary of the region intersects the northern tropics at approximately 20° north latitude and continues somewhat south.
A series of authors have referred the territory of the Saharo-Arabian Region to the Paleotropical Kingdom (e.g., Engler 1899, 1924; Newbigin 1936; Good 1947, 1974; Schmithüsen 1961, Mattick 1964; Walter and Straka 1970; Tolmatchev 1974) but others, with rather considerable evidence, include it (or part of it) in the Holarctic Kingdom (Drude 1890; Krasnov 1899; Hayek 1926; Eig 1931; Rikli 1934; Wulff 1944; Szafer 1952; Gaus-

HOLARCTIC KINGDOM 133

sen 1954; Ozenda 1958, 1964; Turrill 1959; Lavrenko 1962; Zohary 1963, 1973; Ehrendorfer 1971; Quézel 1978).

While the northern and eastern boundaries of the Saharo-Arabian Region are relatively clearly expressed (their delimitation in Asia being based almost completely on Zohary 1973), the problem of its southern boundary in Africa is difficult to resolve (Quézel 1978). As frequently mentioned in the literature, the boundary between the Holarctic and the paleotropic floras in the Sahara is rather obscure. As Quézel (1965, 1978:481) has shown, whereas a large portion of the Sahara belongs, for the most part, to the Saharo-Arabian Region, toward the south the question becomes complicated by the extension or even predominance of the elements of tropical origin. This is basically connected to the large climatic fluctuations that took place in the Sahara during the entire Quaternary (see Quézel 1965, 1978).

On his map of floristic regions of the world, Gaussen (1954) placed the boundary between the Holarctic and tropical floras in large part south of 20° north latitude, with the exception of eastern Mauritania, where the tropical flora extends a tongue northward from 20° north latitude. Hayek (1926) extends the boundary still further south. I accept the boundary of the Holarctic flora which in principle agrees with the boundary of Gaussen (1954), Clayton and Hepper (1974), White (1976), and Wickens (1976).[73]

The flora of the Saharo-Arabian Region is not rich and consists of no more than 1,500 species,[74] but the number of endemic species is at least 310 (and probably somewhat greater).[75] However, the number of endemic genera is not great and some of them overlap more or less extensively into the Mediterranean (Quézel 1978) and Irano-Turanian Regions. The following genera are endemic or nearly so:

Caryophyllaceae: *Xerotia* (1, Arabian Peninsula)
Chenopodiaceae: *Fredolia* (1, Morocco, Algeria), *Nucularia* (1, Sahara), *Traganopsis* (1, Morocco)
Amaranthaceae: *Saltia* (1, southern part of Arabian Peninsula)
Brassicaceae: *Ammosperma* (2, Sahara), *Eremophyton* (1, Sahara), *Foleyola* (1, Sahara), *Lonchophora* (1, Sahara), *Muricaria* (1, Sahara), *Nasturtiopsis* (1–2, Sahara, Egypt, Arabian Peninsula, Palestine), *Oudneya* (1, Sahara), *Pseuderucaria* (3, Morocco to Palestine), *Quezelia* (1, Sahara: Tibesti), *Reboudia* (2, Morocco, Algeria, Libya, Egypt, Palestine), *Schimpera* (1, Egypt, Palestine, Arabian Peninsula, Syria, Iraq, Iran), *Schouwia* (2, Sahara, Arabian Peninsula), *Stigmatella* (1, Palestine), *Zilla* (3, northern Africa, Arabian Peninsula, Palestine, Syria, Iraq)
Apiaceae: *Adenosciadium* (1, southeastern part of Arabian Peninsula), *Ammodaucus* (1, Sahara; close to *Daucus*)
Gentianaceae: *Monodiella* (1, Sahara: Tibesti)
Boraginaceae: *Echiochilon* (6, Sahara, Arabian Peninsula)

Scrophulariaceae: *Omania* (1, Arabian Peninsula)
Lamiaceae: *Physoleucas* (1, Arabian Peninsula)
Asteraceae: *Lifago* (1, Sahara; very close to *Filago*), *Perralderia* (4, northwestern Africa), *Tibestina* (1, Sahara), *Warionia* (1, northwestern Sahara)
Hyacinthaceae: *Battandiera* (1, northwestern Sahara)

There is also series of endemic genera shared with the Mediterranean Region—for example, *Enarthrocarpus* (5), *Randonia* (3), and *Anacyclus*. *Traganum* (2) is a shared endemic with the Canary Islands. Other series of endemics are shared with the Macaronesian–Mediterranean–Saharo-Arabian, the Saharo-Arabian–Irano-Turanian, and the Saharo-Arabian–Sudanian areas. The genus *Hammada* (12) is an endemic common to the Mediterranean Saharo-Arabian and Irano-Turanian regions, as well as to the northwestern deserts of the Indian Peninsula. As in the Mediterranean flora, there are genera in common with South Africa.

Various desert and semidesert formations and woodlands dominate the vegetative cover of the Saharo-Arabian Region (for literature, see Zohary 1973; White 1983).

1. **Saharan Province** (Provinz der Grossen Sahara, Engler 1908, 1912, 1924; Western Province, Zohary 1973). This province includes the Sahara from the shores of the Atlantic Ocean to the Libyan Desert inclusively, with the exception of parts of the Mediterranean shore that have a Mediterranean flora.[76]

In the vast territory of this province, great expanses exist that are completely or almost completely devoid of vegetation. Only in deep valleys with ground water and in oases, and especially in the montane regions, does there develop a more or less rich vegetation composed of several floristically different types, In the north, one observes a gradual transition to the Mediterranean flora, and in the south a similar transition to the tropical. In some places, as at Tibesti (Quézel 1958), tropical elements can form true enclaves of the Sahelian flora (see under the Sudano-Zambezian Region).

There are several endemic Saharan genera (including the gentianaceous genus *Monodiella* on Tibesti) and a number of endemic species, although the total percentage of endemics is much less than in Mediterranean Africa (see Quézel 1978:504). Some remarkable endemics are localized in the Sahara mountains, including *Cupressus duprezziana, Myrtus nivellei,* and *Olea laperrinii.*

2. **Egyptian-Arabian Province** (Bezirk der ägyptisch-arabischen Wüste, Hayek 1926). This province includes Egypt eastward from the Nile Valley, the Sinai Peninsula, the extratropical part of the Arabian Peninsula, the greater part of southern Palestine, part of Jordan, the southern part of the

HOLARCTIC KINGDOM 135

Syrian Desert, and lower Mesopotamia, where the boundary continues not very far north of Balad, Kuwait, and the Bahrein Islands.[77]
According to Zohary (1973:69), a number of typically Saharan genera have their terminus in Arabia and in the adjacent parts of the Syrian Desert. Examples are species of *Eremobium, Maresia, Reboudia, Schouwia, Zilla, Reaumuria, Retama, Fagonia, Zygophyllum, Aristida*, and a number of others.

8. Irano-Turanian Region

Eig 1931; Zohary 1950, 1962, 1963, 1973, p.p.; Davis 1965; Takhtajan 1969, 1974; Davis and Hedge 1971, p.p.; Hedge 1976; Wickens 1976; Quézel 1978;[78] Région Oriental proprement dite, Boissier 1867; Steppenregion, Grisebach 1872, 1884, p.p.; Western and Central Asiatic Region, Good 1947, 1974; Brice 1966; Turanian Region, Turrill 1959; West-und Zentralasien, Mattick 1964; Orientalisch-Turanische Region, Meusel et al. 1965.

In this vast region—the borders of which were first rather exactly outlined by Boissier (1867) in his classic work, "Flora Orientalis"—one includes central and eastern Anatolia, a greater part of Syria, part of southern and eastern Palestine, a small part of the Sinai Peninsula, part of Jordan, the northern part of the Syrian Desert, upper Mesopotamia, a large part of the Armenian Highlands, the arid and semiarid areas of southern and eastern Transcaucasia, Hyrcania (Talish and neighboring areas along the Caspian shore of Iran), the Iranian Plateau without the tropical deserts, the southern spurs of the Hindu Kush Range, the southern slopes and spurs of the western Himalayas to the west from 83° east longitude, and the entire vast arid territory from southeastern European Russia and eastern Transcaucasia up to the Gobi, inclusive.

The flora of the Irano-Turanian Region is characterized by a rather high number of endemic genera and very high species endemism (probably no less than 25%). The richest flora is that of the Iranian Plateau, and the most impoverished flora is that of eastern Central Asia. Among the endemic and subendemic genera are the following:

Ranunculaceae: *Alexeya* (1, northwestern Himalayas: Kashmir; very close to *Paraquilegia*), *Paropyrum* (1, Dzungarian Altai, Tien Shan, Pamiro-Alai; very close to *Paraquilegia*)
Berberidaceae: *Bongardia* (1, encountered also in the eastern Mediterranean)
Fumariaceae: *Cryptocapnos* (1, southeastern Afghanistan), *Fumariola* (1, Pamiro-Alai), *Roborowskia* (1, Pamiro-Alai, Kashgaria)
Caryophyllaceae: *Acanthophyllum*, including *Allochrusa* (70, in the north en-

countered also in Siberia), *Ankyropetalum* (3, Palestine, Syria, Iraq, Iran; encountered also in the eastern Mediterranean), *Diaphanoptera* (2, Khorasan—Paropamisus; closely related to *Acanthophyllum*), *Kabulia* (1, eastern Afghanistan), *Kughitangia* (2, Pamiro-Alai), *Ochotonophila* (2, central and northern Afghanistan), *Pentastemenodiscus* (1, central Afghanistan), *Phrynellia* (*Phryna;* 1, Anatolia), *Schischkinella* (1, Anatolia), *Scleranthopsis* (1, Afghanistan), *Thylacospermum* (1, from the Dzungarian Alatau and central Tien Shan to Tibet), *Tytthostemma* (1, eastern Iran to the Tien Shan and Pamiro-Alai)

Chenopodiaceae: *Agriophyllum* (5), *Alexandra* (1), *Anthochlamys* (5), *Arthrophytum* (7), *Bienertia* (1), *Borsczowia* (1), *Cyathobasis* (1, Anatolia), *Gamanthus* (4), *Girgensohnia* (4), *Halanthium* (6, in the Caucasus to Daghestan), *Halarchon* (1, Afghanistan), *Halimocnemis* (13), *Halocharis* (12), *Halostachys* (1), *Halotis* (2), *Horaninovia* (7), *Iljinia* (1), *Kalidium* (5), *Kalidiopsis* (1, Anatolia), *Kirilowia* (2), *Londesia* (1), *Microgynoecium* (1, Central Asia), *Micropeplis* (1, close to *Halogeton*), *Nanophyton* (3), *Ofaiston* (1), *Panderia* (2), *Physandra* (1, western Tien Shan), *Piptoptera* (1, Kyzyl Kum, Kara Kum), *Raphidophyton* (1, Kazakhstan, western Tien Shan: Karatau), *Sympegma* (1)

Polygonaceae: *Pteropyrum* (5, Iraq, Iran, Afghanistan, Pakistan, Turkmenistan)

Plumbaginaceae: *Acantholimon* (120; encountered in the eastern Mediterranean area), *Aeoniopsis* (1, Afghanistan), *Cephalorrhizum* (4), *Chaetolimon* (3), *Dictyolimon* (4, Afghanistan to western Himalayas), *Ikonnikovia* (1, northern Tien Shan)

Hamamelidaceae: *Parrotia* (1, Hyrcania, but with a small exclave in the eastern part of the Alazan Valley near the town Kutkashen), *Parrotiopsis* (1, Afghanistan: Nuristan; Pakistan: Swat, Kurram Valley; northwestern India: Kashmir)

Brassicaceae: *Acanthocardamum* (1, southwestern Iran), *Alyssopsis* (1, Talish, Iran), *Anchonium* (2), *Atelanthera* (1), *Botschantzevia* (1, Kazakhstan—Karatau Range), *Brossardia* (1, northern Iraq, western Iran), *Buchingera* (1, southern Transcaucasia, Iran, Kyzylkum, Kopet Dagh, Tien Shan, Pamiro-Alai, Afghanistan, Pakistan), *Calymmatium* (2, western Pamirs, eastern Afghanistan), *Camelinopsis* (1, northern Iraq, northern Iran), *Catenularia* (1, Soviet Central Asia: southern Pamiro-Alai—low mountains of southern Tajikistan), *Chalcanthus* (1, Iran, Kopet Dagh, Tien Shan, Pamiro-Alai, Afghanistan), *Chartoloma* (1, Soviet Central Asia), *Cithareloma* (3, Soviet Central Asia, Afghanistan), *Clastopus* (2, Iraq, Iran), *Cryptospora* (3, Iran, Afghanistan, Soviet Central Asia), *Cymatocarpus* (3, southern Transcaucasia, Iran, from Kopet Dagh to Pamiro-Alai, Afghanistan), *Cyphocardamum* (1, eastern Afghanistan), *Desideria* (4, Karakorum, Kashmir, Nepal, eastern Pamirs; characterized by a calyx with fused segments!), *Didymophysa* (2, eastern Anatolia, northern Iraq, southern Transcaucasia, Iran, western Tien Shan, Pamiro-Alai, Afghanistan, Pakistan; encountered in the eastern part of the Greater Caucasus), *Dielsiocharis* (1, Iran, Turkmenistan: Kopet Dagh), *Dilophia* (5, Central Asia), *Diptychocarpus* (1, Iran, Afghanistan, Pakistan, Soviet Central Asia), *Douepia* (1, Pakistan), *Drabopsis* (1, all areas), *Elburzia* (1, northwestern Iran; related to *Petrocallis*), *Eremoblastus* (1, the lower reaches of the rivers Volga and Ural), *Eurycarpus* (1, western Tibet), *Fortuynia* (2, Iran, Afghanistan, Pakistan), *Galitzkya* (3, northern Kazakhstan, Mongolia, western China), *Graellsia* (4, Anatolia, Iraq, Iran, Kopet Dagh, Pamiro-Alai, Afghanistan, Pakistan), *Gynophorea* (1, central Afghanistan), *Hedinia* (3, southeastern Altai, Mongolian Altai, Tien Shan, Pamiro-Alai, northwestern Himalayas, Tibet), *Irania* (5, Iraq, Iran, Afghanistan), *Iskandera* (2, Pamiro-Alai), *Lachnoloma* (1, Iran, Soviet Central Asia), *Leiospora* (6, Soviet Central Asia), *Micrantha* (1, Iran), *Microstigma* (3, Inner and Outer Mongolia), *Moriera* (2, Iran, central Kopet Dagh, Afghanistan; very close to *Aethionema*), *Nasturtiicarpa* (1, Afghanistan), *Octoceras* (1, Iran, Afghanistan, Pakistan, Soviet Central Asia), *Oreoloma* (3, Mongolia and adjacent China), *Pachypterygium* (4; very close to *Isatis*), *Parlatoria* (2, Anatolia; northern Iraq, Syria, Iran), *Parryopsis* (1, Tibet), *Peltariopsis* (2, southeastern Anatolia, southern Transcaucasia, northern Iraq, Iran), *Phaeonychium* (2, Soviet Central Asia; Pamiro-Alai, Afghanistan, Kashmir, western Tibet), *Physocardamum* (1, eastern Anatolia), *Physoptychis* (2, eastern Anatolia, northern Iraq, southern Transcaucasia, Iran), *Prionotrichon* (4, Kopet Dagh, Pre-Balkhash deserts, Dzungarian Alatau, Pamiro-Alai, Afghanistan), *Pseudoanastatica* (1, southern Transcaucasia, Iran; very close to *Clypeola*), *Pseudocamelina* (7, Iraq, Iran), *Pseudoclausia* (10, from western Kopet Dagh to Tien Shan and Pamiro-Alai, Iran, Afghanistan), *Pseudofortuynia* (1, southwestern Iran), *Pterygostemon* (1, eastern Kazakhstan), *Pugionium* (3, Mongolia, Inner Mongolia, Ordos), *Pycnoplinthus* (1, Tibet, Himalayas), *Pyramidium* (1, southwestern and southeastern Afghanistan), *Rhammatophyllum* (3, Kazakhstan), *Robeschia* (1, from Sinai Peninsula to Pakistan), *Sameraria* (15, all areas), *Sisymbriopsis* (2, Tien Shan and Pamiro-Alai, Kashgaria, Tibet), *Sophiopsis* (5, Kazakhstan, Soviet Central Asia, Kashgaria, Pakistan, Mongolia), *Spirorrhynchus* (1, Iran, Pakistan, Soviet Central Asia), *Spryginia* (7, from eastern Kopet Dagh to Pamiro-Alai and Afghanistan), *Strausiella* (1, western Iran), *Streptoloma* (2, Kazakhstan, Soviet Central Asia, Iran, Afghanistan), *Stubendorffia* (8, Soviet Central Asia), *Taphrospermum* (2, Central Asia), *Tauscheria* (2, Central Asia), *Tetracme* (8, Iran, Afghanistan, Pakistan, Soviet Central Asia), *Texiera* (1, Lebanon, Syria, Iraq, Turkey), *Trichochiton* (2, Soviet Central Asia, Afghanistan), *Tschihatschewia* (1, Anatolia), *Veselskya*

(1, Afghanistan), *Vvedenskiella* (2, Kashgaria, Kashmir), *Winklera* (2, Pamiro-Alai, Afghanistan, Pakistan), *Zerdana* (1, Iran: southern Zagros) Resedaceae: *Homalodiscus* (3, Iran, Oman Mountains, Kopet Dagh; very close to *Ochradenus*) Frankeniaceae: *Hypericopsis* (1, southern Iran: Fars; very close to *Frankenia*) Sapotaceae: *Reptonia* (2, Oman Mountains, Afghanistan, Pakistan) Primulaceae: *Dionysia* (about 30, southeastern Anatolia, Mardin and Hakkari, Iraq, Iran, Oman Mountains, Turkmenistan, Afghanistan, Pakistan: Baluchistan, western Pamiro-Alai), *Kaufmannia* (2, northern Tien Shan, Dzungarian Alatau) Thymelaeaceae: *Stelleropsis* (18, western Asia and Soviet Central Asia; very close to *Stellera*) Crassulaceae: *Pseudosedum* (40, Iran, Turkmenistan, Afghanistan, Pakistan: Chitral, Tien Shan), *Sempervivella* (4, Afghanistan, northwestern Himalayas) Rosaceae: *Hulthemia* (1–2, from western Iran to Dzungaria; very close to *Rosa*), *Potaninia* (1, Mongolia), *Spiraeanthus* (1, Soviet Central Asia) Fabaceae: *Ammodendron* (3–4, Iran, Afghanistan, Central Asia), *Ammopiptanthus* (1–2, Soviet Central Asia, Mongolia and adjacent China), *Calispepla* (1, western Gissar Mountains), *Chesneya* including *Chesniella* (about 25, from Syrian Desert and Anatolia to Mongolia), *Eremosparton* (3, southeastern part of European Russia and Central Asia), *Eversmannia* (1–2, southeast part of the European Russia, northern Iran, Afghanistan, Soviet Central Asia), *Oreophysa* (1, northern Iran), *Physocardamum* (1, eastern Anatolia), *Pseudolotus* (1, Afghanistan), *Smirnowia* (1, southern Turan; scarcely distinct from *Sphaerophysa*), *Sphaerophysa* (2, from Syria and Anatolia to northern China), *Vavilovia* (1, Anatolia, Lebanon, northern Iraq, southern Transcaucasia, northern and northwestern Iran; in the Caucasus as far as the Greater Caucasus Range; close to *Pisum*) Sapindaceae: *Stocksia* (1, southern Iran—Kermanshah, eastern Iran, northwestern and southeastern Afghanistan, Pakistan) Zygophyllaceae: *Malacocarpus* (1, Soviet Central Asia, Afghanistan; close to *Peganum*), *Tetraena* (1, Mongolia) Apiaceae: *Actinanthus* (1, western Asia), *Actinolema* (2, Anatolia, Syria, northern Iraq, southern Transcaucasia, Iran), *Alococarpum* (1, northwestern Iran), *Aphanopleura* (4 southern Transcaucasia, Iran, Afghanistan), *Artedia* (1, western Asia), *Calyptrosciadium* (1, northwestern Afghanistan), *Cephalopodum* (1, Pamiro-Alai), *Coriandropsis* (1, Kurdistan), *Crematosciadium* (1, northeastern Afghanistan), *Crenosciadium* (1, Anatolia), *Cryptodiscus* (4, Iran to Central Asia), *Cymbocarpum* (5, Anatolia, southern and eastern Transcaucasia, Talish, Iran; encountered in the eastern Mediterranean area), *Dicyclophora* (1, southwestern Iran), *Diplotaenia* (1, southeastern Anatolia, northern and western Iran),

Dorema (about 16, southern Transcaucasia, Iran, Soviet Central Asia, Afghanistan, Pakistan), *Elaeosticta* (about 22, Iran, Afghanistan, Soviet Central Asia), *Eremodaucus* (1, Caucasus to Afghanistan and Central Asia), *Ergocarpon* (1, eastern Iraq, Zagros), *Fergania* (1, northern slopes of the Alai and Turkestan ranges), *Grammosciadium* (8), *Hyalolaena* (4), *Hymenolyma* (2, Soviet Central Asia), *Johrenia* (20), *Kalakia* (1, Elburz), *Komarovia* (1, western Pamiro-Alai), *Korovinia* (3, Soviet Central Asia), *Korshinskia* (3, western Tien Shan—Pamiro-Alai), *Kosopoljanskia* (1, Syrdarian Karatau—Talassian Alatau), *Krasnovia* (1, Soviet Central Asia), *Lipskya* (1, western Pamiro-Alai), *Lisaea* (2), *Mastigiosciadium* (1, central Afghanistan), *Mediasia* (1, western Tien Shan, Pamiro-Alai), *Mogoltavia* (1, Mogoltau—Pamiro-Alai), *Oedibasis* (3, Soviet Central Asia, Kazakhstan), *Oliviera* (1), *Ormopterum* (2, Soviet Central Asia, Pakistan), *Ormosciadium (1), Paulia* (2, western Pamiro-Alai, northern Afghanistan), *Pastinacopsis* (1, northern Tien Shan), *Pilopleura* (2, western Dzungarian Alatau, Tien Shan), *Pinacantha* (1, northeastern Afghanistan), *Polylophium* (1, western Asia), *Psammogeton* (4, Iraq, Iran, Afghanistan, Soviet Central Asia, Pakistan, northwestern India), *Pyramidoptera* (1, Central Afghanistan), *Rhabdosciadium* (5, eastern Anatolia, Kurdistan, Iran), *Schrenkia* (10, western Tien Shan—Pamiro-Alai; one species reaches Tarbagatai and southern Altai), *Schtschurowskia* (1, western Tien Shan—Pamiro-Alai), *Sclerotiaria* (1, Kirgisian Alatau), *Semenovia* (about 10, Soviet Central Asia, Afghanistan, Pakistan, northwestern India), *Seselopsis* (1, northern Tien Shan), *Sphaenolobium* (3, western Tien Shan), *Spongiosyndesmus* (1, northern Afghanistan), *Stenotaenia* (6, Anatolia, southern Transcaucasia, Iran), *Stewartiella* (1, Pakistan: Baluchistan), *Synelcosciadium* (1, Syria), *Szovitsia* (1, northeastern Anatolia, Transcaucasia, northwestern Iran), *Thecocarpus* (1, Iran) *Tricholaser* (1, northeastern Afghanistan), *Trigonosciadium* (3, Anatolia, Iraq, western Iran), *Turgeniopsis* (1, western Asia), *Vvedenskya* (1, Pamiro-Alai), *Zeravschania* (1, western Pamiro-Alai), *Zosima* (10, western and Central Asia, Pakistan, northwestern Himalayas; encountered in Daghestan and Ciscaucasia)
Apocynaceae: *Poacynum* (3, Soviet Central Asia)
Rubiaceae: *Aitchisonia* (1, Afghanistan: Kurram; Pakistan), *Leptunis* (2–3, southern Transcaucasia, Iran to Afghanistan and Soviet Central Asia), *Microphysa* (1, Soviet Central Asia), *Warburgina* (1, Syria, Palestine)
Boraginaceae: *Caccinia* (5), *Choriantha* (1, Iraq: Kurdistan), *Craniospermum* (4, Soviet Central Asia), *Heliocarya* (1, Iran; close to *Caccinia*), *Heterocaryum* (4, southeastern European part of the USSR, Anatolia, Syria, Iraq, southern Transcaucasia, Iran, Afghanistan, Pakistan, Soviet Central Asia), *Lepechiniella* (10, northern Iran, Afghanistan, Pakistan, northwestern Himalayas, Soviet Central Asia, Mongolia; close to *Cynoglossum*),

Mattiastrum (25, eastern Anatolia, Iraq, Iran, Afghanistan, Pakistan, Soviet Central Asia, Kashmir), *Microula* (2, Tibet), *Oreogenia* (*Lasiocaryum;* 7, Afghanistan, Soviet Central Asia), *Phyllocara* (1, Anatolia, Syria, Sinai Peninsula, Iraq, western Iran), *Stephanocaryum* (2, Soviet Central Asia: western Tien Shan, eastern Fergana), *Suchtelenia* (2–3, Transcaucasia, Iran, western Kazakhstan, Central Asia), *Tianschaniella* (1, Tien Shan), *Trachelanthus* (4, Palestine, Iraq, Iran, Soviet Central Asia)

Scrophulariaceae: *Bungea* (2, from Anatolia and southern Transcaucasia to Soviet Central Asia), *Leptorhabdos* (1, eastern Transcaucasia, Iran, Afghanistan, Pakistan, Central Asia), *Nathaliella* (1, Soviet Central Asia: Alai Range), *Spirostegia* (1, Soviet Central Asia: southern Pamiro-Alai)

Lamiaceae: *Chamaesphacos* (1, very close to *Thuspeinanta*), *Dorystoechas* (1, Anatolia), *Drepanocaryum* (1, Soviet Central Asia, Pakistan; very close to *Nepeta*), *Eremostachys* (about 60, from Anatolia, Syria and Palestine to Soviet Central Asia and the Punjab; encountered in the eastern Mediterranean area), *Gontscharovia* (1, Soviet Central Asia, northwestern Himalayas, very close to *Micromeria*), *Hymenocrater* (9, southern Transcaucasia, Iran, Turkmenistan, Afghanistan, Pakistan), *Hypogomphia* (3, Afghanistan, Soviet Central Asia), *Lagochilus* (about 40, Iran, Afghanistan, Central Asia, northwestern Himalayas), *Metastachydium* (1, Soviet Central Asia: central and northern Tien Shan), *Pentapleura* (1, Kurdistan), *Perovskia* (7, northeastern Iran, Afghanistan, Pakistan, Soviet Central Asia, Tibet, northwestern Himalayas), *Pseudomarrubium* (4, Soviet Central Asia: Syrdarinian Karatau), *Stachyopsis* (4, Soviet Central Asia, Dzungarian Alatau and adjoining mountain regions of China), *Thuspeinanta* (2, Iran, Afghanistan, Pakistan, Punjab, Soviet Central Asia), *Zataria* (1, Iran, Afghanistan, Pakistan)

Campanulaceae: *Cryptocodon* (1, Soviet Central Asia: western Tien Shan—Kukhistan), *Cylindrocarpa* (1, Soviet Central Asia: western Tien Shan—Kukhistan), *Michauxia* (7, Anatolia, Syria, Lebanon, Iraq, southern and southeastern Transcaucasia, Iran), *Ostrowskia* (1, Soviet Central Asia: western Tien Shan and Pamiro-Alai, Afghanistan), *Sergia* (2, Soviet Central Asia: western Tien Shan, Pamiro-Alai), *Zeugandra* (1, western Iran: Kermanshakh, between Kermanshakh and Shakhabad and in the mountains above Shakhabad; related to *Campanula*)

Asteraceae: *Acanthocephalus* (2, Soviet Central Asia, Afghanistan, Pakistan: Chitral), *Acantholepis* (1), *Amblyocarpum* (1, Hyrcania), *Anura* (1, Soviet Central Asia: Nuratau), *Brachanthemum* (5, central Kazakhstan to Soviet Central Asia), *Callicephalus* (1, northeastern Anatolia, southern Transcaucasia, northwestern and northern Iran), *Cancrinia* (4, Soviet Central Asia), *Cancriniella* (1, southern Kazakhstan), *Chamaegeron* (3, Kazakhstan, Soviet Central Asia, Iran, Afghanistan, Pakistan), *Chrysophthalmum*

(2, southern and eastern Anatolia, northern Syria, northern Iraq), *Chrysopappus* (1, Kurdistan), *Codonocephalum* (3, northern Iraq, Iran, Afghanistan, Pakistan, Soviet Central Asia: Turkmenistan, Tien Shan), *Cousinia* (about 600, from central Anatolia, Syria, and Lebanon to Dzungarian Gobi and northwestern Himalayas; encountered in Europe only in the Lower Volga region), *Cousiniopsis* (1, Soviet Central Asia: southern Turan—Pamiro-Alai, Afghanistan), *Cymbolaena* (1), *Dipterocome* (1, Syrian Desert, southern Transcaucasia, Iran), *Epilasia* (4; very close to *Scorzonera*), *Garhadiolus* (2), *Grantia* (6, Arabian Peninsula, Iran), *Gundelia* (1, inner and southern Anatolia, Syrian Desert, northern Iraq, southern Transcaucasia, Iran, western Afghanistan, Kopet Dagh; westward as far as western Syria and Cyprus), *Handelia* (1, Soviet Central Asia), *Heteracia* (2), *Hyalea* (1–2; close to *Centaurea*), *Karelinia* (1, southeastern European part of the USSR, northeastern Iran to Mongolia), *Kaschgaria* (2, Soviet Central Asia; close to *Tanacetum*), *Karvandarina* (1, Baluchistan; related to *Jurinea*), *Lachniphyllum* (2–3), *Lamyropappus* (1, Soviet Central Asia), *Lepidolopha* (6, western Tien Shan—Pamiro-Alai), *Lepidolopsis* (1, Iran, Afghanistan, Soviet Central Asia), *Lipskyella* (1, Soviet Central Asia), *Microcephala* (3, Iran, Soviet Central Asia), *Modestia* (2, Soviet Central Asia: Karategin—Darvaz), *Myopordon* (2, southern and southwestern Iran), *Nikitinia* (1, Kopet Dagh), *Olgaea* (about 15, Soviet Central Asia, Pakistan, China: Sinkiang), *Perplexia* (2, Iran, Soviet Central Asia), *Plagiobasis* (3, Soviet Central Asia, Afghanistan), *Polychrysum* (1, Soviet Central Asia: Gissaro—Darvaz, northern Afghanistan), *Postia* (4, Syria, southwestern Iran), *Pseudohandelia* (1, Iran, Afghanistan, Soviet Central Asia; close to *Tanacetum*), *Pterachaenia* (1, Afghanistan, Pakistan), *Russowia* (1, Soviet Central Asia), *Schischkinia* (1), *Schmalhausenia* (1, Soviet Central Asia: western Tien Shan), *Sclerorhachis* (2, northeastern Iran, Afghanistan), *Siebera* (2, inner and southern Anatolia, Syria, Lebanon, northern Iraq, western and northern Iran, Afghanistan, southern Soviet Central Asia), *Stilpnolepis* (1, Mongolia), *Stizolophus* (2, southern Transcaucasia, Iran, Turkmenistan; close to *Centaurea*), *Syreitschikovia* (2, Soviet Central Asia), *Tanacetopsis* (about 15, Soviet Central Asia; very close to *Cancrinia*), *Thevenotia* (2, Iraq, Iran, Afghanistan, Soviet Central Asia), *Tomanthea* (10, Anatolia, Iraq, southern Transcaucasia, Iran), *Trichanthemis* (about 8, Soviet Central Asia), *Tugarinovia* (1, Mongolia), *Uechritzia* (3, eastern Anatolia, Soviet Central Asia, Himalayas), *Ugamia* (1, Soviet Central Asia: western Tien Shan), *Wendelboa* (1, northern Pakistan: Chitral; very close to *Taraxacum*), *Xylanthemum* (6, northeastern Iran, Afghanistan, Soviet Central Asia)

Poaceae: *Henrardia* (3, from Anatolia and Syria to Tien Shan and northwestern Himalayas), *Heteranthelium* (1, from Syria, Lebanon and Palestine

to Pakistan and Soviet Central Asia), *Littledalea* (4, Central Asia), *Orinus* (2, Kashmir, southern Tibet, Sinkiang), *Pilgerochloa* (1, Anatolia), *Psammochloa* (1, Central Asia), *Rhizocephalus* (1, Anatolia, Syria, Palestine, southern Transcaucasia, Iran, Soviet Central Asia), *Sinochasea* (1, China: Sinkiang)
Hyacinthaceae: *Alrawia* (2, northeastern Iraq, western Iran)
Amaryllidaceae: *Ungernia* (8, Iran, Soviet Central Asia, Afghanistan)
Ixioliriaceae: *Ixiolirion* (3–4, from Anatolia to Pakistan and Chinese Dzungaria; encountered in western Siberia)
Araceae: *Eminium* (6–9, western Asia, Soviet Central Asia, encountered also in the eastern Mediterranean area)

The Irano-Turanian Region is divided into subregions—the Western Asiatic and the Central Asiatic—but the boundary between them is not always very easy to describe.

8A. WESTERN ASIATIC SUBREGION

Takhtajan 1974

The Western Asiatic Subregion[79] includes the entire western part of the Irano-Turanian Region up to the Prebalkhash area, central Tien Shan, and the western Himalayas. It includes the floristically richest part of the Irano-Turanian Region, the vegetation of which is distinguished by its great variety. In Hyrcania, in the Zagros Mountains in Iran, in western Tien Shan, and in the Western Himalayan Province, one still finds relict stands of mesophytic forests. In many places considerable areas are covered with the xerophytic oak forests. The Mediterranean element is strongly expressed, especially in the flora of the oak forests and woodlands.

The Western Asiatic Subregion, especially the Iranian Plateau, represents the basic center of formation of the Irano-Turanian flora. Here one finds many endemic species of genera, such as *Delphinium, Silene, Acanthophyllum, Gypsophila, Calligonum, Atraphaxis, Acantholimon, Limonium, Althaea, Alcea, Euphorbia, Haplophyllum, Alyssum, Aethionema, Erysimum, Isatis, Dionysia, Prunus, Chesneya, Astragalus, Hedysarum, Onobrychis, Trigonella, Ferula, Onosma, Nepeta, Phlomis, Eremostachys, Salvia, Stachys, Thymus, Scrophularia, Verbascum, Heliotropium, Convolvulus, Asperula, Galium, Achillea, Anthemis, Tanacetum, Artemisia, Centaurea* and related genera, *Cousinia, Echinops, Helichrysum, Jurinea, Scorzonera, Tragopogon, Eremurus, Tulipa, Allium,* and *Iris.*

1. Mesopotamian Province (Zohary 1962, 1963, 1973; Takhtajan 1970, 1974; Sous-Région Mésopotamienne, Boissier 1867; Bezirk von Mesopotamien, Hayek 1926; Domaine Mésopotamien, Eig 1931; Mesopotamian Subregion of the Irano-Turanian Region, Guest 1966). According to Zohary (1973:90), the central part of this province is the Syrian Desert, extending northward to the "Fertile Crescent" in southeastern Anatolia. The province comprises some of the highlands of eastern Edom and of Sinai, the Judean Desert, northern Iraq up to the foothills of the Kurdistan Mountains, and also some spots of Khuzestan in southwestern Iran. Though there are a considerable number of endemic species, endemism is much lower than in the majority of other provinces of the Irano-Turanian Region.

The vegetation of the Mesopotamian Province consists mainly of steppe and semidesert communities on grey, calcareous desert soil, often gravelly and stony (Guest 1966).

2. **Central Anatolian Province** (Meusel et al. 1965; Kamelin 1973; Central Anatolian Subprovince of the Armeno-Iranian Province, Takhtajan [in Takhtajan and Fedorov] 1972). This province occupies the inner arid and semiarid parts of Anatolia, including the Anatolian Plateau. Its eastern boundary, which Davis (1965, 1971) calls "the Anatolian Diagonal," runs from the vicinity of Bayburt and Gümüsane to the southwest through the vicinity of Erzincan, and extends further along the Anti-Taurus. This boundary, which Davis considers "a remarkable floral break through the middle of Inner Anatolia" (1971:19), separates the Central Anatolian Province from the Armeno-Iranian Province, to which it is close in many respects.

The central Anatolian flora is characterized by rather high species endemism (probably about 30%) and contains a series of monotypic endemic genera—for example, *Kalidiopsis*, a genus close to *Kalidium*, which is a member of the halophytic vegetation of the southwestern shores of lake Tuz Gölü in the center of the Anatolian Plateau; *Cyathobasis*, a genus close to *Girgensohnia*, which is absent in the flora of Anatolia; and *Crenosciadium*, a genus close to *Opopanax*, occupying a small area between lakes Eğridir Gölü and Beysehir Gölü. Endemics of the Central Anatolian Province are linked with both the Irano-Turanian and the Mediterranean floras. The strongest Irano-Turanian affinities are shown by the halophytic flora around the salt lakes of the Konya Ovasi Plain, especially in the Chenopodiaceae, and also the Plumbaginaceae (e.g., *Limonium anatolicum*; see Davis 1971). But as Davis points out, the major part of the central Anatolian basin's endemic flora shows few such connections with the flora of Iran or that of the deserts east of the Caspian. These endemics are often related to species still growing in peripheral (often more or less Mediterranean) areas. They include *Delphinium venulosum, Consolida stenocarpa,* and *Salvia halophila.* Other endemics (e.g., *Silene salsuginea, Limonium globuliferum, Astragalus ovalis,* and *Verbascum helianthemoides*) apparently have no close relatives, and their geographic relations are less clear. Other centers of endemism are found on areas of

gypseous chalk surrounding Cankiri and especially Sivas, where such remarkable endemics as *Salvia vermifolia* grow (Davis 1971).
On the whole, the composition of the endemic flora of the Central Anatolian Province indicates that it is to some extent transitional in character between a typical Irano-Turanian flora and the flora of the eastern Mediterranean. One may assume that, before the broad invasion of the Irano-Turanian plants, the flora of the Anatolian Plateau was basically of the Mediterranean type, while in its northern part it was nearer to the Euxinian flora. In the west and in the south, the Anatolian flora is still closely related to the eastern Mediterranean, but in the north it is related to the Euxine. Many genera characteristic of the Iranian Plateau (*Acantholimon, Acanthophyllum, Calligonum, Ferula, Eremostachys, Cousinia* et al.) are represented in the Central Anatolian Province by relatively few species, although there is a large diversity in such genera as *Aethionema, Alyssum, Isatis,* pinnate-leaved *Salvia, Achillea,* and others (Davis, 1965), and also the genus *Centaurea* s.l.

3. **Armeno-Iranian Province** (Engler 1899, 1903, 1924, p.p.; Alekhin 1944; Good 1947, 1974; Efremov [in Armand et al.] 1956; Takhtajan [in Takhtajan and Fedorov] 1972, p.p.; Takhtajan 1974, p.p.; Domaine Iranien, Eig 1931; Iranian Province, Grossheim 1936, 1948; Irano-Anatolian Province, Zohary 1962, 1963, 1973; Davis 1965, p.p.).[80] The vast area of this province embraces the eastern part of the Anatolian Plateau (east of the "Anatolian Diagonal"), most of the Armenian Highlands, dry regions of southern Transcaucasia, Zuvand, Kopet Dagh, the greater part of Iran (excluding the south and southwest coastal tropical regions, and regions of the relict Hyrcanian forests), and part of Afghanistan. The limits of this phytochorion were roughly defined by Boissier (1867), who called it "Sousrégion des plateaux."
The Armeno-Iranian Province has a rich, distinctive flora very high in generic and specific endemism. There are more than 20 endemic genera (see Hedge and Wendelbo 1978) as well as endemic subgenera and sections, and a large number of endemic species. Endemic taxa are particularly abundant in the families Asteraceae, Apiaceae, Boraginaceae, Brassicaceae, Chenopodiaceae, Fabaceae, Lamiaceae, Liliaceae s.l., Plumbaginaceae, Rubiaceae, and Scrophulariaceae.
Floristically, the Armeno-Iranian Province is rather heterogenous and can be divided into several subprovinces, some of which may even deserve treatment as separate provinces. But it would be premature to subdivide this chorion into separate provinces until a more complete analysis of the flora is made. This will be possible only after the completion of K. H. Rechinger's "Flora Iranica."
Within the limits of the Armeno-Iranian Province, the Armenian Subprovince (Province of Armenian Highlands, Grossheim and Sosnovsky 1928;

Armenian province, Magakyan 1941, p.p.; Sector of Armenian Highlands, Zohary 1973, p.p.) stands somewhat apart. It occupies a large part of the Armenian Highlands, including the Ardahan and Kars plateaus, and a large area the border of which lies in the west between Sivas and Erzincan and then heads south, where it bends around the Malatya Mountains and goes up to the basin of Lake Van. The eastern border bends around Lake Van from the west and extends northwestward, passing somewhere between Kağizman and Iğdir, and on to Mount Aragatz (Alagöz). There are many endemics shared with the adjacent phytochoria, with which this subprovince intergrades. Nevertheless, the Armenian Subprovince has many endemic species of its own, which we list the following:

Ranunculus sintenisii, Dianthus nodiflorus, D. robustus, D. sessiliflorus, Gypsophila briquetiana, G. tuberculosa, Hypericum scabroides, H. uniglandulosum, Isatis erzurumica, Reseda armena, Alcea karsiana, A. sophiae, Lathyrus karsianus, many species of *Astragalus, Malabaila lasiocarpa, Trigonosciadium intermedium, Onosma arcuatum, Nonea karsensis, Verbascum hajastanicum, Scrophularia versicolor, Micromeria elliptica, Campanula ledebouriana, Inula macrocephala, I. discoidea, Tanacetum mucroniferum, Cousinia bicolor, C. brachyptera, C. euphratica, C. woronowii, Cirsium sommieri, Centaurea sessilis, C. kurdica, C. fenzlii, C. tomentella, Echinops melitensis, Uechtritzia armena, Asphodeline tenuiflora, Ornithogalum tempskyanum, Bellevalia coelestis, Tulipa kaghyzmanica, T. mucronata, Allium sosnovskyanum, A. stenopetalum,* and *Crocus karsianus.*

Characteristic of the Armenian Subprovince are various types of grass steppes (including typical *Stipa* steppes); tragacanthic (thorn-cushion) steppes with the dwarf, spiny, cushionlike shrublets of *Astragalus, Acantholimon, Onobrychis cornuta, Acanthophyllum,* and so on; different kinds of woodlands (especially juniper woodlands); and high mountain vegetation (subalpine and alpine meadows and alpine carpets of dwarf perennial herbs with leaves crowded at the base, forming a rosette, and comparatively large flowers, like *Campanula tridentata* and *Gentiana verna*). In the northern part of the subprovince some areas contain more or less preserved remains of pine forests (*Pinus sylvestris* subsp. *kochiana*). Remains of oak forests, usually degraded and parklike, are common. In the northernmost part of the subprovince, oak forests are composed of *Quercus macranthera,* whereas in the southern parts the oak forests (and oak woodlands) consist of *Quercus brantii* (the most common species), *Q. robur* subsp. *pedunculiflora, Q. libani* and *Q. infectoria* subsp. *boissieri.* In the basin of Lake Van, one also finds *Q. petraea* subsp. *pinnatiloba.*

The next subprovince is the Atropatenian (Takhtajan 1941, Takhtajan [in Takhtajan and Fedorov] 1972).[81] It embraces: arid and semiarid parts of southern Transcaucasia; the basin of Lake Sevan; the Diabar Depression in Zuvand; the eastern parts of Turkish Armenia, including the basin of Lake

Van and the northern part of the Kurdish Alps (in Hakkari), which form a southern fringe of the Armenian Highlands; Iranian Azerbaijan and some of the adjacent areas in Iran. The western border of the Atropatenian Subprovince starts with the east and southeast slopes of Mount Aragatz, runs in southwest and bends around the western side of Mount Nimrud (Nemrut Dagh) and Lake Van, and goes along the Kurdish Alps and then east and southeast, bending round the Lake Urmia (Rizaieh) from the south. In the east the Atropatenian flora reaches the Tehran area and the south slopes of the central part of the Elburz Mountains, including Mt. Demavend. In the north and northeast the Aropatenian Subprovince forms a more or less clear border with the Caucasian and especially with the Hyrcanian Province, but in the east, south, and southeast its borders are not quite clear. Many Atropatenian elements reach the Shiraz area in the south.

The Atropatenian Subprovince is characterized by high species endemism. It has some endemic genera (*Elburzia, Kalakia, Szovitsia*). Genera with the largest number of endemic Atropatenian species are *Astragalus* and *Allium*, then *Nepeta* and *Salvia*, followed by *Acantholimon, Cousinia* and *Isatis*. These are followed by *Centaurea* s.l., *Euphorbia, Helichrysum, Iris* (especially subsection *Oncocyclus*), *Onosma, Convolvulus, Scrophularia*, and *Salsola*. The Atropatenian Subprovince is one of the most active centers of speciation in all of Western Asia. There are numerous endemic species, including:

Delphinium carduchorum, Papaver bipinnatum, Corydalis persica, Acanthophyllum (Allochrusa) bungei, Acanthophyllum (Allochrusa) versicolor, Dianthus cancescens, D. crossopetalus, D. grossheimii, Silene eremetica, S. prilipkoana, Telephium oligospermum, Anabasis eugeniae, Salsola cana, Seidlitzia florida, Acantholimon araxanum, A. hohenackeri, A. karelinii, Hypericum formosissimum, Dionysia teucrioides, Aethionema diastrophis, Atropatenia rostrata, Crambe armena, Cymatocarpus grossheimii, Erysimum buschii, E. feodorovii, E. nachyczevanicum, E. nanum, E. wagifii, Isatis ornithorhynchus, Lepidium lyratum, Neurotropis armena, N. szovitsiana, Peltariopsis drabicarpa, P. grossheimii, P. planisiliqua, Pseudocamelina szowitsii, Sameraria odontophora, Sterigmostemum acanthocarpum, Euphorbia coniosperma, E. grossheimii, Potentilla porphyrantha, Prunus (Amygdalus) nairica, P. (Amygdalus) urumiensis, Pyrus raddeana, Astragalus karakuschensis, A. latifolius, A. mesites, A. paradoxus, A. strictifolius, A. szovitsii, Colutea komarovii, Hedysarum atropatanum, Onobrychis buhseana, O. heterophylla, O. subacaulis, Polygala hohenackeriana, Haplophyllum kowalenskyi, H. laeviusculum, H. schelkovnikovii, H. tenue, Aphanopleura trachysperma, Dorema glabrum, Echinophora orientalis, Elaeosticta glaucescens, Ferula persica, Peucedanum pauciradiatum, Prangos uloptera, Seseli leptocladum, Szovitsia callicarpa, Valerianella amblyotis, Jaubertia szovitsii, Galium bullatum, G. hyrcanicum, Rubia rigidifolia, Heliotropium gracillimum, H. gypsaceum, H. schahpurense, H. szovitsii, Rindera media, Onosma gaubae, Nonea anchusoides, Solenanthus formosus, Convolvulus gracillimus, Scrophularia atropatana,

S. nachitschevanica, Verbascum spp., *Veronica microcarpa, Marrubium parviflorum, M. persicum, Nepeta erivanensis, N. kochii, N. meyeri, N. trautvetteri, Salvia brachyantha, S. dracocephaloides, S. fominii, S. limbata, Sideritis balansae, Ziziphora rigida, Campanula bornmuelleri, C. karakushensis, C. massalskyi, C. radula, Anthemis grossheimii, Achillea cuneatiloba, Tanacetum canescens, T. tabrisianum, T. uniflorum,* numerous species of *Cousinia, Onopordum armenum, Cirsium congestum, Carduus nervosus, Jurinea armeniaca, J. pulchella, Serratula coriacea, Oligochaeta divaricata, Centaurea carduchorum, C. erivanensis, C. gracillima, C. pseudoscabiosa, C. schischkinii, C. xanthocephala, C. vanensis, Tomanthea daralaghezica, T. phaeopappa, Scorzonera armeniaca, S. candavanica, Tragopogon marginatus, T. sosnovskyi, Lactuca azerbaijanica, Cephalorrhynchus takhtadzhianii, Asphodeline szovitsii, Ornithogalum schelkownikowii, Scilla atropatana, Bellevalia makuensis, Muscari longipes, Gagea improvisa, Tulipa florenskyi, Nectaroscordum tripedale, Allium akaka, A. derderianum, A. dictyoprasum, A. leonidii, A. mariae, A. woronowii, Iris barnumae, I. pseudocaucasica,* and *Triticum araraticum.*

The vegetation of the Atropatenian Subprovince is basically xerophytic in character, the xerophytic nature gradually increasing from west to east and from north to south. Most characteristic for this subprovince are various phryganoid formations, trangacanthic steppes, different kinds of shrublands, and open woodlands ("steppe forests"), especially juniper, almond and pistacia-almond woodlands. The alpine vegetation is extremely xerophytic and contains many elements of mountain tragacanthic steppes including some dwarf spiny cushion shrublets and perennials.

In the east, the Atropatenian Subprovince intergrades with the Khorassan Subprovince (Takhtajan [in Takhtajan and Fedorov] 1972; Khorassan district of the Iranian province, Popov 1950; Turkmeno-Iranian province, Korovin 1962, p.p.; Turkmeno-Iranian mountain province, Lavrenko 1965; Turkmenische Provinz, Meusel et al. 1965; Khorassan-Kopet Dagh provinces, Kamelin 1970, 1973; Cherneva 1974). The Khorassan Subprovince is very clearly linked with the Atropatenian Subprovince, with which it has much in common. It includes the Turkmeno-Khorassan Mountains (the massif of the Big Balchan and Kopet Dagh in the north and Nishapur Mountain in the south), the eastern Elburz, and the Gorgano-Meshed line of valleys that is situated between them. Popov extended his "Khorassan district" eastward to the Tedzhen River valley. There is an endemic genus *Diaphanoptera.* It is very closely related to subgenus *Allochrusa* of the genus *Acanthophyllum.* This subprovince is closely connected with the Atropatenian Subprovince through the south slopes of Elburz (Alborz).

There are many species endemic to the Khorassan Subprovince (see Kamelin 1970, 1973; Hedge and Wendelbo 1978), including *Ranunculus pulsatillifolius, Corydalis chionophila, Ficus kopetdagensis, Acantholimon khorassanicum, A. raddeanum, Dionysia kossinskyi, Peltaria turkmena, Thlaspi stenocarpum,*

Stroganowia litwinowii, Alcea freyniana, Cotoneaster discolor, C. turcomanicus, C. tytthocarpus, Pyrus turcomanica, Colutea atabaevii, Haplophyllum obtusifolium, Ferula turkomanica, Crucianella sintenisii, Heliotropium litwinowii, H. mesinanum, Mattiastrum gorganicum, M. gracile, M. turcomanicum, Veronica spp., *Crepis khorassanica, C. turcomanica, Atropa komarovii, Mandragora turkomanica, Salvia khorassanica,* many species of *Cousinia, Helichrysum kopetdagenese, Tanacetum turcomanicum, Jurinea antoninae, J. antonowii, J. sintenisii, Perplexia (Jurinea) microcephala, Tragopogon kopetdagensis, T. tomentosulus, Cephalorrhynchus kossynskyi, Hyacinthella litwinowii, H. transcaspica, Eremurus kopetdaghensis, E. subalbiflorus, Fritillaria raddeana, Tulipa hoogiana, T. wilsoniana, Allium bodeanum, A. cristophii, A. giganteum, A. helicophyllum, A. kopetdagense, A. vavilovii, Iris fosterana, Crocus michelsonii.* There are also many endemics common to the Atropatenian and Khorassanian Subprovinces, including *Gypsophila aretioides, Acanthophyllum macronatum, Euphorbia marschalliana,* and *Ferula oopoda.*

Different types of shrublands and open woodlands (mainly the juniper and pistacia-almond "steppe forests") and phryganoid formations are characteristic of this subprovince. There are significant areas of psammophytic and halophytic vegetation in which the typical Turanian flora predominates.

The Atropatenian Subprovince also has much in common with the Kurdo-Zagrosian Subprovince (Kurdisch-südwestiranische Gebiet, Rechinger 1951*b*; Kurdisch-südwestiranische Provinz, Meusel et al. 1965; Kurdo-Zagrosian sector, Zohary 1973). This subprovince embraces part of Turkish Armenia in southeast Anatolia, the mountains of Iraqi Kurdistan (Amadiya, Rowanduz, and Sulaimaniya districts), and the northwest and central parts of the Zagros system in the broader sense (including Iranian Kurdistan and parts of southern Iranian Azerbaijan) to the south up to approximately the latitude of the towns Kazerun and Shiraz.

The rich flora of the Kurdo-Zagrosian Subprovince includes several endemic genera (*Brossardia, Zeugandra, Choriantha,* and *Alrawia;* see Hedge and Wendelbo 1978) and numerous endemic species. There are a number of endemics common to the Kurdo-Zagrosian and Atropatenian subprovinces, including *Michauxia laevigata*. In the Iraqi part of the subprovince there occur isolated localities (enclaves) of some east Mediterranean species, including *Pinus brutia, Cercis siliquastrum,* and *Fontanesia phillyreoides* (see Zohary 1973). Relict localities of *Zelkova carpinifolia* occur in Anatolian and Iranian Kurdistan (see Browicz 1978, map 100). The vegetation is characterized by phryganoid formations and mountain steppes (especially the tragacanthic steppes), as well as by oak forests (especially forests of *Quercus brantii, Q. infectoria,* and *Q. libani*) and by different kinds of open woodlands.

The Fars-Kermanian Subprovince (Fars-Kermanian district of the Iranian province, Popov 1950) includes the southeastern part of the Zagros system to the south and east of Shiraz, the Yazd and Kerman Mountains, and the

Bandar Abbas Mountains. The boundaries of the Fars-Kermanian Subprovince are not very clear, especially in the northwest, where there are gradual transitions between the floras of the Fars-Kermanian and the Kurdo-Zagrosian subprovinces. The genus *Zerdana* and a number of species are endemics common to these two subprovinces. The flora of the south and southwest parts of the Fars-Kermanian Subprovince gradually turns into the tropical Omano-Sindian flora. *Stocksia* (Sapindaceae) and *Cometes surratensis* (Caryophyllaceae) are endemic to the borders between the Fars-Kermanian and Omano-Sindian floras (see Hedge and Wendelbo 1978). Among the endemics of the Fars-Kermanian Subprovince, the monotypic genus *Hypericopsis* (Frankeniaceae) is the most remarkable; it occurs in the Fars and Shiraz areas. There is also an endemic, monotypic brassicaceous genus *Acanthocardamum* (closely related to *Lepidium*) and many endemic species, including species of *Acantholimon* and *Dionysia, Isatis pachycarpa, Matthiola flavida, Pseudocamelina aphragmodes, Sterigmostemum laevicaule, Rosularia modesta, Cotoneaster persica, Heliotropium borasdjunense, H. gaubae, Alkanna leptophylla, Caccinia kotschyi, Nonea hypoleia, Convolvulus argyracanthus, C. zargarianus, Nepeta assurgens, Cousinia fragilis,* and *Allium lalesarensis*.

The Fars-Kermanian Subprovince is almost without forests. The preserved relict "islands" of various kinds of usually open woodlands convey some idea about the climax vegetation of the area. A good example is the relict island preserved on the top of the Mount Kuli-e Genou (2,300 m high) just north of Bandar Abbas. According to Wendelbo (personal communication), it is a comparatively dense stand of *Juniperus excelsa* subsp. *polycarpos, Olea aucheri, Acer monspessulanum* subsp. *persicum, Prunus scoparius,* the endemic *P. wendelboi* (*Amygdalus wendelboi* Freibag), and *Pistacia khinjuk*. *Dionysia revoluta* and *Rosularia modesta* are two other interesting species in this woodland. The lower part of the mountain has an Omano-Sindian flora. There is a small enclave of a predominantly Fars-Kermanian flora in Oman (in Jabal-Akhadar) with *Juniperus excelsa* subsp. *polycarpos, Olea aucheri,* and *Dionysia* (see Wendelbo 1961).

The Central Iranian Subprovince (Zentraliranishce Provinz, Meusel et al. 1965; Central Iranian sector, Zohary 1973) occupies the Central Plateau, including two great basins of interior drainage (the large saline Dasht-e-Kavir and an almost empty, sandy Dasht-e-Lut), and part of southern Afghanistan. According to Zohary (1973), "this sector harbours the most typical flora of Iran's steppes and deserts and constitutes the bulk of the Irano-Turanian territory in Iran." There is an endemic central Iranian genus *Heliocarya* (in the low mountains around Esfahan) and a considerable number of endemic species.

The vegetation of the Central Iranian Subprovince has a predominantly desert character. Vast areas are occupied by *Artemisia* communities, alternating with halophytic and psammophytic vegetation. In some high mountains one finds preserved remnants of pistacia-almond open woodlands ("steppe-forests"). The remnants of a *Prunus scoparia* and *Pistacia khinjuk* community are often very characteristic, and in Wendelbo's opinion (personal communication), this community must have been very important before grazing was introduced by man. *Pteropyrum aucheri* communities are very common and very characteristic. The large enclaves of the Turanian (Aralo-Caspian) flora are typical in the desert and semidesert areas, with species such as *Agriophyllum latifolium, Horaninovia ulicina, Chrozophora gracilis, Lachnoloma lehmannii, Smirnowia turkestana,* and *Chamaesphacos illicifolius* found mostly on sand dunes (Wendelbo, personal communication).

4. **Hyrcanian Province** (Grossheim and Sosnovsky 1928; Grossheim 1936, 1948; Prozorovsky and Maleev 1947; Meusel et al. 1965; Takhtajan 1970, 1974; Lenkoran Province, Kuznetsov 1909). This province includes the relict forests of the southwestern and southern Caspian coastal plains of Lenkoran (in southeastern Transcaucasia) and of Gilan and Mazanderan (in Iran), the northeastern slopes of the Talish Range, and the northern slopes of the Elburz Mountains and its eastern extension. Thus it is shaped like an arc, beginning at a little more than 48° and ending at a little more than 56° east longitude. In the east the Hyrcanian flora gradually diminishes. Some of its restricted exclaves are encountered in the territory of the Khorassan Subprovince (of the Armeno-Iranian Province) in the direction of Mt. Bujnurd in Iran. The Hyrcanian Province is one of the most distinct and clearly outlined provinces in the Irano-Turanian Region.

The flora of the Hyrcanian Province differs very greatly in composition from those of the neighboring, typical Irano-Turanian floras. Circumboreal taxa dominate in the flora of the Hyrcanian Province, particularly Euxinian-Caucasian elements. Thus the Hyrcanian Province may be described as an enclave of the circumboreal (or, more accurately, predominantly circumboreal) flora in the territory of the Irano-Turanian Region. The Hyrcanian Province is a relict island of an ancient mesophilous forest of Tertiary flora, surrounded on all sides by a typical Irano-Turanian flora. Such relict islands occur also in other parts of the Irano-Turanian Region, particularly in the western Himalayas. Since I use the "rule of integrity of boundaries" throughout this book, I include such enclaves in those phytochoria in the territory of which they are located.

Although the Hyrcanian flora is dominated quantitatively by circumboreal elements, other floristic elements also play significant roles. It is sufficient to point out that three of the most important forest-forming species—namely, the subendemic *Parrotia persica, Quercus castaneifolia,* and *Acer velutinum*—are completely absent from the circumboreal flora. *Parrotia persica* is close to the monotypic, western Himalayan genus *Parrotiopsis*. For a long time it was considered an endemic Hyrcanian genus, but recently Safarov (1977) has found a small population of it in the eastern part of the Alazan Valley, 15–20 km southwest of the city of Kutkashen (Azerbaijan). The main part of the distribution of *Parrotia* is nevertheless found in the Hyrcanian Province, where it extends from Talish to Gulistan in Gorgan, so it may definitely be considered a Hyrcanian subendemic. *Quercus castaneifolia* stands very close to *Q. afares*, which grows in the Algerian Atlas Mountains. It is widely distributed in the Hyrcanian Province, but in the east small exclaves of it are found in the Khorassan Subprovince, and other disjunct areas are found in the southeastern part of the Greater Caucasus. It also is a typical subendemic. As regards *Acer velutinum* (*A. insigne*), it is close to the Himalayan species *A. caesium* (from Kashmir to Nepal). Despite its exclaves in the Greater Caucasus, *A. velutinum* is so widely distributed in the Hyrcanian Province (extending eastward to almost 56° east longitude), that this species may also be considered subendemic. One of the most interesting Hyrcanian endemics is *Gleditsia caspia*. Its closest relatives are found entirely outside the circumboreal flora, being *G. japonica* and related species.

The Hyrcanian endemics and subendemics include *Ranunculus dolosus, Epimedium pinnatum* subsp. *pinnatum*, probably *Buxus hyrcana* (close to *B. sempervirens*, and a subspecies of it), *Alnus subcordata, Saponaria brodeana, Primula heterochroma, Cyclamen elegans, Populus caspica, Alcea hyrcana, Daphne rechingeri, Pyrus boissierana* (subendemic), *P. grossheimii, P. hyrcana, Rubus hyrcanus, R. persicus, Geranium montanum, Chaerophyllum meyeri, Ilex hyrcana* (very close to *I. aquifolium* and probably a subspecies of it), *Lindelofia kandavanensis, Solanum kieseritzkii, Scrophularia hyrcana, S. megalantha, S. rostrata, Scutellaria tournefortii, Stachys persica, S. talyschensis, Origanum hyrcanum, Amblyocarpum inuloides, Cousinia hablizii, Fritillaria grandiflora, F. kotschyana, Lilium ledebourii, Ornithogalum bungei, Ruscus hyrcanus* (very close to *R. aculeatus*), *Crocus caspius, C. hyrcanus*, and *Allium lenkoranicum*.

There are many Mediterranean elements in the Hyrcanian Province, but these are in general different from the Mediterranean elements in the Euxine and Caucasian provinces. In particular, the Hyrcanian Province is characterized by *Cupressus sempervirens* (eastward to Gorgan; see Browicz 1978, map 1), which beyond the Mediterranean Region is encountered in southern Iran, but is absent from northern Anatolia and the Caucasus. The Hyrcanian Province also contains *Myrtus communis*, a Mediterranean and Irano-Turanian species (eastward to Afghanistan), which is absent in the Caucasus but present in places in northern Anatolia.

In contrast with the Euxine and the Caucasian provinces, the Hyrcanian Province is characterized by the complete absence of *Rhododendron* and species of *Abies, Picea*, and *Pinus*, although *Albizia julibrissin* is found here. To the west of Hyrcania it is found only in northeastern Anatolia, where it was probably naturalized very early.

The vegetation of the Hyrcanian Province is basically forest. In the lowlands the coastal forests are characterized by *Alnus subcordata, Pterocarya pterocarpa, Populus caspica,* and *Gleditsia caspia.* In the lower and middle mountain belts the forest-forming species are *Quercus castaneifolia, Parrotia persica, Carpinus betulus, Zelkova carpinifolia,* and several other species. Around 1,200 m above sea level (and in some places around 600–900 m) there are beech forests of *Fagus orientalis,* which usually reach heights of 1,800–2,000 m above sea level, and in places even up to 2,400 m. Oak forests of *Quercus macranthera* (a species also distributed in the Caucasus and in northern Anatolia) reach heights of 2,500 m above sea level in spots and extend eastward to 56° east longitude. Above these one finds a belt of subalpine vegetation.

5. **Turanian or Aralo-Caspian Province** (Engler 1899, 1903, 1924; Graebner 1910; Fedchenko 1925; Grossheim and Sosnovsky 1928; Grossheim 1936, 1948; Alekhin 1944; Takhtajan 1970, 1974).[82] This province includes the deserts and semideserts of eastern Transcaucasia; the Precaspian Lowlands from the lower reaches of the Terek through the lower reaches of the Volga up to the Ural River; and wide expanses extending from the Ural River and the eastern shores of the Caspian Sea to the basin of the Lake Ala Kul, including the Mugodzhary Hills, the Ust Urt Plateau, the Turan Lowland (centered on the Aral Sea), the Kara Kum and Kyzyl Kum sand deserts, the Muyunkum Sands, Bet-Pak-Dala, the Sary-Ishikotrau Sands, and the foothills of the piedmont plains along the Kopet Dagh, and extending along the Pamiro-Alai and along the western Tien Shan, the intermontane valleys of Mirzachul, the Karshinsky Steppe, and other plains in southern Uzbekistan.

The flora of the Turanian Province displays much in common with the flora of the Armeno-Iranian Province, where the probable center of origin of many of its elements is located. It also has a great deal in common with the central Asian flora, especially in the Chenopodiaceae, which is perhaps the most characteristic family of the Turanian flora. It contains a series of endemic or almost endemic genera (e.g., *Alexandra, Rhaphidophyton, Piptoptera,* and *Smirnowia*), and a multitude of endemic species. The vegetative cover of the province is primarily desert, semidesert, and steppe formations.

Several subprovinces could be recognized, and some authors have even described them as independent provinces (see especially Prozorovsky and Maleev 1947; Lavrenko 1962). However, it seems best that floristic classification within the Turanian Province await new phytochorionomic investigations.

6. **Turkestanian Province** (Provinz von Turkestan, Engler 1882; Takhtajan 1970, 1974; Provinz des Turkestanischen Gebirglandes, Engler 1899,

1903, 1924; southern Turkestanian Province, Prozorovsky [in Prozorovsky and Maleev] 1947; southern Turkestanian Montane Province, Korovin 1958, 1961, 1962; Montane Central Asian Province, Kamelin 1973).[83] This province includes the greater part of the mountains and uplands of Central Asia[84] up to the central Tien Shan and the Pamirs to the east, a greater part of Badghys (both the Soviet and Afghanian parts), the Karabil Uplands, Afghanian Turkistan, Badakhshan (both the Soviet and Afghanian parts), the Paropamisus Range, the Safed Koh Range, western Hindu Kush, the Koh-l-Baba Range, and to the east up to Nuristan and the Kabul River Valley (which are, however, assigned to the Western Himalayan Province).

The Turkestanian Province is in many ways a transitional province, closely allied in the west with the Armeno-Iranian flora, and in the east with the Central Asiatic flora on one side and with the northwestern Himalayan flora on the other. However, the province has not only an exceptionally large number of endemic species but no less than 50 endemic genera as well. The flora of this province is the subject of a large literature (see Korovin 1962; Kamelin 1973).

On foothills and low mountains, considerable areas are occupied by various types of desert vegetation, especially by ephemerals. On the mountain slopes, formations of deciduous shrubs are characteristic. In the majority of cases these are secondary communities that have come into existence as a result of deforestation. On the mountains, considerable areas are occupied by peculiar grasslands of the steppe type, which are characterized by the abundance of large apiaceous plants (including many species of *Ferula*), ephemerals and geophytes, as well as by woodlands or, more rarely, by juniper forests.

Broad-leaved forests occur only on separate massives. Persian walnut, *Juglans regia*, forms forests in the western part of Tien Shan and on the Gissar Range. Along with the pure stands of walnut, there are maple-walnut and apple-walnut forests, which are characteristic for more sunny slopes. The parklike maple forests of *Acer semenovii* and *A. turkestanicum* are one of the most characteristic forest types.

Only comparatively small areas are occupied by spruce forests of *Picea schrenkiana* (*P. morinda* subsp. *tianschanica*) and fir forests of *Abies semenovii* (*A. sibirica* subsp. *semenovii*). The latter usually does not form pure stands and grows together with spruce, apple, and walnut.

The high mountain vegetation consists of meadows (mainly on the Tien Shan mountains), steppes, and formations of cushionlike subshrubs.

7. **Northern Baluchistanian Province** (Belutschistanische Provinz, Meusel et al. 1965, p.p.; Baluchistanian Province, Kamelin 1973). This province occupies part of the eastern outskirts of the Iranian Uplands, bordered from the east by the arc of the Sulaiman Range, which includes also

the Takht-i-Sulaiman Massif ("the throne of Solomon"). This province also includes the Quetta-Pishin Plateau and Toba Kakar Hills. However, its boundaries and chorionomic status are not completely clear. Many characteristic xerophytic elements of the Irano-Turanian flora are found here, at the southeastern limits of their distribution. There are many Armeno-Iranian and Turkestanian elements, and on the eastern slopes of the Sulaiman Mountains there are enclaves of Western Himalayan forest flora. But here the forest is parklike and gradually disappears to the southwest. The Western Himalayan element permeates also to the Quetta Basin where, for example, one finds *Pinus wallichiana* and *P. gerardiana*.

8. **Western Himalayan Province** (Hooker 1904, 1907; Turrill 1953; Meusel et al. 1965; Takhtajan 1970, 1974; West Himalaya, Clarke 1898; Western Himalayan Region, Chatterjee 1940, 1962; Maheshwari et al. 1965; Bezirk der Gebirgswälder des Westlichen Himalaya, Hayek 1926). This province includes Nuristan (Kafiristan); valleys of the rivers Kabul, Kurram, Kunar, Swat, and Gilgit; Waziristan; and the southern slopes and offspurs of the western Himalayas, westward approximately from 83° east longitude at an average of over 1,000 m above sea level (Kashmir, Simla, Mussoorie, and Naini Tal belong to this province). In contrast to the other provinces of the Irano-Turanian Region, a considerable part of the Western Himalayan Province is characterized by a monsoon climate, which imparts a definite character to its plant life.

Among the endemics, the most interesting is the monotypic genus *Parrotiopsis*. This province has much in common not only with the flora of the Iranian Highlands but also its western part with the Mediterranean flora, a number of elements of which (e.g., *Myrtus communis*) are found here at the eastern limits of their ranges. Also very characteristic is the endemic *Cedrus deodara*, distributed in Nuristan, on Mt. Gardez (2,700 m above sea level), in Swat, Kurram, southern Waziristan, Kashmir, and in the Himalayas to Kumaun (Garhwal) and western Nepal. Other characteristic conifers are *Abies spectabilis*, *A. pindrow*, the endemic *Picea smithiana* (Nuristan, Swat, Kurram, southern Waziristan and the Himalayas up to central Nepal), *Pinus wallichiana*, the almost endemic *P. gerardiana* (Nuristan, Mt. Gardez, Kurram, Quetta, northern and southern Waziristan, northwestern Himalayas), and *Cupressus torulosa* (Kashmir to central Nepal). In the lower belts of subtropical forests, the evergreen oaks are characteristic (*Quercus incana*, an endemic *Q. dilatata*, *Q. semecarpifolia*, and *Q. baloot*). *Juglans regia* subsp. *kamaonia* (*J. kamaonia* Dode) is also characteristic. In contrast to the eastern Himalayas, the number of *Rhododendron* species is not great here. The most characteristic are *R. arboreum* and, in the high mountains, *R. campanulatum*, *R. anthopogon*, and *R. barbatum*. The endemic species here are *R. afghanicum* and *R. colletianum*. Representatives of the Magnoliaceae are completely absent, but the Lauraceae are represented (species of *Litsea*, *Neolitsea*, and *Machilus*) as is *Schisandra*. Species endemism is rather high, and besides those just mentioned, there is a series of interesting endemic and subendemic woody plants, such as *Buxus wallichiana*, *Sarcococca saligna*, *Alnus nitida*, *Corylus jacquemontii*, *Sorbus cashmiriana*, *S. lanata*, *Staphylea emodi*, *Aesculus indica*, *Abelia triflora*, *Lonicera purpurascens*, *Fraxinus micrantha*, *Olea ferruginea*, and *Syringa emodi*. There are numerous endemic herbaceous plants. The Western Himalayan Province has many species in common with (or that are close relatives to) those found in other parts of the Irano-Turanian region (e.g., *Biebersteinia odora*, *Bupleurum lanceolatum*, *Ferula* spp., *Heracleum canescens*, *Prangos pabularia*, species of *Artemisia*, *Scorzonera*, *Tanacetum*, and many others), but also has species in common with the eastern Asiatic and especially with the eastern Himalayan flora. It may be said that the flora of this province occupies a kind of transitional position and is a link between the ancient Mediterranean (Tethyan) and eastern Asiatic floras. Yet it becomes evident to anyone who visits the western Himalayas that the Irano-Turanian element predominates here.

8B. Central Asiatic Subregion

Grubov 1959, p.p., 1963, p.p.; Lavrenko 1962; Takhtajan 1974; Zentralasiatisches Gebiet, Engler 1882, p.p., 1903, p.p., 1924, p.p.; Diels 1908, p.p.; Roi 1941; Turrill 1959, p.p.; Schmithüsen 1961, p.p.; Tolmatchev 1974; Turanian Subregion, Wickens 1976, p.p.

As treated here, the boundaries of this subregion are close to those defined by Lavrenko (1962). It includes a vast territory of desert and steppes, stretching from the central Tien Shans and the Pamirs to the Greater Khingan Mountains. According to Grubov (1963:13), the Central Asiatic Subregion "occupies basically the territory of the inner drainage of the Asian continent and itself represents an independent kingdom of unique cold deserts, desert-steppes and high mountain steppes." Its flora is comparatively depauperate (scarcely more than 5,000 species; Grubov 1963), but endemism is high, with a particularly large number of endemic genera, including some that are systematically rather isolated.

Owing to the aridity and very great extremes of temperature, the plant world of the Central Asiatic Subregion is limited and comparatively uniform. Mountain steppes predominate (changing, in places, into forest steppes), as well as high mountain meadows, but various desert and semidesert formations are especially characteristic. One also encounters islands of spruce, spruce-fir, larch, and impoverished broad-leaved forests.

There are several provinces within this subregion (see especially Grubov 1959, 1963; Korovin 1962).

1. **Central Tien Shan Province** (Provinz des Tien-Shan, Engler 1882, p.p.; Central Tien Shan Province, Prozorovsky [in Prozorovsky and Maleev] 1947; Korovin 1962; Takhtajan 1970, 1974; Central Tien Shan–Zaalai Province, Kamelin 1973). The exact boundaries of this province are still not completely clear, and Korovin and Kamelin have sketched them somewhat differently. Nevertheless, the following clearly belong here: the central ranges of the Tien Shan; probably part of the Alai Range and the Alai Valley; and, in the north, undoubtedly the Issyk-Kul Basin and its adjacent slopes. In the south the boundary goes along the crest of the Transalai (Zaalai) Range; in the west it follows along the Fergana Range and further along the Susamyr Tau to the Talassian Ala-Tau; in the north it travels along the Kirgizian Range and the Kungei Ala-Tau.

Despite the comparatively large size of its territory, the flora of this province contains probably no more than 1,800 species. But it does contain one endemic genus, *Tianschaniella*, and comparatively many endemic species, including *Ammopiptanthus nanus*, *Zygophyllum kaschgaricum*, and *Salsola* spp.

The dominant vegetation types are steppe communities, *Kobresia* meadows (most frequently of *K. capilliformis*), and sedge swamps of various species of *Carex*). Several authors call the central Tien Shan "a country of grasses and sedges," which is fully justified. Arboreal and shrubby vegetation is found here only in fragments, and occupies a small area. In remote areas of the central Tien Shan—especially in the Terskei-Ala-Tau and the Kungei-Ala-Tau—there are separate, small islands of spruce forests. In deep gorges one also encounters stands of *Populus* spp., *Betula tianschanica*, species of *Lonicera*, and occasionally *Prunus armeniaca* (*Armeniaca vulgaris*), *P. padus*, and others (Korovin 1962).

2. **Dzungaro–Tien Shan Province** (Prozorovsky [in Prozorovsky and Maleev] 1947; Korovin 1962; Takhtajan 1970, 1974; Dzungaro-Turanian Province, Grubov 1959, 1963, p.p.). This province includes the Kirgizian Range and the Talassian Ala-Tau (north slopes), Transilian (Zailiisky) Ala-Tau, the southern slopes of the Tarbagatai Range, the Dzungarian Ala-Tau, and the eastern Tien Shan.

The flora of this province is comparatively young and contains a large number of boreal elements. It includes endemic genera such as *Pastinacopsis*, *Sclerotiaria*, and *Seselopsis*, and about 50 endemic species. Vegetation cover consists of sagebrush-grass semideserts, feather-grass steppes, broad-leaved forests, spruce fir taiga, subalpine meadows, and "carpets" of *Kobresia*.

3. **Mongolian Province** (Popov 1940, 1950; Grubov 1959, 1963; Grubov [in Grubov and Fedorov] 1964; Meusel et al. 1965; Takhtajan 1970,

1974). In a series of works, Grubov plotted rather precisely the borders of this unique floristic province, which includes: the greater part of the mountain ranges of the Mongolian Altai; the Basin of the Great Lakes; the Valley of Lakes, Middle Khalkha, and eastern Mongolia; the Gobi Altai; the Gobi Desert, Ala Shan Desert, Ordos Plateau, Tsaidam Basin, Takla Makan Desert, Tarim and Lob Nor Basin; and northwestern Kansu, north of Lanchow.

The flora of this province is very distinctive and, as Grubov frequently stressed, fairly old. Here one finds an entire series of taxonomically and geographically isolated taxa, including *Ephedra przewalskii, Gymnocarpos przewalskii* (a second species found from the Canary Islands to Baluchistan), *Potaninia mongolica,* the genus *Ammopiptanthus,* the rather isolated, monotypic genus *Tetraena, Zygophyllum xanthoxylon,* and *Nitraria sphaerocarpa.* The leading families of the Mongolian flora are Chenopodiaceae, Asteraceae, Poaceae, Fabaceae, and Brassicaceae. The dominant formations are various desert and semidesert communities and steppes. Their most characteristic elements are *Anabasis brevifolia, Stipa gobica* and *S. glareosa, Salsola passerina* and *S. laricifolia, Artemisia* spp., *Sympegma regelii, Brachanthemum gobicum, Ajania* spp., *Haloxylon ammodendron, Kalidium gracile, Reaumuria songarica, R. kaschgarica,* and *R. trigyna, Allium mongolicum* and *A. polyrhizum, Ephedra przewalskii, Zygophyllum xanthoxylon* and *Z. kaschgaricum,* and *Nitraria sphaerocarpa* (Grubov 1963, personal communication 1983).

4. Tibetan Province (Provinz der Tibetanischen Hochwüste, Engler 1899, 1903, 1924; Tibetan Province, Grubov 1959, 1963; Grubov [in Grubov and Fedorov] 1964; Pamiro-Tibetan Province, Korovin 1962; Lavrenko 1962). According to Grubov (1959, 1963), this province includes the Pamirs and Tibet (Chan Tang, Veitzan, and southern Tibet). The western boundary has a rather complicated configuration, passing approximately through the line of 73° east longitude. The Transalai (Zaalai) Range forms the northern boundary, and the eastern Hindu Kush mountain region forms the southern. Further east, the southern boundary passes along the Karakorum Range and then along the Himalayas. In the eastern part, the boundary extends in an arc to the northeast, bending to the south around the Bayan Khara Shan Mountains. Thus one properly includes in the Tibetan Province the Pamirs (eastern Pamirs), eastern slopes of the Karakoram, and a greater part of the Tibetan Plateau. Within the boundaries of the USSR, this province is just barely represented, by the eastern Pamirs.

The flora of the Tibetan Province is not rich, and consists of probably no more than 1,000 species. This is the youngest flora in the entire Central Asiatic Subregion, and its history really begins only after the Quaternary glaciation.

The vegetation of most of the Tibetan Plateau consists mainly of species of Irano-Turanian—and particularly of Central Asiatic—origin. To them belong such species as *Eurotia ceratoides, E. compacta, Kochia prostrata, Micro-*

ula tibetica, Myricaria prostrata, Reaumuria kashgarica, Rheum spiciforme, Thylacospermum caespitosum, Chamaerhodos sabulosa, Oxytropis aciphylla, Thermopsis alpina, species of *Astragalus, Tanacetum tibeticum, T. gracilis, Artemisia wellbyi, Ptilagrostis concinna, Stipa purpurea, Agropyron thoroldianum,* species of *Helictotrichon* and *Festuca, Elymus junceus, E. lanuginosus, Cleistogenes thoroldii, Ephedra distachya,* and *E. gerardiana.* The vegetation of the central part of the Tibetan Plateau (Chang Tang) is extremely scattered and consists predominantly of halophytes. The main life forms are cushionlike perennials (*Artemisia wellbyi, Astragalus malcolmii, A. arnoldii, Acantholimon diapensioides, Thylacospermum caespitosum, Saussurea tridactyla, S. wellbyi, Thermopsis alpina*), perennial grasses (*Stipa purpurea, Ptilagrostis concinna*), and subshrubs (*Eurotia ceratoides, E. compacta, Tanacetum tibeticum, Ephedra distachya, E. gerardiana, Myricaria prostrata, Reaumuria kaschgarica*). Only in the lowland places and along the rivers and lakes can one see closed communities—namely, formations of *Kobresia* (mainly *K. myosuroides*) and *Carex* (mainly *C. sabulosa*). The only arborescent plant of Chang Tang is *Juniperus squamata*. In the eastern part of the province, the dominant formations are high mountain meadow steppes of *Kobresia* (most frequently *K. stenocarpa*) and *Carex* (especially *C. sabulosa*). *Polygonum sphaerostachyum* and *P. viviparum* are very frequent also. In addition, there are many cushionlike plants (*Thylacospermum caespitosum, Arenaria musciformis, Stellaria decumbens, Androsace tapete*) and low hemicryptophytes and geophytes.

Southern Tibet, which is situated between the Transhimalayas and Himalayas, is the warmest part of Tibet. The dominant vegetation is an open community ("gravel desert") of dwarf perennials—mainly of *Oxytropis aciphylla, Astragalus tibeticus, Thermopsis alpina, Artemisia salsoloides, Stellera chamaejasme*. In the eastern part of southern Tibet, the following xerophilous shrubs occur: *Sophora viciifolia, Berberis* spp., *Lonicera spinosa, Ceratostigma griffithii, Buddleja tibetica, B. lindleyana, Elaeagnus pungens, Hippophaë rhamnoides,* and also *Rosa sericea, Cotoneaster acutifolia, Dasyphora fruticosa* s.l., and *Crataegus* spp. On the upper parts of the mountain slopes stands of *Juniperus pseudosabina* occur. In the southernmost part of Tibet, along the northern slopes of the Himalayas, solonchaks are found (*Suaeda* spp., *Salsola* spp., *Atriplex* spp., *Iris oxypetala*, etc.; (see Ward 1936, 1942; Grubov 1963).

C. MADREAN SUBKINGDOM

The flora of southwestern North America and of northern and central Mexico differs so strongly from the boreal and Tethyan floras that it is well worthy of the chorionomic rank of a separate subkingdom. The name "Madrean" is derived from the Sierra Madre Occidental. Despite the presence of some common genera (such as *Arbutus, Cercis, Cupressus, Pistacia, Salvia, Viburnum*, etc.) and the striking physiognomic and ecological resemblances between some Mediterranean and Madrean plant communities (especially between Mediterranean "maquis" and Californian "chaparral"), the Madrean flora arose and evolved independently from that of the Tethyan Subkingdom. There is only one region in this subkingdom—the Madrean.

9. Madrean Region

Madrean Region, Takhtajan 1969, 1970, 1974.

This region includes most of the southwestern United States (north as far as southeastern Oregon and the Snake River Plains of Idaho) and northern and central Mexico (south to and including the Sierra Madre del Sur, but excluding the Balsas Depression).

The Madrean region has a very rich flora, including 3 small endemic families (Simmondsiaceae, Fouquieriaceae, and Pterostemonaceae) and well over 250 endemic genera (nearly 100 in the Asteraceae alone). A fourth family, the Crossosomataceae, is nearly endemic to the region. Seven of its 10 species are endemic, and only one is wholly extralimital.

Although no large family is nearly endemic to the Madrean region, the Onagraceae, Polemoniaceae, and Hydrophyllaceae have their major center of diversity here. The more widespread families Cactaceae, Brassicaceae, Fabaceae (sens. lat.), Apiaceae, Boraginaceae, Scrophulariaceae, and Asteraceae are also especially well represented.

Species endemism in the Madrean region is very high, probably well over 50%. The following are among the endemic and nearly endemic genera of vascular plants:

Lauraceae: *Umbellularia* (1, southwestern Oregon, California)
Saururaceae: *Anemopsis* (1)
Papaveraceae: *Arctomecon* (3, southwestern U.S.), *Canbya* (2; 1 Great Basin, 1 Mohavean district of the Sonoran Subprovince), *Dendromecon* (2, California and northwestern Baja California), *Eschscholzia* (10; 1 species north to the Columbia River), *Platystemon* (1, southern Oregon to northern Baja California, east to Utah and Arizona), *Romneya* (2, southern California, northern Baja California), *Stylomecon* (1, central California, northern Baja California)
Achatocarpaceae: *Phaulothamnus* (1, northern Mexico)
Nyctaginaceae: *Acleisanthes* (about 10, southern California to southern Arizona, western and southern Texas, Mexican Highlands), *Anulocaulis* (5, southwestern U.S., Mexico), *Cyphomeris* (2, southwestern U.S., Mex-

ican Highlands), *Hermidium* (1, California to Nevada and Utah), *Selinocarpus* (about 8, southwestern U.S., Mexico)
Cactaceae: *Ancistrocactus* (4, Chihuahuan and Tamaulipan), *Ariocarpus* (6, southern Texas, Mexican Highlands), *Bartschella* (1, Baja Californian); *Bergerocactus* (1, Californian and Baja Californian, close to *Cereus*); *Carnegiea* (1, southwestern California, southern Arizona, Sonora; close to *Cereus*), *Cochemiea* (5, Baja California and adjacent islands), *Echinocactus* (about 12, Sonoran and Mexican Highlands), *Echinofossulocactus* (about 30, mainly Mexican Highlands), *Epithelantha* (2, Chihuahuan), *Ferocactus* (about 25, Sonoran and Mexican Highlands), *Lophophora* (2, Chihuahuan and Tamaulipan subprovinces, south to the Mexican Altiplano), *Mamillopsis* (2, Sierra Madre Occidental), *Neolloydia* (about 15, from California and Arizona to Utah, Texas, and the Mexican Highlands); *Pelecyphora* (2, Chihuahuan and Sierra Madre Oriental), *Sclerocactus* (8, Great Basin and Sonoran), *Thelocactus* (about 20, southern Texas and the Mexican Highlands)
Portulacaceae: *Talinopsis* (1, Sonoran)
Caryophyllaceae: *Achyronichia* (2, southeastern California, Baja California, Arizona, Sonora), *Scopulophila* (1, Mohavean district of the Sonoran Subprovince)
Amaranthaceae: *Acanthochiton* (1, Sonoran, close to *Amaranthus*); *Dicraurus* (1, Chihuahuan)
Chenopodiaceae: *Aphanisma* (1, California and Baja California), *Meiomeria* (1, Chihuahuan), *Zuckia* (1, Great Basin)
Polygonaceae: *Dedeckera* (1, Mohavean district of the Sonoran Subprovince), *Gilmania* (1, Mohavean district of the Sonoran Subprovince), *Goodmania* (1, California), *Harfordia* (1, Baja California), *Hollisteria* (1, California), *Nemacaulis* (1, California and Sonora); *Pterostegia* (1, California and Sonora)
Simmondsiaceae: *Simmondsia* (1, southern California, Baja California, Arizona, Sonora)
Ericaceae: *Ornithostaphylos* (1, Baja California and adjacent California; very close to *Arctostaphylos*), *Xylococcus* (1, Baja California and adjacent California)
Fouquieriaceae: *Fouquieria,* including *Idria* (11, southern California, Baja California and several adjacent islands, Arizona, western Texas, Sonora, south into Sinaloa)
Cucurbitaceae: *Brandegea* (4, Sonora and California), *Sicyosperma* (1, southern Arizona and northern Sonora), *Tumamoca* (1, Arizona south into central Sonora), *Vaseyanthus* (2, Sonora)
Capparaceae: *Isomeris* (1, Californian to Baja Californian, Cedros Island; close to *Cleome*), *Koeberlinia* (1, Sonora; a taxonomically isolated genus),

Oxystylis (1, Mohavean district of the Sonoran Subprovince), *Wislizenia* (1 or more, Californian and Sonoran)

Brassicaceae: *Dithyrea* (5, southwestern U.S., south into Baja California and southern Sonora), *Dryopetalon* (4, Sonoran), *Glaucococarpum* (1, Great Basin), *Heterodraba* (1, southwestern Oregon to northwestern Baja California), *Lyrocarpa* (2, Sonoran), *Nerisyrenia* (5, Sonoran and Mexican Highlands), *Synthlipsis* (3, Sonoran and Mexican Highlands), *Tropidocarpus* (2, California)

Sterculiaceae: *Fremontodendron* (2, southern California, northern Baja California, western Arizona), *Nephropetalum* (1, Tamaulipan)

Euphorbiaceae: *Tetracoccus* (4, California, northern Baja California, western Nevada)

Crassulaceae: *Dudleya* (about 40, Californian and Sonoran), *Graptopetalum* (about 10, Sonoran), *Lenophyllum* (1, Tamaulipan), *Parvisedum* (4, Californian; close to *Sedum*)

Saxifragaceae: *Jepsonia* (3, California, northern Baja California and adjacent islands)

Rosaceae: *Adenostoma* (2, from northern California to Baja Californian), *Chamaebatia* (2, California and Baja California), *Chamaebatiaria* (1, mainly Great Basin), *Coleogyne* (1, Great Basin and Sonoran), *Cowania* (about 5, Great Basin and Sonoran), *Fallugia* (1, southern California to Nevada, Utah, Texas, Mexico; close to *Cowania*), *Heteromeles* (1, from northern California to adjacent northern Baja California and Santa Catalina and San Clemente islands; close to *Photinia*), *Lindleya* (1), *Lyonothamnus* (1, islands of southern California: Santa Catalina, San Clemente, Santa Rosa, and Santa Cruz), *Purpusia* (1, California to Nevada and Arizona); *Vauquelinia* (3, Sonoran)

Crossosomataceae: *Apacheria* (1, Sonoran, in the Chiricahua Mountains of Arizona), *Crossosoma* (2, Californian and Sonoran), *Forsellesia* (7, only 5 endemic)

Onagraceae: *Burragea* (1, Baja California; close to *Gongylocarpus*), *Heterogaura* (1, southwestern Oregon to southern California), *Semeiandra* (1, Sierra Madre Occidental), *Xylonagra* (1, Baja California)

Fabaceae sens. lat.: *Genistidium* (1, Chihuahuan), *Hesperothamnus* (6, Mexican Highlands), *Olneya* (1, southeastern California, Baja California, western and southern Arizona, Sonora), *Parryella* (1, Great Basin and Sonoran), *Peteria* (4), *Psorothamnus* (9), *Pickeringia* (1, southern California, Santa Cruz Island, northwestern Baja California), *Sphinotospermum* (1, Sonoran)

Rutaceae: *Choisya* (6, Sonoran and Mexican Highlands), *Cneoridium* (1, southern California, San Clemente Island, northern Baja California), *Sargentia* (1, Sonoran)

Simaroubaceae: *Holacantha* (2, Sonoran; close to *Castela*)
Zygophyllaceae: *Morkilla* (2, Mexican Highlands) *Sericodes* (1, Chihuahuan), *Viscainoa* (1, Baja California, many of the islands of the Gulf of California, and Sonoran coast of the Gulf of California)
Anacardiaceae: *Pachycormus* (1, Baja California)
Malpighiaceae: *Echinopterys* (3, mainly Mexican Highlands)
Celastraceae: *Acanthothamnus* (1, Mexican Highlands), *Canotia* (1, Sonoran and Great Basin; a very isolated genus within the family), *Mortonia* (8, Sonoran and Mexican Highlands), *Orthosphenia* (1, Sonoran), *Rzedowskia* (1, Mexican Highlands)
Rhamnaceae: *Adolphia* (2, mainly Sonoran)
Hydrangeaceae: *Carpenteria* (1, California), *Fendlera* (4, mainly Sonoran), *Fendlerella* (3)
Pterostemonaceae: *Pterostemon* (2, Mexican Highlands)
Apiaceae: *Apiastrum* (1, California and Baja California), *Coulterophytum* (5, Mexican Highlands), *Neogoesia* (3, Mexican Highlands)
Apocynaceae: *Cycladenia* (1, California), *Streptotrachelus* (1, Mexican Highlands)
Asclepiadaceae: *Himantostemma* (1, Sonoran Subprovince), *Mellichampia* (2, Mexican Highlands), *Microdactylon* (1, Mexican Highlands), *Rothrockia* (2, Baja Californian and Sonoran subprovinces)
Oleaceae: *Hesperelaea* (1, Guadelupe Island, off the coast of Baja California)
Loasaceae: *Cevallia* (1, Sonoran and Mexican Highlands), *Petalonyx* (5, mainly Sonoran)
Convolvulaceae: *Petrogenia* (1, Chihuahuan)
Hydrophyllaceae: *Emmenanthe* (1, mainly Sonoran), *Eriodictyon* (8, mainly Sonoran), *Eucrypta* (2); *Lemmonia* (1, Californian and Sonoran), *Pholistoma* (3, Californian and Sonoran), *Tricardia* (1, Great Basin and Sonoran), *Turricula* (1, southern Californian and northern Baja California)
Lennoaceae: *Ammobroma* (1, southeastern Californian south to central Baja California, Santa Catalina Island, Arizona)
Solanaceae: *Margaranthus* (1, Arizona, New Mexico, southern and western Texas, south into Mexico), *Oryctes* (1, mainly Great Basin)
Buddleiaceae: *Emorya* (1, mainly Tamaulipan and Chihuahuan)
Scrophulariaceae: *Clevelandia* (1, Baja California), *Gentrya* (1, Sierra Madre Occidental), *Leucophyllum* (12, Sonoran and Mexican Highlands), *Mohavea* (2, southern California and Baja California, east to Nevada and Arizona), *Ophiocephalus* (1, northern Baja California)
Bignoniaceae: *Chilopsis* (1, Sonoran)
Acanthaceae: *Berginia* (3, Sonoran and Mexican Highlands), *Carlowrightia* (20, mainly Sonoran and Mexican Highlands, but south to Costa Rica), *Holographis* (4, Sonoran and Mexican Highlands), *Mexacanthus* (1, Mexican Highlands; close to *Anisocanthus*)

Verbenaceae: *Burroughsia* (2, California, south into Baja California)
Lamiaceae: *Acanthomintha* (3, southern California south into adjacent Baja California), *Pogogyne* (5, southwestern Oregon to northwestern Baja California), *Poliomintha* (about 4, southern California, Baja California, Arizona to Utah, Texas, northeastern Mexico), *Salazaria* (1, southern California to Utah, western Texas, and northeastern Mexico)
Campanulaceae s.l.: *Nemacladus* (about 10), *Parishella* (1, southern California), *Pseudonemacladus* (1, Mexican Highlands)
Asteraceae: *Acamptopappus* (2, mainly Sonoran Subprovince), *Achaenipodium* (1, Trans-Mexican Volcanic Belt), *Achyraechaenia* (1, California, north to Oregon and south to northern Baja California and San Clemente, Santa Rosa, and Santa Cruz islands), *Adenopappus* (1, Mexican Highlands), *Adenothamnus* (1, Baja Californian), *Agiabampoa* (1, Sierra Madre Occidental), *Alvordia* (3, Baja Californian), *Amauria* (2, Baja Californian), *Amphipappus* (1, Mohavean District of the Sonoran Subprovince), *Anisocoma* (1, mainly in the Mohavean district of the Sonoran Subprovince), *Arnicastrum* (1, Sierra Madre Occidental; an interesting genus that may link *Arnica* to the Heliantheae), *Atrichoseris* (1, mainly in the Mohavean district of the Sonoran Subprovince), *Baeriopsis* (1, Guadelupe Island, off the coast of Baja California), *Baileya* (3, mainly Sonoran and southern Great Basinj), *Barroetia* (about 5, Mexican Highlands), *Bartelettia* (1, Chihuahuan), *Bebbia* (2, Sonoran), *Benitoa* (1, Californian), *Blepharizonia* (1, Californian), *Bolanosa* (1, Mexican Highlands), *Calycadenia* (11, mainly Californian), *Calycoseris* (2, Sonoran), *Carphochaete* (5, Sonoran and Mexican Highlands), *Carterothamnus* (1, Baja Californian), *Chaetadelpha* (1, Great Basin; close to *Lygodesmia*); *Chaetopappa* (about 15), *Chamaechaenactis* (1, Great Basin), *Chromolepis* (1, Mexican Highlands), *Chrysactinia* (4, Sonoran and Mexican Highlands), *Clappia* (1, Tamaulipan), *Coreocarpus* (11, mainly Sonoran; 1 Sierra Madre del Sur), *Corethrogyne* (3, mainly Californian), *Coulterella* (1, Baja Californian), *Cymophora* (3, Mexican Highlands), *Dichaetophora* (1, Sonoran), *Dicoria* (4, Sonoran), *Dicranocarpus* (1, Chihuahuan), *Dimeresia* (1, Great Basin), *Dyscritothamnus* (2, Mexican Highlands), *Eastwoodia* (1, Californian), *Eatonella* (2, mainly Californian and Great Basin), *Enceliopsis* (4, mainly Great Basin and adjacent Sonoran), *Eryngiophyllum* (1, Sierra Madre Occidental), *Eutetras* (2, Mexican Highlands), *Faxonia* (1, Baja Californian), *Geraea* (2, mainly Sonoran), *Glyptopleura* (2, Great Basin, and Mohavean district of the Sonoran Subprovince), *Greenmaniella* (1, Sierra Madre Oriental), *Guardiola* (10, Sonoran and Mexican Highlands), *Haplocalymma* (2, Mexican Highlands), *Haploesthes* (2, mainly Sonoran), *Hecastocleis* (1, Mohavean district of the Sonoran Subprovince, plus adjacent Great Basin), *Hemizonia* (31, mainly Californian), *Hofmeisteria* (about 10, Sonoran and Mexican High-

lands), *Holocarpha* (4, Californian), *Holozonia* (1, Californian), *Hymenoclea* (2, Californian and Sonoran), *Hymenothrix* (4, Californian and Sonoran), *Iostephane* (3, mainly Mexican Highlands), *Jaliscoa* (2, Mexican Highlands), *Lagophylla* (5, mainly Californian, but one species more widespread), *Lepidospartum* (3, Californian, Chihuahuan, and Mohavean district of the Sonoran Subprovince), *Lessingia* (about 10, Californian and Sonoran), *Malperia* (1, southern Californian, northern Baja Californian), *Marshalljohnstonia* (1, Sierra Madre Occidental), *Monolopia* (4, Californian), *Monoptilon* (2, Californian and Sonoran), *Munzothamnus* (1, San Clemente Island, California), *Nicolletia* (5, mainly Sonoran), *Olivaea* (1, Mexican Highlands), *Oxypappus* (2, Mexican Highlands), *Parthenice* (1, Sonoran), *Pelucha* (1, Chihuahuan and Baja Californian), *Pericome* (2, mainly Sonoran), *Peucephyllum* (1, Sonoran), *Pionocarpus* (1, Sierra Madre Occidental), *Plummera* (2, Sonoran), *Psathyrotes* (4, mainly Sonoran), *Pseudobahia* (3, Californian), *Pseudoclappia* (1, Chihuahuan), *Rafinesquia* (2, Californian and Sonoran), *Rhysolepis* (2, Mexican Highlands), *Sartwellia* (3, Sonoran), *Selloa* (1, Trans-Mexican Volcanic Belt), *Stenocarpha* (1, Sierra Madre Occidental), *Stephanodoria* (1, Sierra Madre Oriental), *Strotheria* (1, Chihuahuan; close to *Dyssodia*), *Stylocline* (6), *Syntrichopappus* (2, Sonoran Subprovince), *Trichocoronis* (2, Sonoran), *Trichoptilium* (1, Sonoran), *Urbinella* (1, Mexican Highlands), *Venegasia* (1, California and Baja California), *Viguethia* (1, Sierra Madre Oriental), *Whitneya* (1, California)

Tecophilaeaceae: *Odontostomum* (1, California)

Hyacinthaceae: *Chlorogalum* (5, southern Oregon, California, northern Baja California)

Alliaceae: *Behria* (1, Baja California), *Bloomeria* (2, California, northern Baja California), *Muilla* (5, California, northern Baja California), *Triteleiopsis* (1, Baja Californian and Sonoran subprovinces; close to *Brodiaea*)

Hesperocallidaceae: *Hesperocallis* (1, California and Arizona)

Agavaceae: *Beschorneria* (about 10, mainly Mexican Highlands), *Hesperaloë* (2, Sonoran)

Anthericaceae: *Eremocrinum* (1, Arizona and Utah)

Dracaenaceae: *Dasylirion* (about 15, Sonoran), *Nolina* (about 25, Sonoran and Californian)

Poaceae: *Allolepis* (1, Chihuahuan), *Neostapfia* (1, Californian; close to *Anthochloa*), *Orcuttia* (5, California and northern Baja California), *Reederochloa* (1, Chihuahuan), *Swallenia* (1, Sonoran Subprovince), *Vaseyochloa* (1, Tamaulipan)

Arecaceae: *Washingtonia* (2, southeastern California, northern Baja California, southwestern Arizona, Sonora)

As shown by the foregoing list, the Sonoran Province is linked to all the other provinces of the Madrean Region by endemic genera common to two or more provinces. Moreover, there are a great many species, among both endemic and more widespread genera, that occur in two or more of the provinces but are restricted to the Madrean Region, or are largely Madrean but also encroach into the Rocky Mountain Region. The Great Basin Province does not have nearly so many endemic genera as the other provinces in the Madrean Region.

In contrast to other North American floristic regions, the Madrean Region supports a basically dryland flora. At lower elevations it is mostly desert and semidesert, or (in California) Mediterranean-type sclerophyllous woodland or chaparral, with some grassland as well. At upper elevations, or in otherwise more favorable (moister) sites, it supports a usually rather open forest or woodland of conifers (especially pines) and small to middle-sized broad-leaved trees (especially oaks). Only the highest elevations are above timberline.

The Madrean flora has a long evolutionary history. The tropical or subtropical, more or less humid climate of the Cretaceous period began to deteriorate in the Eocene, and by the end of the Oligocene the lower temperature and especially the increasing aridity made life difficult in most of the Madrean Region for plants adapted to moist, tropical or subtropical conditions. Thus there was a strong selective pressure toward the evolution of species adapted to drier and (at least toward the north) cooler conditions than had prevailed in the Cretaceous and early Tertiary.

The Madrean Region consists of 4 provinces—the Great Basin Province, the Californian Province, the Sonoran Province, and the Mexican Highlands Province. Each of these can again be divided, and we here lay particular stress on the 4 Sonoran subprovinces (Baja Californian, Sonoran, Chihuahuan, and Tamaulipan) and the 5 subprovinces of the Mexican Highlands (Sierra Madre Occidental, Trans-Mexican Volcanic Belt, Mexican Altiplano, Sierra Madre Oriental, and Sierra Madre del Sur).

1. The Great Basin Province (Gleason and Cronquist 1964; Ornduff 1974; Unterprovinz des Great Basin, Engler 1903, 1924; Bezirk des Great Basin, Hayek 1926). The Great Basin Province occupies nearly all of the hydrographic Great Basin of western United States (excluding the slopes of the mountains that mark its western border) and floristically allied territory. The latter consists mainly of two large units—the Snake River Plains in southern Idaho; and the major part of the geologic Colorado Plateau in eastern Utah, northern Arizona, northwestern New Mexico, and westernmost Colorado. The Uinta Mountains in northeastern Utah are excluded and assigned to the Rocky Mountain Floristic Province, but the Uinta Basin, just to the south of the mountains, is an integral part of the Colorado Plateau portion of the Great Basin Province.

Some of the boundaries of the Great Basin Province are fairly sharp, but others are more vague (see discussion under Rocky Mountain and Vancouverian provinces). The eastern base of the Sierra Nevada in California makes a good line, which is followed for some distance by a major highway. The eastern base of the Warner Mountains in northeastern California and adjacent Oregon forms a more arbitrary boundary, especially since there is some sagebrush country to the west of these mountains. Farther north, in Oregon, the western boundary is marked by the coniferous forest that extends for some distance eastward from the Cascade Mountains. The base of the Blue Mountains and associated ranges of central Oregon forms a fairly good northern boundary, as do the mountains on the north side of the Snake River Plains in Idaho. The eastern boundary in Colorado is rather vague, especially inasmuch as the geologic Colorado Plateau extends well into the southern Rocky Mountain portion of the Rocky Mountain Province. A considerable portion of the southern boundary is marked by an abrupt drop in elevation to the south, so that the transition to the adjoining Sonoran Province is accomplished within a few kilometers. In central and eastern Arizona, the top of this declivity is called the Mogollon Rim.

The Great Basin Province is largely cool or cold desert, in contrast to the warm desert of the Sonoran Province to the south. The elevation in much but not all of the area is over 1,300 m. The climate is continental, with warm or hot summers and rather cold winters. Spring tends to be late, the summers are dry, and there is typically a rather long, dry autumn. Most of the precipitation comes in the winter, often as snow, which accumulates especially in the mountains.

Most of the Great Basin proper has an anticline-syncline or tilted fault-block structure, with numerous narrow mountain ranges trending north-south, separated by broad, dry valleys. There are many local interior-drainage basins with a central playa or alkaline lake, in addition to the major areas that drain into Great Salt Lake in Utah, Carson Sinks in western Nevada, and Harney Lake in southeastern Oregon. The Colorado Plateau, in contrast, mostly drains to the Gulf of Mexico through the Colorado River. It is a land of mesas, cliffs, steep-sided canyons, and flat-topped mountains, reflecting massive uplift and erosion of flat-lying rocks. The La Sal, Abajo, and Henry mountains in southeastern Utah are geologic outliers of the Southern Rocky Mountains, arising in the Colorado Plateau. The flora of the LaSals in particular shows a strong Southern Rocky Mountain influence. *Erigeron mancus,* an alpine LaSal endemic, exemplifies both the influence and the distinction: it is obviously related to and derived from *E. pinnatisectus* of the Southern Rocky Mountains. San Francisco Mountain in north central Arizona is another outlier of the Southern Rocky Mountains. It is of particular interest because it helped inspire C. Hart Merrian (1890, 1893, 1898) to develop his concept of North American life zones almost a century ago.

The Great Basin flora evidently took shape during the Miocene, and its boundaries may have been fairly stable since that time. Increasing aridity forced the evolution of a dryland flora, but the cold winters restricted immigration from the Sonoran Province to the south.

Despite the numerous very local species, the percentage of endemism in the flora of the Great Basin Province is not so high as in the other Madrean provinces, probably amounting to no more than 25%. Most of the species that are widespread within the province also extend into the Rocky Mountain Region or into the Sonoran or Californian Province, or even beyond. The situation is exemplified by *Artemisia tridentata*, which not only occurs throughout the province but also permeates the southern part of the Rocky Mountain Province and has a total range that extends into parts of the Vancouverian, Californian, Sonoran, and North American Prairies provinces.

The most characteristic plant of the Great Basin Province is sagebrush (*Artemisia tridentata* and closely related species), a low to middle-sized shrub, seldom as much as 2 m tall. Sagebrush dominates large areas in the valleys, foothills, and plains, giving way to species of *Atriplex* and other members of the Chenopodiaceae in alkaline or very dry soils. It also extends well up into the mountains, sometimes even to above timberline. Species of *Argopyron* (sens. lat.), *Poa*, *Festuca*, *Stipa*, and other grasses commonly grow intermingled with the sagebrush and may even be locally dominant. The amount of grass in these sagebrush-grass communities has been considerably reduced by overgrazing during the twentieth century.

Immediately above the sagebrush zone, there is typically an open juniper or pinyon-juniper woodland, with sagebrush still an important member of the community. The juniper is most commonly *J. osteosperma* (*J. utahensis*, the Utah juniper). The pinyon pine is *Pinus monophylla* (single-leaf pinyon) in the Great Basin proper, and *P. edulis* (two-leaf pinyon) on the Colorado Plateau.

In the more eastern and southern parts of the province, especially on the Colorado Plateau and the mountains arising from it, there are considerable patches of *Quercus gambelii* or *Pinus ponderosa* (in different communities), partly in place of the pinyon-juniper woodland and partly above it. The Gambel oak, a low tree or scarcely more than a large shrub, often forms a dense chaparral. Conversely, ponderosa pine forms open forests. It is more characteristically a tree of the Rocky Mountain Region, and in the Great Basin Province it usually does not get so large as elsewhere, but it forms a fine forest on the Kaibab Plateau (a portion of the Colorado Plateau) north of the Grand Canyon in Arizona. Both of these communities require a more favorable moisture-balance than the pinyon-juniper woodland.

At middle and upper elevations in the Great Basin Province, the plant communities and the flora look more and more like those of the southern part of the Rocky Mountain Province. *Pseudotsuga menziesii* var. *glauca*, *Picea engelmannii* or *Picea pungens*, and *Abies lasiocarpa* form patches of forest in favorable sites, with the *Picea* and *Abies* usually at somewhat higher elevations than the *Pseudotsuga*. *Populus tremuloides* forms large clonal groves just as in the Southern Rocky Mountains. These tend to expand at the margins during moist years and die back during dry ones. Spruce and fir or even Douglas fir may be the timberline trees, but often there is an open timberline forest composed largely of *Pinus flexilis* (limber pine), *P. longaeva* (Intermountain bristlecone pine), or sometimes *P. albicaulis*. *Pinus longaeva* has only recently been distinguished from *P. aristata* of the Southern Rocky Mountains, but it appears to be fully distinct. It is especially noteworthy in attaining the greatest age of any known tree—up to about five thousand years.

Aside from the aspen and Gambel oak, broad-leaved trees do not form a large element in the flora of the Great Basin Province. *Populus fremontii* grows along rivers and dry washes toward the south, sometimes even forming a gallery forest, as along the Colorado River. *Populus angustifolia* grows in some of the canyons in the mountains. *Acer grandidentatum* and *A. glabrum* are small trees or arborescent shrubs in some of the canyons and in favorable sites on the more open mountain slopes, especially northward. *Acer negundo* grows here and there along streams in some of the canyons, especially in Utah.

Above timberline the flora of the Great Basin Province consists largely of species that also occur in some part of the Rocky Mountain Region, or of local species of more or less widespread genera. Some of the genera that contribute to the alpine flora are well distributed in Eurasia as well as in North America. Among these are *Ranunculus, Arenaria, Silene, Arabis, Draba, Saxifraga, Potentilla, Astragalus, Carex,* and *Poa*. Other genera are characteristically western cordilleran, although they may also have a wider range. Among these are *Phlox, Mertensia, Castilleja, Erigeron, Haplopappus,* and *Townsendia*. *Eriogonum* and *Penstemon* center in the Great Basin Province (as noted later), but *Senecio* is cosmopolitan. Relatively few species in the Great Basin Province are circumboreal arctic-alpines. Among these are *Saxifraga bronchialis* sens. lat., *S. caespitosa* sens. lat., and *S. cernua* (all in the LaSal Mountains), *Saxifraga adscendens* (Ruby Mountains), *Silene acaulis, Carex nardina, Poa alpina,* and *P. arctica*.

By far the largest genus in the Great Basin Province is *Astragalus*, with about 175 species. More than 60 of these are endemic. Many are very local and occur in specialized habitats that probably did not exist as long as ten thousand years ago. The Great Basin Province is the center of diversity for the North American branch of this vast genus, but there is an even greater center in the Irano-Turanian Region in Eurasia. Except for a few boreal taxa, the American and Eurasian species of *Astragalus* belong to different phylads of the genus. Most of the American species have chromosome numbers based on $x = 11-13$, whereas the Eurasian ones have $x = 8$. In spite of the apparent antiquity of the geographic and phyletic separation of the two groups, they show a remarkable parallelism in morphologic and ecologic diversification.

The next largest genera in the Great Basin Province are *Eriogonum* and *Penstemon*, each with well over 100 species in the province. Probably more than 50 species of *Eriogonum* are endemic, as are nearly 50 of *Penstemon*. *Eriogonum* is widespread especially in western United States, and *Penstemon* in temperate North America as a whole, but both genera center in the Great Basin Province. Neither is easily traced in some other source. Both must be considered autochthonous in the Great Basin.

The subgenus *Oreocarya* of *Cryptantha* also centers in the Great Basin Province, especially in the Colorado Plateau portion. Some 25 species (out of 40 in the province and about 50 in all) are endemic.

Cymopterus is another genus with a Great Basin center: 28 of its approximately 40 species occur in the Great Basin Province, and 19 are endemic, or nearly so. The related genus *Lomatium* is also well represented, but is even better developed in the portion of the Rocky Mountain Region that lies in the United States.

The shrubby genus *Chrysothamnus*, although not large in number of species, reaches its greatest abundance and diversity in the Great Basin Province. Several species of *Chrysothamnus* are conspicuous members of the sagebrush community.

Several other genera that are widespread in western North America (or beyond) also have significant numbers of endemic species in the Great Basin Province. Among these are *Erigeron* (18 endemic species, out of more than 50 in the province; more than 200 species in the genus as a whole); *Phacelia* (a dozen endemic species out of 50; about 150 in the genus); *Castilleja* (8 endemic species out of 30; about 200 in the genus); and *Gilia* (7 endemic species out of about 30; about 60 in the genus). *Gilia* and *Phacelia* are also especially well represented in California and in the Mohave desert.

It seems that almost every identifiable topographic, petrologic, or edaphic unit in the Great Basin Province has its own set of endemic species. For example, a 30-kilometer segment of the Bear River Range in northern Utah has 4 local species—*Primula maguirei, Musineon lineare, Penstemon compactus,* and *Erigeron cronquistii*. A fifth species, *Draba maguirei*, occurs only in this same area and the nearby Wellsville Mountains. Tuffaceous slopes in Leslie Gulch, a draw about 10 km long in Malheur County, Oregon, support three very local species—*Mentzelia packardiae, Ivesia rhypara* and *Senecio ertterae*. And so it goes.

The hanging gardens along the Colorado River and its major tributaries are a distinctive feature that stand out sharply from the surrounding rocks and desert vegetation. These hanging gardens are associated with perched water tables. Water percolates through the sandstone cliffs until it reaches a relatively impervious layer of clay. Erosion of the sandstone is accelerated in these moist zones, forming ledges and hollows that support a comparatively lush community. Some species, such as *Primula specuicola, Mimulus east-*

woodiae, Cirsium rydbergii, and *Zigadenus vaginatus,* are endemic to the hanging gardens. Others, such as *Ostrya knowltonii,* also occur irregularly in more or less similar habitats through Arizona and New Mexico to western Texas. A few, such as *Adiantum capillus-veneris,* are widespread in moister regions elsewhere, but in the Great Basin Province find a suitable habitat mainly in the hanging gardens.

Given the restricted distribution and often precise adaptation of many of the endemic species of the Great Basin Province to a narrowly defined niche, it is difficult to conceive that all or even most of them have a long history of repeated migration in association with the climatic upheavals of the Pleistocene. Many of them must have originated in situ during the past several thousand years.

The Snake River Plains form the only large segment of the province with but few endemic species. *Phlox aculeata* is confined to the western half of these plains, and *Allium aasease* is even more restricted, being known only from Gem and Ada counties in western Idaho. *Oenothera psammophila* grows on the sand dunes near St. Anthony, toward the upper (eastern) end of the plains. It is closely related to the more widespread *Oe. caespitosa.* *Hackelia cronquistii* grows on the low hills at the west end of the plains in Oregon. A few other taxa from the Snake River Plains that have been described as distinct species are now generally treated as subspecies or varieties of more widespread species. In general, the Snake River Plains carry a somewhat attenuated extension of the flora of the Great Basin (not the Colorado Plateau).

Cercocarpos ledifolius (mountain mahogany) is an interesting species of Rosaceae that occurs at various elevations on dry mountain slopes throughout most of the Great Basin Province and adjoining parts of the Sonoran and Rocky Mountain provinces. It is an intricately branced small tree or arborescent shrub with small, firm, evergreen leaves. The very hard, tough wood is heavier than water.

Over much of its length, the southern border of the Great Basin Province is marked by the disappearance of the sagebrush community and its replacement by the *Larrea-Ambrosia* community of the Sonoran Province. Two common and conspicuous species notably transgress this boundary. *Yucca brevifolia* (Joshua tree) is an ungainly coarse shrub or small tree that has been said to cast about as much shade as a barbed-wire fence. It characteristically grows along the ecotone between the *Artemisia* and the *Larrea-Ambrosia* communities, extending a short distance into both, often farther into the *Larrea-Ambrosia.* It is at most a codominant with the smaller shrubs that characterize these other communities. *Coleogyne ramosissima* (black brush) is a small-leaved, evergreen shrub not unlike sagebrush in aspect. It often forms a zone interposed between the other two communities, and is difficult to assign to either province.

Like the Rocky Mountain Province, the Great Basin Province is marked by prominent altitudinal zonation of the vegetation. The shrub-dominated community of the lowlands gives way at successively higher elevations to open woodland or chaparral brushland, then open or patchy forest and, at the highest elevations in some of the mountains, to alpine tundra.

Water relations are of critical importance throughout the province. A north slope may be forested or wooded, whereas an adjoining south slope is more open. Cold air drainage and restricted insolation in the canyons cause local reversals of zonation, with the trees extending to lower elevations in the canyons than on the ridges. One may climb through a spruce-fir forest in a north-facing hollow and come out on the summit to find it covered with sagebrush.

2. Californian Province (Engler 1882,[85] Harshberger 1911; Howell 1957, p.p.; Munz 1959; Thorne 1963; p.p.; Gleason and Cronquist 1964; Stebbins and Major 1965, p.p.; Ornduff 1974, p.p.; Raven 1977, p.p.; Raven and Axelrod 1978, p.p.).

The Californian Province occupies a major part of the state of California and a small part of northwestern Baja California Norte (Mexico). Many phytogeographers also include a part of southwestern Oregon, but this area is transitional to the Vancouverian Province, in which it is here included.

The long Central Valley of California forms the heartland of the California Province. This is surrounded on all sides by mountains of varying height. The more northerly of these mountains (Sierra Nevada, southern Cascade Mountains, mountains of the Klamath area, and the western part of the North Coast Ranges of California) are here assigned to the Vancouverian Province, in the Rocky Mountain Region. The inner North Coast Ranges, the South Coast Ranges, the Transverse Ranges in southern California, and the western foothills and lower slopes of the Sierra Nevada and Cascade Mountains all belong to the California Province. The lower limit of the ponderosa pine forests in the Cascade Mountains and Sierra Nevada (about 1,200 m in the southern Sierra Nevada) may be the most useful boundary line in that area, but many characteristically Californian species extend well up into the ponderosa pine zone, and some go even higher. The higher parts of the mountains of southern California and northern Baja California carry attenuated versions of the Sierran forests, surrounded and considerably infiltrated by Californian and/or Sonoran elements.

We limit the Californian Province more narrowly (in accordance with Gleason and Cronquist 1964) than do some authors, such as Raven and Axelrod (1978), so that it conforms essentially to the area in which the vegetation and flora are shaped by a Mediterranean climate. Following such a concept, we consider the Californian Province to be basically Madrean, with some influence and infiltration from the Vancouverian Province and, to a lesser extent, from the Rocky Mountain Province and other parts of the Boreal Subkingdom.

The present flora of the Californian Province is largely descended from the Madro-Tertiary flora that evolved in the southwestern United States and northern Mexico under the influence of gradually increasing aridity. Sclerophyllous shrubs and small trees formed an important element in the vegetation of California at least as long ago as the Miocene epoch. These Miocene and Pliocene sclerophylls are thought to have been vegetatively active for much of the year, however, instead of being dormant in the summer like most of the modern species. The present Mediterranean climate, with prolonged summer drought, appears to have developed only during the past million or so years, at least partly in relation to the rise of the Sierra Nevada. Intense speciation, mainly from preexisting Madrean elements, occurred during this climatic change.

Raven and Axelrod (1978) calculated that the Californian Province harbors 4,452 native species of vascular plants, of which 2,125 (47.7%) are endemic. They defined the province more broadly than we, including not only the Klamath area and all of the North Coast Ranges, but also the southern Cascade Mountains and the Sierra Nevada. The number of native species in the province as here defined must be less than 4,000, but the percentage of endemics is, if anything, higher than Raven and Axelrod calculated for the larger, more heterogeneous area. Thus we may say that roughly half of the native species in the province are nearly or quite endemic to it.

In addition to the many endemic species, a considerable number of genera and a few small families are endemic to the Californian Province or have their principal center of diversity there. Among the more notable of these are: Limnanthaceae (10 species, 9 in California); subtribe Madiinae of the tribe Heliantheae in the Asteraceae (nearly 100 species in all, the vast majority endemic); tribe Gileae in the Polemoniaceae (about 170 species in all, 100 in the province, 60 or more endemic); *Mimulus* (100 or more species in all, about 75 in the province, many endemic); *Caulanthus* (15 species in all, 11 in the province, 7 endemic); *Streptanthus* (30+ species in all, 20+ in the province, all endemic); *Arctostaphylos* (40+ species, nearly all in the province, many endemic); *Dudleya* (about 50 in all, more than half in the province, most of these endemic); *Lotus* subg. *Hosackia* (about 40 in all, about 30 in the province, many endemic); *Clarkia* (33 in all, 29 in the province, many endemic); *Ceanothus* (50+ in all, 44 in the province, 37 endemic); *Cryptantha* subg. *Krynitzkia* (50+ species, about 25 in the province, some endemic); *Collinsia* (18 species, 14 in the province, most endemic); *Downingia* (14 in all, 10 in the province, 5 endemic); *Brodiaea* (15 species, all in the province, 13 endemic); *Triteleia* (14 in all, 13 in the province, 12 endemic). *Eriogonum* and *Astragalus,* which have major centers in the Great Basin

HOLARCTIC KINGDOM 173

Province, also have considerable numbers of species in the California Province. The presence of about 10 species of the widespread genus *Cupressus* may also be noted.

Endemic species include: *Ophioglossum californicum*, species of *Cheilanthes* (including *C. californica*), *Pinus coulteri, P. muricata, P. radiata, P. sabiniana, P. torreyana, Pseudotsuga macrocarpa*, species of *Cupressus* (including *C. macrocarpa*), *Eschscholzia lemmonii, Meconella californica, Papaver californicum, Romneya coulteri, Quercus agrifolia, Q. douglasii, Q. lobata, Q. tomentella, Aphanisma blitoides, Atriplex californica, Chorizanthe* spp., many species of *Eriogonum, Hollisteria lanata, Paeonia californica, Crossosoma californicum*, many species of *Arctostaphylos, Lavatera assurgentiflora, Dirca occidentalis*, species of *Dudleya, Jepsonia parryi, Adenostoma sparsifolium, Chamaebatia australis, Lyonothamnus floribundus*, many species of *Astragulus, Lupinus* spp., *Pickeringia montana, Clarkia delicata, C. rubicunda* and several others, *Heterogaura heterandra, Cneoridium dumosum, Adolphia californica, Ceanothus arboreus* and some other species of this genus, *Mentzelia* spp., *Gilia* spp., *Leptodactylon californicum, Phacelia* spp., *Amsinckia douglasiana, Cryptantha* spp., *Castilleja* spp., *Galvezia speciosa, Mimulus* spp., all three species of *Acanthomintha, Lepechinia* spp., *Monardella* spp., *Pogogyne abramsii, P. nudiuscula, Pycnanthemum californicum, Salvia* spp., *Satureja chandleri, Stachys* spp., *Downingia cuspidata, Githopsis* spp., *Nemacladus* spp. *Erigeron sanctarum*, species of *Haplopappus, Perezia microsephala, Layia* spp., *Senecio lyonii, Calochortus catalinae, C. clavatus, C. pulchellus, C. umbellatus* and few other species of this genus, *Chlorogalum pomeridianum, Fritillaria biflora* and several other species of this genus, *Allium* spp., *Bloomeria crocea, B. clevelandii*, species of *Brodiaea, Muilla maritima, Triteleia clementina, T. versicolor, Nolina interrata, Smilax californica, Discanthelium californicum, Orcuttia californica, Poa douglasii, P. napensis*, and *P. tenerrima*.

A phytogeographical relationship between the western United States and temperate or warm-temperate South America has attracted attention and comment from many botanists (cf. Raven 1963). There are more than a hundred examples of the occurrence of the same or closely related species in these two areas, but not elsewhere. The common feature in the large majority of these disjunctions is that the Californian Province provides part or all of the North American area, and central and/or northern Chile provides part or all of the South American area. These taxa with disjunct distribution all have small disseminules. Nearly all of them are self-compatible (many self-pollinated), or at least one member of the pair is so. Most of them are plants of open habitats, in which establishment can be relatively easy once the seed arrives. Many of them are annual. It is doubtless significant that no similar pattern of distribution exists among vertebrates.

Long-distance dispersal, by birds or whatever method, is the obvious explanation for the pattern. The varying degrees of relationship between the members of a given pair suggest repeated rather than concurrent dispersal events. The past existence of climatically suitable stepping stones seems highly unlikely, especially in view of the fact that the Mediterranean climate of both California and Chile appears to be geologically rather recent. A postulated Californian-Chilean landbridge in the past is purely fanciful and without geological foundation.

In most cases the Californian distribution appears to be primary and the Chilean one secondary. *Limonium californicum,* a California salt-marsh endemic, provides a contrary example. It is homostylic and self-compatible, in contrast to *L. guaicuru* of Chile, which is heterostylic and self-incompatible.

Some examples of species that occur in both California and Chile are *Carpobrotus chilensis, Cardionema ramosissima, Paronychia franciscana, Lepidium nitidum, Lotus subpinnatus, Trifolium depauperatum, T. macraei, T. microdon, Sanicula crassicaulis, S. graveolens, Microcala quadrangularis, Pectocarya ferocula, P. pusilla* (perhaps only recently introduced in Chile), *Plagiobothrys myosotoides, P. scouleri, Orthocarpus attenuatus, Amblyopappus pusillus, Madia sativa,* and *Psilocarphus brevissimus.* Some species-pairs, with the Californian species listed first, are *Chorizanthe coriacea* and *C. chilensis; Helianthemum scoparium* and *H. spartioides; Acaena californica* and *A. trifida; Clarkia davyi* and *C. tenella; Collomia grandiflora* and *C. cavanilesii; C. linearis* and *C. biflora; Linanthus pygmaeus* and *L. pusillus; Pectocarya peninsularis* and *P. dimorpha; Plagiobothrys greenei* and *P. gracilis; Downingia humilis* and *D. pusilla; Legenere limosa* and *L. valdiviana; Lasthenia glaberrima* and *L. kunthii; Psilocarphus tenellus* and *P. berteri;* and *Poa douglasii* and *P. commungii.*

There is a degree of altitudinal zonation in the Californian Province. Most of the floor of the Central Valley and other lowlands was originally a treeless grassland, but some oaks extend out into the valley floor, especially toward the north. Climbing from the valley floor, one at first encounters scattered oaks in a sea of grass. This very open oak woodland merges upwards, in relatively favorable sites, with an oak woodland or oak-pine woodland, in which the trees are less widely spaced. In the Coast Ranges, especially northward, the oak woodland merges upward with a mixed broad sclerophyll forest. In drier, less favorable sites in the foothills and lower mountains, one finds chaparral instead of oak woodland. All of these zones are also represented in southern California, to the south of the Great Valley and its surrounding mountains.

It has been widely believed that Central Valley was originally (in terms of human time) a great prairie dominated by perennial bunchgrasses. *Stipa pulchra* was the most abundant single species. *Stipa cernua* and species of *Aristida, Elymus, Festuca, Koeleria, Melica,* and *Poa* were also common. The interstices between the bunchgrasses contained some annual grasses and a wide variety of annual and perennial herbs in other families. The annual grasses included *Aristida oligantha, Deschampsia danthonoides,* and species of *Orcuttia* and *Vulpia,* among others. The Papaveraceae (notably *Eschscholzia*), Caryophyllaceae, Fabaceae (notably *Lupinus*), Onagraceae (notably *Clarkia*), Apiaceae, Polemoniaceae, Lamiaceae, Scrophulariaceae (notably *Orthocarpus* and *Mimulus*), Liliaceae sens. lat. (notably *Allium, Brodiaea* and *Calochortus*), and especially the Asteraceae (many genera and species) were also well represented. In most years the land looked like a lush flower garden for a few weeks in the spring. A Californian botanist, reminiscing in the mid-1940s, about the magnificent floral displays in the 1920s and early 1930s, before so much of the land had been converted to other uses, was told by a much older botanist, "you should have seen it in the 90s."

The Central Valley is now largely devoted to irrigated farms and other appurtenances of civilization. The native grasses in what remains of the grassland have been largely replaced by Mediterranean annuals such as *Bromus rubens* (the most abundant species), *B. hordeaceus, Avena barbata, A. fatua,* and *Taeniatherum caput-medusae.* Many of the other native species, especially the annuals, still survive in considerable quantity. Some of them, such as *Eschscholzia californica* (the California poppy), have become popular ornamentals. Weedy Mediterranean annual dicotyledons such as *Erodium cicutarium* are now also abundant in the Central Valley and elsewhere in California.

More recently (Wester 1981), it has been suggested that perennial grasses were not an important component of the pristine community on the dry floor of the Central Valley, but reached their best development on the slightly more mesic surrounding hills. In this view the showy flowers of the valley floor were accompanied then, as now, mainly by annual rather than perennial grasses. These grasses were of course the native species rather than the presently abundant Mediterranean weeds. This newer concept has not yet been subjected to critical analysis by other ecologists.

The vernal pools are one of the most important special habitats in the Central Valley. Water accumulates in them during the winter and evaporates during the spring and summer. The vernal pools differ from the playas of desert regions (such as those in the Great Basin and Sonoran provinces) in their usually smaller size, associated with smaller catch-basins, and—more importantly—in the fact that plant growth annually follows the retreating margins of the pool to its center. There is no permanently barren central playa. At least a hundred species are especially adapted to growth around the margin or on the floor of these vernal pools. Most are annual. Many are endemic to the Californian Province, but others (especially those that grow around the margins) have a wider range and occur in seasonally moist habitats elsewhere. Some of the more characteristic species of the floors are *Isoetes howellii, Eryngium aristulatum, E. pinnatisectum, E. vaseyi, Navarretia leucocephala, Plagiobothrys austiniae, P. humistratus, P. hystriculus, P. trachycarpus, P. undulatus, Mimulus tricolor, Downingia bella, D. concolor, D. cuspidata,*

D. elegans, D. insignis, D. ornatissima, D. pulchella, D. pusilla, D. yina, Legenere limosa, Evax caulescens, Psilocarphus brevissimus, Orcuttia californica, O. pilosa, and *O. tenuis.* Some of the more characteristic species from around the margins are *Trifolium barbigerum, T. cyathiferum, T. depauperatum, T. fucatum, Limnanthes douglasii, Plagiobothrys acanthocarpus, P. distantiflorus, Pogogyne zizyphoroides, Orthocarpus campestris, Blennosperma nanum, Lasthenia burkei, L. chrysantha, L. fremontii, L. glaberrima, L. platycarpha, Layia chrysanthemoides, Machaerocarpus californicus, Juncus uncialis,* and *Alopecurus howellii*. Only a few foreign species have been able to invade the vernal pool habitat.

The Central Valley is largely surrounded by oak woodland. There are about 15 species of *Quercus* in the province, and several of them are common and widespread. The most common trees in most of the oak woodland are *Quercus lobata* (valley oak), *Q. douglasii* (blue oak), and *Pinus sabiniana* (digger pine). Blue oak and digger pine often grow together, but they do not usually grow with valley oak. Both of these oaks are deciduous. Several evergreen oaks, most notably *Q. agrifolia* (coast live oak), and *Q. wislizenii* (interior live oak) are also common in the oak woodland, and *Q. agrifolia* is an integral part of the broad sclerophyll forest in the Coast Ranges as well. *Quercus engelmannii* (a semi-evergreen species) is common in southern California. Favorable sites in the oak woodland (often called the foothill woodland) often harbor a few other deciduous trees, such as *Aesculus californica* and (toward the south only) *Juglans californica.*

A mixed broad sclerophyll forest occupies the more mesic sites in the Coast Ranges, and occurs also in a few places in the foothills of the Sierra Nevada, where the Californian Province abuts on the ponderosa pine forests of the Sierran district of the Vancouverian Province. The most important dominant species are *Quercus agrifolia, Q. chrysolepis, Arbutus menziesii, Lithocarpus densiflora,* and *Umbellularia californica. Pinus coulteri* often grows with *Quercus chrysolepis* in these otherwise hardwood forests. All of these dominant species except the pine sprout freely after fire. Fires are not so frequent here as in the chaparral community, but they still play an important role.

In the North Coast Ranges of California and in the Klamath area in southern Oregon, the broad sclerophyll forest passes into the mainly coniferous forest of the Vancouverian Province. Douglas fir may grow intermingled with the broad sclerophyll trees. In some places the broad sclerophyll forest abuts on the coast redwood forest (Vancouverian Province), occurring above the fog belt in which the redwood grows. Some elements of the broad sclerophyll forest, such as madrone, extend north irregularly in relatively dry habitats all the way to Puget Sound.

Chaparral covers many of the hills and lower mountains of California, occupying sites too dry for the oak woodland. The community is adapted to drought and fire, passing through repeated cycles of burning and regrowth. It is a rare stand that escapes fire for more than fifty years.

The drier and more exposed sites in the chaparral community are commonly dominated by *Adenostoma fasciculatum* (chamise), an evergreen shrub with small, narrow leaves—that is, a narrow sclerophyll. Ephemeral annuals and short-lived perennial herbs germinate after fire in the chamise chaparral and form a vigorous herbaceous carpet that persists for a few years until the shrub cover becomes too dense. The seeds of some of these species may lie dormant in the ground for years until the return of conditions favorable to them. A similar herbaceous growth occurs after fire in the other chaparral communities.

Progressively more mesic sites in the chaparral community are typically dominated by species of *Ceanothus, Arctostaphylos,* and shrubby oaks. These are all broad sclerophylls. Many but not all of the species sprout after fire. The taxonomy of both *Ceanothus* and *Arctostaphylos* is complex, and interspecific hybrids are frequent. The explosive speciation of the past million years has not yet produced clear breaks between some of the taxa.

The most mesic (but still rather dry) sites in the chaparral community are characterized by small oaks, notably *Quercus dumosa* (scrub oak) and a shrubby phase of *Q. wislizenii*. A wide variety of other shrubs occur with the oaks. *Toxicodendron diversilobum* (poison oak, a member of the Anacardiaceae), an allergenic shrub or vine that is widely distributed in California, Oregon, and Washington, is a prominent member of this as well as several other communities.

The Coast Ranges and the foothills of the Sierra Nevada contain scattered outcrops of serpentine, most of them smaller than the massive blocks in the Klamath area to the north (discussed under the Vancouverian Province). An important ecological feature of serpentine is the virtual absence of available calcium. As elsewhere in dry climates, serpentine tends to be more sparsely vegetated than other substrates, and it carries a specialized flora. Some species of *Allium* grow indiscriminately on serpentine or in other barren habitats where the competition is minimal, but many species in a wide range of the other genera grow preferentially on serpentine or are restricted to it. Some species are limited to a single outcrop and may have a total range less than 100 m long. Others grow on two or more outcrops that are not necessarily adjacent, and some seem to find a large proportion of the serpentine exposures over a wider range. Among the many Californian species associated with serpentine are *Quercus durata, Arenaria howellii, Chorizanthe brewerii, C. uniaristata, Eriogonum argillosum, E. covilleanum, Arabis mcdonaldiana, Streptanthus barbiger, S. breweri, S. batrachopus, S. howellii, S. insignis, S. polygaloides, Thelypodium flavescens, Astragalus breweri, Lupinus spectabilis, Linum adenophyllum, L. bicarpellatum, L. californicum, L. clevelandii, Lomatium howellii, Phacelia greenei, Cryptantha mariposae, Castilleja neglecta, Cordylanthus nidularius, Mimulus brachiatus, Haplopappus ophitidis, Layia discoidea, Madia hallii, Senecio clevelandii, S. greenei, Calochortus umbellatus, Fritillaria falcata, F. glauca,* and *F. purdyi.*

Several rather small species of closed-cone conifers (*Pinus* and *Cupressus*) form distinctive local communities in inhospitable (but not especially dry) habitats here and there in the Californian Province, especially along the coast. They are all fire-trees, in which the cones tend to persist on the tree for a number of years, opening after fire. Some of them occur on serpentine or on other soils deficient in one or more nutrients. All are thought to be relicts from a more mesic Tertiary coniferous forest. Their ranges may well have expanded—and contracted also—in association with Pleistocene reversals in climate.

The closed-cone pines include *Pinus attenuata*, *P. muricata*, *P. radiata* (Monterey pine, now widely planted in Mediterranean climates), *P. remorata*, *P. torreyana*, and *P. contorta* var. *contorta*. *Pinus contorta* is also widespread as a fire-tree in the Rocky Mountain Region, in more slender and erect varieties known as lodgepole pine. The other species are nearly or quite endemic to the Californian Province.

The taxonomy of *Cupressus* is complex, and some taxa treated as species by some authors are reduced to infraspecific rank by others. In one view the genus contains about 25 species, all in the Northern Hemisphere. Ten of these are native to California, and most of the 10 are nearly or quite endemic to the Californian Province. These are *Cupressus abramsiana*, *C. bakeri*, *C. forbesii*, *C. goveniana*, *C. macnabiana*, *C. macrocarpa*, *C. nevadensis*, *C. pygmaea*, *C. sargentii*, and *C. stephensonii*. All of these species have restricted distributions (most of them discontinuous), and some have very narrow limits indeed. The famous and picturesque Monterey cypress (*C. macrocarpa*) is known from two groves along the windswept coast of Monterey County, and the Gowen cypress (*C. goveniana*) is also known from only two groves, likewise in Monterey County, but a few kilometers inland.

3. Sonoran Province (Thorne 1963; Gleason and Cronquist 1964; Chaparal-Provinz and Sonora-Provinz, Engler 1899, 1903, 1924).[86] The Sonoran Province occupies much of northern Mexico and a considerable fringe of southwestern United States, as far north as southern Nevada and a tiny corner of southwestern Utah. The portion in the United States is continuous from southeastern California to southern Texas, but the portion in Mexico is partly divided into segments by the Gulf of California, the Sierra Madre Occidental, and the Sierra Madre Oriental. These two Sierra Madres form parts of the Mexican Highlands Province.

The province has four well-marked subprovinces. From west to east these are Baja California (excluding the northwest corner), the Sonoran Desert, the Chihuahuan Desert, and the Tamaulipan Thorn-scrub. A low-lying portion of the Sonoran Desert near the Colorado River in southeastern California is often called the Colorado Desert. The Mohave Desert forms a distinctive northwestern district of the Sonoran Desert. The district is here broadly interpreted to include Death Valley and some desert mountain ranges that border on the Great Basin Province, as well as the Mohave Desert proper.

No one species or genus characterizes the Sonoran Province as a whole, but two genera collectively may almost do so. *Larrea* is dominant in much of the Sonoran (including Mohave) and Chihuahuan deserts, and is also common in a large part of Baja California, but not in the Tamaulipan Thorn-scrub. *Prosopis* is a dominant element in the Tamaulipan Thorn-scrub and is also common in the Chihuahuan Desert and parts of Baja California and the more southern parts of the Sonoran Desert, but it is not significant in the Mohave district.

Both *Larrea* and *Prosopis* also have a wider range. *Larrea* has 1 species (*L. tridentata*, creosote bush) in North America, and 4 in deserts of South America. The North American species is mainly confined to the Sonoran Province, but has outlying stations (even local communities) in the drier parts of the Mexican Highlands Province. *Prosopis* has about 40 species, in warm, dry parts of the Old World as well as of the New. In North America it encroaches into the southern part of the North American Prairies Province, and it occurs irregularly southward into South America, as well as being common in the West Indies. It is a prominent element in the thorn-scrub vegetation of Paraguay and northern Argentina.

Both *Larrea* and *Prosopis,* as represented in North America, have their antecedents in South America. The time of migration is debatable, but it may well have been relatively recent. The North American *Larrea tridentata* is so closely related to *L. divaricata* of Argentina that the two have often been regarded as conspecific. The North American *Prosopis reptans* var. *cinerascens* is paralleled by *P. reptans* var. *reptans* in northern Argentina, and the common Sonoran species *P. glandulosa* was long confused with the very closely related *P. juliflora* of the Caribbean Region and *P. chilensis* of subtropical South America.

Cacti are an important part of the vegetation in the Sonoran Province, much more so than in the Californian and Great Basin provinces to the north. Baja California alone has some 90 native cacti, about 60 of them endemic. *Carnegiea gigantea* (saguaro), the largest of all cacti, occurs mainly in southern Arizona and northern Sonora. Several species in other genera, found mainly in Mexico, get nearly as large. *Opuntia*, with jointed stems, is well represented throughout the province. Species with flattened stem segments are called prickly pear, and species with terete segments are called cholla. Some of the chollas disjoint very easily when brushed against, and are known as jumping cholla.

Cercidium (palo verde) is also a common and conspicuous plant in most of the Sonoran Province, although it scarcely reaches the Mohave Desert. The several species are robust, spiny shrubs or small trees with notably green twigs (whence the common name). They are leafless for most of the year, and the leaves when present are relatively small and insignificant. *Cercidium* is widespread in dry places in the American tropics and is a common member of the desert flora in northern Argentina.

The taxonomically isolated genus *Fouquieria* has its home in the Sonoran and Mexican Highlands provinces. One of the species, *F. splendens* (the ocotillo), is fairly widespread in the Sonoran Province (excluding the Mohave and Tamaulipan areas), but the others have more limited ranges. The famous *Fouquieria* (*Idria*) *columnaris*, the boojum tree, occurs mainly in the southern part of Baja California Norte, with a few outlying stations across the Gulf of California in Sonora.

The Asteraceae are by far the largest family in the Sonoran Province, as they are in many other parts of the world outside the moist tropics. They do not dominate the landscape, however, as do the legumes, cacti, and *Larrea*. One species, *Ambrosia dumosa*, is codominant with *Larrea* in the Mohave Desert, but it is smaller and less conspicuous than the *Larrea*.

Monocotyledons make up less than 15% of the Sonoran flora, in contrast to nearly 25% for the world as a whole. The large genus *Carex*, which has so many species throughout the Boreal Subkingdom, has only a handful here. The grasses are somewhat better represented, and there is a considerable amount of desert grassland or oak woodland in the more nearly mesic, upland parts of the province. *Agave*, *Yucca* and other firm-leaved succulents of the family Agavaceae are well developed in the less extreme habitats, as they are also in the Mexican Highlands Province. Bulbous plants, so common among the monocotyledons, are in general not well adapted to the conditions of the Sonoran Province.

As in all dryland regions, the nature of the parent rock has a profound and continuing influence on the soils in the Sonoran Province. Limestone, sandstone, shale, and the various sorts of crystalline rocks provide different substrates with different (though not mutually exclusive) floras.

Gypsum provides one of the most notable special habitats in the Chihuahuan subprovince and adjacent parts of the Tamaulipan subprovince and the Sierra Madre Oriental (Mexican Highlands Province). It also occurs here and there in other parts of the Mexican Highlands, in addition to the Sierra Madre Oriental. Some species are wholly confined to gypseous habitats, some are more common there than elsewhere, some are indifferent, and many are excluded. Among the gypsophilous species of the Chihuahuan subprovince and nearby areas are *Notholaena bryopoda, Anulocaulis eriosolenus, A. gypsogenus, A. leisolenus, A. reflexus, Selinocarpus purpusianus, Atriplex reptans, Drymaria elata, D. lyropetala, Nerisyrenia castillonii, N. gracilis, N. incana, N. linearifolia, Frankenia gypsophila, F. jamesii, F. johnstonii, Fouquieria shrevei, Astragalus gypsodes, Dalea filiciformis, Petalonyx crenatus, Nama canescens, N. carnosum, N. purpusii, N. stevensii, N. stewartii, Phacelia gypsogenia, Aster gypsophilus, Dicranocarpus parviflorus, Flaveria anomala, F. oppositifolia, Gaillardia gypsophila, G. multiceps, G. powellii, G. henricksonii, Haploesthes greggii, Haplopappus johnstonii, "Machaeranthera" gypsophila, "Machaeranthera" re-*

stiformis, Sartwellia flaveriae, S. mexicana, S. puberula, Strotheria gypsophila, Thelesperma ramosius, T. scabridulum, Muhlenbergia bryopoda, and *Sporobolus nealleyi.*

The floras of Baja California (exclusive of the essentially Californian northwest corner), the Sonoran Desert, and the Chihuahuan Desert are all about the same size, in the range of 2,500–3,000 native species. No figures are available for the Tamaulipan Thorn-scrub, but it may be comparable. Most of the species are of course common to two or more of the subprovinces, so that the total flora may be in the range of 5,000–6,000 species. Endemism for the province as a whole has not been carefully calculated, but it is probably more than 25%.

The floras of Baja California and the Chihuahuan Desert are now fairly well understood. The percentage of endemism in these two is strikingly different, in the range of 20–25% for the former, and 8–10% for the latter. The relatively high endemism in Baja California may reflect the presence of the Gulf of California as a physical barrier to migration, and also the fact that more than half of the peninsula is frost-free. In contrast, all of the Chihuahuan Desert and most of the Sonoran (including the Mohave) Desert and the Tamaulipan Thorn-scrub are subject to some frost. The elevation of the Chihuahuan Desert gradually increases southward into Mexico, so that the southern part is not so much warmer than the northern part as might otherwise be expected.

Baja California is not completely cut off from the rest of the Sonoran Province. The Sonoran Desert flora is continuous around the head of the Gulf of California and extends south for more than 300 km in a fringe along the northeastern side of the peninsula. This strip is largely dominated by *Larrea* and is here only arbitrarily included in the Baja Californian subprovince.

Larrea is also a significant element in most of the rest of Baja California, but other genera are more prominent. Cacti, many of them arborescent, are common and conspicuous throughout. Sarcocaulous shrubs or small trees are often dominant, and coarse succulents such as *Agave* and *Yucca* are common. Dominance at a particular site may be shared by species of a dozen or more genera, including *Opuntia, Pachycereus, Atriplex, Fouquieria, Jatropha, Pedilanthus, Cercidium, Bursera, Larrea,* and *Yucca.* Low annuals in the Asteraceae and other families make a brief show at the end of the rainy season in a good year. The flora may be highly diversified, but much of the ground is bare, especially after the annuals wither and die.

Southward the flora of Baja California takes on a progressively more tropical aspect. Cacti are abundant to the southern tip of the peninsula; species of *Ferocactus, Lemaireocereus, Mammillaria,* and *Pachycereus* are prominent elements. Common woody plants include species of *Acacia, Pithecel-*

lobium, Bursera, Cyrtocarpa, Sapindus, Erythea, and other genera. Some of these reach tree size, but the community is always open, never forming a dense forest.

Although *Larrea* is the most pervasive element in the Sonoran and Chihuahuan subprovinces, it is of course not always the only dominant. Woody species of *Yucca* stand up above the *Larrea* in parts of both of these areas. The largest of these species, *Yucca brevifolia,* forms grotesque elfin "forests" along the western and northern borders of the Mohave Desert, where it also overlaps into the sagebrush zone of the Great Basin Province. Other sorts of large shrubs or even arborescent cacti rise above the *Larrea* in other parts of these deserts, except the Mohave.

Chaparral and oak woodlands resembling those of California occur in some of the uplands of the Sonoran and Chihuahuan subprovinces, especially in bands encircling the isolated mountain ranges of central to southeastern Arizona and southwestern New Mexico. The resemblance is mainly in broad aspect and in some of the dominant genera (*Quercus, Arctostaphylos, Ceanothus*), however. Many of the other genera and most of the species are different, although *Quercus turbinella, Arctostaphylos pungens,* and *Ceanothus greggii* are common members of the Arizona–New Mexico chaparral that *do* extend west to California. Vegetative activity of the plants is of course attuned to the seasonal distribution of rainfall.

Above the chaparral and oak woodland in these mountains there may be a zone of junipers and pinyon pine, which in turn gives way to a more characteristic coniferous forest. The highest peaks extend above timberline and have a more or less alpine vegetation. Toward the northern part of these subprovinces many of the species in the upper vegetational zones are the same as those of comparable habitats in the Rocky Mountains and Great Basin. There are, for example, many Rocky Mountain species in the mountains of trans-Pecos Texas.

The Tamaulipan subprovince is not so dry as the other parts of the Sonoran Province. The principal vegetation is more nearly a thorn-scrub than a truly desert type. Species of *Acacia, Prosopis* (especially *P. glandulosa*) and *Cercidium* are among the most important dominants. Others include *Aloysia gratissima, Castela texana, Celtis pallida, Karwinskia humboldtiana, Ziziphus obtusifolia,* and so on.

Fairly large parts of the Tamaulipan Thorn-scrub might be considered a mesquite-grassland, with *Prosopis glandulosa* and/or *P. reptans* var. *cinerascens* as the dominant woody plants. The common grasses include some species that are widespread in the North American Prairies Province, such as *Bouteloua hirsuta,* and others that have more limited ranges and reach their best development in and near the Tamaulipan Thorn-scrub, such as *Aristida roemeriana, Trichachne hitchcockii,* and *Tridens texanus.*

As in all desert regions, the vegetation in most of the Sonoran Province is sparse, with much open ground between individual plants. A large part of the area is dominated by widely spaced shrubs not more than 1.5 m tall. Such open habitats are also well suited to annual plants, and in a good year these are abundant, diversified, and conspicuous for a short time at the end of the rainy season.

Whittaker (1975:156) points out that "in the warm semideserts there is no such convergence of dominant form as in the cool semideserts and other biomes; evolution here has produced [a] divergence of plant forms." This generalization applies best to the less extreme desert habitats, which are too dry for real trees and too dry to permit a continuous crown cover or ground cover, but not so dry as to impose rigid limitations on non-tree forms of growth. It is well illustrated in parts of Baja California, where microphyllous desert shrubs, pachycaulous arborescent shrubs, cacti of all sizes (including tree-cacti) and plants of diverse other specialized forms (but not ordinary trees) grow intermingled. Some of the less extreme parts of the Sonoran and Chihuahuan deserts present more or less similar mixtures. Still, one can drive through mile after mile of the Mohave Desert and some other parts of the Sonoran Desert and see little but two small-leaved evergreen shrubs, one (*Larrea tridentata*) sclerophyllous, the other (*Ambrosia dumosa*) with soft leaves. Large stretches of the Chihuahuan Desert are even more monotonous, dominated by *Larrea* alone.

4. Province of the Mexican Highlands (Provinz des mexikanischen Hochlandes, Engler 1899, 1903, 1924; Mexican Highlands, Good 1947, 1974; Aztekische Provinz, Engler 1882; Kingdom of the Mexican Mountains, Schouw 1823). The topographic, ecologic, and floristic diversity of central Mexico is so great that no scheme of classification can do it justice. Alexander von Humboldt said that the vegetative cover of the area might well be considered a microcosm of that of the whole world. Doing our best, we here define the Mexican Highlands Province to include the major mountain systems of Mexico, from the southern slopes of the Sierra Madre del Sur on the south, to the northern ends of the Sierra Madre Occidental and Sierra Madre Oriental on the north. The Altiplano of central Mexico, south of the Chihuahuan Desert and north of the Trans-Mexican Volcanic Belt, is also included. The Balsas Depression, lying between the Sierra Madre del Sur and the volcanic belt, is excluded and is considered to form a peninsula of the Central American Province, joining with the rest of that province at the west. The mountains of Chiapas have many species in common with or closely related to those of the Mexican Highlands Province, but these mountains are wholly surrounded by tropical vegetation and are here considered to be northern enclaves in the Central American Province.

It is useful to consider the Mexican Highlands Province as consisting of five subprovinces—the Sierra Madre Occidental, the Trans-Mexican Volcanic Belt, the Mexican Altiplano, the Sierra Madre Oriental, and the Sierra Madre del Sur.

The Sierra Madre Occidental and the Sierra Madre Oriental rise only gradually from opposite sides of the Altiplano. On the seaward side, however, each of these long ranges drops off abruptly from a usual altitude of 2,500–3,000 m nearly to sea level. The Sierra Madre Occidental, in particular, has great cliffs from which waterfalls may plunge 300 m or more. These two major ranges are geologically very different. The Sierra Madre Oriental consists very largely of marine limestone, whereas the Sierra Madre Occidental is built up mainly of massive felsitic flows stacked one on another.

The climate in the Mexican Highlands Province is on the dry side, but not really arid. The rainy season is in the summer, mainly in July, August, and September, tapering off in October. The big burst of flowering comes at or shortly after the end of the rainy season. Virtually all of the province is subject to some frost in the winter. In contrast, the Central American Province (Caribbean Region) to the south is frost-free except at the highest elevations.

There are no good figures for endemism in the province as a whole. It must be relatively high, surely well over 25%. Some of the endemics are more or less widespread within the province, but each of the subprovinces except the Altiplano also has many local endemic species. A traveler from one subprovince to another will encounter many unfamiliar species in familiar genera.

The conditions of the Mexican Highlands Province are ideal for the large family Asteraceae, which reaches its best development in open, sunny, fairly warm but not really tropical habitats the world over. Species of *Viguiera, Verbesina, Eupatorium, Stevia,* and *Senecio* are ubiquitous in the province. The large and archaic subtribe Verbesininae of the archaic tribe Heliantheae centers in the Mexican Highlands Province, and is especially well represented in the Sierra Madre Occidental. The more advanced subtribe Tagetinae is also well developed here.

The not-so-large family Crassulaceae also does well under such conditions. The Trans-Mexican Volcanic Belt alone has 28 species of *Sedum,* half of them endemic.

Agave and other genera of Agavaceae likewise thrive in the Mexican Highlands, as do the Cactaceae. Tree-cacti of various genera allied to *Cereus* are common especially in the lower, warmer, more southern parts of the province, often in strange contrast to the surrounding small deciduous trees or shrubs. The archaic genus *Pereskia,* with well-developed leaves, is also common in the warmer parts of the province and is even used as a hedgeplant. *Pereskia* is basically Caribbean, ranging south to northern South America and east to the West Indies.

The Mexican Highlands Province is one of the great centers of diversity for *Pinus:* more than 25 of the ±95 species of the genus occur here. Various kinds occur at all elevations up to timberline, wherever the climate is moist enough to permit the growth of trees but not moist enough to permit a more mesophitic forest to flourish. Most of them are resistant to fire, and some have closed cones. Typically they form open forests, alone or with various species of *Quercus.*

The principal pine at elevations of 3,000–4,000 m is *Pinus hartwegii.* Toward the upper part of its altitudinal range it is the only tree in the forest, and in the Trans-Mexican Volcanic Belt it reaches the highest elevation of any pine in the world.

At more moderate elevations in the volcanic belt and in the Sierra Madre Occidental, *P. engelmannii, P. montezumae,* and *P. teocote* are important species, and *P. arizonica* is common at upper elevations in the northern part of the Sierra Madre Occidental. In the Sierra Madre Oriental, *P. montezumae, P. patula, P. pseudostrobus,* and *P. tenuifolia* are among the most common species. At lower elevations and toward the southern part of the province, *P. ayacahuite, P. michoacana, P. oocarpa,* and *P. pseudostrobus* are among the important species. These latter pines also extend well south of the province, reaching Guatemala and even (in the case of *P. pseudostrobus*) Nicaragua.

In the most favored, mesic sites in the mountains, at elevations of 1,200–3,300 m, there are fairly dense groves of *Abies religiosa,* either pure or mixed with other trees. In the Sierra Madre Occidental and the Sierra Madre Oriental these other trees may include the Rocky Mountain taxon *Pseudotsuga menziesii* var. *glauca.*

Quercus is certainly one of the largest as well as one of the most abundant genera in Mexico. Rzedowski (1978:263) considers that there are more than 150 and perhaps as many as 200 species in Mexico as a whole. Most of these occur in the Mexican Highlands Province. No list of only a few species can begin to do justice to the oak communities of Mexico.

The oaks grow in the same sort of places as the pines, often intermingled with them. It is not obvious why oaks are favoured in some places and pines in other, apparently similar places.

Some of the Mexican species of *Quercus,* particularly those with small, hard leaves, are evergreen, but most of them are more or less deciduous. The leafless period is typically short, however—often only a month or two in the driest part of the year.

No topographic feature marks the transition from the Chihuahuan Desert (on the north) to the Mexican Altiplano (on the south). We here consider the boundary to be marked by the disappearance of *Larrea* as a dominant element, and its replacement by a more mixed, less xeromorphic community in which *Acacia,* tree-*Opuntia,* and tree-*Ipomoea* are conspicuous. These species of *Opuntia* and *Ipomoea* are trees only by grace of a loose interpretation and by contrast with other species of their respective genera. They are treelike in having a single stout trunk toward the base, but they are usually only about 2–5 m tall. This same community extends onto parts of the flanks of the bordering mountains, especially the Sierra Madre Occidental. Occasional hills arising in the Altiplano carry an open oak woodland comparable to that of the nearby mountains.

The major peaks of the Trans-Mexican Volcanic Belt rise far above timberline, and some are snowcapped. The timberline, which is here at about 3,900–4,000 m, differs from timberline in more northern mountains in that there is no Krummholz or band of stunted trees at the upper limit of the forest. At elevations above 3,000 m, the pine is *Pinus hartwegii*, a stately tree, nearly as large at timberline as it is farther down the mountain. The stand is typically open, with a ground cover of coarse bunchgrasses, notably *Festuca tolucensis, Calamagrostis tolucensis,* and *Muhlenbergia quadridentata*. The bunchgrass community continues up the mountain for some distance beyond the pines, sometimes to 4,200 m or more, but with some substitution of one species for another.

A number of herbaceous or shrubby dicotyledons are adapted to conditions near and above timberline on these volcanoes. Some of them, such as species of *Ranunculus, Arenaria, Cerastium, Alchemilla, Trifolium, Geranium,* and *Achillea,* belong to boreal genera that are not well represented at lower altitudes in Mexico. Others are alpine representatives of genera that are common enough below. In the Asteraceae alone, there are at least 7 endemic alpine species—namely, *Cirsium nivale, Gnaphalium sarmentosum, G. vulcanicum, Senecio calcarius, S. gerberaefolius, S. procumbens,* and *S. roseus*.

The east side of the Sierra Madre Oriental differs from most of the rest of the Mexican Highlands in being relatively moist, especially toward the south. South of 22° north latitude, the lower slopes are essentially tropical and support an evergreen forest like that of the coastal lowland. Higher up, at elevations of roughly 1,000–2,000 m, there is an irregular and interrupted strip of warm-temperate to subtropical mesophytic forest with variously deciduous, semi-evergreen, and evergreen species. Above that there is a zone of pine and/or oak forest comparable in appearance (but not in specific composition) to that of the Sierra Madre Occidental. Patches of the mesophytic forest occur on the Atlantic slope of other mountain ranges at intervals all the way to Guatemala.

This mid-level strip of forest is of particular ecologic and phytogeographic interest. The relative abundance of moisture in the summer, and its scarcity in the winter, provide a very rough approximation of a temperate-zone deciduous forest climate with summer growth and winter dormancy of trees. Yet the winter is very mild, and many tropical species can survive it. Consequently the forest presents a heterogeneous, varying mixture of species and genera not ordinarily seen growing together elsewhere.

A considerable number of species characteristic of the Appalachian Floristic Province occur in this mixed forest (Watson 1890; Fernald 1931; McVaugh 1943; Miranda and Sharp 1950; Dressler 1954; Rzedowski 1965). Among these are *Pinus strobus, Hamamelis virginiana, Liquidambar styraciflua, Platanus occidentalis, Laportea canadensis, Fagus grandifolia, Carpinus caroliniana, Ostrya virginiana, Carya illinoensis, C. ovata, Prunus serotina, Cercis canadensis, Nyssa silvatica, Cornus florida, Rhamnus caroliniana, Mitchella repens, Epifagus virginiana,* and several species of *Quercus, Alnus,* and other genera. Such endemic Mexican species as *Taxus globosa, Magnolia dealbata, M. schiedeana, Illicum mexicanum, Fagus mexicana, Myrica pringlei, Carya mexicana, Tilia longipes, Acer skutchii* and many others are closely related vicariants of Appalachian species.

Tropical elements growing in these mixed forests include species of *Beilschmiedia, Phoebe, Ternstroemia, Rapanea, Eugenia, Turpinia, Meliosma,* and even various palms and tree-ferns (*Cyathea*), as well as herbaceous and shrubby species belonging to a number of chiefly tropical families. Bromeliad epiphytes are frequently abundant. Even more than the other Appalachian genera, *Liquidambar* is likely to grow with the more or less tropical plants here, and it is often not fully deciduous. These species amply illustrate the individualistic concept of the plant association, as expounded, for example, by Gleason (1926).

It is noteworthy that the Sierra Madre Oriental was uplifted only during the Pliocene and Pleistocene. Thus the habitat in which these Appalachian elements now occur did not exist in the Miocene and earlier times. The mountains farther south in Mexico and Guatemala which support some of these same Appalachian species have an earlier (likely Miocene) origin, but they were also much uplifted during the Pliocene.

Conditions permitting the floristic exchange between eastern Mexico and eastern United States probably began early in the Pleistocene, in association with the climatic changes marked by continental glaciation farther north. Presumably each return of glaciation reflected a return of climatic conditions favorable for renewed exchange. The present disjunction in range results from the decrease in rainfall in Texas and northeastern Mexico since the height of the Wisconsin glaciation (Dressler 1954:91–94).

Pleistocene migration between the eastern United States and Mexico was of course not all in one direction. The most obvious movement was from the United States to Mexico, but *Cacalia rugelia* provides a reverse example. This local endemic of upper altitudes in the Blue Ridge (Appalachian Province) has no close relatives in the United States but is obviously related to some species of the Mexican Highlands.

The dominant vegetation of most of the province ranges from open woodland or semidesert grassland or thorn-scrub at lower elevations to rather open forest with pine and oak as the principal dominants at upper elevations. The thorn-scrub in some of the drier areas is not very different in appearance from that of the Tamaulipan area, with *Acacia* and *Prosopis* playing a dominant role. There are also bits and pieces of more typical desert, outliers of the Chihuahuan Desert to the north. Toward the south, especially in the Sierra Madre del Sur and on the south side of the Trans-Mexican Volcanic Belt (bordering on the Balsas Depression), there is a considerable amount of open, rather low deciduous forest, with a preponderance of basically tropical genera such as *Bursera* and *Cordia*. This more or less tropical deciduous forest and associated thorn-scrub represent an infiltration of central American elements into the Mexican Highlands.

NOTES

1. Diels (1908) introduced the term "Holarctic Kingdom" ("Holarktisches Florenreich"), having borrowed it from the zoogeographer Heilprin (1887). Heilprin used this term for both the Neoarctic and Paleoarctic Faunistic Regions. In Engler's system this kingdom is called the "Nordliches extratropiches oder boreales Florenreich."

2. The Circumboreal Region corresponds to two regions in Meusel et al. (1965)—the Circumarctic and the Circumboreal—and to two analogous regions in Tolmatchev (1974)—the Arctic and Boreal (see general bibliography). In Engler's system the territory of the Circumboreal Region is divided into three regions: the Arctic, the Subarctic (or region of conifers), and the Mideuropean. Division of this territory into three or even into two independent regions does not appear justified to me.

3. Here (and throughout the book) numbers in parentheses are number of species in a genus.

4. See note 7 below.

5. Engler termed this province the "Atlantische Provinz"; Braun-Blanquet called it "Domaine Atlantique"; and Gaussen named it "Domaine atlantico-européen."

6. Several authors exclude from the Atlantic-European Province a part (Braun-Blanquet 1923b, 1928), and even a large part (Roisin 1969) of the Jutland Peninsula. Others (e.g., Walter and Straka 1970) include in it the entire peninsula and even Fyn Island. In addition, two later authors also include in this province the southeast shores of Norway and a considerable part of the southwest coast of Sweden (where the boundary moves south of Göteborg).

7. The flora of the Faroe Islands is very depauperate, and the Atlantic element is poorly represented here. Nevertheless, we find on these islands such characteristic species of the Atlantic-European Province as: *Hymenophyllum wilsonii, Blechnum spicant, Hypericum pulchrum, Erica cinerea, Anagallis tenella, Polygala serpyllifolia, Galium saxatile, Narthecium ossifragum, Scilla verna, Carex binervis,* and so on.

8. The Central European Province conforms approximately to the "Dominio centroeuropo" of Giacomini (in Giacomini and Fenaroli 1958).

9. According to Favarger (1972), 67% of the flora of the Pyrénées is the same as that of the western Alps.

10. In 1882, Engler established a series of provinces for the territory which encompassed the Central European and part of the Eastern European floras—namely, "Provinz der Alpenländer"; "Provinz der Apenninen"; "Provinz der Karpathen"; "Provinz der europäischen Mittelgebirge"; "Danubische Provinz"; and "Sarmatische Provinz," including central Russia to the east. Later, Engler (1899, 1924) eliminated his "Danubische Provinz" and included it in his newly created "Pontische Provinz" (the territory of which he extended to the "Nordkaspische Steppe"). Finally, his "Provinz der westpontischen Gebirgsländer" included several regions of the Central European Province. This produced a rather complex and intricate mosaic of phytochoria, including several very artificial provinces (especially "Pontische Provinz"). Unfortunately, the system of phytochoria elaborated by Meusel et al. (1965)—in which the "Mitteleuropäische Region" (including the Atlantische, and also the Sarmatische and Pontische provinces) is subdivided into 13 independent provinces— cannot be considered a step forward. Therefore, I definitely prefer the concept of Braun-Blanquet, and accept his "Mitteleuropäische Province" with only slight changes in its borders (mainly in the eastern boundary).

11. This includes the northern slopes of the Pontus Range and the Artvin area lying to the southeast of it, situated along the lower course of the Coruh River and its tributaries. But Maleev (1938) completely excluded from the Euxine Province the Oltu area, which is situated in the basin of the Oltu River.

12. The boundaries of the Euxine Province in northern Anatolia and in European Turkey are mapped in Davis (1971).

13. Maleev (1938) extended the northern boundary of the Euxine Province in Bulgaria somewhat south of Varna.

14. Grossheim (1948:189) included in the Caucasian Province also my Megri District (Takhtajan 1941), which I however assign to the Armeno-Iranian Province.

15. In my opinion, the Eastern European Province includes two provinces of Braun-Blanquet (1928, 1964)—the "Zentral-Russische Provinz," and part of the "Sarmatische Provinz."

16. Engler also included in this province Iceland, the Faroe Islands, and the Shetland Islands, with which it is difficult to agree.

17. Krylov, having created this province, named it "the Province of the Western Siberian Lowlands," and Schischkin called it "the Province of Western Siberian Forests." It unites two provinces of Kuznetzov (1912)—the Forest Province, and the Steppe Province of Western Siberia. And it agrees with the largest part of the "West Siberian Subprovince" (without the Altai) and the "Provinz Subarktisches Asien oder Sibirien" of Engler (1899, 1924). Earlier Engler (1882) designated this province "Nordsibirische Provinz." He brought the eastern boundary of his "Western Siberian Subprovince" to the Yenisey, which fully agrees with contemporary views.

18. In creating this province, Kuznetzov called it "the Alpine Province of the Altai and the Sayan" and defined it with considerably narrower boundaries than did Krylov and subsequent authors.

19. Krylov, who created this province, called it "the Province of the Middle Siberian Plateau." He understood it in a very broad sense and included a considerable

part of northeastern Siberia in it. In contrast, Vasiliev and Shumilova understood it in a considerably narrower sense than we do. Hence Vasiliev recognized the Leno-Viliuisko-Aldan (Central-Yakutian) and the Aldan as independent provinces.

20. The Transbaikalian Province is usually understood to correspond partly to the Dahurian Province of Krylov (1919) and, to a lesser extent, to the Baikal-Olekminsk Province of Vasiliev (1956). The boundaries of the province are most correctly outlined by Peshkova (1972).

21. Here the Northeastern Siberian Province is somewhat broader than the Northeastern Siberian Province of Vasiliev (1956), and it agrees partly with the Yano-Kolymsk Province of Shumilova (1962). It corresponds closely to the four provinces of Yurtsev (1974)—namely, the Verkhoyansk, the Northern-Okhotsk, the Aniuisk, and the Anadyro-Koryaksk—which may better be described as subprovinces.

22. The Okhotsk-Kamchatka Province corresponds to Kuznetzov's (1912) Forest Province of the Okhotsk Shores and, in part, to the Chukotsk-Okhotsk Province of Krylov (1919).

23. The "line of Miyabe" is floristic line separating the southern Kurile Islands from the northern one and passing between the islands of Iturup and Urup (Hara 1959).

24. Engler called the Canadian Province "Provinz Subarktisches Amerika" and divided it into four subprovinces: "Südliches Alaska," "Peace und Athabasca Riverland," "Nordliches Ontario," and "Quebec and Labrador". Dice called it the "Hudsonian Province," and Good called it the "Province of Canada and Alaska." Gleason and Cronquist called it the "Northern Conifer Province", in which they included also some parts of northern Eurasia, dominated by coniferous forests, that are here treated as separate provinces. The Canadian Province corresponds to Scoggan's "Boreal Forest Floral Region," plus his "Acadian Forest Floral Region" and a large part of his "Great Lakes–St. Lawrence Forest Floral Region." Both of these latter regions show considerable influence of the Appalachian Floristic Province (in the North American Atlantic Region), as well as having a few more or less distinctive elements of their own (see also Scoggan 1966, 1978).

25. The "Ostasiatisches Gebiet" of Diels coincides approximately with: the "Chinenisch-japanisches Gebiet" of Grisebach (1872, 1884); the "Ostasiatisches Florenreich" of Drude (1890); the region of "Temperiertes Ostasien" of Engler (1912), but without Kamchatka and the Aleutian Islands; the "Chinesisch-japanisches Gebiet" of Hayek (1926); the "Région Sino-japonaise" of Chevalier and Emberger (1937); and the "Sino-Japanese Region" of Good (1947, 1974). It is important to point out that my understanding of the boundaries of the Eastern Asiatic Region is somewhat broader than is that of these authors.

26. The Manchurian Province is similar to the Manchurian Floristic Region of Komarov (1897, 1901), but with considerable expansion of its western boundary. Good (1974) called this chorion "Manchuria and South-Eastern Siberia," but did not specify its boundaries. Sometimes this province also is called the "Amurian" (see Fedorov [in Grubov and Fedorov] 1964).

27. In the Russian geographical literature, Dahuria is understood to consist of the forest steppe and the steppe territory of southeastern Transbaikalia and northeastern Mongolia.

28. Schischkin (1947) called this province "Sakhalino-Kurilian," and Good (1947, 1974) called it the "North Japan and South Sakhalin."
29. Kudo (1927) and Hara (1959) drew the southern boundary of the Eastern Asiatic flora on Sakhalin along the so-called Schmidt Line, joining Due on the west shore with Poronaisk in Terpeniya Bay (formerly Taraika Bay).
30. As in other cases, Good did not give the boundaries of the "Korean and South Japan Province" that he created.
31. Hara (1959) drew the southern boundary of the "Hokkaido (Yezo) floral region" (in which he also includes the Kurile Islands and southern Sakhalin) along the Isikari Depression. Both these depressions appear to be floristic boundaries, but the majority of the most typical representatives of the Japanese flora (including *Fagus crenata* and *Ilex leucoclada*; see Horikawa 1972) do not go beyond the Kuromatsunai Depression. In particular, Maekawa (1974:18, 73) drew the southern boundary on his "Yezo-Mutsu Region" along the Kuromatsunai Depression.
32. Engler referred his "Provinz der Bonin-Inseln" to the "Monsugebiet" of the Paleotropical Floristic Kingdom.
33. In Tuyama's list of endemic taxa, a number of varieties are mentioned which I would prefer to consider as subspecies.
34. Engler called this province the "Provinz der Luschu- oder Riu-Kiu-Inseln" and included it in the "Monsungebiet" of the Paleotropical Floristic Kingdom. Hara (1959), who described this chorion in detail, called it the "Ryukyu floral region." He also included in it the Sakishima Islands, which are, however, floristically closer to Taiwan. Maekawa (1974) drew the boundaries of his "Ryukyu Region" in a similar manner.
35. In his system, Good (1947, 1974) joined Taiwan and the Ryukyu Islands into one province, which can hardly be considered proper.
36. As shown in a series of works (especially Li and Keng 1950), the flora of the Hengchun Peninsula and of the islands of Lanyu (Botel Tobago) and Lutao possesses a tropical character.
37. Engler (1924) called it the "Provinz des Nördlichen Chine (Nördlich vom Tsin-ling-shan)." Earlier, Engler (1903, 1912) recognized a much larger province, when in addition to Northern China he also included Korea ("Provinz des Nördlichen Chine und Koreas").
38. It corresponds approximately with the Chinese part of the "Mittelchinensisch-mitteljapanisches Übergangsgebiet" of Handel-Mazzetti (1926, 1931), but without southeastern China.
39. In the opinion of K'ien Ch'ung-Shu, Wu Cheng-i, and Ch'en Ch'ang-tao (1957), *Ginkgo biloba* is still preserved to this very day in the region of Tyan'mushan as a wild plant. This opinion is held by several other Chinese botanists as well.
40. The Sikang-Yünnan Province agrees partly (especially in its eastern boundary) with the "Gebiet des Hochlandes und Hochgebirge von Yünnan und West-Setschwan" of Handel-Mazzetti (1926, 1931) and with the Yünnan-Szechwan Sub-province of the Himalayan-Chinese Province of Fedorov (in Grubov and Fedorov 1964). But it also includes the northern mountains of northeastern Burma, of Laos, and of northwestern Tonkin. This vast province probably should be divided into several independent provinces, which would require detailed study. Nevertheless, it

now is possible to clearly discern the salient subprovinces—namely the Sikang-Szechwan Subprovince and the Yünnan Uplands. These correspond to the districts that were similarly named by Fedorov. However, it is impossible to designate the boundary between them at this time.

41. As Ching Zen-chan (1958) showed, the fern flora of southwestern China, especially Yünnan, is one of the richest in the world.

42. The Northern Burmese Province corresponds to the North Burmese Subregion of Ward (1944:554), who places it in his Sino-Himalayan Region. It also corresponds to parts of the Assamo-Upper Burmese Province of Fedorov (in Grubov and Federov 1964), which he assigns to the Paleotropical Kingdom.

43. Ward (1944) wrote that nothing is known about approximately four-fifths of the area. Since that time, little has changed, and considerably more than half of the territory remains totally unknown, botanically speaking.

44. Tropical vegetation is found mainly in deep river valleys and occupies no more than 10% of the entire territory. But as Ward (1946b) points out, many trees usually found in tropical evergreen rain forests reach the zone of subtropical forests, sometimes attaining heights of 5,000 feet above sea level.

45. This chorion was already created by Clarke (1898), but with the rank of region ("Eastern Himalayan Region"). Both Clarke and especially Hooker ([1904] 1907) incorrectly outlined the western border of the chorion. Stearn (1960) was the first to defined it more exactly.

46. Both the Tsangpo and the Dihang are parts of the Brahmaputra River, individual parts of which possess various names.

47. The Khasi-Manipur Province corresponds partly with the "Naga-Khasia Endemic Centre" of Clayton and Panigrahi (1974), which they created on the basis of a chorological analysis of grasses of India, with the use of computer techniques.

48. In a letter to Joseph Hooker, C. B. Clarke stated that the flora of the Naga Hills was considerably different from the flora of that altitude in the Khasi Hills, even though less than 160 kilometers apart. At the same time, considerable similarity is observed with the flora of Sikkim, and especially of Darjeeling, which is located almost 800 km away and is separated by the wide valley of the Brahmaputra River (Bor 1942b). Thus Clarke (1898) felt that the flora of the Naga Hills is more Himalayan than the Khasi flora, and Rao (1974) agrees. However, the floras of the Khasi and the Naga hills share many endemics. As a whole, the flora of the Naga Hills gives the impression of being a mixed, transitional flora from Khasi-Manipur to the Eastern Himalayan Province, but considering the proximity of the Naga Hills to the Shillong Plateau and to Manipur, as well as the considerable distance from the Himalayas, I am nevertheless inclined to think that they should be included in the Khasi-Manipur Province. This does not, however, exclude the possibility that detailed investigations of the flora of this poorly studied land may necessitate the creation of a separate province for the Naga Hills.

49. I have not included those rather numerous Eastern Asiatic taxa which reached the Nilgiri Plateau in southern India and the mountains of Malesia as a result of events during the glacial epoch.

50. The greatest concentration of plants with eastern Asiatic relationships is in the southern part of the Appalachian Uplands, especially in the southern parts of the Blue Ridge system (Braun 1955).

51. On the map, "Forests and Floristic Regions of Canada," Scoggan (1966)

assigned his "Acadian" and "Great Lakes-Saint Lawrence" regions to the Appalachian Province.

52. Many Appalachian elements actually extend further west, reaching North Dakota.

53. The Tethyan Subkingdom agrees approximately with the Afro-Eurasian part of Popov's (1927, 1929) Ancient Mediterranean Region and with Quézel's (1978) Mesogean Subkingdom.

54. In his list of endemic Macaronesian genera of the Crassulaceae, Bramwell (1976) also includes *Monanthes* (about 13 species), but one of them also grows in Morocco.

55. The monotypic *Drusa glandulosa,* previously considered a Canary Island endemic, was recently discovered in the northeast of Somalia (Lavranos 1975).

56. Macaronesia also was included in the Mediterranean Region by Lavrenko (1962). But he accepted the eastern boundary of the region in its classical interpretation.

57. Some authors doubt that *Hermodactylus* is a native in Lebanon and Palestine.

58. Engler referred this province to the "Nordafrikanisch-indisches Wüstengebiet." Hayek (1926) did likewise, considering this phytochorion at the rank of district ("Sudmarokkanischer Bezirk"). In contrast with Engler, Hayek referred all of this region to the Holarctic Kingdom, which was doubtless correct. Chevalier (Chevalier and Cuenot 1932) included this territory in the Mediterranean Region (in the "Mauritanian Mediterranean Subregion"), and Quézel (1965, 1978) and many other contemporary authors referred it to the Mediterranean Region.

59. Engler, Rikli, and many others, including myself, put part of northern Egypt (the maritime belt in the northwestern part of the country) in the Mediterranean Region. However, there are good reasons to agree with Zohary (1973:162), who maintains that "Egypt has no Mediterranean territory." There is a relatively large percentage of Mediterranean plants in Egypt, but as Zohary shows, they are scattered among segetal, ruderal, hydrophytic, halophytic, and psammolittoral communities, and some even penetrate into desert oases. In Egypt, there are no obvious Mediterranean floristic complexes and communities, and the characteristic elements of the sclerophyll forest and maquis are absent.

60. Engler also included the Balearic Islands in this province, but I believe it is better to consider them as an independent province.

61. Rivas-Martinez (1973) and Rivas-Martinez et al. (1977) divided the territory of the Iberian Peninsula into 11 provinces (including the Atlantic, the Orocantabrian, and the Pyreneeian), which is not sufficiently substantiated floristically. In any case, the floristic geography of the Iberian Peninsula has still not reached the stage where one can present a detailed chorionomic system. With further study, the provinces of Rivas-Martinez will most likely be reduced to subprovinces, and in some cases perhaps even to districts. In addition, his system is not purely floristic but, as he himself indicates, represents a "synthesis" of data from floristics, phytocoenology, history, ecology, geology, and geography. Engler and Gilg (1919) and Engler (1924) divided his Iberian Province into the following subprovinces: (1) sub-Pyreneeian Iberia, excluding the Pyrénées themselves, (2) central Iberia, (3) eastern Iberia, (4) the Balearic Islands, (5) west Atlantic Iberia, and (6) north Atlantic Iberia. This division is certainly out of date, but there is nothing better for the time being.

62. Engler (1882, 1924) included the Balearic Islands in his Iberian Province, and

Meusel et al. (1965) placed them in their "Südostiberisch-Balearische Provinz." Rivas-Martinez (1973) and Rivas Martinez et al. (1977) placed them in their "Provincia Valenciano-Catalano-Provenzal-Balear."

63. *Spiroceratium* has been considered by some authors as a separate genus (e.g., Bolós 1958), but is now usually united with the *Pimpinella*.

64. The boundaries of the Ligurio-Tyrrhenian flora in France are based mainly on Flahault (1937) and Gaussen (1938), and within Italy on Giacomini (in Giacomini and Fenaroli 1958).

65. The Adriatic Province corresponds to the Adriatic Subprovince of the "Mittlere Mediterranprovinz" of Engler (1903, 1924), with the "Zircumadriatische Pflanzengeografische Zone" of Adamović (1933), and with the "Zirkumadriastische Provinz" of Meusel et al. (1965).

66. Later, Engler (1903, 1924) included the territory of the East Mediterranean Province in his broad "Mittlere Mediterranprovinz," which also includes the Adriatic, the Euxine, and the Crimean-Novorossiysk provinces of later authors.

67. Thus the East Mediterranean Province includes the Aegean Province of Maleev (1938:191), who included in it western Anatolia, the islands of the Aegean Sea, the southern part of the Balkan Peninsula, and the entire series of provinces created by Meusel et al. (1965)—namely, Westhellenische Provinz, Ägaische Provinz, Westanatolische Provinz, Südanatolische Provinz, Palästinisch-Libanische Provinz, and part of Mazedonisch-Thrazische Provinz.

68. The Crimean-Novorossyisk Province is synonymous with the Crimean (Krymsky) Province of Grossheim and Sosnovsky (1928) and with the Tauride (Tavrichesky) Province of Grossheim (1948).

69. According to Kuznetzov (1909) and Grossheim and Sosnovsky (1928), in western Transcaucasia the boundary takes the following form: in the north it goes somewhat north to Anapa to Krasnodar, then turns sharply to the south to an intersection with the main range and arrives at the sea further along the course of the Tuapse River.

70. The Crimean form of this pine is usually considered a separate species, *P. stankewiczii* (Sukacz.) Fomin, but it may be best considered a subspecies. It is sometimes referred to *P. pityusa* Stev., but this species is so close to *P. brutia* that at best it represents a subspecies of it.

71. The Saharo-Arabian Region more or less corresponds to the Saharo-Arabian parts of the "Région du Dattier" of Boissier (1867) and of the "Wüstenregion" of Grisebach (1872, 1884); the territory of "Sahara und Arabien" of Drude (1890); the "Nordafrikanisch-indisches Wüstengebiet" of Engler (1899, 1903, 1908, 1924), of Hayek (1926), and of Mattick (1964); the "Région Saharo-sindienne" of Eig (1931) and of many other authors; "the Northern Palaeotropical Desert Region" of Newbigin (1936); and "the North African and Indian Desert Region" of Good (1947, 1974) and of Turrill (1959).

72. Although a considerable number of Saharo-Arabian elements also penetrate into some parts of southern Iran, they are nowhere centered within a particular area (Zohary 1973:239).

73. Earlier (Takhtajan 1974), I accepted the boundary of the Saharo-Arabian Region as created by Zohary (1973).

74. For the entire Sahara, including its southern tropical region, Ozenda (1958) estimated 1,200 species in all.

75. According to Quézel (1978), there are 190 endemic species for the entire Saharan flora (Arabia excluded), which represents 11.6% of the entire flora. And according to Zohary (1973), in the eastern part of the region there are probably 150 endemic species. But it should be kept in mind that Zohary excluded from the Saharo-Arabian Region the uplands of the Ahaggar and of the Tibesti, which reduces the number of endemics.

76. Engler put the eastern boundary of his "Provinz der Grossen Sahara" at the Nile (compare also Graebner 1910, 1929). Based on lichenological data, Reichert (1936) placed the boundary between the western and eastern parts of the Sahara "somewhere near Tunisia." According to Zohary (1973:95), the dividing line between his western and eastern province occurs somewhere in the Libyan Desert.

77. Hayek (1926) includes in this phytochorion only Egypt east of the source of the Nile, the Sinai, and the Arabian Peninsula, and refers the major part to his "Persisch-Indische Provinz." I base the boundaries of this province principally on Zohary (1973).

78. Zohary, Davis and Hedge, Wickens, and a series of other authors also include in the Irano-Turanian Region the "Mauritanian Steppes Province" in northern Africa (i.e., the belt of northwestern African "Hauts Plateaux"). But Ozenda (1964) and Quézel (1978) are correct in denying the existence of a separate, independent territory of the Irano-Turanian flora in northwestern Africa, although they recognize the appearance of many species in common.

79. The Western Asiatic Subregion corresponds to the Asiatic part of Zohary's (1963, 1973) "Western Irano-Turanian Subregion," but with inclusion also of the Hyrcanian Province, which I refer to the Irano-Turanian Region.

80. Engler called the Armeno-Iranian Province the "Armeno-iranische Mediterranprovinz." It corresponds to the "Armenisch-iranische Steppen-region" of Drude (1890).

81. A. V. Prozorovsky and V. P. Maleev (1947) award this chorion the rank of a province.

82. The Turanian Province includes the three provinces of Meusel et al. (1965): the Araxian; the southern Turanian; and the Aralo-Caspian in the narrow sense. For Central Asia, Korovin (1958, 1961, 1962) divided the territory of the Turanian or Aralo-Caspian flora into two provinces—the Turanian Desert Province and the Central Kasakhstanian Province. The boundary of the Turanian flora in Central Asia which is used in the present work follows that of Korovin.

83. In the limits of their Southern Turkestan Province, Prozorovsky (in Prozorovsky and Maleev 1947) recognized a separate Paropamisus Subprovince, and in the boundaries of his Turkmeno-Iranian Province, Korovin (1962) created the separate Badghys District, in which he also included the Karabil upland.

84. The Soviet part of Central Asia is called Middle Asia (Srednyaya Aziya) by the Russian authors.

85. Engler called this province "Californische Küstenprovinz." Later (1902, 1903, 1924), he included it in his "West-amerikanische Wüsten- und Steppenprovinz."

86. Thorne introduces the name Sonoran Province in a table, without indicating

its limits. In his "Chaparal-Provinz," Engler included much of New Mexico and Texas, and in his "Sonora-Provinz" he included Sonora, most of the Californian Peninsula, southern Arizona, and a small part of New Mexico. The phraseology varies in successive versions.

II
PALEOTROPICAL KINGDOM (PALEOTROPIS)

The Paleotropical Kingdom includes the tropics of the Old World, with the exception of Australia. It contains all the tropical islands of the Pacific Ocean, with the exception of several islands along the shores of America.

The extremely rich flora of the Paleotropical Kingdom contains about 40 endemic families, of which we mention only Matoniaceae, Dipteridaceae, Stangeriaceae, Welwitschiaceae, Degeneriaceae, Rafflesiaceae s.str., Nepenthaceae, Didiereaceae, Didymelaceae, Ancistrocladaceae, Dioncophyllaceae, Scytopetalaceae, Medusagynaceae, Scyphostegiaceae, Sarcolaenaceae, Sphaerocephalaceae, Huaceae, Pandaceae, Crypteroniaceae, Duabangaceae, Strephonemataceae, Psiloxylaceae, Dirachmaceae, Phellinaceae, Lophopyxidaceae, Salvadoraceae, Medusandraceae, Mastixiaceae, Hoplestigmataceae, and Lowiaceae. The number of genera and especially of species is difficult to calculate.

Extending from the great vastness of Africa to Polynesia, the Paleotropical Kingdom is very diverse, and I recognize the following subkingdoms within its borders: African, Madagascan, Indomalesian, Polynesian, and Neocaledonian.

A. AFRICAN SUBKINGDOM

Good 1947, 1974; Mattick 1964; Takhtajan 1969, 1970, 1974; Werger 1978a.

This subkingdom includes the larger part of the African continent, the tropical deserts of the Arabian Peninsula, and tropical deserts of Iran, Pakistan, and northwestern India. The African Subkingdom may be divided into the following regions: Guineo-Congolian; Uzambara-Zululand; Sudano-Zambezian; Karoo-Namib; and St. Helena and Ascension Region.

10. Guineo-Congolian Region

Waldregion von Guinea und dem Congogebiet, Drude 1890; Région congo-guinéene, Trochain 1952; Région guineo-congolaise, Monod 1957; Lebrun 1958, 1961; Schmithüsen 1961; Aubréville 1962; Léonard 1965; White 1965, 1971; White (in Chapman and White) 1970; White 1978, 1979, 1983; White and Werger 1978; Troupin 1966; Takhtajan 1970, 1974; Knapp 1973; Clayton and Hepper 1974; Tolmatchev 1974; Wickens 1976; Werger 1978a; Brenan 1978; Denys 1980.

The Guineo-Congolian Region[1] extends from southwestern Gambia and southwestern Senegal to northwestern Angola, and includes the Congo River Basin. Eastward it reaches the southwestern corner of the Sudan Republic, southwestern Uganda, western Kenya, and northwestern Tanzania, tentatively including the basin of Lake Victoria. There is a gap in the coastal strip between southeastern Ghana and southwestern Dahomey.

The flora of this region is rich. It has several endemic families and subfamilies: Dioncophyllaceae, Scytopetalaceae, Huaceae, Strephonemataceae, Lecythidaceae-Napoleonoideae, Medusandraceae, and Hoplestigmataceae. There are many endemic genera (especially in Cameroon and Gabon), including *Afroguatteria, Dennettia, Gilbertiella, Letestudoxa, Pseudartabotrys* and *Toussaintia* (Annonaceae), *Scyphocephalium* (Myristicaceae), *Fleurydora* (Ochnaceae), *Habropetalum* (Dioncophyllaceae), *Lecomptedoxa, Letestua* and *Tulestea* (Sapotaceae), *Amphimas, Angylocalyx, Augouardia, Camoensia, Eurypetalum, Gilbertodendron, Neochevalierodendron, Paraberlinia, Sinderopsis* and *Tessmannia* (Fabaceae), *Butumia, Letestuella* and *Stonesia* (Podostemaceae), *Crateranthus* (Lecythidaceae), *Oriciopsis* (Rutaceae), *Gymnostemon* (Simaroubaceae), *Chonopetalum* and *Laccodiscus* (Sapindaceae), *Lepidobotrys* (Oxalidaceae), *Octoknema* (Olacaceae), *Coleactina, Globulostylis* and *Temnopteryx* (Rubiaceae), *Djaloniella* (Gentianaceae), *Dinkladeodoxa* (Bignoniaceae), *Bafutia* (Asteraceae), *Moldenkea* (Amaryllidaceae), and *Hypseochloa* (Poaceae). According to White (1983:74), there are about 8,000 species, of which more than 80% are endemic.

The characteristic vegetation types of the Guineo-Congolian Region are tropical rain forest and in some parts—especially in the Congo—forest-savanna mosaic (see White 1983).

This region is divided into the following three provinces.

1. **Upper Guinea Province** (Domaine Éburnéo-libérien, Lebrun 1961; Domaine libéro-ivoréen, Aubréville 1962; Upper Guinea centre of endemism, White 1979; Upper Guinea Domain, Denys 1980). This province comprises the northeastern part of the region and stretches along the Atlantic coast of Africa from southwestern Gambia to eastern Ghana (approximately to 0° longitude). It contains a number of endemic genera, including *Habropetalum, Dinkladeodoxa*, the Liberian monotypic genus *Maschalocephalus* (the only Old World genus of the American family Rapateaceae), and a considerable number of endemic species, among them *Pitcairnia feliciana* (the only Old World member of the American family Bromeliaceae). The dominant vegetation type is tropical rain forest.

2. **Nigerian-Cameroonian Province** (Domaine nigero-camerounais, Lebrun 1961; Nigerian-Camerounian Domain, Werger 1978*a*, p.p.; Lower Guinea centre of endemism, White 1979; Lower Guinea Domain, Denys 1980). This province occupies southwestern Dahomey, southern Nigeria, western Cameroon, Equatorial Guinea, the greater part of Gabon, and western Congo. It contains several endemic genera and many endemic species. Tropical rain forests and forest-savanna mosaics are the main vegetation types of the province.

3. **Congolian Province** (Domaine congolais, Aubréville 1962; Congo Domain, Clayton and Hepper 1974; Wickens 1976; Brenan 1978; Congolian Domain, Werger 1978*a;* Congo Basin Domain, Denys 1980). This vast province comprises all of the remaining parts of the Guineo-Congolian Region. The flora of this province, especially the flora of eastern Nigeria and Cameroon, is very rich and contains many endemic genera and numerous endemic species. The Cameroons system, the islands of the Gulf of Guinea (Fernando Po, Principé, São Tomé, and Annobon), and the Virunga Volcanoes and Mt. Ruwenzori harbor rich montane and afroalpine floras with a high percentage of endemism (Morton 1972; Brenan 1978). The montane flora of Cameroons Mountain and the Bamenda Highlands as well as that of the neighboring islands has striking affinities with that of the mountains of East Africa and of Europe (Hooker 1864; Keay 1955). Among the plants with boreal and Mediterranean affinities are *Cerastium octandrum, Cardamine hirsuta, Radiola linoides, Umbilicus botryoides*, and *Succisa trichocephala*. A number of plants (including *Thalictrum rhynchocarpum, Clematis simensis, Hypericum lanceolatum, Viola abyssinica, Agauria salicifolia, Philippia mannii, Crassula alba, Adenocarpus mannii,* and *Arundinaria alpina*) are conspecific with the East African types. There are even more boreal and Mediterranean elements on Mt. Ruwenzori and on the Virunga Volcanoes in East Africa. Tropical rain forests and forest-savanna mosaic are the main vegetation types of this province. For the high mountains, montane forest, montane wood-

land, and the ericaceous belt are characteristic, as well as the alpine belt on the highest mountains. On the Virunga Volcanoes and Mt. Ruwenzori there occur dense *Senecio* woodlands, *Helichrysum* scrub, *Alchemilla* scrub, tussock grassland, and *Carex* bogs and related communities (see Hedberg 1964).

11. Uzambara-Zululand Region

Domaine forestier oriental, Monod 1957; Domaine oriental, Aubréville 1962; Domaine côtier oriental, Troupin 1966; Uzambara-Zululand Domain, White (in Chapman and White) 1970; Wickens 1976; Brenan 1978; Indian Ocean Coastal Belt, Werger 1978a; White and Moll 1978; Region of the Indian Ocean Coastal Belt, Denys 1980.

This region comprises a relatively narrow strip along or near the eastern coast of Africa, from the extreme southeastern corner of Somalia southward to the neighborhood of Port Elizabeth in the eastern Cape Province. Monod (1957) treated the Indian Ocean Coastal Belt as an outlying oriental "domaine" of the Guineo-Congolian Region, and a number of authors more or less accepted this view (see Aubréville 1962; Léonard 1965; White 1965; White [in Chapman and White] 1970; Troupin 1966; Clayton and Hepper 1974; Wickens 1976; Brenan 1978). But on the basis of a comprehensive study of the phytochoria of Africa, White (1979) later concluded that the floristic relationships between the Indian Coastal Belt and the Guineo-Congolian Region were much exaggerated. According to White (1979:27) there are "no grounds for treating the Indian Ocean Coastal Belt, either in whole or in part, as belonging to the Guineo-Congolian Region." Factor analysis confirms White's conclusion (Denys 1980). Both White (1979) and Denys (1980) consider this area a separate entity.

The flora of the Uzambara-Zululand Region is rich and contains many endemics. There are two endemic families—Stangeriaceae and Rhynchocalycaceae—and about 20 endemic genera (including *Stangeria*, *Englerodendron* and *Schefflerodendron*) and many endemic species. According to White (1979), the degree of specific endemism is possibly as high as 40%. There are some linking taxa with Madagascar (e.g., species of *Ludia*, *Aphloia*, *Macphersonia*, and *Mascarenhasia*), as well as with the Somalo-Ethiopian and Zambesian provinces and with the Guineo-Congolian, Karoo-Namib, and Cape regions.

According to White and Moll (1978:564), "the floras of the northern and southern parts of the Coastal Belt, although part of the same continuum, are so different that it is necessary to recognize two major phytochoria—the Zanzibar-Inhambane and the Tongoland-Pondoland transitional and mosaic Regions." But it is necessary to note that the term "region," as used by White

(1976, 1979), is not entirely equivalent to region in the Englerian hierarchical chorionomic system. I prefer to consider these phytochoria as two provinces of one region.

1. Zanzibar-Inhambane Province (Zanzibar-Inhambane Regional Mosaic, White 1976, 1978, 1979, 1983; White and Moll 1978; Werger 1978*a*). According to White and Moll (1978) and White (1983:184), this phytochorion occupies a coastal belt from southern Somalia (1° N) to the mouth of the Limpopo River (25° S), and varies in width from 50 to 200 km, except where it penetrates inland along certain broad river valleys. It lies mostly below 200 m. There are 4 endemic genera: *Cephalosphaera* (Myristicaceae), *Grandidiera* (Flacourtiaceae), *Englerodendron* and *Stuhlmannia* (Fabaceae). The endemic genera are confined or almost confined to the East Usambara Mountains. There are about 3,000 species, of which at least several hundred (including 92 forest trees) are endemic (White 1983). The endemics are centered mainly on the Shimba Hills and East Usambara Mountains. Impoverishment to the south is rapid. The prevalent vegetation types are tropical forest, bushlands, and thickets (White and Moll 1978; White 1983).

2. Tongoland-Pondoland Province (Tongoland-Pondoland Regional Mosaic, White 1976, 1978, 1979, 1983; White and Moll 1978; Werger 1978*a;* Tongoland-Pondoland Region, Goldblatt 1978). According to White and Moll (1978) and White (1983), this province stretches from the mouth of the Limpopo River (25° S) to about Port Elizabeth (34° S). In the north it is up to 240 km wide, but locally in the south, where mountains come close to the sea, its width is no more than 8 km (White and Moll 1978:572; White 1983:197). The flora of this province comprises about 3,000 species. Endemism is more pronounced than in the Zanzibar-Inhambane Province. There are two endemic monotypic families—Stangeriaceae and Rhynchocalycaceae—and about 15 endemic genera (Goldblatt 1978), including *Stangeria* (Stangeriaceae), *Bachmannia* (Capparaceae), *Umtiza* (Fabaceae), *Galpinia* and *Rhynchocalyx* (Rhynchocalycaceae), *Harpephyllum* and *Loxostylis* (Anacardiaceae), *Pseudosalacia* (Celastraceae), *Ephippiocarpa* (Apocynaceae), *Anastrabe* (Scrophulariaceae), *Mackaya* (Acanthaceae), and *Jubaeopsis* (Arecaceae). Specific endemism is high. According to White and Moll (1978), more than 200 larger woody species—about 40% of the total—are endemic. Of 35 species of the genus *Encephalartos,* 18 are Tongoland-Pondoland endemics. Some genera, including *Bersama, Diospyros* (section Royena), *Euclea* and *Rhoicissus,* have their centre of variation in this phytochorion. Among the endemics are a number of succulent tree euphorbias and *Aloe bainesii.* White and Moll also point out that some Tongoland-Pondoland endemic species are geographically widely separated from the other species of their genera. Some of the above facts lead them to conclude that the

Tongoland-Pondoland area "has served as a refuge for genera which were formerly more widespread on the African mainland" (White and Moll 1978:573). The vegetation is a mosaic of subtropical forest and coastal grassland that interfingers with elements of the "afromontane" flora (White 1976, 1983; White and Moll 1978; Goldblatt 1978).

12. Sudano-Zambezian Region

Lebrun 1947, 1958; Schmithüsen 1961; White 1965; White (in Chapman and White) 1970; Knapp 1973; Clayton and Hepper 1974; Tolmatchev 1974; Wickens 1976; Brenan 1978; Werger 1978*a*; Werger and Coetzee 1978; Région soudano-angolan, Trochain 1952; Monod 1957; Takhtajan 1969, 1974; Quézel 1978 (see Holarctic bibliography).

This region comprises the huge area from the Atlantic coasts of southern Mauritania, Senegal, and northeastern Guinea to the Sudan Republic (including a greater part of it), northeastern and eastern tropical Africa, the island of Socotra, tropical parts of the Arabian Peninsula, and tropical deserts of Iran, Pakistan, and northwestern India. To the north it is bounded by the deserts and semideserts of the Saharo-Arabian and Irano-Turanian Regions (in the Tethyan Subkingdom), in the west it borders the Guineo-Congolian Region, while in southern Africa it extends to the deserts and semideserts of the Karoo-Namib Region and the sclerophyllous fynbos of the Cape Region (in the Cape Kingdom).

The Sudano-Zambezian Region has three small endemic families (Barbeyaceae, Dirachmaceae, and Kirkiaceae), a number of endemic genera, and many endemic species. The characteristic genera of the region include *Acacia, Combretum, Terminalia, Brachystegia, Hyparrhenia,* and so on (Wickens 1976).

The main vegetation types of the region are open woodland, savanna, and grassland (see White 1983).

The Sudano-Zambezian Region is divided into four subregions: Zambezian, Sahelo-Sudanian, Eritreo-Arabian, and Omano-Sindian.

12A. ZAMBEZIAN SUBREGION

Angolo-Tanzanian Subregion, Takhtajan 1970, 1974; Zambezian Region, Goldblatt 1978; Denys 1980; White 1983.

The Zambezian Subregion is a vast area that extends from 3° to 26° south latitude and from the Atlantic Ocean almost to the Indian Ocean. It includes southern Zaire (Shaba), most of Angola, the whole of Zambia, Malawi, and

Zimbabwe, most of Tanzania and Mozambique, the northeastern part of Namibia, the greater part of Botswana (including Central Kalahari), and most of the eastern part of the Republic of South Africa, together with Lesotho and most of Swaziland. According to Werger (1978b), its border with the Karoo-Namib Region is floristically quite clearly defined and follows major physiographic lines. In southwestern Angola, about halfway between Moçamedes and Benguala, and in northern Namibia, the border runs gradually further inland along the eastern edge of the escarpment. In Namibia the border generally follows the escarpment until the vicinity of Windhoek. South of Windhoek the border between the Karoo-Namib and Sudano-Zambezian regions is formed by the eastern frontier of the dune area of the southern Kalahari and runs further south, crossing the Orange River just upstream of the Orange-Vaal confluence near Luckhoff (Werger 1978b:309).

This subregion has a rich flora, which contains a number of endemic genera, including *Viridivia* (Passifloraceae), *Colophospermum* (Fabaceae), *Ageratinastrum* and *Rastrophyllum* (Asteraceae), *Triceratella* (Commelinaceae), *Richardsiella* (Poaceae), and so on. The majority of endemic genera occur in Angola (including *Mischogyne* of Annonaceae, *Aizoanthemum* of Aizoaceae, *Sedopsis* of Portulacaceae, *Carrisoa* and *Caulocarpus* of Fabaceae, *Aframmi*, *Pseudoselinum* and *Spuriodaucus* of Apiaceae) and in the Drakensberg system (including *Platylophus* of Cunoniaceae, *Seemannaralia* of Araliaceae, *Gonioma* of Apocynaceae, *Strobilopsis* of Scrophulariaceae, *Heteromma* of Asteraceae, and *Rhodohypoxis* and *Saniella* of Hypoxidaceae). There are many endemic species, including *Dianthus chimanimaniens* (Zimbabwe), *Hypericum oligandrum* (Zambia), *Doryalis spinosa* (Malawi), *Homalium chasei* (Zimbabwe), *Glyphaea tomentosa* (Mozambique), *Dombeya brachystemma* (Zambia), *D. calantha* (Malawi), *D. leachii* (Mozambique), *Monotes discolor* (Zambia and northeastern Botswana), *Bombax mosambicense* (Mozambique), *Clutia brassii* (Malawi), *Rhus monticola* (Malawi), *Sorindeia rhodesica* (Zambia), *Helichrysum whyteanum* (Malawi), and *Volkiella*, *Mosdenia* (Transvaal), and others. The Drakensberg system is characterized by the highest percentage of local endemics.

Over most of the plateau, species of *Brachystegia*, *Julbernardia*, and *Isoberlinia* are characteristic, whereas in the lowland plains *Colophospermum mopane* is widely distributed (Wickens 1976). For the higher parts of the Drakensberg system, Holarctic or predominantly Holarctic genera are characteristic, such as *Hypericum*, *Cerastium*, *Silene*, *Geum*, *Trifolium*, *Valeriana*, *Myosotis*, *Ajuga*, *Caucalis*, *Bromus*, *Aira*, *Koeleria*, *Agrostis*, *Poa*, *Festuca*, and so forth.

The Zambezian Subregion is a transitional area, ranging from typically tropical parts to more temperate areas with occasional frosts in winter. The vegetation consists mainly of woodlands, savannas, and grasslands (see White 1983). As Brenan (1978:458) indicates, one of the remarkable features of this phytochorion is the prevalence of huge areas of woodland and savanna-woodland in which the genera *Brachystegia* and *Julbernardia* are conspicuous or dominant—that is, the so-called miombo, a rich vegetation type with a great number of both woody and herbaceous species. Brenan (1978) notes that many of the local endemics so frequent in Zambia, Angola, and southern Zaire occur in miombo. The tree savannas, which are also well represented in this phytochorion, usually occur at lower elevations and lower rainfall; there is a prevalence of *Acacia* and *Combretum* (Brenan 1978:459). Palm savannas, scrubs, montane forests, and subalpine and alpine formations occur locally (see Werger 1978b). The high mountain vegetation is developed in the Drakensberg system and higher parts of Lesotho. The vegetation of the subalpine belt of the Drakensberg consists mainly of bunchgrass land, especially *Themeda triandra*, while the subalpine belt of Lesotho is almost entirely covered by *Themeda-Festuca* grassland. The vegetation of the alpine belt consists of climax heath communities dominated chiefly by low, woody species of *Erica* and *Helichrysum* interpersed with grassland that is dominated by species of *Merxmuellera, Festuca,* and *Pentaschistis* (Killick 1978).

The Zambezian Subregion contains one province.

1. Zambezian Province (Domaine Zambesien, Lebrun 1947, p.p.; White 1965; White [in Chapman and White] 1970; White 1971, 1976; Troupin 1966; Clayton and Hepper 1974; Wickens 1976; Brenan 1978; Werger 1978a, 1978b; Werger and Coetzee 1978; Angola-Zambesia Domaine, Monod 1957).

12B. SAHELO-SUDANIAN SUBREGION

Sous-région soudanais, Chevalier and Emberger 1937; Sahelo-Sudanian Subregion, Takhtajan 1969, 1970, 1974; Sudanian Region, White 1976, 1978, 1983; Denys 1980.

This subregion extends from the Atlantic coasts of southern Mauritania and Senegal to the Sudan Republic (including most of it), northern Uganda, and western Kenya, forming a belt bounded on the north by the Saharo-Arabian Region (in the Tethyan Subkingdom). According to White (1983: 103), the very few endemic genera include *Butyrospermum* (Sapotaceae), *Pseudocedrela* (Meliaceae), and *Haematostaphis* (Anacardiaceae), and there are possibly no more than 2,750 species, of which about one-third are endemic. Important genera include *Acacia, Commiphora, Isoberlinia, Anogeissus, Terminalia, Hyparrhenia,* and *Andropogon*. The dominant vegetation types are woodland and grassland. It has two provinces.

1. Sahelian Province (Domaine sahélian, Trochain 1952; Sahelian Domain, Clayton and Hepper 1974; Wickens 1976; Brenan 1978). This province occupies the northern part of the Sahelo-Sudanian Subregion and forms a narrow belt that extends from the Atlantic coasts of southern Mauritania and Senegal eastward to the Red Sea coast of the Sudan Republic. South of Lake Chad the border of the province juts out southward, occupying part of northern Cameroon to approximately 10° north latitude; along the eastern border of the Sudan Republic, it reaches the Red Sea coast. The southern boundary of this province largely corresponds to the southernmost extent of the Pleistocene sand invasion (Wickens 1976). Specific endemism is very low. Certain widespread species of *Acacia* (*A. senegal, A. seyal,* and *A. nubica*) are characteristic, as is the absence of *Terminalia* (Brenan 1978). Semidesert grassland and shrubland are characteristic for the northern belt, as is *Acacia* wooded grassland and deciduous bushland for the southern belt of the province.

2. **Sudanian Province** (Sudanische Parksteppenprovinz, Engler 1908, 1912, 1924, p.p.; Hayek 1926, p.p.; Domaine soudanien, Trochain 1952; Sudanian Domain, White 1965; White [in Chapman and White] 1970; Clayton and Hepper 1974; Wickens 1976; Brenan 1978). This province extends from the Atlantic coast of Senegal eastward to the frontier of the Sudan Republic with Ethiopia. The western border consists of a short stretch along the shores of Senegal with Cape Almadi in the center. In the east the border of the Province turns southward and extends to the basin of Lake Victoria, including the northwestern corner of Zaire, northern Uganda, and two small parts of western Kenya. Specific endemism is much higher than in the Sahelian Province. Important genera include *Anogeissus, Terminalia, Hyparrhenia, Andropogon,* and *Isoberlinia*. *Khaya senegalensis* is also characteristic of this Province (Wickens 1976). The main vegetation type of the Sudanian Province is woodland, which in the southern part is characterized by abundant *Isoberlinia doka*.

12C. Eritreo-Arabian Subregion

Gruenberg-Fertig 1954; Meher-Homji 1965; Eritreo-Arabian Province, Reichert 1921; Zohary 1963, 1973; Somalo-Ethiopian Subregion, Takhtajan 1969, 1970; Northeast African Highland and Steppe Region, Good 1947, 1974; Domaine somalo-éthiopien, Lebrun 1947; Monod 1957; Somalia-Masai regional centre of endemism, White 1983.

This subregion comprises some small isolated parts of the Sudan Republic (including the southernmost part of the coastal zone and the southeastern part of the country), Ethiopia (except some parts along its western border), Somalia, two small areas near the eastern border of Uganda, most of Kenya, a considerable part of Tanzania, the southwestern and southern parts of the

Arabian Peninsula, and the island of Socotra. A small exclave of the Eritreo-Arabian flora is found in Jebel Elba and its surroundings (at the twenty-second parallel) in southeastern Egypt (Zohary 1973).

The Eritreo-Arabian Subregion includes two endemic monotypic families—Barbeyaceae, which is distributed in Ethiopia, Eritrea, Somalia and adjacent parts of the Arabian Peninsula; and Dirachmaceae, which occurs on Socotra. There are about 50 endemic genera (mostly oligo- and monotypic), including *Poskea* (Globulariaceae) which occurs in Somalia and on Socotra, and the palm *Wissmannia*, which occurs in Somalia and southern Arabian Peninsula. There are many endemic species. According to White (1983:111), the following nonendemic genera have important concentrations of endemic species: *Acacia* (30), *Aloe, Boscia* (7), *Boswellia* (6), *Cadaba* (10), *Ceropegia, Commicarpus* (7), *Commiphora* (60), *Crotalaria* (30), *Euphorbia, Farsetia* (8), *Indigofera* (20), *Ipomoea* (20), *Jatropha* (6), *Maerua* (10), *Moringa* (9), *Neuracanthus* (8), *Otostegia* (5), *Psilotrichum* (7) and *Terminalia* (5). Of the 120 or so Stapelieae known from the Sudano-Zambezian Region, all but about 8 are endemic. According to Brenan (1978:458), the "Horn of Africa" has at least a dozen endemic species of *Acacia*—more than any other part of Africa.

The northern part of this subregion includes a large area of afromontane and afroalpine flora, which has strong affinities both with the other high mountain floras of Africa and those of the boreal and Mediterranean regions. No less than 80% of this flora is endemic to the high mountains of tropical East Africa and Ethiopia, indicating that is has long been isolated from other high mountain and temperate floras (Hedberg 1961, 1970). White (1965, 1978, 1983) proposed the recognition of an archipelago-like Afromontane Region, comprising "islands" of afromontane and afroalpine flora very widely distributed on the African mainland. This concept has been accepted by a number of authors (Troupin 1966; Wickens 1976; Werger 1978*a;* Brenan 1978; Goldblatt 1978; Denys 1980), and some (Werger 1978*a;* Killick 1978) even recognize two regions (Afromontane and Afroalpine). I prefer to follow Engler (1908), Lebrun (1947), Monod (1957), and others, who regard the flora of the African mountains as parts of the various phytochoria in which they occur.

The Eritreo-Arabian Subregion is divided into three provinces.

1. Somalo-Ethiopian Province (Nordafrikanische Hochland und Steppenprovinz, Engler 1903, 1924; Domaine somalo-éthiopien, Lebrun 1947; Domaine oriental, Lebrun 1947; Oriental Domain, White 1965; Afroorien-tal Domain, Hepper [in Clayton and Hepper] 1974; Wickens 1976; Brenan 1978). This province includes the entire African part of the Eritreo-Arabian Subregion. There are a number of endemic genera, including:

Calyptrotheca (Portulacaceae), *Allmaniopsis, Dasysphaera, Chionothrix, Neo-

PALEOTROPICAL KINGDOM 207

centema, Pleuropterantha, Pseudodigera, Sericomopsis, and *Volkensinia* (Amaranthaceae), *Gyroptera* and *Lagenantha* (Chenopodiaceae, close to *Salsola*), *Loewia* (Turneraceae), *Cephalopentandra* and *Myrmecosicyos* (Cucurbitaceae), *Puccionia* (Brassicaceae), *Rumicarpus* (Tiliaceae), *Afrovivella* (Crassulaceae), *Arthrocarpum, Cordeauxia, Dicraeopetalum, Spathionema* and *Vatovaea* (Fabaceae), *Kelleronia* (Zygophyllaceae), *Scassellatia* (Anacardiaceae), *Bottegoa, Hypseloderma* and *Sennia* (Sapindaceae), *Erythroselinum* and *Cynosciadium* (close to *Pimpinella,* Apiaceae), *Lamellisepalum* (Rhamnaceae), *Dibrachionostylus, Paolia,* and *Pentanopsis* (Rubiaceae), *Kanahia* and *Socotora* (Asclepiadaceae), *Simenia* (Dipsacaceae), *Pseudolithos* and *Rhytidocaulon* (Asclepiadaceae), *Cladostigma* and *Sabaudiella* (Convolvulaceae), *Erythrochlamys, Hyperaspis* and *Capitanya* (Lamiaceae), *Xylocalyx* (Scrophulariaceae), *Chamaeacanthus, Golaea* and *Lindauea* (Acanthaceae), *Drakebrockmania, Keniochloa, Leptagrostis, Odontelytrum,* and *Pseudozoysia* (Poaceae).

This province also contains many endemic species, particularly within such genera as *Acacia, Commiphora, Combretum,* the cactiform *Euphorbia, Grewia,* and so on (Wickens 1976; Brenan 1978). There are many afromontane and afroalpine taxa, especially in Ethiopia, which include:

Podocarpus falcatus, Juniperus procera, Ocotea bullata, s.l., *Anemone thomsonii, Ranunculus stagnalis, Trichocladus elliptica, Parietaria debilis, Myrica salicifolia, Montia fontana, Cerastium afromontanum, C. octandrum, Sagina abyssinica, S. afroalpina, Stellaria mannii, Uebilinia spathulaefolia, Rumex bequetii, Hypericum peplidifolium, H. revolutum, Ritchiea albersii, Arabidopsis thaliana, Arabis alpina, Cardamine hirsuta, C. obliqua, Oreophyton falcatum, Subularia monticola, Agauria salicifolia, Blaeria spicata, Erica arborea, Philippia* spp., *Aningueria adolfi-friederici, Maesa lanceolata, Mitrogyna rubrostipulata, Rapanea melanophloes, Anagallis serpens, Cola greenwayi, Macaranga kilimandscharica, Suregada procera, Gnidia glauca, Crassula granvikii, C. schimperi, Sedum crassularia, S. epidendrum, Umbilicus botryoides, Alchemilla* spp., *Hagenia abyssinica, Prunus africana, Acacia abyssinica, Albizia schimperana, Parochaetus communis, Trifolium* spp., *Epilobium stereophyllum, Fagaropsis angolensis, Turraea holstii, T. robusta, Geranium arabicum, Schefflera abyssinica, S. myriantha, Caucalis melanantha, Haplosciadium abyssinicum, Heracleum abyssinicum, Hydrocotyle monticola, Pimpinella oreophila, Schimperella verrucosa, Apodytes dimidiata, Ilex mitis, Catha edulis, Thesium kilimandscharicum, Galium glaciale, Galiniera coffeoides, Sebaea brachyphylla, Swertia* spp., *Dipsacus pinnatifidus, Scabiosa columbaria, Myosotis* spp., *Salvia* spp., *Satureja* spp., *Thymus* spp., *Callitriche stagnalis, Nuxia congesta, Bartsia* spp., *Halleria lucida, Hebenstretia dentata, Limosella africana, Veronica* spp., *Lobelia rhynochopetalum, Wahlenbergia pusilla, Anthemis tigreensis, Cineraria grandiflora, Conyza variegata, Cotula abyssinica, Dichrocephala alpina, Erigeron alpinus, Gnaphalium schulzii, Haplocarpha* spp., *Helichrysum* spp., *Nannoseris schimperi, Senecio* spp. (including a few species of the subgenus *Dendrosenecio* in the southern part of the

Province), *Kniphofia foliosa, Hesperantha petitiana, Romulea columnae, R. fischeri, Cyanotis barbata, Eriocaulon schimperi, Juncus capitatus, Luzula abyssinica, L. johnstonii, Carex* spp., *Scirpus* spp., *Agrostis* spp., *Aira caryophyllea, Andropogon amethystinus, Anthoxanthum* spp., *Arundinaria alpina, Bromus cognatus, Deschampsia* spp., *Festuca abyssinica, Helictotrichon elongatum, Keniochloa* spp., *Koeleria capensis, Pentaschistus mannii,* and *Poa* spp.

In northern Ethiopia species of such genera as *Primula, Saxifraga,* and *Rosa* occur (Hedberg 1957, 1961, 1964, 1965). The dominant vegetation types are *Acacia-Commiphora* deciduous bushland and thicket, semidesert grassland and shrubland, evergreen and semievergreen bushland and thicket, montane forest, ericaceous scrub (the most important dominants of which are *Erica arborea,* and especially *Philippia keniensis* subsp. *abyssinica;* Hedberg 1961), and alpine vegetation (*Dendrosenecio* woodland, *Helichrysum* scrub, *Alchemilla* scrub, tussock grassland, and *Carex* bogs and related communities; see Hedberg 1964).

2. **South Arabian Province** (South Arabian Domain, Clayton and Hepper 1974; Wickens 1976; Brenan 1978; Yemen and South Arabia, Good 1974; Sous-domaine sud-arabique, Monod 1957; Yemen Province, Takhtajan 1970). This province comprises the southern and southwestern parts of the Arabian Peninsula, including Hadhramaut and Asir (where it extends almost as far north as Mecca). The flora of the province is predominantly African and is the richest in the Arabian Peninsula, containing about 225 endemic species. According to Zohary (1973:244), the vegetation of this area is most complex and is closely reminiscent of that of mountainous East Africa, especially Somalia, both in its floristic makeup and in its altitudinal zonation. On eroded plains, *Acacia-Maerua* bushland is dominant; in drained valleys, *Ziziphus spina-christi* and *Cissus quadrangularis* are most characteristic; on the slopes up to 1,500 m above sea level, the *Acacia-Commiphora* deciduous bushland is characteristic. *Ficus sycomorus* and *F. salicifolius* often grow near water courses. Species of *Combretum* and *Terminalia* also enter this type of vegetation. Around and above 1,500 m, in the Hijaz and Asir mountains, there is an evergreen montane forest or scrub that is characterized or dominated by *Olea europaea* subsp. *africana* and *Tarchonanthus camphoratus*. The high zone of the ranges that run parallel to the coast is inhabited by a coniferous forest of *Juniperus procera*. The high mountains (2,500–3,000 m and over) show islands of afroalpine flora and vegetation on top. The most characteristic species here are *Helichrysum abyssinicum, Lactuca yemensis, Pittosporum abyssinicum, Senecio harazianus, S. hadiensis, Alchemilla cryptantha, Rubus petitianus, Potentilla reptans, Rosa abyssinica, Gerbera piloselloides, Crassula* spp., and many grasses and herbs, which form a kind of open, heathlike dwarf-shrub formation (Zohary 1973).

3. **Socotran Province** (Good 1947, 1974; Takhtajan 1970). This province comprises the island of Socotra. The flora of this island is unique and contains an endemic family, Dirachmaceae, and a number of endemic genera, including *Haya* and *Lochia* (Caryophyllaceae), *Dendrosicyos* (Cucurbitaceae), *Lachnocapsa* (Brassicaceae), *Dirachma* (Dirachmaceae), *Nirarathamnos* (Apiaceae), *Placopoda* (Rubiaceae), *Mitolepis* and *Socotranthus* (Asclepiadaceae), *Angkalanthus, Ballochia,* and *Trichocalyx* (Acanthaceae). The majority of its species also occur on the African mainland, but a considerable number are confined to the island. Among the most remarkable endemic species are *Dendrosicyos socotranus, Punica protopunica, Dirachma socotrana,* and *Dracaena cinnabari.*

Socotra is covered by semidesert dwarf shrubland, succulent shrubland, semidesert grassland with shrubs and trees, and in some parts by upland grassland and evergreen thicket mosaic as well. The vegetation of the limestone plateau is the most variable. On the slopes of the Hamadera Hills and on the southern slopes of the Hagghier Massif there is a remarkable formation dominated by *Dracaena cinnabari.* Another spectacular community is the succulent shrubland of the limestone cliffs and valley slopes, especially the rugged slopes of the north. Characteristic species include *Dendrosicyos socotranus, Adenium socotranum, Euphorbia arbuscula, E. spiralis, Dorstenia gigas, Kleinia scottii, Kalanchoe robusta,* and *Aloe perryi* (White 1983:255; see also Popov 1957; Gwynne 1968).

12D. OMANO-SINDIAN SUBREGION

This subregion comprises northeastern Oman and the easternmost part of the United Arabian Emirates (including the entire coastal belt of the Oman Gulf), tropical deserts of Iran along the coasts of the Persian Gulf and Gulf of Oman, tropical deserts of Mekran, a large area of the plains of the Indus River Basin (including the province of Sind in southern Pakistan and the plains of Punjab in Pakistan and India), the Thar Desert (the Indian Desert), Hariana, all of western Rajasthan (west of the Aravalli Range and Yamuna River), and northern Gujarat.

The Omano-Sindian Subregion is floristically richer than the Sahelo-Sudanian, but it is considerably impoverished in comparison with the Eritreo-Arabian Subregion. The dominant vegetation types of this subregion are open woodland and desert.

1. **Province of Oman.** This province comprises the Arabian part of the Omano-Sindian Subregion. The flora of this province is comparatively rich and contains a number of endemics, including *Delphinium penicillatum, Her-*

niaria mascatensis, Pteropyrum scoparium, Rumex limoniastrum, Limonium arabicum, Ochradenus (Homalodiscus) aucheri, Reptonia mascatensis, Primula aucheri, Tephrosia haussknechtii, Rhus aucheri, Polygala mascatensis, Convolvulus mascatensis, C. ulicinus, Salvia macilenta, Teucrium mascatense, Verbascum akdarensis, Barleria aucheriana, Iphiona horrida, and *Phagnalon arabicum.*

Although this province is separated from the other part of the subregion by the Gulf of Oman, there are many species growing on both sides of the gulf, including *Juniperus macropoda, Cometes surratensis, Viola cinerea, Capparis elliptica, Cleome oxypetala, Dionysia mira, Euphorbia larica, Ebenus stellata, Taverniera glabra, Tephrosia persica, Trigonella uncata, Caucalis stocksiana, Pycnocycla aucheriana, Galium ceratopodum, Jaubertia aucheri, J. hymenostephana, Oldenlandia retrorsa, Nerium mascatense, Lonicera aucheri, Echiochilon persicum, Heliotropium lasiocarpum, Lactuca auriculata, Aristida pogonoptila, Sporobolus arabicus,* and *S. minutiflorus.*

According to Mandaville (1975), the broad vegetation zones of Central Jabal Al-Akhdar are *Acacia* desert parkland (dominant species *Acacia tortilis*), mountain wadi associations (dominant species *Ziziphus spina-christi, Acacia* spp., *Ficus salicifolia*), Euphorbia larica slopes (dominant species *Euphorbia larica, Jaubertia* spp.), *Reptonia-Olea* woodland (*Reptonia mascatensis* and *Olea europaea* subsp. *africana*), and *Juniperus* summit zone (*Juniperus macropoda* and *Cymbopogon* spp., together with *Dodonaea viscosa, Euryops pinifolius,* and *Teucrium mascatense).*

2. South Iranian Province (Südiranisches Gebiet, Rechinger 1951*b*; Garmsirische Provinz, Meusel et al. 1965). This province comprises tropical parts of Fars and Mekran (Makran) in southern Iran and the tropical part of Mekran in southern Baluchistan (southwestern Pakistan). It extends at lower altitudes in a belt of varying width bordering the Persian Gulf and Gulf of Oman. This territory, which is rather continuous in its southernmost and southwestern parts, becomes interrupted further north and sends numerous branches northward and eastward along the wadis that traverse the southern ridges of the Zagros and Mekran ranges (Zohary 1973). It is therefore very difficult to draw any accurate line between the territories of the Omano-Sindian and Irano-Turanian floras in Iran and Pakistan.

The flora of this province is the richest within the Omano-Sindian Subregion and contains a number of endemic and subendemic taxa. There are several endemic or subendemic genera including *Zhumeria* (Lamiaceae), which is of restricted range in the Bandar Abbas area (see Hedge and Wendelbo 1978). Examples of endemic species are *Diceratella canescens, D. floccosa, Grewia makranica, Argyrolobium kotschyi, Crotalaria furfuracea, C. iranica, Pycnocycla nodiflora, Fagonia acerosa, F. subinermis, Monsonia commixta, Otostegia aucheri, Salvia mirzayanii, Schweinfurthia papilionacea, Achillea eriophora, Echinops kotschyi, Grantia aucheri* and two other allied species, and *Platychaeta* spp. (see Zohary 1973; Hedge and Wendelbo 1978).

The dominant vegetation is tropical desert, which consists of various communities, such as the following: association of *Periploca aphylla—Prunus (Amygdalus) arabica*; association of *Euphorbia larica—Jaubertia aucheri*; association of *Euphorbia larica—Convolvulus acanthocladus*; association of *Euphorbia larica—Sphaerocoma aucheri*; association of *Convolvulus acanthocladus, Helianthemum sessiliflorum, Gymnocarpos decander, Heliotropium persicum, Convolvulus spinosus, Cymbopogon schoenanthus, Eremopogon foveolatus, Aristida* spp., *Forskaelea tenacissima*, and *Jaubertia calycoptera*; and association of *Acacia flava—Prosopis spicigera* (see Zohary 1963).

3. Sindian Province (The province of Sindh, Hooker and Thomson 1855; Sindische Provinz, Meusel et al. 1965; The Indus-Plain Province, Hooker [1904] 1907; Chatterjee 1940, 1962; Maheshwari et al. 1965). This province includes the Province of Sind in Pakistan, the plains of Punjab in Pakistan and India, the Thar Desert, Hariana, western Rajasthan, and northern Gujarat. In the southeast, the boundary of the Sindian Province passes somewhat east of Mehsana (northern Gujarat), extending further through the vicinity of Godhra; east of Baroda and Dabhoi, it turns southwest and extends to the Gulf of Cambay, somewhat south of Broach.

Despite the enormous area of this province, its flora is not rich, containing no endemic genera and only a few endemic species, among them *Moringa concanensis, Abutilon* spp., and *Tephrosia falciformis*. The most common trees and shrubs are *Acacia senegal, A. leucophloea, A. jacquemontii, A. catechu, A. nilotica, Anogeissus pendula, Prosopis cineraria, Salvadora oleoides, Ziziphus nummularia, Capparis decidua, Boswellia serrata, Maytenus emarginatus, Commiphora wightii, Grewia tenax, Euphorbia caducifolia, E. nivulia, Calotropis procera, Balanites aegyptiaca, Moringa concanensis, Mimosa hamata, Leptadenia pyrotechnica, Calligonum polygonoides, Crotalaria burhia, Lycium barbarum, Dipterygium glaucum, Fagonia arabica, Sericostoma pauciflorum, Cordia gharaf, Bauhinia racemosa, Securinega (Flueggea) leucopyrus*, and *Tamarix* spp. *Nannorrhops ritchieana* occurs in some places. The common grasses are species of *Panicum, Lasiurus, Cenchrus, Aristida, Cymbopogon, Eleusine*, and so on. The dominant vegetation types are *Acacia senegal* forest, tropical desert thorn forest, *Ziziphus nummularia* scrub, *Euphorbia* scrub, *Salvadora-Cassia auriculata* scrub, and desert dune scrub.

13. Karoo-Namib Region

Monod 1957; White 1965; White (in Chapman and White) 1970; White 1976, 1978, 1979, 1983; Troupin 1966; Volk 1966; Takhtajan 1969, 1974; Knapp 1973; Clayton and Hepper 1974; Wickens 1976; Brenan 1978; Goldblatt 1978; Werger 1978*a*, 1978*b*.

This region includes the large desert and semidesert areas in the southwestern part of southern Africa, south and west of the Zambezian Province and to the north of the Cape Region. It has a rich and very distinct flora, characterized by very high degree of endemism, which indicates its great antiquity. There is one endemic family (Welwitschiaceae), and there are at least 80 endemic genera and numerous endemic species. The majority of endemic genera belongs to the Aizoaceae. The families Asteraceae, Scrophulariaceae, Liliaceae s.l., and Poaceae (particularly the tribe Stipeae) are also very developed here and have many endemics. Endemic genera include the most remarkable monotypic genus *Welwitschia*, which extends almost continuously from San Nikolau, at 14°20' south latitude in southern Angola, to the Kuiseb River just south of the Tropic of Capricorn; its eastern limit in Namibia is near the small town Welwitschia (White 1983:143). *Welwitschia* represents a very ancient gymnospermous stock. Other endemic genera are *Hypertelis* and *Plinthus* (Aizoaceae), most of the Aizoaceae-Mesembryanthemoideae, *Ceraria* (Portulacaceae), *Phaeoptilum* (Nyctaginaceae), *Arthraerua*, *Calicorema*, and *Leucosphaera* (Amaranthaceae), *Grielum* (Rosaceae), *Adenolobus* and *Xerocladia* (Fabaceae), *Augea* and *Sisyndite* (Zygophyllaceae), *Nymania* (Meliaceae), *Sarcocaulon* (Geraniaceae), *Ectadium* and *Microlema* (Asclepiadaceae), *Didelta* (Asteraceae), and *Kaokochloa, Leucophrys, Monelytrum*, and *Phymaspermum* (Poaceae).

There is very high percentage of endemic species in this region. According to White (1983:137), the following genera have important concentrations of endemic species: *Aloe, Anacampseros, Babiana, Chrysocoma, Cotyledon, Crassula, Eriocephalus, Euphorbia, Gasteria, Haworthia, Hermannia, Pentzia, Pteronia, Sarcocaulon, Stipagrostis, Tetragonia*, and *Zygophyllum*. Especially characteristic are numerous species of Aizoaceae-Mesembryanthemoideae (*Mesembryanthemum* and genera close to it, about 1,500 endemic species), *Tetragonia, Adenia*, Crassulaceae, *Acacia, Euphorbia*, Zygophyllaceae, *Oxalis, Pelargonium*, Asclepiadaceae-Stapelieae (6 endemic genera and about 160 endemic species; White 1983), many dwarf shrubby Asteraceae, many species of the Iridaceae, the Liliaceae s.l., the Amaryllidaceae, and grasses. Cape elements contribute significantly to the Karoo-Namib flora, and small relict enclaves of Cape flora are characteristic in the winter rainfall parts of Namaqualand and southern Karoo (Goldblatt 1978; Werger 1978*b*).

According to Werger (1978*b*), "in the arid Karoo-Namib Region the vegetation shows a remarkably high degree of structural and physiognomic diversity and includes some most peculiar formations." Typical of this region is a succulent or narrow-leaved scrub formation or a sand desert with taller, woody plants on the mountain slopes. Open, steppelike grasslands are common on sandy areas. Patches of evergreen and deciduous forests and woodland are confined to the banks of perennial streams or occur along major watercourses (Werger 1978*b*:243).

The Karoo-Namib Region is subdivided into four provinces.

1. **Namib Province** (Domaine du Namib, Monod 1957; Namib Domain, Werger 1978b). This province occupies the extremely arid coastal strip from about 150 km south of Lobito (Angola) to about 30 km south of Alexander Bay at the mouth of the Orange River (Werger 1978b). There are a considerable number of endemics or subendemics in this province, such as *Welwitschia mirabilis* (partly in Namaland Province), *Trianthema heteroensis, Arthraerua horrida, Acanthosicyos horridus* (a remarkable thorny shrub with long tendrils and very long thick roots growing on sand dunes), *Sarcocaulon mossamedense* (a thorny xerophyte with succulent stems), and a number of species of the genus *Stipagrostis* (see De Winter 1971; Werger 1978b).

In some areas, such as the vicinity of the Swakop River, *Welwitschia mirabilis* occurs in peculiar formations in which it is the only conspicuous feature of the vegetation. Stands of *Welwitschia* occur particularly in slight depressions, sometimes with *Maerua welwitschii, Salvadora persica,* and *Acacia tortilis* subsp. *heteracantha* as companion species. Grasses and some other herbs and dwarf shrubs may also be prominent in these stands. Further northward *Welwitschia* occurs locally in the grassland of Namib fringes and does not determine the vegetation physiognomically (Werger 1978b).

2. **Namaland Province** (Namaqualand Desert province, Pole-Evans 1922, p.p.; Domaine du Namaqualand, Monod 1957; Namaland Domain, Werger 1978a, 1978b). According to Werger (1978b), the Namaland Province covers the narrow escarpment belt inland of the Namib Province, but southward it broadens gradually, and south of Windhoek it comprises extensive areas on the plateau, including the southern Kalahari (the "Southern Kalahari Subdomain" of Werger 1978a), which is floristically distinct from the central Kalahari (Bremekamp 1935) and to some extent transitional (White 1965). The southern border of the province runs for the most part a few kilometers south of the Orange River, from the vicinity of Vioolsdrif via the Pfadder area to the vicinity of Upington. The Namaland Province contains many endemics or subendemics, such as *Leucosphaera bainesii,* species of *Euphorbia, Zygophyllum dregeanum* and *Z. macrocarpum, Curroria decidua, Barleria lichtensteiniana, Berkheya chamaepeuce, Stachys burchelliana, Panicum arbusculum* (see Volk 1964, 1966; Werger 1978b). Various kinds of open shrub and dwarf shrub communities are typical vegetation. Several species of *Commiphora, Euphorbia* and *Rhigosum,* and *Aloe dichotoma* are common. In the southern Kalahari, which is distinct geomorphologically and floristically from the remainder of the Namaland Province, psammophilous vegetation is the most characteristic (Werger 1978b).

3. **Western Cape Province** (Western Cape Domain, Werger, 1978a, 1978b). According to Werger (1978a:237), this province occupies the coastal strip and escarpment mountains in South Africa south of the Namib Province. From the Richterveld it extends northward into southwest Africa as far north

as the vicinity of Aus and occurs in a wedge of rugged, mountainous country between the Namib and Namaland provinces. In the south the Little Karoo, lying between the Cape mountain ridges, the Langeberg, and the Swartberg ranges, is included in the province.

The Western Cape Province contains many endemic or subendemic species, including many members of the Aizoaceae and Crassulaceae. It is rich in succulent species, mainly belonging to the Aizoaceae, Crassulaceae and Euphorbiaceae, but also to the Portulacaceae, Chenopodiaceae, Geraniaceae, Zygophyllaceae, Asteraceae and Liliaceae s.l. This province includes the highly succulent dwarf-shrub vegetation along the western coast, on the western escarpment mountains, and on the western edge of the plateau, while in the south it is fringed by Cape fynbos (Werger 1978*b*).

4. Karoo Province (Karoo Province, Pole-Evans 1922; Domain de Karroos, Monod 1957; Karoo Domain, Werger 1978*a*, 1978*b*). Werger (1978*b*) includes in this province the summer rainfall area east of the Western Cape Province and south of the Namaland Province. In the center of the South African plateau it borders on the Zambesian Province of the Sudano-Zambesian Region.

According to Werger (1978*b*:238), in the Karoo Province the Asteraceae are represented by a great number of species, most of which are strongly xeromorphic, dwarf shrubs. As he mentions, endemics and near-endemics are numerous, and include *Nestlera humilis, Pteronia glauca, Thesium hystrix, Zygophyllum gilfillani, Euphorbia aequoris,* and many others. The vegetation is typically an open dwarf-shrub formation. The dwarf shrubs have mostly narrow, ericoid, or pubescent leaves. Succulents are not so common, except in the areas transitional to the Western Cape Province. But grasses are conspicuous, particularly toward the eastern margin. On hillsides, scattered larger shrubs or small trees emerge from the dwarf-shrub and grass layer, and along the riverbeds a woodland or gallery forest is common (Werger 1978*b*:286–287).

14. St. Helena and Ascension Region

Good 1947, 1974; Schmithüsen 1961; Mattick 1964; Takhtajan 1969, 1970, 1974; Gebiet der südatlantischen Inseln, Engler 1899, 1903, 1924.

Despite the small size of these two volcanic islands, their flora is so distinctive that they are usually treated as a separate region. At the time of its discovery, Ascension Island was almost devoid of vegetation, and at the present time it has only three endemic species of flowering plants (*Euphorbia origanoides, Hedyotis adscensionis, Sporobolus durus*). In contrast, the island of

St. Helena was "entirely covered with forests, the trees drooping over the tremendous precipices that overhang the sea" (Hooker 1896:17). These were later destroyed by goats and other domestic animals, which the Portuguese introduced in 1513. The flora of the island shows the strongest relationship with the flora of Africa, especially southern Africa.

Despite the fact that the original flora of this region is not rich in number of genera and species, it is distinguished by high species endemism. Generic endemism is not very high, and only 9 of its genera are endemic, 7 being monotypic. The endemic genera are confined to St. Helena. They show the following relationships:

Trochetiopsis (2 species, Sterculiaceae) is very close to the Mascarene genus *Trochetia* (6 species in Mauritius and Réunion [Marais 1981]).
Nesiota (Rhamnaceae) is closest to the South African genus *Phylica* (about 150 species in southern Africa, on Madagascar and the island of Tristan da Chunha), which is in turn related to South American genera.
Mellissia (Solanaceae) is probably a derivative of *Physalis* (Carlquist 1974).
Trimeris (Lobeliaceae) is very close to *Lobelia*, to the Hawaiian genus *Clermontia*, and to the neotropical genus *Centropogon* (see Mabberley 1974).
Commidendrum (4 species, 1 extinct) and *Melanodendron* (Asteraceae) are probably derived from *Conyza*-like ancestors, whereas the third endemic genus of the Asteraceae, *Petrobium*, is probably a derivative of *Bidens* (Carlquist 1974). Monotypic arborescent genera *Pladaroxylon* (*P. leucadendron*) and *Lachanodes* (*L. arborea*) belong to the "senecionoid" stock.

Of the 39 species of flowering plants on St. Helena, 38 (97%) are endemic (Turrill 1948), but 11 are now extinct (Melville 1979). Examples of endemic species (except the species of endemic genera) are *Frankenia portulacifolia* (shrubby), *Acalypha rubra* (now extinct, closely related to a Mascarene species, *A. reticulata*), *Pelargonium cotyledonis*, *Plantago robusta*, *Wahlenbergia* spp., *Sium helenianum*, *Psiadia rotundifolia* (now extinct, other species of this genus in Africa, Madagascar, Mascarenes, and Socotra), and *Senecio prenanthiflorus*. There is also an endemic tree-fern, *Dicksonia arborescens*. The affinities of the flora are predominantly with Africa, and particularly with southern Africa (Hooker 1896; Turrill 1948; Melville 1979), but also with the Madagascan Region.

The remnant of the indigenous flora of St. Helena "is almost confined to a few patches towards the summit of Diana Peak, the central ridge, 2,700 feet above the sea" (Hooker 1896:17). The indigenous flora of Ascension, which lies about 1,280 km north of St. Helena, is situated high up on the flank of Green Mountain. It is "probably only a small remnant of a much larger flora that had been almost completely overwhelmed by volcanic eruptions and latterly had been further depleted by human interference" (Melville

1979:367). According to Melville (1979), the surviving indigenous flora is very sparse and includes *Euphorbia origanoides, Aristida adscensionis, Sporobolus durus, Hedyotis adscensionis* and *Wahlenbergia linifolia*. He notes that *Euphorbia origanoides* is nearer to the West African *E. trinervia*, and that *Hedyotis adscensionis* is nearer to African species than to the St. Helena *H. arborea*. The *Wahlenbergia* is also on St. Helena, and the fern *Asplenium ascensionis* has recently been found on St. Helena (Melville 1979). In the past there was also a report of the presence of *Commidendrum rugosum*, but no herbarium specimen vouchers exist (Melville 1979).

Despite the exceptionally depauperate nature of the flora of this region, Engler (1899, 1903, 1924) found it possible to divide it into two provinces— the Ascension Island Province, and the Province of St. Helena.

B. MADAGASCAN SUBKINGDOM

Takhtajan 1969, 1970, 1974.

15. Madagascan Region

Engler 1882, 1903, 1912; Engler and Gilg 1919; Graebner 1910; Perrier de la Bâthie 1936; Lebrun 1947; Humbert 1955; Turrill 1959; Schmithüsen 1961; Good 1974; Mattick 1964; White 1965; White (in Chapman and White) 1970; Takhtajan 1969, 1970, 1974; Tolmatchev 1974; Clayton and Hepper 1974; Wickens 1976; Malagassisches Florenreich, Drude 1890; Malagasy province, Krasnov 1899; Provinz der ostafrikanischen Inseln, Hayek 1926; Empire floral de Madagascar et des iles Mascareignes, Chevalier and Emberger 1937; Group malgache: Région occidentale et Région orientale, Monod 1957.

This region includes Madagascar, the Comoros Islands, the Aldabra Islands, the Seychelles, the Amirante Islands, the Mascarene Islands and numerous small islands distributed between them.[2]

The flora of the Madagascan Region is extraordinarily distinctive and is characterized by exceptionally high endemism at the family, genus, and species levels. It includes 12 endemic families (Didymelaceae, Didiereaceae, Barbeuiaceae, Diegodendraceae, Asteropeiaceae, Medusagynaceae, Sarcolaenaceae, Sphaerosepalaceae, Rousseaceae, Physenaceae, Psiloxylaceae, and Geosiridaceae). There are well over 400 endemic genera of vascular plants, of which Madagascar itself accounts for 350 (Good 1974:147). Of the approximately 10,000 species of flowering plants, over 80% are endemic (Humbert 1959; Koechlin et al. 1974; Rauh 1979).

The Orchidaceae are the largest family (over 900 species) of the Madagascan flora, followed in decreasing order by the Rubiaceae, Asteraceae, Fabaceae, Poaceae, Acanthaceae, Cyperaceae, Sterculiaceae, Melastomataceae, Asclepiadaceae, and Euphorbiaceae (Koechlin et al. 1974). In terms of numbers of endemic genera, the Rubiaceae occupy first place, followed in descending order by the Acanthaceae, Arecaceae, Orchidaceae, Fabaceae, Sapindaceae, Poaceae, Asteraceae, and Euphorbiaceae. There are many endemic genera of vascular plants, including:

Pteridaceae: *Ochropteris* (1)
Oleandraceae: *Psammiosorus* (1)
Winteraceae: *Takhtajania* (1, a very isolated genus within the family, which constitutes a separate subfamily; see Leroy 1978)
Annonaceae: *Ambavia* (2), *Feneriva* (1), *Pseudannona* (2)
Canellaceae: *Cinnamosma* (3)
Myristicaceae: *Brochoneura* (3), *Haematodendron* (1), *Mauloutchia* (1)
Monimiaceae: *Decarydendron* (1), *Ephippiandra* (1), *Hedycaryopsis* (1), *Monimia* (4), *Phanerogonocarpus* (2), *Schrameckia* (1), *Tambourissa* (26)
Lauraceae: *Ravensara* (18; related to *Cryptocarya*)
Chloranthaceae: *Ascarinopsis* (1; very close to *Ascarina* and possibly only a section of it)
Balanophoraceae: *Ditepalanthus* (1)
Menispermaceae: *Burasaia* (4), *Orthogynium* (1), *Rameya* (2), *Rhaptonema* (6), *Spirospermum* (1), *Strychnopsis* (1)
Barbeuiaceae: *Barbeuia* (1)
Didiereaceae: *Alluaudia* (6), *Alluaudiopsis* (2), *Decaryia* (1), *Didierea* (2)
Portulacaceae: *Talinella* (2)
Didymelaceae: *Didymeles* (2)
Hamamelidaceae: *Dicoryphe* (15)
Dilleniaceae: *Neowormia* (1)
Asteropeiaceae: *Asteropeia* (7)
Diegodendraceae: *Diegodendron* (1)
Medusagynaceae: *Medusagyne* (1)
Clusiaceae: *Eliaea* (1), *Ochrocarpos* (1), *Paramammea* (1)
Sapotaceae: *Calvaria* (15), *Capurodendron* (22), *Faucherea* (11), *Labramia* (8), *Northea* (5), *Tsebona* (1)
Myrsinaceae: *Badula* (about 20), *Monoporus* (8), *Oncostemum* (about 100)
Flacourtiaceae: *Antinisa* (3), *Bembicia* (1), *Bembiciopsis* (1), *Bivinia* (1; close to *Calantica*), *Prockiopsis* (1), *Tisonia* (14; isolated within the family)
Passifloraceae: *Adenia*
Turneraceae: *Mathurina* (1)
Cucurbitaceae: *Ampelosicyos* (3), *Lemuriosicyos* (1), *Seyrigia* (5), *Tricyclandra* (1), *Trochomeriopsis* (1), *Xerosicyos* (4), *Zambitsia* (1), *Zygosicyos* (1)

Tiliaceae: *Pseudocorchorus* (6)
Dipterocarpaceae: *Vateriopsis* (1, very close to *Vateria*)
Sarcolaenaceae: *Leptolaena* (12), *Rhodolaena* (5), *Sarcolaena* (10), *Schizolaena* (7), *Xerochlamys* (16)
Sphaerosepalaceae: *Dialyceras* (3), *Rhopalocarpus* (15)
Sterculiaceae: *Astiria* (1), *Cheirolaena* (1), *Ruizia* (3), *Trochetia* (6)
Malvaceae: *Helicteropsis* (2), *Humbertianthus* (1), *Humbertiella* (3), *Jumelleanthus* (1), *Macrostelia* (3), *Megistostegium* (3), *Perrierophytum* (9)
Moraceae: *Ampalis* (2), *Maillardia* (4), *Pachytrophe* (1)
Urticaceae: *Neopilea* (1)
Euphorbiaceae: *Cordemoya* (1), *Lautembergia* (4), *Voatamalo* (2), *Wielandia* (1)
Thymelaeaceae: *Stephanodaphne* (9)
Rousseaceae: *Roussea* (1)
Chrysobalanaceae: *Grangeria* (1)
Podostemaceae: *Endocaulos* (1), *Palaeodicraeia* (1), *Thelethylax* (2)
Rhizophoraceae: *Macarisia* (2)
Lythraceae: *Capuronia* (1), *Tetrataxis* (1)
Combretaceae: *Calopyxis* (22)
Psiloxylaceae: *Psiloxylon* (1)
Melastomataceae: *Amphorocalyx* (5), *Dionycha* (3), *Gravesia* (100), *Rhodosepala* (1), *Rousseauxia* (13), *Veprecella* (20)
Fabaceae: *Apaloxylon* (2), *Bathiaea* (1), *Baudouinia* (6), *Brandzeia* (1), *Bremontiera* (1), *Brenierea* (1), *Chadsia* (18), *Elignocarpus* (1), *Gagnebina* (1), *Lemuropisum* (1), *Mendoravia* (1), *Phylloxylon* (4), *Tetrapterocarpon* (1)
Rutaceae: *Humblotiodendron* (2), *Ivodea* (6)
Simaroubaceae: *Pleiokirkia* (1), *Perriera* (2)
Meliaceae: *Astrotrichilia* (14), *Calodecaryia* (1), *Humbertioturraea* (6), *Malleastrum* (12), *Neobeguea* (3)
Ptaeroxylaceae: *Cedrelopsis* (3)
Anacardiaceae: *Faguetia* (1), *Micronychia* (5), *Operculicarya* (3)
Sapindaceae: *Beguea* (1), *Chouxia* (1), *Conchopetalum* (2), *Crossonephelis* (1), *Doratoxylon* (5), *Hornea* (1), *Molinaea* (10), *Neotina* (2), *Plagioscyphus* (10), *Pseudopteris* (2), *Tina* (16), *Tinopsis* (10), *Tsingya* (1)
Physenaceae: *Physena* (2)
Balsaminaceae: *Impatientella* (1)
Malpighiaceae: *Digoniopterys* (1), *Microsteira* (28), *Philgamia* (3), *Rhynchophora* (1)
Trigoniaceae: *Humbertiodendron* (1)
Icacinaceae: *Grisollea* (2)
Celastraceae: *Brexiella* (2), *Evonymopsis* (4), *Hartogiopsis* (1), *Herya* (1), *Polycardia* (9), *Ptelidium* (2), *Salvadoropsis* (1)

PALEOTROPICAL KINGDOM

Olacaceae: *Phanerodiscus* (3)
Loranthaceae: *Socratina* (1)
Rhamnaceae: *Bathiorhamnus* (3)
Proteaceae: *Dilobeia* (1)
Escalloniaceae: *Forgesia* (1)
Montiniaceae: *Kaliphora* (1), *Melanophylla* (9)
Araliaceae: *Cuphocarpus* (1), *Geopanax* (1), *Indokingia* (1), *Sciadopanax* (1)
Apiaceae: *Anisopoda* (1), *Phellolophium* (1)
Loganiaceae: *Androya* (1)
Rubiaceae: *Breonia* (16), *Canephora* (5), *Carphalea* (9), *Cremocarpon* (1), *Danais* (40), *Fernelia* (2), *Homollea* (3), *Homolliella* (1), *Myonima* (4), *Neoschimpera* (1), *Paracephaelis* (1), *Peponidium* (20), *Pyrostria* (10), *Schismatoclada* (20), *Schizenterospermum* (4), *Scyphochlamys* (1), *Trigonopyren* (9)
Apocynaceae: *Cabucala* (16), *Craspedospermum* (1), *Hazunta* (8), *Oistanthera* (1), *Plectaneia* (12), *Roupellina* (1)
Asclepiadaceae: *Camptocarpus* (5), *Cryptostegia* (12), *Harpanema* (1), *Pentopetia* (10), *Stapelianthus* (2), *Stephanotis* (5), *Tanulepis* (5), *Trichosandra* (1)
Gentianaceae: *Gentianothamnus* (1), *Tachiadenus* (10)
Oleaceae: *Comoranthus* (2), *Linociera* (2), *Noronhia* (40)
Convolvulaceae: *Humbertia* (1; sometimes placed in a family of its own)
Schrophulariaceae: *Allocalyx* (1), *Bryodes* (3), *Hydrotriche* (1)
Bignoniaceae: *Colea* (20), *Kigelianthe* (3), *Ophiocolea* (5), *Perichlaena* (1), *Phyllarthron* (13), *Phylloctenium* (2), *Rhodocolea* (6)
Pedaliaceae: *Uncarina* (5)
Gesneriaceae: *Colpogyne* (1)
Acanthaceae: *Ambongia* (1), *Boutonia* (1), *Corymbostachys* (1), *Periestes* (2), *Stenandriopsis* (10)
Verbenaceae: *Acharitea* (1), *Adelosa* (1)
Lamiaceae: *Mahya* (1), *Perrierastrum* (1)
Campanulaceae: *Berenice* (1, close to *Cephalostigma*), *Heterochaenia* (3)
Lobeliaceae: *Dialypetalum* (5)
Asteraceae: *Aphelexis* (10), *Apodocephala* (4), *Centauropsis* (10), *Cylindorcline* (1), *Eriothrix* (1), *Oliganthes* (9), *Syncephalum* (7)
Hydrocharitaceae: *Calypsogyne* (1; incertae sedis?)
Triuridaceae: *Seychellaria* (1)
Geosiridaceae: *Geosiris* (1)
Hyacinthaceae: *Rhodocodon* (8)
Asphodelaceae: *Lomatophyllum* (13; close to *Aloë*)
Herreriaceae: *Herreriopsis* (1)
Dioscoreaceae: *Avetra* (1)
Orchidaceae: *Aeranthes* (30), *Ambrella* (1), *Arnottia* (4), *Benthamia* (25),

Bonniera (2), *Cymbidiella* (5), *Cryptopus* (3), *Gymnochilus* (2), *Lemurella* (4), *Oeonia* (7), *Oeoniella* (3), *Perrierella* (1), *Sobennikoffia* (4)
Strelitziaceae: *Ravenala* (1)
Commelinaceae: *Pseudoparis* (2)
Eriocaulaceae: *Moldenkeanthus* (1; closely related to the neotropical genera *Leiothrix* and *Paepalanthus*)
Poaceae: *Cyphochlaena* (2), *Decaryochloa* (1), *Hickelia* (2), *Hitchcockella* (1), *Lecomptella* (1), *Nastus* (13), *Neostapfiella* (3), *Perrierbambus* (2), *Poecilostachys* (20), *Pseudocoix* (1), *Pseudostreptogyne* (1), *Viguierella* (1)
Arecaceae: *Acanthophoenix* (1), *Antongilia* (1), *Beccariophoenix* (1), *Bismarckia* (1; close to the African genus *Medemia*), *Chrysalidocarpus* (20), *Deckenia* (1), *Dictyosperma* (1), *Dypsis* (21), *Hyophorbe* (= *Mascarena;* 5; a distinct genus of chamaedoreoid palms, the remainder of which are American; Moore 1973, 1978), *Latania* (3), *Lodoicea* (1; belongs to the *Borassus* alliance, which includes also *Borassodendron* and *Latania;* Moore 1973), *Louvelia* (3; related to the South American genera *Ceroxylon* and *Juania*), *Marojejya* (1), *Masoala* (1), *Neodypsis* (14), *Neophloga* (about 30), *Nephrosperma* (1), *Phoenicophorium* (1), *Phloga* (1), *Ranevea* (1), *Ravenea* (9; related to the South American genera *Ceroxylon* and *Juania*), *Roscheria* (1), *Sindroa* (1), *Tectiphiala* (1), *Verschaffeltia* (1), *Vonitra* (4)
Araceae: *Arophyton* (3), *Calephyton* (1), *Protarum* (1)

A series of endemic genera is common to both Madagascar and other islands of the Madagascan Region (Good 1974). They include *Ochropteris* (Madagascar, Mascarenes), *Monimia* (Madagascar, Mascarenes), *Tambourissa* (Madagascar, Mascarenes), *Maillardia* (Madagascar, Reunion), *Rameya* (Madagascar, Comores), *Dicoryphe* (Madagascar, Comores), *Oncostemum* (Madagascar, Mascarenes), *Badula* (Madagascar, Reunion, Mauritius, Rodriquez), *Lautembergia* (Madagascar, Mauritius), *Stephanodaphne* (Madagascar, Comores), *Grangeria* (Madagascar, Mauritius), *Brandzeia* (Madagascar, Seychelles), *Gagnebina* (Madagascar, Mauritius), *Phylloxylon* (Madagascar, Mauritius), *Humblotiodendron* (Madagascar, Comores), *Molinaea* (Madagascar, Mascarenes), *Grisollea* (Madagascar, Seychelles), *Breonia* (Madagascar, Comores, Mauritius), *Danais* (Madagascar, Mascarenes), *Peponidium* (Madagascar, Comores), *Pyrostria* (Madagascar, Mauritius, Rodriguez), *Trigonopyren* (Madagascar, Comores), *Hazunta* (Madagascar, Comores, Seychelles), *Camptocarpus* (Madagascar, Mauritius), *Tanulepis* (Madagascar, Rodriguez), *Comoranthus* (Madagascar, Comores), *Noronhia* (Madagascar, Comores, Mauritius), *Colea* (Madagascar, Mascarenes, Seychelles), *Ophiocolea* (Madagascar, Comoros Islands), *Phyllarthron* (Madagascar, Comores Islands), *Periestes* (Madagascar, Comoros Islands), *Seychellaria* (Madagascar, Seychelles), *Lomatophyllum* (Madagascar, Mascarenes), *Cryptopus* (Madagascar, Mascarenes), *Sobennikoffia* (Madagascar, Mascarenes), *Chrysalidocarpus* (Madagascar, Comores Islands, indicated also for Pemba Island), and *Ravenea* (Madagascar, Comores Islands).

Thus Madagascar appears to have the greatest number of endemic genera in common with the Comoros Islands and also with the Mascarenes. Madagascar and the Seychelles have considerably fewer endemic genera in common.

Despite the closeness of Madagascar and of the Comoros Islands to the eastern shores of Africa, the African element in their floras is not as great as might be supposed, and constitutes scarcely more than one-fourth of the entire composition of the flora (about 27% for Madagascar, according to Pierrier de la Bâthie 1936).[3] This is explained by the fact that the separation of Madagascar from Africa had already begun at the end of the Paleozoic era, and effective migration between them (and probably between Madagascar and India as well) ceased no later than 70–75 million years ago (i.e., in the late Cretaceous), and possibly even somewhere in the middle of the Cretaceous period. The prolonged closeness of Madagascar with the Indostan Peninsula (and with the island of Sri Lanka)—and an even more prolonged migratory connection between them through the chains of archipelagos—explains the appearance of shared taxa that are not found in Africa. There are an especially great number of such taxa in the Seychelles Islands, located closest to Sri Lanka and the Indostan Peninsula. The following are among those genera shared with the Indian Region (and also with other regions of the Indomalesian Subkingdom): *Cycas, Dillenia, Wormia, Erythrospermum, Vateriopsis*[4] *Nepenthes, Geniostoma, Timonius, Trichopus, Ochlandra, Lepironia, Cephalostachyum, Pothos,* and so on.

In the flora of the Madagascan Region, one observes both the most remote and often the most enigmatic relationships. In this group are the Madagascan species of *Trachypteris* (the second species occurring in South America and in the Galapagos); the taxonomically very isolated Madagascan representatives of the Winteraceae (*Takhtajania perrieri*); representatives of *Phenax* (2 species in Madagascar, and the rest in Central and South America); *Hibbertia* (Madagascar, New Guinea, Australia, New Caledonia); *Rheedia* (1 species on Madagascar, and all the remaining species in Central and South America); *Keraudrenia* (1 species on Madagascar, and the rest in Australia); *Rulingia* (Madagascar, Australia); *Stillingia* (1–2 species on the Mascarene Islands, 1 species in eastern Malesia and on Fiji, and the remaining species in America); *Weinmannia* (Madagascar, Mascarene Islands, Malesia, islands of the Pacific Ocean, New Zealand, America from Mexico to Chile); *Macadamia* (1 species on Madagascar, one species on Celebes, the remainder in eastern Australia and New Caledonia); *Oplonia* (5 species on Madagascar, one species in Peru, and 8 species in the West Indies); *Abrotanella* (Rodrigues Island, New Guinea, Australia, New Zealand, Auckland Islands, South America); *Astelia* (Mascarene Islands, New Guinea, Australia, Tasmania, in the Pacific Ocean up to the Hawaiian Islands); *Colmia* (Mascarene Islands and New Caledonia); *Agrostophyllum* (Seychelles Islands to Malesia and Polynesia); *Lophoschoenus* (Seychelles Islands, Borneo, New Caledonia); *Thoracostachyum* (2 species on the Seychelles Islands, the remaining 3 in Malesia and Polynesia); and others (see Dejardin et al. 1973).

Several endemic genera are found to have extraordinarily remote geographical relationships. This particularly includes the monotypic genera *Ascarinopsis* in the Chloranthaceae (very close to *Ascarina*, 8 species of which are distributed in Malesia, Polynesia, and New Zealand); *Humbertiodendron* of the Trigoniaceae (the remaining 3 genera of which are distributed in Malesia and tropical America); *Herreriopsis* of the Herreriaceae, consisting only of 2 genera (the second genus, *Herreria*, consists of 8 species found in South America); and *Ravenala* of the Strelitziaceae (very close to *Phenakospermum*, found in South America).

Despite the relatively small size of its territory, the Madagascan Region is chorologically very diverse. Consequently, it is presently divided into a somewhat greater number of provinces than was done in the floristic system of Engler, who placed all of Madagascar into one province.

1. Eastern Madagascan Province (Flore du vent, Perrier de la Bâthie 1921; Région malgache orientale, Humbert, 1927, 1955; Perrier de la Bâthie 1936; Rauh 1973, 1979; Koechlin et al. 1974; Cornet and Guillaumet 1976; White 1983; Domaine de l'Est, Humbert 1927, 1955; Domaine du Centre, Humbert 1927, 1955; Domaine du Sambirano, Humbert 1927, 1955; these provinces have also been accepted by Perrier de la Bâthie 1936, Koechlin et al. 1974, and Cornet and Guillaumet 1976). This province includes the eastern part of the island of Madagascar, except for its northern extreme (north of Vohemar); the high plateau, including the center of the island and the mountains; and the northwestern part of the island ("Domaine du Sambirano"), together with the island Nossi-Bé. There are 6 endemic families— Didymelaceae, Diegodendraceae, Asteropeiaceae, Sarcolaenaceae, Sphaerosepalaceae and Geosiridaceae—and a monotypic subfamily—Takhtajanoideae of the family Winteraceae—in this province. There are approximately 1,000 genera, of which about 160 (16%) are endemic (White 1983:235), including *Cinnamosma, Brochoneura, Mauloutchia, Ravensara, Dicoryphe, Neowormia, Plagioscyphus, Dilobeia, Melanophylla, Cymbidiella*, and all the endemic Madagascan palm genera. Even the small Nossi-Bé Island has its own endemic genus, *Crossonephelis*. According to White (1983:235), there are about 6,100 species, of which about 4,800 (78.7%) are endemic.

The main types of vegetation are lowland rain forest very rich in species, with a series of Myristicaceae, *Anthostema madagascariensis, Tambourisa gracilis* and *Weinmannia* spp., and many epiphytes (including *Rhipsalis madagascariensis* and *Angraecum sesquipedale*); montane forest rich in tree ferns, epiphytes (especially orchids), ferns, and mosses; mossy forest (in the high mountains, over 2,000 m above the sea level) with abundant epiphytes, mostly mosses and lichens (the upper border of this forest is formed by ericaceous bush *Philippia* and grassland); and evergreen sclerophyllous forest of the western slopes of the central plateau between 800 and 1,000 m (*Uapaca bojeri*, members of endemic family Sarcolaenaceae, and *Asteropeia* spp.). The Sambirano Subprovince of northwest Madagascar is a transition zone between the tropical rain forest and the east and the sclerophyllous forest of the western slopes of the central plateau (Rauh 1973, 1979; Koechlin et al. 1974). Much of the rain forest of this province has been destroyed for cultivation.

2. **Western Madagascan Province** (Région occidentale, Perrier de la Bâthie 1936; Rauh 1973, 1979; Domaine de l'Ouest, Humbert 1955; Koechlin et al. 1974; Cornet and Guillaumet 1976). This province includes the western side of the island (except its southwestern part) and the northern Diego-Suarez region, which is separated by the Sambirano Subprovince of the Eastern Madagascan Province. There are many endemic genera and numerous endemic species. The dominant types of vegetation are dry deciduous forest and grassland. Species of *Dalbergia* and *Commiphora*, *Hildegardia erythrosiphon*, and many woody succulents (including *Pachypodium* spp., *Adenia* spp., *Ampelocissus elephantina* and *Aloë* spp.) are characteristic. Forests on the calcareous plateau have the most distinctive flora (including the famous tropical ornamental plant *Delonix regia*) (Koechlin et al. 1974).

3. **Southern and Southwestern Madagascan Province** (Région du Sud-Ouest ou meridionale, Perrier de la Bâthie 1921, p.p.; Southern and southwestern region, Rauh 1979; Domaine subdésertique du Sud-Ouest, Humbert 1927; Domaine du Sud-Ouest, Perrier de la Bâthie 1936; Domaine du Sud, Humbert 1955; Koechlin et al. 1974; Cornet and Guillaumet 1976; White 1983). This province occupies the southwestern and southern coastal zone of the island, which is characterized by a very high percentage of endemic taxa. There is an almost endemic family—Didiereaceae, many endemic genera (including *Seyrigia, Xerosicyos, Stapelianthus*), and numerous endemic species. The most characteristic formation is *Euphorbia-Didierea* thorn-scrub in which aphyllous *Euphorbia* species and the very thorny Didiereaceae (especially *Didierea* itself) predominate. There are also some trees of 12–15 m height, such as *Adansonia fony, Gyrocarpus americanus* subsp. *capuronianus, Tetrapterocarpon geayi, Givotia madagascarensis,* and *Zanthoxylum seyrigii* (Koechlin et al. 1974). One of the most interesting plants of southern Madagascar is an endemic, very bristly dwarf terrestrial *Rhipsalis horrida,* which forms cushions only a few centimeters high (Barthlott 1983:246). According to Barthlott (1983), this remarkable species is derived

from an epiphytic neotropical ancestor and has secondarily adapted to terrestrial life. Another Madagascan endemic species is *Rhipsalis fasciculata*, which is more widely distributed.

4. Comoro Province. Besides the volcanic Comoro Islands, this province probably includes the coralline Aldabra Islands and a number scattered, low islands of the western Indian Ocean (including Cosmoledo, Glorieusa, Providence, and Farquhar), which are either raised coral-reef limestone islands or sand cays on sea-level reefs. The flora of the Comoros is very similar to the flora of Madagascar, and especially to the flora of the lowland rain forests. The Comoros contain two endemic monotypic genera—*Cremocarpon* and *Ranevea*. According to Voeltzkow (see Renvoize 1979) there are 416 indigenous species in the Comoros, of which 136 are endemic. *Ocotea comoriensis, Khaya comorensis, Philippia comorensis* and *Diospyros comorensis* are among the most characteristic species. Natural vegetation of the Comoros survives only in the mountains. The most extensive forests occur on the southern and eastern slopes of Karthala. The principal species are *Ocotea comoriensis* and *Khaya comorensis*.

5. Mascarene Province (Engler 1882, 1903, 1924; Good 1947, 1974; Takhtajan 1970). This province consists of the Mascarene Islands (Réunion, Mauritius, and Rodriguez), and probably also the islands of Cargados Carajos, which have only 17 indigenous species.

The flora of this province has one endemic family, Psiloxylaceae, and it is characterized by high generic and specific endemism. The endemic genera include *Pseudannona* (Mauritius), *Badula, Mathurina* (Rodriguez), *Astiria* (Mauritius), *Cheirolaena* (Mauritius), *Ruizia* (Réunion), *Cordemoya, Forgesia* (Réunion), *Roussea* (Mauritius), *Bremontiera* (Réunion), *Tetrataxis* (Mauritius), *Psiloxylon* (Réunion, Mauritius), *Doratoxylon, Hornea* (Mauritius), *Herya* (Réunion), *Fernelia, Mahya* (Réunion), *Myonima* (Réunion, Mauritius), *Scyphochlamys* (Rodriguez), *Oistanthera* (Mauritius), *Trichosandra* (Mauritius), *Allocalyx* (Réunion), *Bryodes, Cylindrocline* (Mauritius), *Berenice* (Réunion), *Heterochaenia, Eriothrix* (Réunion), *Calypsogyne* (Mauritius), *Aeranthes, Arnottia, Benthamia, Bonniera, Gymnochilus, Oeonia, Oeoniella, Nastus, Pseudostreptogyne* (Réunion), *Acanthophoenix, Dictyosperma, Hyophorbe* (Réunion, Mauritius), *Latania*, and *Tectiphiala* (Mauritius). Species endemism probably approaches 50%. Of 38 genera, 8 are endemic to Mauritius. In addition, the Indomalesian element is very well developed in the flora of these islands.

The indigenous vegetation of Rodriguez has been destroyed, and only individual plants remain in some places. The original vegetation was apparently a low forest with *Sideroxylon galeatum, Elaeodendron orientale, Mathurina penduliflora, Diospyros diversifolia, Terminalia bentzoe, Foetidia rod-

riguesiana, and the palm *Dictyosperma album*. Palm stands of *Latania verschaffeltii* and *Hyophorbe verschaffeltii*—together with the drought-resisting *Pandanus heterocarpus*—were probably characteristic of the coral plain on the drier eastern coast (White 1983:258; Wiehe 1949; Vaughan 1968). The natural vegetation of Mauritius has disappeared from most of the island. According to White (1983:257), from the accounts of early explorers it appears that palm stands consisting of *Latania lontaroides, Dictyosperma album,* and *Hyophorbe* spp. occurred where rainfall is less than 1,000 mm per year. The moister forests were probably dominated by *Diospyros tesselaria* (endemic) and *Elaeodendron orientale* (also in Réunion and Rodriguez), associated with *Foetidia mauritiana, Stadmannia oppositifolia, Hornea mauritiana* (endemic), and *Terminalia bentzoe* (also in Réunion and Rodriguez). The upland communities are somewhat better preserved (see Vaughan and Wiehe 1937).

According to Rivals (1952, 1968) and White (1983:258), there are the following plant communities on Réunion:

1. Coastal. Mangrove, halophytic, and *Ipomoea pes-caprae* and *Scaevola* communities.
2. Dry, megathermic forest (now destroyed). Below 400 m on the dry side of the island. Important species were *Elaeodendron orientale, Terminalia bentzoe, Mimusops petiolaris, Diospyros melanida* and *Ocotea obtusata*.
3. A complex of moister forests formerly ascending from sea level to 1,800 (2,000) m on the wet side of the island and occurring between 400 and 1,200 m on the dry side. Characteristic species include *Calophyllum tacamahaca, Grangeria borbonica,* and *Pittosporum senacia*. At 1,700 m on the windward slopes, this forest is replaced by 4–5 m high elfin thicket dominated by *Forgesia borbonica*.
4. *Acacia heterophylla* scrub forest, which occurs in rain-shadow areas between 1,200 and 2,000 m.
5. Ericoid communities. Between 2,000 and 2,500 m, the vegetation of slopes is dominated by *Philippia montana*, and above 2,500 m it is dominated by *Stoebe passerinoides*.

6. Seychelles Province (Engler 1882, 1903, 1924; Good 1947, 1974; Takhtajan 1970). This province includes the granitic Seychelles proper and the low sand-cay islands of the Amirantes. It has one endemic family, the Medusagynaceae (Seychelles Islands), and the following endemic genera: *Neowormia, Medusagyne, Vateriopsis* (very close to *Vateria*), *Wielandia, Geopanax, Indokingia, Neoschimpera, Deckenia, Lodoicea, Nephrosperma, Phoenicophorium, Roscheria, Vershaffeltia,* and *Protarum,* as well as 72 endemic species of flowering plants (including species of the endemic genera; Procter 1974, 1984*b*). All are confined to the granitic islands. There are about 766 species of flowering plants in the Seychelles and about 85 ferns and fern allies

(Procter 1984*b*). The most remarkable endemic species is *Medusagyne oppositifolia*—a very rare plant which is found in open exposed communities, distinctly xerophytic in character, in deep clefts between granite masses (Procter 1984*a*).

Vesey-FitzGerald (1940) and Jeffrey (1968) reconstructed the original vegetation as follows (see White 1983:257):

1. Coastal formations, including mangrove and *Ipomoea pes-caprae* formations.
2. Lowland rain forest. Up to 300 m on Mahé and Silhouette. Dominated by *Imbricaria seychellarum* and *Calophyllum inophyllum*, accompanied by *Dillenia ferruginea*, *Intsia bijuga*, and *Vateriopsis* (*Vateria*) *seyshellarum* (Mahé only). The tall lowland forests of the granitic islands have long since disappeared, and native woodland now exists only in inaccessible inland and upland localities (Stoddart 1984).
3. Intermediate forest. From 300 to 500 m on Mahé and Silhouette. Dominated by *Dillenia ferruginea* and *Northea seychellana*, accompanied by *Soulamea terminalioides*, *Colea seychellarum*, *Campnosperma seychellarum*, *Riseleya griffithii*, *Aphloia theiformis*, and *Pandanus hornei*. Most of the endemic species occur in this formation.
4. Mossy montane forest. Above 550 m on Mahé. Dominated by *Nothea sechellana*, accompanied by *Roscheria melanochaetes*, *Timonius seychellensis*, and *Nepenthes pervillei*.
5. Dryer forest. In dryer parts of Mahé, Silhouette, and Praslin. Dominated by *Dillenia ferruginea*. Among the associates are *Diospyros seychellarum*, *Dodonaea viscosa*, *Memecylan eleagni*, and several endemic palms, including the famous *Lodoicea maldivica*, the double coconut (Praslin and Curieuse only), *Verschaffeltia splendida*, and *Deckenia nobilis* (see also Procter 1984*a*).

C. INDOMALESIAN SUBKINGDOM

Good 1947, 1974; Mattick 1964; Takhtajan 1969, 1970, 1974.

Despite the fragmentation of the vast territory of the Indomalesian Subkingdom, which consists of two gigantic peninsulas—Indostan and Indochina—and a multitude of islands (from the Maldive Islands in the west to the islands of Samoa in the east), its flora is characterized by many common elements of various taxonomic ranks.

Endemism in this phytochorion is unusually high. It contains 16 endemic families of vascular plants (Matoniaceae, Cheiropleuriaceae, Degeneriaceae,

Barclayaceae, Pentaphylacaceae, Scyphostegiaceae, Plagiopteraceae, Anisophylleaceae, Crypteroniaceae, Lophopyxidaceae, Mastixiaceae, Aralidiaceae, Leeaceae, Pentaphragmataceae, Trichopodaceae, and Lowiaceae), and a very large number of endemic genera and species, the quantity of which is difficult to estimate, even very approximately, at the present time. No other phytochorion contains so many ancient, archaic forms of flowering plants as does the Indomalesian Subkingdom.

This subkingdom is divided into four regions: Indian, Indochinese, Malesian, and Fijian.

16. Indian Region

Good 1947, 1974; Mattick 1964; Takhtajan 1969, 1970, 1974; Rao 1972; Indostan Region, Tolmatchev 1974.

The Indian Region includes almost all of Indostan (with the exception of the northwestern parts included in the Sudano-Zambezian Region), the Gangetic Plain, tropical foothills of the Himalayas, Sri Lanka, the Laccadive Islands, the Maldive Islands, and the Chagos Archipelago.

The flora of the Indian Region has no endemic families, but there are a number of endemic genera (Good 1947, 1974; Rao 1972, 1979), probably little more than 120. Of these, the following deserve mention:

Monimiaceae: *Hortonia* (2, Sri Lanka)
Amaranthaceae: *Indobanalia* (1, southern mountains of Western Ghats)
Dilleniaceae: *Schumacheria* (3, Sri Lanka)
Clusiaceae: *Poeciloneuron* (2, western peninsular India)
Flacourtiaceae: *Chlorocarpa* (1, Sri Lanka)
Cucurbitaceae: *Dicoelospermum* (1, southern and western India)
Tiliaceae: *Erinocarpus* (1, western peninsular India)
Dipterocarpaceae: *Stemonoporus* (15, Sri Lanka), *Vateria* (2, southern India, Sri Lanka)
Bombacaceae: *Cullenia* (3, Sri Lanka, and the western part of the State of Madras, India)
Malvaceae: *Dicellostyles* (1, Sri Lanka), *Julostylis* (1, Sri Lanka)
Euphorbiaceae: *Copianthus* (1, India; very isolated within the family), *Mischodon* (1, western and southern India, Sri Lanka)
Podostemaceae: *Dalzellia* (1, southern India), *Farmeria* (2, southern India, Sri Lanka), *Griffithella* (1, western peninsular India), *Hydrobryopsis* (1, Western Ghats and southern India), *Indotristicha* (1, southern India; close to *Tristicha*); *Willisia* (1, western peninsular India)
Rhizophoraceae: *Blepharistemma* (1, western peninsular India)

Myrtaceae: *Meteoromyrtus* (1, western peninsular India; close to *Eugenia*)
Melastomataceae; *Kendrickia* (1, western India, Sri Lanka)
Fabaceae: *Eleiotis* (2, India), *Humboldtia* (6, southern and southwestern India, Sri Lanka); *Wagatea* (1, western peninsular India; closely related to *Caesalpinia*)
Rutaceae: *Chloroxylon* (1, western peninsular India, Sri Lanka), *Pamburus* (1, southern India, Sri Lanka)
Anacardiaceae: *Nothopegia* s. str. (6, India, Sri Lanka)
Sapindaceae: *Glenniea* (1, Sri Lanka), *Otonephelium* (1, western peninsular India; close to *Nephelium*)
Loranthaceae: *Helicanthes* (1, western India)
Apiaceae: *Polyzygus* (1, western peninsular India), *Vanasushava* (1, western peninsular India)
Vitaceae: *Puria* (1, western India; close to *Cissus*)
Rubiaceae: *Byrsophyllum* (2, India, Sri Lanka), *Fergusonia* (1, southern India, Sri Lanka), *Leucocodon* (1, Sri Lanka), *Nargedia* (1, Sri Lanka), *Schizostigma* (1, Sri Lanka), *Scyphostachys* (2, Sri Lanka)
Apocynaceae: *Petchia* (1, Sri Lanka), *Walidda* (1, Sri Lanka; very close to *Wrightia*)
Asclepiadaceae: *Baeolepis* (1, western peninsular India), *Decalepis* (1, peninsular India), *Frerea* (1, western peninsular India), *Oianthus* (4, India), *Seshagiria* (1, western peninsular India), *Utleria* (1, southern India)
Gentianaceae: *Hoppea* (2, India, Sri Lanka)
Scrophulariaceae: *Bonnayodes* (1, Sri Lanka)
Gesneriaceae: *Championia* (1, Sri Lanka), *Jerdonia* (1, western peninsular India)
Acanthaceae: *Carvia* (1, peninsular India; close to *Strobilanthes*), *Didyplosandra* (3–7, peninsular India, Sri Lanka; close to *Strobilanthes*), *Gantelbua* (1, peninsular India), *Indoneesiella* (2, India), *Kanjarum* (1, southern India), *Leptacanthus* (5, peninsular India, Sri Lanka; close to *Strobilanthes*), *Mackenziea* (9, India, Sri Lanka; close to *Strobilanthes*), *Meyenia* (1, India, Sri Lanka; close to *Thunbergia*), *Nilgirianthus* (20, peninsular India; close to *Strobilanthes*), *Phlebophyllum* (8, peninsular India; close to *Strobilanthes*), *Plaesianthera* (1, Sri Lanka), *Pleocaulus* (3, peninsular India; close to *Strobilanthes*), *Pseudostenosiphonium* (9, Sri Lanka), *Santaupaua* (1, peninsular India), *Stenisiphonium* (6, India, Sri Lanka), *Supushpa* (1, western peninsular India), *Taeniandra* (1, peninsular India; close to *Strobilanthes*), *Thelepaepale* (1, peninsular India, Sri Lanka; close to *Strobilanthes*), *Xenacanthus* (4, peninsular India)
Asteraceae: *Cyathocline* (3, India), *Glossocardia* (2, India), *Lamprachaenium* (1, western peninsular India), *Nanothamnus* (1, Konkan hills in Maharashtra)

Triuridaceae: *Hyalisma* (1, India)
Burmanniaceae: *Haplothismia* (1, western peninsular India)
Orchidaceae: *Cottonia* (2, southwestern India, Sri Lanka), *Diplocentrum* (2, India, Sri Lanka), *Ipsea* (3, western and southern India, Sri Lanka), *Papilionanthe* (1, India, Sri Lanka), *Sirhookera* (2, southern India, Sri Lanka)
Zingiberaceae: *Cyphostigma* s. str. (1, Sri Lanka)
Cyperaceae: *Ascopholis* (1, southern India)
Poaceae: *Bhidea* (1, western India), *Coelachyropsis* (1, southern India, Sri Lanka), *Danthonidium* (1, western peninsular India), *Dichaetaria* (1, southern India, Sri Lanka), *Hubbardia* (1, western Ghats), *Indochloa* (2, India), *Indopoa* (1, western Ghats; very close to *Tripogon*), *Limnopoa* (1, southern India; close to *Coelachne*), *Lopholepis* (1, southern India, Sri Lanka), *Pogonachne* (1, Maharashtra State), *Pommereulla* (1, southern India, Sri Lanka), *Pseudodichanthium* (1, Maharashtra State), *Trilobachne* (1, western peninsular India), *Triplopogon* (1, western peninsular India), *Woodrowia* (1, western peninsular India), *Zenkeria* (3, southern India, Sri Lanka)
Arecaceae: *Loxococcus* (1, Sri Lanka)
Araceae: *Anaphyllum* (2, southern India), *Lagenandra* (6, southern India, Sri Lanka), *Theriophonum* (6, peninsular India, Sri Lanka)
Incertae sedis: *Pouslowia* (1, southern India, Sri Lanka)

In comparison with other regions of the Indomalesian Subkingdom, the Indian Region is characterized by the absence of endemic genera of archaic families of flowering plants, except for *Hortonia* and *Schumacheria* on Sri Lanka. Not only are there no endemic genera of the order Magnoliales,[5] but also none of the Ranunculales, Hamamelidales, Saxifragales, Rosales, or even Caryophyllales. Nor are there any endemic genera of the most archaic families of the monocots, and the palms are represented by only one endemic genus. However, one finds relatively many endemic genera of such specialized families as the Podostemaceae, the Acanthaceae, and the Poaceae. The explanation for this peculiar poverty of endemic genera in the Indian Region may lie in the geologic past of Indostan, which collided with Asia and became a part of it only in the Eocene, probably about 45 million years ago (Molnar and Tapponnier 1977). In contrast to Madagascar, Indostan did not become a sufficiently efficient and stable area of evolutionary processes; thus nothing evolved in Indostan similar to those exceptionally peculiar taxa which characterize the Madagascan Region.

1. Sri Lanka Province (Ceylon province, Hooker and Thomson 1855; Hooker [1904] 1907; Good 1947, 1974; Turrill 1953; Meusel et al. 1965;

Takhtajan 1970). This province includes the island of Sri Lanka and the Maldive Islands. The flora of the Sri Lanka Province is characterized by a considerable number of endemic genera and species. It has a more clearly marked Malesian character than does the flora of the Indian Peninsula (Good 1974). Species endemism is at least 30% (Willis 1922). The coral Maldive Islands are characterized by a very poor flora and lack any endemic species.

2. **Malabar Province** (Hooker and Thomson 1855; Hooker [1904] 1907; Good 1947, 1974; Turrill 1953; Takhtajan 1970; Malabar Region, Chatterjee 1940, 1962; Puri 1960; Maheshwari et al. 1965; Subramanyam and Nayar 1974). This province includes the western coastal plains of peninsular India from the southern vicinities of Broach to Cape Comorin; the western side of the Western Ghats together with the Nilgiri, Anaimalai, Palni, and Cardamom hills. It also includes the coral Laccadive Islands. There are a considerable number of endemic genera and numerous endemic species (about 1,500 in the western Ghats, according to Subramanyam and Nayar 1974). Of the endemic species, I mention only:

Gymnocranthera canarica, Myristica malabarica, Knema attenuata, Actinodaphne lanata, Apollonias arnottii, Litsea nigrescens, Apama barberi, A. siliquosa, Daphniphyllum neilgherrense, Tetracera akara, Calophyllum apetalum, Garcinia cambogia, Poeciloneuron indicum, P. pauciflorum, Begonia aliciae, Isonandra lanceolata, Palaquium ellipticum, Antistrophe serratifolia, Vaccinium leschenaultii, Ardisia blatteri, Rapanea daphnoides, R. wightiana, Elaeocarpus tuberculatus, Erinocarpus nimmonii, Dipterocarpus indicus, Hopea utilis, Vateria indica, Aporosa lindleyana, Baccaurea coutrallensis, Emblica fischeri, Pseudoglochidion anamalayanum, Pittosporum dalycaulon, Humboldtia oligocantha, Inga cynometroides, Ormosia travancorica, Wagatea spicata, Griffithella hookeriana, Indotristicha ramosissima, Willisia selaginoides, Meteoromyrtus wynaadensis, Blepharistemma membranifolia, Aglaia anamallayana, Amoora lawii, Dysoxylum malabaricum, Gluta travancorica, many species of *Impatiens, Nothopegia travancorica, Aralia malabarica, Schefflera capitata, Polyzygus tuberosus, Apodytes beddomei, Nothopodytes foetida, Glyptopetalum lawsonii, Erythropalum populifolium, Olax wightiana, Helicia nilagirica, H. travancorica, Acranthera grandiflora, Octotropis travancorica, Beaumontia jerdoniana, Chilocarpus malabaricus, Ellertonia rheedii, Baeolepis nervosa, Ligustrum travancoricum, L. walkeri, Linociera malabarica, Myxopyrum serrulatum, Pedicularis perottetii, Christisonia bicolor, C. calcarata, C. cytinoides, C. lawii, Didymocarpus ovalifolia, Jerdonia indica, Calacanthus grandiflorus, Campanula wightii, Adenoon indicum, Lamprochaenium microcephalum, Nanothamnus sericeus, Haplothismia exannulata, Chiloschista pusilla, Diplocentrum congestum, D. recurvum, Cymbopogon travancorensis, Hubbardia heptaneuron, Indochloa oligocantha, Ochlandra travancorica, Oxytenanthera monostigma, Arenga wightii, Bentinkia coddapanna, Pinanga dicksonii,* and about 9 species of *Calamus.*

The Malabar Province can be divided into 4 subprovinces (or phytogeographical regions, according to Subramanyam and Nayar 1974)—namely, (1) Konkan Subprovince (Concan Province, Hooker and Thomson 1855); (2) Karnataka-Kerala Subprovince (Malabar Province, Hooker and Thomson 1855); (3) Nilgiri Subprovince (Nilgiri Phytogeographical Region, Subramanyam and Nayar 1974), and (4) Anaimalai-Palni-Cardamom Subprovince (Anaimalai-Palni-Cardamom Phytogeographical Region, Subramanyam and Nayar 1974).

The Konkan Subprovince occupies the northern part of the Malabar Province. Its southern boundary runs approximately along the valley of the Kalinadi River. There are a few endemic genera and a number of endemic species, including *Nanothamnus sericeus*. The main types of vegetation are the montane evergreen forest in the Western Ghats of Maharashtra and moist deciduous forest on the windward side of the Western Ghats.

The Karanataka-Kerala Subprovince stretches from the River Kalinadi to the Cape Comorin. Floristically, it is the richest part of the province and accounts for the majority of its endemic taxa. The main types of vegetation are tropical evergreen rain forest and moist deciduous forest.

The Nilgiri Subprovince occupies the Nilgiris or Blue Mountains, situated between 11°12′ and 11°43′ north latitude and 76°14′ and 77°1′ east longitude and rising abruptly from the surrounding areas. The Nilgiri (as well as the Anaimalai, Palni, and Cardamom hills) is a unique tropical island of subtropical and temperate flora which shows a pronounced relationship with the flora of the Himalayas and of the Khasi and Naga hills. There are no endemic genera, but the subprovince has a number of endemic species and subspecies, including *Actinodaphne lanata, Berberis nilghirensis, Daphniphyllum neilgherrense, Hypericum mysorense, Rhododendron nilagiricum, Vaccinium neilgherense*. The most characteristic vegetation type of the Nilgiri is the so-called shola—an evergreen forest with thick undergrowth. The most conspicuous shrubs and trees of the shola are *Hydnocarpus alpina, Michelia nilgirica, Berberis tinctoria, Mahonia leschenaultii, Garcinia cambogia, Gordonia obtusa, Ternstroemia gymnanthera, Ilex denticulata, I. wightiana, Euonymus crenulatus, Microtropis ramiflora, Cinnamomum wightii, Meliosma wightii, M. microcarpa, Osyris wightiana, Pentapanax leschenaultii, Schefflera racemosa*, and *Macaranga indica*. The undergrowth consists mainly of *Clematis wightiana, Viola serpens, Polygala arillata, Parthenocissus neilgheriensis* and *Osbeckia leschenaultiana*. There is a rich growth of orchids. In the open downs, there are many herbaceous and shrubby species such as *Anemone rivularis, Ranunculus reniformis, Cardamine hirsuta, Viola* spp., *Polygala* spp., *Hypericum mysorense, Impatiens* spp., *Crotalaria* spp., *Indigofera pulchella, Smithia gracilis, Rubus moluccanus, Parnassia mysorensis, Rhodomyrtus tomentosa, Bupleurum mucronatum, Heracleum rigens, H. hookerianum, Galium asperifolium, Campanula fulgens, C. wightii, Swertia* spp. and many others (Subramanyam and Nayar 1974).

3. **Deccan Province** (Hooker and Thomson 1855; Hooker [1904] 1907; Good 1947, 1974; Turrill 1953; Takhtajan 1970; Deccan Region, Chatterjee 1940, 1962; Puri 1960; Maheshwari et al. 1965). This province includes most of the peninsular India eastward from the Malabar Province and to the south from the Indo-Gangetic alluvial plain. The Vindhya Range, Kaimur Range, and Rajmahal Hills are also parts of this province. The chorionomic boundaries of the Deccan Province are not well defined, and there are some transitional areas between the adjacent provinces. It is especially difficult to indicate the exact boundary between the Deccan and Malabar provinces. As Hooker ([1904] 1907:35) put it, "in the large river valleys and those of the higher hills, types of the Malabar Flora penetrate far to the east." In contrast to the Malabar flora, the flora of the Deccan Province has fewer Malesian elements, and in some more arid areas, it contains a considerable admixture of the characteristic Omano-Sindian elements.

Throughout the Deccan Province, the most characteristic vegetation formations are moist and dry deciduous or monsoon forests and thorn forest. The most characteristic trees of these forests are *Tectona grandis* (which occurs at intervals over the whole area of the Deccan Province), *Shorea robusta* (sal tree of Indian forestry), *Hardwickia binata, Boswellia serrata, Butea frondosa, Anogeissus pendula, Santalum album, Dalbergia sissoo, Terminalia alata, T. chebula, Bassia latifolia, Acacia catechu, Anogeissus latifolia, Cedrela toona, Soymida febrifuga, Euphorbia neriifolia, Garuga pinnata, Heterophragma roxburghii, Lagerstroemia parviflora, Gmelina arborea, Adina cordifolia, Ougeinia dalbergioides, Schrebera swietenioides, Emblica officinalis, Borassus flabellifer,* and *Phoenix sylvestris.*

To the east of the Maikal Range, and especially on the Chota Nagpur plateau, forests are very rich floristically. The dominant species is *Shorea robusta.* The most common associates are *Terminalia alata, Buchanania lanzan, Cleistanthus collinus, Anogeissus latifolia, Tectona grandis, Pterocarpus marsupium, Terminalia chebuba, Aegle marmelos, Lagerstroemia parviflora, Bassia latifolia, Syzygium cumini, Dillenia pentagyna,* and so on.

On the higher levels of the Maikal Range and on the Chota Nagpur, the forests of shola character somewhat resemble the sholas of the Western Ghats. For these forests *Pterospermum acerifolium, Phoenix robusta,* and *Clematis nutans* are characteristic.

A considerable part of Deccan is covered by thorn forests, with *Acacia* spp., *Euphorbia* spp., *Balanites roxburghii, Cordia myxa, Ziziphus mauritiana,* and *Calotropis procera. Phoenix sylvestris* grows in relatively damper places. Vast areas of Deccan are at present short-grass savannahs and semideserts.

The flora of the Coromandel Subprovince (Hooker [1904] 1907; Turrill 1953) or the Cornatic Province (Hooker and Thomson 1855; Razi 1955) differs somewhat from that of the rest of the Deccan Province. This subprovince (or province?) occupies the strip of lowland between the Eastern Ghats and the sea, and stretches from Orissa to Tirunelveli: "Thickets of thorny evergreens and deciduous trees and shrubs abound, belonging to the genera *Flacourtia, Randia, Scutia, Diospyros, Mimusops, Garcinia, Sapindus, Pterospermum,* etc." (Hooker [1904] 1907:37). *Strychnos nux-vomica* is also characteristic. The evergreen forest characterized by *Eugenia bracteata, Memecylon umbellatum, Carallia integerrima, Linociera malabarica,* and *Mimusops hexandra* is limited to the coastal areas from north of Madras to Point Calimere. For the extreme southern coast, *Acacia planifrons, Cocculus leaba, Capparis aphylla, Cassia ovata* and *C. angustifolia* are characteristic (Mani 1974*a*). In the coast areas there are mangrove swamps with *Avicennia officinalis, Lumnitzera racemosa, Bruguiera* spp., *Rhizophora* spp., and Chenopodiaceae in the mouth of the rivers. The sandy beaches have *Hydrophylax maritima, Ipomoea pes-caprae, Sesamum prostratum,* and *Spinifex squarrosus*.

4. **Upper Gangetic Plain Province** (Hooker and Thomson 1855; Razi 1955; Gangetic Plain Province, Hooker [1904] 1907, p.p.; Gangetic Plain Region, Chatterjee 1940, 1962, p.p.). This province extends from the Aravalli Hills and the Yamuna River eastward approximately to the Kosi River (i.e., a little above the bend of the Ganges at the Rajmahal Hills). It includes tropical foothills of the Siwalik Range and the entire lowlands of the upper courses of the Ganges.

The flora of this province lacks endemic genera and contains comparatively few endemic species. The forests have been largely destroyed in most of the area. The comparatively best preserved forests are confined to the Siwalik Hills, where they are broken by tracts of savanna and scrubs. The dominant species is *Shorea robusta*. For these forests, characteristic species include *Anogeissus latifolia, Buchanania lanzan, Dendrocalamus strictus, Terminalia bellirica, Bauhinia variegata, Acacia* spp., *Emblica officinalis, Erythrina suberosa, Bombax ceiba, Syzygium* spp., *Cassia fistula, Woodfordia fruticosa,* and *Indigofera* spp. Considerable areas of the Upper Gangetic Plain are occupied by more or less halophytic formations. In many places, *Acacia* spp. and other small trees and shrubs are scattered here and there. Such grasses as *Heteropogon contortus, Bothriochloa* spp., and *Themeda* spp. are characteristic. On the most saline soils, *Salvadora persica* is almost the only tree that succeeds.

5. **Bengal Province** (Hooker and Thomson 1855; Razi 1955; Bengal and Sundarban Subprovinces of the Gangetic Plain Province, Hooker [1904] 1907; Chatterjee 1940, 1962; Turrill 1953). This province occupies the entire lowlands of the lower courses of the Ganges and Brahmaputra (including nearly the whole territory of Bangladesh), along with lowlands of Orissa northward from the Mahanadi River, the tropical parts of Tripura, Manipur, Nagaland, Assam, and a part of northern Arakan in Burma. The very rich flora and luxuriant evergreen vegetation of the Bengal Province "contrasts

favorably with the upper valley of the Ganges" (Hooker [1904] 1907:25). There are a few endemic genera and many endemic species. The dominant type of vegetation is tropical semi-evergreen forest. The Sundarban Subprovince is characterized by very rich estuarian floras, which "contain more local species than do any other botanical regions of India" (Hooker [1904] 1907:26).

17. Indo-Chinese Region

Chevalier and Emberger 1937; Vidal 1960; Takhtajan 1969, 1970, 1974; Tolmatchev 1974; Continental South-east Asiatic Region, Good 1947, 1974; Südostasiatisches Florengebiet, Mattick 1964; Indo-Chinese Subregion of the Indo-Malesian Region, Fedorov (in Grubov and Fedorov) 1964.

This region includes the southeastern and eastern border parts of Bangladesh, with adjoining tropical regions of India on the east; the entire tropical part of Burma; Thailand (with the exception of some northernmost parts); Indochina (with the exception of some northernmost parts); the Andaman Islands; the tropical regions of southwestern and southern China; and the island of Hainan.

There is only one endemic family, Plagiopteraceae, but the region contains over 250 endemic genera, including the following:

Pinaceae: *Ducampopinus* (1, Indochina; very close to *Pinus*)
Magnoliaceae: *Kmeria* (2, southern China, Thailand and Indochina)
Annonaceae: *Enicosanthellum* (2, southern China, Vietnam)
Hamamelidaceae: *Chunia* (1, Hainan), *Mytilaria* (1, Indochina)
Juglandaceae: *Annamocarya* (1, southern China, Indochina; very close to *Carya*, and treated by some authors as a section)
Theaceae: *Paranneslea* (1, Indochina)
Ochnaceae: *Indosinia* (1, Indochina)
Sapotaceae: *Aisandra* (2, southern Indochina), *Eberhardtia* (4, southern China, northern Indochina), *Sinosideroxylon* (3, southeastern China, Hong Kong, northern Vietnam)
Flacourtiaceae: *Dankia* (1, Indochina)
Capparaceae: *Hypselandra* (1, Burma), *Neothorelia* (1, Indochina), *Poilanedora* (1, Annam; taxonomically very isolated), *Tirania* (1, Indochina)
Plagiopteraceae: *Plagiopteron* (1, Lower Burma)
Tiliaceae: *Hainania* (1, Hainan), *Sicrea* (1, Thailand, Indochina)
Malvaceae: *Cenocentrum* (1, Indochina)
Urticaceae: *Meniscogyne* (2, Indochina), *Petelotiella* (1, Tonkin)
Euphorbiaceae: *Deutzianthus* (1, Indochina), *Glyphostylus* (1, Thailand, In-

dochina; close to *Excoecaria*), *Oligoceras* (1, Indochina), *Poilaniella* (1, southern China, Indochina), *Sphyranthera* (1, Andaman Islands), *Thyrsanthera* (1, Indochina)
Thymelaeaceae: *Rhamnoneuron* (1, Indochina)
Podostemaceae: *Diplobryum* (1, southern Vietnam), *Polypleurella* (1, Thailand)
Melastomataceae: *Sporoxeia* (1, Burma, southern China), *Scorpiothyrsus* (1, Indochina, Hainan), *Stapfiophyton* (4, southern China)
Fabaceae: *Afgekia* (3, southern China, Burma, Thailand), *Antheroporum* (2, Cochin China, Thailand), *Endomallus* (1, Laos and Vietnam; close to *Cajanus*)
Connaraceae: *Schellenbergia* (1, Lower Burma)
Rutaceae: *Thoreldora* (1, Indochina), *Tractopevodia* (1, Burma)
Anacardiaceae: *Allospondias* (3, southern China, Burma, Thailand, Laos, Vietnam; close to *Spondias*)
Podoaceae: *Campylopetalum* (1, Thailand)
Sapindaceae: *Arfeuillea* (1, Thailand, Indochina), *Boniodendron* (1, Indochina), *Cnemidiscus* (1, southern Indochina), *Phyllotrichum* (1, Indochina), *Sapindopsis* (1, Hainan)
Icacinaceae: *Pittosporopsis* (1, Yünnan, Burma, Thailand, Laos, Vietnam)
Celastraceae: *Annulodiscus* (1, Indochina)
Rhamnaceae: *Chaydaia* (2, southern China, northern Indochina)
Vitaceae: *Acareosperma* (1, Indochina)
Rubiaceae: *Alleizettella* (2, Indochina, Hong Kong), *Leptomischus* (1, Indochina), *Mouretia* (1, Indochina), *Notodontia* (2, northern Vietnam), *Paedicalyx* (1, Hainan, northern Indochina), *Pubistylus* (1, Andaman Islands), *Symphyllarion* (1, Indochina), *Thysanospermum* (1, Hong Kong), *Xanthonneopsis* (1, Indochina), *Xanthophytopsis* (2, southern China, northern Indochina)
Apocynaceae: *Aganonerion* (1, Indochina), *Argyronerium* (1, Indochina), *Boussigonia* (2, Indochina), *Hanghomia* (1, Indochina), *Muantum* (1, Lower Burma, peninsular Thailand), *Parabarium* (20, southern China, Indochina), *Spirolobium* (1, Indochina), *Xylinabariopsis* (2, Indochina)
Asclepiadaceae: *Atherolepis* (3, Burma, Thailand), *Costantina* (1, Indochina), *Graphistemma* (1, Hong Kong), *Gymnemopsis* (2), *Harmandiella* (1), *Merrillanthus* (1, Hainan), *Spirella* (2, Indochina), *Zygostelma* (1, Thailand)
Solanaceae: *Atrichodendron* (1, Indochina)
Scrophulariaceae: *Geoffraya* (2, Indochina), *Petitmenginia* (2, southern China, Indochina), *Pseudostriga* (1, Indochina), *Trichotaenia* (2, Indochina)
Bignoniaceae: *Hexaneurocarpon* (1, Indochina), *Spathodeopsis* (2, Indochina)

Gesneriaceae: *Cathayanthe* (1, Hainan), *Dasydesmus* (1, southern China), *Raphiocarpus* (1, southern China), *Trisepalum* (3, Burma)
Acanthaceae: *Antheliacanthus* (1, Thailand), *Chroësthes* (1, southern China, Indochina), *Dossifluga* (1, Thailand), *Graphandra* (1, Thailand), *Larsenia* (1, Thailand), *Parajusticia* (1, Indochina), *Plegmatolemma* (2, Thailand), *Psiloësthes* (1, Indochina), *Thysanostigma* (1, Thailand)
Verbenaceae: *Dimetra* (1, Thailand), *Paravitex* (1, Thailand), *Tsoongia* (1, southern China, Indochina)
Lamiaceae: *Wenchengia* (1, Hainan)
Campanulaceae: *Numaeacampa* (1, Indochina)
Asteraceae: *Aëtheocephalus* (1, Indochina), *Camchaya* (4, Thailand, Indochina), *Colobogyne* (1, Indochina), *Jodocephalus* (3, Thailand, Indochina)
Convallariaceae: *Antherolophus* (1, Indochina; close to *Aspidistra*), *Colania* (1, Indochina; close to *Aspidistra*)
Orchidaceae: *Cephalantheropsis* (1), *Schoenomorphus* (1, Indochina), *Smitinandia* (Thailand, Indochina)
Zingiberaceae: *Gagnepainia* (3, Indochina), *Pommereschea* (2, Burma), *Siliquamomum* (1, Indochina)
Commelinaceae: *Aëtheolirion* (1, Thailand), *Spatholirion* (2, southern China, Indochina, Thailand)
Poaceae: *Brousmichea* (1, Indochina), *Kerriochloa* (1, Thailand), *Ratzeburgia* (1, Burma), *Thyrsostachys* (2, Assam, Burma, Thailand)
Arecaceae: *Zalacella* (1, Indochina)
Araceae: *Pseudodracontium* (7, Thailand, Indochina), *Pycnospatha* (2, Thailand, Laos)

The number of endemic species of the Indochinese flora is very great, and at present it is impossible even to approximate it.

The natural vegetation of the Indochinese Region is better preserved and is considerably richer than that in the Indian Region. In contrast with the Indian Region, the Indochinese Region is characterized by montane forests of evergreen species of *Lithocarpus, Castanopsis,* and *Quercus.*

1. South Burmese Province (Meusel et al. 1965; Lower Burmese Province, Takhtajan 1970; Burmese Province, Hooker [1904] 1907, p.p.; Lower Burma Region, Chatterjee 1940, p.p.). This province includes the tropical parts of the eastern extremity of India, the southeastern part of Bangladesh, and Lower Burma and its neighboring islands, with the exception of Tenasserim. The South Burmese Province is by no means homogeneous. The flora of its western and southern parts is markedly different from the flora of the central and eastern areas. The province is characterized by very high generic and specific endemism. The endemic genera include *Hypsel-*

andra, Plagiopteron, Schellenbergia, Tractopevodia, Trisepalum, Pommereschea, and *Ratzeburgia.*

2. **Andamanese Province** (Andaman Region, Maheshwari et al. 1965, p.p.). In spite of the fact that the forests of the Andaman Islands are floristically very close to the Burmese (Turrill 1953; Puri 1960), it is expedient to describe these islands as a separate province. This decision is based not only on the considerable distance of the Andamanese Province from Burma but also on the appearance in it of the endemic genus *Pubistylus* and a considerable number of endemic species. These include:
Myristica andamanica, Miliusa tectona, Orophea hexandra, Polyalthia parkinsonii, Trivalvaria dubia, Cryptocarya andamanica, Dillenia andamanica, Ardisia andamanica, Maesa andamanica, Dipterocarpus kerrii, Hopea andamanica, Blachia andamanica, Drypetes andamanica, Glochidion andamanicum, Macaranga andamanica, Mallotus andamanicus, Dichapetalum andamanicum, Linostoma andamanica, Mezonevron andamanicum, Planchonia andamanica, Mangifera andamanica, Canarium manii, Ailanthus kurzii, Lagerstroemia hypoleuca, Amoora manii, Dysoxylum andamanicum, Memecylon andamanicum, Hippocratea andamanica, Gouania andamanica, Linociera parkinsonii, Lasianthus andamanicus, Pubistylus andamanensis, Peristrophe andamanica, Strobilanthes glandulosus (Thothathri 1962).
Malesian elements form a noticeable part of the flora of the Andaman Islands. For example, such typical Malesian species as *Canarium denticulatum* are encountered on the Andaman Islands. This role increases considerably in the flora of the neighboring Nicobar Islands, relating them to the Malesian Region.

3. **South Chinese Province** (Fedorov [in Grubov and Fedorov] 1964; Takhtajan 1970).[6] This province includes the southern tropical regions of Yünnan, a small part of northern Vietnam, the Luichow Peninsula, the island of Hainan, and the coastal belt of continental China from Nanning to Macao. It also seems necessary to include in this province the eastern extremity of Burma, and some northern parts of Thailand and Laos. The South Chinese Province is characterized by the widespread occurrence of Holarctic elements; in several areas, it contains rather substantial enclaves of the Holarctic flora. There are a number of endemic genera, including *Chunia, Eberhardtia, Sinosideroxylon, Hainania, Sapindopsis, Chaydaia, Graphistemma, Paedicalyx, Thysanospermum, Xanthophytopsis, Cathayanthe, Dasydesmus, Raphiocarpus,* and *Wenchengia.* Endemic species are numerous.

4. **Thailandian Province** (Takhtajan 1970; Siamische Provinz, Meusel et al. 1965).[7] This province includes a larger part of Thailand, the western regions of Laos, and probably Tenasserim (Burma). Endemic genera include

Polypleurella, Campylopetalum, Zygostelma, Antheliacanthus, Dossifluga, Graphandra, Larsenia, Plegmatolemma, Thysanostigma, Dimetra, Paravitex, and *Aëtheolirion.* In the lowlands of peninsular Thailand typical rain forests occur (Vidal 1979).

5. **North Indochinese Province.** This province includes tropical parts of North Laos and North Vietnam (including north Annam), except those parts which belong to the South Chinese Province. Floristically, it is closely related to the South Chinese Province but also has a number of its own endemic taxa. There are a few endemic genera, including *Petelotiella* and *Notodontia*. One of the most remarkable endemic species of this province is *Platanus kerrii,* which differs from all other species of the genus by its pinnately veined, merely toothed leaves.

6. **Annamese Province** (Takhtajan 1970; Secteur annamitique, Vidal 1960). This province includes central Vietnam (central and southern Annam) and neighboring regions of Laos. There are many endemic representatives of the Orchidaceae, Fagaceae, Euphorbiaceae, Rubiaceae, Annonaceae, and the Celastrales, as well as some endemic species of the Theaceae, Styracaceae, Rosaceae, Ericaceae, Malpighiaceae, Pinaceae, and Cycadaceae (Schmid 1974). Endemic genera include *Ducampopinus* and *Poilanedora.*

7. **South Indochinese Province** (Meusel et al. 1966).[8] Besides Cambodia, this province includes the southern part of Vietnam (Cochinchine), a part of southernmost Laos, and a small part of Thailand. The vegetation consists of various types of lowland tropical formations, including savannah forests and grasslands (especially in southeastern Vietnam, where dune vegetation is also characteristic), but in the lowlands of south eastern Thailand and southwestern Cambodia, typical rain forests occur (Vidal 1979).

This subdivision of the Indochinese Region is of a very preliminary nature, and will undoubtedly undergo major changes as future research is completed.

18. Malesian Region

Diels 1908, p.p.; Hayek 1926, p.p.; Wulff 1944; Good 1947, 1974; Turrill 1959; Balgooy 1971, p.p.

The Malesian Region includes the following: the Cocos Islands, Christmas Island, the Nicobar Islands, the southern part of the Malay Peninsula, the entire Malesian Archipelago (the Great Sunda Islands and the Lesser Sunda

Islands, the Philippines, the Molucca Islands, and also numerous groups of much smaller islands), New Guinea, the Aru Islands, the Admiralty Islands, the Bismark Archipelago, the Solomon Islands, the Louisiade Archipelago, and numerous small islands. In the north the Malesian Region reaches the southern extremity of Taiwan (the Henchun Peninsula) and to the islands of Lanyu (Botel Tobago or Hungt'ou) and Lutao (Hoshaotao).

The Malesian Region has four endemic families—the Matoniaceae, Scyphostegiaceae, Duabangaceae, and Lophopyxidaceae—and perhaps about 400 endemic genera (Good 1974:151). The largest number of endemic genera are in New Guinea (at least 140), followed by Borneo, the Philippines, Malacca, Sumatra, Java, and the Solomon Islands (Steenis 1950c). Endemic genera deserving special mention are *Matonia, Aromadendron, Elmerillia, Thottea, Rafflesia, Tetramerista, Dryobalanops, Scyphostegia, Lophopyxis,* and *Stenomeris.*

The extremely rich flora of the Malesian Region, estimated to comprise approximately 25,000 species of flowering plants (Steenis 1971), is characterized by exceptionally large numbers of ancient, archaic forms of flowering plants. Due to the strategic location of the Malesian Region as a center where very important migration routes cross each other, the flora is of great importance for the resolution of many major problems concerning the historical biogeography of the higher plants. Malesia is indeed "one of the most interesting parts of the world" (Good 1974:149). Unfortunately, the remarkable flora of this region of the world, especially the flora of New Guinea, is still inadequately studied.

Warburg (1900) distinguished "the Malesische Florengebietsgruppe" and the "Papuanische Florengebietsgruppe," and referred the Molucca Islands to Malesia. As later investigations showed—and despite the strong similarity of their flora—Malesia and Papualand also have considerable differences. As Merrill (1936:261) said, within the limits of Malesia *sensu amplo,*

> there developed two great centers of origin and dispersal; one was the Borneo-Java-Sumatra-Malay Peninsula part of ancient Sundaland, and the other the New Guinea part of ancient Papualand. Plants and animals extended their ranges from these two centers: from the first, most of them went north into the Asiatic continent and northeast into the Philippines; and from the second they went north through Gilolo, the Moluccas, and Celebes into the Philippines south into eastern Australia, and east into Polynesia.

Malesia and Papualand are sometimes considered to be two independent floristic regions (Mattick 1964; Takhtajan 1969, 1970, 1974; Tolmatchev 1974), but at present, it seems more expedient to consider these two phytochoria as subregions. The boundary between them corresponds to

Zollinger's floristic line (Steenis 1950c, 1979). Zollinger (1857) runs his dividing line between Mindanao and Celebes and along the Makassar Straits, and extends it southward toward the east of the Lesser Sunda Islands, thus separating the Papuan Subregion (Steenis's East Malesian Province) from the rest of Malesia. Hallier (1912, see Steenis 1979:103) later came to similar conclusions. Wallace (1863) also draws his line between Mindanao and Celebes and along the Makassar Straits, but he extends it southward between Bali and Lombok, which evidently is not supported by phytogeographical data. Beginning with Wegener himself (1924) and a phytogeographer (Lam 1934), a number of authors—both geologists and biogeographers—have attempted to explain biogeographical dividing lines (especially a line between Borneo and Celebes) by continental movements—that is, by the collision between Gondwanaland and Laurasia, and particularly between Sula Peninsula and Celebes (see especially Schuster 1972; Raven and Axelrod 1974, Whitmore 1981, 1982). But there are some discrepancies connected with the enigmatic geological history of Celebes. There is some geological evidence to suggest that only the eastern arm of Celebes was formerly part of Gondwanaland (Audley-Charles 1981), which creates serious difficulties in attempts to explain floristic or faunistic lines by geologic events, independently from the history of biota.

Nearly all of the Malesian islands are mountainous, and some of them have high mountains, such as Kinabalu on Borneo and the Papuan Snow Range in New Guinea. High elevations in these islands harbour representatives of a number of predominantly temperate genera of both the Northern and Southern hemispheres. Of the northern genera may be mentioned *Berberis, Anemone, Ranunculus, Thalictrum, Stellaria, Hypericum, Viola, Cardamine, Primula, Astilbe, Rubus, Potentilla, Geranium, Lonicera, Valerianella, Galium, Gentiana, Myosotis, Veronica, Euphrasia, Aster, Cirsium, Disporum, Agrostis, Deschampsia,* and *Poa.* The southern genera include *Haloragis, Gunnera, Nertera, Thelymitra, Caladenia, Microlaena,* and others. The associated phytogeographical problems have interested a number of botanists and have been the subject of major studies by Stapf (1894) and especially by van Steenis (1934, 1935, 1962, 1965, 1972). By comparing distributional areas of species of the strictly microtherm Malesian mountain genera—about 900 species in all—van Steenis concluded that there have been three main paths by which many temperate species migrated into the Archipelago: by way of the Malay Peninsula, Sumatra, Java, and the Lesser Sunda Islands; by Taiwan and the Philippines; and from Australia by way of New Guinea. Temperate species could reach these tropical islands during the Pleistocene epoch, when the vegetation zones of higher elevations were considerably lower, on the order of 1,500 m (Flenley 1979; Whitmore 1981; Axelrod and Raven 1982:934): "The stepping stones required for dispersal would have been very much closer" (Whitmore 1981:39).

PALEOTROPICAL KINGDOM   241

18A. MALESIAN SUBREGION

Malesische Florengebietsgruppe, Warburg 1900, p.p.; Indonesische Gebiet, Schmithüsen 1961; Malesian (Malayan) Region, Takhtajan 1969, 1970, 1974; Malayan Region, Tolmatchev 1974.

1. **Malay Province** (The Malay Peninsula, Good 1947, 1974; Malayan District, Thorne 1963). This province includes the southern part of the Malay Peninsula, where the northern boundary of the province proceeds (as was established by Kloss 1922) along the line leading from the mouth of the Kedah River (close to Alor Star in the state of Kedah, Malaysia), northeastward to Songkhla (or Singgora) on the eastern shore of the peninsular part of Thailand (see also Steenis 1950b; Keng 1970). The Malay Province also includes the island of Singapore, the Riau Archipelago, and probably the Anambas Islands.

The Malay Province is characterized by relatively high generic endemism. The endemics include *Hexapora* (1), *Maingaya* (1), *Andresia* (1), *Pernettyopsis* (1), *Leptonychiopsis* (1), *Kostermansia* (1), *Burkilliodendron* (1), *Perilimnastes* (1), *Hederopsis* (1), *Wardenia* (1), *Pycnorhachis* (1), *Aleisanthia* (2), *Becheria* (1, also on the island of St. Barbe), *Klossia* (1), *Mesoptera* (1), *Perakanthus* (2), *Codonoboea* (4), *Micraeschynanthus* (1), *Orchadocarpa* (1), *Stenothyrsus* (1), *Acrymia* (1), and *Calospatha* (2). According to Steenis (1950c:LXXI), it is "a typical Malayan flora intimately allied to the floras of Sumatra and Borneo but differing strongly from that of Indochina" (see also Keng 1970).

2. **Kalimantan (Bornean) Province** (Takhtajan 1970; Borneo, Good 1947, 1974; Bornean District, Thorne 1963). This province includes Borneo (Kalimantan), the Bunguran (Natuna) Islands, Palawan Island, the Calamian Islands, the Tawitawi Islands, Jolo Island, the Pangutaran Islands, and Laut Island. The Kalimantan Province has the highest genus and species endemism within the Malayan Subregion. It has much in common with the flora of the Malay Province, especially with its southeastern part. A few genera—such as *Stemmatodaphne, Ashtonia, Leucostegane, Campimia*, and *Racemobambos*—are confined to the Malay Province and the Kalimantan Province.

3. **Philippinean Province** (Provinz der Philippinen, Engler 1882; Provinz der Philippinen und Süd-Formosa, Engler 1899, 1903, 1924; Philippines, Good 1947, 1974; Philippinean Province, Thorne 1963; Takhtajan 1970). This province includes the Philippine Archipelago (with the exclusion of Palawan Island and the Calamian Islands); Basilan Island (the northernmost island in the Sulu Archipelago); a series of small islands close to Mindanao; the Babuyan Islands and the Batan Islands located north of Luzon; and a southern extremity of the island of Taiwan (the Hengchun

Peninsula);[9] and two small islands to the east of southern Taiwan, Lanyu and Lutao.

Both generic and species endemism in the Philippinean Province is very high. It contains over 30 endemic genera, and specific endemism approximates 68%; in the primary forests it is approximately 84% (Merrill 1946). The Philippines' location is an interface among several floristic entities. In its high mountains, the eastern Asiatic flora exerts a very strong influence. And in both low and high altitudes there is a peculiar infiltration of strictly Australian types, several of which reach as far north as the small islands between Luzon and Taiwan (Merrill 1946).

4. Sumatran Province (Takhtajan 1970; Sumatran District, Thorne 1963). This province includes the island of Sumatra and its surrounding islands, including Belitung Island. In the Sumatran Province I also include the Nicobar Islands; a Malesian element plays a considerably greater part in its flora than in the flora of the neighboring Andaman Islands (Thothathri 1962), which belong to the Indochinese Region.

5. South Malesian Province (Steenis 1950c, 1979; Meusel et al. 1965; South Malesian Subprovince, Balgooy 1960; Javan District and Lesser Sunda Province, Thorne 1963; Javan Province and Province of Sunda Islands, Takhtajan 1970). This province includes the islands of Java, Madura, Kangean, Bali, Lombok, Sumbawa, Sumba, Flores, Timor, as well as small islands lying close to them, the islands of Damar and Babar, and the Tenimber Islands. It also includes Christmas Island, located 300 km south of western Java.

The South Malesian Province is characterized by the lack of an autochthonous flora and by low generic and specific endemism (Steenis 1979). "The very high percentage of wide-ranging genera and species; the absence of many rain forest genera common in the other Malesian islands and the negligible endemism characterize the floristic province, South Malesia" (Steenis 1979:104). This province contains only four monotypic endemic genera: *Sclerachne* and *Teyleria*, both Acanthaceae; *Septogarcinia* (Clusiaceae); and *Grisseea* (Apocynaceae). In comparison with other provinces, specific endemism is low. Interestingly, despite the fact that Java and its neighboring islands are located rather closer to Australia than to the Philippines, Australian elements play a smaller role in the flora of the South Malesian Province; according to Kalkman (1955), only 2.8% belong to the Australian element.

18B. Papuan Subregion

Balgooy 1976; Papuanische Florengebietsgruppe and Ostmalesien, Warburg 1900, p.p.; East Malesian Province, Steenis 1950c; Papuan Region, Takhtajan 1970; Tolmatchev 1974.

1. **Celebesian (Sulawesian) Province** (Thorne 1963). This province includes the island of Celebes (Sulawesi) along with its neighboring islands, the Sangihe Islands and Sula Islands.

2. **Moluccan Province** (Thorne 1963, p.p.;[10] Takhtajan 1970). This province includes the Moluccan Islands and the Banda Islands. There is an endemic palm genus, *Siphokentia* (2).

3. **Papuan Province** (Papuasische Provinz, Engler 1899, p.p., 1903, p.p., 1924, p.p.; Thorne 1963; Takhtajan 1970; New Guineal and Aru, Good 1947, 1974). This province includes New Guinea, the Aru Islands, Misool Island, Salawati, Waigeo, the Schouten Islands, the Trobriand Islands, Murua Island, the D'Entrecasteaux Islands, and the Louisiade Archipelago. The flora of this province contains "an impressive number of endemic genera" out of a total of 1,463 genera (Balgooy 1976). According to even the most conservative estimate, it contains at least 8,500 endemic species, out of a total of more than 9,000 species (Good 1960, 1974). It is thus the richest province in the Indomalesian Subkingdom. Despite its closeness to Australia, the flora of the Papuan Province has a clearly pronounced Indomalesian character (Balgooy 1976). As Good (1960, 1974) pointed out, only about 60 genera have an Australian relationship, while 495 exhibit an Indomalesian relationship (for discussion, see Steenis 1979; Axelrod and Raven 1982).

4. **Bismarckian Province** (Thorne 1963, 1969; Bismarckian Province and the Province of the Solomon Islands, Takhtajan 1970). This province includes the Bismarck Archipelago, the Admiralty Islands, and the Solomon Islands. The flora of this province is a notably attenuated version of the Papuan flora (Good 1969; Thorne 1969; Balgooy 1971). There is only one monotypic, endemic genus in the Bismarck Archipelago (the genus *Clymenia*, which is very close to *Citrus*). On the Solomon Islands there are four such genera (*Cassidispermum, Whitmorea, Allowoodsonia, Kajewskiella*). In comparison with the Bismarck Archipelago, the Solomon Islands are floristically much more distinctive, more "individual." There is every reason to consider the Solomon Islands, on the one hand, and the Bismarck Archipelago and the Admiralty Islands, on the other, as two distinct subprovinces ("districts," according to Thorne 1963).

19. Fijian Region

Takhtajan 1969, 1970, 1974; Smith 1973, 1979; Melanesische Provinz, Engler 1903, 1912; Engler und Gilg 1919; East Melanesian Province, Balgooy 1971.

The Fijian Region includes the New Hebrides, the Santa Cruz Islands,[11] the Fiji Islands, the Rotuma Islands, the Wallis Islands, the Horn Islands, the Samoa Islands, the Tonga Islands, and Niue Island. The Fijian Region's only endemic family is the Degeneriaceae (Fiji). There are about 12 endemic genera, belonging to four families:

Degeneriaceae: *Degeneria* (1, Fiji)
Sterculiaceae: *Pimia* (1, Fiji)
Rubiaceae: *Gillespiea* (1, Fiji), *Hedstromia* (1, Fiji; close to *Psychotria*), *Readea* (3, Fiji; close to *Psychotria*), *Sarcopygme* (5, Samoa; close to *Morinda*), *Squamellaria* (2, Fiji), *Sukunia* (1, Fiji; very close to *Gardenia*)
Arecaceae: *Balaka* (20, Fiji, Samoa), *Carpoxylon* (1, New Hebrides), *Goniocladus* (1, Fiji), *Neoveitchia* (1, Fiji)

The greatest concentration of endemic genera and species is found on the Fiji Islands. The flora is close to the Malesian, and has especially much in common with the floras of New Guinea and of the Solomon Islands.

1. New Hebridean Province (Takhtajan 1970; New Hebridian District, Thorne 1963). This province includes the New Hebrides (from the islands of Torres in the northwest to Aneityum Island in the southeast) and the Banks Islands. These are relatively young volcanic islands. While they began to develop in the late Eocene, their current surface was formed only in post-Pliocene and even in post-Pleistocene times. Because they were never physically joined with current or past continents, their flora (and fauna) developed entirely by the accidental transfer of diaspores across the sea expanses surrounding them, and primarily from Malesian sources (Chew 1975). Thus the flora of this province is relatively depauperate and "disharmonic," is deprived of relict elements, and has only one endemic genus (*Carpoxylon*). The flora of the New Hebrides includes 534 genera and 1,120 species (Braithwaite 1975; Chew 1975). Species endemism probably does not exceed 15% (Chew 1975).

2. Fijian Province (Takhtajan 1970; Fijian Province, Thorne 1963, p.p.; Fijian District, Thorne 1963). This province includes the Fiji Islands, Rotuma Island, Uvea Island, the Horn Islands, the Samoa Islands, the Tonga Islands, and Niue Island. The Degeneriaceae are the one endemic family in the Fijian Province. There are also 11 endemic genera. Compared to the flora of the New Hebridean Province, Fijian Province—especially that of the Fiji Islands—is considerably older and richer, contains a series of relicts unknown to the New Hebrides and Banks Islands, and is characterized by higher species endemism (over 70%; Smith 1955*b*).

PALEOTROPICAL KINGDOM 245

D. POLYNESIAN SUBKINGDOM

Good 1947, p.p., 1974, p.p.; Mattick 1964, p.p.; Takhtajan 1969, 1970, 1974.

The Polynesian Subkingdom includes areas with purely island floras that are basically of post-Pliocene or even post-Pleistocene age. There are no endemic families, but high generic and species endemism is characteristic. Historically, the Polynesian flora seems to be a derivative of the Indomalesian: "These floras have largely been derived in ancient times from the larger islands in the west, particularly New Guinea and its adjacent islands groups and the Philippines" (Merrill 1946:205). At the same time, Polynesia and Micronesia—like Papuasia and the Philippines—contain some definitely Australian types, and in the east, particularly Hawaii, some plants that can only be considered to be derived from the American forms. But even in Hawaii the great majority of the indigenous genera are definitely Malesian.

20. Polynesian Region

Engler 1882; Takhtajan 1969, 1970, 1974; Tolmatchev 1974, p.p.; Region of Melanesia and Micronesia, p.p., and Region of Polynesia, Good 1947, 1974; Polynesien, p.p., and Melanesien und Micronesien, Mattick 1964; Polynesian Province, Thorne 1963.

The Polynesian Region includes the Caroline Islands, the Mariana Islands, Marcus Island, Wake Island, the Marshall Islands, Nauru Island, Banaba (Ocean) Island, the Gilbert Islands, Howland Island, Baker Island, the Ellice Islands, the Phoenix Islands, the Tokelau Islands, the Line Islands, the Cook Islands, the Society Islands, the Tubuai Islands, Rapa Island, the Tuamotu Archipelago, the Marquesas Islands, the Mangareva Island, Pitcairn Island, Henderson Island, Ducie Island, Easter Island, and Sala-Y-Gomez.
The following are among the relatively few endemic genera:

Annonaceae: *Guamia* (1, Mariana Islands)
Tiliaceae: *Tahitia* (1, Society Islands; very close to *Berrya*)
Malvaceae: *Lebronnecia* (1, Marquesas Islands, close to the Hawaiian genus *Kokia*)
Urticaceae: *Metatrophis* (1, Society Islands; its systematic position is not completely clear—several authors assign it to the Moraceae)
Araliaceae: *Bonnierella* (2, Society Islands; close to *Polyscias*), *Reynoldsia* (14)
Gesneriaceae: *Cyrtandroidea* (1, Marquesas Islands)

Campanulaceae: *Apetahia* (3, Society Islands, Marquesas Islands, Rapa Island), *Sclerotheca* (4, Cook Islands, Society Islands, Rapa Island)
Asteraceae: *Fitchia* (7)
Arecaceae: *Pelagodoxa* (1, Marquesas Islands)

1. **Micronesian Province** (Takhtajan 1970; Micronesian District, Thorne 1963). This province includes the Caroline Islands, the Mariana Islands, Marcus Island, Wake Island, the Marshall Islands, Nauru Island, Banaba (Ocean) Island, the Gilbert Islands, Howland Island, Baker Island, the Ellice Islands, the Phoenix Islands, and the Tokelau Islands. According to Fosberg et al. (1979), 78 species are endemic to the Marianas, 267 species are endemic to the Carolines, and 24 species are endemic to the two groups collectively. There is only one endemic genus (*Guamia*).

2. **Polynesian Province** (Engler 1899, 1903, p.p., 1924, p.p.; Thorne 1963, p.p.; Takhtajan 1970; Polynesian District, Thorne 1963). This province includes the Line Islands, the Cook Islands, the Society Islands, the Tubuai Islands, Rapa Island, the Tuamotu Archipelago, the Marquesas Islands, the Mangareva Islands, Pitcairn Island, Henderson Island, Ducie Island, Easter Island, and Sala-Y-Gomez. There are a few endemic genera and many endemic species.

21. Hawaiian Region

Gebiet der Sandwich-Inseln, Engler 1882, 1903, 1924; Hayek 1926; Hawaiian Region, Good 1947, 1974; Schmithüsen 1961; Mattick 1964; Takhtajan 1969, 1970, 1974; Tolmatchev 1974; Hawaiian Province, Balgooy 1960, 1971; Thorne 1963.

The Hawaiian Region is the most spatially isolated of all the floristic regions of the world. It is also one of the smallest areas, consisting of the Hawaiian Islands and the Johnston Atoll. However, the Hawaiian flora is so unique that most authors assign it the rank of a region.

The flora of the Hawaiian Islands is a typical case of the "disharmonic" island flora, which originated due to chance immigration of propagules from various sources. It is interesting that, despite its geographic proximity to the American continent, plants with American relatives play a comparatively small role in the composition of its flora. Among the seed plants, more than twice as many have an Indomalesian origin, and among the ferns, more than four times as many have an Indomalesian origin (Fosberg 1948). Immigrants from the Australian flora also play a definite role in the floristic composition of the Hawaiian Region. Among the seed plants the Australian element is about equal to the American element (18.3% of seed plants are of American origin and 16.5% are of Australian; Fosberg 1948). In the Hawaiian flora, many widely distributed tropical genera and even families (including conifers and all families of the order Magnoliales and related orders) are absent. The Orchidaceae are surprisingly poorly represented, and the number of genera of vascular plants does not exceed 230 (226, according to Balgooy 1971). However, the Hawaiian flora is characterized by exceptionally high generic and species endemism. The endemic genera are the following (based on Stone 1967, with changes):[12]

Aspleniaceae: *Diellia* (5), *Sadleria* (6)
Caryophyllaceae: *Schiedea*, including *Alsinidendron* (24; belongs to the *Alsineae*)
Amaranthaceae: *Nototrichium* (4)
Violaceae: *Isodendrion* (14, very close to *Melicytus*, of the Fiji Islands, New Zealand, and Norfolk Island)
Begoniaceae: *Hillebrandia* (1)
Malvaceae: *Hibiscadelphus* (4; very close to *Hibiscus*), *Kokia* (5; close to *Gossypium*)
Urticaceae: *Neraudia* (5), *Touchardia* (1)
Euphorbiaceae: *Neowawraea* (1; close to *Margaritaria*)
Rutaceae: *Platydesma* (4)
Hydrangeaceae: *Broussaisia* (1–2; close to *Dichroa*)
Araliaceae: *Munroidendron* (1)
Loganiaceae: *Labordia* (23; close to *Geniostoma*)
Rubiaceae: *Bobea* (5; very close to *Timonius*)
Apocynaceae: *Pteralyxia* (2–3)
Convolvulaceae: *Perispermum* (1; very close to *Banamia*, and perhaps does not deserve generic status)
Solanaceae: *Nothocestrum* (5)
Lamiaceae: *Haplostachys* (5), *Stenogyne* (28)
Lobeliaceae: *Brighamia* (2; phylogenetic relations not completely clear), *Clermontia* (over 40), *Cyanea* (about 65), *Delissea* (9; close to *Cyanea* and *Rollandia*), *Rollandia* (14), *Trematocarpus (Trematolobelia)* (3–4; closest to *Sclerotheca*, of the Cook Islands and Society Islands)
Asteraceae: *Argyroxiphium* (about 7; closest to the American *Madiinae*), *Dubautia*, including *Railliardia* (31; close to *Argyroxiphium*), *Hesperomannia* (3; related to American genera), *Lipochaeta* (24), *Remya* (2), *Wilkesia* (1; very close to *Argyroxiphium*)
Poaceae: *Dissochondrus* (1; very close to *Setaria*)

The Hawaiian flora also possesses an entire series of endemic subgenera and especially sections, the most remarkable of which are the endemic sections of *Santalum*. Species endemism exceeds 97% (about 2,700 endemic species), "the highest rate of endemism of any island, island group, or continental area of comparable size in the world" (Fosberg 1979:99). The Hawaiian Region consists of only one province.

1. **Hawaiian Province** (Thorne 1963).

E. NEOCALEDONIAN SUBKINGDOM

Takhtajan 1969, 1970, 1974.

The Neocaledonian flora is so highly distinctive that it surely deserves the rank of a separate subkingdom. This subkingdom contains only one region.

22. Neocaledonian Region

Good 1947, 1974; Balgooy 1960, 1971; Mattick 1964, p.p.; Takhtajan 1969, 1970, 1974; Tolmatchev 1974; Neocaledonian Subregion, Thorne 1965.

The Neocaledonian Region includes the island of New Caledonia, the Loyalty Islands, and the Isle of Pines.[13] The region is characterized by several endemic families (Stromatopteridaceae, Austrotaxaceae, Amborellaceae, Paracryphiaceae, Oncothecaceae, Strasburgeriaceae and the Phellinaceae), as well as over 130 endemic genera of vascular plants, including the following:

Stromatopteridaceae: *Stromatopteris* (1)
Aspidiaceae: *Cionidium* (1)
Austrotaxaceae: *Austrotaxus* (1)
Podocarpaceae: *Parasitaxus* (1; the only parasitic conifer, close to *Podocarpus*)
Cupressaceae: *Neocallitropsis* (1; close to *Callitris*)
Winteraceae: *Exospermum* (2), *Zygogynum* (6)
Amborellaceae: *Amborella* (1)
Monimiaceae: *Nemuaron* (2), *Canaca* (1), *Carnegieodoxa* (1)
Balanophoraceae: *Hachettea* (1; very isolated genus within the family)
Moraceae: *Sparattosyce* (2; close to *Ficus*)
Myricaceae: *Canacomyrica* (1; occupies an isolated position in the family)
Dilleniaceae: *Trisema* (7; very close to *Hibbertia*)
Paracryphiaceae: *Paracryphia* (1–2)
Oncothecaceae: *Oncotheca* (2)
Strasburgeriaceae: *Strasburgeria* (1)

PALEOTROPICAL KINGDOM 249

Clusiaceae: *Montrouziera* (5)
Epacridaceae: *Cyathopsis* (1)
Sapotaceae: *Achradotypus* (6), *Blabeia* (1), *Corbassona* (2), *Leptostyllis* (8), *Ochrothallus* (3; close to *Chrysophyllum*), *Pichonia* (1), *Pycnandra* (11–12), *Pyriluma* (2), *Rhamnoluma* (3), *Sebertia* (2), *Trouettea* (1)
Capparaceae: *Oceanopapaver* (1; occupies an isolated position in the family and is sometimes placed in its own family, the Oceanopapaveraceae)
Sterculiaceae: *Acropogon* (3)
Bombacaceae: *Maxwellia* (1)
Euphorbiaceae: *Bocquillonia* (6), *Cocconerion* (2), *Dendrophyllanthus* (1), *Lasiochlamys* (1), *Neoguillauminia* (1), *Ramelia* (1)
Thymelaeaceae: *Deltaria* (1), *Solmsia* (2)
Cunoniaceae: *Codia* (13), *Pancheria* (25)
Myrtaceae: *Archirhodomyrtus* (4), *Arillastrum* (1), *Cloëzia* (8), *Myrtastrum* (1), *Pleurocalyptus* (1), *Purpureostemon* (1), *Stereocaryum* (3), *Uromyrtus* (10)
Fabaceae: *Arthroclianthus* (20), *Nephrodesmus* (7)
Rutaceae: *Boronella* (4), *Bouzetia* (1), *Comptonella* (2), *Cupheanthus* (5), *Dutaillyea* (5), *Myrtopsis* (8), *Oxanthera* (4; close to *Citrus*), *Platyspermation* (1), *Sarcomelicope* (2), *Zieridium* (3)
Meliaceae: *Anthocarapa* (2)
Anacardiaceae: *Montagueia* (1)
Sapindaceae: *Gongrodiscus* (1), *Loxodiscus* (1), *Podonephelium* (4), *Storthocalyx* (4)
Phellinaceae: *Phelline* (10)
Icacinaceae: *Anisomallon* (1), *Gastrolepis* (1), *Sarcanthidion* (1)
Celastraceae: *Dicarpellum* (5), *Lecardia* (1), *Peripterygia* (1), *Salaciopsis* (5)
Santalaceae: *Amphorogyne* (3), *Daenikera* (1)
Proteaceae: *Beauprea* (12), *Beaupreopsis* (1), *Garnieria* (1), *Sleumerodendron* (1)
Alseuosmiaceae: *Memecylanthus* (1), *Periomphale* (2)
Araliaceae: *Apiopetalum* (3), *Botryomeryta* (1), *Dizygotheca* (17), *Enochoria* (1), *Eremopanax* (10), *Myodocarpus* (12), *Nesodoxa* (1), *Octotheca* (2), *Pseudosciadium* (1), *Schizomeryta* (1), *Strobilopanax* (3)
Rubiaceae: *Atractocarpus* (10), *Bonatia* (1), *Captaincookia* (1), *Holostyla* (2), *Merismostigma* (1), *Morierina* (2), *Neofranciella* (1), *Normandia* (1)
Apocynaceae: *Cerberiopsis* (3), *Podochrosia* (1)
Gesneriaceae: *Depanthus* (1)
Verbenaceae: *Oxera* (25), *Neorapinia* (1)
Campynemataceae: *Campynemanthe* (1)
Orchidaceae: *Eriaxis* (3), *Megastylis* (7), *Pachyplectron* (2)
Poaceae: *Greslania* (4)
Arecaceae: *Actinokentia* (2), *Alloschmidia* (1), *Basselinia* (9), *Brongniarti-*

kentia (2), *Burretiokentia* (2), *Campecarpus* (1), *Chambeyronia* (2), *Clinosperma* (1), *Cyphokentia* (1), *Dolicokentia* (1), *Kentiopsis* (1), *Lavoxia* (1), *Mackeea* (1), *Moratia* (1), *Pritchardiopsis* (1), *Veillonia* (1)
Incertae sedis: *Serresia* (1)

No other area on earth with a size comparable to New Caledonia possesses such a great number of endemic families and genera. In terms of percentage of endemic genera, only the Hawaiian Islands and the Juan-Fernandez Islands may be more or less comparable with New Caledonia (about 16% here). It is interesting that many of the endemic genera are not monotypic; the number of species of several endemic genera reaches into several tens. This shows that intensive processes of speciation have occurred here for a long time. Moreover, the flora of New Caledonia has a considerable number of archaic genera, and one encounters here 6 out of the 12 vesselless woody genera of flowering plants (*Amborella, Belliolum, Bubbia, Drimys, Exospermum,* and *Zygogynum*), three of which (*Amborella, Exospermum,* and *Zygogynum*) are endemic.

In generic composition the flora of New Caledonia has much in common with that of the Indomalesian flora (especially with the New Hebrides, Fiji, New Guinea, and the Solomon Islands), and with the flora of southeastern Australia.[14] The Indomalesian relationship is most strongly pronounced in the families Rubiaceae, Euphorbiaceae, Sapotaceae, Moraceae, and Fagaceae, and among monocotyledons the Orchidaceae, Arecaceae, and Pandanaceae. It is interesting that the subsection *Bipartitae* of *Nothofagus,* and also the section *Antholoma* of the genus *Sloanea,* are found both in New Caledonia and New Guinea (Balgooy 1971). In terms of generic composition, a curious connection exists with the New Hebrides and with Fiji. Thus 4 genera (*Chambeyronia, Cyclophyllum, Dizygotheca,* and *Strobilopanax*) are limited in their distribution to New Caledonia and the New Hebrides, and 6 others (*Acicalyptus, Acmopyle, Buraeavia, Mooria, Piliocalyx,* and *Storckiella*) are limited to New Caledonia and Fiji. Two other genera (*Alpandia* and *Guillainia*), besides being found in New Caledonia, are found only in the New Hebrides and on New Guinea. No less interesting is the distribution of the genus *Kermadecia* (Proteaceae), 4 species of which are endemic to New Caledonia, 1 to the New Hebrides, and 2 to Fiji.

The relationship with the purely Australian flora is most obvious in the Proteaceae, and especially in the Myrtaceae. The endemic Indomalesian genus *Helicia* is completely absent, but the largely Australian genus *Grevillea* is relatively well represented (4 species). The majority of New Caledonian genera of the Proteaceae are either endemic to New Caledonia or show affinities with Australia. Thus *Macadamia* has 6 endemic species on New Caledonia, 5 in Australia (Queensland and New South Wales), one on Celebes, and 1 on Madagascar, while *Stenocarpus* has 12 endemic species on New Caledonia, 4 in northern and eastern Australia, 1 in Indonesia (Aru), and 1 on New Guinea. *Macadamia* and *Stenocarpus* thus indicate a relationship of the flora of New Caledonia with both the Indomalesian flora and with the Australian. In contrast, *Knightia* exhibits completely different relationships, being represented by 2 species on New Caledonia and 1 in New Zealand.

A connection with the Australian flora is well pronounced in the subfamily Leptospermoideae of the Myrtaceae—especially in genera such as *Xanthostemon, Callistemon, Metrosideros,* and *Baeckea*—but it is interesting that in the flora of New Caledonia there is not one species of *Eucalyptus,* the dispersal of which apparently began only after the complete isolation of New Caledonia. Connections with the Australian flora, especially with New South Wales, are also evident in representatives of the Dilleniaceae, Rutaceae, Epacridaceae, Goodeniaceae, and others. It is estimated that 15 genera are distributed only in New Caledonia and Australia (mainly Queensland).

Interestingly, those elements of the flora with connections to the Indomalesian (particularly the New Guinean) flora are found in rain forests. Yet those elements with Australian relationships are characteristically found in dry shrub formations and savannas. The first dominates in the northern part of the island, while the second is more characteristic of its southern part (Compton 1917).

Although New Caledonia is separated from the surrounding archipelagos, especially from the New Hebrides, by deep seas, and is connected with the Three Kings Islands and the northern Island of New Zealand by the submerged Norfolk Range, the similarity of its flora with that of New Zealand is less than might be expected. There are only 18 genera common to New Caledonia and New Zealand, whereas New Caledonia has 474 genera in common with Queensland, and 482 in common with New Guinea (Thorne 1969). Nevertheless, the genera *Xeronema, Knightia,* and *Libocedrus* s. str. are limited in their distribution to New Caledonia and New Zealand, and the more temperate New Caledonian alpine flora has as much in common with the present flora of New Zealand as it does with the alpine flora of New Guinea (Thorne 1965). As Thorne indicated, the following genera provide excellent examples of such connections: *Agathis, Podocarpus, Dacrydium, Libocedrus, Uncinia, Astelia, Cordyline, Ascarina, Knightia, Muehlenbeckia, Hedycaria, Quintinia, Pittosporum, Weinmannia, Nothofagus, Corynocarpus, Metrosideros, Cyathodes, Meryta, Schefflera, Dracophyllum,* and *Geniostoma*. In addition, as the same author notes, in Tertiary times this relationship was even stronger, as seen from the discovery in Tertiary deposits in New Zealand remains of *Araucaria,* the group "brassii" of the genus *Nothofagus,* and several genera of the Proteaceae represented in the present flora of New Caledonia. To this I would add the interesting fact that on Norfolk Island, a part of the submerged Norfolk Island Range, one finds the endemic *Araucaria heterophylla,* very close to the New Caledonian endemic *A. columnaris.* Norfolk Island also has several other elements in common with New Caledonia—

for example, *Geitonoplesium cymosum*. Lord Howe Island has even more in common with New Caledonia. Yet in its major features the flora of both Lord Howe and Norfolk Island is more closely linked to that of the Neozeylandic Region (in the Holantarctic Kingdom).

The most richly represented families in the flora of New Caledonia are the Rubiaceae, Orchidaceae, Myrtaceae, and Euphorbiaceae, followed by the Apocynaceae, Araliaceae, and Cunoniaceae. Although the legumes, grasses, sedges, and composites are well represented in the majority of floras, in New Caledonia they play a more modest role. These families, particularly the grasses and composites, are represented in New Caledonia mainly by alien species, but in the serpentine areas the percentage of alien species is very low, and there the paucity of species of these families is especially noteworthy. This interesting situation to which Compton (1917) also called attention, may be explained by the separation of New Caledonia from other land masses before such comparatively young families as the composites and grasses had attained an adequate distribution. Thorne (1969) also noted the poor representation of younger taxa in the New Caledonian flora.

New Caledonia probably contains about 2,700 species of seed plants, of which approximately 2,500 (over 90%) are endemic to this region. A series of families (e.g., the Proteaceae) are represented by unique, endemic species. Nonendemic species of the flora of New Caledonia occur in mangrove growths and coastal vegetation. Their penetration into the territory of New Caledonia can be explained by water and air routes. Thus New Caledonia is characterized by extraordinarily high endemism, exceeding that of the flora of Madagascar and comparable only to the endemism of the Hawaiian Region. A considerable amount of varied relief contributed to the intensive process of species formation, as did the specific characteristics of the substrate, especially the large areas of serpentine and peridotite. A rather high percentage of the nonendemic species of New Caledonia are also found in the Fijian Region.

Forests occupy only about 10% of the territory of New Caledonia and usually are very disturbed. On some of offshore islands they are disturbed less than on the mainland. No true lowland rain forest occurs in the region. The ultrabasic outcrops in the drier parts of New Caledonia are covered by a highly distinctive edaphic formation—the so-called maquis de terrains minier—dominated by *Melaleuca leucadendron* (niaouli), ferns, and various shrubs, but without grasses. Soils of these peculiar New Caledonian maquis-like scrubs are characterized by extreme poverty of phosphate and potash, and richness of iron and other minerals. The vegetation of New Caledonia is especially rich in conifers. Many of them are confined to the forests and woodlands adjoining the maquis of the ultrabasic outcrops. The most distinctive of these conifers is *Araucaria cookii*, which is especially well developed on some offshore islands.

PALEOTROPICAL KINGDOM

The Neocaledonian Region contains only one province.

1. **Neocaledonian Province** (Engler 1882; Thorne 1963).

NOTES

1. Engler (1882) called the Guineo-Congolian Region the "West Afrikanisches Wald-gebiet," and Good (1947, 1974) called it the "West African Rain-forest Region." Later, Engler (1899, 1903, 1907, 1924) reduced this region to the rank of a province ("West Afrikanische Waldprovinz").
2. Engler and several other authors called the Madagascan Region the "Madagassisches Gebiet," Humbert (1955) called it the "Région malgache," and Turrill (1959) termed it the "Mascarene Region."
3. According to the somewhat out-of-date figures of Perrier de la Bâthie (1936), Africa and Madagascar have 170 species in common, and 43 endemic Madagascar genera have an African relationship.
4. The Seychelles Islands endemic *Vateriopsis seychellarum* is sometimes included in *Vateria*.
5. Provided one does not accept the highly doubtful genus *Phoenicanthus* (Annonaceae), which for good reasons can be considered a synonym of *Orophea*.
6. The South Chinese Province includes the "Secteur tonkinois" (Tonkin eastward from the Red River) and "Secteur lao-birman" of Vidal (1960).
7. The Thailand Province corresponds partially to the "Secteur lao-siamois" of Vidal (1960).
8. The South Indochinese Province corresponds partially to "Secteur Sud-Indochinois" ("Cambodge, Cochinchine et probablement Siam peninsulaire") of Vidal (1960).
9. See Li and Keng (1950) about the floristic relations of the Hengchun Peninsula.
10. Thorne (1963) also includes in his "Moluccan Province" the island of Tenimber (Timor Laut), but it is located west of the so-called Line of Zollinger. Steenis (1950) refers this island to his "South Malaysian Province." Resolution of the phytochorionomic position of Tenimber requires detailed study.
11. Although the Santa Cruz Islands are transitional toward Papuan flora (especially toward the flora of Solomon Islands), I follow Smith (1979) and include these islands in the Fijian Region.
12. In his list of endemic genera, Stone (1967) also cited two endemic genera of red algae and one endemic moss, which I omit.
13. Good (1974) also includes in the Neocaledonian Region Lord Howe Island and Norfolk Island. I assign these to the Neozeylandic Region (in the Holantarctic Kingdom).
14. Relations with the Indomalesian and Australian floras are approximately equal, but the Malesian influence is somewhat stronger: "On the whole there are slightly more Malesian or Malesia-centered Paleotropical genera than Australian or Australia-centered Paleotropical genera" (Balgooy 1971:95).

III
NEOTROPICAL KINGDOM (NEOTROPIS)

The Neotropical Kingdom occupies the southernmost, tropical part of the Florida Peninsula, parts of the lowlands and shores of Mexico, all of Central America, the islands of the Antilles, the greater part of South America (with the exception of its southern parts, which belong to the Holantarctic Kingdom), and a series of tropical islands adjoining both the North and South American continents.

The Neotropical Kingdom contains the following endemic and partly subendemic families: Hymenophyllopsidaceae, Stylocerataceae, Marcgraviaceae, Caryocaraceae, Pelliceriaceae, Quiinaceae, Lissocarpaceae, Lacistemataceae, Tovariaceae, Brunelliaceae, Rhabdodendraceae, Alzateaceae, Tepuianthaceae, Eremolepidaceae, Dulongiaceae, Columelliaceae, Desfontainiaceae, Plocospermataceae, Dialypetalanthaceae, Saccifoliaceae, Nolanaceae, Duckeodendraceae, Goetzeaceae, Heliconiaceae, Thurniaceae, Abolbodaceae, and the Cyclanthaceae.

The neotropical flora has a common origin with the paleotropical and it may be assumed, at least for the flowering plants, that its initial nucleus had its roots in the Paleotropical Kingdom. There are many families with a pantropical distribution—that is, their distribution includes both the Old and New World tropics (sometimes extending beyond the boundaries of the tropics). Examples include the Annonaceae, Canellaceae, Myristicaceae, Hernandiaceae, Lauraceae, Piperaceae, Peperomiaceae, Balanophoraceae, Moraceae, Urticaceae, Dilleniaceae, Tetrameristaceae, Ochnaceae, Sauvagesiaceae, Sapotaceae, Passifloraceae, Sterculiaceae, Bombacaceae, Chrysobalanaceae,

Rhizophoraceae, Myrtaceae, Melastomataceae, Meliaceae, Anacardiaceae, Sapindaceae, Malpighiaceae, Celastraceae, Proteaceae, Bignoniaceae, Orchidaceae, and Arecaceae.

A considerable number of genera also are found in the tropics of both hemispheres—probably no less than 450. Evidently, over the course of considerable time a direct migration occurred between the tropics of the Old and New Worlds, encompassing not only migration from the Old World to the New but also in the opposite direction. One of the most remarkable examples of the past connection of the flora of tropical West Africa with the flora of tropical America is the distribution of the Bromeliaceae. Almost all species of this family occur in tropical America, with the exception of *Pitcairnia feliciana,* an endemic of the Upper Guinea Floristic Province of tropical Africa (in the Guinea-Congolian Subkingdom). An analogous distribution of another neotropical family is that of the Rapateaceae, the majority of the genera of which are distributed in tropical South America, with the exception of the monotypic *Maschalocephalus,* which is also endemic to the Upper Guinea Province.

The number of such examples could be multiplied. However, as is well known, the separation of South America from Africa occurred long enough ago that the neotropical flora developed independently for a very long time, evolving about 25 endemic families and a great multitude of endemic genera and species. According to Good (1974:156), the total number of genera unknown outside tropical America appears to approach 3,000. Gentry (1982) estimated that there are 3,660 neotropical genera.

The Neotropical Kingdom is divided into five regions.

23. Caribbean Region

Caribbean Region, Good 1947, 1974; Schmithüsen 1961; Mattick 1964; Takhtajan 1969, 1970, 1974; Tolmatchev 1974; Dominio Caribe, Cabrera and Willink 1973.

The Caribbean Region includes the tropical lowland plains and coasts of Mexico, the southernmost tropical part of the Florida Peninsula, the Florida Keys, the Bahama Islands and the island of Bermuda, the Greater and Lesser Antilles, all of Central America from Mexico to Panama (inclusive), the shores of Ecuador, Colombia, and western Venezuela, the Revillagigedo Islands, the Galápagos Islands, and Cocos Island.

The flora of this region is very rich. It includes one endemic family the Plocospermataceae (Mexico and Guatemala). The number of endemic genera probably exceeds 600 (Good 1974), and the number of endemic species is very great.

1. Central American Province (Provinz des tropischen Zentral-Amerika und Süd-Kalifornien, Engler 1899, p.p., 1903, p.p., 1924, p.p.; Südäquatoriale andine Provinz, Engler 1899, 1903, p.p., 1924, p.p.; Provincia Pacifica and Provincia Guajira, Cabrera and Willink 1973). This province includes the tropical lowland plains and shores of Mexico and all remaining Central America south of Mexico, and also the shores of Ecuador, Colombia, and western Venezuela, beginning from the province of Guayas in Ecuador to the Paraguana Peninsula in Venezuela. In this province I also place the volcanic Revillagigedo Islands and the small Cocos Island.

The flora of the Revillagigedo Islands contains about 125 species and subspecies of vascular plants, of which 37 are endemic (Johnston 1931). It is closest to the flora of tropical Mexico, Central America, and Baja California. The flora of Cocos Island is rather poor (about 100 species of vascular plants). It consists primarily of widely distributed species, and therefore I include it in this province on a purely conditional basis.

2. West Indian Province (Engler 1882, 1903, 1924; Gleason and Cronquist 1964; Caribische Provinz, Hayek 1926, p.p.; Caribbean Province, Takhtajan 1970; Provincia Caribe, Cabrera and Willink 1973). This province includes the southern tropical part of the Florida Peninsula (approximately south of the latitude of Miami), the Greater and Lesser Antilles, the Bahamas, and Bermuda.

The West Indian Province has about 200 endemic genera (Good 1974), and the number of endemic species is very great (probably about 50%), especially in the Greater Antilles. Cuba, which is the largest island, has about 6,000 species, 47% of which are endemic, 27% of them being found on serpentine soils (Berazaín 1977; Howard 1979). Among the most noteworthy endemic genera, I mention only the monotypic genus *Microcycas*, endemic to western Cuba.

Each of the four islands of the Greater Antilles (Cuba, Jamaica, Haiti, and Puerto Rico) has its own set of endemic species. Furthermore, Cuba has partially different sets of endemics in the mountains at the two ends. The islands of the Lesser Antilles have much smaller and less distinctive floras, with a progressively stronger South American (Amazonian) influence toward the south.

3. Galapageian Province (Good 1947, 1974; Thorne 1963; Takhtajan 1970; Cabrera and Willink, 1973; Gebiet der Galapagos-Inzeln, Engler 1882, 1903, 1924; Bezirk der Galapagos-Inseln, Hayek 1926). This Province includes only the Galápagos Islands.

The flora of these islands is relatively poor, with only 529–543 species, subspecies, or varieties of indigenous vascular plants, including a number of extinct taxa (according to Porter 1979); it is also rather poorer in endemic taxa than earlier supposed. Therefore, while in Engler's floristic system the Galapagos were treated as a separate, independent region, at the present time its chorionomic rank is reduced to the level of province, and sometimes even lower ("no more than a district within the Neotropic," according to Balgooy 1971:119). The province contains 4 endemic genera of the Asteraceae: *Darwiniothamnus* (1; very close to *Erigeron*); *Macraea* (1; close to the American species of the pantropical genus *Wedelia*); *Lecocarpus* (3; like *Macraea*, it is referred to the Heliantheae); and *Scalesia* (8; also referred to the Heliantheae). It also has 2 endemic genera of Cactaceae (*Brachycereus* and *Jasminocereus*) and 1 endemic monotypic genus of the Cucurbitaceae (*Sisyocaulis*). Endemic species and subspecies number 211–231 and constitute 40–43% of the indigenous flora (Porter 1979). Cabrera and Willink (1973) include the Galápagos Province ("Provincia de les Galapagos") in their "Dominio Caribe," which probably is the correct solution.

24. Region of the Guayana Highlands

Region of Venezuela and Guiana, Good 1947, p.p., 1974, p.p.; Mattick 1964, p.p.; Region of the Guayana Highlands, Takhtajan 1970, 1974; Dominio Guayano, Cabrera and Willink 1973; Orinoco Region, Tolmatchev 1974, p.p.; Nordbrasilianisch-guianenische Provinz, Engler 1882, p.p.

This small but unique—and doubtless ancient—floristic region with a "spectacularly distinctive flora" (Maguire 1979:244) occupies the Guayana Highland of northern South America east to Andes, (i.e., parts of southern Venezuela and adjoining regions of Colombia, Brazil, and the Guianas).

The rich flora of this region contains over 8,000 species, including no less than 4,000 endemics. The endemic species are concentrated in the high mountains and outwash lowlands, where endemism reaches 90 to 95% (Maguire 1970). The region contains three endemic families—the Hymenophyllopsidaceae, Tepuianthaceae, and Saccifoliaceae—and also one recently described subfamily, the Pakaraimoideae of the Dipterocarpaceae.

The presence of representatives of the Dipterocarpaceae in the Neotropical Kingdom is of special interest because until very recently this family was considered to be completely paleotropical. In 1951 an enigmatic plant was discovered in the Upper Mazaruni River Basin in the savannah on the Pakaraima Plateau, and also not far to the west, in Venezuela (the Cerro Guaiquinima on the Rio Paragua). It has now been described as a new genus and a new species, *Pakaraimaea dipterocarpacea* Maguire et Ashton. Multifaceted study of this new monotypic genus led to the unexpected conclusion that it belongs to the Dipterocarpaceae, but is a rather isolated taxon within this family and constitutes a special, neotropical subfamily, the Pakaraimoideae (Maguire et al. 1977). Data from comparative morphology led to the conclusion that *Pakaraimaea* is closer to the African genera *Monotes* and *Marquesia* than to the Asiatic dipterocarps. This is completely understandable in light of the newest global tectonic representations of the past union of South America and Africa.

The flora of the Guayana Highlands contains about 100 endemic genera (Maguire 1970). Along with clearly neotropical relationships, several of these genera—like *Pakaraimaea*—reveal African relationships, while others indicate Indomalesian relationships. The last include the comparatively recently described genus, *Pentamerista*, referred to the Tetrameristaceae (Maguire 1972*a*), which was long considered completely paleotropical (with the single Malesian genus, *Tetramerista*). The ancient flora of the Guayana Highlands, containing many relicts, holds exceptionally great interest for the history of floras, partly for the verification of past floristic connections with paleotropical regions.

The Region of the Guayana Highlands probably contains only one province.

1. **Guayana Province** (Province of Guayana, Maguire 1970, 1972*b*, 1979; Cabrera and Willink 1973). Maguire (1979) divides his "province of the Roraima Formation" into three subprovinces: Pakaraima–Gran Sabana Subprovince; the Rió Caroni–Río Negro Subprovince, and the Trans Río Negro–Colombian Guayana Subprovince.

25. Amazonian Region

Good 1947, 1974; Schmithüsen 1961; Mattick 1964; Takhtajan 1969, 1970, 1974; Tolmatchev 1974; Région de l'hylaea brasilienne, Chevalier and Emberger, 1937; Dominio Amazónica, Cabrera and Willink 1973, p.p.

This vast region includes the lowlands of the Amazon Basin, the coast of eastern Venezuela and its adjacent islands, Trinidad and Tobago, the greater part of Guyana, all of Surinam, all of French Guiana, and part of the northeastern coast of Brasil, up to Sao Marcos in Maranhao.

The very rich flora of the Amazon Region contains only three endemic families, the Dialypetalanthaceae, Duckeodendraceae, and Rhabdodendraceae, but at least 100 endemic genera and at least 3,000 endemic species (Good 1974).

The most distinguishing characteristic of the vegetation of the Amazon Region is its magnificent tropical rain forests, which A. von Humboldt called Hylaea. Nowhere else in the world do tropical rain forests occupy so great an area nor are they elsewhere distinguished by such diversity of vegetative form as occurs in the Amazon Basin.

1. **Amazonian Province** (Engler 1903, 1914, 1924; Hayek 1926; Cabrera and Willink 1973; Amazonian lowland and montane rain forest, Maguire, lett. comm., 1976, ined.).[1] This is the largest province of the Amazonian Region, containing a very great number of endemic taxa. The dominant vegetation cover is evergreen tropical rain forest, but in places one also encounters woodlands, shrubby thickets, and savannas.

2. **Llanos Province** (Hayek 1926; Provincia Venezolana, Cabrera and Willink 1973; Provincia de la Sabana, Cabrera and Willink 1973; Venezuelan Coastal Andes and Venezuelan-Colombian Llanos, Maguire 1976 ined.). I include in the Llanos Province[2] all of the coast of Venezuela from the Paraguana Peninsula to approximately 63°30' west longitude, the valley of the middle course of the Orinoco River and its left tributaries in the territory of Venezuela, and a considerable part of northeastern Colombia to the east and northeast of Bogotá.[3] The flora of this province is not completely uniform, especially given that—along the eastern cordillera and further through the Cordillera de Merida—a considerable number of Andean elements spread into it.[4]

26. Brazilian Region

Tolmatchev 1974; Région tropicale extra-amazonienne deu Brésil, Chevalier and Emberger 1937; South Brazilian Region, Good 1947, p.p., 1974, p.p.; Ost- und Südbrasilien, Mattick 1964, p.p.; Central Brazilian Region, Takhtajan 1970, 1974; p.p.; Dominio Chaqueno, Cabrera and Willink 1973, p.p.; Südbrasilianische Provinz, Engler 1899, p.p.; 1903, p.p., 1924, p.p.; Provinz der brasilianischen Savannen und Catingas, Hayek 1926, p.p.

This region includes the Brazilian Highlands, the Caatinga (a semidesert tropical woodland in the northeastern Brazilian Highlands), the Gran Chaco and the islands of Saint Paul's Rocks (0°56' N, 29°22' W), Fernando de Noronha (3°50' S, 32°25' W), Trinidade (20°30' S, 29°20' W), and Martin Vaz (20°30' S, 25°51' W). The region has about 400 endemic genera (Good 1974), including *Antonia* and *Diclidanthera*.

1. **Caatinga Province** (Maguire, lett. comm., 1976, ined.; Highlands of Eastern Brazil, Good 1947, 1974); Südbrasilianisch Provinz: Catingas-Unterprovinz, Engler 1903, 1924; Graebner 1910, p.p.; Provinz der brasilianischen Savannen und Catingas: Bezirk der Catingas, Hayek 1926). Following Maguire, I include in this province the northeastern regions of

Brazil, beginning in the north from the environs of Turiaçú and ending approximately at the southern boundary of the State of Espírito Santo. According to Cabrera and Willink (1973), "Provincia de la Caatinga" occupies only part of this very vast territory.

The flora of the Caatinga Province is rich in endemic species. The vegetative cover consists primarily of thorny shrubs, cactus parklands, and different types of savanna, but is also characterized by islands of montane variable-moisture evergreen forests.

2. Province of Uplands of Central Brazil (Uplands of Central Brazil, Good 1947, 1974; Cerrados-Planalto do Brasil, Maguire 1976, ined.). Maguire includes in this province a considerable part of Bolivia and the northeastern parts of Paraguay, as well as part of Brazil. The flora of this province is the richest in the Brazilian Region and is characterized by high endemism. The most dominant vegetation types are savannas with xerophilic shrubs and woodlands ("Cerrados"; see Eiten 1972).

3. Chacoan Province (Good 1947, 1974; Takhtajan 1970; Provincia Chaquena, Cabrera and Willink 1973, p.p.; Argentinische Provinz: Unterprovinz des Gran-Chaco, Engler 1903, 1924; Provinz der brasilianischen Savannen und Catingas: Bezirk des Gran-Chaco, Hayek 1926). This province includes Gran Chaco and Yungas and is characterized by predominantly arid tropical forests and woodlands ("forests of Chaco").

4. Atlantic Province (Takhtajan 1970; Provincia Atlantica, Cabrera and Willink 1973; South Brazilian Region and Eastern coasts, Good 1947, 1974). This province includes the southeastern coast of Brazil and its islands, and is characterized by evergreen tropical rain forests.

5. Paraná Province (Provincia Paranese, Cabrera and Willink 1973). This province includes the area south of Brasília to the west from Serra-do-Mar, and to the north from 30° south latitude, the extreme northeastern part of Argentina, and the eastern part of Paraguay. The Paraná Province contains several endemic or almost endemic genera and a rather considerable number of endemic species, among which perhaps the most noteworthy is *Araucaria angustifolia*, closely related to the Chilean *A. araucana*. The following are also characteristic of the Paraná Province:

Dicksonia sellowiana, Podocarpus lambertii, P. sellowii, Drimys brasiliensis, Cinnamomum (Phoebe) porosum, Nectandra membranacea, Ficus spp., *Gallesia guararema, Luehea divaricata, Centrolobium robustum, Dalbergia nigra, Enterolobium contortisiliquum, Holocalyx balansae, H. glaziovii, Hymanaea stilbocarpa, Inga edulis, Lonchocarpus* spp., *Machaerium* spp., *Melanoxylon brauna, Myrocarpus frondosus, Myroxylon peruiferum, Parapiptadenia rigida, Peltophorum dubium, Piptadenia* spp., *Pithecellobium quaranticum, P. hassleri, Plathymenia foliosa, Schizolobium excelsum,* various taxa of the Myrtaceae, *Cariniana estrellensis, Balfourodendron riedelianum, Cabralea cangarena, C. oblongifolia, Cedrela fissilis, C. glaziovii, Vochysia* spp., *Ilex paraguayensis, Aspidosperma polyneuron, Patagonula americana, Tabebuia ipe, T. pulcherrima,* the palms *Acrocomia totai, Arecastrum romanzoffianum, Euterpe edulis, Syagrus romanzoffianus* and bamboos of the genera *Chusquea, Guadua,* and *Merostachys.*

The most characteristic vegetation types of the Paraña Province are the araucarian and deciduous mesophytic subtropical forests, and also tall-grass grasslands ("Campos Limpos, geschlossene Grassfluren in den Hochlagen Südbrasiliens," Hueck and Seibert 1972); these last consist of *Aristida pallens, A. venusta, Paspalum stellatum, Anthaenantia lanata, Andropogon* spp., *Elionurus, Trachypogon,* and *Chloris.*

27. Andean Region

Engler 1882, 1903, 1912; Engler and Gilg 1919; Graebner 1910; Hayek 1926, p.p.; Good 1947, p.p., 1974, p.p.; Mattick 1964, p.p.; Takhtajan 1969, p.p., 1970, p.p., 1974, p.p.; Tolmatchev 1974, p.p.; Dominio Adino-Patagónico, Cabrera and Willink 1973.

This long region, drawn out from south to north, includes the western coastal mountain ranges and the shores of South America from northwestern Venezuela to northern Chile. It contains endemic families Styloceratceae, Hypseocharitaceae, Rhynchothecaceae, Ledocarpaceae, Columelliaceae, and Desfontainiaceae, and has probably several hundred endemic genera, as well as numerous endemic species.

The flora of the Andean Region holds special interest from the point of view of the origin and path of migration of its different elements. Here I can only mention the results of the classical studies of Diels and other authors, which are well summarized by Wulff (1944). In the flora of the Andean Region (along with indigenous elements with clearly neotropical origins), a highly important role is played by Holantarctic elements which, via mountain chains, reached Ecuador and even northeastern Colombia and northwestern Venezuela (Sierra-de-Perija and the Cordillera de Merida). Typical examples of Holantarctic genera that extend into the northern part of the Andean Region are *Colobanthus, Azorella,* and *Ourisia.* Of the Holantarctic species spreading far to the north of the Andean Region, *Caltha sagittata* (referable to the Holantarctic section Psychrophila) deserves mention. Also noteworthy is the role of Holarctic elements spreading into the Andean Region from the north. From the north, *Quercus* extends to northern Colombia; *Pedicularis* to Ecuador; *Berberis, Hydrangea,* and *Viburnum* to southern Chile; and *Ribes* to Tierra del Fuego. Along with these genera—which have migrated gradually, and over a considerable period of time—there is also a series of most interesting cases of long-distance dispersal by means of chance agents (see Raven 1963; Thorne 1972; and especially Raven and Axelrod 1974, which contains an extensive literature review). The Andean Region can probably be divided into two provinces.

1. **Northern Andean Province** (Northern Andes, Smith and Johnston 1945).

2. **Central Andean Province** (Southern Andes, Smith and Johnston 1945, p.p.). The northern and partly also the southern boundary of the Central Andean Province is not completely clear and will be clarified only after special chorionomic investigations based on the chorology of an adequate number of Andean plants.

NOTES

1. Engler named the Amazonian Province the "Provinz des Amazonenstromes oder Hylaea," and Hayek called it "Provinz der Hylaea." I consider the boundaries of this province to be closest to those given in the floristic map of Maguire, who kindly sent his map to me.
2. The term "Llanos" is used here as the name of a geographic region, not as the name of a special type of savanna vegetation.
3. In the understanding of the boundary of the Llanos Province accepted here, it only partially corresponds to Hayek's "Provinz der Llanos."
4. Thus in his floristic map, Maguire referred a considerable part of northern Venezuela to his "Andean complexes."

IV
CAPE KINGDOM (CAPENSIS)

The Cape Kingdom is the smallest of the world's floristic kingdoms, but because of its exceptionally distinctive flora and its independent historical development, phytogeographers unanimously separate it from the rest of Africa. The Cape Kingdom has only one region.

28. Cape Region

Gebiet des südwestlichen Kaplandes, Engler 1899, 1903, 1924; Gebiet des Kaplandes, Graebner 1910; Pole-Evans 1922; Hayek 1926; Good 1947, 1974; Mattick 1964; Takhtajan 1969, 1970, 1974; Tolmatchev 1974; Wickens 1976; Taylor 1978; Goldblatt 1978; White 1983.

The Cape Region includes the southern extremity of Africa from the mouth of the Olifants River in the northwest to the vicinity of Port Elizabeth in the east. According to Goldblatt (1978:375), the border of this region extends from Nieuwoudtville in the north, following the eastern slopes of the Cedarberg, and then east from Karoopoort along the north slope of the Witteberg, Swartberg, Baviaans Kloof, and Groot Winterhoek mountains, ending at Port Elizabeth. All territory south and west of this line to the coast forms the Cape Region.

The flora of this geographically small region is unusually rich, including 8,550 species, of which 73%—or 6,252 species—are endemic (Goldblatt 1978).[1] It includes 8 endemic or subendemic families: Grubbiaceae, Roridulaceae, Bruniaceae, Geissolomataceae, Greyiaceae, Penaeaceae, Retziaceae, and Stilbaceae. About 200 genera or more (including genera of endemic families) are endemic to the region, and a considerable number of nonendemic genera are centered here. The majority of the endemic genera are monotypic or oligotypic, with the systematically most isolated genera being monotypic as a rule (Weimarck 1941). The Endemic genera include: *Mystropetalon* (Balanophoraceae; a very isolated genus in the family), *Discocapnos* (Fumariaceae), 26 genera of the Aizoaceae, 19 genera of the Ericaceae, *Grubbia* (Grubbiaceae), *Roridula* (Roridulaceae), 4 monotypic or oligotypic genera of the Brassicaceae, *Lachnostylis* (Euphorbiaceae), *Cryptadenia* (Thymelaeaceae), *Platylophus* (Cunoniaceae), 11 genera of the Bruniaceae, *Geissoloma* (Geissolomataceae), *Greyia* (Greyiaceae), 7 genera of Penaeaceae, 11 genera of the Fabaceae, 11 genera of the Rutaceae, *Heeria* and *Laurophyllus* (Anacardiaceae), *Smelophyllum* (Sapindaceae), *Maurocenia* (Celastraceae), *Thesidium* (Santalaceae), 10 genera of the Proteaceae, of which the largest, *Leucadendron*, has about 8 species (one of which extends beyond the borders of the Cape Region, to the Drakensberg Mountains), *Glia*, *Hermas*, *Rhyticarpus*, and *Thunbergiella* (Apiaceae), *Orphium* (Gentianaceae), *Oncinema* (Asclepiadaceae), *Carpacoce* (Rubiaceae), *Echiostachys* (Boraginaceae), *Retzia* (Retziaceae), *Agathelpis, Gosela, Ixianthes*, and *Microdon* (Scrophulariaceae), *Charadrophila* (Gesneriaceae), 5 genera of the Stilbaceae, 9 genera of the Campanulaceae, about 36 genera of the Asteraceae, *Baeometra, Dipidax*, and *Neodregea* (Colchicaceae), 10 genera of Iridaceae, *Amphisiphon* and *Androsiphon* (Hyacinthaceae), *Amaryllis, Carpolyza*, and *Cybistetes* (Amaryllidaceae), *Dilatris, Lanaria*, and *Wachendorfia* (Haemodoraceae), *Pauridia* (Hypoxidaceae), 11 genera of the Orchidaceae, 5 genera of the Cyperaceae, 8 genera of the Restionaceae, and 5 genera of the Poaceae (see Goldblatt 1978).

According to White (1983:132), 70 genera have their greatest concentration of species here. Among them are *Agathosma* (130 endemic species), *Aspalathus* (240), *Cliffortia* (70), *Crassula* (145), *Erica* (520), *Ficinia* (50), *Metalasia* (30), *Muraltia* (100), *Phylica* (140), *Protea* (85), and *Restio* (40).

Some of the very characteristic members of the Cape flora are species of *Erica, Phylica, Cliffortia, Muraltia, Roridula, Metalasia*, and *Stoebe*; genera of the tribe Diosmeae of the Rutaceae; and numerous taxa of the Proteaceae, represented by species of *Protea* and an entire series of endemic genera, including the almost endemic genus, *Leucadendron*. One species of *Leucadendron, L. argenteum* (the "Silver Tree"), is among the most characteristic plants of this region. The following genera also have many species here: *Pelargonium, Helichrysum, Senecio, Eriospermum, Gasteria, Haworthia, Lachenalia, Romulea, Moraea, Aristea, Geisorhiza, Hesperantha, Ixia*, and *Babiana*.

The Cape flora appears to be an inexhaustible source of ornamental plants, especially of bulbous and tuberous monocots.

The characteristic vegetation of the Cape Region is a maquis-like formation ("fynbos") with numerous shrubs and small trees with sclerophyllous and frequently ericoid leaves. In general, the physiognomic and ecological similarity of fynbos with Mediterranean maquis is quite great; it is sometimes called "Cape Maquis." According to White (1983:134), in typical fynbos, true trees are virtually absent. The only species with well-defined boles are *Leucadendron argenteum*, *Widdringtonia cedarbergensis*, and *W. schwarzii*. Two other species, *Widdringtonia cupressoides* and *Olea capensis*, which often occur as trees elsewhere, in fynbos are usually of bushy habit and less than 7 m tall. Fynbos is exceptionally well expressed in the mountains close to Capetown. The following are characteristic of the upper layer of fynbos: species of *Protea*, *Leucadendron*, *Leucospermum*, and *Gymnosporia*, and also *Olea africana* and several other species. Characteristic of the second layer are numerous small shrubs of the Ericaceae, Fabaceae, Rutaceae, Campanulaceae, Asteraceae, and also *Gymnosporia buxifolia* and *Asparagus capensis*. Many semishrubs, large herbaceous perennials, and monocotyledonous geophytes in particular are also present. In the winter months bulbous and tuberous geophytes cover the area with their bright flowers. Grasses and the Cyperaceae are not numerous, but representatives of the Restionaceae are very common. Annuals are usually not numerous. The sclerophyll vegetation includes many endemic species.

Resident South African botanists emphasize the importance of the low-nutrient sandstone substrate of much of the province in fostering a Mediterranean aspect of the vegetation. They also note that summer fog in the mountain valleys tends to ameliorate the seasonal drought.

Within the Cape Region, one also finds isolated stands of evergreen forests, which are especially pronounced in the coastal region between the cities of George to the west and Humansdorp to the east. The upper layer in these forests consists largely of *Podocarpus falcatus*, *P. latifolius*, *Olea laurifolia*, *Olinia cymosa*, and frequently other species as well (including the deciduous *Celtis kraussiana*). In the lower layer one most often encounters *Ocotea bullata*, *Apodytes dimidiata*, *Curtisia dentata*, *Trichocladus crinitus*, *Platylophus trifoliatus*, *Halleria lucida*, *Gonioma camassi*, and so on. Also characteristic are the lianas *Secamone alpini* and *Rhoicissus capensis*, and the epiphytes *Peperomia retusa* and also *Vittaria isoëtifolia* and other ferns. The herbaceous cover is characterized by the large, ranunculaceous *Knowltonia capensis*. In damp places one encounters large-leaved ferns, including *Cyathea dregei*, *Alsophila capensis*, *Lonchitis pubescens*, and *Marrattia fraxinea*. In the mountains 230 km north of Capetown, remains of former forests are preserved: they are small groves of pure stands of *Widdringtonia cupressoides*, which is endemic for this area. Another species of this genus, *W. schwarzii*, forms parklike groves in montane ravines of the inner regions not far from Knysna (Adamson 1938; Adamson [in Haden-Guest et al.], Acocks 1953; Walter 1968; Knapp 1973). Until the seventeenth century (i.e., until colonization of South Africa by Europeans), forests occupied a considerably greater area, and the sclerophyll shrubs were in comparison less widely distributed (Acocks 1953).

According to White (1983:135), along the inner margin of the Cape Region there is a narrow band of arid fynbos which forms the transition from typical Cape to typical Karoo vegetation. Ericaceae are absent, and Proteaceae and Restionaceae, though conspicuous (especially the latter), are few in species. Typical Karoo genera are well represented. Succulents, including *Euphorbia mauritanica*, are often present.

In the past, the Cape flora occupied a much larger territory in South Africa than at present, but due to the increasing dryness of the climate it continues to decrease steadily, giving way to the flora of Karoo (Marloth 1908; Wulff 1944).

In the complex and far from completely understood history of the flora of the Cape Region, particular interest exists about its relationships with other floras of the Southern Hemisphere, extending even to the time when Gondwanaland was represented by one continent or only beginning to break up. In this regard, the most interesting families are the Proteaceae and the Restionaceae. Johnson and Briggs (1975) concluded that the initial diversification of the family Proteaceae occurred in that part of Gondwanaland which formed Australasia. Migration in the Proteaceae occurred up until the full separation of Gondwanaland; however, representatives of only two of its subfamilies reached that part of it which now constitutes South Africa.

As Johnson and Briggs note, similar conclusions may also be reached about relationships within the Restionaceae, the other large "Afroaustralian" family. South Africa is very rich in species of the Restionaceae, but has less diversity at the higher taxonomic levels than does Australia, which also possesses such a markedly isolated genus as *Anarthria*, placed by some into a separate family, the Anarthriaceae. Johnson and Briggs (1975) stress that not only is there not one genus in common between Africa and Australia but also the African group as a whole is distinguished from all Australian groups. Evidently the distribution of the Restionaceae occurred in basically the same way as that of the Proteaceae. Thus both the Proteaceae and the Restionaceae underwent their development in the Australasiatic part of Gondwanaland and reached South Africa when direct migration between the two was still possible. Wulff (1944) has already written about this.

However, there is a whole series of such elements in common between the Cape (and, in general, the South African) flora and the floras of Australia, New Caledonia, and temperate South America, and this is difficult if not impossible to explain solely by continental drift. Here I mean such genera as *Cunonia* (1 species in South Africa, 16 species in New Caledonia); *Metrosideros* (South Africa-eastern Malesia, Australia, New Zealand, Polynesia); *Bulbinella* (about 5 species in South Africa and 6 species in New Zealand, on Campbell Island and on the Auckland Islands); *Caesia* (2 species in South Africa, 1 species on Madagascar, and 7 species in Australia, Tasmania, and New Guinea); and *Tetraria* (38 species in South Africa, 1 species in East Africa, 1 species on Kalimantan (Borneo), and 4 species in Australia). It is even more difficult to accept a "predrift" distribution of *Todea barbara*, which is common to South Africa, Australia, Tasmania, and New Zealand. Most likely the dispersal of these genera occurred in the later stages of continental drift—beginning probably from the middle of the late Cretaceous—through connections of their archipelagos.

Although species of the taxa *Podocarpus, Widdringtonia,* Cunoniaceae, *Acaena, Roridula,* Proteaceae, *Metrosideros, Phylica, Tetraria,* Restionaceae, and some other components of the Cape flora were undoubtedly of southern origin, many others—including species of *Celtis, Pittosporum, Cliffortia, Ilex, Olea, Muraltia, Lobostemon, Stoebe,* and representatives of the Ericoideae (in the Ericaceae) and of the Diosmeae (in the Rutaceae)—had a northern origin (Levyns 1964). Thus as Diels (1908) had already noted, the Cape flora (like the flora of New Zealand) exhibits a duality: on the one hand it shows significant connections with the contemporary tropical African flora, and on the other it represents totally original elements—derivatives of the ancient flora of the Southern Hemisphere.

Only one province can be recognized in the Cape Region.

1. **Cape Province.**

NOTES

1. The small Cape Peninsula, on which the city of Capetown is located, has 2,622 species assignable to 702 genera (according to the now somewhat outdated information of Adamson and Salter 1950).

V
AUSTRALIAN KINGDOM (AUSTRALIS)

The flora of Australia, the smallest and most isolated continent, is very distinctive and is characterized by high endemism. Its composition includes an entire series of endemic families: Platyzomataceae, Austrobaileyaceae, Idiospermaceae, Gyrostemonaceae, Baueraceae, Davidsoniaceae, Cephalotaceae, Eremosynaceae, Stylobasiaceae, Emblingiaceae, Akaniaceae, Tremandraceae, Tetracarpaeaceae, Brunoniaceae, Blandfordiaceae, Doryanthaceae, Dasypogonaceae, and Xanthorrhoeaceae. In addition, the Australian Kingdom appears as the center of origin of such families as the Eupomatiaceae, Pittosporaceae, Epacridaceae, Stackhousiaceae, Myoporaceae, and Goodeniaceae (Burbidge 1960), as well as of a series of subfamilies, tribes, and many genera. The following families play the greatest role in the flora of Australia: Poaceae, Fabaceae, Asteraceae, Orchidaceae, Euphorbiaceae, Cyperaceae, Rutaceae, Myrtaceae (especially the subfamily Leptospermoideae), and Proteaceae. Also very characteristic for the Australian flora are representatives of the following families: Chenopodiaceae, Epacridaceae, Cunoniaceae, Pittosporaceae, Stackhousiaceae, Myoporaceae, Stylidiaceae, and Goodeniaceae. At the same time, one is struck by the absence of many widely distributed groups such as *Equisetum*, the Papaveraceae, the Myricaceae, the Saxifragaceae s. str., the subfamily Maloideae of the Rosaceae, bamboo, and such families as the Begoniaceae, the Valerianaceae, and the Maranthaceae. An exceptionally great role is played in the vegetation of Australia by numerous species of *Acacia*, and especially by *Eucalyptus*, as well as by *Casuarina*, *Melaleuca*, *Leptospermum*, and numerous species of the Proteaceae (especially *Banksia* spp.).

The Australian Kingdom contains about 550 endemic genera of vascular plants, of which the largest are *Hakea* (Proteaceae), *Eremophila* (Myoporaceae), *Pultenaea* (Fabaceae), *Daviesia* (Fabaceae), *Boronia* (Rutaceae), *Prostanthera* (Lamiaceae), *Dampiera* (Goodeniaceae), *Dryandra* (Proteaceae), *Calytrix* (Myrtaceae), *Verticordia* (Myrtaceae), *Maireana* (formerly in *Kochia*, Chenopodiaceae), *Bossiaea* (Fabaceae), *Conospermum* (Proteaceae), *Darwinia* (Myrtaceae), *Isopogon* (Proteaceae), *Spyridium* (Rhamnaceae), *Angianthus* (Asteraceae), *Mirbelia* (Fabaceae), and *Thomasia* (Sterculiaceae).[1] About 80% of the species are endemic and belong mainly to endemic and near-endemic genera (Beadle 1981*b*).

Just as some taxa of the Australian flora—such as the Annonaceae, Lauraceae, Menispermaceae, Nepenthaceae, Euphorbiaceae, Sapotaceae, Flacourtiaceae, Tiliaceae, Malvaceae, Ulmaceae, Urticaceae, Moraceae (including *Antiaris* in Arnhem Land), Euphorbiaceae, Myrtaceae-Myrtoideae, Fabaceae, Rutaceae, Meliaceae, Burseraceae, Anacardiaceae, Sapindaceae, Celastraceae, Santalaceae, Rhamnaceae, Vitaceae, Rubiaceae, Apocynaceae, Asclepiadaceae, Verbenaceae, Orchidaceae, Zingiberaceae, Arecaceae, and so on—show more or less obvious tropical connections, others—such as the Proteaceae, Epacridaceae, Restionaceae, Centrolepidaceae, Stylidiaceae, Donatiaceae, and Goodeniaceae—show a connection with the floras of the Southern Hemisphere.

More or less direct overland migration between Australia and East Antarctica (via the South Tasmanian Rise)—and between West Antarctica and South America—seems to have been possible until approximately the close of the Eocene (Raven 1979; see also Crook 1981). This explains the appearance of many common taxa, not only of families (Araucariaceae, Podocarpaceae, Winteraceae, Berberidopsidaceae, Proteaceae, Restionaceae, and other families characteristic of the Southern Hemisphere) but also of many genera (e.g., *Nothofagus, Hebe, Donatia, Drapetes*), many of which are also a part of the New Zealand flora. In the majority of cases these common elements represent relicts of an ancient Holantarctic flora, which in the Australian Kingdom are best preserved in the mountains of southeastern Australia and Tasmania. Yet the Australia flora has only 1 genus in common with the flora of the Cape Kingdom (*Caesia*), only 2 genera in common with Madagascar (*Keraudrenia* and *Rulingia*), and only a few in common with the whole of South Africa and Madagascar (including *Helipterum, Wurmbea, Chrysithrix,* and *Restio*). This may be because a direct migratory connection between Africa and Australia was broken much earlier than in the case of Antarctica and Australia.

The initial nucleus of the flora of the Australian Kingdom arose as a result of the transformation of elements of the ancient Holantarctic flora. The origin of the xerophilous flora of Australia resulted from the increase in aridity during the Miocene. According to Galloway and Kemp (1981:76),

"around the Middle Miocene the sea surface cooled and precipitation on land declined especially in the center where grass appears in the fossil record. A sudden extra cooling of the sea surface in the latest Miocene may be associated with a period of increased dryness in the south also."
The Australian Kingdom is divided into three floristic regions.

29. Northeast Australian Region

Good 1947, 1974; Schmithüsen 1961; Mattick 1964; Takhtajan 1969, 1970, 1974.

This large and rather heterogenous region consists of the northern, eastern, and southeastern parts of Australia, along with the coastal islands, including Tasmania. The Northeast Australian Region contains a number of endemic families (Platyzomataceae, Austrobaileyaceae, Idiospermaceae, Baueraceae, Davidsoniaceae, Tetracarpaeaceae, Akaniaceae, Blandfordiaceae, and Doryanthaceae) and over 150 endemic genera, including such taxonomically very isolated genera such as *Bowenia* and *Lepidozamia* (Zamiaceae), *Microcachrys* and *Microstrobos* (Podocarpaceae), *Athrotaxis* (Taxodiaceae), *Diselma* (Cupressaceae)*Prionotes* (Epacridaceae), *Blepharocarya* (Anacardiaceae), *Irvingbaileya* (Icacinaceae), *Bellendena* (Proteaceae), *Isophysis* (Iridaceae), *Drymophila* (Philesiaceae), and *Petermannia* (Philesiaceae).

1. **North Australian Province** (Engler 1912, 1924; Good 1947, 1974; Takhtajan 1970; Northern Province, Gardner 1944, 1959; Aplin 1975; Beard 1978, 1981). This province includes northwestern part of Australia (the Fortescue-De Gray region, which centers round the Hamersley Range, and the Dampier Archipelago); Dampier Land; the Kimberley Plateau and the Arnhem Land, with adjacent islands; and the islands and shores of the Gulf of Carpentaria, with the adjacent zone of tropical dry forests and woodlands.

The North Australian Province has only a few endemic genera, including *Ondinea* (Nymphaeaceae), *Pachynema* (Dilleniaceae) and *Hygrochloa* (Poaceae). This province—especially the Kimberley district—contains a great number of Malesian elements, which finds their highest expression in the low-lying moist soils (Gardner 1944). In Kimberley district occurs an endemic representative of the Afro-Madagascan genus *Adansonia* (*A. gregorii*), which according to Beadle (1981*a*:186) possibly entered Australia by oceanic dispersal in relatively recent times.

Most of the province is covered by *Eucalyptus* forests and woodlands. *Acacia* is poorly represented but is locally dominant on some stony places. In the dryer south, the *Eucalyptus* woodlands become very open, merging into savanna. The swampy areas are mainly covered by Cyperaceae. For permanently moist soil, *Mitrasacme, Byblis,* and *Utricularia* are characteristic.

2. Queensland Province (Good 1947, 1974; Takhtajan 1970). This province includes eastern parts of Cape York Peninsula and eastern Australia (including the Great Dividing Range) south to the north coast of New South Wales. It contains four endemic families (Platyzomataceae, Austrobaileyaceae, Idiospermaceae, and Akaniaceae) and has a rather large number of endemic genera, including:

Pteridoblechnum, Bowenia, Fitzalania, Haplostichanthus, Piptocalyx, Valvanthera, Neostrearia, Ostrearia, Baileyoxylon, Peripentadenia, Neoroepera, Oreodendron, Irvingbaileya, Neorites, Austromuellera, Buckinghamia, Carnarvonia, Cardwellia, Darlingia, Hollandaea, Opisthiolepis, Musgravea, Placospermum, Athertonia, Motherwellia, Hexaspora, Schistocarpaea, Kreysigia, Mobilabium, and *Normanbya.*

In terms of number of species, this province is the richest in the Australian Kingdom. Although basically it is completely Australian, there is also a strongly expressed Malesian (especially Papuan) influence. The Cape York Peninsula is especially rich in Malesian elements (including species of *Cananga, Mitrephora, Balanophora, Nepenthes, Ternstroemia, Garcinia, Agapetes, Palaquium, Xanthophyllum, Peripterygium, Areca, Borassus,* and *Caryota*). According to Burbidge (1960:154), in a number of cases genera with Malesian affinities have their southernmost occurrence in the "MacPherson-Macleay Overlap area" (which covers a corner of southeast Queensland and the high-rainfall areas of the north coast of New South Wales).

Most of the province is covered by *Eucalyptus* woodland and forests. In southeastern Queensland and on the north coast of New South Wales, the tall *Eucalyptus* forests are well developed. Patches of tropical rain forest occur, especially in the northeastern and eastern coasts, while *Acacia* and *Casuarina* communities occur locally in relatively drier areas. Small patches of *Nothofagus moorei* forests occur in a few isolated localities in eastern highlands.

3. Southeast Australian Province (Diels 1906; Good 1947, 1974; Takhtajan 1970). This province includes a considerable part of New South Wales and Victoria, and also a very small southeastern part of the state of South Australia. According to Nelson (1981), this phytochorion has diffuse and imprecise boundaries.

The Southeast Australian Province contains a series of endemic genera, including *Streptothamnus* (2), which is very closely related to the Chilean genus *Berberidopsis,* and the monotypic genera *Wittsteinia* and *Parantennaria.* In the southeastern part of the province, especially in the Australian Alps, there are rather many elements in common with the neighboring Tasmanian Province. These include a series of endemic genera (including the conifer,

Microstrobos) and rather numerous endemic species (including *Eucalyptus ovata*).

The province is characterized by humid and dry *Eucalyptus* forests, by sclerophyll woodlands and savannas and, in places, by humid montane forests of the Australian-Tasmanian type. *Nothofagus cunninghamii* forest occurs in some small areas in Victoria. For the high mountain vegetation, subalpine woodlands and alpine herbfields are characteristic (Costin 1981). On low-nutrient sandstone substrate the vegetation often has a Mediterranean, maquis-like aspect.

4. Tasmanian Province (Engler 1882; Hayek 1926; Good 1947, 1974; Burbidge 1960; Takhtajan 1970). This province includes Tasmania, along with its adjacent islands. It contains one arguably distinct endemic, monotypic family, the Tetracarpaeaceae, and 13 endemic genera: *Diselma, Microcachrys, Prionotes, Anodopetalum, Tetracarpaea, Agastachys, Bellendena, Cenarrhenes, Acradenia, Nablonium, Pterygopappus, Milligania,* and *Isophysis* (*Hewardia*). With the exception of *Milligania* (4), which is close to *Astelia*, all of them are monotypic. The genus *Campynema* (2) is an endemic in common with Tasmania and New Caledonia. It is close to the monotypic New Caledonian genus, *Campynemanthe*.

Tasmania shares a number of endemic genera with the Australian continent, including *Athrotaxis, Microstrobos, Spyridium, Hakea, Telopea, Bedfordia,* and *Blandfordia*. Species endemism in the Tasmanian flora is about 20% (Rodway 1923; Burbidge 1960; Kirkpatrick and Brown 1984). The endemic element is most strongly represented in alpine and rain forest ecosystems (Kirkpatrick and Brown 1984). As Hooker (1855–1860) has already shown, in all its basic characteristics, the flora of Tasmania is similar to the flora of Victoria—especially to the mountainous parts of that state. The presence of no less than 13 endemic genera and a considerable number of endemic species (about 246, according to Specht et al. 1974; 293 endemic taxa at the species level or below, according to Kirkpatrick and Brown 1984) allows Tasmania to be described as a separate province.

Among the endemic species, the greatest role in the vegetation is played by the following: *Phyllocladus aspleniifolius, Microstrobos hookerianus* (a second species in southeastern Australia), *Microcachrys tetragona, Dacrydium franklinii, Athrotaxis cupressoides* and *A. selaginoides* (a third species in Australia), *Diselma archeri, Nothofagus gunnii, Eucalyptus coccifera, E. urnigera,* and *Anodopetalum biglandulosum* (Schweinfurth 1962). The Tasmanian flora contains a large number of taxa that are closely related to taxa in various cool temperate floras (see Moore 1972), as well as a large number of genera centered in the temperate regions of the Northern Hemisphere, including *Ranunculus, Epilobium, Euphrasia, Veronica, Mentha,* and *Carex* (Curtis 1967; Nelson 1981).

Most characteristic for the vegetation of Tasmania are forests (different types of eucalypt forests and mountain forests of *Athrotaxis* and other conifers) and woodland. The highest parts of the island have well-developed characteristic alpine and subalpine shrub-dominated communities, including the distinctive cushion-heaths and conifers. These communities are dominated by shrubs of endemic genera such as *Bellendena* and *Diselma*, and by endemic species of other genera such as *Archeria, Richea, Orites, Phyllocladus, Podocarpus,* and *Leptospermum* (Costin 1981; Nelson 1981). As Schweinfurth (1962) showed, the southwestern part of Tasmania ("Tasmania *sensu stricto*," as he called it), which contains the largest number of endemic species, is strongly distinguished from the northeastern part of the island.

30. Southwest Australian Region

Good 1947, 1974; Schmithüsen 1961; Mattick 1964; Takhtajan 1969, 1970, 1974.

This region occupies a comparatively small territory in southwestern Australia, including the Darling and Sterling Ranges and the entire coast from the Edel Peninsula in the north to approximately the midpoint between the environs of the cities of Esperance and Eyre in the east.

The region contains four endemic families (Cephalotaceae, Eremosynaceae, Stylobasiaceae, and Emblingiaceae),[2] and about 125 endemic genera (Good 1974), including *Actinostrobus, Dipteranthemum, Cosmelia, Sollya, Andersonia, Acidonia, Dryandra, Franklandia, Stirlingia, Synaphea, Beaufortia, Corynanthera, Chamaelaucium, Hypocalymma, Scholtzia, Diplolaena, Nuytsia, Hemiandra, Pentaptilon, Kingia, Anigozanthos, Phlebocarya, Tribonanthes, Conostylis, Johnsonia, Baxteria, Dasypogon, Rhizantella, Evandra, Reedia, Anarthria, Ecdeiocolea, Lyginia,* and *Onychosepalum.* Certain genera, although not endemic to southwestern Australia, have considerable concentrations of species in that area (e.g., *Banksia, Adenanthos, Isopogon, Calytrix,* and *Daviesia;* Nelson 1981).

According to somewhat outdated calculations, of 2,841 species in the native flora of southwestern Australia, 2,472 are endemic (Beard 1969). Thus, both as to the absolute number of endemic species and as to the percentage of species endemism (87%, on Beard's figures), the Southwest Australian Region is one of the most distinctive in all the world. Of the numerous endemic species, I mention only *Macrozamia riedlei, Pilostyles hamiltonii,* and *Eucalyptus diversicolor.*

Within the Southwest Australian Region, the autochthonous Australian element is most richly developed:

Its real home today is the large triangular-crescentic area bordered by the coastline from Shark Bay to Israelite Bay, and extending inland to the 175 mm (7 in.) winter isohyet. Within this area autochthonian flora finds its highest expression in the sandy areas, either on the sand heaths, or in the sandy swamplands of the south-western littoral (Gardner 1944:26).

The autochthonous xeromorphic flora exhibits here the greatest diversity. The most representative families are Casuarinaceae, Epacridaceae, Droseraceae, Myrtaceae, Proteaceae, Rhamnaceae, Verbenaceae, Goodeniaceae, Asteraceae, Dasypogonaceae, Xanthorrhoeaceae, and Haemodoraceae. The degree of endemism is particularly high in the high shrublands and heath formations found to the north of the Hill River and to the east of the Fitzgerald River (Aplin 1975:73).

One of the most characteristic features of this region is the predominance of sclerophyllous, xeromorphic shrubs and the paucity of grasses (the monotypic *Diplopogon* is only endemic genus). The main vegetation types are various *Eucalyptus* forests and woodlands, sand heaths, and swamp formations. The region consists of only one province.

1. **Southwest Australian Province** (Diels 1906; Gardner 1944, 1959; Burbidge 1960; Beard 1969, 1978, 1981; Aplin 1975; Nelson 1981).

31. Central Australian or Eremaean Region

Central Australian Region, Good 1947, 1974; Mattick 1964; Takhtajan 1969, 1970, 1974.

This region extends from the drier parts of the western Australian coast around Shark Bay to the Darling River in New South Wales and includes: the Great Sandy Desert (extending to the shores of the Indian Ocean between the Edel Peninsula and Roebuck Bay); the Gibson Desert; the Great Victoria Desert; the Nullarbor Plain; almost all the territory of the state of South Australia, along with the adjacent islands (with the exception of the very small southeastern part, referred to the Northeast Australian Region); the deserts of the Northern Territory; a considerable part of the Great Artesian Basin in the state of Queensland; and the northwestern part of the state of New South Wales.

The flora of this region has no endemic families, but contains 85 endemic genera (Burbidge 1960), of which over half belong to three families: Asteraceae (20), Chenopodiaceae (15), and the Brassicaceae (12). Among the endemic and subendemic genera, I mention only *Babbagia, Dysphania, Malococera, Threlkeldia, Enchylaena, Uldinia, Embadium, Rutidosis, Astrebla, Neurachne, Plectrachne, Triodia,* and *Zygochloa* (the latter is closely related to the

coastal *Spinifex*). A large genus, *Eremophila*, is mainly restricted to this region. Species endemism is very high, probably greater than 90%.
Characteristic vegetation includes *Acacia, Eucalyptus,* and *Casuarina* woodlands and shrublands; different types of tree or, more often, shrubby savannas (in places, in combination with desert woodlands); and grasslands (both the perennial hummock grasslands with a predominance of sclerophyllous grasses *Plectrachne* spp., *Triodia* spp., *Zygochloa paradoxa,* etc., and the ephemeral grasslands that appear only after rains). In drier areas there is a sparser desert vegetation, which in saline places, especially in the southern part of Australia, consists of various kind of halophytic shrublands dominated chiefly by species of *Atriplex* and *Maireana*. Sandy areas are either devoid of vegetation or are covered with *Zygochloa paradoxa* or with scattered trees of *Callitris, Casuarina, Exocarpos,* dwarf eucalypts, acacias, *Eucarya, Codonocarpus cotonifolius, Crotalaria cunninghamii,* and so on.

In some more favorable areas, plants of rain-forest type have persisted. The richest of these refugia is in central Australia, in Palm Valley and the Macdonell Ranges of the Northern Territory (Chippendale 1959; quoted in Beadle 1981*b*). The most remarkable plants of this refugium are a narrow endemic cycad, *Macrozamia macdonellii,* and a palm, *Livistona mariae*.

1. **Eremaean Province** (Diels 1906; Engler 1912, 1924; Hayek 1926; Gardner 1944, 1959; Aplin 1975; Beard 1978, 1981). This province, which includes the territory of the entire Central Australian Region, can probably be divided into three separate subprovinces, or even provinces: (1) the Province of Northern and Eastern Savannas; (2) the Eremaean Province, in the narrow sense (central deserts); and (3) the Southern Australian Province (Takhtajan 1970).

NOTES

1. See Burbidge's (1960) list of endemic seed plants, where 566 names are given. Since 1960, several new genera have been added, including the Queensland genus, *Idiospermum*.
2. Many authors include *Stylobasium* (2 species) in the Chrysobalanaceae.

VI
HOLANTARCTIC KINGDOM (HOLANTARCTIS)

I interpret the Holantarctic Kingdom[1] more broadly than it is usually understood, and I include in it not only the cold and temperate zones of the Southern Hemisphere but also part of the subtropical zone.

In terms of richness of its flora, the Holantarctic Floristic Kingdom is considerably poorer than the Holarctic. Nevertheless, the Holantarctic Kingdom includes about a dozen small, monotypic or oligotypic, endemic families (Thyrsopteridaceae, Lactoridaceae, Gomortegaceae, Hectorellaceae, Halophytaceae, Malesherbiaceae, Francoaceae, Aextoxicaceae, Vivianiceae, Misodendraceae, Tribelaceae, and Griseliniaceae) and a considerable number of endemic genera, many of which are characterized by very disjunct distributions.

Despite the fact that the territories included in the Holantarctic Kingdom are located at quite large distances from each other (e.g., the Juan Fernandez Islands and Tierra Del Fuego are separated from New Zealand and Lord Howe Island by a colossal expanse of sea), they share many genera in common, as well as close and even identical species. The following list of endemic Holantarctic genera common to two or more of its regions clearly indicates the floristic connections among different parts of this floristic kingdom. Endemic and almost endemic Holantarctic taxa with disjunct distributions include:

Blechnum penna-marina: subantarctic South America, South Georgia Island, Antarctic islands, eastern Australia, Tasmania, New Zealand and adjacent islands

Laurelia (2 spp.): Chile, New Zealand

Caltha, sect. *Psychrophila* (6 spp.): Andes to Ecuador, Tierra del Fuego, Falkland Islands, southeastern Australia, Tasmania, New Zealand

Ranunculus biternatus: Patagonia, southern Chile, Tierra del Fuego, Falkland Islands, South Georgia Island, Antarctic Islands, Macquarie Island

R. acaulis: southern Chile, Falkland Islands, New Zealand, Stewart Island, Chatham Island, Auckland Islands

Colobanthus (20 spp.): Andes, subantarctic South America, Falkland Islands, Antarctic Peninsula and adjacent islands, South Georgia Island, Antarctic Islands, mountains of southeastern Australia and Tasmania, New Zealand and adjacent islands

Acaena adscendens: Antarctic Islands, Macquarie Island

Geum, subgen. *Oncostylus* (9 spp.): subantarctic South America, Tasmania, New Zealand and adjacent islands

G. parviflorum: Chile (to Tierra del Fuego), New Zealand

Eucryphia (5–6 spp.): Chile, southeastern Australia, Tasmania

Tillaea moschata: Chile, Tierra del Fuego, Falkland Islands, Antarctic Islands, New Zealand and adjacent islands

Epilobium sect. *Sparsiflorae* (10 spp.): Tierra del Fuego (1 sp.), Tasmania (1 sp.), New Zealand and adjacent islands

Griselinia (7 spp.): Chile, southeastern Brazil, New Zealand, Stewart Island

Pseudopanax (10 spp.): temperate South America, New Zealand

Azorella (70 spp.): from the northern Andes to subantarctic South America, Falkland Islands, Antarctic islands, New Zealand

A. selago: Tierra del Fuego, Antarctic islands, Macquarie Island

Mida (2 spp.): Juan Fernandez Islands, New Zealand (North Island)

Jovellana (7 spp.): Chile, New Zealand

Ourisia '(20 spp.): Andes of South America, Tasmania (1 sp.), New Zealand, Stewart Island

Tetrachondra (2 spp.): Patagonia, New Zealand

Callitriche antarctica: Falkland Islands, South Georgia Island, Antarctic islands, Auckland Islands, Campbell Island, Antipodes

Coprosma pumila: Kerguelen Island, southeastern Australia, Tasmania, Macquarie Island, Campbell Island, Auckland Islands, Antipodes, Stewart Island, New Zealand

Hebe elliptica: Tierra del Fuego, western Patagonia north to 45°53' south latitude, Falkland Islands, New Zealand and surrounding islands

Hypsela (5 spp.): Andes, eastern Australia, New Zealand

Phyllachne (4 spp.): subantarctic South America, Tasmania, New Zealand, Stewart Island, Auckland Islands, Campbell Island

Selliera (5 spp.): subantarctic South America, Tasmania, New Zealand, Stewart Island

Donatia (2 spp.): Chile, Australia, Tasmania, New Zealand, Stewart Island

Cotula subg. *Leptinella:* Tierra del Fuego, Antarctic islands, Australia, Tasmania, New Zealand
C. plumosa: Antarctic islands, Macquarie Island, Campbell Island, Auckland Islands, Antipodes
Taraxacum magellanicum: subantarctic South America, New Zealand, Stewart Island, Chatham Islands
Luzuriaga (3 spp.): Peru, Chile, Patagonia, Tierra del Fuego, Falkland Islands, New Zealand, Stewart Island
Rostkovia (2 spp.): subantarctic South America, Falkland Islands, South Georgia Island, Campbell Island, New Zealand (*R. magellanica*), Tristan da Cunha (*R. tristanensis*), and isolated in Ecuador
Marsippospermum (3–4 spp.): subantarctic South America, Falkland Islands, Campbell Island, Auckland Islands, New Zealand (South Island)
Juncus scheuchzerioides: subantarctic South America, South Georgia Island, Antarctic islands, Macquarie Island, Campbell Island, Auckland Islands, Antipodes
Carex trifida: Tierra del Fuego, western Patagonia (north to 43°35' south latitude), Falkland Islands, Macquarie Island, New Zealand (South Island), Auckland Islands, Campbell Island, the Snares
Agrostis magellanica (= *A. antarctica*): subantarctic South America, Falkland Islands, Crozet Islands, Kerguelen Islands, Prince Edward Islands, Macquarie Island
Deschampsia antarctica: subantarctic South America (Andes from 34°10' south latitude to Tierra del Fuego), Falkland Islands, South Georgia Island, South Shetland Islands, Antarctic Peninsula and adjacent islands, South Orkney Islands, South Sandwich Islands, Prince Edward Islands, Crozet islands, Kerguelen Islands, Heard Island
Festuca erecta: Tierra del Fuego, Falkland Islands, South Georgia Islands, Kerguelen Islands, Macquarie Island

To this list one can add a considerable number of lichens and mosses that are common to subantarctic America and New Zealand. There is also a series of genera and sections the disjunct distribution of which extends far beyond the limits of the Holantarctic Kingdom, embracing not only southeastern Australia but also Malesia, while in America some members of the series extend northward to Central America and even to the Hawaiian Islands. The series includes the section *Edwardsia* of the genus *Sophora* and the genera *Oreomyrrhis, Hebe, Nertera, Abrotanella,* and *Oreobolus.*

The appearance of many taxa common to the Juan Fernandez Islands and temperate South America, to the Antarctic islands and New Zealand and its adjacent islands, and also to Tasmania and the mountains of southeastern Australia and, in part, South Africa has long caused botanists to posit an ancient center of temperate flora in the Southern Hemisphere. Hooker (1853) had already noted the "botanical affinity" between extratropical America, the Antarctic islands, New Zealand, and Tasmania. In the introduction to his "Flora Tasmaniae" (1860:civ), he concluded that "the many bonds of affinity between the three southern Floras, the Antarctic, Australian, and South African, indicate that these may all have been members of one great vegetation, which may once have covered as large a southern area as the European now does a Northern." Today, the existence of an ancient temperate Holantarctic flora that once was united and became fractionated is no longer challenged.

According to current opinion, the formation and expansion of the Holantarctic flora continued through that age when today's isolated parts of the Holantarctic Kingdom were joined together and constituted portions of the gigantic southern continent, Gondwanaland. The breakup of this land mass began in the early Cretaceous, but occurred mainly in the first half of the Tertiary, when the flowering plants were already the dominant group of plants on earth. During the Pleistocene glaciation, many elements of the Holantarctic flora moved further northward, especially in South America where, in the high mountain flora of the Andes, they reached Ecuador and Colombia. These include, for example, the genera *Colobanthus, Acaena, Azorella,* and *Ourisia*. Simultaneously, thanks to the lowering of the snowline, a series of Holarctic plants could spread along the lower mountains of the Isthmus of Panama and reach southern Chile. But the Holarctic element had already penetrated Holantarctica in the Tertiary, which explains the appearance in subantarctic South America and New Zealand of well-isolated endemic species and even supraspecific taxa (including sections) of such genera as *Caltha, Ranunculus, Berberis, Stellaria, Rumex, Draba, Geum, Hydrangea, Saxifraga, Viburnum, Gentiana, Valeriana, Veronica, Euphrasia, Pedicularis, Plantago, Juncus, Luzula,* and *Poa*.

Both the geological and the biogeographical data indicate that Africa (along with Madagascar) was the first to separate from Gondwanaland. During the end of the early Cretaceous, Africa had already separated considerably from South America and Antarctica. Therefore, the Holantarctic element is more weakly represented in South Africa, on Madagascar, and on the Mascarenes than in other temperate and subtropical countries of the Southern Hemisphere. Nevertheless, a large number of taxa exist in common among Africa, Madagascar, the Reunion Islands, and other parts of the previous continent of Gondwanaland. After a wide separation of Australia from Antarctica, the development of Australia's biota followed an independent path, with the Holantarctic element being preserved mainly in the mountains of southeastern Australia and Tasmania. At the same time, the Holantarctic element continued to predominate in New Zealand and its adjacent islands. The Holantarctic Kingdom includes the following four regions: Fernándezian; Chile-Patagonian; South Subantarctic Islands; and Neozeylandic.

32. Fernándezian Region

Engler 1882, 1903, 1924; Good 1947, 1974; Schmithüsen 1961; Mattick 1964; Takhtajan 1969, 1970, 1974.

The Fernándezian Region includes the Juan Fernandez Islands (Alexander Selkirk, Robinson Crusoe, and Santa Clara) and the Desventuradas Islands (San Ambrosio, San Felix; 26°20' south latitude, 80° west latitude). The islands are volcanic peaks of the submersed Juan Fernandez Range. The flora of the region contains only two endemic families, the Thyrsopteridaceae and Lactoridaceae, and over 20 endemic genera:

Thyrsopteridaceae: *Thyrsopteris* (1)
Lactoridaceae: *Lactoris* (1; endemic to Robinson Crusoe Island and almost extinct)
Myrtaceae: *Nothomyrcia* (1)
Boraginaceae: *Nesocaryum* (1), *Selkirkia* (1)
Lamiaceae: *Cuminia* (1)
Asteraceae-Senecioneae: *Robinsonía* (7), *Rhetinodendron* (1), *Symphochaeta* (1)
Asteraceae-Centaureinae: *Centaurodendron* (2), *Yunquea* (1); these very closely related genera probably are derivitives of some continental species of *Centaurea* s.l., probably from the section *Plectocephalus,* but under the conditions of island life acquired the arborescent habit (see Skottsberg 1937; Carlquist 1974).
Asteraceae-Cichorieae: *Dendroseris* (including *Hesperoseris, Phoenicoseris,* and *Rea,* 9; Carlquist 1974), *Lycapsus* (1, Desventuradas Islands), *Thamnoseris* (1, Desventuradas Islands; probably related to *Dendroseris)*
Arecaceae: *Juania* (1; close to *Ceroxylon* and to the Madagascan and Comoros Islands genus, *Revenea)*
Poaceae; *Megalachne* (1), *Pantathera* (1), *Podophorus* (1)

About 70% of the vascular plant flora of 195 species (Skottsberg 1945*b*), is endemic. There are 98 species of flowering plants and 17 species of ferns endemic to this region, and the majority of them are threatened. Among them are species of *Ophioglossum, Trichomanes,* and *Hymenophyllum, Dicksonia berteriana* (very close to the New Zealand *D. lanata),* species of *Dryopteris,*

Polystichum, Arthropteris, Asplenium, Blechnum, Pellaea, and *Pteris, Ranunculus caprarum* (close to two New Zealand species), *Sophora fernandeziana* and *S. masafuerana* (close to Chilean and New Zealand species), the already extinct *Santalum fernandezianum,* the arborescent *Plantago fernandezia, Cladium scirpoideum,* arborescent species of *Chenopodium,* and species of *Peperomia, Cardamine, Escallonia, Acaena, Gunnera, Uncinia, Carex,* and *Polypogon.*

Floristically, the Juan Fernandez Islands are closest to the Chile-Patagonian Region, especially to the Middle Chilean Province, and consequently must be included in the Holantarctic Kingdom—not in the Neotropical, as in almost all other floristic systems. Glaciation influenced the flora of the Juan Fernandez Islands and the Desventuradas Islands rather less than the flora of Chile, so that these islands have the character of a refuge (Skottsberg 1945*b*). The two groups of islands are markedly different from each other floristically. This is explained by the fact that the richest forest flora is preserved on Robinson Crusoe Island, whereas Alexander Selkirk Island is characterized by a well-developed alpine flora. Only a third of all species of the region are found on both islands (36 ferns and 27 flowering plants).

The flora of the Fernándezian Region includes a series of isolated taxa, such as *Thyrsopteris elegans, Lactoris fernandeziana* and several arborescent genera of the Apiaceae and the Asteraceae, and a series of genera and species related to genera and species of the Neozeylandic Region, Oceania, and the Hawaiian Islands. To the latter we can refer to regrettably already extinct *Santalum fernandezianum* and species of the genera *Arthropteris, Blechnum, Peperomia, Boehmeria, Fagara, Haloragis, Coprosma,* and *Plantago.* The region also contains a small group with Andean affinities, represented by the endemic genera *Selkirkia, Cuminia,* and *Juania;* endemic species of *Berberis;* and endemic, arborescent species of *Nicotiana* (Skottsberg 1945*b*). But as expected, the most significant element of the Juan Fernandez flora is the Chilean, although as Skottsberg (1945*b*:151) remarked, Fernandezian species "largely are endemic and as a rule very distinct." These include species of *Drimys, Escallonia, Azara, Myrceugenia, Ugni, Gunnera, Colletia, Rhaphithamnus, Hesperogreigia, Ochagavia,* and *Chusquea.* Finally, a significant subantarctic-Magellanian element exists on Alexander Selkirk Island, but the majority of these species are not endemic (Skottsberg 1945*b*).

According to Skottsberg (1945*b*:151), the humid parts of Robinson Crusoe Island were formerly densely forested from the coast to the highest ridges, but they are now preserved in a more or less original state only in the relatively less accessible places. The forests are evergreen and subtropical-temperate, consisting principally of *Myrceugenia fernandeziana,* which is as a rule accompanied by *Fagara mayu,* an endemic race of *Drimys winteri, Sophora fernandeziana,* and *Coprosma pyrifolia.* Woody lianas are absent, but 2 twining ferns (*Arthropteris altescandens* and *Blechnum schottii*) are common. There are many epiphytic ferns, mosses, liverworts, and lichens. In the undergrowth one encounters the tree-ferns *Dicksonia berteriana* and *Blechnum cycadifolium*, and also the gigantic *Gunnera peltata*. Along streams stretch thickets of *Boehmeria excelsa*. Herbaceous cover consists mainly of broad-leaved ferns and a few herbs.

Above some 350–400 m above sea level, the forest becomes less monotonous. New trees appear: *Azara fernandeziana, Coprosma hookeri, Rhaphithamnus venustus,* and the palm *Juania australis*. There is an abundance of large ferns, among which the most characteristic is the endemic *Thyrsopteris elegans*. Extremely rarely one encounters the small shrub *Lactoris fernandeziana,* the most interesting plant of the Fernándezian Region. There are many representatives of the Hymenophyllaceae. Higher up, in the fog zone, almost pure stands of *Dicksonia berteriana* are characteristic.

On Alexander Selkirk Island, the forest is approximately similar to the type found on Robinson Crusoe Island, but it is drier and consists fundamentally of *Myrceugenia schulzei* with an admixture of *Drimys winteri, Coprosma pyrifolia, Rhaphithamnus venustus,* and *Fagara externa*. Most of the species just mentioned for Robinson Crusoe Island are absent on Alexander Selkirk Island or are replaced by vicarious species.

On the foot of vertical cliffs or over the narrow ridges of Robinson Crusoe Island, a most remarkable formation developed an *Eryngium-Robinsonia* scrub, with a predominance of *Robinsonia gayana, Eryngium bupleuroides,* and species of *Dendroseris*. Much rarer are *Centaurodendron dracaenoides, Rheitinodendron berteroi, Selkirkia berteroi,* and the remarkable arboreous plantain, *Plantago fernandezia*. Rare forest trees grow in places with more well-developed soil cover. On sheer rocks one finds the endemic bromeliad, *Ochagavia elegans*.

Grasslands with alternating thickets of shrubs cover the forestless western part of Robinson Crusoe Island and the entire basal region of Alexander Selkirk Island. The alpine vegetation is represented by thick mats of trailing herbs and shrubs, tufted perennials, and cushion plants "reproducing the Magellanian heath." The genera *Abrotanella, Acaena, Galium, Erigeron,* and *Luzula* are represented here by endemic species (Skottsberg 1922, 1945*b*, 1956).

The flora of the Desventuradas Islands consists of only 19 vascular plants assignable to 16 genera of 13 families (Skottsberg 1937, 1945*b*). Three monotypic genera (*Nesocaryum, Lycapsus,* and *Thamnoseris*) and 12 species are endemic. On San Felix Island, only 7 species occur, 3 of which are also encountered on San Ambrosio Island.

The region consists of only one province.

1. **Fernándezian Province** (Thorne 1963).

33. Chile-Patagonian Region

The Chile-Patagonian Region[2] includes the extratropical parts of South America, from the subtropical "monta" of Argentina to Tierra del Fuego inclusive, as well as the Falkland islands (Islas Malvinas), Isla de los Estados, Islas Diego Ramirez, South Georgia Island, the South Sandwich Islands, the South Shetland Islands, the South Orkney Islands, and part of the Antarctic Peninsula with several adjoining islands (including Adelaide Island).

The Chile-Patagonian Region is characterized by a rather diverse flora, consisting basically of derivatives of ancient Holantarctic elements, but with a considerable admixture of taxa of Holarctic origin, and in the north also of neotropical elements. The flora of the region includes 8 endemic or subendemic families (Gomortegaceae, Halophytaceae, Malesherbiaceae, Tribelaceae, Francoaceae, Aextoxicaceae, Vivianiaceae, and the Misodendraceae), as well as a multitude of endemic or almost endemic genera, which are encountered in the majority of cases in Chile. Of the endemic and subendemic genera we mention only the following:

Podocarpaceae: *Saxegothaea* (1, southern Chile, western Patagonia)
Cupressaceae: *Austrocedrus* (1), *Fitzroya* (1, Chile, northern Patagonia), *Pilgerodendron* (1, southern Chile)
Monimiaceae: *Peumus* (1, Chile)
Gomortegaceae: *Gomortega* (1, Chile)
Lardizabalaceae: *Boquila* (1, Chile), *Lardizabala* (2, Chile)
Ranunculaceae: *Barneoudia* (3, Chile, Argentina), *Hamadryas* (5, Magellanian Province)
Phytolaccaceae: *Anisomeria* (2, Chile)
Cactaceae: *Austrocactus* (8, Chile, Argentina), *Copiapoa* (15, Chile), *Eriosyce* (1, Chile), *Eulychnia* (5, Chile), *Maihuenia* (5, Chile, Argentina), *Pterocactus* (6, Argentina)
Portulacaceae: *Calandriniopsis* (4, Chile), *Lenzia* (1, Chile), *Monocosmia* (1, Chile, Patagonia)
Halophytaceae: *Halophytum* (1, subarid regions of Argentina from La Rioja to Santa Cruz)
Caryophyllaceae: *Microphyes* (2, Chile), *Philippiella* (1, Patagonia), *Reicheëlla* (1, Chile)
Chenopodiaceae: *Holmbergia* (1, Uruguay, Argentina; very close to the Australian genus, *Rhagodia*)
Epacridaceae: *Lebetanthus* (1, Patagonia, Tierra del Fuego)
Berberidopsidaceae: *Berberidopsis* (1, Chile; very close to the Australian genus *Streptothamnus*)
Malesherbiaceae: *Malesherbia* (27, mainly Chile, but encountered in Argentina and Peru)

Frankeniaceae: *Niederleinia* (3, Patagonia)
Brassicaceae: *Agallis* (1, Chile), *Decaptera* (1, Chile), *Hexaptera* (13), *Onuris* (6, Chile, Patagonia), *Schizopetalon* (8, Chile), *Werdermannia* (3-4, northern Chile)
Elaeocarpaceae: *Crinodendron* (3, Chile and Argentina)
Malvaceae: *Cristaria* (40)
Euphorbiaceae: *Adenopeltis* (2), *Avellanita* (1, Chile)
Aextoxicaceae: *Aextoxicon* (1, Chile)
Thymelaeaceae: *Ovidia* (4)
Cunoniaceae: *Caldcluvia* (1, Chile)
Saxifragaceae: *Saxifragella* (2, Magellanian Province); *Saxifragodes* (1, Magellanian Province)
Francoaceae: *Francoa* (1, Chile), *Tetilla* (1, Chile)
Rosaceae: *Quillaja* (4)
Onagraceae: *Oenotheridium* (1, Chile)
Myrtaceae: *Tepualia* (1, Chile)
Fabaceae: *Gourliea* (1, southern Peru, Chile, Argentina), *Zuccagnia* (1, Chile)
Rutaceae: *Pitavia* (1, Chile)
Zygophyllaceae: *Pintoa* (1, Chile), *Metharme* (1, northern Chile)
Sapindaceae: *Bridgesia* (1, Chile), *Valenzuelia* (2, Chile, Argentina)
Ledocarpaceae: *Wendtia* (3, Chile, Argentina)
Vivianiaceae: *Araeoandra* (1, Chile), *Caesarea* (1, southern Brazil and Uruguay), *Cissarobryon* (1, Chile), *Viviania* (3, Chile and Argentina)
Tropaeolaceae: *Magallana* (2)
Malpighiaceae: *Dinemagonum* (3, Chile), *Tricomaria* (1, Argentina)
Santalaceae: *Myoschilos* (1, Chile), *Nanodea* (1)
Misodendraceae: *Misodendrum* (11, in the temperate forest region from 33° south latitude to the Straits of Magellan; parasitic on *Nothofagus*)
Rhamnaceae: *Talguenea* (1, Chile)
Escalloniaceae: *Valdivia* (1, Chile)
Tribelaceae: *Tribeles* (1)
Apiaceae: *Bustillosia* (1, Chile), *Domeykoa* (4, northern Chile), *Gymnophyton* (6, Andes of Chile and of Argentina), *Huanaca* (2), *Laretia* (2, Andes of Chile), *Mulinum* (20, southern Andes)
Rubiaceae: *Cruckshanksia* (7, Chile)
Apocynaceae: *Elytropus* (1, Chile)
Loasaceae: *Scyphanthus* (2, Chile)
Solanaceae: *Benthamiella* (10, Argentina, Patagonia), *Combera* (2, Andes of Chile and Argentina), *Phrodus* (4, Chile), *Schizanthus* (8, Chile), *Trechonaetes* (3, Chile and Argentina), *Vestia* (1, Chile)
Nolanaceae: *Alona* (6, Chile; a second genus, *Nolana*, with approximately 60 species, is distributed from Peru to Patagonia with 1 species in the Galapagos Islands)

Scrophulariaceae: *Melosperma* (1, Chile), *Monttea* (3, Chile)
Bignoniaceae: *Argylia* (10, Chile, Argentina, north to southern Peru), *Campsidium* (1, Chile, Argentina)
Gesneriaceae: *Asteranthera* (1, Chile), *Hygea* (1, Chile), *Mitraria* (1, southern Chile), *Sarmienta* (1, southern Chile)
Verbenaceae; *Thryothamnus* (1, Chile), *Urbania* (2, Chile)
Lamiaceae: *Kurtzamra* (1), *Oreosphacus* (1, Chile)
Lobeliaceae: *Cyphocarpus* (2, Chile)
Calyceraceae: *Calycera* (15), *Moschopsis* (8, Chile, Patagonia)
Asteraceae: *Aylacophora* (1, Patagonia), *Calopappus* (2, Andes of central Chile), *Chiliotrichum* (2), *Closia* (10, Chile), *Dolichlasium* (1, central Andes of Argentina), *Doniophyton* (2, Andes of Chile and Argentina, and Patagonia), *Eriachaenium* (1, Tierra del Fuego), *Gypothamnium* (1, northern Chile, *Huarpea* (1, Andes of La Rioja, Argentina), *Leuceria* (46, Andes from southern Peru to Patagonia), *Leunisia* (1, central Chile), *Leophopappus* (5, Andes from southern Peru to northern Chile and northwestern Argentina), *Macrachaenium* (2, Patagonia, Tierra del Fuego), *Marticorenia* (1, central Chile), *Moscharia* (2, central Chile), *Oxyphyllum* (1, northern Chile), *Pachylaena* (2, Andes of Chile and Argentina), *Plazia* (2, Andes from Peru to Chile and Argentina), *Pleocarphus* (1, northern Chile), *Polyachyrus* (7, from Peru to central Chile), *Triptilion* (12, central Chile, 1 in Patagonia), *Urmenetea* (1, Andes of northern Chile and northwestern Argentina)
Juncaginaceae: *Tetroncium* (1, Tierra del Fuego, western Patagonia northward to 40° south latitude, Andean Patagonia, Falkland Islands)
Iridaceae: *Chamelum* (3, Chile, Argentina), *Solenomelus* (3, Chile)
Tecophilaeaceae: *Conanthera* (5, Chile), *Tecophilaea* (2, Andes of Chile), *Zephyra* (1, Chile)
Alstroemeriaceae: *Leontochir* (1, Chile), *Schickendantzia* (1–2, Argentina)
Alliaceae: *Ancrumia* (1, Chile), *Erinna* (1, Chile), *Garaventia* (1, Chile), *Gethyum* (1, Chile), *Gilliesia* (3–4, Chile), *Leucocoryne,* including *Latace* (15, Chile), *Miersia* (5, Bolivia, Chile), *Solaria* (2, Chile), *Speea* (1, Chile), *Stemmatium* (*Stephanolirion;* 1, Chile: Atacama), *Tristagma* (5, Chile, Argentina, Patagonia)
Amaryllidaceae: *Hieronymiella* (3, Argentina), *Placea* (6, Chile)
Anthericaceae: *Bottionea* (1–2, Chile)
Luzuriagaceae: *Lapageria* (1, southern Chile)
Philesiaceae: *Philesia* (1, southern Chile)
Corsiaceae: *Arachnitis* (1, Chile), *Tapeinia* (2, southern Chile, Patagonia)
Orchidaceae: *Asarca* (20)
Bromeliaceae: *Fascicularia* (5, Chile)
Juncaceae: *Patosia* (2, Chile, Argentina)
Poaceae: *Chaetotropis* (6), *Ortachne* (2, Chile)

Arecaceae: *Jubaea* (1, Chile)

There are very many endemic species, including *Araucaria araucana*, species of *Podocarpus, Caltha, Berberis, Nothofagus, Gaultheria, Eucryphia, Escallonia, Hydrangea, Fuchsia, Gunnera, Griselinia, Azorella, Tropaeolum, Cissus, Embothrium, Lomatia, Fabiana, Solanum, Calceolaria, Lobelia, Baccharis, Dasyphyllum, Alstroemeria,* and *Astelia.* The majority of endemics is concentrated in Chile, where the greatest number of ancient Holantarctic elements are preserved.

1. Northern Chilean Province (Engler 1882, p.p.).[3] This province includes the semideserts of northern Chile with the characteristic vegetation formations "Zwergstraugh- und Sukkulentenformation" (*Adesmia, Fuchsia, Verbena, Oxalis gigantea, Carica, Cassia, Trichocereus, Eulynchia, Copiapoa,* and *Opuntia*) and "Ephemere Kräuterfluren" (Schmithüsen 1956; Hueck and Seibert 1972). In the north, the boundary passes somewhat south of 25° south latitude, proceeding along the shores to Cape San Pedro. Besides the plants just mentioned, the Northern Chilean Province is characterized by species of *Aristolochia, Anisomeria, Neoporteria, Calandrinia, Reicheëlla, Cristaria, Malvastrum, Gourliea, Prosopis, Bridgesia, Balbisia, Tropaeolum, Skythanthus, Cruckshanksia, Cordia, Heliotropium, Schizanthus, Solanum, Calceolaria, Argylia, Plantago, Lobelia, Bahia, Gypothamnium, Oxyphyllum, Scilla, Alstroemeria, Leucocoryne, Tecophilaea, Hippeastrum, Placea, Sisyrinchium,* and *Dioscorea.*

2. Middle Chilean Province (Chilenische Übergangsprovinz von 30°30′–37° s. B., Engler 1903, 1924; Provincia Chileana, Cabrera and Willink 1973; northern part of the Provincia Subantarctica, Cabrera and Willink 1973). This province includes the greater part of Chile from the southern part of the Province of Coquimbo (30°30′ south latitude) in the north to the southern area of the Province of Aysén (approximately 47–48° south latitude), adjacent islands, and also several contiguous parts of Argentina.

The flora of the Middle Chilean Province is characterized by a considerable number of endemic genera and species, and in fact contains the overwhelming share of the endemic taxa of the Chile-Patagonian Region. Some of the endemic or subendemic species are various conifers, *Laurelia aromatica, Aristolochia chilensis, Boquila trifoliata, Lardizabala biternata,* species of *Berberis, Nothofagus, Gaultheria, Acaena* and *Rubus, Weinmannia trichosperma,* species of *Eucryphia, Escallonia, Hydrangea, Ribes, Myrceugenia, Fuchsia, Gunnera,* and *Griselinia, Pseudopanax laetevirens,* and species of *Azorella, Tropaeolum, Cissus,* and *Embothrium.*

Characteristic of the coastal zone of the Middle Chilean Province is the so-called ancient Chilean floristic element (Skottsberg 1916), which is historically related to the neotropical flora. It includes *Cryptocarya rubra, Peumus boldus, Puya* spp., *Acacia caven, Lithraea caustica, Passiflora pinnatistipula, Crinodendron patagua, Baccharis* spp., and others. In the valleys of the Chilean provinces of Valparaiso and Santiago (31–35° south latitude), there previously occurred rich stands of the Chilean palm, *Jubaea chilensis,* but its area is at present greatly diminished.

The role of the Holantarctic element increases with increased elevation. The coastline and the slopes of the Andes are covered with forest communities and xerophilous shrubs, deciduous *Nothofagus obliqua* (whose southern border is Llanquihue), and species with leathery, evergreen leaves, such as *Quillaja saponaria, Lithraea caustica,* and *Peumus boldus.* But south of the Province of Colchagua, this flora is replaced by open forests that consist mainly of deciduous species of *Nothofagus* (*N. glauca, N. leonii,* and *N. alessandri*), according to Stein (1956). On the highest slopes of the Andes, one more frequently encounters the evergreen *N. dombeyi* (beginning in the Province of Calchagua); while beginning in the Province of Talca, one finds the deciduous *N. procera* (the southern boundary of which is Valdivia).

Two other deciduous species (*N. pumilio* and *N. antarctica*) are usually encountered from 1,300 m above sea level to the upper boundary of forests (beginning in the Province of Nuble). In the central plain of this region, from 37° south latitude (northward from Yungay), the deciduous *N. obliqua* replaces the formation of *Acacia caven* (currently, *N. obliqua* is encountered only as isolated trees in cultivated fields). The forests of the foothills of the Andes in the province of Nuble, Bio-Bio, and Malleco consist mainly of *N. obliqua, N. procera,* and *N. dombeyi,* improving in quality and quantity southward. *N. glauca, N. leonii,* and *N. alessandri* do not extend farther south than the Province of Maule, and thus are endemic for the Middle Chilean Province. But all the remaining species of *Nothofagus* encountered in this region range far to the south—in fact, to Tierra del Fuego (Stein 1956).

The forest flora of the Middle Chilean Province is rich in conifers: one encounters 9 species belonging to 7 genera. The most remarkable representative of the conifers is *Araucaria araucana,* which has a restricted distribution in the Andes from the volcano Antuko (37°30′ south latitude) to the volcano Lanin (39°30′ south latitude), at elevations from about 1,200 m above sea level to the upper boundary of the forest, thus extending into the territory of Argentina. It is also found on the slopes of the coastal cordillera from 37°20′ to 38°40′ south latitude. *Araucaria* forms pure stands or mixtures with *Nothofagus dombeyi* and *N. pumilio.* The most characteristic conifer of Chile, *Fitzroya cupressoides,* occurs as an element of the rain forest and is found from 39°45′ to 43°29′ south latitude, extending north to the shores of the Cordillera Valdivia (to heights of about 3,000 m above sea level). In the south it is often associated with *Pilgerodendron uviferum,* distributed from Valdivia (40° south latitude) to Tierra del Fuego inclusive. *Austrocedrus*

(*Libocedrus*) *chilensis, Podocarpus andinus,* and *P. salignus* are encountered mainly north of the Province of Osorno. *Saxegothaea conspicua* is distributed from 35°20' to 45° south latitude (Stein 1956).

3. **Pampean Province** (Pampasprovinz, Engler 1882; Hayek 1926; Cabrera and Willink 1973; Argentinische Provinz: Pampas-Zone, Engler 1899, 1903, 1924). This province includes the southern extremity of Brazil (southern parts of the State of Rio Grande do Sul), all of Uruguay, and considerable territory in the northeastern and central areas of Argentina; southward, it extends to Montevideo.

The flora of this province is dominated by species of *Stipa, Piptochaetium, Aristida, Melica, Briza, Bromus, Eragrostis,* etc. Also characteristic are species of *Paspalum, Panicum, Bothriochloa,* and *Schizachyrium,* especially in the northern areas. Along with grasses, representatives of many other families are usually present, including species of *Margyricarpus, Baccharis, Heimia, Berroa, Vicia, Chaptalia, Oxalis,* and *Adesmia.* The most characteristic vegetation types of this Province are Pampas and other grasslands.

4. **Patagonian Province** (Argentinisch-patagonische Provinz, Engler 1882, p.p.; Andin-patagonische Provinz plus Argentinische Provinz, Unterprovinz (Zone) des Espinale, Engler 1899, 1903, 1924).[4] This province includes the undulating plateau descending from the southern Andes (where it averages only about 800 to 1,000 m above sea level) down to the Atlantic Ocean. In the north it reaches the left shores of the lower and middle course of the Rio Negro and rises to the middle course of the Rio Colorado. To the south the province extends to the northern part of Tierra del Fuego.

Owing to the severity of the climate and comparatively uniform environmental conditions, the flora of the Patagonian Province is rather depauperate. Almost all Patagonian endemics are related to species of the Magellanian flora, but the greatest part of the flora of the Patagonian Province is directly related to the montane flora of central Chile. This includes such species as *Berberis heterophylla, Adesmia canescens, Anarthrophyllum rigidum, Azorella caespitosa, Schinus dependens, Quinchamalium majus, Monnina angustifolia, Verbena mendocina, Alstroemeria patagonica,* and *Carex gayana.* Through the Andean chain, which serves as a gigantic "transtropical high mountain bridge," the flora of Patagonia is connected with the temperate floras of the Northern Hemisphere (Beetle 1943). Yet the Patagonian flora is also connected with the desert floras of South and North America, as was shown long ago (Gray and Hooker 1880; Johnston 1940). Many of these species are halophytes; *Larrea divaricata* and *Koeberlinia spinosa* can serve as examples.

The vegetation consists of open communities that are characterized by various shrubs, small shrubs, a variety of perennial and annual herbs, and grasses. The shrubs commonly belong to the Asteraceae (e.g., *Baccharis marginalis, Picris echinoides*), or to the Fabaceae (*Prosopis strombulifera, Cassia aphylla*). But there are also representatives of other families, such as *Atamisquea emarginata, Mulinum spinosum,* and *Monttea aphylla*. Shrubs such as *Mulinum spinosum* form thorny cushions.

5. Magellanian Province (Skottsberg 1960;[5] Magellano-Antarctic Region, Tolmatchev 1974, p.p.). This province includes the southern subantarctic part of Chile with all of its adjoining islands southward from 47–48° south latitude (although along the Andes, the Magellanian elements extend northward almost to 44° south latitude, and individual species are distributed rather further, but usually not farther than 36° south latitude); subantarctic parts of western and Andean Patagonia (where the Magellanian flora begins from 51°30′ south latitude); the Falkland Islands; South Georgia Island; the South Sandwich Islands; the South Orkney Islands; the South Shetland Islands; the western shores of the Antarctic Peninsula and islands along its south to the small island of Neny, on the eastern shore of Marguerity Bay. The island of Neny is the southern limit of the distribution of vascular plants.

The flora of this province is extraordinarily poor; the ecological characteristics of the vegetation are reminiscent of the Arctic Province of the Circumboreal Region. Here one finds only a small number of endemic or almost endemic genera: *Hamadryas, Saxifragella, Saxifragodes, Nanodea, Eriachaenium, Macrachaenium, Nassauvia, Tetroncium,* and *Tapeinia*. The number of endemic species is significantly less than in the other provinces of the Chile-Patagonian Region.

Within the province, the flora of the Falkland Islands is of special interest. It includes 163 species of vascular plants, of which 14 (8.6%) are endemic (Moore 1968, 1979). Alboff (1902, 1904) concluded that despite their remoteness from Tierra del Fuego, the Falkland Islands belong entirely to the Tierra del Fuegean flora. As Alboff (1904:18) wrote,

> In my opinion, the Falkland Islands as is well known, are completely deprived of woody vegetation because of their exposed position in the middle of the ocean, which subjects them to the blowing of the wind from all directions of the compass. They represent only a characteristic formation of the Tierra del Fuegan vegetation—a formation, which we already described under the title "balsambogs," or dry peats. It seems to us that no essential difference exists between the vegetation of the Falkland Islands and the vegetational cover of those islands deprived of forest, which are scattered in the canal of Beagle.

Subsequent investigations completely supported the conclusions of Alboff (see Skottsberg 1913, 1945*a*; Moore 1968). It is interesting that the fern *Grammitis kerguelensis* is the only species encountered both on the Falkland Islands and on subantarctic islands. The following species are endemic to the

Falkland Islands: *Arabis macloviana, Calandrinia feltonii, Chevreulia lycopodioides, Erigeron incertus, Gnaphalium affine, Hamadryas argentea, Leucheria suaveolens, Lilaeopsis macloviana, Nassauvia gaudichaudii, N. serpens, Nastanthus falklandicus, Senecio littoralis, S. vaginatus,* and *Sisyrinchium filifolium.*
Although the Falkland Islands were not subjected to Andean glaciation, they are totally devoid of arborescent vegetation. According to Alboff, this lack is explained by the constantly blowing strong winds. The flora has only two small shrubs—*Chiliotrichum diffusum* and *Hebe elliptica.* The greater part of the islands is covered with grasslands, with *Cortaderia pilosa* and *Poa flabellata* forming the major components. The latter is characteristic of the littoral belt, especially for the rocky sites and pebbly areas. Characteristic of the relatively dry sites are heaths with *Empetrum rubrum, Pernettya pumila,* and *Gaultheria antarctica.* On the heaths, one often encounters great cushions of species of *Azorella,* especially *A. lycopodioides* and *A. filamentosa.* Here and there one finds small moss bogs, which are dominated by *Astelia pumila, Rostkovia magellanica,* or *Juncus scheuchzerioides* (Skottsberg 1913, 1945a; Moore 1968).

The flora of South Georgia Island is much poorer. It contains only 25 species (Greene 1964, 1969), among which there is 1 endemic (*Uncinia smithii*). The remaining species are *Lycopodium megellanicum, Ophioglossum crotalophoroides, Hymenophyllum falklandicum, Blechnum penna-marina, Cystopteris fragilis, Polystichum mohrioides, Grammitis kerguelensis, Ranunculus biternatus, Colobanthus quitensis, C. subulatus* (encountered also on Tierra del Fuego and very close to *C. hookeri,* growing on the Auckland Islands and on Campbell Island in the Neozeylandic Region), *Cerastium fontanum, Montia fontana, Acaena decumbens* (= *A. magellanica*) and *A. tenera, Callitriche antarctica, Galium antarcticum, Juncus inconspicuus* and its close relative *J. scheuchzerioides, Rostkovia magellanica* (occurring also on the Falkland Islands, in temperate South America, and in the Neozeylandic Region), *Deschampsia antarctica, Poa flabellata, Festuca erecta, Phleum alpinum,* and *Alopecurus antarcticus* (Greene 1964, 1969; Walton and Greene 1971). It is interesting that mosses exhibit the highest endemism. *Poa flabellata* and *Acaena tenera* play a basic role in the vegetative cover of South Georgia Island, and are found mainly in the coastal belt. According to Skottsberg (1921), the vascular plant flora of South Georgia Island arrived by wind and by birds from Tierra del Fuego and the Falkland Islands. The moss flora, containing 2 endemic genera (*Skottsbergia* and *Pseudodistichum*), represents a remnant of the ancient Antarctic flora. But the presence on South Georgia Island of an endemic species of *Uncinia* probably indicates the preservation of a small number of representatives of the more ancient flora of vascular plants. The scanty vegetation of South Georgia Island grows in sheltered valleys and along the seashores.

The flora of the South Shetland Islands (located at a distance of 770 km southeast of Cape Horn and about 160 km north of the Trinity Peninsula, which is the northernmost part of the Antarctic Peninsula) is reduced to only 2 species, *Colobanthus quitensis* and *Deschampsia antarctica* (Lindsay 1971). The analogous floristic composition is observed on the Orkney Islands and on the Antarctic Peninsula, where these two species are encountered only on its western shore and coastal islands, extending from Almond Point (Palmer Coast) in the north to Adelaide Island, Lagotellerie Island, and Neny Island to the northern coast of Marguerity Bay (Greene and Holtum 1971). The flora of the South Sandwich Islands is the most depauperate, consisting of only 1 species, *Deschampsia antarctica,* known only from one locality on Kandemas Island (Greene and Holtum 1971; Corner 1971).

Paleobotanical data suggest a rather rich past plant world on the Antarctic Peninsula and its nearby islands. Especially interesting are the fossil plant remains discovered on Seymour Island, located at the northern tip of the Antarctic Peninsula (about 64° south latitude). Here one finds representatives of the Schizaeaceae, Cyatheaceae, Polypodiaceae, Araucariaceae (pollen of *Agathis* and *Araucaria*), Podocarpaceae, Winteraceae (*Drimys*), Monimiaceae (*Laurelia* and *Mollinedia*), Lauraceae (*Lauriphyllum*), Fagaceae (pollen of *Nothofagus*), Myricaceae (*Myrica*), Brassicaceae (pollen), Cunoniaceae (*Caldcluvia*), Fabaceae (*Leguminosites*), Myrtaceae (pollen), Onagraceae (pollen of the *Fuchsia* type), Aquifoliaceae, Proteaceae (*Knightia* and species of *Lomatia*), and Cyperaceae (*Scirpitis,* possibly related to *Schoenoplectus;* Dusén 1908; Florin 1940; Cranwell 1959; Wace 1965). Conifers and *Nothofagus* predominate (Cranwell 1959). The composition of this flora indicates some similarity with the Valdivian forests of western Chile between 41° and 48° south latitude (Skottsberg 1916; Godley 1960; Wace 1965).

34. Region of the South Subantarctic Islands

Good 1947, 1974; Schmithüsen 1961; Takhtajan 1969, 1970, 1974; Kerguelen und benachbarte Inselgruppen, Mattick 1964.[6]

This transitional region between the Neozeylandic and the Chile-Patagonian regions includes the islands of Tristan da Cunha along with Gough Island, the Prince Edward Islands, the Crozet Islands, Amsterdam Island, Saint Paul Island, the Kerguelen Islands, and Heard Island (including McDonald Island, a small island located close to it).

The flora of this region is very depauperate, but the paleobotanical data indicate that in the past it was considerably richer, and that coniferous and *Nothofagus* forests covered large expanses. Two endemic, monotypic genera are characteristic of the region: *Pringlea* of the Brassicaceae (Prince Edward Islands, Crozet Islands, Kerguelen Islands and Heard Island), and *Lyallia* of the Hectorellaceae, which is very close to the Caryophyllaceae (Kerguelen Islands). The flora of Tristan da Cunha is relatively richer (44 species).

Paleobotanical data, although scanty, indicate that in Tertiary times the subantarctic oceanic islands supported a fairly rich flora, including trees. In the Tertiary, the Kerguelen Islands had forests of *Araucaria* and of Podocarpaceae. It is interesting that on the Kerguelen Islands no remains of *Nothofagus* have been detected, and there is reason to suppose that flowering plants then held a generally subordinate position (Cookson 1947; Wace 1965).

1. Tristan-Goughian Province (Gebiet von Tristan D'Acunha, Engler 1882). This province includes the Tristan da Cunha Islands and Gough Island. Despite being closer to South Africa (2,900 km) than to South America (3,200 km), the islands of Tristan da Cunha and Gough Island are much more closely allied to latter (Moore 1979).

The flora of the Tristan da Cunha Islands consists of 51 genera and 70 species of vascular plants, of which 37% of the flowering plants and 42% of the ferns are endemic (Wace and Dickson 1965; Christophersen 1968; Moore 1979). The flora of Gough Island has about 60 species of vascular plants (of which 32 species are flowering plants), but among them there is not one absolutely indisputable endemic (Wace 1961). *Lycopodium diaphanum* on Tristan da Cunha is close to *L. contiguum* in South America. Among the ferns, the following have a South American relationship: *Hymenophyllum aeruginosum* (close to the Chilean species *H. ferrugineum*), *Eriosorus cheilanthoides*, *Vittaria vittarioides*, *Asplenium alvarezence* (closely related to *A. magellanicum*), and *Dryopteris aquilina* (almost inseparable from the Chilean species *D. spectabilis;* Tryon 1966). Yet *Asplenium platybasis* is also found on the island of Saint Helena and is close to the African species, *A. friesiorum*.

Among the flowering plants, the African element is represented by *Pelargonium acugnaticum* (very close to *P. grossularioides*) and by the dominant tree species of the islands, *Phylica arborea*, which also is encountered on Amsterdam Island and on the Mascarene Islands (the center of diversity of the genus is located in southern Africa and on Madagascar). But many more numerous floristic ties exist with the flora of temperate South America—specifically with the flora of the Magellanian Province, and also with the flora of other provinces of the Holantarctic Kingdom. The endemic *Ranunculus carolii* is close to the Holantarctic species, *R. biternatus* (and also to *R. crassipes*), and *Peperomia berteroana* is common to both Juan Fernandez and the Tristan da Cunha islands. The endemics *Acaena sarmentosa* and *A. stangii* are close to the polymorphic species *A. sanguisorbae*, which is distributed in Australia, Tasmania, and New Zealand. This group also includes *A. insularis*, endemic to Amsterdam Island (Christophersen 1968). The endemic *Hydrocotyle capitata* is related to *H. marchantioides*, distributed in Patagonia and southern Chile. Described from Tristan da Cunha, *Apium australe* is found not only on Gough Island (under the name *A. goughense*) but also in Chile (north to 35° south latitude), western and Andean Patagonia, southeastern Australia, Tasmania, New Zealand and adjacent islands, and on Lord Howe Island. *Nertera depressa*, besides being found on Tristan da Cunha, is distributed on Chiloe Island, Tierra del Fuego, Andean Patagonia (north to 41°10' south latitude), the Falkland Islands, Australia, New Guinea, New Zealand, Auckland Islands, and on Stewart and Campbell islands. Two other endemic species, *N. assurgens* and *N. holmboei*, are close to *N. depressa*. *Chevreulia sarmentosa* (Asteraceae) occurs in South America from Paraguay to Bolivia, and also on Tristan da Cunha (Christophersen 1968). Also of great interest is the endemic *Rostkovia tristanensis*, which is close to *R. magellanica*.

2. **Kerguelenian Province** (Skottsberg 1960; Gebiet der Kerguelen, Engler 1882, 1903, 1924). This province includes the Prince Edward Islands, Crozet Islands, the Kerguelen Islands (including the numerous surrounding small islands), Heard Island (along with McDonald Island), Saint Paul Island, and Amsterdam Island. The flora of the province consists of 45 species of vascular plants, including 2 species of *Lycopodium* (*L. magellanicum* and *L. saururus*) and 10 fern species (including *Elaphoglossum randii*, endemic to the Prince Edward Islands). Flowering plant endemics are *Ranunculus moscleyi* (Kerguelen Islands), *Lyallia kerguelensis* (Kerguelen Islands), *Pringlea antiscorbutica*, *Plantago pentasperma* (Amsterdam Island), *P. stauntonii* (Saint Paul Island), *Uncinia dikei* (Prince Edward Islands), *Poa novareae* (Amsterdam Island and Saint Paul Island), and so on. The highest endemism occurs on the islands of Amsterdam and Saint Paul, where 7 of the 17 species are endemic, in contrast to the Kerguelen Islands, where only 2 of the 29 vascular plants species are endemic, and the Prince Edward Islands, where there is only 1 endemic—a fern—out of the 19 species. The Crozet Islands and Heard Island have no endemics.

Despite its location between southern Africa and Australia, the Kerguelenian Province is floristically closest to Tierra del Fuego, with which it has 17 species in common. Next in terms of closeness is the flora of New Zealand, with which Kerguelenia has 11 species of flowering plants in common. Most characteristic for the vegetation of the Kerguelenian Province are Kerguelenian cabbage (*Pringlea antiscorbutica*), mats formed by pure growths of *Cotula plumosa*, and large cushions of *Azorella selago*.

35. Neozeylandic Region

Engler 1882, 1903, 1924; Oliver 1925; Hayek 1926; Cockayne 1928; Wulff 1944; Good 1947, 1974; Turrill 1959; Takhtajan 1970, 1974; Tolmatchev 1974.

This region includes New Zealand (North and South islands), Stewart Island, the Snares Islands, the Three Kings Islands, Lord Howe Island and Ball's Pyramid, Norfolk Island, the Kermadec Islands, Chatham Islands, the Antipodes Islands, Campbell Island, and the Auckland Islands. The region contains no endemic family, but has about 45 endemic genera:

Loxsomaceae: *Loxsoma* (1, North Island)
Winteraceae: *Pseudowintera* (3, New Zealand, Stewart Island)
Balanophoraceae: *Dactylanthus* (1, North Island; remotely related to the New Caledonian genus *Hachettea*)
Hectorellaceae: *Hectorella* (1, South Island)
Myrsinaceae: *Elingamita* (1, North Island)
Passifloraceae: *Tetrapathaea* (1, New Zealand)
Brassicaceae: *Ischnocarpus* (1, South Island), *Notothlaspi* (2, South Island), *Pachycladon* (1, South Island)
Tiliaceae: *Entelea* (1, New Zealand, Three Kings Islands)
Malvaceae: *Hoheria* (5, New Zealand)
Myrtaceae: *Lophomyrtus* (2, New Zealand), *Neomyrtus* (1, New Zealand, Stewart Island)
Fabaceae: *Carmichaelia* (39, with 38 in New Zealand and 1 species on Lord Howe Island), *Chordospartium* (1, South Island), *Corallospartium* (1, South Island), *Notospartium* (3, South Island), *Streblorrhiza* (1, Philip Island near Norfolk; now extinct)
Loranthaceae: *Ileostylus* (1, New Zealand, Stewart Island), *Tupeia* (1, New Zealand)
Escalloniaceae: *Ixerba* (1, North Island; systematic position of the genus is disputable)
Alseuosmiaceae: *Alseuosmia* (8, with 7 species on North Island and 1 species on both islands of New Zealand)
Araliaceae: *Stilbocarpa*, including *Kirkophytum* (2, the Snares, Stewart Island and adjacent islets, Antipodes, Campbell Island, Auckland Islands, and Macquarie Island)
Apiaceae: *Coxella* (1, Chatham Islands; very close to *Aciphylla*), *Lignocarpa* (2, New Zealand), *Scandia* (2, New Zealand)
Boraginaceae: *Myosotidium* (1, Chatham Islands)
Scrophulariaceae: *Parahebe* (15, New Zealand, Stewart Island; very close to *Hebe*)
Gesneriaceae: *Negria* (1, Lord Howe Island), *Rhabdothamnus* (1, North Island; close to *Negria* and to *Coronanthera*, which is represented by 10 species in New Caledonia and 1 species in Queensland)
Verbenaceae: *Teucridium* (1, New Zealand)
Stylidiaceae: *Oreostylidium* (1, New Zealand, Stewart Island)

Asteraceae: *Brachyglottis* (1, New Zealand, Three Kings Islands; very close to *Senecio*), *Haastia* (3, South Island), *Kirkianella* (1, New Zealand; close to *Crepis*), *Leucogenes* (2, New Zealand, Stewart Island; close to *Helichrysum*), *Lordhowea* (1, Lord Howe Island), *Pachystegia* (1, South Island; close to *Olearia*)
Phormiaceae: *Phormium* (2; 1 species in New Zealand, on Stewart Island and the Chatham Islands, and 1 on Norfolk Island)
Orchidaceae: *Aporostylis* (1, New Zealand, Stewart Island, Chatham Island, Auckland Islands, Campbell Island, Antipodes)
Restionaceae: *Sporadanthus* (1, North Island, Chatham Islands)
Arecaceae: *Hedyscepe* (1, Lord Howe Island; close to *Rhopalostylis*), *Howeia* (2, Lord Howe Island), *Lepidorrhachis* (1, Lord Howe Island), *Rhopalostylis* (3; 1 species in New Zealand, on the Chatham Islands, Norfolk Island, and Kermadec Islands)

Species endemism of the flora of the Neozeylandic Region is very high. Among all the ferns it constitutes 40%, but in the genera *Dicksonia* and *Cyathea* it is approximately 75%. The most remarkable endemic fern species include *Cardiomanes* (*Trichomanes*) *reniforme, Leptopteris superba,* and *Mecodium* (*Hymenophyllum*) *dilatatum.* It is interesting that the fern flora of New Zealand reveals much in common with that of temperate Australia and Tasmania (about 45% of the species in common) and surprisingly little with the flora of subantarctic America. Of these few common elements we mention *Sphaerocionium* (*Hymenophyllum*) *ferrugineum* (the darkest parts of the rain forests of New Zealand, Chile, and the Juan Fernandez Islands); *Grammitis billardieri* (New Zealand, especially the South Island, and eastern Australia, southern Chile, the Falkland Islands, and the Kerguelen Islands); *Blechnum penna-marina* (New Zealand, especially the southern Alps, eastern Australia, southern Chile, Tristan da Cunha, the Kerguelen Islands, Crozet Islands, Marion Island, which is one of the Prince Edward Islands, and also the islands of Saint Paul and Amsterdam); and *Polystichum mohrioides* (Auckland Islands, Campbell Island, Falkland Islands, South Georgia Island, Marion Island, Amsterdam Island, and southern Chile).

Exceptionally high species endemism occurs in the coniferous flora—almost 100%. Among these endemics are *Araucaria heterophylla* (Norfolk Island and its neighboring Philip Island); *Agathis australis* (lowland forests of the North Island, almost from the North Cape to 38° south latitude); 7 species of *Podocarpus;* 6 species of *Dacrydium;* 3 species of *Phyllocladus;* and 2 species of *Libocedrus.* The greater part of the species is found in New Zealand; some (most of the species of *Podocarpus* and *Dacrydium*) also grow on Stewart Island. Some of the New Zealand species of *Podocarpus* and *Dacrydium* have vicariant species in Chile.

The flowering plant flora of the Neozeylandic Region is also distinguished by high species endemism (apparently no less than 80%). Genera with the most endemic species are *Ranunculus, Clematis, Colobanthus, Gaultheria, Dracophyllum, Myrsine, Pimelea, Pittosporum, Alseuosmia, Acaena, Carmichaelia, Metrosideros, Epilobium, Schizeilema, Aciphylla, Anisotome, Gentiana, Coprosma, Myosotis, Euphrasia, Ourisia, Parahebe, Hebe, Celmisia, Olearia, Cotula, Abrotanella, Raoulia, Helichrysum, Senecio, Astelia, Luzula, Uncinia,* and *Carex.* These include some endemic genera (e.g., *Alseuosmia, Carmichaelia,* and *Parahebe*).

The following are basic elements of the Neozeylandic flora: Holantarctic, in the broad sense; paleoneozeylandic; paleotropic (Indomalesian, Polynesian, and Neocaledonian); Australian; and cosmopolitan.

As the paleobotanical data indicate, the Holantarctic element thrived especially in the late Cretaceous and in the Paleogene. At that time the following genera in particular were well-represented: *Araucaria* (one species, which has survived only on Norfolk Island); *Athrotaxis* (at the present time only in Australia and Tasmania); and *Casuarina* (now completely absent from the Neozeylandic Region). *Nothofagus* and the Proteaceae were more diversely represented. It is interesting that among the microfossils in McMurdo in eastern Antarctica (deposits from the late Cretaceous to the Oligocene) pollen grains of conifers, palms, the Proteaceae, and *Nothofagus* were found (Cranwell et al. 1960). But already in the late Cretaceous, New Zealand had begun to separate from Antarctica, and in the Miocene its northern part entered into a zone of subtropical climate. The paleotropical element gradually began to inhabit New Zealand from the north through volcanic archipelagos, and in the Miocene and later along the now almost entirely submerged Lord Howe and Norfolk rises. As a result, tropical genera such as the following penetrated into New Zealand: *Macropiper, Metrosideros, Elaeocarpus, Homalanthus, Dysoxylum, Alectryon, Parsonsia, Avicennia, Elytranthe,* palms, *Freycinetia,* and many orchids. The Holantarctic element again began to extend its area in the Pliocene and especially in the Pleistocene. In the Pleistocene many Holarctic elements penetrated the region.

The separation of New Zealand from Australia occurred before formation of the characteristic Australian flora. Therefore the flora of New Zealand has no genera such as *Acacia* and *Eucalyptus,* and the Proteaceae, which are richly represented in Australia, have in New Zealand only 1 species of *Knightia* (2 other species in New Caledonia) and 1 of *Persoonia* (about 60 remaining species in Australia). Therefore, the elements that Australia and Tasmania have in common with New Zealand are much more nearly Holantarctic than properly Australian. But in the fern flora, which is more ancient than the flora of flowering plants, the elements that New Zealand has in common with Australia and Tasmania constitute 40%.

1. Lord Howean Province. This province includes the small island of Lord Howe (9.6 km long, maximum width 2.9 km) and Ball's Pyramid Island (almost devoid of vegetation). Its flora consists of 180 genera and 226 species of vascular plants (Green 1970). The majority of genera are represented by 1, rarely by 2, and even more rarely by 3 species, which indicates the purely relictual character of the flora. Among the flowering plants, only *Coprosma* (Rubiaceae) is represented by 4 species. Endemism is exceptionally high: there are 5 endemic genera (*Negria, Lordhowea, Lepidorrhachis, Hedyscepe,* and *Howeia*) and 70 endemic species and subspecies.

The majority of endemic taxa, including the genus *Negria*, are related to New Zealand taxa, and somewhat fewer are related to the flora of Australia (11 species) and of Polynesia (10 species). *Carmichaelia*, a leafless xerophyte with green phylloclades and a genus very characteristic of New Zealand, is represented on Lord Howe Island by 1 endemic species (*C. exsul*), the only species of this genus known outside New Zealand (Green 1979). Also endemic on Lord Howe Island are *Sophora howinsula* (= *S. tetraptera* subsp. *howinsula*), which is closest to the New Zealand species *S. tetraptera*, and *Melicytus novae-zelandiae* subsp. *centurionis*, which is very close to the New Zealand typical subspecies (Green 1970). According to Green (1979), 2 endemic species of *Olearia* on Lord Howe are related more to the New Zealand species than to the Australian representatives of this genus. He also mentions that a new subspecies of the New Zealand grass *Chionochloa conspicua* has been found on the flanks of Mt. Lidgbird; it is a member of a genus known only from New Zealand, except for one species in Australia. Furthermore, species of *Pittosporum, Planchonella, Dracophyllum, Melicope, Pimelea, Coprosma, Senecio,* and *Uncinia* that are endemic to Lord Howe Island are related to New Zealand taxa. The orchid *Bulbophyllum tuberculatum* is encountered both on Lord Howe Island and in New Zealand (North Island).

Despite the general numerical predominance of the Australian element, the most characteristic Australian genera (*Eucalyptus, Acacia,* and *Casuarina*) and the Proteaceae are absent from Lord Howe Island. The connection with New Zealand is deeper and more ancient. However, *Bubbia*—represented on Lord Howe by the endemic species, *B. howeana* (type of the genus)—is otherwise absent from the Neozeylandic Region. The remaining species of the genus (about 30) grow in New Guinea, Queensland, and New Caledonia. Completely mysterious is the presence of a member of the South African genus *Dietes* (Iridaceae)—namely, the endemic *D. robinsoniana*, which is quite distinct from the other species and therefore could not have been introduced by man (Green 1979). Analogous disjunct distributions are observed in the flora of New Zealand (*Pelargonium inodorum* and others).

Although the presence of Australian and cosmopolitan elements characteristic mainly of coastal formations (Akhmetiev 1972), may be explained by the powerful eastern Australian ocean current and by the wind, this explanation cannot be applied to the New Zealand and New Caledonian plants. They may be explained instead by the past existence of archipelagic "steppingstone" migration routes and in part even by direct overland connections (Wallace 1880; Oliver 1911; Paramonov 1963), especially inasmuch as, according to geological data, Lord Howe Island was considerably larger in the Pleistocene than it is at present (see Akhmetiev 1972).

It is interesting that the zoologist Paramonov (1963) concluded that the fauna of Lord Howe Island is closer to that of New Zealand than to that of Australia. He considers Lord Howe to be part of a larger land mass—a kind of "microcontinent" that he called "Howeania." Howeania connected Lord Howe and Ball's Pyramid with the New Zealand Plateau, but was joined with Australia only before the dispersal of typical Australian genera. In Paramonov's opinion, New Caledonia and Norfolk Island constituted another land mass, which was separated from Howeania in the south only by a comparatively narrow strait. Even at the present time Lord Howe Island is joined to New Zealand by a submerged rise or ridge, the "Lord Howe Rise"; in the north, this ridge approaches close to the Tropic of Capricorn.

2. **Norfolkian Province.** This province consists of Norfolk Island and a small island lying south of it, Philip Island. The native flora of the province consists of 174 species of vascular plants (Hoogland [in Turner, Smithers, and Hoogland] 1968). Here grow *Psilotum nudum;* the endemic *Tmesipteris forsteri; Lycopodium cernuum;* 40 species of ferns, including 14 endemic species such as *Blechnum norfolkianum;* the endemic "Norfolk Island pine," *Araucaria heterophylla;* and 128 species of flowering plants, of which 36 are endemic to the province. These last include *Clematis cocculifolia, Boehmeria australis, Achyranthes arborescens, Melicytus latifolius, Capparis nobilis, Zehneria baueriana, Ungeria floribunda* (a second species of the genus occurs in eastern Australia), *Hibiscus insularis* (known from Philip Island, where it exists today in the wild only as 3 or 4 bushes; Green 1979), *Wickstroemia australis, Euphorbia norfolkiana, Pittosporum bracteolatum, Planchonella costata, Evodia littoralis, Dysoxylum patersonianum, Meryta angustifolia, Melodinus baueri, Myoporum obscurum, Coprosma baueri* and *C. pilosa, Korthalsella disticha, Cordyline obtecta,* 5 species of Orchidaceae, *Carex neesiana, Agropyron kingianum, Rhopalostylis baueri,* and *Freycinetia baueriana.*

According to Green (1979:50), "the connection with New Zealand is perhaps even stronger in the case of Norfolk Island." The flora of the Norfolkian Province has 5 species in common with New Zealand—namely, *Euphorbia glauca* (New Zealand, Stewart Island, Chatham Islands); *Muehlenbeckia australis* (Three Kings Islands, New Zealand, Stewart Island, Chatham Islands); *Ileostylus micranthus* (New Zealand, Stewart Island); *Phormium tenax* (New Zealand, Stewart Island, Chatham Islands, Auckland Islands); and *Nestegis apetala* (northernmost part of New Zealand). In addition, it has an entire series of vicariant species and subspecies, in particular:

Melicytus ramiflorus subsp. *oblongifolius—M. ramiflorus* subsp. *ramiflorus* (Kermadec Islands, New Zealand, Stewart Island)
Coprosma baueri—C. repens (Kermadec Islands, Three Kings Islands, New Zealand)
Planchonella costata—P. novo-zelandica (New Zealand)
Rapanea crassifolia—R. kermadecensis (Kermadec Islands)
Pennantia endlicheri—P. corymbosa (New Zealand)
Cordyline obtecta—C. australis (New Zealand, Stewart Island)
Rhopalostylis baueri—R. sapida (New Zealand, Chatham Islands), *R. cheesemanii* (Kermadec Islands)
Freycinetia baueriana subsp. *baueriana—F. baueriana* subsp. *banksii* (New Zealand)

3. Kermedecian Province (Engler 1914; Cockayne 1928). This province includes the Kermadec Islands, with a total area of 33 km². The flora of this province includes about 120 species of vascular plants, of which 16 are endemic: *Cyathea kermadecensis* and *C. milnei, Ascarina lanceolata, Boehmeria dealbata, Rapanea kermadecensis, Neopanax kermadecense, Homalanthus polyandrus, Metrosideros kermadecensis, Coprosma acutifolia, C. petiolata, Hebe breviracemosa, Scaevola gracilis, Erechtites kermadecensis, Carex kermadecensis, C. ventosa,* and *Rhopalostylis cheesemanii*.

4. Northern Neozeylandic Province (Northern Mainland Province, Cockayne 1921, 1928; Bezirk von Northern Mainland, Hayek 1926). This province includes the North Island of New Zealand northward from a line going from the Mokau River; it proceeds further to the north of the Mamaku Plateau and reaches the Bay of Plenty somewhat south of Tauranga. As Cockayne (1928:379) notes, this boundary is to some degree artificial. The province also includes the Three Kings Islands in the north, and all remaining tiny islands in the east.

The flora of the Northern Neozeylandic Province is the richest in the Neozeylandic Region and is characterized by the considerable representation of paleotropical—principally Malesian—elements (especially in the fern flora). On this basis, many authors exclude this province from the Neozeylandic Region. Thus Engler (1903, 1924) includes the North Island of New Zealand with eastern Australia, Lord Howe Island, Norfolk Island, the Kermadec Islands, New Caledonia, and the Chatham Islands in his "Araucarien-Provinz." The influence of the Australian flora is shown especially strongly in the Orchidaceae and partly in the Cyperaceae, but the majority of the most typical Australian genera of dicots are absent here. In contrast, Turrill (1959) and Good (1974) with complete justification included North Island in the New Zealand Region, along with the Kermadec Islands and Chatham Island. The flora of the Northern Neozeylandic Province is in fact not tropical but subtropical, and paleotropical elements, although numerous, are not dominant.

Endemism in the Northern Neozeylandic flora is rather high. Among the endemic taxa are the monotypic genera *Loxsoma, Elingamita, Ixerba,* and numerous species: *Thelypteris gongylodes, Davallia tasmanii* (Three Kings Islands), *Agathis australis, Dacrydium kirkii, Libocedrus plumosa, Beilschmiedia tarairi, Litsea calicaris, Paratrophis smithii, Hoheria populnea, Dracophyllum lessonianum, D. patens, D. viride, D. matthewsii, D. pyramidale, Planchonella novo-zelandica* (a species very close to the Norfolkian representatives of this genus, and also to the Fijian species, *P. vitiensis*), *Rapanea dentata, Ackama rosifolia, Weinmannia silvicola, Corokia cotoneaster, C. buddleoides, Pittosporum pimeleoides, P. fairchildii, P. ellipticum* and several other species of *Pittosporum, Alseuosmia ligustrifolia, A. linariifolia, A. atriplicifolia, A. palaeiformis, Metrosideros albiflora, Phebalium nudum* (the only New Zealand representative of this Australian genus), *Meryta sinclairi, Pseudopanax discolor, Pomaderris kumeraho* and several other species of this genus, *Mida salicifolia, Elytranthe adamsii, Persoonia toru* (the only representative endemic to this province of this typically Australian genus), *Olea apetala, Coprosma spathulata, C. arborea, C. macrocarpa, C. dodonaeifolia, Myosotis matthewsii, Hebe insularis* (endemic to the Three Kings Islands), *H. pubescens, H. ligustrifolia, H. obtusata, H. bollonsii, Tecomanthe speciosa* (endemic to The Three Kings Islands, with the remaining representatives of the genus found mainly in New Guinea), *Utricularia protrusa, U. novaezelandiae, Avicennia resinifera, Pratia physaloides, Lagenophora lanata, Olearia albida, Cassinia amoena, Xeronema callistemon* (the other species of this genus grows in New Caledonia), *Astelia trinervia, A. banksii, Collospermum microspermum, Cordyline pumilio, C. kaspar, Yoania australis,* and *Hydatella inconspicua.*

5. Central Neozeylandic Province (Central Mainland Province, Cockayne 1921, 1928; Bezirk von Central Mainland, Hayek 1926). This province includes the territory of North Island south from the Northern Neozeylandic Province, Kapiti Island, and part of South Island, to a line extending from Greymouth to Amuri Bluff. Endemism is rather high. It has 2 endemic, monotypic genera—*Chordospartium* and *Pachystegia*—and a large number of endemic species. In contrast to the Northern Neozeylandic Province, there are no endemic species of ferns or conifers. In comparison with the Northern Province, the Central Neozeylandic Province is characterized by a considerable increase of Holantarctic and paleoneozeylandic elements, and by a corresponding reduction of the influence of the Malesian flora.

6. **Southern Neozeylandic Province** (Southern Mainland Province, Cockayne 1921, 1928; Bezirk von Southern Mainland, Hayek 1926). This province includes the entire southern part of South Island, along with Stewart Island and other small islands (including Ruapuke Island and Solander Island, which is located northwest of Stewart Island). The flora of this province has a more temperate character than does that of the Central Neozeylandic Province. Within this province, Stewart Island forms a special district, for which the following endemic species are characteristic: *Ranunculus kirkii, Aciphylla traillii, Anisotome flabellata, Gentiana gibbsii, Celmisia glabrescens, C. rigida, Abrotanella muscosa, A. filiformis, Raoulia goyenii, Carex longiculmis, Danthonia pungens,* and *Poa guthriesmithiana.*

7. **Chathamian Province** (Cockayne 1921, 1928; Bezirk der Chatham-Inseln, Hayek 1926). This province consists of the Chatham Islands. It has 2 monotypic, endemic genera, *Coxella* and *Myosotidium.* Among its 260 species, 40 are endemic, including *Hymenanthera chathamica, Cyathodes robusta, Dracophyllum paludosum, D. arboreum, Rapanea chathamica, Corokia macrocarpa, Geranium traversii, Pseudopanax chathamicus, Aciphylla traversii, Gentiana chathamica, Coprosma chathamica, Hebe dieffenbachii, H. barkeri, H. chathamica, Olearia chathamica, O. semidentata, O. traversii, Cotula featherstonii, C. renwickii, C. potentillina, Senecio radiolatus, S. huntii, Sonchus grandifolius, Astelia chathamica, Carex chathamica, C. ventosa,* and *Agropyron coxii.*

8. **Province of New Zealand Subantarctic Islands** (Subantarctic Province, Cockayne 1921, 1928; Provinz der subantarktischen Inseln, Hayek 1926: Skottsberg 1960; Antipodean District, Thorne 1963). This province includes the Snares, the Auckland Islands, Campbell Island, the Antipodes, and also Macquarie Island, which Engler (1903, 1924) separated as a separate province; in the past, it probably also included the Bounty Islands.[7] The higher plant flora of the subantarctic islands of New Zealand consists of 190 species (Cockayne 1928), of which about 50 are endemic to the province. They are:
Ranunculus pinguis, R. subscaposus, Urtica aucklandica, Stellaria decipiens, Colobanthus hookeri, Dracophyllum scoparium, Geum parviflorum subsp. *albiflorum, Acaena minor, Epilobium confertifolium, E. antipodum, Stilbocarpa polaris, Schizeilema reniforme, Anisotome latifolia, A. antipoda, A. acutifolia, Gentiana cerina, G. concinna, G. antarctica, G. antipoda, Coprosma cuneata, Myosotis antarctica, M. capitata, Hebe benthamii, Plantago aucklandica, P. triantha, Callitriche aucklandica, Pleurophyllum speciosum, P. criniferum, P. hookeri, Celmisia vernicosa, Cotula lanata, C. plumosa, Abrotanella spathulata, A. rosulata, Senecio antipodus, Bulbinella rossii, Uncinia hookeri, Hierochloë*

brunonis, Chionochloa antarctica, Deschampsia gracillima, D. penicellata, Poa ramosissima, P. hamiltonii, P. aucklandica, P. incrassata, Puccinellia antipoda, and *P. macquarensis.*

The Auckland Islands have the richest flora in the province containing 187 species (Johnson and Campbell 1975), including the following endemic species and subspecies: *Urtica aucklandica, Geum parviflorum* subsp. *albiflorum, Gentiana cerina* subsp. *cerina, G. concinna, Plantago aucklandica, Callitriche aucklandica, Poa breviglumis* subsp. *maorii,* and *P. incrassata.* Campbell Island has 115 species, the Antipodes have 60 species, and Macquarie Island has 35 species in all.

NOTES

1. In the literature, the Holantarctic Kingdom usually is referred to by the term Antarctic (Antartcis), suggested by Drude (1890). Following Szafer (1952), I prefer to call it the Holantarctic.

2. I interpret the Chile-Patagonian Region rather broadly: besides the "Patagonian Region" and the "Pampas Region" of Good (1947, 1974), it also includes the southern part of his "Andean Region" and extends southward to the Antarctic Peninsula and southeastward to the South Sandwich Islands.

3. In his later work, Engler included northern Chile (southward to 30°30' south latitude) in his "Nordliche und mittlere hochandine Provinz" (Engler 1899, 1903, 1924). In my opinion, the Northern Chilean Province corresponds to "Das Vegetations gebiet der Subtropischen Zwergstrauchformationen des Kleinen Nordens" and "Das Vegatationsgebiet der an Frühlingshygrophyten reichen Strauch- und Zwergstrauchformationen des Kleinen Nordens" ("Gebiet von La Serena") of Schmithüsen (1956).

4. The Patagonian Province, as understood here, includes all territory of the "Provincia Patagonica," the "Provincia del Monte," and the greater part of the "Provincia del Espinal" of Cabrera and Willink (1973), or the "Unterprovinz des Espinale" of the "Argentinische Provinz" of Engler (1903, 1924). These three provinces could probably correctly be described as subprovinces. However, one cannot exclude the possibility that further study would lead to the necessity of the rank of province for these choria. The Patagonian Province as understood here does not agree with the "Patagonische Provinz" of Hayek (1926), in that it does not include the southern part of Tierra del Fuego and in the north it includes a rather significant part of his "Andines Gebiet."

5. Skottsberg (1960) divided his "Magellanian Province" into (1) "Patagonian-Fuegian District," (2) "Andine Patagonian-Fuegian District," and (3) "Falkland and S. Georgia District."

6. In the first version of his floristic system, Engler (1882) divided the subantarctic islands into three independent regions: "Gebiet der Kerguelen," "Gebiet der

Amsterdan-Inseln," and "Gebiet von Tristan d'Acunha." Later, Engler (1899, 1903, 1924) united two previous regions into one ("Gebiet von Tristan d'Acunha, St. Paul und Amsterdam-Inseln").

7. It is assumed that the higher vegetation on the Bounty Islands was annihilated by numerous seals and penguins, which even polished the rocks with their bodies.

APPENDIX

LIST OF FAMILIES OF LIVING VASCULAR PLANTS

Division I. Lycopodiophyta

Class A. Lycopodiopsida

Order 1. Lycopodiales

1. Lycopodiaceae (*Lycopodium* and *Phylloglossum*). 2 genera/250 species. Nearly cosmop. (*Lycopodium*), but the monotypic *Phylloglossum* is endemic to western and southern Australia, Tasmania, and New Zealand.

Class B. Isoëtopsida

Order 1. Selaginellales

1. Selaginellaceae (*Selaginella*). 1/700. Nearly cosmop., but the majority of species occur in the rain forests of tropical regions.

Order 2. Isoëtales

1. Isoëtaceae (*Isoëtes*, incl. *Stylites*). 1/150 or more. Widely distributed in both hemispheres, but absent from oceanic islands (except New Zealand).

Division II. Psilotophyta

Class A. Psilotopsida

Order 1. Psilotales

1. Psilotaceae (*Psilotum*). 1/2. Trop. and partly subtrop. regions of both hemispheres.
2. Tmesipteridaceae (*Tmesipteris*). 1/10. New Hebrides, Samoa and other Polynesian islands, Norfolk, New Caledonia, Australia, Tasmania, New Zealand, Stewart Island, Auckland Islands, Chatham Islands, and Kermadec Islands; to the north it reaches Philippines.

Division III. Equisetophyta

Class A. Equisetopsida

Order 1. Equisetales

1. Equisetaceae (*Equisetum*). 1/15. Widely distributed, except in tropical Africa, Australia, and New Zealand.

Division IV. Polypodiophyta

Class A. Ophioglossopsida

Order 1. Ophioglossales

1. Ophioglossaceae (*Botrychium, Ophioglossum, Helminthostachys*). 3/60. Cosmop.; the monotypic *Helminthostachys* extends from India and Sri Lanka to Taiwan and New Caledonia.

Class B. Marattiopsida

Order 1. Marattiales

1. Marattiaceae. 7/150 or more. Mainly in tropical humid forests of both hemispheres, but extend to subtropical regions also. The majority of genera are centered in the tropics of southeastern Asia. *Danaea* (20 or more species) is the only genus endemic to tropical America.

Class C. Polypodiopsida

Subclass A. Osmundidae

Order 1. Osmundales

1. Osmundaceae (*Osmunda, Todea, Leptopteris*). 3/20 (30–37). Mainly in tropical and temperate regions of both hemispheres. The monotypic *Todea* occurs in South Africa, Australia, and New Zealand.

Order 2. Plagiogyriales

1. Plagiogyriaceae (*Plagiogyria*). 1/50. From the Himalayas to Korea, Japan, the Ryukyus, Taiwan, and the Philippines, east and southward to New Guinea and northeastern Australia, and in America from Mexico to Bolivia and southeastern Brazil.

Subclass B. Schizaeidae

Order 1. Schizaeales

1. Schizaeaceae (incl. Anemiaceae, Lygodiaceae, and Mohriaceae). 4/145 or more. Trop. and subtrop., with a few species in temperate regions of East Asia, North America and South Africa. *Mohria* (2) is endemic to southern and southeastern Africa, Madagascar, and the Mascarenes.

Order 2. Pteridales

1. Pteridaceae (incl. Actiniopteridaceae, Cryptogrammaceae, Hemionitidaceae, Negripteridaceae, Sinopteridaceae, and Taenitidaceae). 35/700. Widely distributed, but concentrated mainly in trop. and subtrop. regions. The largest genus, *Pteris* (250), is nearly cosmop.
2. Platyzomataceae (*Platyzoma*). 1/1. Northeastern Australia.
3. Adiantaceae (*Adiantum*). 1/200. Nearly cosmop.; most numerous in South America.
4. Vittariaceae. 6/105. Pantropical (*Antrophyum* and *Vittaria*), neotropical (*Hecistopteris* and *Anetium*), and paleotropical (*Monogramma* and *Rheopteris*). *Rheopteris* is endemic to New Guinea.
5. Parkeriaceae (*Ceratopteris*). 1/3. Trop. and subtrop.; nearly pantropical.

Subclass C. Marsileidae

Order 1. Marsileales

1. Marsileaceae. 3/60. Trop. and temperate regions. *Regnellidium* (1) is endemic to southern Brazil and adjacent Argentina.

Subclass D. Gleicheniidae

Order 1. Gleicheniales

1. Stromatopteridaceae (*Stromatopteris*). 1/1. New Caledonia.
2. Gleicheniaceae (*Gleichenia* and *Dicranopteris*). 2/120. Pantropical, subtrop. Asia, and the Southern Hemisphere.

Order 2. Matoniales

1. Matoniaceae (*Matonia* and *Phanerosorus*). 2/4. Malay Peninsula, Sumatra, Borneo, Moluccas (Amboina), and New Guinea.

Subclass E. Polypodiidae

Order 1. Polypodiales

1. Dipteridaceae (*Dipteris*). 1/8. From northeastern India, southern China, and Taiwan across the Malay Archipelago to New Guinea, New Caledonia, and Polynesia.
2. Cheiropleuriaceae (*Cheiropleuria*). 1/1. From Indochina to New Guinea.
3. Polypodiaceae (incl. Grammitidaceae, Loxogrammaceae, and Platyceriaceae). 40/1000 or more. Widely distributed, but mainly in tropical regions. *Polypodium* (150) is nearly cosmop.

Subclass F. Hymenophyllidae

Order 1. Hymenophyllales

1. Hymenophyllaceae (*Hymenophyllum* and *Trichomanes*). 2/600. Mainly trop. regions and the Southern Hemisphere; a few species in temperate regions of the Northern Hemisphere.

Order 2. Loxsomales

1. Loxsomaceae (*Loxsoma* and *Loxsomopsis*). 2/2–3. North Island of New Zealand (*Loxsoma*) and tropical America (mountains of Costa Rica, Colombia, Ecuador, Peru, and Bolivia; *Loxsomopsis*).

Order 3. Hymenophyllopsidales

1. Hymenophyllopsidaceae (*Hymenophyllopsis*). 1/7. Northern South America (the Roraima formation of the states of Amazonas and Bolivar in Venezuela).

Order 4. Dicksoniales

1. Thyrsopteridaceae (*Thyrsopteris*). 1/1. Juan Fernandez Islands.
2. Culcitaceae (*Culcita*). 1/7. Trop. America, the Azores, Madeira, Tenerife, Portugal, and southern Spain (subgenus *Culcita*), and from Malesia and eastern Australia to Samoa (subgenus *Calochlaena*).
3. Dicksoniaceae (*Cibotium, Dicksonia,* and *Cystodium*). 3/30. Trop. and subtrop. Asia, Australia, Tasmania, New Caledonia, Samoa, New Zealand, Stewart Island, Chatham Islands,

Hawaii, and Central and South America. *Dicksonia arborescens* is endemic to St. Helena.
4. Lophosoriaceae (*Lophosoria*). 1/1. America, from Mexico and the West Indies to the Juan Fernandez Islands and Patagonia.
5. Metaxyaceae (*Metaxya*). 1/1. Tropical America.
6. Cyatheaceae (*Sphaeropteris, Alsophila, Nephelea, Trichipteris, Cyathea,* and *Cnemidaria*). 6/500. Trop. and subtrop. regions, south to South Africa, Madagascar, New Zealand, Kermadec Islands, Stewart Island, Chatham Islands, Auckland Islands, and temperate South America, a few passing the Tropic of Cancer.
7. Dennstaedtiaceae (incl. Hypolepidiaceae). 8/160. Trop., and subtrop., and south-temperate regions. The monotypic and very polymorphic genus *Pteridium* is nearly cosmop. *Coptidipteris* (1) is endemic to eastern Asia, *Leptolepia* (2) to New Guinea, Queensland, and New Zealand, and *Oenotrichia* (2) to New Caledonia.
8. Lindsaeaceae. 9/260. Trop., subtrop., and south-temperate regions. *Humblotiella* and *Sambirania* are endemic to Madagascar.

Order 5. Aspidiales

1. Thelypteridaceae. 1 (Tryon and Tryon 1982), 3 (K. Iwatsuki 1964), or 32 (Pichi-Sermolli 1977) /800. Nearly cosmop., but with relatively few species in boreal and south-temperate regions.
2. Aspleniaceae. 12–13/700. Widely distributed, but mainly in tropical and subtropical regions. The largest genus, *Asplenium* (650), is nearly cosmop.
3. Aspidiaceae (Dryopteridaceae; incl. Athyriaceae, Onocleaceae, Peranemataceae, and Woodsiaceae). 50–60/1550. Very widely distributed, but concentrated mainly in trop. and subtrop. regions.
4. Lomariopsidaceae (incl. Bolbitidaceae and Elaphoglossaceae). 5/480. Trop., subtrop., and south-temperate regions.
5. Oleandraceae. 3/55. *Oleandra* (40) is pantropical, *Arthropteris* (15) occurs in trop. Africa and adjacent Arabia east to Sri Lanka, southern China, Taiwan, Vietnam, and the Philippines, south to eastern Australia, New Zealand, and the Kermadec Islands, and eastward through the Pacific islands to Tahiti and the Juan Fernandez Islands. Monotypic *Psammiosorus* is endemic to Madagascar and very closely related to *Arthropteris*.

6. Davalliaceae (incl. Gymnogrammitidaceae and Nephrolepidaceae). 9/145. South Africa (Natal), Madagascar, Canary Islands, southwestern Europe, tropical and subtropical Asia to Polynesia, Australia and New Caledonia, New Zealand, and the Auckland Islands. *Nephrolepis* (20) has 6 species in America.
7. Blechnaceae (incl. Stenochlaenaceae). 9/175. Trop. and subtrop. regions of both hemispheres, with a few species ranging to the temperate regions. The largest genus, *Blechnum* (150), is nearly cosmop.

Subclass G. Salviniidae

Order 1. Salviniales

1. Salviniaceae (*Salvinia*). 1/10 Trop., subtrop., and warm temperate regions.
2. Azollaceae (*Azolla*). 1/6. Trop., subtrop., and warm temperate regions.

Division V. Ginkgophyta

Class A. Ginkgoopsida

Order 1. Ginkgoales

1. Ginkgoaceae (*Ginkgo*). 1/1. Southeastern China, along the northwestern border of Chê kiang in western Tien mu shan in the hilly country of the Lower Yangtze Valley.

Division VI. Pinophyta

Class A. Pinopsida

Order 1. Cephalotaxales

1. Cephalotaxaceae (*Cephalotaxus*). 1/6–7. From Khasi and the Naga Hills and eastern Himalayas to Vietnam, Taiwan, Korea, and Japan.

Order 2. Podocarpales

1. Podocarpaceae. 8/135–140. Mostly the Southern Hemisphere, extending north to Japan, Central America, and the West Indies.
2. Phyllocladaceae (*Phyllocladus*). 1/5. The Philippines, Borneo, Sulawesi, New Guinea, Tasmania, and New Zealand.

Order 3. Taxales

1. Austrotaxaceae (*Austrotaxus*). 1/1. New Caledonia

2. Taxaceae (*Amentotaxus, Pseudotaxus, Taxus, Torreya*). 4/19. The Northern Hemisphere south to Sumatra and Sulawesi (*Taxus celebica*) and southern Mexico. *Amentotaxus* (4) is distributed in Burma, continental China, Taiwan, and Vietnam; *Pseudotaxus* (1) is endemic to eastern China.

Order 4. Araucariales

1. Araucariaceae. 2/33, *Agathis* (13): Malesia, New Guinea, New Britain, Solomon Islands, Australia (Queensland), northern New Zealand, New Caledonia, New Hebrides, and Fiji; *Araucaria:* southern South America (mountains of southeastern Brazil and northeastern Argentina, Chile, and western Argentina), New Guinea, eastern Australia, Norfolk Island, New Caledonia, Loyalty Islands, and the Isle of Pines.

Order 5. Pinales

1. Pinaceae. 11/250. The Northern Hemisphere south to Borneo, Sumatra, Central America, and the West Indies. The tropical Asiatic species *Pinus merkusii* is the only member of the family which crosses the equator.

Order 6. Sciadopityales

1. Sciadopityaceae (*Sciadopitys*). 1/1. Japan.

Order 7. Cupressales

1. Taxodiaceae. 9/16. North America and Mexico, Japan, northeastern Burma, continental China and Taiwan, Vietnam, and Tasmania.
2. Cupressaceae. 19/130. Nearly cosmop.

Division VII. Cycadophyta

Class A. Cycadopsida

Order 1. Cycadales

1. Cycadaceae (*Cycas*). 1/20. From Madagascar and the Mascarenes to Sri Lanka, southern India, Madras State, eastern Assam, Burma, East Asia (north to Kyushu Island), Southeast Asia, New Guinea, and northern and northeastern Australia, and in the western Pacific eastward to Fiji, Samoa, and Tonga.
2. Stangeriaceae (*Stangeria*). 1/1. Southeastern Africa (Tongaland-Pondoland Region).
3. Zamiaceae. 8/80. Trop. America, the West Indies, trop. and South Africa, Australia.

Division VIII. Gnetophyta

Class A. Ephedropsida

Order 1. Ephedrales

1. Ephedraceae (*Ephedra*). 1/40. North and South America (especially in western North America) and Eurasia (especially in the Mediterranean and Western and Central Asia).

Class B. Welwitschiopsida

Order 1. Welwitschiales

1. Welwitschiaceae (*Welwitschia*). 1/1. Deserts of Angola and southwestern Africa, mainly in the coastal desert of Namib.

Class C. Gnetopsida

Order 1. Gnetales

1. Gnetaceae (*Gnetum*). 1/30. Trop. America, trop. West Africa, northeastern India, Southeast Asia, New Guinea, Solomon Islands, Caroline Islands, and Fiji.

Division IX. Magnoliophyta

Class A. Magnoliopsida

Subclass A. Magnoliidae

Superorder Magnolianae

Order 1. Magnoliales

1. Degeneriaceae (*Degeneria*). 1/1. Fiji.
2. Himantandraceae (*Galbulimima*). 1/2. New Guinea, New Britain, the Moluccas, and northeastern Australia (Queensland).
3. Magnoliaceae. 12/230. Southern India and Sri Lanka, eastern Himalayas, Assam, Burma, East and Southeast Asia, New Guinea, southeastern North America, Central America, the West Indies, South America; centered mainly in East and Southeast Asia.

Order 2. Eupomatiales

1. Eupomatiaceae (*Eupomatia*). 1/2. Eastern New Guinea and eastern Australia, from tropical Queensland to temperate Victoria.

Order 3. Annonales

1. Annonaceae. 130/2300. Almost wholly confined to the tropics, but the genus *Asimina* (8) occurs in eastern North America as far north as Southern Ontario.
2. Canellaceae. 6/20. Eastern and southern trop. Africa (*Warburgia*), Madagascar (*Cinnamosma*), and trop. South America and the West Indies.
3. Myristicaceae. 16/300. Pantropical, but especially in trop. Asia.

Order 4. Winterales

1. Winteraceae. 9/100. Madagascar (*Takhtajania*), Malayan Archipelago (except Sumatra, Java, and Timor), New Guinea, Solomon Islands, eastern Australia and Tasmania, New Zealand, Lord Howe Island, New Caledonia, Fiji, America from southern Mexico to the Straits of Magellan, and the Juan Fernandez Islands. *Takhtajania* is a very isolated and advanced genus within the family and constitutes a separate subfamily (Takhtajanioideae) or perhaps even a monotypic family.

Order 5. Illiciales

1. Illiciaceae (*Illicium*). 1/40. Bhutan, Khasi Hills, Burma, East and Southeast Asia, southeastern North America, eastern Mexico, and the West Indies.
2. Schisandraceae (*Schisandra* and *Kadsura*). 2/47. India and Sri Lanka, Himalayas from Simla to Bhutan, Assam, northern Burma, East and Southeast Asia, Malesia, and 1 species of *Schisandra* in southeastern North America.

Order 6. Austrobaileyales

1. Austrobaileyaceae (*Austrobaileya*). 1/1. Northern Queensland.

Order 7. Laurales

1. Amborellaceae (*Amborella*). 1/1. New Caledonia.
2. Trimeniaceae. 2/5–9. Sulawesi, the Moluccas, New Guinea, New Britain, Bougainville, eastern Australia, New Caledonia, Fiji, Samoa, and the Marquesas Islands.
3. Monimiaceae. 24/330. Trop. and subtrop. regions, especially of the Southern Hemisphere, where it extends to temperate regions.

4. Atherospermataceae. 5/13. New Guinea, Australia, Tasmania, New Caledonia (*Nemuaron*), and New Zealand and Chile (*Laurelia*).
5. Siparunaceae (*Siparuna, Bracteanthus, Glossocalyx*). 3/170. Trop. America, the West Indies, trop. West Africa (*Glossocalyx*).
6. Gomortegaceae (*Gomortega*). 1/1. Central Chile.
7. Hernandiaceae (*Hernandia* and *Illigera*). 2/42. Trop. Africa, Madagascar, Mauricius, Reunion, South and Southeast Asia (north to southern China, Ryukyu Islands, and Bonin Islands), New Guinea, New Caledonia, Fiji and other Pacific islands, northeastern Australia, and trop. America.
8. Calycanthaceae (*Calycanthus, Sinocalycanthus, Chimonanthus*). 3/7. China and North America (*Calycanthus*).
9. Idiospermaceae (*Idiospermum*). 1/1. Northern Queensland.
10. Lauraceae (incl. Cassythaceae). 30–45/2500–3000. Trop. and subtrop. The main centers are in Southeast Asia and trop. America. Some genera extends north and south into temperate areas.
11. Gyrocarpaceae. 2/26. *Gyrocarpus:* pantropical; *Sparattanthelium:* only trop. America.

Order 8. Lactoridales

1. Lactoridaceae (*Lactoris*). 1/1. Juan Fernandez Islands.

Order 9. Chloranthales

1. Chloranthaceae. 5/70–75. Madagascar (*Ascarinopsis*, close to *Ascarina*), tropical Himalayas, South, East, and Southeast Asia, New Guinea and Melanesia to the Marquesas on the east, New Hebrides, Fiji, New Caledonia, New Zealand, and trop. America.

Order 10. Piperales

1. Saururaceae. 5/7. From Himalayas to Japan, Taiwan and the Philippines, Indochina, North America and Mexico.
2. Piperaceae. 4–5/2100. Pantropical.
3. Peperomiaceae. 4/1000. Pantropical. The oligotypic genus *Verhuellia* and monotypic genera *Manekia* and *Piperanthera* are endemic to West Indies.

Order 11. Aristolochiales

1. Aristolochiaceae. 7/about 625. Tropical, subtropical, and

warm temperate regions of both hemispheres, except Australia.

Superorder Rafflesianae

Order 12. Hydnorales

1. Hydnoraceae. 2/20. Africa from Ethiopia to the Cape and Madagascar (*Hydnora*), and America from Paraguay and northern Argentina to Patagonia (*Prosopanche*).

Order 13. Rafflesiales (Cytinales)

1. Rafflesiaceae. 3/15. Northeastern India (*Sapria himalayana*) and Southeast Asia.
2. Apodanthaceae (*Apodanthes, Pilostyles, Berlinianche*). 3/35. America from the southern United States to the Straits of Magellan, South Africa, western Asia (Asia Minor, Iraq and Iran), and western Australia. *Apodanthes* is endemic to trop. South America and *Berlinianche* is endemic to trop. Africa.
3. Mitrastemonaceae (*Mitrastemon*). 1/5. East and Southeast Asia, Mexico, and Central America.
4. Cytinaceae. 2/10. Trop. and South Africa and Madagascar, Mediterranian Region, Asia Minor, western Caucasus (*Cytinus*), and Mexico and Salvador (*Bdallophytum*).

Order 14. Balanophorales (excl. Cynomoriaceae)

1. Balanophoraceae. 18/45 (110). Pantropical; only a few species in subtrop. regions. *Mystropetalon* (1–3), which is one of the most isolated genus within the family, is endemic to South Africa. *Hachettea* (1) is endemic to New Caledonia, and a related genus, *Dactylanthus* (1), is endemic to New Zealand.

Superorder Nepenthanae

Order 15. Nepenthales

1. Nepenthaceae (*Nepenthes*). 1/72. Madagascar, Seychelles, Sri Lanka, northeastern India (Assam), Southeast Asia, New Guinea, northern Australia, New Caledonia. The Seychelles species *N. pervillei* is sometimes regarded as a separate genus, *Anurosperma*.

Superorder Nymphaeanae

Order 16. Nymphaeales

1. Cabombaceae (*Cabomba* and *Brasenia*). 2/7–8. America from the northern United States to the West Indies and Argentina, trop. Africa, eastern Himalayas (Bhutan), Assam (Khasi Hills), East Asia, eastern Australia.
2. Nymphaeaceae. 5/60–80. Nearly cosmopolitan in fresh water habitats.
3. Barclayaceae (*Hydrostemma*). 1/2–3. Southeast Asia (Burma, Thailand, Malay Peninsula, Singapore, and Borneo) and New Guinea.

Order 17. Ceratophyllales

1. Ceratophyllaceae (*Ceratophyllum*). 1/6. Worldwide in fresh water.

Superorder Nelumbonanae

Order 18. Nelumbonales

1. Nelumbonaceae (*Nelumbo*). 1/2. Warmer parts of Asia and Australia and the eastern United States.

Subclass B. Ranunculidae

Superorder Ranunculanae

Order 19. Ranunculales

Suborder Lardizabalineae

1. Lardizabalaceae. 8/35. From the western Himalayas to northern Vietnam, continental China, Hainan, Taiwan, Korea, Japan, and disjunct in central Chile (*Lardizabala* and *Boquila*).
2. Sargentodoxaceae (*Sargentodoxa*). 1/1. Continental China, northern Laos, and northern Vietnam.

Suborder Menispermineae

3. Menispermaceae. 67/400 or more. Trop. and subtrop., with only few species in temperate areas. The genus *Menispermum* (2) is widespread in eastern Siberia, Transbaicalia, East Asia, and in North America from southern Canada to Florida and in Mexico.

Suborder Ranunculineae

4. Ranunculaceae. 45/2000. Worldwide, but centered in

temperate and cold regions of the Northern and Southern Hemispheres, chiefly in north-temperate regions.
5. Circaeasteraceae (incl. Kingdoniaceae?). 2/2. Himalayas from Garhwal to Bhutan, southeastern Tibet, and continental China.

Suborder Berberidineae

6. Hydrastidaceae (*Hydrastis*). 1/1. Eastern North America.
7. Nandinaceae (*Nandina*). 1/1. Continental China and Japan.
8. Berberidaceae (incl. Leonticaceae and Podophyllaceae). 13/650. North Africa, Europe, trop., subtrop., and temperate Asia, and South America.

Order 20. Glaucidiales

1. Glaucidiaceae (*Glaucidium*). 1/1. Japan (Hokkaido and Honshu).

Order 21. Papaverales

1. Papaveraceae. 25/250. Mainly in temperate and subtrop. parts of the Northern Hemisphere; best developed in the Mediterranean, Irano-Turanian, and Eastern Asiatic regions, and in the southwestern United States.
2. Hypecoaceae (*Hypecoum*). 1/15. From the western Mediterranean to Mongolia and northern China.
3. Fumariaceae. 15/470. Mainly in north-temperate regions, but a few species of *Corydalis* occur on mountains in East Africa, and a few small genera in South Africa.

Superorder Paeonianae

Order 22. Paeoniales

1. Paeoniaceae (*Paeonia*). 1/35. Mediterranean Region, Europe, subtrop., temperate, and partly cold regions of Asia, and western North America (only 2 species).

Subclass C. Caryophyllidae

Superorder Caryophyllanae

Order 23. Caryophyllales

1. Phytolaccaceae (incl. Agdestidaceae and Petiveriaceae). 16/75. Trop. and subtrop. regions, mostly in America. *Anisomeria* (2) and *Ercilla* (1) are endemic to Chile, *Monococ-*

cus (1) is endemic to Australia and New Caledonia, and *Lophiocarpus* (4) is endemic to South Africa.
2. Achatocarpaceae (*Achatocarpus* and *Phaulothamnus*). 2/10. From Texas and northwestern Mexico to Paraguay and Argentina.
3. Barbeuiaceae (*Barbeuia*). 1/1. Madagascar.
4. Nyctaginaceae. 30/300. Trop. and subtrop., mainly in America. *Phaeoptilum* (1) is endemic to southwestern Africa.
5. Aizoaceae (incl. Mesembryanthemaceae). 130/2000. Drier parts of trop. and subtrop. regions, but centered mainly in South and southwestern Africa. There is also a small center in western and southern Australia.
6. Tetragoniaceae (*Tetragonia*). 1/60. Mostly in the Southern Hemisphere, especially South Africa.
7. Molluginaceae. 13/100. Trop. and subtrop. regions, especially in South Africa. *Macarthuria* (4) is endemic to southwestern and southeastern Australia.
8. Stegnospermataceae (*Stegnosperma*). 1/3. Baja California to Central America and the West Indies.
9. Portulacaceae. 19/575. Nearly cosmop., but centered in America (western North America and the Andes) and in South Africa.
10. Hectorellaceae (*Hectorella* and *Lyallia*). 2/2. Southern New Zealand and Kerguelen Islands.
11. Basellaceae. 4/20. Mostly in trop. and subtrop. America, but also in trop. Africa, Madagascar, India, New Guinea, and some Pacific islands.
12. Halophytaceae (*Halophytum*). 1/1. Patagonia.
13. Cactaceae. 105 (50–220)/2200. Mostly drier parts of North Central and South America. *Rhipsalis* is also in Africa (from Sierra Leone and Ethiopia to Angola and southeastern Africa), Madagascar and adjacent islands, the Mascarenes, Seychelles, and Sri Lanka.
14. Didiereaceae. 4/11. Madagascar.
15. Caryophyllaceae. 80/2000 or more. Cosmop., but mainly in temperate regions of the Northern Hemisphere; best developed in the Mediterranean and Irano-Turanian regions.
16. Amaranthaceae. 65/900. Widespread in trop., subtrop., and temperate regions; trop. species centered in Africa and America.
17. Chenopodiaceae. 105/1600. Widespread in subtrop. and temperate regions, especially in saline habitats, particularly around the Mediterranean and Red Sea, in the Irano-Tura-

nian Region, South Africa, alkaline prairies of the United States, the Pampas of Argentina, and deserts of Australia.

Order 24. Polygonales

1. Polygonaceae. 45/1100. Widespread, but chiefly in northern temperate regions.

Superorder Plumbaginanae

Order 25. Plumbaginales

1. Plumbaginaceae (incl. Limoniaceae). 20/775. Nearly cosmop., particularly in dry and saline habitats. Best developed in the Mediterranean and Irano-Turanian regions.

Subclass D. Hamamelididae

Superorder Trochodendranae

Order 26. Trochodendrales

1. Trochodendraceae (*Trochodendron*). 1/1. From Korean Peninsula and Japan to Taiwan.
2. Tetracentraceae (*Tetracentron*). 1/1. Eastern Himalayas (eastern Nepal to NEFA), northern Burma, and southwestern and central China.

Order 27. Cercidiphyllales

1. Cercidiphyllaceae (*Cercidiphyllum*). 1/2–3. Kunashir Island, Japan, continental China.

Order 28. Eupteleales

1. Eupteleaceae (*Euptelea*). 1/2. Eastern Himalayas, Assam, Mishmi Hills, southwestern and central China, and Japan.

Superorder Eucommianae

Order 29. Eucommiales

1. Eucommiaceae (*Eucommia*). 1/1. Continental China.

Superorder Hamamelidanae

Order 30. Hamamelidales

1. Hamamelidaceae (incl. Altingiaceae and Rhodoleiaceae). 28/100. Eastern North America, Central America, trop. and South Africa, Madagascar, eastern Mediterranean, southeastern Transcaucasia and northern Iran, western and eastern

Himalayas, Assam, Manipur, East and Southeast Asia, New Guinea, and northeastern Australia.
2. Platanaceae (*Platanus*). 1/10. From Balkan Peninsula to western Himalayas, Indochina, and America from Canada to Mexico.

Order 31. Daphniphyllales

1. Daphniphyllaceae (*Daphniphyllum*). 1/9–10. Southwestern India (Western Ghats), Sri Lanka, Himalayas (Mussoorie to Bhutan), southern Tibet, Assam, northern Burma, East and Southeast Asia.

Order 32. Balanopales

1. Balanopaceae (*Balanops*). 1/9. Trop. Australia, New Caledonia, New Hebrides, and Fiji.

Order 33. Didymelales

1. Didymelaceae (*Didymeles*). 1/2. Madagascar.

Order 34. Myrothamnales

1. Myrothamnaceae (*Myrothamnus*). 1/2. Trop. and South Africa and Madagascar.

Order 35. Buxales

1. Buxaceae. 4/80–100. Trop. and South Africa, Madagascar, northeastern Africa, Socotra, from the western Mediterranean to northern Iran, Sri Lanka, southern and southwestern India, Punjab, Himalayas (to Afghanistan on the west), Assam, Burma, East and Southeast Asia, southeastern United States, the West Indies, and Central America.
2. Styloceratacae (*Styloceras*). 1/3. Andes of Colombia, Equador, Peru, and Bolivia.
3. Simmondsiaceae (*Simmondsia*). 1/1. Southernmost California and adjacent Sonora, western Arizona, and northern Baja California.

Order 36. Casuarinales

1. Casuarinaceae (*Casuarina*). 1/65. Mainly Australia and Tasmania, but some species occur on Pacific islands (especially New Caledonia) and in Malesia, and some in continental Southeast Asia and as far north as Burma.

Order 37. Fagales

Suborder Fagineae

1. Fagaceae. 7–8/900. Widely distributed in trop., subtrop., and temperate regions, except trop. and South Africa and a greater part of South America.

Suborder Betulineae

2. Betulaceae (incl. Carpinaceae and Corylaceae). 6/150. Widely distributed in temperate regions of the Northern Hemisphere, but some species of *Alnus* extend to Chile and Argentina.

Superorder Juglandanae

Order 38. Myricales

1. Myricaceae. 3/50. The largest genus, *Myrica*, is widespread in subtrop., temp., and cold regions, except Australia; *Comptonia* (1) is endemic to eastern North America; and *Canacomyrica* (1) is endemic to New Caledonia.

Order 39. Juglandales

1. Rhoipteleaceae (*Rhoiptelea*). 1/1. Southwestern China and northern Vietnam.
2. Juglandaceae. 8/58. Widespread in subtrop. and temperate regions of the Northern Hemisphere, with a few species extending into western South America, Malesia, and New Guinea.

Subclass E. Dilleniidae

Superorder Dillenianae

Order 40. Dilleniales

1. Dilleniaceae. 11/350. Trop. and subtrop. regions, but centered in Asia and Australasia, especially in Australia.

Superorder Theanae

Order 41. Actinidiales

1. Actinidiaceae (incl. Saurauiaceae) (*Saurauia, Actinidia, Clematoclethra*). 3/300. Trop. and subtrop. regions, but some species of *Actinidia* extend to the temperate parts of East Asia.

Order 42. Paracryphiales

1. Paracryphiaceae (*Paracryphia*). 1/1. New Caledonia.

Order 43. Theales

1. Stachyuraceae (*Stachyurus*). 1/16. Himalayas (Nepal to Bhutan), southern Tibet, Assam, northern Burma, continental China, Taiwan, Japan, and northern Indochina.
2. Theaceae (incl. Sladeniaceae and Ternstroemiaceae). 28/550. Trop. and subtrop. regions, but some species in temperate areas of East Asia and eastern North America.
3. Oncothecaceae (*Oncotheca*). 1/2. New Caledonia.
4. Marcgraviaceae. 5/100. Trop. America.
5. Pentaphylacaceae (*Pentaphylax*). 1/2. Southern China, Vietnam, Malay Peninsula, and northern Sumatra.
6. Tetrameristaceae. 2/4. Malayan Peninsula, Borneo and Sumatra (*Tetramerista*), and Guayana Highlands of northern South America (*Pentamerista*).
7. Symplocaceae (*Symplocos*). 1/250–300. Trop. and subtrop. regions of South, East, and Southeast Asia, Solomon Islands, New Guinea, Lord Howe Island, Australia from northeastern Queensland to the northern coast of New South Wales, New Caledonia, Fiji, and in America from the State of Washington to southern Brazil.
8. Caryocaraceae (*Anthodiscus* and *Caryocar*). 2/23. Trop. America from Costa Rica to Paraguay, but chiefly in Guayana and the Amazon Basin.
9. Asteropeiaceae (*Asteropeia*). 1/7. Madagascar.
10. Pelliceriaceae (*Pelliceria*). 1/1. Trop. America along the Pacific coast from Costa Rica to Ecuador.
11. Bonnetiaceae (incl. Kielmeyeroideae). 11/100. Trop. America, with 1 genus (*Ploiarium*, 3) in Southeast Asia, the Moluccas, and New Guinea.
12. Clusiaceae or Guttiferae (incl. Hypericaceae). 35/900. Widespread in trop., subtrop., and temperate regions, but chiefly in humid tropics.

Order 44. Medusagynales

1. Medusagynaceae (*Medusagyne*). 1/1. Seychelles.

Order 45. Ochnales

1. Ochnaceae. 32/500. Trop. and subtrop. regions, especially Southeast Asia, West Africa, and South America.

2. Lophiraceae (*Lophira*). 1/2. Trop. Africa.
3. Sauvagesiaceae (Luxemburgiaceae). 7/75. Trop. regions of Africa, Asia, New Guinea, and especially South America.
4. Diegodendraceae (*Diegodendron*). 1/1. Madagascar.
5. Strasburgeriaceae (*Strasburgeria*). 1/1. New Caledonia.
6. Quiinaceae. 4/50. Trop. America, chiefly the Amazon Basin.
7. Scytopetalaceae. 5/20. Trop. West Africa.

Order 46. Ancistrocladales

1. Ancistrocladaceae (*Ancistrocladus*). 1/20. Trop. Africa, South Asia, and continental Southeast Asia.

Order 47. Elatinales

1. Elatinaceae. 2/33. Trop., subtrop., and temperate regions. *Elatine* is mostly aquatic and almost cosmop., with the majority of species in temperate areas. *Bergia* is less often aquatic and mostly tropical and subtropical in distribution.

Superorder Lecythidanae

Order 48. Lecythidales

1. Lecythidaceae (incl. Asteranthaceae and Barringtoniaceae). 25/400. From trop. Africa, Madagascar, and the Comoro Islands to tropical Asia and eastward to the Pacific islands, from the Ryukyu Islands and Guam to New Guinea and Australia, and in trop. America.

Superorder Sarracenianae

Order 49. Sarraceniales

1. Sarraceniaceae. 3/15. North America (*Darlingtonia* and *Sarracenia*) and Guayana Highlands of northern South America (*Heliamphora*).

Superorder Ericanae

Order 50. Ericales

1. Clethraceae (*Clethra*). 1/65. East and Southeast Asia, southeastern United States, Central America, and tropical South America, with a single endemic species on Madeira Island.
2. Ericaceae (incl. Vacciniaceae, Pyrolaceae, and Monotropaceae, and excl. *Wittsteinia*). 140/3500. Widespread in subtrop., temperate, and cold regions and in trop. mountains,

with large concentrations in the Himalayas, New Guinea, and southern Africa; absent from most of Australia.

3. Epacridaceae (excl. *Wittsteinia*). 30/400. Mainly in Australia and Tasmania, but extending to New Zealand and New Caledonia, and with a few species in southern South America.
4. Cyrillaceae. 3/14. Central America, northern South America, the West Indies, and the coastal plain of the southeastern United States.
5. Empetraceae. 3/5–6. *Empetrum* (2–3) is widespread in cool temperate regions of the Northern Hemisphere and in southern South America and some nearby islands. *Corema* (2) occurs in eastern North America and southwest Europe (the Iberian Peninsula and the Azores). *Ceratiola* (1) occurs in the coastal plain of the southeastern United States.
6. Grubbiaceae (*Grubbia*). 1/3. Cape Province of South Africa.

Order 51. Diapensiales

1. Diapensiaceae. 6/20. Temperate and cold regions of the Northern Hemisphere.

Order 52. Ebenales

1. Styracaceae (excl. *Afrostyrax*). 12/150. Eastern Mediterranean (*Styrax officinale*), eastern Himalayas (Nepal to NEFA), Assam, eastern India, Burma, East and Southeast Asia and adjacent islands, and America from the United States to northern Argentina. Best developed in Asia, especially in China and Indochina.
2. Ebenaceae. 2/500. *Diospyros* is pantropical, with the greatest concentration in Malesia; only a few species in western Asia, Japan, and the southeastern United States. *Euclea* (14) is confined to East and South Africa.
3. Lissocarpaceae (*Lissocarpa*). 1/2. Trop. South America.

Order 53. Sapotales

1. Sapotaceae (incl. Sarcospermataceae). (35)50–60(75)/800. Pantropical, with a relatively few species extending into temperate regions.

Order 54. Primulales

1. Theophrastaceae. 4–5/110. America from Mexico, southern Florida, and the Bahama Islands to northern Paraguay, and also on the Hawaiian Islands.

APPENDIX

2. Myrsinaceae. 32–34/1000. Widely distributed in trop. and subtrop. regions, and extending to southern Korea, northern China, Japan, and Mexico in the Northern Hemisphere, and to New Zealand, Australia, South Africa, and northern Argentina in the Southern Hemisphere.
3. Aegicerataceae (*Aegiceras*). 1/2. In mangroves of South and Southeast Asia, New Guinea, Bismark Archipelago, Solomon Islands, Lord Howe Island, and northeastern Australia.
4. Primulaceae (incl. Coridaceae). 28–30/1000. Almost cosmop., but chiefly in temperate and cold regions of the Northern Hemisphere.

Superorder Violanae

Order 55. Violales

1. Berberidopsidaceae. 3 ?/3. According to J. F. Veldkamp (Blumea 30:21–29, 1984) *Berberidopsis* has two species—*B. carallina* Hook. f. in Chile, costal cordilleras, and *B. beckleri* (F. v. M.) Veldk. (= *Streptothamnus beckleri* F. v. M.) in eastern Australia (Queensland and New South Wales), but the Australian species differs markedly from the Chilean species (flowers solitary and auxillary, sculpturing of pollen grains striate, placentas 5, mesotesta without stone-cells, petioles not pulvinate, leaf blades heterophyllous, beneath with scattered glandular dots) and most probably deserves a generic status. A monotypic genus, *Streptothamnus,* occurs in eastern Australia, Queensland, and New South Wales.
2. Flacourtiaceae (excl. *Berberidopsis, Streptothamnus,* and *Physena*). 85–90/1250–1300. Trop. and subtrop. regions, with some species in temperate areas of East Asia, South Africa, and North and South America.
3. Lacistemataceae (*Lacistema* and *Lozania*). 2/27. West Indies and trop. America from Mexico to Paraguay.
4. Peridiscaceae (*Peridiscus* and *Whittonia*). 2/2. Trop. South America.
5. Violaceae. 22/900. Cosmop., but chiefly temperate.
6. Dipentodontaceae (*Dipentodon*). 1/1. Eastern Himalayas, northeastern India, Burma, southeastern Tibet, and southwestern China.
7. Dioncophyllaceae. 3/3. Trop. West Africa.
8. Scyphostegiaceae (*Scyphostegia*). 1/1. Borneo.
9. Passifloraceae (incl. Paropsiaceae and excl. *Physena*). 16/650–

700. Widespread in trop., subtrop., and warm-temperate regions, but best developed in trop. America and Africa.

10. Turneraceae. 9/120. Trop. and subtrop. regions of America and Africa, Madagascar, and also the Island of Rodriguez (*Mathurina*).
11. Malesherbiaceae (*Malesherbia*). 1/35. Western South America in the Andes from southern Peru to northern Chile and western Argentina. Most species grow in northern Chile.
12. Achariaceae. 3/3. South Africa.

Order 56. Caricales

1. Caricaceae. 4/30–55. Trop. and subtrop. America from Mexico and the West Indies to northern Chile and Argentina; also in trop. West Africa (*Cylicomorpha*, 2).

Order 57. Cucurbitales

1. Cucurbitaceae. 90/700. Widespread in trop. and subtrop. regions, with relatively few species in temperate and cold areas.

Order 58. Begoniales (Datiscales)

1. Datiscaceae. 3/4. *Octomeles* (1): Malay Archipelago, New Guinea, New Britain, and Solomon Islands; *Tetrameles* (1): Sri Lanka, Andaman Islands, Indochina, and Malesia; *Datisca* (2): from the eastern Mediterranean to the western Himalayas, and in western North America.
2. Begoniaceae. 5/950–1000. Widespread in trop. and subtrop. regions. Best developed in northern South America and trop. Asia, and absent in Australia. *Hillebrandia* (1): Hawaiian Islands; *Semibegoniella* (2): Ecuador; *Begoniella* (5): Colombia; and *Symbegonia* (13): New Guinea. A few species of *Begonia* extend to northern China.

Order 59. Salicales

1. Salicaceae. 3/400. Mostly in temperate and cold regions of the Northern Hemisphere, but extending to southern South America, South Africa, and Southeast Asia; lacking in New Guinea and Australia. The monotypic genus *Chosenia* occurs in eastern Siberia, Sakhalin, northern Japan, and northeastern China.

Order 60. Tamaricales

1. Tamaricaceae. 3/85–90. Africa, Europe, and Asia, mainly in the Mediterranean and Irano-Turanian regions, but also in Norway, southwestern, northwestern, and northeastern Africa, the Arabian Peninsula, India, Sri Lanka, Burma, China, Korea, and Japan.
2. Frankeniaceae. 4/90. Saline habitats in America, Africa, Eurasia, and Australia, but centered in the Mediterranian Region and in western Asia.

Order 61. Fouquieriales

1. Fouquieriaceae (*Fouquiria*)]. 1/11. Arid parts of the southwestern United States and Mexico.

Order 62. Capparales

1. Capparaceae (incl. Cleomaceae, Oxystylidaceae, Pentadiplandraceae, Koeberliniaceae, and *Oceanopapaver;* excl. *Canotia, Emblingia,* and *Physena*). 45/850. Widespread in trop., subtrop., and warm-temperate regions. The monotypic genus *Koeberlinia* (southern U.S. and Mexico) is rather isolated within the family.
2. Brassicaceae or Cruciferae. 376–380/3200. Cosmop., but mainly in temperate and cold regions, with the greatest concentration of genera and species in the Mediterranean and Irano- Turanian regions and a lesser concentration in western North America, temperate South America, South Africa, and Australia.
3. Tovariaceae (*Tovaria*). 1/2. Trop. America and the West Indies.
4. Resedaceae. 6/75. Centered in the Mediterranean Region, extending into Canary Islands, parts of northern Europe, the Arabian Peninsula, western and Central Asia, and northwestern India; a few species occur in the southwestern United States, northern Mexico, and South Africa.

Order 63. Batales

1. Gyrostemonaceae. 5/16. Australia.
2. Bataceae (*Batis*). 1/2. Coasts of trop. and subtrop. America and also in the Galapagos, the Hawaiian Islands, New Guinea, and northeastern Australia.

Order 64. Moringales

1. Moringaceae (*Moringa*). 1/14. From North and northeastern Africa to the Arabian Peninsula and across southern Iran to India; also in southwestern Africa and Madagascar.

Superorder Malvanae

Order 65. Bixales (Cistales)

1. Bixaceae (*Bixa*). 1/3–4. Trop. America and the West Indies.
2. Cochlospermaceae. 2/20–38. *Cochlospermum* (15–30): trop. and subtrop. America, trop. Africa, and from Southeast Asia to northern Australia; *Amoreuxia* (4–7): America from the southern United States to Peru.
3. Cistaceae. 8/200. Mostly in north-temperate and warm temperate regions; especially abundant in the Mediterranean Region and eastern United States. A few species occur in the West Indies and in South America.

Order 66. Malvales

1. Elaeocarpaceae (excl. *Muntingia*). 9/390. Madagascar and Mauritius, Socotra, eastern Himalayas, India, Burma, East and Southeast Asia, Moluccas, New Guinea, New Caledonia, New Hebrides, Fiji, Samoa, Tonga, and some other Pacific islands, eastern Australia and Tasmania, New Zealand, from Mexico to trop. South America, the West Indies, and temperate South America from Peru to Chile.
2. Plagiopteraceae (*Plagiopteron*). 1/1. Lower Burma.
3. Tiliaceae (incl. *Goethalsia* and *Muntingia*). 50/450. Widely distributed throughout the trop. regions, especially in South America, Africa, and Southeast Asia. The genus *Tilia* extends into the temperate regions of the Northern Hemisphere.
4. Dipterocarpaceae. 15/570. Mainly tropical regions of the Old World, centered in the rain forests of Malesia. There are only 2 genera (*Marquesia* and *Monotes*) in trop. Africa, and 1 monotypic genus (*Pakaraimaea*) in northern South America (Guayana Highlands).
5. Sarcolaenaceae. 8/39. Eastern Madagascar.
6. Sphaerosepalaceae (*Rhopalocarpus* and *Dialyceras*). 2/14. Madagascar.
7. Sterculiaceae (incl. Byttneriaceae and the genus *Maxwellia*). 60–65/1000. Pantropical, extending into subtropical and temperate regions.
8. Huaceae (*Hua* and *Afrostyrax*). 2/3. Trop. Africa.

9. Bombacaceae. 25–30/200. Pantropical, centered in the rain forests of South America, especially in Brazil.
10. Malvaceae. 75–85/1500–1600. Cosmop., but best developed in trop. regions.

Superorder Urticanae

Order 67. Urticales

Suborder Ulmineae

1. Ulmaceae (incl. Celtidaceae). 15/200. Widely distributed in trop. and temperate regions, especially in the Northern Hemisphere.

Suborder Urticineae

2. Moraceae. 53/1400 or more. Widely distributed in trop. and subtrop. regions, but some species occur in temperate regions of both hemispheres.
3. Cannabaceae. 2/3. Temperate regions of the Northern Hemisphere.
4. Cecropiaceae. 6/200. Mainly trop. America, but also trop. Africa (*Musanga* and *Myrianthus*) and trop. Asia (*Poikilospermum*).
5. Urticaceae. 45/850 or more. Widely distributed in trop. and subtrop. regions, but some species occur in temperate and cold regions.

Order 68. Barbeyales

1. Barbeyaceae (*Barbeya*). 1/1. Northeastern Africa and adjacent part of southwestern Arabia.

Superorder Euphorbianae

Order 69. Euphorbiales

1. Euphorbiaceae (incl. *Picrodendron*). 300/7500. Nearly cosmop., except the arctic regions, but best developed in trop. and subtrop. regions.
2. Pandaceae. 4/28. Trop. Africa, trop. Asia, and New Guinea.
3. Dichapetalaceae. 4/200. Pantropical, but *Dichapetalum* is extending to South Africa.
4. Aextoxicaceae (*Aextoxicon*). 1/1. Chile.

Order 70. Thymelaeales

1. Thymelaeaceae. 50/500. Nearly cosmop., but best developed

in trop. Africa and Australia, with lesser centers in the Mediterranean Region and in western, East, and Southeast Asia.

Subclass Rosidae

Superorder Rosanae

Order 71. Cunoniales

1. Cunoniaceae. 24/350 (excl. *Aphanopetalum*?). Chiefly temperate and subtrop. regions of the Southern Hemisphere between 13° and 35° S, mainly Australia, New Caledonia, and New Guinea; a few (*Weinmannia* spp.) north to the Philippines, southern Mexico, and the West Indies.
2. Baueraceae (*Bauera*). 1/3. Temperate eastern Australia and Tasmania.
3. Eucryphiaceae (*Eucryphia*). 1/16. Southeastern Australia, Tasmania, and Chile.
4. Davidsoniaceae (*Davidsonia*). 1/1. Northeastern Australia (Queensland and New South Wales).
5. Brunelliaceae (*Brunellia*). 1/50. Trop. America (from southern Mexico, Costa Rica, and the West Indies to Venezuela and the Andes of Colombia, Bolivia, and Peru).

Order 72. Bruniales

1. Bruniaceae. 12/75. South Africa, almost entirely the Cape Province, with only a single species in Natal.

Order 73. Geissolomatales

1. Geissolomataceae (*Geissoloma*). 1/1. Southwestern Cape Province of South Africa.

Order 74. Saxifragales

1. Penthoraceae (*Penthorum*). 1/3. East Asia, Indochina, and Atlantic North America.
2. Crassulaceae. 30/1500. Widely distributed (except Polynesia and Australia), mostly in South Africa.
3. Cephalotaceae (*Cephalotus*). 1/1. Southwestern Australia.
4. Saxifragaceae. 30/600. Mostly north-temperate regions, with a few species in south-temperate regions and in trop. mountains.
5. Grossulariaceae (*Ribes*, incl. *Grossularia*). 1/150. Temperate

Eurasia, northwestern Africa, North and Central America, and Pacific South America to Fuegia.
6. Greyiaceae (*Greyia*). 1/3. Southeastern South Africa.
? 7. Rousseaceae (*Roussea*). 1/1. Mauritius.
8. Vahliaceae (*Bistella*). 1/4. Trop. and South Africa, southwestern Asia, and northwestern India.
9. Eremosynaceae (*Eremosyne*). 1/1. Southwestern Australia.
10. Francoaceae (*Francoa* and *Tetilla*). 2/2. Chile.
11. Parnassiaceae (*Parnassia*). 1/50. Temperate and cold regions of the Northern Hemisphere, especially in East Asia and northwestern North America.
12. Lepuropetalaceae (*Lepuropetalon*). 1/1. From the southeastern United States and Mexico to Chile.

Order 75. Droserales

1. Droseraceae. 4/83. Cosmop., with concentration in Australia and New Zealand. *Drosophyllum* is endemic to the Iberian Peninsula.

Order 76. Rosales

1. Rosaceae (subfamilies Spiraeoideae, Quillajeoideae, Rosoideae, Dichotomanthoideae, Maloideae, Prunoideae, Prinsepioideae). 110–120/3000–3350. Cosmop., but best developed in temperate and subtrop. regions of the Northern Hemisphere.
2. Chrysobalanaceae. 14/480. Throughout the lowland tropics and both hemispheres, but mostly in Central and South America and the West Indies, extending into subtrop. regions of North America, South Africa, and East Asia.
3. Neuradaceae. 3/10. Arid regions of North Africa, Arabian Peninsula, Syria, Irak, northern Iran, Afghanistan, Pakistan, and India (*Neurada*), and South and southwestern Africa (*Grielum* and *Neuradopsis*).

Order 77. Crossosomatales

1. Crossosomataceae (*Crossosoma, Forsellesia, Apacheria*). 3/10–11. Arid parts of the western United States and adjacent Mexico.

Order 78. Podostemales

1. Podostemaceae (incl. Tristichaceae). 45–50/130–200. Pantropical, especially in Asia and America.

Superorder Myrtanae

Order 79. Rhizophorales

1. Anisophylleaceae (incl. Polygonanthaceae). 4/36. Mostly in trop. Africa and Asia, with 1 species of *Anisophyllea* and the genus *Polygonanthus* (2) in trop. South America.
2. Rhizophoraceae. 13/130. Pantropical, mostly in the Old World.

Order 80. Myrtales

1. Crypteroniaceae (*Axinandra, Crypteronia, Dactylocladus*). 3/10. India (Assam and Bengal), Sri Lanka, Andaman Islands, Malay Peninsula, Philippine Islands, and the Malay Archipelago.
2. Lythraceae (excl. *Alzatea* and *Rhynchocalyx*). 25/500–550. Mainly trop. and subtrop., especially in America; relatively few species in temperate and cold regions.
3. Punicaceae (*Punica*). 1/2. From the Balkan Peninsula to the western Himalayas (*P. granatum*) and on Socotra (*P. protopunica*).
4. Rhynchocalycaceae (*Rhynchocalyx*). 1/1. South Africa.
5. Duabangaceae.* 1/2. Southeast Asia and New Guinea.
6. Sonneratiaceae (*Sonneratia*). 1/5. From coasts of East Africa and Madagascar to Micronesia, New Hebrides and the Solomons, northern Australia and New Caledonia.
7. Alzateaceae (*Alzatea*). 1/1. Trop. America.
8. Onagraceae. 17/650–700. Widespread in trop., subtrop., and temperate regions, especially in the New World, where it is centered in southwestern North America.
9. Combretaceae (incl. Strephonemataceae?). 20/475–500. Mainly trop. regions, with only a few species extending into subtrop. areas.
10. Oliniaceae (*Olinia*). 1/10. Trop. East Africa and South Africa, with 1 species on St. Helena Island.
11. Penaeaceae. 7/25. Cape Province of South Africa.
12. Psiloxylaceae (*Psiloxylon*). 1/1. Mauritius, Réunion.
13. Myrtaceae (incl. Heteropyxidaceae and Kaniaceae). 144/3000. Trop. and subtrop. regions, centered in America and eastern and southwestern Australia.

*Duabangaceae, fam. nov., affinis Sonneratiaceae, sed petala obovato-rotundata, undulatocrispata, antherae recurvae, grana pollinis triporata, stylus crassus, stigma capitatum, 4–8-lobum, capsula loculicide 4–8-valvis, testa utrinque producta, parenchyma ligni paratracheale, radii heterocellulares. Typus: *Duabanga* Buch.-Ham.

14. Melastomataceae (incl. Memecylaceae). 245/3500–4000. Mainly trop. and subtrop. regions, centered in South America.

Order 81. Trapales

1. Trapaceae (*Trapa*). 1/1, or 3, or 15–30. Trop., subtrop., and temperate regions of the Old World, except Australia.

Order 82. Haloragales

1. Haloragaceae (incl. Myriophyllaceae and excl. Gunneraceae and Hippuridaceae). 8/100. Widely distributed, but best developed in the Southern Hemisphere, especially in Australia.

Superorder Fabanae

Order 83. Fabales

1. Fabaceae or Leguminosae (incl. Caesalpiniaceae and Mimosaceae). 630/18,000. Cosmop.

Superorder Rutanae

Order 84. Connarales

1. Connaraceae. 17/350. Pantropical; best developed in Africa, Southeast Asia, and trop. America.

Order 85. Rutales

Suborder Rutineae

1. Rutaceae (incl. Flindersiaceae). 150/1500–1600. Widely distributed in trop., subtrop., and warm-temperate regions, especially in South Africa and Australia.
2. Rhabdodendraceae (*Rhabdodendron*). 1/3. Trop. South America.
3. Tetradiclidaceae.* 1/1. North Africa and the Irano-Turanian Region.
4. Cneoraceae. 2/3. Cuba, Canary Islands, and western and central Mediterranean. *Neochamaelea* (1) is endemic to the Canary Islands.
5. Simaroubaceae (incl. Irvingiaceae). 28–30/210. Trop. and subtrop., with a few species in warm-temperate regions.
6. Surianaceae (*Suriana*). 1/1. Trop. coasts.
7. Zygophyllaceae (excl. *Balanites, Nitraria, Peganum,* and

---

*Tetradiclidaceae, fam. nov. (subfam. Tetradiclidoideae Engler, in Engler and Prantl, Die natürlichen Pflanzenfamilien III, 4:355, 1896). Typus: *Tetradiclis* Steven.

Tetradiclis). 22/220. Mostly trop. and subtrop.; fewer in warm temperate regions.
8. Nitrariaceae (*Nitraria*). 1/10. From North Africa and southeastern Europe to Central Asia and southwestern Australia.
9. Balanitaceae (*Balanites*). 1/25. Trop. and North Africa, western Asia to India and Burma.
10. Peganaceae (*Peganum* and *Malacocarpus*). 2/6. Southern Europe, North Africa, arid and semiarid parts of Asia and North America (southwestern Texas and northern Mexico).
11. Meliaceae (incl. Aitoniaceae). 52/550. Trop. and subtrop.; a few species in warm temperate regions.
12. Kirkiaceae (*Kirkia*). 1/8. Trop. and South Africa.
13. Tepuianthaceae (*Tepuianthus*). 1/5. Northern South America (incl. Guayana Highlands).
14. Ptaeroxylaceae. 2/4–5. South Africa (*Ptaeroxylon*, 1) and Madagascar (*Cedrelopsis*, 3–4).
15. Burseraceae. 17/500–550. Pantropical, especially in trop. America, northeastern Africa, and Malesia.
16. Anacardiaceae (incl. *Blepharocarya* and Julianiaceae). 77–79/600. Mainly trop. and subtrop. regions, but with some species in warm temperate regions.
17. Podoaceae (Dobineaceae) (*Dobinea* and *Campylopetalum*). 2/3. Central and eastern Himalayas, northeastern India (Naga and Mishmi Hills), southwestern China, and Thailand (*Campylopetalum*).

Order 86. Leitneriales

1. Leitneriaceae (*Leitneria*). 1/1. Coastal plain of the southeastern United States, from southern Missouri to Texas and Florida.

Order 87. Coriariales

1. Coriariaceae (*Coriaria*). 1/20. Western Mediterranean, temperate and subtrop. Himalayas to Japan and New Guinea, New Zealand, the South Pacific Islands, and America from Mexico to Chile.

Order 88. Sapindales (Acerales)

1. Staphyleaceae. 5/60. North-temperate regions, trop. Asia, West Indies, Central and South America. *Tapiscia* (1) is endemic to China.

2. Sapindaceae. 150/2000. Trop. and subtrop. regions; relatively few species in warm temperate regions.
3. Aceraceae (*Acer* and *Dipteronia*). 2/120 (110–160). North-temperate regions and Southeast Asia. *Dipteronia* (2) is endemic to continental China.
4. Hippocastanaceae. 2/15. Balkan Peninsula, from western Himalayas to Japan and northern Indochina, temperate North America (*Aesculus*), and from Mexico to northern South America (*Billia*).
5. Stylobasiaceae (*Stylobasium*). 1/2. Southwestern Australia in dry sandy areas.
6. Emblingiaceae (*Emblingia*). 1/1. Western and southwestern Australia.
7. Bretschneideraceae (*Bretschneidera*). 1/1. Western and southwestern China.
8. Melianthaceae (*Bersama* and *Melianthus*). 2/15. Trop. and South Africa.
9. Akaniaceae (*Akania*). 1/1. Eastern Australia from the southern coast of Queensland to northeastern New South Wales.
10. Sabiaceae (incl. Meliosmaceae). 3/160. Peninsular India, Himalayas, East Asia (north to Korea and Japan), Southeast Asia, Solomon Islands, and trop. America (north to Mexico).
11. Physenaceae (*Physena*). 1/2. Madagascar.

Order 89. Linales

1. Hugoniaceae. 6/60. From trop. Africa and Madagascar to New Caledonia and trop. South America. *Indorouchera* is confined to Indochina and western Malesia; *Rouchera* occurs in trop. South America.
2. Ctenolophonaceae (*Ctenolophon*). 1/3. Trop. Africa and western Malesia.
3. Linaceae. 6/250. Widely distributed, especially in temperate and subtrop. regions.
4. Ixonanthaceae (*Allantospermum, Cyrillopsis, Ixonanthes, Ochtocosmus, Phyllocosmus*). 5/35. Trop. Africa, Madagascar, Himalayas, northeastern India, southern China, Southeast Asia, New Guinea, and trop. America.
5. Humiriaceae. 8/50. Trop. South America (north to Costa Rica) with 1 species of *Sacoglottis* in trop. West Africa.
6. Erythroxylaceae (incl. Nectaropetalaceae; *Aneulophus, Erythroxylum, Nectaropetalum, Pinacopodium*). 4/260. Pantropical, but most abundant in the Andes, in the Amazon Basin of South America, and in Madagascar.

Order 90. Geraniales

1. Oxalidaceae (incl. Averrhoaceae and Lepidobotryaceae). 7/900. Mostly trop. and subtrop. regions, with a few species widespread in temperate regions.
2. Hypseocharitaceae (*Hypseocharis*). 1/8. Andes of South America.
3. Geraniaceae. 5/750. Almost cosmop., but mainly in temperate and subtrop. regions.
4. Vivianiaceae (*Viviania*). 1/30. Southern Brazil and Chile.
5. Rhynchothecaceae (*Rhynchotheca*). 1/1. Andes of South America.
6. Ledocarpaceae (*Balbisia* and *Wendtia*). 2/11. Western South America, mostly in the Andes.
7. Biebersteiniaceae (*Biebersteinia*). 1/4–5. From Greece to western Tibet.
8. Dirachmaceae (*Dirachma*). 1/1. Socotra.

Order 91. Balsaminales

1. Balsaminaceae. 4/600. Mainly trop. Asia and Africa, with a few species in temperate regions of Eurasia, Africa, and North America.

Order 92. Tropaeolales

1. Tropaeolaceae. 3/92. Mountains of America from Mexico to central Chile and Argentina (*Magellana* and *Trophaeastrum* are endemic to Patagonia).

Order 93. Limnanthales

1. Limnanthaceae (*Floerkea* and *Limnanthes*). 2/11. North America, chiefly California.

Order 94. Polygalales

1. Malpighiaceae. 60/1200. Trop. and subtrop. regions, especially in South America.
2. Trigoniaceae. 4/35. Trop. America (*Trigonia* and *Euphronia*), Madagascar (*Humbertiodendron*), and western Malesia (*Trigoniastrum*).
3. Vochysiaceae. 6/200. Mostly trop. America and the West Indies, but *Erismadelphus* is in trop. West Africa.
4. Polygalaceae (incl. Diclidantheraceae and Xanthophyllaceae). 15/900. Nearly cosmop., except for the Arctic, Polynesia, and New Zealand.

5. Krameriaceae (*Krameria*). 1/15. From the southwestern United States to Argentina and Chile.
6. Tremandraceae. 3/43. Australia (especially western Australia) and Tasmania.

Superorder Celastranae

Order 95. Celastrales

1. Aquifoliaceae (*Ilex*, incl. *Byronia*, and *Nemopanthus*). 2/400. *Ilex* is nearly cosmop.; *Nemopanthus* is found only in eastern North America.
2. Phellinaceae (*Phelline*). 1/10. New Caledonia.
3. Icacinaceae (incl. Phytocrenaceae). 58/400. Pantropical, with relatively few species in temperate regions.
4. Sphenostemonaceae (*Sphenostemon*). 1/7. Eastern Malesia (Sulawesi, Moluccas, New Guinea), Queensland, and New Caledonia.
5. Cardiopteridaceae (*Peripterygium*). 1/3. Northeastern India (Assam and Bengal), Burma, Southeast Asia, New Guinea, and Queensland.
6. Celastraceae (incl. Canotiaceae, Goupiaceae, Hippocrateaceae, Siphonodontaceae, Tripterygiaceae, and ? *Pottingeria*; excl. *Forsellesia*). 57/860. Widely distributed, but chiefly in trop. and subtrop. regions.
7. Lophopyxidaceae (*Lophopyxis*). 1/2. Malay Peninsula, Borneo, eastern Malesia, Palau and Solomon islands.
8. Stackhousiaceae (*Macgregoria*, *Stackhousia* and *Tripterococcus*). 3/27. Malesia, Micronesia, New Guinea, Australia, Tasmania, and New Zealand.
9. Salvadoraceae (*Azima*, *Dobera*, and *Salvadora*). 3/12. From Africa and Madagascar to Pakistan, India, Sri Lanka, Burma, and Southeast Asia.
10. Corynocarpaceae (*Corynocarpus*). 1/4–5. Aru Islands, New Guinea, Polynesia, Queensland, New Hebrides, New Caledonia, and New Zealand.

Order 96. Santalales

1. Olacaceae (incl. Aptandraceae, Erythropalaceae, Octoknemaceae, and Schoepfiaceae). 25/250. Trop. and subtrop. regions of Africa, Asia, Australia, and South America.
2. Opiliaceae. 8/50. Widespread in trop. and subtrop. regions, especially in Asia. *Agonandra* (12) is the only American genus (from Mexico to Argentina).

3. Medusandraceae (*Medusandra*). 1/1. Equatorial West Africa.
4. Santalaceae. 35/400. Nearly cosmop., but mostly trop. and subtrop. regions. Best developed in relatively dry areas.
5. Eremolepidaceae (*Antidaphne, Eremolepis, Eubrachion*). 3/11. Trop. America and West Indies.
6. Misodendraceae (*Misodendrum*). 1/11. Temperate South America from the Straits of Magellan north to about 33° south latitude.
7. Loranthaceae. 65/850. Mainly trop. and subtrop. regions, especially in the Southern Hemisphere.
8. Viscaceae. 8/450. Cosmop., but mostly in trop. and subtrop. regions.

Superorder Proteanae

Order 97. Proteales

1. Proteaceae. 75/1050. Trop. and South Africa, Madagascar, trop. Asia, Malesia, Australia, Tasmania, New Hebrides, New Caledonia and other Pacific Islands, New Zealand, and Central and South America south to Chile.

Superorder Rhamnanae

Order 98. Rhamnales

1. Rhamnaceae. 58/900. Nearly cosmop., but mostly in trop. and subtrop. regions.

Order 99. Elaeagnales

1. Elaeagnaceae (*Elaeagnus, Hippophaë,* and *Shepherdia*). 3/50. Subtrop. and temperate regions of the Northern Hemisphere and in trop. Asia, with 1 species of *Elaeagnus* in Queensland.

Superorder Vitanae

Order 100. Vitales

1. Vitaceae. 12/700. Trop. and subtrop. regions, with relatively few species in temperate regions (mostly species of *Vitis, Ampelopsis,* and *Parthenocissus*).
2. Leeaceae (*Leea*). 1/70. Paleotropical.

Superorder Cornanae

Order 101. Hydrangeales

1. Tetracarpaeaceae (*Tetracarpaea*). 1/1. Tasmania.
2. Escalloniaceae (incl. *Abrophyllum,* ? *Brexia* ? *Carpodetus,*

Corokia, ? *Ixerba*, and ? *Polyosma;* excl. *Pottingeria* ?). 11–14/185. The Southern Hemisphere, mostly South America and Australia.
3. Iteaceae (*Itea* and *Choristylis*). 2/22. East and South Africa, Himalayas, Assam, East and Southeast Asia, North America.
4. Dulongiaceae (Phyllonomaceae; *Phyllonoma*). 1/8. Trop. America from Mexico to Peru.
5. Tribelaceae (*Tribeles*). 1/1. Temperate South America.
6. Hydrangeaceae (incl. Philadelphaceae). 16/250. Temperate and subtrop. regions of the Northern Hemisphere, mostly East Asia (especially China) and North America; a few species in Southeast Asia and in the Andes, from Mexico to southern Chile.
7. Kirengeshomaceae (*Kirengeshoma*). 1/1. Japan, Korea, and continental China.
8. Pterostemonaceae (*Pterostemon*). 1/2. Mexico.
9. Griseliniaceae (*Griselinia*). 1/6. New Zealand, Chile, Paraguay, and southeastern Brazil.
10. Melanophyllaceae (*Melanophylla*). 1/8. Madagascar.
11. Montiniaceae (*Grevea, Montinia, Kaliphora*). 3/5. East and South Africa and Madagascar; *Kaliphora* is endemic to Madagascar.
12. Columelliaceae (*Columellia*). 1/4. Andes from Colombia to Bolivia.
13. Desfontainiaceae (*Desfontainia*). 1/1. Andes from Costa Rica to Cabo de Hornos.
14. Alseuosmiaceae (*Alseuosmia, Crispiloba,* and *Wittsteinia*). 3/12 or more. New Guinea, Australia, New Zealand, and New Caledonia. A monotypic genus, *Crispiloba* (van Steenis 1984), is endemic to Queensland.
15. Roridulaceae. 1/2. Cape Province of South Africa.

Order 102. Aralidiales

1. Aralidiaceae (*Aralidium*). 1/1. Southern Thailand, peninsular Malaysia, Singapore, Sumatra, Anambus Islands, and Borneo.

Order 103. Cornales

1. Davidiaceae (*Davidia*). 1/1. Southwestern and central China.
2. Nyssaceae (*Camptotheca* and *Nyssa*). 2/6–7. Eastern Himalayas, Assam (Khasi Hills), China, Southeast Asia, eastern North America.
3. Cornaceae (*Swida* and *Cornus;* the latter often separated into

several segregates, such as *Afrocrania, Chamaepericlimenum, Benthamidia,* and *Dendrobenthamia*). 2–6/55. North-temperate regions and the Arctic, southern China, Central America, Peru, Bolivia, and trop. East Africa (only *Afrocrania*).
4. Curtisiaceae (*Curtisia*). 1/1. South Africa and southeastern trop. Africa.
5. Mastixiaceae (*Mastixia*). 1/13. India, eastern Himalayas, Sri Lanka, southwestern and southern China, Indochina, Malesia, Solomon Islands, and New Guinea.
6. Aucubaceae (*Aucuba*). 1/6. Eastern Himalayas, northern Burma, continental China, Taiwan, Ryukyu Islands, southern Korea, and Japan.
7. Garryaceae (*Garrya*). 1/13. North America from Washington to Panama, with 1 species in the Greater Antilles.
8. Alangiaceae (*Alangium*). 1/20. Trop. Africa, Madagascar, East and trop. Asia, New Guinea and adjacent islands, eastern Australia, New Caledonia, and Fiji.

Order 104. Helwingiales

1. Helwingiaceae (*Helwingia*). 1/4–5. Temperate eastern Himalayas (central Nepal to Bhutan), northeastern India (Assam and Manipur), northern Burma, continental China, northern Vietnam, Taiwan, Ryukyu Islands, and Japan.

Order 105. Toricelliales

1. Toricelliaceae (*Toricellia*). 1/3. Eastern Himalayas (western Nepal to Bhutan), northern Burma, and continental China.

Order 106. Apiales (Araliales)

1. Araliaceae. 65–70/750–800. Trop. and subtrop. regions, with relatively few species in temperate areas.
2. Apiaceae or Umbelliferae. 300/2500–3000. Cosmop., but chiefly in north-temperate regions.

Order 107. Pittosporales

1. Pittosporaceae. 9/200 or more. Trop. and subtrop. regions of the Old World, mostly Australia.
2. Byblidaceae (*Byblis*). 1/2. Western and northern Australia and southernmost New Guinea.

Order 108. Gunnerales

1. Gunneraceae (*Gunnera*). 1/50. Trop. and South Africa,

APPENDIX

Madagascar, Malesia (excl. the Malay Peninsula), New Guinea, Tasmania, New Zealand, Hawaii, Mexico to Chile, and the Juan Fernandez Islands.

Order 109. Dipsacales

1. Caprifoliaceae (incl. Carlemanniaceae). 13/260. Mostly north-temperate and boreal regions, with a few species in trop. mountains; best developed in East Asia.
2. Viburnaceae (*Viburnum*). 1/225. Temperate and subtrop. regions of the Northern Hemisphere, especially East Asia and eastern North America.
3. Sambucaceae (*Sambucus*). 1/20–40. Nearly cosmop., but only 1 species in Africa (mountains of East Africa).
4. Adoxaceae (*Adoxa, Sinadoxa,* and *Tetradoxa*). 3/3. North-temperate regions (*Sinadoxa* and *Tetradoxa* only in continental China).
5. Valerianaceae. 12/400. Nearly cosmop., but best developed in north-temperate regions (especially in the Mediterranean Region and in western Asia), and in the Andes of South America, being absent from Australia, New Zealand, and much of Africa.
6. Triplostegiaceae (*Triplostegia*). 1/2. Temperate and alpine Himalayas (Garhwal to Bhutan), Assam, southeastern Tibet, Burma, continental China, Taiwan, Sulawesi, and New Guinea.
7. Morinaceae (*Morina, Acanthocalyx, Cryptothladia*). 3/13. From the Balkans (Greece, Bulgaria, and Romania) and Lebanon and Israel to continental China.
8. Dipsacaceae. 9/300. Eurasia and Africa, especially in Mediterranean Region and western Asia.

Subclass G. Lamiidae*

Superorder Gentiananae

Order 110. Gentianales

1. Loganiaceae (incl. Antoniaceae, Spigeliaceae, Strychnaceae, and the genus *Polypremum*; excl. *Desfontainia*). 20/350. Trop. and subtrop. regions, with relatively few species in temperate areas.

---

*Here I accept Professor F. Ehrendorfer's (Strasburger's Lehrbuch der Botanik, 32 Auflage, Gustav Fischer Verlag, Stuttgart, 1983) proposal to subdivide the subclass Asteridae into smaller subclasses Lamiidae and Asteridae. But unlike Ehrendorfer, I prefer to place the order Dipsacales in Rosidae rather than in Lamiidae.

2. Potaliaceae. 3/50. South America from Colombia to Amapa in Brazil (*Potalia*, 1), trop. Africa, Comoro Islands, and Madagascar (*Anthocleista*, 14), and trop. Asia, northeast Queensland, and in the Pacific from the Marianas to the Marquesas and the Tubai Islands and New Caledonia (*Fagraea*, 35).
3. Plocospermataceae (*Plocosperma*, incl. *Lithophytum*). 1/2. Southern Mexico and Guatemala.
4. Rubiaceae (incl. Henriqueziaceae). 500/6500–7000. Cosmop., but mainly in trop. and subtrop. regions with some species in temperate and cold regions (even in the Arctic and Antarctic).
5. Theligonaceae (*Theligonum*). 1/3. From the Canary Islands across the Mediterranean to East Asia.
6. Dialypetalanthaceae (*Dialypetalanthus*). 1/1. Brazil.
7. Apocynaceae. 200/2000. Trop. and subtrop. regions, with relatively few representatives in temperate areas.
8. Asclepiadaceae (incl. Periplocaceae). 250/2000. Trop. and subtrop. regions, especially in Africa, with relatively few representatives in temperate areas.
9. Gentianaceae. 80/1000. Cosmop., but mainly in temperate and subtrop. regions and also in trop. mountains.
10. Saccifoliaceae (*Saccifolium*). 1/1. Southern Venezuela.
11. Menyanthaceae. 5/35–40. Cosmop.

Order 111. Oleales

1. Oleaceae (incl. Nyctanthaceae). 29/600. Nearly cosmop., but best developed in Southeast Asia and Australasia.

Superorder Loasanae

Order 112. Loasales

1. Loasaceae. 13/280. Temperate and trop. regions of North and South America.

Superorder Lamianae

Order 113. Convolvulales

1. Convolvulaceae (incl. Dichondraceae and Humbertiaceae). 58/1660. Nearly cosmop., but best developed in trop. and subtrop. regions.
2. Cuscutaceae (*Cuscuta*). 1/170. Nearly cosmop., but best developed in America, especially in warmer regions.

Order 114. Polemoniales

1. Polemoniaceae (incl. Cobaeaceae). 15/300. Mostly north-temperate regions, especially in North America, particularly the west, but extending south to northern South America, Peru, and Chile, and north to Alaska.

Order 115. Boraginales

1. Hydrophyllaceae (excl. *Lithophytum*). 18/250. Best developed in America, especially in the western United States, with a few representatives in Africa, trop. Asia, Australia, and the Hawaiian Islands.
2. Lennoaceae (*Ammobroma*, *Lennoa*, *Pholisma*). 3/4–5. From the southwestern United States across Mexico to Colombia and Venezuela.
3. Ehretiaceae (incl. Cordiaceae). 13/400. Trop. and subtrop. regions; best developed in Central and South America.
4. Hoplestigmataceae (*Hoplestigma*). 1/2. Western trop. Africa, ranging from the Cameroons south to Gabon.
5. Boraginaceae (incl. Heliotropiaceae). 100/2000. Widespread in trop., subtrop., and temperate regions, but best developed in the Mediterranean and Irano-Turanian regions.
6. Wellstediaceae (*Wellstedia*). 1/2. Southwestern and northeastern Africa and Socotra.

Order 116. Solanales

1. Solanaceae (incl. Salpiglossidaceae). 90–95/2900. Nearly cosmop., but best developed in trop. South America.
2. Nolanaceae (*Alona* and *Nolana*). 2/60. Southern Peru and northern Chile; 1 species on Galapagos Islands.
3. Duckeodendraceae (*Duckeodendron*). 1/1. Amazon Basin in Brazil.
4. Sclerophylacaceae (*Sclerophylax*). 1/12. Argentina, Paraguay, and Uruguay.
5. Goetzeaceae (excl. *Lithophytum*). 4/6. West Indies.

Order 117. Scrophulariales (Bignoniales)

1. Buddlejaceae (incl. *Peltanthera* and excl. *Polypremum*). 7/120. Trop. and subtrop. regions of America, Africa, and Asia.
2. Scrophulariaceae (incl. Ellisiophyllaceae, Orobanchaceae, Selaginaceae, and the genera *Jerdonia* and *Oftia*). 220/4000. Cosmop., but most abundant in temperate regions (especially in the Northern Hemisphere) and in trop. mountains.

3. Globulariaceae (*Globularia* and *Poskea*). 2/30. Macaronesia, the Mediterranean Region, Europe, western Asia, and also Somalia and Socotra (*Poskea*, 2).
4. Retziaceae (*Retzia*). 1/1. The Cape of Good Hope.
5. Bignoniaceae (incl. *Paulownia* and *Wightia*, and possibly also *Brandisia*). 120/650–700. Mainly trop. regions, especially in South America, with a few species in subtrop. and temperate Asia.
6. Pedaliaceae. 12/120. From trop. and South Africa and Madagascar to Malesia, New Guinea, and northern Australia and adjacent islands.
7. Trapellaceae (*Trapella*). 1/1–2. East Asia.
8. Martyniaceae. 4/16. America, mainly trop. regions.
9. Gesneriaceae. 125/2000. Pantropical, with only a few species in the Pyrénées and the Balkan Peninsula.
10. Plantaginaceae (*Plantago*, *Littorella*, and *Bougueria*). 3/255. Temperate regions and in trop. mountains.
11. Lentibulariaceae. 4/180. Cosmop.
12. Myoporaceae (*Bontia*, *Myoporum*, *Eremophila*; excl. *Oftia*). 3/150. Mainly Australia, but *Myoporum* (32) also extends to New Guinea, East Asia, New Zealand, Pacific islands, and Mauritius, and *Bontia* (1) occurs in the West Indies and northern South America.
13. Acanthaceae (incl. Nelsoniaceae and Thunbergiaceae). 254/2700. Mainly trop., with only a few species in temperate regions.
14. Mendonciaceae (*Mendoncia* and *Gilletiella*). 2/60. South America, trop. Africa, and Madagascar.

Order 118. Lamiales

1. Verbenaceae (incl. Avicenniaceae, Chloanthaceae or Dicrastylidaceae, Phrymaceae). 100/3000. Trop. and subtrop., with only few species in temperate regions.
2. Stilbaceae. 5/12. South Africa.
3. Lamiaceae or Labiatae (incl. Tetrachondraceae). 200/3500. Cosmop., but best developed in the Mediterranean and Irano-Turanian regions.
4. Callitrichaceae (*Callitriche*). 1/17. Cosmop., but mainly in temperate regions.

Order 119. Hippuridales

1. Hippuridaceae (*Hippuris*). 1/1. Temperate and cold regions of the Northern Hemisphere.

Order 120. Hydrostachyales
  1. Hydrostachyaceae (*Hydrostachys*). 1/25. Trop. and South Africa, Madagascar.
Order 121. Cynomoriales
  1. Cynomoriaceae (*Cynomorium*). 1/2. Canary Islands and the Mediterranean and western Asia to Mongolia.

Subclass H. Asteridae
Superorder Asteranae
Order 122. Campanulales
  1. Pentaphragmataceae (*Pentaphragma*). 1/30. Lower Burma, southern China, Indochina, Malesia, and New Guinea.
  2. Campanulaceae (incl. Cyphiaceae and the genus *Berenice*). 40–45/900 or more. Worldwide, but best developed in temperate regions of the Northern Hemisphere. There are a number of small endemic genera in South Africa and adjacent islands of the Indian Ocean, including a monotypic *Berenice* (very close to *Cephalostigma*) on Réunion, but in general the Southern Hemisphere is extremely poor in the Campanulaceae. South America has only certain species of *Wahlenbergia, Lagousia,* and *Cephalostigma*. In Australia, Tasmania, and New Zealand there are only several species of *Wahlenbergia*. *Cephalostigma* is the only genus of the family confined to the tropics.
  3. Lobeliaceae (incl. *Cyphocarpus, Nemacladus,* and *Parishella*). 33/1300. Worldwide, but best developed in the Americas and in tropical regions of other continents, with a few species in temperate areas.
  4. Sphenocleaceae (*Sphenoclea*). 1/2. One paleotropical, 1 in west Africa.
  5. Stylidiaceae. 5/164. Mainly extratropical parts of Australia, but also in Tasmania, New Zealand, South and especially Southeast Asia, and southernmost South America.
  6. Donatiaceae (*Donatia*). 1/2. New Zealand and Tasmania and southernmost South America.
Order 123. Goodeniales
  1. Goodeniaceae. 15/300. Mostly in Australia (particularly western Australia) and Tasmania, with a few species in New Zealand, Japan, trop. Asia, Africa, Madagascar, trop. America, and Pacific islands.

2. Brunoniaceae (*Brunonia*). 1/1. Australia and Tasmania.

Order 124. Calycerales

1. Calyceraceae. 6/60. Central and South America, mostly in extratropical regions.

Order 125. Asterales

1. Asteraceae or Compositae. 1150–1300/20000–21000. Cosmop., but best developed in temperate and subtrop. regions.

Class B. Liliopsida

Subclass A. Alismatidae

Superorder Alismatanae

Order 1. Alismatales

Suborder Butomineae

1. Butomaceae (*Butomus*). 1/1. Temperate Eurasia.

Suborder Alismatineae

2. Limnocharitaceae (*Hydrocleys* incl. *Ostenia, Limnocharis,* and *Butomopsis* [*Tenagocharis*]). 3/12 Trop. and subtrop. regions of both hemispheres.
3. Alismataceae. 11–13/100. Cosmop., but best developed in the Northern Hemisphere, especially in North America.

Order 2. Hydrocharitales

1. Hydrocharitaceae. 15/100–110. Widely distributed, except in cold and arid areas.

Order 3. Aponogetonales

1. Aponogetonaceae (*Aponogeton*). 1/45. Trop. regions of the Old World and South Africa. Most species are found in Africa, Madagascar, and Comores.

Order 4. Juncaginales (Scheuchzeriales)

1. Scheuchzeriaceae (*Scheuchzeria*). 1/1–2. Cold regions of the Northern Hemisphere; especially common in sphagnum bogs.
2. Juncaginaceae (incl. Maundiaceae and Triglochinaceae). 3–4/

15. Widespread in temperate and cold regions of both hemispheres, usually in marshy habitats. Two monotypic genera—*Cycnogeton* (very closely related to *Triglochin*) and *Maundia*—are endemic to Australia and Tasmania. *Tetroncium* (1) is endemic to antarctic South America.
3. Lilaeaceae (*Lilaea*). 1/1. Mountains of Pacific America from British Columbia to Chile and Argentina.

Order 5. Potamogetonales

1. Potamogetonaceae. 2/100. *Potamogeton* is cosmop., but *Groenlandia* (1) is found only in western Europe, North Africa, and southwestern Asia.
2. Ruppiaceae (*Ruppia*). 1/7. Nearly cosmop. Usually in brackish water in coastal areas, but a few species are found inland in fresh water in South America and New Zealand.

Order 6. Posidoniales

1. Posidoniaceae (*Posidonia*). 1/5. Mediterranean Sea and southern Australian and Tasmanian coasts.

Order 7. Zosterales

1. Zosteraceae. 3/20. Temperate seas of Northern and Southern Hemispheres, with a few species extending into tropical seas. *Phyllospadix* (5) occurs on the coasts of Japan and Pacific North America. *Heterozostera* (1) occurs on coasts of Australia and Chile.

Order 8. Cymodoceales

1. Zannichelliaceae. 4/7–8. Nearly cosmop., in fresh and brackish water. *Zannichellia* (1) is nearly cosmop.; *Althenia* (1 or 2) is found in the western Mediterranean and North Africa; *Lepilaena* (4) in Australia and New Zealand; and *Vleisia* (1) in Cape Province of South Africa.
2. Cymodoceaceae. 5/20. Mainly tropical and subtrop. seas, with a few species in warm temperate waters. *Amphibolis* (2) is limited to the temperate waters of southern Australia, and *Cymodocea nodosa* to the Mediterranean Sea.

Order 9. Najadales

1. Najadaceae (*Najas*). 1/40–50. Nearly cosmop., in fresh or brackish water.

Subclass B. Liliidae

Superorder Triuridanae

Order 10. Triuridales

1. Triuridaceae. 7/80. Trop. America, Africa, and Asia.

Superorder Lilianae

Order 11. Liliales

1. Melanthiaceae (incl. Petrosaviaceae and Nartheciaceae). 20/150. Mainly in extratropical regions of the Northern Hemisphere, especially in East Asia and North America; only a few members of the family occur in trop. Asia and South America.
2. Uvulariaceae (Uvularieae, Tricyrtideae, ? Scoliopeae). 8/55. Himalayas, East and Southeast Asia, North America, New Guinea, and Australia.
3. Colchicaceae (Anguillarieae, Iphigenieae, Colchiceae). 12–13/175. Temperate Eurasia and North America, the Mediterranean and Irano-Turanian regions, Himalayas, China, trop. Asia, trop. and South Africa, Madagascar, Arabian Peninsula, New Guinea, Australia, and New Zealand. *Wurmbea* (incl. *Anguillaria*) occurs in South Africa and Australia; and *Iphigenia* is distributed in South Africa, trop. Asia, trop. Australia, and New Zealand.
4. Campynemataceae. 2/3. Tasmania (*Campynema*) and New Caledonia (*Campynemanthe*).
5. Iridaceae. 70–75/1800. Very widely distributed in trop., subtrop., and temperate regions, especially in South Africa, the eastern Mediterranean, western and Central Asia, and Central and South America.
6. Geosiridaceae (*Geosiris*). 1/1. Madagascar and an adjacent Saint Mary Island.
7. Burmanniaceae (incl. Thismiaceae). 18–20/130. Trop. and subtrop. regions of both hemispheres, north to Japan and the eastern United States, south to Mozambique, southern Australia, Tasmania, New Zealand, and Paraguay.
8. Corsiaceae. 2/27. New Guinea, Solomon Islands, northern Australia (*Corsia*), and Chile (*Arachnitis*, 1).
9. Tecophilaeaceae (excl. *Lanaria*?). 6/18–20. South and trop. Africa and Madagascar (*Cyanella* and *Walleria*), Chile (*Conanthera, Tecophilaea* and *Zephyra*), and California (*Odontostomum*).

10. Cyanastraceae (*Cyanastrum*). 1/6. Trop. Africa.
11. Calochortaceae (*Calochortus*). 1/60. From British Columbia (Canada) to Guatemala; best developed in California.
12. Liliaceae (Lilieae, Tulipeae, Gageeae, and ? Medeoleae). 10/470. Temperate and subtrop. regions of the Northern Hemisphere.

Order 12. Asparagales

1. Convallariaceae (incl. Aspidistraceae, Ophiopogonaceae, Peliosanthaceae and Polygonataceae). 17/230. Mainly the Northern Hemisphere, but best developed in East Asia, the Himalayas, and North America.
2. Ruscaceae (*Danaë, Ruscus,* and *Semele*). 3/13. The Azores, Madeira, and the Canary Islands, Mediterranean Region, western, central, southeastern Europe, northern Asia Minor, western and southeastern Transcaucasia, and northern Iran.
3. Asparagaceae (*Asparagus*). 1/300. Old World, mostly in dry places, especially in Africa.
4. Herreriaceae. 2/10. Subtrop. and trop. South America (*Herreria*) and Madagascar (*Herreriopsis,* 1).
5. Dracaenaceae (incl. Asteliaceae, Nolinaceae, and Sansevieriaceae). 10/285. Trop. and subtrop. regions of both hemispheres (mainly Africa, Southeast Asia, New Zealand, and America) and temperate regions of the Southern Hemisphere. *Cohnia* occurs on Mascarenes (2) and in New Caledonia (1). *Milligania* (4) is endemic to Tasmania. Nolineae occur in the southwestern United States and Mexico.

Order 13. Amaryllidales (Agavales)

1. Hyacinthaceae (Scilleae, Hyacintheae, Massonieae, Chlorogaleae, and Bowieae). 35/825. Widely distributed in both hemispheres; best developed in South Africa and in the Mediterranean and Irano-Turanian regions.
2. Alliaceae (incl. Agapanthaceae, Gilliesiaceae, and Milulaceae). 30/650 or more. Widely distributed except in tropical regions, Australia, and New Zealand. Best developed in temperate regions of the Northern Hemisphere. There are many endemic genera in Peru and Chile (all genera of the tribe Gilliesieae and the genera *Garaventia* [*Steinmannia*] and *Leucocoryne* of the tribe Allieae). The genus *Milula* (1), which is the only alliaceous genus with a dense cylindrical spicate inflorescence subtended by one spath-like bract, occurs in central Nepal and Tibet.

3. Hesperocallidaceae (*Hesperocallis* and possibly *Leucocrinum*). 1–2/1–2. *Hesperocallis* in California and Arizona; *Leucocrinum* in western North America from Oregon and northern California to Colorado and New Mexico.
4. Funkiaceae (*Hosta*). 1/40. East Asia.
5. Agavaceae (incl. Yuccaceae). 9/400. Trop. and subtrop. regions of North, Central, and northern South America and in the West Indies. Best developed in Mexico and the southwestern and southern United States.
6. Hemerocallidaceae (*Hemerocallis*). 1/15. Central Europe (1), Siberia, and East Asia.
7. Phormiaceae (incl. Dianellaceae; *Phormium*, ? *Xeronema*, *Dianella*, *Rhuacophila*, *Stypandra*, *Excremis*). 6/30. Mainly in the Southern Hemisphere. *Phormium* (2): New Zealand, Stewart Island, Chatham Island, Auckland Islands, Norfolk Island; *Xeronema* (2): New Caledonia, Poor Knights Islands, and Hen (Taranga) Island; *Dianella* (20): from eastern trop. Africa, Madagascar, and the Mascarenes across Sri Lanka and India to southern China, Indochina, Malesia, New Guinea, Australia, Tasmania, New Zealand, Norfolk Island, New Caledonia, Fiji, Polynesia (to Hawaii), and northern and western South America; *Rhuacophila* (1): Southeast Asia to New Caledonia, but excluding Australia; *Stypandra* (5): Australia and Tasmania; *Excremis* (1): mountains of Venezuela, Colombia, and Peru.
8. Blandfordiaceae. 1/4. Eastern Australia (Queensland, New South Wales, and eastern and southern Victoria) and Tasmania.
9. Amaryllidaceae. 65/900. Mainly trop. and subtrop. regions of both hemispheres, especially Central and South America, trop. and South Africa, and the Mediterranean Region.

Order 14. Doryanthales

1. Doryanthaceae (*Doryanthes*). 1/2–3. Australia (eastern Queensland and coastal New South Wales).

Order 15. Asphodelales

1. Asphodelaceae (Asphodeleae, Aloeeae, and Kniphofieae). 17/850. Europe, Macaronesia, the Mediterranean and Irano-Turanian regions, western Himalayas, southwestern China, South and trop. Africa, Madagascar, Mascarenes, Arabian Peninsula, and Socotra. *Bulbinella* (20) occurs both in South Africa and in New Zealand and on Auckland and Campbell

islands. The predominantly African genus *Bulbine* (incl. *Bulbinopsis*) includes 3 Australian species. In the New World there is only 1 monotypic genus *Glyphosperma,* endemic to northern Mexico.

2. Anthericaceae (Anthericeae, Thysanoteae, Hodgsonioleae, Simethideae, and Johnsonieae). 32/630. Trop. and South Africa, Madagascar, India, Australia, Tasmania, South America, with a few species in North America, Europe, and East Asia, Malesia, New Guinea, New Zealand, and New Caledonia. *Caesia* (12) occurs both in South Africa and Madagascar and in Australia (9). A number of monotypic and oligotypic genera (including *Herpolirion* and *Sowerbaea*) are endemic to Australia. Only 3 genera (*Anemarrhena, Diuranthera,* and *Terauchia*) are endemic to East Asia. *Paradisea* (2) is endemic to mountains of Europe, *Bottionea* (1) to Chile, *Pasithea* (1) to Peru and Chile, and *Eremocrinum* (1) to the western United States.

3. Eriospermaceae (*Eriospermum*). 1/80. Africa, about 50 in South Africa, centered in the southwestern Cape Province.

4. Ixioliriaceae (*Ixiolirion*). 1/5. Palestine and the Irano-Turanian Region.

5. Xanthorrhoeaceae (*Xanthorrhoea*). 1/15. Australia and Tasmania.

6. Dasypogonaceae (incl. Calectasiaceae, Kingiaceae, and Lomandraceae). 8/55. Australia, Tasmania, New Guinea, New Britain, and New Caledonia. Mostly concentrated in Australia, with only a few species in New Guinea and adjacent islands and 1 species in New Caledonia. The largest genus, *Lomandra,* contains about 42 species, 40 of which are endemic to Australia (mostly in the south and east); only *L. multiflora* occurs both in Queensland and adjacent southern New Guinea, while *L. banksii* occurs in Queensland, adjacent southern New Guinea, and also in New Caledonia. *Acanthocarpus* (1), *Chamaexeros* (3), and *Dasypogon* (2) are endemic to southwestern Australia, and *Romnalda* (1) is endemic to New Guinea, New Britain, and Japen Islands (Geelvink Bay).

7. Hanguanaceae (*Hanguana*). 1/1. Sri Lanka, Thailand, Malay Archipelago, Palau Island, northern Australia.

8. Aphyllanthaceae (*Aphyllanthes*). 1/1. Western Mediterranean.

Order 16. Alstroemeriales

1. Alstroemeriaceae. 4/200. From Mexico and the West Indies

to southern South America, mostly in the Andes. *Leontochir* (1) is endemic to Chile and *Schickendantzia* (2) to Argentina.

Order 17. Smilacales

1. Philesiaceae (incl. Geitonoplesiaceae, Luzuriagaceae, Petermanniaceae, and the genus *Drymophila*). 8/13. Southeast Africa (*Behnia*, 1), eastern Malesia, New Guinea, Solomon Islands, New Hebrides, New Caledonia, Fiji, Australia, New Zealand and Stewart Island, temperate South America (Peru, Chile, and western Argentina), and the Falkland Islands. *Philesia* (1) and *Lapageria* (1) occur in Chile. *Luzuriaga* (3) occurs in Peru, Chile, Argentina, and the Falkland Islands (2), and in New Zealand and Stewart Island (1, *L. parviflora*). *Drymophila* (2), which is related to *Eustrephus*, is endemic to eastern Australia.
2. Smilacaceae (incl. Rhipogonaceae). 3/380. Mainly trop. and subtrop. regions, with a few species in warm-temperate regions. *Rhipogonum* (7) occurs in New Guinea, northern and eastern Australia, and New Zealand. *Heterosmilax* (11) is distributed in Assam, Burma, continental China, Taiwan, Thailand, Ryukyu Archipelago, and western Malesia.

Order 18. Dioscoreales

1. Dioscoreaceae (incl. Stenomeridaceae). 5/700. Pantropical, with some temperate representatives. *Stenomeris* (2) occurs in Malesia; *Avetra* (1) on the eastern coast of Madagascar; *Rajania* (25) in the West Indies; *Tamus* (4) in western Europe, the Macaronesian and Mediterranean Regions, and the Caucasus and northern Iran. The largest genus *Dioscorea* is pantropical with a few temperate species.
2. Trichopodaceae (*Trichopus*). 1/1. Southern India, Sri Lanka, and the Malay Peninsula.
3. Taccaceae (*Tacca*). 1/10. Mainly trop. regions of the Old World, with 1 species in northern trop. South America.
4. Stemonaceae (incl. Croomiaceae). 4/30. Mainly East, South, and Southeast Asia and northeastern Australia, but 1 species of *Croomia* occurs in the southeastern United States.
5. Trilliaceae (*Daiswa, Kinugasa, Paris,* and *Trillium*). 4/60. Europe, Mediterranean, Caucasus, Siberia, East Asia, Himalayas, northeastern India, northern Burma, northern Indochina, and North America.

Order 19. Haemodorales

1. Hypoxidaceae (excl. *Campynema* and *Campynemanthe*). 9/150. Africa (mainly South Africa), the Mascarenes, Seychelles, trop. and subtrop. regions of Asia (north to eastern Himalayas, southwestern China and Japan), New Guinea, Australia, Tasmania, New Zealand, America from the Atlantic United States to Uruguay, and the West Indies.
2. Haemodoraceae (incl. Conostylidaceae and ?*Lanaria*). 16–17/90. South Africa (*Barberetta, Dilatris, Lanaria* and *Wachendorfia*), New Guinea, Australia and Tasmania, and also North America (*Lachnanthes*, 1 and *Lophiola*, 2), and South and Central America (*Xiphidium, Hagenbachia, Pyrrorhiza* and *Schiekia*).

Order 20. Orchidales

1. Orchidaceae (incl. Apostasiaceae and Cypripediaceae). 750/20,000–25,000. Cosmop., but mainly in trop. regions, especially in Southeast Asia and trop. America. The most primitive genera *Apostasia* (7) and *Neuwiedia* (9) occur in the eastern Himalayas, Assam, Sri Lanka, Burma, Thailand, Indochina, southern China, southern Japan, Malesia, New Guinea and adjacent islands, and northeastern Queensland (only 1 species of *Apostasia*).

Superorder Pontederianae

Order 21. Pontederiales

1. Pontederiaceae. 9/34. Pantropical, but best developed in South America.

Order 22. Philydrales

1. Philydraceae. 4/5. East and Southeast Asia, Guam Island, Andaman Islands, New Guinea, and Australia (all 4 genera occur in Australia).

Superorder Bromelianae

Order 23. Bromeliales

1. Bromeliaceae. 46/2100. Trop. and partly subtrop. and warm temperate regions of America, with only 1 species of *Pitcairnia* in trop. West Africa (Guinea).

Order 24. Velloziales

1. Velloziaceae. 6/260. Trop. and South Africa, Madagascar, Arabian Peninsula, and arid regions of trop. America (especially in Brazil).

Superorder Zingiberanae

Order 25. Zingiberales (Musales)

1. Strelitziaceae. 3/7. Trop. South America (*Phenakospermum*), eastern South Africa (*Strelitzia*), and Madagascar (*Ravenala*).
2. Musaceae (*Ensete* and *Musa*). 2/42. From trop. Africa to northeastern Queensland and the Pacific.
3. Heliconiaceae (*Heliconia*). 1/150. Trop. and subtrop. regions of Central and South America with 1 species (*H. indica*) widespread in the islands of the southwestern Pacific.
4. Lowiaceae (*Orchidantha*). 1/7–8. Southern China, Indochina, Malay Peninsula, and Borneo.
5. Zingiberaceae. 45–47/1000–1300. South, East and Southeast Asia, Malesia, and New Guinea, with a few species in northeastern Australia.
6. Costaceae. 4/200. Pantropical, but best developed in Central and South America.
7. Cannaceae (*Canna*). 1/55–60. Trop. and subtrop. parts of America, north to South Carolina and Florida, south to northern regions of Chile and Argentina.
8. Marantaceae. 30/400. Almost pantropical (except Australia), but most abundant in trop. America.

Superorder Hydatellanae

Order 26. Hydatellales

1. Hydatellaceae (*Hydatella* and *Trithuria*). 2/7. Australia, Tasmania, and New Zealand.

Superorder Juncanae

Order 27. Juncales

1. Juncaceae. 8–9/400. The 6 or 7 mono- or oligotypic genera are wholly confined to the Southern Hemisphere, but 2 large genera—*Juncus* and *Luzula*—are best developed in the Northern Hemisphere, mostly in temperate and cold regions. The only shrubly genus, *Prionium* (1), is endemic to South Africa.

2. Thurniaceae (*Thurnia*). 1/3. Venezuela, Guayana, and parts of the Amazon Valley.

Order 28. Cyperales

1. Cyperaceae. 120/5600. Cosmop., but most abundant in temperate and cold regions.

Superorder Commelinanae

Order 29. Commelinales

1. Commelinaceae (incl. Cartonemataceae). 47/700. Mainly trop. and subtrop. regions, with a few species in temperate regions of East Asia, southern North America, and Australia.
2. Mayacaceae (*Mayaca*). 1/4 or 10. The southeastern United States, West Indies, and Central and South America, with 1 species in Africa (Angola, Zaire, and Zambia).
3. Xyridaceae (*Xyris* and *Achlyphila*). 2/250. Trop. and subtropical regions, mostly in America, but also in Africa, South and Southeast Asia, New Caledonia, Australia, and Tasmania. *Achlyphila* (1) is endemic to Venezuela.
4. Abolbodaceae (*Abolboda* and *Orectanthe*). 2/20. Trop. South America; *Orectanthe* (1) is endemic to Venezuela.
5. Rapateaceae. 16/100. Trop. South America, with a single monotypic genus, *Maschalocephalus,* in trop. West Africa. Best developed in the Guayana Highland Region of northern South America.
6. Eriocaulaceae. 13/1200. Mainly in trop. regions, especially in the New World. A few species of the genus *Eriocaulon* extend to temperate regions of East Asia and North America.

Order 30. Restionales

1. Flagellariaceae (*Flagellaria*). 1/3. Trop. and subtrop. regions of the Old World, mainly on Pacific islands.
2. Joinvilleaceae (*Joinvillea*). 1/2. Malesia, New Hebrides, Atoll Caroline, Solomon Islands, New Caledonia, Fiji, western Samoa, and the Hawaiian Islands.
3. Restionaceae (incl. Anarthriaceae and Ecdeiocoleaceae). 39/340. Temperate South America (Chile and Argentina), Cape Province of South Africa, trop. Africa (Malawi), Madagascar, Indochina and Malay Peninsula, New Guinea, Australia and Tasmania, and Chatam Island. Best developed in Cape Province, extratropical regions of Australia, and Tasmania.

4. Centrolepidaceae (excl. Hydatellaceae; *Aphelia, Centrolepis,* and *Gaimardia*). 3/35. Temperate South America (Falkland Islands and Tierra Fuego), southern China (Hainan), Indochina, Malesia, New Guinea, Australia, Tasmania, and New Zealand. Best developed in Australia and Tasmania.

Order 31. Poales

    1. Poaceae or Gramineae. 880/10,000. Cosmop.

Subclass C. Arecidae

Superorder Arecanae

Order 32. Arecales

    1. Arecaceae or Palmae. 210/3000. Pantropical, with a few species in warm temperate regions.

Order 33. Cyclanthales

    1. Cyclanthaceae. 11/180. Central and trop. South America and the West Indies.

Superorder Pandananae

Order 34. Pandanales

    1. Pandanaceae. 3/800. Tropics of the Old World. Best developed in Malesia, Melanesia, and Madagascar; relatively few species occur in temperate regions (Japan, China, and New Zealand). *Sararanga* (2) occurs on the Philippine Islands, New Guinea, Solomon Islands, and Admiralty Islands.

Superorder Aranae

Order 35. Arales

    1. Araceae (incl. Acoraceae and Pistiaceae). 110/2500. Pantropical, with a few species in temperate regions.
    2. Lemnaceae. 6/43. Cosmop.

Superorder Typhanae

Order 36. Typhales

    1. Typhaceae (incl. Sparganiaceae). 2/30. Nearly cosmop., but mainly in the Northern Hemisphere.

BIBLIOGRAPHY

GENERAL

Alekhin, V. V. 1944. Geography of Plants (in Russian). Moscow.
Aubréville, A. 1970. Vocabulaire de biogéographie appliquée aux régions tropicales. Adansonia, sér. 2, 10:439–497.
Axelrod, D. 1975. Plate tectonics and problems of angiosperm history. Mém. Mus. National d'hist. Naturelle, sér. A, 88:72–86.
Axelrod, D. I., and P. H. Raven. 1972. Evolutionary biogeography viewed from plate tectonic theory. Challenging Biological Problems: Directions toward Their Solution, ed. J. A. Behnke, 218–236. New York.
Balgooy, M. M. J. van. 1960. Preliminary plant-geographical analysis of the Pacific. Blumea 10, 2:385–430.
——. 1969. A study on the diversity of island floras. Blumea 17:139–178.
——. 1971. Plant-geography of the Pacific. Blumea Suppl. 6:1–222.
Barber, H. N., H. E. Dadswell, and H. D. Ingle. 1959. Transport of driftwood from South America to Tasmania and Macquarie Island. Nature 184:203–204.
Braun-Blanquet, J. 1919. Essai sur les notions "d'élément" et de "territoire" phytogéographiques. Arch. phys. nat. Genève, sér. 5, 1:497–512.
——. 1923. L'origine et la développement des flores dans le Massif central de France. Paris and Zurich.
——. 1928. Pflanzensoziologie. Grundzüge der Vegetationskunde. I. Berlin.
——. 1964. Pflanzensoziologie. Grundzüge der Vegetationskunde. 3 Aufl. Vienna and New York.
Candolle, A. P. de. 1808. Icones plantarum Galliae rariorum. I. Paris.
Candolle, Alph. de. 1855. Géographie botanique raisonnée. I–II. Paris and Genève.
Carlquist. S. 1974. Island Biology. New York.

Chevalier, A., and L. Cuenot. 1932. Biogeographie. In Traité de géographie physique, E. de Martonne. Paris.
Chevalier, A., and L. Emberger. 1937. Les régions botaniques terrestres. Encyclopédie française 5:5.64–1 to 5.66–7. Paris.
Clayton, W. D., and F. H. Hepper. 1974. Computer-aided chorology of West African grasses. Kew Bull. 29:213–234.
Clayton, W. D., and G. Panigrahi. 1974. Computer-aided chorology of Indian grasses. Kew Bull. 29:669–686.
Croizat, L. 1952. Manual of Phytogeography. The Hague.
———. 1958. Panbiogeography. I–III. Caracas.
Darlington, P. J. 1957. Zoogeography: The Geographical Distribution of Animals. New York.
Diels, L. 1908. Pflanzengeographie. Berlin.
———. 1918. Pflanzengeographie. 2 Aufl. Berlin.
Diels, L., and F. Mattick. 1958. Pflanzengeographie. Berlin.
Drude, O. 1890. Handbuch der Pflanzengeographie. Stuttgart.
Dupont, P. 1962. La flore atlantique européene. Toulouse.
Ehrendorfer, F. 1971. Die Florenreiche der Erde und ihre Floren— und Vegetationsgebiete. In Lehrbuch der Botanik, E. Strasburger, 765–774. 30 Aufl. Jena.
Engler, A. 1879, 1882. Versuch einer Entwicklungsgeschichte der Pflanzenwelt, insbesondere der Florengebiete, seit der Tertiärperiode. I–II. Leipzig.
———. 1899. Die Entwickelung der Pflanzengeographie in den letzten hundert Jahren. A. V. Humboldt-Gentenarschrieft. Berlin.
———. 1902. Die pflanzengeographische Gliederung Nordamerikas. Notizbl. Bot. Gart. Mus. Berlin, App. IX, 1–94.
———. 1903. Syllabus der Pflanzenfamilien. 3 Aufl. Berlin.
———. 1905. Über floristische Verwandtschaft zwischen den tropischen Afrika und Amerika. Sitsb. Preuss. Akad. Wiss. Berlin 6:180–231.
———. 1908. Pflanzengeographische Gliederung von Afrika. Sitzb. Preuss. Akad. Wiss. Berlin 38:781–837.
———. 1912. Syllabus der Pflanzenfamilien. 7 Aufl. Berlin.
———. 1924. Übersicht über die Florenreiche und Florengebiete der Erde. In Syllabus der Pflanzenfamilien, A. Engler and E. Gilg, 9–10 Aufl. Berlin.
Engler, A., and L. Diels. 1936. Syllabus der Pflanzenfamilien. 11 Aufl. Berlin.
Gaussen, H. 1954. Géographie des plantes. 2d ed. Paris.
Gaussen, H., and C. Leredde. 1949. Les endémiques pyrénéo-cantabriques dans la region centrale des Pyrénées. Bull. Soc. Bot. Fr. 96:57–83.
Geptner, V. G. 1936. General Zoogeography (in Russian). Moscow-Leningrad.
Gleason, H. A., and A. Cronquist. 1964. The Natural Geography of plants. New York.
Good, R. 1947. The Geography of the Flowering plants. London.
———1974. The Geography of the Flowering plants. 3d ed. London.
Graebner, P. 1910. Lehrbuch der allgemeinen Pflanzengeographie. Leipzig.
———. 1929. Lehrbuch der allgemeinen Pflanzengeographie. Leipzig.
Grisebach, A. 1872. Die Vegetation der Erde nach ihrer klimatischen Anordnung. Bd. I and II. Leipzig.
———. 1884. Die Vegetation der Erde nach ihrer klimatischen Anordnung. Bd. I

and II. 2 Aufl. Leipzig.
Hayek, A. 1926. Allgemeine Pflanzengeographie. Berlin.
Heilprin, A. 1887. The Geographical and Geological Distribution of Animals. New York.
Hemsley, W. B. 1885 (1884). Report on the botany of the Bermudas and various other islands. Part 2. In Report on the Scientific Results of the Voyage of N.M.S. "Challenger" during the Years 1873–76. Botany 1, 2:133–283. London.
Hooker, J. D. 1882. On geographical distribution. Brit. Assoc. Rep. 1881:727–738.
Howden, H. F. 1974. Problems in interpreting dispersal of terrestrial organisms as related to Continental Drift. Biotropica 6, 1:1–6.
Iwatsuki, K. 1964. Taxonomic studies of Pteridophyta VIII. 10. Classification of the genus *Thelypteris*. Acta Phytotax. Geobot. 21:35–38.
Krasnov, A. H. 1899. Geography of Plants (in Russian). Kharkov.
Kuznetsov, N. I. 1909. Principles for the division of the Caucasus into botanico-geographic provinces (in Russian). Memoirs of the St. Petersburg Academy of Sciences 24(1):1–174.
Lavrenko, E. M. 1964a. Geography of Plants (in Russian). Great Soviet Cyclopedia. 2d ed. 10:475–478. Moscow.
———. 1964b. Botanico-geographic dominions and areas of plants (in Russian). In Physico-geographic Atlas of the World, 288. Moscow.
Mattick, F. 1964. Übersicht über die Florenreiche und Florengebiete der Erde. In Syllabus der Pflanzenfamilien, A. Engler, 626–630. Bd. 2. 12 Aufl. Berlin.
Mayr, E. 1969. Principles of Systematic Zoology. New York.
Neil, W. T. 1969. The Geography of Life. New York and London.
Newbigin, M. I. 1936. Plant and Animal Geography. London.
Ozenda, P. 1964. Biogéographie végétale. Paris.
Peters, J. A. 1971. A new approach in the analysis of biogeographical data. Smithsonian Contrib. Zool., 107.
Pichi Sermolli, R. E. G. 1977. Tentamen Pteridophytorum genera in taxonomicum ordinem redigendi. Webbia 31:313–512.
Popov, M. G. 1940. Contributions toward a monograph of the genus *Eremostachys* (in Russian). New Memoirs of the Moscow Society of Naturalists 19:1–166.
———. 1963. Foundations of Florogenetics (in Russian). Moscow.
Raven, P. H., and D. J. Axelrod. 1974. Angiosperm biogeography and past continental movements. Ann. Missouri Bot. Gard. 61, 3:539–673.
Ridley, H. N. 1930. The Dispersal of Plants throughout the World. Kent.
Rikli, M. 1913. Die Florenreiche. In Handwörterbuch der Naturwissenschaften 4:776–857. Jena.
———. 1934. Geographie der Pflanzen. In Handwörterbuch der Naturwissenschaften 4:907–1002. Jena.
Schmidt, R. P. 1954. Faunal realms, regions and provinces. Quart. Rev. Biol. 29:322–331.
Schmithüsen, J. 1961. Allgemeine Vegetationsgeographie. Berlin.
Schouw, J. F. 1823. Grundzüge einer allgemeinen Pflanzengeographie. Berlin.
Smith, A. C. 1970. The Pacific as a key to flowering plant history. H. L. Lyon Arb. Lecture 1:1–27.
———. 1973. Angiosperm evolution and the relationship of the floras of Africa and

America. In Tropical Forest Ecosystems in Africa and South America: A Comparative Review, eds. B. J. Meggers, E. S. Ayensu, and W. D. Dickworth, 49–61. Washington.
Steenis, C.G.G.J. van. 1962. The land-bridge theory in botany with particular reference to tropical plants. Blumea 11:235–372.
———. 1963. Transpacific floristic affinities, particularly in the tropical zone. In Pacific Basin Biogeography, ed. J. L. Gressit, 219–231. Honolulu.
Stoyanov, N. 1950. Textbook of Plant Geography (in Bulgarian). Sofia.
Szafer, W. 1952. Zarys ogolnej geograffii roslin, Warszawa.
Takhtajan, A. 1969. Flowering Plants, Origin and Dispersal. Edinburgh.
Takhtajan, A. L. 1970. Origin and Dispersal of the Flowering Plants (in Russian). Leningrad.
———. 1974. Floristic division of the world (in Russian). In Life of Plants 1:117–153. Moscow.
———. 1978. Floristic Regions of the World (in Russian). Soviet Sciences Press. Leningrad Branch.
Thiselton-Dyer, W. T. 1909. Geographical distribution of plants. In Darwin and Modern Science, ed. A. C. Seward, 298–318. Cambridge.
Thorne, R. F. 1963. Biotic distribution patterns in the tropical Pacific. In Pacific Basin Biogeography, ed. J. L. Gressitt, 311–354. Honolulu.
———. 1973*a*. Major disjunctions in the geographic ranges of seed plants. Quart. Rev. Biol. 47, 4:365–411.
———. 1973*b*. Floristic relationships between tropical Africa and tropical America. In Tropical Forest Ecosystems in Africa and South America: A Comparative Review, eds. D. J. Meggers, E. S. Ayensu, and W. D. Dickworth, 27–47. Washington.
Tolmatchev, A. I. 1970. On some quantitative inter relation among the floras of the world (in Russian). Bulletin of Leningrad State University (Series Biology) 15, 3:3–74.
———. 1974. Introduction to the Geography of Plants (in Russian). Leningrad.
Tryon, R. M., and A. F. Tryon. 1982. Ferns and allied plants. New York, Heidelberg Springer-Verlag.
Turrill, W. B. 1953. Pioneer Plant Geography. The Hague.
———. 1958. The evolution of floras with special reference to those of the Balkan peninsula. J. Linn. Soc. London (Bot.) 56, 365:136–152.
———. 1959. Plant geography. In Vistas in Botany, eds. R. C. Rollins and G. Taylor, 2:172–229. London.
Udvardy, M. D. F. 1975. A classification of the biogeographical provinces of the world. JUGN Occasional paper 18. Morges.
Vavilov, N. I. 1935. Botanico-Geographical Foundations of Selection (in Russian). Moscow and Leningrad.
Walter, H. 1962, 1968. Die Vegetation der Erde in ökologischer Betrachtung. Bd. I und II. Jena.
Walter, H., and H. Straka. 1970. Arealkunde. Floristisch-historische Geobotanik. 2 Aufl. Stuttgart.
White, F. 1971. The taxonomic and ecological basis of chorology. Mitt. Bot. Staatssamml. München 10:91–112.

———. 1983. The vegetation of Africa. UNESCO, Paris.
Wulff, E. V. 1934. An Attempt to divide the Earth into Vegetation Regions on the Basis of the Quantitative Distribution of Species (in Russian). Leningrad.
———. 1944. Historical Geography of Plants: History of Floras of the Earth (in Russian). Moscow and Leningrad.

I. HOLARCTIC KINGDOM (HOLARCTIS)

Adamovič, L. 1909. Die Vegetationsverhältnisse der Balkanländer (Mösische Länder). Leipzig.
———. 1933. Die pflanzengeographische Stellung und Gliederung Italiens. Jena.
Alexandrova, V. D. 1977. Geobotanical subdivision of the Arctic and Antarctic (in Russian). Leningrad.
Armand, D. L., B. F. Dobrynin, Yu. K. Efremov, L. Ya. Siman, E. M. Myrsaev, and L. I. Sprygina. 1956. Extralimital Asia. Physical Geography (in Russian). Moscow.
Axelrod, D. I. 1958. Evolution of the Madro-Tertiary geoflora. Bot. Rev. 24:433–509.
———. 1975. Evolution and biogeography of Madrean-Tethyan sclerophyll vegetation. Ann. Missouri Bot. Gard. 62:289–334.
———. 1985. Rise of the grassland biome of central North America. Bot. Review 51:163–201.
Barbero M., A. Benabid, C. Peyre, and P. Quézel. 1980. Sur la presence au Maroc de *Laurus azorica* (Seub) Franco. Ann. Jard. bot. Madrid 37, 2:467–472.
Barbour, M. G., and J. Major, eds. 1977. Terrestrial Vegetation of California. New York-London-Sydney-Toronto.
Böcher, T. W. 1978. Phytogeography of Greenland: Survey and outlook (in Russian with English summary). In The Arctic Floristic Region, ed. B. A. Yurtsev, 127–142. Leningrad.
Boissier, E. 1867–1888. Flora Orientalis. I–V. Geneva.
Bolós, O. 1958. Grupos corologicos de la flora balear. Publ. Inst. Biol. Apl. (Barcelona) 27:49–71.
Bor, N. L. 1942a. The relict vegetation of the Shillong Plateau. Assam. Ind. For. Records 3, 6:152–195.
———. 1942b. Some remarks upon the geology and flora of the Naga and Khasi Hills. In 150th Anniv. Vol. Royal Bot. Gard. Calcutta, 129–137.
Boulos, L. 1968. The genus *Sonchus* and allied genera in the Canary Islands. Cuad. Bot. 3:19–26.
Bramwell, D. 1972. Endemism in the flora of the Canary Islands. In Taxonomy, Phytogeography and Evolution, ed. D. H. Valentine, 141–159. London and New York.
———. 1974. Los bosques de Canarias, su historia y desarrollo. Revista el Museo Canario 35:13–27.
———. 1976. The endemic flora of the Canary Islands; distribution, relationships and phytogeography. In Biogeography and Ecology in the Canary Islands, ed. G. Kunkel, 207–240. The Hague.

Bramwell, D., and Z. Bramwell. 1974. Wild Flowers of the Canary Islands. London and Burford.
Braun, E. L. 1955. The phytogeography of unglaciated Eastern United States and its interpretation. Bot. Rev. 21, 6:297–375.
Braun-Blanquet, J. 1923a. Über die Genesis der Alpenflora. Verh. Naturf. Ges. Basel 1:243–261.
———. 1923b. L'origine et la développement des flores dans le Massif central de France. Paris and Zurich.
———. 1937. Sur l'origine des éléments de la flore méditerranééne. Stat. Inst. Géobot. Médit. Alpine, Montpellier 56:8–31.
Brice, W. C. 1966. South-West Asia. London.
Browicz, K. 1978. Chorology of trees and shrubs in south-west Asia, vol. 1 (with assistance from Jerzy Zielinski). Kórnik near Poznan.
Cabrera, A. L., and A. Willink. 1973. Biogeografia de America Latina. Ser. Biol. Monogr. 13. Org. American States. Washington, D.C.
Campbell, D. H., and I. L. Wiggins. 1947. Origins of the flora of California. Stanford Univ. Publ. Biol. Sci. 10:1–20.
Candolle, A. P. de. 1808. Rapports sur deux voyages botaniques et agronomique dans les départments de l'Quest et du Sud-Quest de la France. Paris.
Casaseca, B. 1969. Les enclavas mediterráneous en la Espana atlántico-centrooeuropea. Publ. Univ. Sevilla, V Simp. Fl. Europaea, 49–52.
Chaney, R. W. 1940. Tertiary forests and continental history. Bull. Geol. Soc. Amer. 51:469–488.
Chatterjee, D. 1940. Studies on the endemic flora of India and Burma, J. Roy. Asiatic Soc. Bengal 5, 1:19–67.
———. 1962. Floristic patterns of Indian vegetation. Proc. Summer School of Bot., Darjeeling 1960:32–42. New Delhi.
Cherneva, O. V. 1974. A concise analysis of the geographic distribution of species of *Cousina* (in Russian). Bot. Zhurn. 59 (2):183–191.
Chevalier, A. 1935. Les Iles du Cap Vert: Flore de l'Archipel. Rev. Bot. Appl. 15:733–1030.
Ching Zehn-chan. 1958. Geography and Floristic Composition of the Pteridophyte Flora of China (in Russian). Reports of Foreign Scholars at the Delegates Conference of the All-Union Botanical Society, 56–70. Leningrad.
Chodat, L. 1924. Contributions à la géo-botanique de Majorque. Univ. Genève, Inst. Bot. 10, 6:3–116.
Chopik, V. I. 1976. High Mountain Flora of Ukrainian Carpathians (in Ukrainian). Kiev.
Ciferri, R. 1962. La laurisilva canaria: una paleoflora vivente. Ricerca scient. 32 (1):111–134.
Clarke, C. B. 1898. Subareas of British Empire, illustrated by the detailed distribution of Cyperaceae in that Empire. J. Linn. Soc. Lond. 34:1–146.
Clausen, R. T. 1959. *Sedum* of the Trans-Mexican Volcanic Belt. Ithaca, New York: Cornell Univ. Press.
Constance, L. 1963. Amphitropical relationships of the herbaceous flora of the Pacific Coast of North and South America: Introduction and historical review. Quart. Rev. Biol. 38:109–116.

Contandriopoulos, J. 1962. Récherches sur la flore endémique de la Corse et sur ses origines. Ann. Fac. Sc. Marseille 32:1–354.
Cronquist, A. 1978. The biota of the Intermountain region in geohistorical context. In Intermountain Biogeography: A Symposium Great Basin Naturalist Memoirs, 2:3–15.
Czeczott, H. 1926. The Atlantic element in the flora of Poland. Bull. Acad. Pol. Sci. Pettr., ser. B:361–407.
———. 1932. The distribution of some species in Asia Minor and the problem of Pontide. Mitt. Köngl. Naturwiss. Inst. Sofia 10:43–68.
———. 1938–39. A contribution to the knowledge of the flora and vegetation of Turkey. Fed. Rep. Beih., Berlin, 107.
Daubenmire, R. E. 1969. Ecologic plant geography of the Pacific Northwest. Madroño 20:11–128.
Davis, P. H. 1965. Phytogeography of Turkey. In Flora of Turkey and the East Aegean Islands, ed. P. H. Davis, 1:16–26. Edinburgh.
———. 1971. Distribution patterns in Anatolia with particular reference to endemism. In Plant Life of South-West Asia, ed. P. H. Davis, 15–27. Edinburgh.
Davis, P. H., and I. C. Hedge. 1971. Floristic links between N.W. Africa and S.W. Asia. Ann. Naturhist. Mus. Wien 75:43–57.
Davy de Virville, A. 1961. Contribution a l'étude de l'endémisme végétal dans l'Archipel des Canaries. Rev. Gén. Bot. 68:201–213.
Deb, D. B. 1958. Endemism and outside influence on the flora of Manipur. J. Bombay Nat. Hist. Soc. 55, 2:313–317.
———. 1961. Monocotyledonous plants of Manipur Territory. Bull. Bot. Surv. India 3:115–138.
Detling, L. E. 1968. Historical background of the flora of the Pacific Northwest. Bull. Mus. Nat. Hist. Univ. Oregon 3: 1–57.
Dice, L. R. 1943. The Biotic Provinces of North America. Ann Arbor: Univ. Michigan Press, 1–78.
Diels, L. 1901. Die Flora von Central-China. Nach der verhändenon Literatur und neu mitgeteilten original Materiale. Engl. Bot. Jb. 29, 2–5:169–659.
———. 1913. Untersuchungen zur Pflanzengeographie von West-China. Engl. Bot. Jb. 49, Beibl. 109:55—88.
———. 1942. Über die Ausstrahlunger des Holarktischen Florenreches an seinem Südrande. Abhandlungen der Preuss. Akad. der Wissensch., Math.-Naturw. Kl., Jg. 1–42, 1. Berlin.
Dobremez, J. F. 1972. Mise au point d'une méthode cartographique d'étude des montagnes tropicales. Le Népal, écologie et phytogéographie. Thesis, Grenoble.
Doronin, Yu. A. 1973. Cretaceous Pine Forests in the Central Russian Hills and the Donetsk Range. Candidate's thesis. Voronezh.
Douglas, G. W., et al. 1981. The Rare Vascular Plants of the Yukon. Syllogeus 28, National Museum of Natural Sciences, National Museum of Canada, Ottawa.
Dressler, R. L. 1954. Some floristic relationships between Mexico and the United States. Rhodora 56:81–96.
Dupont, P. 1962. La flore atlantique européenne. Introduction à l'étude phytogéographique du secteur ibéro-atlantique. Toulouse.
Eig, A. 1931. Les éléments et les groupes phytogéographiques auxiliaires dans la flore

palestinienne. Rep. Sp. Nov. Regni Veg. Beih. 63:1–201.
Engler, A. 1881. Über die morphologischen Verhältnisse und die geographische Verbreitung der Gattung *Rhus*, sowie der mit ihr verwandten lebenden und ausgestorbenen Anacardiaceae. Bot. Jahrb. 1:365–427.
———. 1902. Die pflanzengeographische Gliederung Nordamerikas. Notizbl. Bot. Gart. Mus. Berlin, App. IX, 1–94.
Engler, A., and E. Gilg. 1919. Übersicht über die Florenreiche und Florengebiete der Erde von A. Engler. In Syllabus der Pflanzenfamilien, A. Engler, 8th ed., 352–364. Berlin.
Favarger, C. 1971. Relations entre la flore méditerranéene et celle des enclaves à végétation subméditerranéene d'Europe centrale. Boissiera 19:149–168.
———. 1972. Endemism in the mountane floras of Europe. In Taxonomy, Phytogeography and Evolution, ed. D. H. Valentine, 191–204. London and New York.
Fedchenko, B. A. 1925. An Outline of the Vegetation of Turkestan (in Russian). Leningrad.
Fedorov, An. A. 1957. The flora of southwestern China and its importance for understanding the vegetation world of Eurasia (in Russian). In Ten Years since the Death of V. L. Komarov, 24–50. Moscow and Leningrad.
———. 1958. Concerning floristic relations of eastern Asia with the Caucasus, as indicated by study of the genus *Pyrus* L. (in Russian). In Materials for the History of the Flora and Vegetation of the USSR 3:230–248. Moscow and Leningrad.
———. 1979. Phytochoria of the Europaean part of the USSR (in Russian). In Flora of the Europaean Part of the USSR, ed. An. A. Fedorov, 4:10–27. Leningrad.
Fernald, M. L. 1925. Persistence of plants in unglaciated areas of boreal America. Mem. Amer. Acad. Arts and Sci. 15:239–342.
———. 1931. Specific segregations and identities in some floras of eastern North America and the Old world. Rhodora 33:25–63.
Fischer, C. E. C. 1938. The flora of Lushai hills. Rec. Bot. Surv. India 12, 2:75–161.
Flahault, Ch. 1937. La distribution géographique des végétaux dans la Région méditerranéenne française. Paris.
Font Quer, P. 1927. La flora de las Pitiusas y sus afinidadas con la de la Peninsula Ibérica. Mem. Real Acad. Cienc. Barcelona, 3a ép, 20, 4:109–154.
Gabrielyan, N. Ts. 1978. The Genus *Sorbus* in Western Asia and the Himalayas (in Russian). Erevan.
Gajewski, W. 1937. Les éléments de la flore de la Podolie polonaise. Planta polonica, Contr. Fl. Pologne, 5:1–210.
Gaussen, H. 1938. Carte floristique de la France. Ann. de Géogr. (Paris), 47° année, 237–256.
———. 1965. La division de l'Europe occidentale en ensembles floristiques. Rev. Roum, Biol. (Bot.), Bucarest 10, 1–2:71–76.
Gaussen, H., and C. Leredde. 1949. Les endémiques pyrénéo-cantabriques dans la région centrale de Pyrénées. Bull. Soc. Bot. Fr., 96, 76° sess. extraord., 57–83.
Giacomini, V., and L. Fenaroli. 1958. Conosci l'Italia: la Flora. Touring Club Italiano. Milano.
Gleason, H. A. 1926. The individualistic concept of the plant association. Bull. Torrey Bot. Club 53:7–26.

Golitsin, S. V., and Yu. A. Doronin. 1970. Relict flora and vegetation (in Russian). In Relics of Nature of the Voronezh Oblast. Voronezh.

Gorchakovsky, P. L. 1963. Endemic and relict elements in the flora of the Urals and their origin (in Russian). In Materials for the History of the Flora and Vegetation of the USSR, 4:285–375. Moscow and Leningrad.

———. 1968. Plants of the European broadleaved forests in the eastern limit of their distribution (in Russian). Transactions of the Institute of Ecology of Plants and Animals of the Ural Branch of the Academy of Sciences of the USSR (Sverdlovsk) 59:1–207.

———. 1969. Fundamental Problems in the History of the Phytogeography of the Urals (in Russian). Sverdlovsk.

Graham, A. 1964 [1965]. Origin and evolution of the biota of southeastern North America: Evidence from the fossil plant record. Evolution 18:571–608.

———. 1972. Outline of the origin and historical recognition of floristic affinities between Asia and Eastern North America. In Floristics and Palaeofloristics of Asia and Eastern North America, ed. A. Graham, 1–18. Amsterdam.

Graham, A., 1973. Vegetation and Vegetational History of Northern Latin America. Amsterdam: Elsevier Publ. Co.

Gray, A. 1846. Analogy between the flora of Japan and that of the United States. Amer. J. Sci and Arts II, 2:135–136.

———. 1859. Observations upon the relationship of the Japanese flora to that of N. America. Amer. J. Sci. and Arts, new ser., 6.

———. 1884. Characteristics of the North American flora. Amer. J. Sci. and Arts III, 28:323–340.

Gray, A., and J. D. Hooker. 1880. The vegetation of the Rocky Mountain region and a comparison with that of other parts of the world. Bull. U.S. Geol. Geogr. Surv. Territories, 6.

Greuter, W. 1972. The relict element of the flora of Crete and its evolutionary significance. In Taxonomy, Phytogeography and Evolution, ed. D. H. Valentine, 161–177. London and New York.

Grosset, G. E. 1967. Paths and time of migration of Crimean-Caucasus forest species into the territory of the Russian plain and subsequent changes of their distributions in relation with the evolution of landforms (in Russian). Bulletin of the Moscow Society for the Study of Nature, Biology Section, 72 (5):47–76.

Grossheim, A. A. 1936. Analysis of the Flora of the Caucasus (in Russian). Baku.

———. 1948. Vegetation Cover of the Caucasus (in Russian). Moscow.

Grossheim, A. A., and D. I. Sosnovsky. 1928. A contribution to the botanico-geographical subdivision of the Caucasus area (in Russian). Proceeding of the Tbilisi Polytechnic Institute 3:1–60.

Grubov, V. I. 1959. A Contribution to the Botanico-Geographic subdivision of Central Asia (in Russian). Leningrad.

———. 1963. Introduction (in Russian). In Plants of Central Asia, 1:5–69. Moscow and Leningrad.

Grubov, V. I., and An. A. Fedorov. 1964. Flora and vegetation of China (in Russian). In Physical Geography of China, ed. V. T. Zaichikov, 324–428. Moscow.

Gruenberg-Fertig, I. 1954. On the Sudano-Deccanian element in the flora of Palestine. Palest. J. Bot. 6:234–240.

Guest, E. R. 1966. The vegetation of Iraq and adjacent regions. In Flora of Iraq, ed. E. Guest, 1:55–108. Baghdad.
Guppy, H. B. 1917. Plants, Seeds and Currents in the West Indies and the Azores. London.
Handel-Mazzetti, H. 1926–27. Das nordost-birmanisch-west-yünnanesische Hochgebirgsgebiet. In Vegetationsbilder, eds. G. Karsten and H. Schenck, 17 Reihe, Haft 7/8:37–48. Jena.
———. 1931. Die pflanzengeographischen Gliederung und Stellung Chinas. Engl. Bot. Jahrb. 64:309–323.
Hansen, A. 1969. Checklist of the vascular plants of the Archipelago of Madeira. Bot. Mus. Municipal do Funchal (Funchal-Madeira) 24:5–61.
Hara, H. 1959. An outline of the phytogeography of Japan. In Distribution Maps of Flowering Plants in Japan, eds. H. Hara and H. Kanai, Fasc. 2:1–96. Tokyo.
———. 1972. Corresponding taxa in North America, Japan and the Himalayas. In Taxonomy, Phytogeography and Evolution, ed. D. H. Valentine, 61–72. London and New York.
Harshberger, J. W. 1911. Phytogeographic survey of North America. In Die Vegetation der Erde, A. Engler und O. Drude, XIII:1–92. Leipzig and New York.
Hedge, I. C. 1970. Some remarks on endemism in Afghanistan. Israel J. Bot. 19:401–417.
———. 1976. A systematic and geographical survey of the Old World Cruciferae. In The Biology and Chemistry of the Cruciferae, eds. J. G. Vaughan and B. M. G. Jones, 1–45. London.
Hedge, I. C., and P. Wendelbo. 1978. Patterns of distribution and endemism in Iran. Notes Roy. Bot. Gard. Edinburgh 36:441–464.
Holmgren, N. H. 1972. Plant geography of the Intermountain region. In Intermountain Flora, A. Cronquist et al., 1. New York: Hafner.
Hooker, J. D. 1862. Outlines of the distribution of Arctic plants. Trans. Linn. Soc. Lond. 23:251–348.
———. 1879. The distribution of the North American flora. Proc. Roy. Inst. 8:568–580 (and 1878. Gard. Chron. N. S. 10:140–142).
———. [1904] 1907. A sketch of the flora of British India. London. Reprinted in Imperial Gazetter of India, Oxford (3), 1, 4:157–212.
Horikawa, Y. 1972. Atlas of the Japanese flora. Tokyo.
Horvat, I., V. Glavač, and H. Ellenberg. 1974. Vegetation der Südosteuropas. Jena.
Horvatič, S. 1967. Fitogeografske značajke i raščlanjenje Jugoslavije. In Analitička Flora Jugoslavije 1:23–61. Zagreb.
Hosokawa, T. 1934. Phytogeographical relationship between the Bonin and the Marianne Islands, etc. J. Soc. Trop. Agr. 6:201–209.
Houérou, H. N. 1959. Recherches écologiques et floristiques sur la végétation de la Tunisie Méridionale. I–II. Univ. d'Algér, Inst. Rech. Sahar., Mém. 6.
Howell, J. T. 1957. The California flora and its province. Leafl. West. Bot. 8:133–138.
———. 1972. A statistical estimate of Munz' Supplement to a California flora. Wasmann J. Biol. 30:93–96.
Hultén, E. 1962, 1971. The Circumpolar Plants. I, II. Stockholm.
———. 1962. Flora of the Aleutian Islands, etc. Stockholm.

———. 1963. Phytogeographical connections of the North Atlantic. In North Atlantic Biota and Their History, eds. A. Löve and D. Löve, 45–72. Oxford.
Humphries, C. J. 1979. Endemism and evolution in Macaronesia. In Plants and Islands, ed. D. Bramwell, 171–199. Academic Press. London.
Igoshina, K. N. 1943. Remnants of broadleaved coenoses among the fir-spruce taiga of the central Urals (in Russian). Bot. Zhurn. 28 (4):144–155.
Iljin, M. M. 1941. Tertiary relict elements in the taiga flora of Siberia and their possible origin (in Russian). In Materials for the History of the Flora and Vegetation of the USSR 1:257–292. Moscow and Leningrad.
Jäger, E. 1968. Die pflanzengeographische Ozeanitätsgliederung der Holarktis und die Ozeanitätsbindung der Pflanzenareale. Fed. Rep. 79:157–335.
Jäger, E. J. 1971. Die pflanzengeographische Stellung der Steppen der iberischen Halbinsel. Flora 160:217–256.
Jepson, W. L. 1925. A manual of the flowering plants of California. Berkeley, Calif.
Johnston, I. M. 1941. Gypsophily among Mexican desert plants. J. Arnold Arbor. 22:145–170.
Kamelin, R. V. 1970. Botanico-geographic peculiarities of the flora of Soviet Kopet-Dag (in Russian). Bot. Zhurn. 55 (10):1451–1463.
———. 1973. Florogenetic Analysis of the Native Flora of Mountainous Middle Asia (in Russian). Leningrad.
Kanai, H. 1963. Phytogeographical observations on the japono-hymalayan elements. J. Fac. Sci. Univ. Tokyo, sect. 3, Bot. 8, 8:305–339.
Kanehira, R. 1933. On the ligneous flora of Formosa and its relationship to that of neighboring regions. Lingnan Sci. J. 12:225–239.
Keener, C. S. 1983. Distribution and biohistory of the endemic flora of the mid-Appalachian shale barrens. Bot. Rev. 49:65–115.
K'ien Ch'ung-shu, Wu Cheng-i and Ch'en Ch'ang-tao. 1957. The project for a geobotanical classification of China (in Russian). In Physico-Geographic Classification of China: A Collection of Articles, 1:131–216. Moscow.
Kitanov, B. 1976. Vegetation-geographic conditions in Bulgaria (in Bulgarian). In Botany, K. Popov, B. Kitanov, I. Ganchev, and A. Kotsev, 492–520.
Kleopov, Yu. D. 1941. Basic features of the development of the flora of broadleaved forests of the European part of the USSR (in Russian). In Materials for the History of the Flora and Vegetation of the USSR 1:183–256. Moscow and Leningrad.
Knapp, R. 1973. Die Vegetation von Afrika. Stuttgart. Fischer.
Kobayashi, S. 1978. A list of vascular plants occurring in the Ogasawara (Bonin) Islands. Ogasawara Research 1:1–33.
Kolakovsky, A. A. 1962. The Vegetation World of Colchis (in Russian). Moscow.
Komarov, V. L. 1897. Botanico-geographic regions of the Amur Basin (in Russian). Transactions of the Saint Petersburg Society of Naturalists 28 (1):1–13.
———. 1901. Flora of Manchuria (in Russian). Vol. I. St. Petersburg.
———. 1908. Introduction to the floras of China and Mongolia (in Russian). Transactions of the St. Petersburg Botanical Garden 29 (1):1–176.
Korovin, E. P. 1958. Phytogeographic regions (in Russian). In Middle Asia, 351–362. Moscow.
———. 1961. A contribution to the botanico-geographic subdivision of Middle Asia (in Russian). Transactions of Tashkent State University, New Series 186.

———. 1962. Vegetation of Middle Asia and Southern Kazakhstan (in Russian). 2. Tashkent.
Korzhinsky, S. I. 1899. Vegetation of Russia (in Russian). Encyclopediac Dictionary of Brokgaus and Efron 54:42–49. St. Petersburg.
Kozlovskaya, N. V., and V. I. Parfenov. 1972. Chorology of the Flora of Belorussia (in Russian). Minsk.
Kozo-Poljansky, B. M. 1931. In the Land of Living Fossils (in Russian). Moscow.
Koz'yakov, S. N. 1962. Eastern boundaries of oaks, lindens and maples on the Zilairian Plateau of the southern Urals (in Russian). Bot. Zhurn. 47 (5):705–709.
Krapivkina, E. D. 1973. Tertiary relics of the black taiga of the Kuznetsk Alatau (in Russian). Bull. of the Tomsk affiliation of the All-Union Botanical Society, Vol. VI. Tomsk.
Krasheninnikov, I. M. 1937. Analysis of the relict flora of the southern Urals in relation to the history of vegetation and the palaeogeography of the Pleistocene (in Russian). Soviet Botany 4:16–45.
———. 1939. Basic pathways of development of the vegetation of the southern Urals, in relation to the palaeogeography of northern Eurasia in the Pleistocene and the Holocene (in Russian). Soviet Botany 6–7:67–99.
Krylov, P. N. 1919. An Outline of the Vegetation of Siberia (in Russian). Tomsk.
Kudo, Y. 1927. Über die Pflanzengeographic der Nordjapans und der Insel Sachalin. Öster. Bot. Z. 76:307–311.
Kulczynski, S. 1924. Das boreale und arktisch-alpine element in der mitteleuropäischen Flora. Bull. Intern. Acad. Polon. Sci. Lettr. Cl. Sci. math.-natur., ser. B, Sci. natur. 1923:127–214.
Kuminova, A. V. 1960. Vegetation cover of Altai (in Russian). Novosibirsk.
———. 1969. Fractional geobotanical subdivision of parts of the Altai-Sayansk geobotanical region (the right banks of the Yenisei; in Russian). In Vegetation of the Right Bank of the Yenisei, 67–135. Novosibirsk.
———. 1973. Characteristic features of the Altai-Sayansk geobotanical region (in Russian). Proceedings of the Tomsk Section of All-Union Botanical Society 6:23–24.
Kunkel, G. G. 1971. On some floristic relationships between the Canary Islands and the neighboring Africa. Mitt. Bot. Staatssaml. München, 10:368–374.
———, ed. 1976. Biogeography and ecology in the Canary Islands. In Monographie Biologicae, 30. The Hague.
Küpfer, P. 1974. Recherches sur les liens parenté entre la flore orophile des Alpes et celle des Pyrénées. Boissiera 23:1–322.
Kurentsova, G. E. 1962. Vegetation of the Prekhankai Plain and Surrounding Foothills (in Russian). Moscow and Leningrad.
Kuznetsov, N. I. 1901. Elements of the Mediterranean region in western Transcaucasia (in Russian). Memoirs of the Russian Geographic Society 23 (3):1–190.
———. 1909. Principles for the division of the Caucasus into botanico-geographic provinces (in Russian). Memoirs of the St. Petersburg Academy of Sciences 24 (1):1–174.
———. 1912. A contribution to the division of Siberia into botanico-geographic provinces (in Russian). Proceedings of the Saint Petersburg Academy of Sciences 14 (6):871–897.

Laasimer, L. R. 1959. The position of Estonia in phytogeographic and geobotanical classification of northern Europe (in Russian). Proceedings of the Academy of Science of the Estonia SSR, 8, Series Biol. 2:95–112.
Lavranos, J. J. 1975. Note on the northern temperate element in the flora of the Ethio-Arabian region. Boissiera 24a:67–69.
Lavrenko, E. M. 1930. Tertiary forest relict centers between the Carpathians and the Altai (in Russian). Journal of the Russian Botanical Society 15 (4):351–363.
———. 1938. History of the study of the flora and vegetation of the USSR based on present day distributions of plants (in Russian). In Vegetation of the USSR, 265–296. Moscow and Leningrad.
———. 1942. Concerning floristic elements and centers of development of the flora of the Eurasian Steppe region (in Russian). Soviet Botany 1–3:39–50.
———. 1950. Fundamental traits of the botanico-geographic division of the USSR and adjacent countries (in Russian). Problems of Botany 1:530–548.
———. 1962. Fundamental traits of the botanical geography of the deserts of Eurasia and northern Africa (in Russian). Komarov Lectures XV:1–169. Moscow and Leningrad.
———. 1965. Provincial division of the central Asiatic and Irano-Turanian Subregion of the Afro-Asiatic Desert Region (in Russian). Bot. Zhurn. 50 (1):3–15.
———. 1970a. Provincial division of the Black Sea-Kazakhstan Subregion of the Steppe Region of Eurasia (in Russian). Bot. Zhurn. 55 (5):609–625.
———. 1970b. Provincial division of the Central Asian Subregion of the Steppe Region of Eurasia (in Russian). Bot. Zhurn. 55 (12):1734–1747.
Lázaro Ibiza, B. 1895. Regiones botánicas de la Peninsula Ibérica. Anal. Soc. Esp. Hist. Nat., 2a, sér. 4:161–208.
Lems, K. 1960. Floristic botany of the Canary Islands. Sarracenia 4:1–94.
Leopold, A. S. 1950. Vegetation zones of Mexico. Ecology 31:507–518.
Leredde, C. 1957. Etude écologique et phytogéographique du Tassili d'Ajjer. Trav. Inst. Rech. Sahar., Alger.
Li, H. L. 1944. The phytogeographical division of China, with special reference to the Araliaceae. Proc. Acad. Nat. Sci. 96:249–277. Philadelphia.
———. 1963. Woody Flora of Taiwan. Narberth, Pennsylvania.
Li, H. L., and H. Keng. 1950. Phytogeographical affinities of southern Taiwan. Taiwania 1:103–128.
Lid, J. 1968. Contributions to the flora of the Canary Islands. Skr. Norske Vid. Akad. Oslo, Math.-Naturv. Kl. 23:1–212.
Liou, T. N. 1934. Essai sur la géographie botanique de Nord et de l'Ouest de la Chine. Contr. Inst. Bot. Nat. Acad. 2:423–451. Peking.
Lipmaa, T. M. 1935. Eesti geobotaanika pöhij coni. Acta et Comm. Univ. Tartuensis 28, 4:1–149.
Litvinov, D. I. 1891. Geobotanical notes about the flora of European Russia (in Russian). Bulletin of the Moscow Society for the Study of Nature, 3.
———. 1927. On some botanico-geographic relations in our flora (in Russian). Leningrad.
Louis, H. 1939. Das natürliche Pflanzenkleid Anatoliens. Stuttgart.
Löve, A. and D. Löve. 1975. Cytotaxonomical Atlas of the Arctic flora. Vaduz.
Lucas, G. Ll., and S. M. Walters, eds. 1976. List of Rare, Threatened and Endemic

Plants for the Countries of Europe. Kew.
McVaugh, R. 1943. The vegetation of the granitic flat-rocks of south-eastern United States. Ecological Monogr. 13:120–166.
Maekawa, F. 1974. Origin and characteristics of Japan's flora. In The Flora and Vegetation of Japan, ed. M. Numata 1–10, 3–86. Tokyo.
Magakyan, A. K. 1941. Vegetation of the Armenian SSR (in Russian). Moscow and Leningrad.
Maheshwari, P., J. C. Sen-Gupta, and C. S. Venkatesh. 1965. Flora of India. The Gazeter of India 1:163–229.
Maire, R., and Th. Monod. 1950. Études sur la flore et la végétation du Tibesti. Mem. Inst. Franç., Afr. Moire 8:7–140.
Maleev, V. P. 1931. Vegetation of the region of the Novorossiysk-Mikhailovsk Pass and its relation to the Crimea (in Russian). In Memoirs of the Nikitski Botanical Garden 13 (2):71–174.
———. 1938. Vegetation of the Black Sea country (Euxine Province of the Mediterranean), its origin and relations (in Russian). Transactions of the Botanical Institute of the Academy of Sciences of the USSR, Series III, 4:135–251.
———. 1941. Tertiary relicts in the flora of the western Caucasus and basic stages in the Quaternary history of its flora and vegetation (in Russian). In Materials for the History of the Flora and Vegetation of the USSR 1:61–144. Moscow and Leningrad.
———. 1947. The Mediterranean forest Region (in Russian). In Geobotanical subdivision of the USSR, 72–86. Moscow and Leningrad.
Malyshev, L. I., ed. 1972. The High Mountain Flora of the Stanovoi Range: Composition, Features, Genesis (in Russian). Novosibirsk.
Mani, M. S., ed. 1974. Ecology and Biogeography in India. The Hague.
Markgraf, F. 1952. Zur Abgrenzung der Mittelmeervegetation.Vegetatio 3: 324–325.
Masamune, G. 1931. A table showing the distribution of all the genera of flowering plants which are indigenous to the Japanese Empire. Ann. Rep. Taihoku Bot. Gard. 1:51–92.
———. 1933. Phytogeographical position of Yakusima, an island situated on the northern part of the Riu-kiu Archipelago. Proc. Fifth Pacific Sci. Congr. Canada, 4.
Merriam, C. H. 1890. Results of a biological survey of the San Francisco Mountain region and desert of the Little Colorado, Arizona, U.S. Dept. Agr. Amer. Fauna, 3.
———. 1893. The geographic distribution of life of North America, with special reference to the Mammalia. Ann. Rep. Regents Smithsonian Inst. for 1891,365–415.
———. 1898. Life zones and crop zones of the United States. U.S. Dept. Agr. Biol. Survey Bull. 10.
Merxmüller, H. 1952. Untersuchungen zur Sippengliederung und Arealbildung in den Alpen. München.
Meusel, H. 1965. Die Reliktvegetation der Kanarischen Inseln in ihren Beziehungen zur Sud- und mitteleuropäischen Flora, S. 117–136. In Gesammelte Vorträge über moderne Probleme der Abstammungslehre, ed. M. Gersh, 1. Jena.
Meusel, H., E. Jäger, and E. Weinert. 1965. Vergleichende Chorologie der Zen-

traleuropeischen Flora. Jena.
Meusel, H., and H. Schubert. 1971. Beiträge zur Pflanzengeographie des Westhimalayas. Flora 160, 2:137–194; 4:373–432; 6:573–606.
Miranda, F., and E. Hernandez X. 1963. Los tipos de vegetacion de Mexico y su clasificacion. Bol. Soc. Bot. Mex. 28:29–179.
Miranda, F., and A. J. Sharp. 1950. Characteristics of the vegetation in certain temperate regions of Eastern Mexico. J. Ecol. 31:313–333.
Miyabe, K., and M. Tatewaki. 1937. On the significance of the Schmidt Line in the plant distribution in Saghalien. Proc. Imp. Acad. Tokyo 13:24–26.
Moggi, G. 1969. Some reflections on the phytogeographical subdivision of Italy. In V simposio de Flora Europea, Sevilla, 229–444.
Monod, T. 1957. Les grandes divisions chorologiques de l'Afrique. Cons. Sci. Afr. Sud. du Sahara (C.S.A.) 24:1–147. London.
Montserrat, P., and L. Villar. 1972. El endemismo iberico. Aspectos ecologicos y fitotopographicos. Bol. Soc. Broter 46:503–527.
Munz, P. A. 1959. A California Flora. Berkeley and Los Angeles: Univ. Calif. Press.
Nakai, T. 1928*a*. The floras of Tsusima and Quelpaert as related to those of Japan and Korea. Proc. Third Pan-Pacific Sci. Congr. 1:893–911.
———. 1928*b*. The vegetation of Dagelet Island, its formation and floral relationship with Korea and Japan. Proc. Third Pan-Pacific Sci. Congr. Tokyo 1:911–914.
———. 1930. The flora of Bonin Islands. Bull. Biogeogr. Soc. Jap. 1:249–278.
Numata, M., ed. 1974. The Flora and Vegetation of Japan. Tokyo.
Ogureeva, G. N. 1962. The position of the Primorski-Krai in botanico-geographic subdivision (in Russian). In Questions on the Natural Subdivision of the Soviet Far East in Relation to Regional Planning, eds. Yu. P. Parmuzin and A. E. Krivolutsky, 134–148.
Ornduff, R. 1974. Introduction to California Plant Life. Berkeley, Los Angeles, and London: Univ. Calif. Press.
Ovchinnikov, P. N. 1957. Basic features of the vegetation and regions of the flora of Tadzhikistan (in Russian). In Flora of the Tadzhik SSR 1:9–20.
Ozenda, P. 1958. Flora du Sahara Septentrional et Central. Paris.
Paczoski, I. 1910. Basic Features of the Development of the Flora of Southwestern Russia (in Russian). Memoirs of the Novorossiysk Society of Naturalists. Kherson.
Patschke, W. 1913. Über die extratropischen ostasiatischen Coniferen and ihre Bedeutung für die pflanzengeographische Gliederung Ostasiens. Bot. Jb. 48:626–776.
Pavlov, N. V. 1929. Introduction to the vegetation cover of the Khangai Mountain-country: Preliminary account of the 1926 botanical expedition into northern Mongolia (in Russian). Materials of the Commission for the Study of Mongolia and Tannu Tuva Region 2:3–72. Leningrad.
Pavlov, V. M. 1972. Botanico-geographic subdivision of the western Tien Shan (in Russian). Bulletin of the Moscow Society for the Study of Nature. Biology Division 77 (6):99–100.
Pawlowski, B. 1970. Remarques sur l'endémisme dans la flore des Alpes et des Carpates. Vegetatio 21:181–243.
Peshkova, G. A. 1966. The Onon-Argunsk Steppes and their place in a system of

botanico-geographic subdivision (in Russian). Proceedings of the Siberian Branch of the Academy of Sciences of the USSR, Series of Biology and Medical Sciences 1:21–27.

———. 1968. Features of the flora and vegetation of extreme southeastern Dahuria, Nerchinsko-Zavod region (in Russian). Bot. Zhurn. 53 (7):990–992.

———. 1972. The Steppe Flora of Baikalian Siberia (in Russian). Moscow.

———. 1976. The question of the botanico-geographic borders of Dahuria (in Russian). Proceedings of the Siberian Branch of the Academy of Sciences of the USSR, Biology Series, 5 (1):39–45.

Polunin, N. 1951. The real Arctic: Suggestions for its delimitation, subdivision and characterization. J. Ecol. 31:308–315.

———. 1959. Circumpolar Arctic Flora. Oxford.

Polunin, O., and A. Huxley. 1965. Flowers of the Mediterranean. London.

Popov, M. G. 1927. Basic traits of the history of development of the flora of Middle Asia (in Russian). Bulletin of Middle Asian State University 15:239–292.

———. 1929. The genus *Cicer* and its species: Toward the problems of the origin of the Mediterranean flora (in Russian). Transactions of Applied Botany, Genetics and Selection 29 (1).

———. 1949. An Outline of the Flora of the Carpathians (in Russian). Moscow.

———. 1950. The use of botanico-geographic methods in the systematica of plants (in Russian). Problems of Botany 1:70–108.

———. 1957. Flora of Middle Siberia (in Russian). Moscow and Leningrad.

Prozorovsky, A. V., and V. P. Maleev. 1947. The Asiatic desert Region (in Russian). In: Geobotanical subdivision of the USSR, 111–146. Moscow and Leningrad.

Quarterman, E. 1950. Major plant communities of Tennessee cedar glades. Ecology 31:234–254.

Quarterman, E., and C. Keever. 1962. Southern mixed hardwood forest: Climax in the southeastern coastal plain, U.S.A. Ecol. Monogr. 32:167–185.

Quézel, P. 1954. Contribution a l'étude de la flore et de la végétation du Hoggar. Trav. Inst. Rech. Sahar., Monogr. régionales 2.

———. 1958. Mission botanique au Tibesti. Inst. Rech. Sahar. Alger Mém. 4.

———. 1964*a*. Contribution à l'étude de l'endémisme ches les Phanérogames sahariens. C. R. Soc. biogéogr. 359:89–103.

———. 1964*b*. L'endémisme dans la flore de l'Algérie. C. R. Soc. biogéogr. 361:137–149.

———. 1965. La végétation du Sahara du Tchad à la Mauritanie. Stuttgart.

———. 1978. Analysis of the flora of Mediterranean and Saharan Africa. Ann. Missouri Bot. Gard. 65:479–534.

Rao, A. C. 1974. The vegetation and phytogeography of Assam-Burma. In Ecology and Biogeography in India, ed. M. S. Mani, 204–246. The Hague.

Rau, M. A. 1974. Vegetation and phytogeography of the Himalaya. In Ecology and Biogeography in India, ed. M. S. Mani, 247–280. The Hague.

Raup, H. M. 1974. Some natural floristic areas in Boreal America. Ecol. Monogr. 17:221–234.

Raven, P. H. 1962. The genus *Epilobium* in the Himalayan Region. Bull. Brit. Mus. (Nat-Hist.), Botany 2, 12:327–382.

———. 1963. Amphitropical relations in the floras of North and South America. Quart. Rev. Biol. 38:151–177.

———. 1977. The California flora. In Terrestrial Vegetation of California, eds. M. G. Barbour and J. Major, 109–137. New York.

Raven, P. H., and D. I. Axelrod. 1974. Angiosperm biogeography and past continental movements. Ann. Missouri Bot. Gard. 61:539–673.

Raven, P. H., and D. I. Axelrod. 1978. Origin and relationships of the California flora. Univ. Calif. Publ. Bot. 72:1–134.

Rechinger, K. H. 1949–50. Grundzuge der Pflanzenverbreitung in der Ägäis. Vegetatio 2:55–119.

———. 1951*a*. Phytogeographia Ägäea. Denkschr. Akad. Wiss. Wien, Math.-Nat. Kl. 105, 3:1–208.

———. 1951*b*. Grundzuge der Pflanzenverbreitung in Iran. Verhandl. Zool. Bot. Ges. Wien 92:181–188.

———. 1965. Der Endemismus in der griechischen Flora. Rev. Roum, Biol., ser. Bot., 10:135–138.

Regel, C. 1959. Vegetationzonen und Vegetationsstufen in der Türkei. Fed. Rep., Beih. 138:230–282.

Reichert, I. 1936. L'Afrique de Nord et sa position phytogéographique au point de vue lichénologique. Bull. Soc. bot. Fr. 83:836–841.

Reverdatto, V. V. 1931. Vegetation of the Siberian Krai: An attempt at a detailed subdivision (in Russian). Proceedings of the Geographic Society 63 (1):43–70.

Rikli, M. 1934. Geographie der Pflanzen. In Handwörterbuch der Naturwissenschaften 4:907:1002. Jena.

———. 1943, 1946, 1948. Das Pflanzenkleid der Mittelmeerländer. I–III. Bern.

Rivas-Martinez, S. 1973. Avance sobre una sintesis corológica de la Peninsula Iberica, Baleares y Canarias. Anal. Inst. Bot. Cavanilles 30:69–87. Madrid.

Rivas-Martinez, S., C. Arnaiz, E. Barreno, and A. Crespo. 1977. Apuntes sobre las provincias corologicas de la Peninsula Iberica e Islas Canarias. Opuscula Botanica Pharmaciae Complutensis 1:1–48.

Roi, H. 1941. Phytogeography of Central Asia. Bull. Fan. Mem. Inst. Biol. Bot. Ser. 11, 5:1–35. Peking.

Roisin, P. 1969. Le domaine phytogéographique atlantique d'Europe. Gembloux.

Rubtsov, N. I., and L. A. Privalova. 1964. Flora of the Crimea and its geographic relations (in Russian). Transactions of the Nikitsky Botanical Garden 37:16–35.

Runemark, H. 1971. Distribution patterns in the Aegean. In Plant Life of South-West Asia, ed. P. H. Davis, 3–14. Edinburgh.

Rzedowski, J. 1962. Contribuciones a la fitogeografia floristica e historica de Mexico. I. Algunas consideraciones acerca del elemento endemico en la flora Mexicana. Bol. Soc. Bot. Mex. 27:52–65.

———. 1965. Relaciones geograficos y posibles origenes de la flora de México. Bol. Soc. bot. Mex. 29:121–177.

———. 1978. Vegetacion de Mexico. Editorial Limusa, Mexico., D.F.

Safarov, I. S. 1977. The new habitat of the Ironwood *Parrotia persica* (DC.) C. A. Mey (family Hamamelidaceae Lindl.) on the Great Caucasus (in Russian). Bot. Zhurn. 62:248–250.

Schmid, E. 1944. Die "atlantische" Flora, eine kritische Betrachtung. Ber. Geobot. Inst. Rübel 1945:124–140.

Schischkin, B. K. 1947. Vegetation of the Soviet Union (in Russian). In Great Soviet Cyclopedia, USSR, 1st ed., 182–215, Moscow.

Schofield, W. B. 1966. Phytogeography of north-western North America: Bryophytes and vascular plants. Madroño 20:155–207.

Scoggan, H. J. 1966. The flora of Canada. Reprinted from the Canada Year Book 1965–1966:1–27. Ottawa.

———. 1978. The flora of Canada. Part 1. National Museum of Natural Science, National Museums of Canada, Ottawa.

Sharp, A. J. 1953. Notes on the flora of Mexico: World distribution of the woody dicotyledonous families and the origin of the modern vegetation. J. Ecol. 41, 2:374–380.

Shreve, F. 1951. Vegetation of the Sonoran Desert. Washington.

Shumilova, L. V. 1962. Botanical Geography of Siberia (in Russian). Tomsk.

Sjörs, H. 1956. Nordisk växtgeografi. Stockholm.

Sochava, V. B. 1964. Classification and cartography of the highest subdivisions of vegetation of the World (in Russian). In Current Problems of Geography, 167–173. Moscow.

Solbrig, O. 1972. The floristic disjunctions between the "Monte" in Argentina and the "Sonoran Desert" in Mexico and the United States. Ann. Missouri Bot. Gard. 59:218–233.

Sprygin, I. I. 1936. Some forest relicts of the Volga uplands (in Russian). Scientific Notes of Kazan State University 96 (6):67–117.

———. 1941. Relict plants of Povolzhie (in Russian). In Materials for the History of the Flora and Vegetation of the USSR 1:293–314. Moscow and Leningrad.

Stearn, W. T. 1946. The floristic regions of the USSR with reference to the genus *Allium*. Herbertia (1944) II:45–63.

———. 1960. *Allium* and *Milula* in the Central and Eastern Himalayas. Bull. Brit. Mus. Nat. Hist. Bot. 2, 6:159–192.

Stebbins, G. L., and J. Major. 1965. Endemism and speciation in the California flora. Ecol. Monogr. 35:1–35.

Stefanov, B. 1924. Mountain formations in the north of Strandsha (in Bulgarian). Ann. of Sofia University, Agronomy Department 5:23–68.

———. 1943. Phytogeographic elements in Bulgaria (in Bulgarian). Symposium of the Bulgarian Academy of Sciences and Research, book 39, 19:1–509.

Stojanoff, N. 1930. Versuch einer Analyse des relicten Elements in der Flora der Balkanhalbinsel. Engl. Bot. Jb. 63, 5:1–368.

Strid, A. 1972. Some evolutionary and phytogeographical problems in the Aegean. In Taxonomy, Phytogeography and Evolution, ed. D. H. Valentine, 289–300. London and New York.

Sunding, P. 1970. Elementer i kanariøyenes flora, og reorier til forklaring av floraens opprinnelse. Blyttia 28, 4:229–259.

———. 1972. The vegetation of Gran Canaria. Skr. Norske Vid. Akad. Oslo, Math.-Nat. Kl., Ny Ser. 29:1–186.

———. 1973. Endemism in the flora of the Cape Verde Islands, with special emphasis on the Macaronesian flora element. Monogr. Biol. Canar. 4:112–117.

———. 1979. Origins of the Macaronesian flora. In Plants and Islands, ed. D. Bramwell, 13–40. London.
Takhtajan, A. L. 1941. Botanico-geographic outline of Armenia (in Russian). Transactions of the Botanical Institute of the Armenian Branch of the Academy of Sciences of the USSR 2:1–180.
———. 1972. Botanico-geographic outline of the Ararat Depression (in Russian). In Flora of Erevan, eds. A. L. Takhtajan and An. A. Fedorov, 7–36. Leningrad.
Tamura, M. 1966. Morphology, ecology and phylogeny of the Ranunculaceae. V. Sci. Repts. College Gen. Ed. Osaka Univ. 14, 2:27–48.
Tatewaki, M. 1963a. Hultenia. J. Fac. Agric. Hokkaido Univ., Sapporo, 53, 2:131–199.
———. 1963b. Phytogeography of the islands of the North Pacific Ocean. In Pacific Basin Biogeography, ed. J. L. Gressitt, 23–28. Honolulu.
Tavares, C. N. 1965. Ilha da Madeira o meio e a Flora. Rev. Fac. Cienc. Lisboa, sér. 2, 13:51–174.
Thorne, R. F. 1969. The California Islands. Ann. Missouri Bot. Gard. 56:391–408.
Tolmatchev, A. I. 1955. Geobotanical Subdivision of Sakhalin Island (in Russian). Moscow and Leningrad.
———. 1956. A contribution to the study of the arctic flora of the USSR (in Russian). Bot. Zhurn. 41:783–796.
———. 1959. The Flora of Sakhalin Island (in Russian). Komarov Lectures XII. Moscow and Leningrad.
———, ed. 1974. Endemic High Mountain Plants of Northern Asia (in Russian). Novosibirsk.
Tumadzhanov, I. I. 1969. Major questions in the botanico-geographic subdivision of mountainous areas (in Russian). Problems of Botany 11: 27–46.
Turin, T. G. 1953. The vegetation of the Azores. J. Ecol. 41: 53–61.
Turrill, W. B. 1929. The Plant Life of the Balkan Peninsula. Oxford.
Tuyama, T. 1953. On the phytogeographical status of the Bonin and Volcano Islands. Proc. Seventh Pacific Sci. Congr. (Auckland and Christchurch, N.Z.) 5:208–212.
Tuyama, T., and S. Asami. 1970. The Nature in the Bonin Islands. Tokyo.
Vasiliev, V. N. 1956. Botanico-geographic subdivision of eastern Siberia (in Russian). Studies of the A. I. Gertsen Leningrad State Pedagogic Institute 116:61–102.
Vinogradov, N. P., S. V. Golitsin, and Yu. A. Doronin. 1960. The white mountain of the Don—A new region of the "Snowy alps" of the Middle Russian hills (in Russian). Bot. Zhurn. 45:524–532.
Vyas, L. N. 1964. Studies on phytogeographical affinities of the flora of North-East Rajasthan. Ind. Forest. 90, 8:535–538.
Walker, E. H. 1976. Flora of Okinawa and the Southern Ryukyu Islands. Washington.
Walter, H. 1956. Das Problem der Zentralanatolischen Steppe. Die Naturwissenschaften 43:97–102.
Warburg, O. 1891. Eine Reise nach der Bonin- und Volcano-Inseln. Verh. Ges. Erdkunde zu Berlin 18:248–266.
Ward, F. Kingdon. 1936. A sketch of the vegetation and geography of Tibet. Proceed. Linnean Soc. London 148 (3):133–160.

———. 1942. An outline of the vegetation and flora of Tibet. In 150th Anniversary Vol. Roy. Bot. Gard. Calcutta, 99–103.
———. 1944, 1945. A sketch of the botany and geography of North Burma. J. Bombay Nat. Hist. Soc. 44:550–574; 45:16–30, 133–148.
———. 1946*a*. Botanical explorations in North Burma. J. Roy. Hort. Soc. 71:318–325.
———. 1946*b*. Additional notes on the botany of North Burma. J. Bombay Nat. Hist. Soc. 46:381–390.
———. 1949. Plant Hunter in Manipur. London.
———. 1956. Return to Irrawady. London.
———. 1957. The great forest belt of northern Burma. Burmese Forester 7, 2:122–131.
———. 1959. A sketch of the flora and vegetation of Mount Victoria in Burma. Acta horti bergiani 22:53–74.
———. 1960. Pilgrimage for Plants. London.
Watson, S. 1890. The relation of the Mexican flora to that of United States (abstract). Proc. Amer. Ass. Advanc. Sci. 39:291–292.
Wendelbo, P. 1961. A monograph of the genus *Dionysia*. Acta Univ. Berg, ser. math.-Rer. nat. 3:6–86.
———. 1971. Some distributional patterns within the Flora Iranica area. In Plant Life of South-West Asia, ed. P. H. Davis, 29–41. Edinburgh.
Wester, L. 1981. Composition of native grasslands in the San Joaquin Valley, California. Madroño 28:231–241.
White, F. 1976. The vegetation map of Africa: The history of a completed project. Boissiera 24:569–666.
Whitmore, T. C., ed. 1981. Wallace's Line and Plate Tectonics. Oxford: Clarendon Press.
Whittaker, R. N. 1975. Communities and Ecosystems. 2d ed. New York: Macmillan.
Wickens, G. E. 1976. The flora of Jebel Marra (Sudan Republic) and its geographical affinities. Kew Bull., Addit. Ser. 5.
Wiggins, I. L. 1980. Flora of Baja California. Stanford, Calif.: Stanford Univ. Press.
Willkomm, M. 1896. Grundzüge der Pflanzenverbreitung auf der iberischen Halbinseln. In Die Vegetation der Erde, A. Engler and O. Drude, I:1–396. Leipzig.
Wilson, E. N. 1919. The Bonin Islands and their ligneous vegetation. J. Arnold Arbor. 1:97–115.
Wood, C. E., Jr. 1971. Some floristic relationships between the southern Appalachians and western North America. In The Distribution History of the Biota of the Southern Appalachians, Part II, ed. P. C. Holdt, 331–404. Flora. Res. Div. Monogr. 2, Virginia Polytechnic Inst. and State Univ. Blacksburg.
Woodson, R. E. 1947. Notes on the "historical factor" in plant geography. Contr. Gray Herb. 165:12–25.
Wu Cheng-yih. 1979. The regionalization of Chinese flora (in Chinese with English summary). Acta Bot. Yunnanica 1 (1):1–22.
Wynne-Edwards, V. C. 1937. Isolated arctic-alpine floras in eastern North America: A discussion of their glacial and recent history. Trans. Roy. Soc. Canada. Ser. 3, 31 (5):33–58.

Young, S. B. 1978. Phytogeography of the North American Arctic (in Russian). In The Arctic Floristic Region, ed. B. A. Yurtsev, 105–126. Leningrad.
Yurtsev, B. A. 1974. Problems of Botanical Geography of Northeastern Asia (in Russian). Leningrad.
Yurtsev, B. A., A. I. Tolmatchev, and O. V. Rebristaya. 1978. The floristic delimitation and subdivision of the Arctic (in Russian). In The Arctic Floristic Region, ed. B. A. Yurtsev, 9–104. Leningrad.
Zohary, M. 1940. Geobotanical analysis of the Syrian Desert. Palest. J. Bot. 2:46–96.
———. 1950. The flora of Iraq and its phytogeographical subdivision. Gvt. Iraq, Director, gen. Agric, Bull. 31:1–201.
———. 1962. Plant Life of Palestine, Israel and Jordan. New York.
———. 1963. On the geobotanical structure of Iran. Bull. Res. Council of Israel. Sect. D. Bot. 11D, Suppl., 1–113.
———. 1973. Geobotanical Foundations of the Middle East. I–II. Stuttgart.
Zohary, M., and G. Orshan. 1965. An outline of the geobotany of Crete. Israel J. Bot. 14, Suppl., 1–49.

II. PALEOTROPIC KINGDOM

Abeywickrama, B. A. 1955. The origin and affinities of the flora of Ceylon. Repr. Proc. 11th Ann. Sess. Ceylon Ass. Advanc. Sci., 1–23.
Ake Assi, L. 1964. Etude floristique de la Côte d'Ivoire. Paris.
Andrews, E. C. 1939. Origin of the Pacific insular floras. Proc. Sixth Pacific Sci. Congr., California 4:613–620.
Aubréville, A. 1962. Position chorologique de Gabon. Flore du Gabon 3:3–11.
Audley-Charles, M. G. 1981. Geological history of the region of Wallace's line. Wallace's Line and Plate Tectonics, ed. T. C. Whitmore, 24–35. Oxford: Clarendon Press.
Axelrod, D. I., and P. H. Raven. 1982. Paleogeography and origin of the New Guinea flora. The Biogeography and Ecology of New Guinea, ed. J. L. Gressitt, 919–941. The Hague.
Bader, F. J. W. 1960. Die Verbreitung boreale und subantarktische Holzgewächse in die Gebirgen des Tropengürteds. Nova Acta Leopoldina 23, 148:1–544.
Balgooy, M. M. J. van. 1960. Preliminary plant-geographical analysis of the Pacific. Blumea 10:385–430.
———. 1969. A study on the diversity of island floras. Blumea 17:139–178.
———. 1971. Plant-geography of the Pacific. Blumea Suppl. 6:1–222.
———. 1976. Phytogeography. In New Guinea vegetation, ed. K. Paijmans, 1–22. Canberra.
Barthlott, W. 1983. Biogeography and evolution in Neo- and Paleotropical Rhipsalinae (Cactaceae). Sonderbd. naturwiss. Ver. Hamburg 7:241–248.
Battistini, R., and R. Richard-Vindard, eds. 1972. Biogeography and Ecology in Madagascar. The Hague.
Baumann-Bodenheim, M. B. 1956. Über die Beziehungen der neu-caledonischen

Flora zu den tropischen un süd-hemisphärische-subtropischen bisextratropischen Flora und die gürtelmässige Gliederung der Vegetations von Neu-Caledonien. Ber. Geobot. Inst. Rübel 1955:64–74. Zurich.

Bews, J. W. 1921. An Introduction to the Flora of Natal and Zululand. Pietermaritzburg.

———. 1922. The South-East African flora: Its origin, migrations, and evolutionary tendencies. Ann. Bot. 36:209–225.

Bhandari, M. M. 1979. Phytogeography of the tropical flora of the Indian Desert. In Tropical Botany, eds. K. Larsen and L. B. Holm-Nielsen, 143–152. London.

Bharucha, F. R., and V. M. Meher-Homji. 1965. On the floral elements of the semi-arid zones of India and their ecological significance. New Phytol. 64, 2:330–342.

Biswas, K. 1943. Systematic and taxonomic studies on the flora of India and Burma. Proc. 30th Ind. Sci. Congr. 2:101–153.

Blasco, F. 1971. Montagnes du Sud de l'Inde. Madras.

Bor, M. L. 1938. The vegetation of the Nilgiris. Ind. Forest. 64:600–609.

Boughey, A. S. 1957. The Origin of the African Flora. Oxford.

———. 1965. Comparisons between the mountane forest floras of the North America, Africa and Asia. Webbia 19, 2:507:517.

Braitwaite, A. F. 1975. The phytogeographical relationships and origin of the New Hebrides fern flora. Phil. Trans. Roy. Soc. Lond., B272:293–313.

Bremekamp, C. E. B. 1935. The origin of the flora of the Central Kalahari. Ann. Transv. Mus. 16:443–455.

Brenan, J. P. M. 1978. Some aspects of the phytogeography of tropical Africa. Ann. Missouri Bot. Gard. 65:437–474.

———. 1979. The flora and vegetation of Tropical Africa. In Tropical Botany, eds. K. Larsen and L. B. Holm-Nielsen, 49–58. London.

Burkill, I. H. 1941–1942. A discussion on the biogeographical divisions of the Indo-Australian archipelago. Proc. Linn. Soc. Lond. 1954:127–138.

Campbell, D. H. 1928. The Australasian element in the Hawaiian flora. Proc. Third Pan-Pacific Sci. Congr. Tokyo 1:938–946.

———. 1933. The flora of the Hawaiian Islands. Quart. Rev. Biol. 8:164–184.

Capuron, R. 1963. Contributions à l'etude de la flore de Madagascar. Adansonia 3:370–400.

Carlquist, S. 1967. The biota of long-distance dispersal. V. Plant dispersal to Pacific islands. Bull. Torrey Bot. Club 94:129–162.

———. 1970. Hawaii: A Natural History. New York.

Champion, H. G. 1936. A preliminary survey of the forest types of India and Burma. Ind. Forest. Records, N.S. 1:1–286.

Chapman, J. D., and E. White. 1970. The Evergreen Forests of Malawi. Oxford.

Chatterjee, D. 1940. Studies on the endemic flora of India and Burma. J. Roy. Asiatic Soc. Bengal. 5, 1:19–67.

———. 1947. Influence of east Mediterranean region flora on that of India. Sci. and Culture 13:9–11.

———. 1962. Floristic patterns of Indian vegetation. Proc. Summer School of Bot., Darjeeling 1960:32–42. New Delhi.

Chevalier, A. 1928. La végétation montagnarde de L'Ouest Afrique et sa genèse. Mém. Soc. Biogéogr. 2.

———. 1932. Le territoire géo-botanique de l'Afrique tropicale nord-occidentale et ses subdivisions. Bull. Soc. Bot. Fr. 80:4–26.

Chevalier, A., and L. Emberger. 1937. Les régions botaniques terrestres. Encyclopédie française 5:5.64–1 to 5.66–7.

Chew, W.-L. 1975. The phanerogamic flora of the New Hebrides and its relationships. Phil. Trans. R. Soc. Lond. B272:315–328.

Clayton, W. D., and F. N. Hepper. 1974. Computer-aided chorology of West African grasses. Kew Bull. 29:213–234.

Clayton, W. D., and G. Panigrahi. 1974. Computer-aided chorology of Indian grasses. Kew Bull. 29:669–686.

Compton, R. H. 1917. New Caledonia and the Isle of Pines. Geogr. J. 49:81–106.

Copeland, E. B. 1948. The origin of the native flora of Polynesia. Pacific Sci. 2:293–296.

Corner, E. J. H. 1958. An introduction to the distribution of *Ficus*. Reinwardtia 4, 3:325–354.

———. 1960. The Malayan Flora. In Proceedings of Centenary and Bicentenary Congress of Biology, 1958 (Singapore), ed. R. D. Purchon, 21–24.

———. 1963. *Ficus* in the Pacific Region. In Pacific Basin Biogeography, ed. J. L. Gressitt, 233–245. Honolulu.

———. 1975. *Ficus* in the Hebrides. Phill. Trans. Roy. Soc. Lond. B272:343–367.

Cornet, A., and J.-L. Guillaumet. 1976. Divisions floristiques et étages de végétation à Madagascar. Cah ORSTOM, sér. Biol. 11, 1:35–42.

Croizat, L. 1968*a*. The biogeography of India: A note on some of its fundamentals. Proc. Symp. Recent. Adv. Tropical Ecol. (Varanasi, India) 2:544–590.

———. 1968*b*. Introduction raisenée à la biogéographie de l'Afrique. Mém. Soc. Broteriana 20:1–451.

———. 1968*c*. The biogeography of the tropical lands and islands east of Suez-Madagascar, with particular reference to the dispersal and formmaking of *Ficus* L. and different other vegetal and animal groups. Atti Ist. Bot. Lab. Critt. Univ. Pavia, ser. 6, 4:1–400.

———. 1969. Riflessioni sulla biogeografia in generale, e su quella della Malesia in particolare. Atti Ist. Bot. Lab. Critt. Univ. Pavia, ser. 6, 5:19–190.

Davis, D. 1969. When did the Seychelles leave India? Nature 220:1225–1226.

Dejardin, J., J.-L. Guillaumet, and G. Mangenot. 1973. Contribution à la connaissance de l'élement non endémique de la flore malgache (végétaux vasculaires). Candollea 28: 325–391.

Denys, E. 1980. A tentative phytogeographical division of tropical Africa based on a mathematical analysis of distribution maps. Bull. Jard. Bot. Not. Belg. 50:465–504.

De Wildeman, E. 1940. De l'origine de certaines éléments de la flore du Congo Belge. Mém. Inst. Roy. Colon. Belge 10:1–355.

De Winter, B. 1971. Floristic relationships between the northern and southern arid areas in Africa. Mitt. Bot. Staatssamml. München 10:424–437.

Diels, L. 1922. Beiträge zur Kenntniss der Vegetation und Flora der Seychellen.

Wiss. Ergebn. Deutsch. Tiefsee Exp. 2, 1–4:409–466.

———. 1930. Ein Beitrag zur Analyse der Hochgebirges-Flora von Neu Guinea. Bot. Jb. 63:324–329.

Durocher-Yvon, F. 1947. Seychelles botanical treasure. La Rev. Agric. de l'île Maurice, 26.

Engler, A. 1891. Über die Hochgebirgeflora des tropischen Africa. Berlin.

———. 1895. Die Pflanzenwelt Ost-Afrikas und der Nachbargebiete. Teil A. Grundzüge der Pflanzenverbreitung in Deutsch-Ost-Afrika und den Nachbargebiete. Berlin.

———. 1904. Plants of the Northern temperate zone in their transition to the high mountains of Tropical Africa. Ann. Bot. 18:523–540.

———. 1907. Syllabus der Pflanzenfamilien. 5. Berlin, 247 S.

———. 1908. Pflanzengeographische Gliederung von Afrika. Sitzb. Preuss. Akad. Wiss. 38:781–835.

———. 1908, 1910, 1915, 1925. Die Pflanzenwelt Afrikas, insbesondere seiner tropischen Gebiet. I, II, III, V (1). Leipzig.

Engler, A., and E. Gilg. 1919. Übersicht über die Florenreiche und Florengebiete der Erde von A. Engler. In Syllabus der Pflanzenfamilien, A. Engler, 8th ed., 352–364. Berlin.

Exell, A. W. 1957. La végétation de l'Afrique tropicale Australe. Données sur la superposition de lignes de distribution nord-sud sur un système fondamental est-ouest. Bull. Soc. Roy. Bot. Belgique 89:101–106.

Exell, A. W., and M. L. Gonçalves. 1973. A statistical analysis of a sample of the flora of Angola. Garcia Orta, ser. bot. 1, 1–2:105–119.

Flenley, J. R. 1979. The Equatorial Rain Forest—A Geological History. Butterworth, London.

Fosberg, F. R. 1948. Derivation of the flora of the Hawaiian islands. In Insects of Hawaii, ed. E. C. Zimmerman, 1:107–119. Honolulu.

———. 1951. The American element in the Hawaiian flora. Pacific Sci. 5:204–206.

———. 1952. Lignes biogéographiques dans l'Ouest de la Pacifique. C. R. Soc. biogéogr. 29:161–166.

———. 1957. The vegetation provinces of the Pacific. In Symposium on Vegetation Provinces of the Pacific 4:15–23. Manila.

———. 1963. Plant dispersal in the Pacific. In Pacific Basin Biogeography, ed. J. L. Gressitt, 273–282. Honolulu.

———. 1974. Phytogeography of atolls and other coral islands. Proc. Second Intern. Coral Reef Symp. 1:389–396. Brisbane.

———. 1979. Tropical floristic botany—concepts and status—with special attention to tropical islands. In Tropical Botany, eds. K. Larsen and L. B. Holm-Nielsen, 89–105. London.

Fosberg, F. R., M.-H. Sachet, and R. Oliver. 1979. A geographical check-list of the Micronesian Dicotyledones. Micronesica 15:41–295.

Gibbs, L. S. 1917. Dutch N.W. Guinea. A contribution to the phytogeography and flora of the Afric Mountains, etc. London.

Gillett, J. B. 1955. The relation between the Highland Floras of Ethiopia and British East Africa. Webbia 11:459–466.

Glassmann, S. F. 1957. The vascular flora of Ponape and its phytogeographical affinities. Proc. Eighth Pacific Sci. Congr. Quezon City 4:201–216.

Goldblatt, P. 1978. An analysis of the flora of southern Africa: Its characteristics, relationships, and origins. Ann. Missouri Bot. Gard. 65:369–436.

Good, R. 1947. The Geography of the Flowering Plants. 1st ed. London.

———. 1950. Madagascar and New Caledonia: A problem in plant geography. Blumea 6:470–479.

———. 1960. On the geographical relationships of the angiosperm flora of New Guinea. Bull. Brit. Mus. (Nat. Hist. Dept.), Bot. 2:205–226.

———. 1963. On the biological and physical relationships between New Guinea and Australia. In Pacific Basin Biogeography, ed. J. L. Gressitt, 301–309. Honolulu.

———. 1969. Some phytogeographical relationships of the angiosperm flora of the Solomon Islands. Phil. Trans. Roy. Soc. Lond. B225:603–608.

———. 1974. The geography of the flowering plants. 4th ed. London.

Green, P. S. 1979. Observation on the phytogeography of the New Hebrides, Lord Howe Island and Norfolk Island. In Plants and Islands, ed. D. Bramwell, 41–53. London.

Gressitt, J. L. 1982. The Biogeography and Ecology of New Guinea. The Hague.

Griffith, A. L. 1946. The vegetation of the Thar Desert of Sind. Ind. Forest. 72:307–309.

Grubov, V. I., and An. A. Fedorov. 1964. Flora and vegetation of China (in Russian). In Physical Geography of China, ed. V. T. Zaichikov, 324–428. Moscow.

Gruenberg-Fertig, I. 1954. On the Sudano-Deccanian element in the flora of Palestine. Palest. J. Bot. 6:234–240.

Guillaumet, J.-L., and G. Mangenot. 1975. Aspects de la spéciation dans la flore malgache. Boissiera 24:119–123.

Guillaumin, A. 1926. Les régions floristiques du Pacifique d'après leur endémisme et la répartition de quelques plantes phanerogames. Proc. Third Pan-Pacific Sci. Congr. Tokyo 1:920–938.

———. 1953. Les caractères floristiques de la Nouvelle Calédonie. Proc. Seventh Pacific Sci. Congr. Auckland 5:120–122.

———. 1964. L'endémisme en Nouvelle Calédonie. C. R. Soc. biogéogr. 356–358:67–74.

Guppy, H. B. 1906. Observations of a naturalist in the Pacific. 2. Plant Dispersal. London.

Gupta, R. K. 1962. Some observations on the plants of the South Indian hill tops (Nilgiri and Palni plateaux) and their distribution in the Himalayas. J. Ind. Bot. Soc. 41, 1:1–15.

Gwynne, M. D. 1968. Socotra. In Conservation of Vegetation in Africa South of the Sahara, eds. I. Hedberg and O. Hedberg. Acta phytogeogr. Suec. 54:179–185.

Hallier, H. 1912. Die Zusammensetzung und Herkunft der Pflanzendecke Indonesiens. In Die Sunda—Expedition des Vereins für Geographie und Statistik zur Frankfurt a/M, ed. J. Elbert, 2:275–302. Frankfurt a/M.

Hauman, L. 1933. Esquisse de la végétation des hautes altitudes sur le Ruwenzori. Bull. Acad. Belg. Cl. Sci. V, 19:602–616, 702–717, 900–917.

Hedberg, O. 1951. Vegetation belts of the East African mountains. Svensk. Bot.

Tidskr. 45:140–202.

———. 1957. Afroalpine vascular plants: A taxonomic revision. Symb. Bot. Upsal. 15:1–411.

———. 1961. The phytogeographical position of the afroalpine flora. Recent Advances Bot. 1:914–919.

———. 1964. Features of afroalpine plant ecology. Acta Phytogeographica Suecica 49:1–144.

———. 1965. Afroalpine flora elements. Webbia 19:519–529.

———. 1970. Evolution of the afroalpine flora. Biotropica 2, 1:16–23.

Hedge, I. C., and P. Wendelbo. 1978. Patterns of distribution and endemism in Iran. Notes Roy. Bot. Gard. Edinb. 36:441–464.

Hepper, F. N. 1965. Preliminary account of the phytogeographical affinities the flora of West Tropical Africa. Webbia 19:593–617.

Hooker, J. D. 1864. On the plants of the temperate regions of the Cameroons Mountains and islands in the Bight of Benin. J. Linn. Soc. Bot. 7:171–240.

———. 1896. Lecture on Insular Floras. London.

———. [1904] 1907. A sketch of the flora of British India. London. Reprinted in Imperial Gazetteer of India Oxford (3), 1, 4:157–212.

Hooker, J. D. A., and T. Thomson. 1855. Flora Indica. London.

Humbert, H. 1927. La destruction d'une flore insulaire par le feu. Prncipaux aspects de la végétation à Madagascar. Mém. Acad. Malgache 5:1–78.

———. 1955. Les territoires phytogéographiques de Madagascar. Leur cartographie. Coll. Intern. Gentre Nat. Rech. Sci. Paris, 439–448.

———. 1959. Origines présumées et affinités de la flore de Madagascar. Mém. Inst. Sci. Madagascar, sér. B, 149–188.

Hutchinson, J. 1946. A Botanist in Southern Africa. London.

Jaeger, P. 1954. Sur la position phytogéographique de Loma (Sierra Leone). VIII. Congrès Intern. Bot. Paris, 48–51.

———. 1965. Sur l'endémisme dans les plateaux soudanais ouest-africaines. C. R. Soc. biogéogr. 365–370:38–48.

Jeffrey, C. 1968. Seychelles. Acta phytogeogr. Suec. 54:275–279.

John, H. St., and A. C. Smith. 1971. The vascular plants of the Horne and Wallis islands. Pacific Sci. 25, 3:313–348.

Kalkman, C. A. 1955. A plant-geographical analysis of the Lesser Sunda Islands. Acta bot. Neerl. 4:200–225.

Kanehira, R. 1935. The phytogeographical relationships between Botel Tobago and the Philippines on the basis of the ligneous flora. Bull. Biogeogr. Soc. Jap. 5:209–211.

———. 1940. On the phytogeography of Micronesia. Proc. Sixth Pacific Sci. Congr. California 4:595–611.

Keay, R. W. J. 1955. Montane vegetation and flora in the British Cameroons. Proc. Linn. Soc. London, Session 165:140–143.

———. 1959. Vegetation Map of Africa South of the Tropic of Cancer. Oxford.

Keng, H. 1970. Size and affinities of the flora of the Malay Peninsula. J. Trop. Geogr. 31:43–56.

Killick, D. J. 1978. The Afroalpine Region. In Biogeography and Ecology of Southern Africa, ed. M. J. A. Werger, 515–560. The Hague.

Kloss, C. B. 1922. Some account of the journey on which the plants were collected. In On Collection of Plants from Peninsular Siam, H. N. Pidley, J. Federation of Malay State Mus. 10:66–80.

Knapp, R. 1973. Die Vegetation von Afrika. Jena.

Koechlin, J., J.-L. Guillaumet, and Ph. Morat. 1974. Flora et végétation de Madagascar. Vaduz.

Krishnaswamy, V. S., and R. S. Gupta. 1952. Rajputana deserts, its vegetation and soil. Ind. Forest. 78:595–601.

Lam, H. J. 1934. Materials toward a study of the flora of the Island of New Guinea. Blumea 1:115–159.

———. 1945. Notes on the historical phytogeography of Celebes. Blumea 5, 3:600–639.

Larsen, K., and L. B. Holm-Nielsen, eds. 1979. Tropical Botany. London.

Lavranos, J. J. 1975. Note on the northern temperate element in the flora of the Ethio-Arabian region. Boissiera 24:67–69.

Lawson, G. W. 1966. Plant Life in Western Africa. Oxford.

Lebrun, J. 1947. La végétation de la plaine alluviale au sud du lac Édouard. Brussels.

———. 1957. Sur les éléments et groupes phytogéographiques de la flore du Ruwenzori. Bull. Jard. Bot. l'Etat. Brux. 27:453–478.

———. 1958. Les orophytes africans. Comm. 6a Sess. Conf. Int. Afr. Occid. 3, Bot., 121–128.

———. 1960. Sur la richesse de la flore de divers territoires africaines. Bull. Séances Acad. Roy. Sci., Nouv. sér. Bruxelles 6, 4:669–690.

———. 1961. Les deux flores d'Afrique tropicale. Acad. Roy. Belg. Cl. Sc. Mém. 8°, 32 (6):1–81.

Legris, P. 1963. La végétation de l'Inde (ecologie et flora). Toulouse.

———. 1969. La Grande Comore, climats et végétation. Pondichery.

Legris, P., and V. M. Meher-Homji. 1968. Floristic elements in the vegetation of India. Proc. Symp. Rec. Adv. Trop. Ecol. 536–543.

Léonard, J. 1965. Contribution à la subdivision phytogéographique de la Région Guinéo-Congolaise d'après repartition geographique d'Euphorbiacées d'Afrique tropicale. Webbia 19:627–649.

Leroy, J. F. 1977. Une sous-famille de Winteraceae endémique à Madagascar: les Takhtajanioideae. Adansonia, sér. 2, 17, 4:383–395.

Leroy, J.-F. 1978. Composition, origin and affinities of the Madagascar vascular flora. Ann. Missouri Bot. Gard. 65:535–589.

Letouzey, R. 1970. Étude phytogéographique du Cameroun. Paris.

Lewis, F. 1926. The altitudinal distribution of the Ceylon endemic flora. Ann. Roy. Bot. Gard. Perad. 10:1–130.

Li, H. L., and H. Keng. 1950. Phytogeographical affinities of southern Taiwan. Taiwania 1:103–128.

Mabberley, D. J. 1974. The pachycaul Lobelias of Africa and St. Helena. Kew Bull. 29:535–584.

Maheshwari, P., J. C. Sen-Gupta, and C. S. Venkatesh. 1965. Flora [of India]. Imperial Gazetteer of India 1:163–229.

Maire, R. 1928. Origine de la flore montagnes de l'Afrique du Nord. Mém. Soc. Biogéogr. 2.

Mandaville, J. P., Jr. 1975. The scientific results of the Oman flora and fauna survey 1975. A Journal of Oman Studies Special Report, published by the Ministry of Information and Culture Sultanate of Oman, 227–267.
Mani, M. S. 1974*a*. The flora [of India]. In Ecology and biogeography in India, ed. M. S. Mani, 159–177. The Hague.
———, ed. 1974*b*. Ecology and Biogeography in India. The Hague.
Marais, W. 1981. Trochetiopsis (Sterculiaceae), a new genus from St. Helena. Kew Bull. 36 (3):645–646.
Marche-Marchad, J. 1955. Le monde végétal en Afrique intertropicale. Paris.
Meher-Homji, V. M. 1962. Phytogeographical Studies of the Semi-arid Regions of India. Ph.D. thesis. Bombay.
———. 1965. On the Sudano-Deccanian flora element. J. Bombay Nat. Hist. Soc. 62, 1:15–18.
———. 1967. Phytogeography of South Indian hill stations. Bull. Torrey Bot. Club 94, 4:230–242.
———. 1970. Some phytogeographic aspects of Rajasthan, India. Vegetation, 21, 4–6:299–320.
———. 1973. Phytogeography of the Indian subcontinent. Progress of Plant Ecology 1:9–88.
Melville, R. 1979. Endangered island floras. In Plants and Islands, ed. D. Bramwell, 361–377. London.
Merrill, E. D. 1910. The Malayan, Australasian and Polynesian elements in the Philippine flora. Ann. Bot. Gard. Buitenzorg, 2 ser., 3d suppl. 1:277–306.
———. 1923*a*. Distribution of the Dipterocarpaceae: Origin and relationships of the Philippine flora and causes of the differences between the floras of Eastern and Western Malaysia. Philippine J. Sci. 23:1–33.
———. 1923*b*. The influence of the Australian flora on the flora of the Philippines. Proc. Pan-Pacific Sci. Congr. 1:323–324.
———. 1923*c*. The correlation of geographical distributions with the geological history of Malaysia. Proc. Pan-Pacific Sci. Congr. 2:1148–1155.
———. 1926. A discussion and bibliography of Philippine flowering plants. Philippine Dept. Agr. Bureau of Sci. Popular Bull. 2:1–239. Manila.
———. 1936. Malaysian phytogeography in relation to the Polynesian flora. In Essays in Geobotany in Honor of W. A. Setchell, ed. T. H. Goodspeed, 247–261. Berkeley, Calif.
———. 1946. Plant Life of the Pacific World. New York.
———. 1953. Some Malaysian phytogeographical problems. Gard., Bull. 9, 1:49–57.
Meusel, H., E. Jäger, and E. Weinert. 1965. Vergleichende Chorologie der Zentraleuropeischen Flora. Jena.
Milne-Redhead, E. 1954. Distributional ranges of flowering plants in Tropical Africa. Proc. Linn. Soc. Lond., 165:25–35.
Molnar, P., and P. Tapponnier. 1977. The collision between India and Eurasia. Scientific American 236 (4):30–41.
Monod, Th. 1957. Les grandes divisions chorologique de l'Afrique. Cons. Sci. Afr. Sud. de Sahara (C.S.A.) 24:1–146. London.
Moore, H. E. 1978. New genera and species of Palmae from New Caledonia. I. Gentes Herb. 11 (4):291–309.

———. 1980. New genera and species of Palmae from New Caledonia. II. Gentes Herb. 12 (1):17–24.
Morton, J. K. 1972. Phytogeography of the West African Mountains. In Taxonomy, Phytogeography and Evolution, ed. D. H. Valentine, 221–236. London and New York.
Ogawa H., K. Yoda, and T. Kira. 1961. A preliminary survey on the vegetation of Thailand. Nature Life South Asia 1:21–157.
Oliver, E. G. H. 1977. An analysis of the Cape flora. In Proceed. Second National Weeds Conf. of South Africa, ed. D. P. Annecke, 1–18. Balkema, Cape Town.
Paijmans, K. 1976. Vegetation. In New Guinea Vegetation, ed. K. Paijmans, 23–105. Canberra.
Perrier de la Bâthie, H. 1921. La végétation malgache. Marseille.
———. 1936. Biogéographie des plantes de Madagascar. Paris.
Pichi-Sermolli, R. E. G. 1957. Una carta geobotanica dell'Africa Orientale (Eritrea, Etiopia, Somalia). Webbia 13:15–132.
Pole-Evans, I. 1922. The main botanical regions of South Africa. Mem. Bot. Surv. S. Afr., Dep. Agr. Pretoria 4:49–53.
Popov, G. B. 1957. The vegetation of Socotra. J. Linn. Soc. Lond., Bot. 55:706–720.
Popov, G. B., and W. Zeller. 1963. Ecological survey report on the survey in the Arabian Peninsula. FAO Progress Report UNSF [DL] ES 6:12–31.
Procter, J. 1974. The endemic flowering plants of the Seychelles: An annotated list, Candollea 29:345–387.
———. 1984a. Vegetation of the granitic islands of the Seychelles. In Biogeography and Ecology of the Seychelles Islands, ed. D. R. Stoddart, 193–207. The Hague.
———. 1984b. Floristics of the granitic islands of the Seychelles. In Biogeography and Ecology of the Seychelles Islands, ed. D. R. Stoddart, 209–220. The Hague.
Puri, G. S. 1960. Indian forest ecology, 2 vols. New Delhi.
Rao, A. S. 1974. The vegetation and phytogeography of Assam-Burma. In Ecology and Biogeography in India, ed. M. S. Mani, 204–246. The Hague.
Rao, C. K. 1972. Angiosperm genera endemic to the Indian forestic region and its neighboring areas. Ind. Forest. 98:560–566.
———. 1979. Angiosperm genera endemic to the Indian Floristic Region and its neighboring areas. II (additions, deletions, and corrections). Ind. Forest. 105, 5:335–341.
Rauh, W. 1973. Über die Zonierung und Differenzierung der Vegetation Madagaskars. Tropische und subtropische Pflanzenwelt, 1. Mainz, Akad. Wiss. Lit. 146 S.
———. 1979. Problems of biological conservation in Madagascar. In Plants and Islands, ed. D. Bramwell, 405–421. London.
Raven, P. H. 1979. Tectonics and Southern Hemisphere biogeography. In Tropical Botany, eds. K. Larsen and L. B. Holm-Nielsen, 3–24. London.
Raven, P. H., and D. I. Axelrod. 1972. Plant tectonics and Australasian paleobiogeography. Science 176:1379–1386.
Razi, B. A. 1955. The phytogeography of Mysore Hill tops. Proc. Nat. Inst. Sci. India 14, 10:87–107; 15, 1:109–144.
Rechinger, K. H. 1951. Grundzuge der Pflanzenverbereitung in Iran. Verh. zool.-bot. Ges. Wien 92:181–188.
Reichert, I. 1921. Die Pilzflora Aegyptens. Engl. Bot. Jb. 56:598–727.

Renvoize, S. A. 1979. The origin of Indian Ocean island floras. In Plants and Islands, ed. D. Bramwell, 107–129. London.
Richards, P. W. 1943. The biogeographical division of the Indo-Australian Archipelago. 6. The ecological segregation of the Indo-Malayan and Australian elements in the vegetation of Borneo. Proc. Linn. Soc. Lond. 154:154–156.
Ridley, H. N. 1937. Origin of the flora of the Malay Peninsula. Blumea Suppl. 1:183–192.
———. 1942. Distribution areas of the Indian flora. 150th Anniv. Vol. Roy. Bot. Gard. Calcutta, 49–52.
Rivals, P. 1952. Études sur la végétation naturelle de l'Ile de la Réunion. Toulouse.
———. 1968. La Réunion. In Conservation of Vegetation in Africa South of the Sahara, eds. I. Hedberg, O. Hedberg, Acta phytogeogr. Suec. 54:272–275.
Roberty, G. 1939–1941. Contribution à l'étude phytogéographique de l'Afrique occidentale française. Candollea 8:83–150.
Rollet, B. 1972. La végétation du Cambodge. Boise et For. de Trop. 144:3–15; 145:23–38; 146:3–20.
Rzedowski, J. 1962. Contributiones à la fitogeografia floristica e historia de México. I. Algunas consideraciones acerca del elemento endémico en la flora mexicana. Bol. Soc. Bot. Méx. 27:52–65.
Schlechter, R. 1905. Pflanzengeographische Gliederung der Insel Neu-Caledonien. Bot. Jb. 35:1–42.
Schmid, M. 1974. Végétation du Viet-Nam. Le massif Sud-Annamitique et les régions limitrophes. Paris.
———. 1975. La flore et la végétation de la partie méridionale de l'Archipel des Nouvelles Hébrides. Phil. Trans. R. Soc. Lond. B272:329–342.
Schuster, R. M. 1972. Continental movements, Wallace's line and Indo-Malayan Australasian dispersal of land plants, some eclectic concepts. Bot. Rev. 38:3–86.
Scrivenor, J. B. 1943. A discussion on the biogeographic division of the Indo-Australian Archipelago, with criticism of the Wallace and Weber Lines and of any other dividing lines and with an attempt to obtain uniformity in the names used for the divisions. Proc. Linn. Soc. Lond. 154:120–165.
Setchell, W. A. 1928. Migration and endemism with reference to Pacific insular floras. Proc. Third Pan-Pacific Sci. Congr. Tokyo 1:869–875.
———. 1935. Pacific insular floras and Pacific palaeogeography. Amer. Naturalist 69:289–310.
Skottsberg, C. 1925. Juan Fernandez and Hawaii. A phytogeographical discussion. B. P. Bishop Mus. Bull. Honolulu 16:1–45.
———. 1936. Antarctic plants in Polynesia. In Essays in Geobotany in Honor of W. A. Setchell, ed. T. H. Goodspeed, 291–311. Berkeley and Los Angeles: Univ. Calif. Press.
———. 1940. The flora of the Hawaiian Islands and the history of the Pacific Basin. Proc. Sixth Pacific Sci. Congr. California 4:685–707.
———. 1956. Derivation of the flora and fauna of Juan Fernandez and Easter Islands. In The Natural History of Juan Fernandez and Easter Island, ed. C. Skottsberg, 1:193–439.
Smith, A. C. 1951. The vegetation and flora of Fiji. Scientific Monthly 73:3–15.

———. 1955a. Botanical studies in Fiji. Smithsonian Inst. Rep. for 1954, 305–315.
———. 1955b. Phanerogam genera with distributions terminating in Fiji. J. Arnold Arbor. 36:273–292.
———. 1979. Flora Vitiensis Nova: A new flora of Fiji. Vol. 1 Lawai, Kauai, Hawaii.
Solem, A. 1958. Biogeography of the New Hebrides. Nature 181: 1253–1255.
Stapf, O. 1894. On the flora of Mount Kinabalu, in North Borneo. Trans. Linn. Soc. Lond. (Bot.) 4 (2):69–264.
Steenis, C.G.G.J. van. 1934. On the origin of the Malaysian mountain flora. Bull. Jard. Bot. Buitenzorg, ser. 3, 13:135–262.
———. 1935. On the origin of the Malaysian mountain flora. 2. Bull. Jard. Bot. Buitenzorg, ser. 3, 13:289–417.
———. 1936. On the origin of the Malaysian mountain flora. 3. Analysis of floristic relationships (first installment). Bull. Jard. Bot. Buitenzorg, ser. 3, 14:56–72.
———, ed. 1950a. Flora Malesiana. Djakarta. Noordhoff-Kolff N.V.
———. 1950b. Flora Malesiana. Introduction. XI–XLIV, Djakarta.
———. 1950c. The delimitation of Malaysia and its main plant geographical divisions. Flora Malesiana 1, 1:LXX–LXXV. Djakarta.
———. 1953. Results of the Archibold expeditions. Papuan *Nothofagus*. J. Arnold Arbor. 34:301–374.
———. 1957. Outline of vegetation types of Indonesia and some adjacent regions. Proc. Pacific Sci. Congr. 8:61–97.
———. 1962. The mountain flora of the Malaysian tropics. Endeavour 21:183–193.
———, ed. 1963. Pacific plant areas. 1. Manila.
———. 1965. Plant geography of the mountain flora of Mt. Kinabalu. Proc. Roy. Soc. B161:7–38.
———. 1971. Plant conservation in Malesia. Bull. Jard. Bot. Nat. Belg. 41:189–202.
———. 1972. The Mountain Flora of Java. E. J. Brill, Leiden.
———. 1979. Plant-geography of east Malesia. Bot. J. Linn. Soc. 79 (2):97–178.
Steenis, C.G.G.J. van, and M. M. J. Balgooy van. 1966. Pacific plant areas. 2. Blumea, Suppl. 5:1–312.
Steenis, C.G.G.J. van, assisted by A. F. Schippers-Lammortse. 1965. Concise plant-geography of Java. In Flora of Java, C. A. Backer and R. C. Bakhuizen van den Brink, Jr. 2:172, Groningen: Nordhoff.
Stoddart, D. R. 1984. Scientific studies in the Seychelles. In Biogeography and Ecology of the Seychelles Islands, ed. D. R. Stoddart, 1–15. The Hague.
Stoddart, D. R., and F. R. Fosberg. 1984. Vegetation and floristics of western Indian Ocean coral islands. In Biogeography and Ecology of the Seychelles Islands, ed. D. R. Stoddart, 221–238. The Hague.
Stone, B. C. 1967. A review of the endemic genera of Hawaiian plants. Bot. Rev. 33, 3:219–259.
Straka, H. 1962. Das Pflanzenkleid Madagascars. Naturwissenschaft. Rundschau 15:178–185.
———. 1963. Das Pflanzenkleid der Maskarenen. Naturwissenschaft. Rundschau 16:100–104.
Subramanyam, K., and M. P. Nayar. 1974. Vegetation and phytogeography of the Western Ghats. In Ecology and Biogeography in India, ed. M. S. Mani, 178–196.

The Hague.
Thorne, R. F. 1963. Biotic distributions patterns in the tropical Pacific. In Pacific Basin Biogeography, ed. J. L. Gressitt, 311–354. Honolulu.
———. 1965. Floristic relationships of New Caledonia. Univ. Iowa Studies Nat. Hist. 20, 7:1–14.
———. 1969. Floristic relationships between New Caledonia and the Solomon Islands. Phil. Trans. Roy. Soc. Lond. B255:595–602.
Thothathri, K. 1962. Contributions to the flora of the Andaman and Nicobar Islands. Bull. Bot. Surv. India 4:281–296.
Trochain, J. L. 1952. Les territoires phytogéographique de l'Afrique Noire Française d'aprés leur pluviométrie. Rec. Trav. Lab. Bot. Géol. Zool. Fac. Sci. Montpellier, sér. Bot. 5:113–124.
———. 1969. Les territoires phytogéographique de l'Afrique Noire francophone d'aprés la trilogie: climat, flore et végétation. C. R. Soc. biogéogr. 402:139–157.
Troupin, G. 1966. Étude phytocénologique du Parc National de l'Akagera et du Rwanda Oriental. Inst. Nat. Rech. Sci. Butare, Publ. 2:1–293.
Turrill, W. B. 1948. On the flora of St. Helena, Kew Bull. 1949:358–362.
Vaughan, R. E. 1968. Mauritius and Rodriguez. In Conservation of Vegetation in Africa South of the Sahara. eds. I. Hedberg and O. Hedberg, Acta phytogeogr. Suec. 54:265–272.
Vaughan, R. E., and P. O. Wiehe. 1937. Studies on the vegetation of Mauritius. I. A preliminary survey of the plant communities. J. Ecol. 25:289–343.
Vesey-FitzGerald, L.D.E.F. 1940. On the vegetation of Seychelles. J. Ecol. 28:465–483.
Vidal, J. E. 1960. La végétation du Laos. Part 2. Toulouse.
———. 1964. Endémisme végétal et systematique en Indochine. C. R. Soc. biogéogr. 41 (362):153–159.
———. 1966. Aspects biogéographiques du Sud-Est asiatique: Laos, Thailande, Cambodge et Sud Vietman. C. R. Soc. biogéogr. 43 (378–379):130–140.
———. 1979. Outline of ecology and vegetation of the Indochinese Peninsula. In Tropical Botany, eds. K. Larsen and L. B. Holm-Nielsen, 109–123. London.
Virot, R. 1956. La végétation canaque. Mém. Mus. Nat. Hist. Nat. Noumea, sér. B, Bot. 7:1–398.
Volk, O. 1964. Die afro-meridional-occidentale Floren-Region in S. W. Afrika. Beitr. Phytologie, Stuttgart:1–16.
Volk, O. H. 1966. Die Florengebiete von SW Afrika. Jb. SW Afrika Wiss. Ges. 20:25–58.
Vyas, L. N. 1964. Studies on phytogeographical affinities of the flora on North-East Rajasthan. Ind. Forest 90, 8:535–538.
———. 1967. Contribution to the flora of North-East Rajasthan. J. Bombay Nat. Hist. Soc. 64, 2:191–231.
Walker, E. H., and R. L. Pendleton. 1957. A survey of the vegetation of South-Eastern Asia. The Indochinese Province of the Pacific Basin. Proc. Eighth Pacific Sci. Congr. (Philippines 1953), 4 (Bot.), 99–114.
Wallace, A. R. 1863. On the physical geography on the Malay Archipelago. J. Roy. Geogr. Soc. 33:217–234.

Warburg, O. 1900. Monsunia. Leipzig.
Wegener, A. 1924. The Origin of Continents and Oceans. London.
Werger, J. J. A. 1978*a*. Biogeographical division of southern Africa. In Biogeography and Ecology of Southern Africa, ed. M. J. A. Werger, 1:147–170. The Hague.
Werger, J. A. 1978*b*. The Karoo-Namib Region. In Biogeography and Ecology of Southern Africa, ed. M. J. A. Werger, 1:231–299. The Hague.
Werger, M. J. A., and B. J. Coetzee. 1978. The Sudano-Zambezian Region. In Biogeography and Ecology of Southern Africa, ed. M. J. A. Werger, 1:301–462. The Hague.
Wiehe, P. O. 1949. The vegetation of Rodrigues Island. Mauritius Inst. Bull. 2:280–304.
White, F. 1965. The savanna woodlands of the Zambezian and Sudanian domains. Webbia 19:651–681.
———. 1975. The vegetation map of Africa: The history of a completed project. Boissiera 24:659–666.
———. 1978. The Afromontane Region. In Biogeography and Ecology of Southern Africa, ed. M. J. A. Werger, 1:463–513. The Hague.
———. 1979. The Guineo-Congolian Region and its relationships to other phytochoria. Bull. Jard. Bot. Nut. Belg. 49:11–55.
———. 1983. The vegetation of Africa. UNESCO, Paris.
White, F., and E. J. Moll. 1978. The Indian Ocean coastal belt. In Biogeography and Ecology of Southern Africa ed. M. J. A. Werger, 1:561–598. The Hague.
White, F., and M. J. A. Werger. 1978. The Guineo-Congolian transition to Southern Africa. In Biogeography and Ecology of Southern Africa, ed. M. J. A. Werger, 1:559–620. The Hague.
Whitmore, T. C. 1973. Plate tectonics and some aspects of Pacific plant geography. New Phytol. 72:1185–1190.
———. 1975. Tropical Rain Forests of the Far East. Oxford.
———. 1981. Palaeoclimate and vegetation history. In Wallace's Line and Plate Tectonics, ed. T. C. Whitmore, 39–42. Oxford: Clarendon Press.
———. 1982. Wallace's line: A result of plate tectonics. Ann. Missouri Bot. Gard. 69:668–675.
Wickens, G. E. 1976. The flora of Jebel Marra (Sudan Republic) and its geographical affinities. Kew Bull. Addit. Ser. 5.
Wild, H. 1968. Phytogeography in South Central Africa. Kirkia 6, 2:197–222.
———. 1975. Phytogeography and the Gondwanaland position of Madagascar. Boissiera 24:107–117.
Willis, J. C. 1922. Age and Area: A Study in Geographical Distribution and Origin of Species. Cambridge: Univ. Press.
Womersley, J. S., and J. B. McAdam. 1957. The Forests and Forest Conditions in the Territory of Palau and New Guinea. Port Moresby.
Zohary, M. 1963. On the geobotanical structure of Iran. Bull. Res. Council of Israel. Sect. D. Bot. 11D, Suppl.:1–113.
———. 1973. Geobotanical Foundations of the Middle East. I–II. Stuttgart.
Zollinger, H. 1857. Observationes botanicae novae. Natuurkundig Tijdschrift voor Nederlandische Indië 14:144–176.

III. NEOTROPICAL KINGDOM

Acosta-Solis, M. 1966. Les divisiones fitogeographicas y las formaciones geobotanicus del Ecuador. Rev. Acad. Colomb. Cienc. Exac. Fisicas y Natur. 12, 48:401–447.
———. 1968. Divisiones fitogeograficas y formaciones geobotanicas del Ecuador. Quito.
Alain, Hno. 1958. La flora de Cuba: Sus principales caracteristicas, su origin probable. Revista Cos. Cub. Bot. 15:36–59, 84–96.
Bates, M. 1965. South America, Flora and Fauna. New York.
Berazaín, R. 1977. Estudio preliminar de la flora serpentinicola de Cuba. Cienc. Bot. 12:11–26.
Bisse, J. 1975. Die floristische Stellung und Gliederung Kubas, Wissensch. Ztschr. d. Friedrich-Schiller Univ. Jena 4:365–371.
Cabrera, A. L. 1971. Fitogeografia de la República Argentina. Bol. Soc. Arg. Bot. 14, 1–2:1–42.
Cabrera, A. L. and A. Willink. 1973. Biogeografia de America Latina. Washington.
Camp, W. H. 1952. Phytophyletic pattern on lands bordering the South Atlantic basin. Bull. Amer. Mus. Nat. Hist. 99, 3:205–216.
Cleef, A. M. 1978. Characteristics of neotropical páramo vegetation and its Subantarctic relations. Erdwiss. Forsch. 1:365–390.
———. 1979. The phytogeographical position of the neotropical vascular Páramo flora with special reference to the Colombia Cordillera Oriental. In Tropical Botany, eds. K. Larsen and L. B. Holm-Nielsen, 175–184. London.
Cuatrecasas, J. 1958. Aspectos de la vegetación natural de Colombia. Rev. Acad. Colomb. Cienc. Exact. Fisicas. Natur. 10, 40:221–268.
Dorst, I. 1967. South America and Central America, a Natural History. New York.
Eiten, G. 1972. The cerrado vegetation of Brazil. Bot. Rev. 38, 2:201–341.
Fittkau, E. J., J. Illies, H. Klinge, G. H. Schwabe, and H. Sioli, eds. 1968. Biogeography, and Ecology in South America. The Hague.
Gentry, A. 1982. Neotropical floristic diversity. Ann. Missouri Bot. Gard. 69:557–593.
Harling, G. 1962. On some Compositae endemic to the Galápagos Islands. Acta horti bergiani 20:63–120.
Hemsley, W. B. 1888. Botany: Biologia Centrali-Americana. London.
Hooker, J. D. 1851. On the vegetation of the Galàpagos Archipelago, as compared with that of some other tropical islands and of the continent of America. Trans. Linn. Soc. Lond. 20:235–262.
Howard, R. A. 1979. Flora of the West Indies. In Tropical Botany, K. Larsen and L. B. Holm-Nielsen, 239–250. London.
Hueck, K., and P. Seibert. 1972. Vegetationskarte von Südamerika. Stuttgart.
Johnston, I. M. 1931. The flora of the Revilla Gigedo Island. Proc. Calif. Acad. Sci., ser. 4, 20, 2:9–104.
Johnston, M. P., and P. H. Raven. 1973. Species number and endemism: The Galápagos Archipelago revisited. Science 179:893–895.
Klotz, G. 1975. Die gemeinsame Erarbeitung einer "Nueva Flora de Cuba," ein Forschungsvorhaben von Botanikern aus Kuba, der DDR und der UdSSR. Wissensch. Ztschr. d. Friedrich-Schiller Univ. 4:351–364.

Kroeber, A. L. 1916. Floral relations among the Galápagos Islands. Univ. Calif. Publ. Bot. 6:199–220.
Labuntsova, M. A. 1969. On the botanico-geographic subdivision of South America (in Russian). Bulletin of the Main Botanical Garden 72:28–33. Moscow.
Lauer, W. 1968. Problemas de la division fitogeografica en America Central. Proc. UNESCO. Mexico Symp. 9:139–155.
Maguire, B. 1970. On the flora of the Guayana Highland. Biotropica 2, 2:85–100.
———. 1972a. Tetrameristaceae in the Botany of the Guayana Highland. Mem. New York Bot. Gard. 23:165–192.
———. 1972b. Guayana as a floristic province: Its relationship within the Neotropics and to the Paleotropics. In Resumenes de los Trabajos I Congreso Latino Americano, Bot. Mex., 55–56.
———. 1979. Guayana, region of the Roraima sandstone formation. In *Tropical Botany*, eds. K. Larsen and L. B. Holm-Nielsen, 223–238. London.
Maguire, B., P. S. Ashton, C. de Zeeum, D. E. Giannasi, and K. J. Niklas. 1977. Pakaraimoideae, Dipterocarpaceae of the Western Hemisphere. Taxon 26, 4:341–385.
Morello, J. 1958. La provincia fitogeografica del Monte. Opera Lilloana 2:1–155.
Onaney, M. 1970. Endemismo en la flore. Atlas Nacional de Cuba. La Habana.
Porter, D. M. 1979. Endemism and evolution in Galápagos Islands vascular plants. In Plants and Islands, ed. D. Bramwell, 225–256. London.
———. 1984. Relationships of the Galapagos flora. Biol. J. Linn. Soc. 21 (1–2):243–251.
Prance, G. T. 1977. The phytogeographic subdivision of Amazonia and their influence on the selection of Biological reserves. In Extinction is Forever, eds. G. T. Prance and T. S. Elias, 195–213, N. Y. Botanical Garden, New York.
———. 1978. The origin and evolution of the Amazon flora. Iterciencia 3 (4):207–222.
———. 1979. Distribution patterns of Lowland neotropical species with relation to history, dispersal and ecology with special reference to Chrysobalanaceae, Caryocaraceae, and Lecythidaceae. In Tropical Botany, eds. K. Larsen and L. B. Holm-Nielsen, 59–86. London.
Raven, P. H. 1963. Amphitropical relations in the flora of North and South America. Quart. Rev. Biol. 29:151–171.
Saakov, S. G. 1970. Endemic palms of Cuba (in Russian). Bot. Zhurn. 55:196–221.
Sampaio, A. J. de 1934. Phytogeographia do Brasil. Sao Paulo.
Seifriz, W. 1940a. The plant life of Cuba. Ecol. Monogr. 13:375–426.
———. 1940b. Die Pflanzengeographie von Cuba. Bot. Hb. 70:441–462.
Sharp, A. J. 1966. Some aspects of Mexican phytogeography. Ciencia Mex. 24:229–232.
Smith, A. C. 1967. The presence of primitive angiosperms in the Amazon Basin and its significance in indicating migrational routes. Atas Simpos Biota Amaz. 4:37–59.
Smith, A. C., and I. M. Johnston. 1945. A phytogeographic sketch of Latin America. In Plants and Plant Science in Latin America, ed. F. Verdoorn, 11–18. Waltham, Mass.
Smith, L. B. 1962. Origins of the flora of Southern Brazil. Contr. U.S. Nat. Herb. 35:215–249.

Stehlé, H. 1945. Le conditions ecologique, la végétation et resources agricoles de l'Archipel des Petities Antiles. In Plants and Plant Science in Latin America, ed. F. Verdoorn 85–100. Waltham, Mass.

Stewart, A. 1911. A botanical survey of the Galápagos Islands. Proc. Calif. Acad. Sci., ser. 4, 1:7–228.

Steyermark, J. A. 1974. The summit vegetation of Cerro Autana. Biotropica 6, 1:7–13.

———. 1979a. Flora of the Cuayana Highland: Endemicity of the generic flora of the summits of the Venezuela Tepuis. Taxon 28:45–54.

———. 1979b. Plant refuge and dispersal centres in Venezuela: their relict and endemic element. In Tropical Botany, eds. K. Larsen and L. B. Holm-Nielsen, 185–221. London.

Svenson, H. K. 1945. A brief review of the Galápagos flora. In Plants and Plant Science in Latin America, ed. F. Verdoorn, 149–150. Waltham, Mass.

———. 1946. Vegetation of the coast of Ecuador and Peru and its relation to the Galápagos Islands. Amer. J. Bot. 33:394–498.

Taylor, N. 1921. Endemism in the Bahama flora. Ann. Bot. 35:523–533.

Toledo-Rizzini, C. 1763. Nota previa sôbre a diviso fitogeográfica (floristico-socióloga) do Brasil. Rev. Brasil. Geogr. 25, 1:3–64.

Tryon, R. 1972. Endemic areas and geographic speciation in tropical American ferns. Biotropica 4, 3:121–131.

———. 1979. Biogeography of the Antillean fern flora. In Plants and Islands, ed. D. Bramwell, 55–68. London.

Verdoorn, F., ed. 1945. Plants and Plant Science in Latin America. Waltham, Mass.

Vuilleumier, B. S. 1971. Pleistocene changes in the fauna and flora of South America. Science 173:771–780.

Weberbauer, A. 1912. Pflanzengeographische Studien in südlichen Peru. Bot. Jb., Beibl. 107:27–46.

———. 1914. Die Vegetationsgliederung des nördlichen Peru um 5° südl. Br. Bot. Jb. 50, Suppl., 72–94.

———. 1945. El mundo vegetal de los Andes Peruanos. Ministerio de Agricultura. Lima.

Wiggins, I. L. 1966. Origins and relationships of the flora of the Galápagos Islands. In The Galápagos, ed. R. I. Bowman, Berkeley, Calif.

Wiggins, I. L., and D. M. Porter. 1971. Flora of the Galápagos Islands. Stanford, Calif.

Williams, L. 1945. The phytogeography of Peru. In Plants and Plant Science in Latin America, ed. F. Verdoorn, 308–312. Waltham, Mass.

IV. CAPE KINGDOM (CAPENSIS)

Acocks, J. P. H. 1953. Veld Types of South Africa. Pretoria.

Adamson, R. S. 1938. The Vegetation of South Africa. London.

———. 1948. Some geographical aspects of the Cape flora. Trans. Roy. Soc. S. Afr. 31, 5:437–464.

———. 1958. The Cape as an ancient African Flora. Advancement of Science 58.
Adamson, R. S., and T. M. Salter. 1950. Flora of the Cape Peninsula. Cape Town and Johannesburg.
Beard, J. S. 1959. The origin of African Proteaceae. J. South. Afr. Bot. 25:231–235.
Bews, J. W. 1921. Some general principles of plant distribution as illustrated by the South African flora. Ann. Bot. 35:1–36.
Bolus, H. 1886. Sketch of the Floral Regions of South Africa. Official Handbook, Cape of Good Hope.
———. 1905. Sketch of the Floral Regions of South Africa. Science in S. Afr.; 1–42.
Diels, L. 1909. Formationen und Florenelemente im nordwestlichen Kapland. Engl. Bot. Jb. 44:91–124.
Engler, A. 1903. Über die Frühlingsflora des Tafelberges bei Kapstadt. Notizbl. Bot. Gart. Mus. Berlin, App. XI, 1–58.
Goldblatt, P. 1978. An analysis of the flora of Southern Africa: its characteristics, relationships and origins. Ann. Missouri Bot. Gard. 65:369–436.
Haden-Guest, S., J. K. Wright, and E. M. Teclaff. 1956. A World Geography of Forest Resources. New York.
Hutchinson, J. A. 1946. A Botanist in Southern Africa. London.
Johnson, L. A. S., and B. G. Briggs. 1975. On the Proteaceae—The evolution and classification of a southern family. Bot. J. Linn. Soc. 70, 2:83–182.
Knapp, R. 1973. Die Vegetation von Afrika. Jena.
Levyns, M. R. 1938. Some evidence bearing on the past history of the Cape Flora. Trans. Roy. Soc. S. Afr. 26:401–424.
———. 1950. The relations of the Cape and the Karroo Floras near Ladismith, Cape. Trans. Roy. Soc. S. Afr. 32:235–246.
———. 1952. Clues to the past in the Cape Flora of today. South African J. Sci. 49:155–164.
———. 1958. The phytogeography of the Proteaceae in Africa. J. South Afr. Bot. 24:1–9.
———. 1964. Migration and origin of the Cape Flora (presidential address). Trans. Roy. Soc. S. Afr. 37:85–107.
Marloth, R. 1908. Das Kapland, insonderheit das Reich der Kapflora, das Waldgebiet und die Karroo, pflanzengeographisch Dargestellt. Wiss. Ergebn. Deutsch. Tiefsee Exped. "Waldivia" 1898–99. Jena.
Nordenstam, B. 1969. Phytogeography of the genus *Euryops* (Compositae): A contribution to the phytogeography of Southern Africa. Opera botanica 23:1–77.
Oliver, E. G. H. 1977. An analysis of the Cape flora. In Proceed. Second National Weeds Conf. of South Africa, ed. D. P. Annecke, 1–18. Balkema, Cape Town.
Phillips, E. P. 1917. A contribution to the flora of the Leribe Plateau and environs, with a discussion on the relationships of the flora of Batutoland, the Kalahari, and the South-Eastern Regions. Ann. S. Afr. Mus. London, 16.
Pole-Evans, I. B. 1918. The plant geography of South Africa. Reprinted from the Official Year Book, 1917. Pretoria.
———. 1922. The main botanical regions of South Africa. Bot. Surv. S. Afr., Mem. 4:49–53.
Taylor, H. C. 1978. Capensis. In Biogeography and Ecology of Southern Africa, ed. M. J. A. Werger, vol. 1:171–229. The Hague: W. Junk.

Weimarck, H. 1941. Phytogeographic groups, centres and intervals within the Cape flora. Lunds Univ. Arsskr. N. F. Avd. 237, 5:1–143.
Werger, J. J. A. 1978. Biogeographycal division of Southern Africa. In Biogeography and Ecology of Southern Africa, ed. M. J. A. Werger, vol. 1:233–299. The Hague: W. Junk.
White, F. 1983. The Vegetation of Africa. UNESCO. Paris.
Wickens, G. E. 1976. The flora of Jebel Marra (Sudan Republic) and its geographical affinities. Kew Bull. Addit. Ser. 5.

V. AUSTRALIAN KINGDOM (AUSTRALIS)

Aplin, T. E. H. 1975. The vegetation of Western Australia. Official Year Book of Western Australia, 14 (new series), 66–81.
Barlow, B. A. 1981. The Australian flora: Its origin and evolution. In Flora of Australia, Vol. I, Introduction, 25–75. Canberra: Australian Government Publishing Service.
Beadle, N. C. W. 1981a. The Vegetation of Australia. Ficher, New York.
———. 1981b. Origins of the Australian angiosperm flora. Ecological Biogeography of Australia, ed. A. Keast, 1:409–469. The Hague.
———. 1981c. The vegetation of the arid zone. In Ecological Biogeography of Australia, ed. A. Keast, 1:697–716. The Hague.
Beard, J. S. 1969. Endemism in the Western Australian flora at the species level. J. Roy. Soc. West. Austral. 52, 1:18–20.
———. 1978. Map of Western Australia showing Botanical Provinces and Districts. Vegmap Publications, 6 Fraser Road, Applecross, W. A.
———. 1981. The history of the phytogeographic region concept in Australia. In Ecological Biogeography of Australia, ed. A. Keast, 1:337–353. The Hague.
Brown, M. J. 1981. Tasmanian flora. In The Vegetation of Tasmania, ed. W. D. Jackson, 104–106. Hobart, Botany Department, University of Tasmania.
Burbidge, N. T. 1960. The phytogeography of the Australian region. Austral. J. Bot. 8, 2:75–212.
Clifford, H. T., and B. K. Simon. 1981. Ecological biogeography of Australian grasses. In Ecological Biogeography of Australia, ed. A. Keast, 1:539–554. The Hague.
Costin, A. B. 1981. Vegetation of high mountains in Australia. In Ecological Biogeography of Australia, ed. A. Keast, 1:719–731. The Hague.
Crook, K. A. W. 1981. The break-up of the Australian-Antarctic segment of Gondwanaland. In Ecological Biogeography of Australia, ed. A. Keast, 1:3–14. The Hague.
Curtis, W. N. 1967. Introduction. In The Endemic Flora of Tasmania, W. N. Curtis and M. Stone, Part 1, 17–19. London.
Diels, L. 1906. Die Pflanzenwelt von West-Australien. In Die Vegetation der Erde, A. Engler and O. Drude, 7:34–72. Leipzig.
———. 1936. The genetic phytogeography of the South-Western Pacific Area, with

particular reference to Australia. In Essays in Geobotany in Honour of William Albert Setchel, 189–194. Berkeley and Los Angeles: Univ. Calif. Press.

Doing, H. 1970. Botanical geography and chorology in Australia. Belmontia 4, 13:81–98.

Galloway, R. W. and E. M. Kemp. 1981. Late Cainozoic environments in Australia. In Ecological Biogeography of Australia, ed. A. Keast, 1:53–80. The Hague.

Gardner, C. A. 1944. The vegetation of Western Australia, J. Roy. Soc. W. Austr. 28:11–86.

———. 1959. The vegetation of Western Australia. In Biogeography and Ecology in Australia, eds. A. Keast, R. L. Crocker, and C. S. Christian, 275–282. The Hague.

Good, R. 1958. The biogeography of Australia. Nature 181:1763–1765.

———. 1963. On the biological and physical relationships between New Guinea and Australia. In Pacific Basin Biogeography, ed. J. L. Gressitt, 301–309. Honolulu.

Hooker, J. D. 1855–1860. Flora Tasmaniae. London.

James, S. H. 1981. Cytoevolutionary patterns, genetic systems and the phytogeography of Australia. In Biogeography and Ecology in Australia, eds. A. Keast et al., 1:763–782. The Hague.

Johnson, L. A. S., and B. G. Briggs. 1975. On the Proteaceae—The evolution and classification of a southern family. Bot. J. Linn. Soc. 70, 2:83–182.

Keast, A., R. L. Crocker, and C. S. Christian, eds. 1959. Biogeography and Ecology in Australia. The Hague.

Kirkpatrick, J. B. 1982. Phytogeographical analysis of Tasmanian alpine flora. J. Biogeogr. 9:255–271.

Kirkpatrick, J. B., and M. J. Brown. 1984. A numerical analysis of Tasmanian higher plant endemism. Bot. J. Linn. Soc. 88:165–183.

Lange, R. T. 1982. Australian Tertiary vegetation: Evidence and interpretation. In A History of Australasian vegetation, ed. J. M. B. Smith, 44–89. Sydney: McGraw-Hill.

Marchant, N. G. 1973. Species diversity in the south-western flora. J. Roy. Soc. W. Austral. 56:23–30.

Moore, D. M. 1972. Connections between cool temperate floras, with particular reference to Southern South America. In Taxonomy, Phytogeography and Evolution, ed. D. H. Valentine, 115–138. London and New York.

Nelson, E. C. 1981. Phytogeography of southern Australia. In Biogeography and Ecology in Australia, eds. A. Keast et al., 1:735–759. The Hague.

Page, C. N., and H. T. Clifford. 1981. Ecological biogeography of Australian conifers and ferns. In Biogeography and Ecology in Australia, eds. A. Keast et al., 1:473–498. The Hague.

Raven, P. H. 1979. Plate tectonics and Southern Hemisphere biogeographs. In Tropical Botany, eds. K. Larsen and L. B. Holm-Nielsen, 3–24. London.

Rodway, L. 1923. The endemic phanerogams of Tasmania. Proc. Pan-Pacific Sci. Congr. Australia, 283–289.

Schweinfurth, U. 1962. Studien zur Pflanzengeographie von Tasmanien. Bonn.

Specht, R. L. 1972. The Vegetation of South Australia. Adelaide.

———. 1981a. Major vegetation formations in Australia. In Biogeography and Ecology in Australia, eds. A. Keast et al., 1:165–297. The Hague.

———. 1981*b*. Biogeography of halophytic angiosperms (saltmarsh, mangrove, and sea-grass). In Biogeography and Ecology in Australia, eds. A. Keast et al., 1:577–589. The Hague.
———. 1981*c*. Evolution of the Australian flora: Some generalizations. In Biogeography and Ecology in Australia, eds. A. Keast et al., 1:785–805. The Hague.
Specht, R. L., E. M. Roe, and V. H. Broughton. 1974. Conservation of major plant communities in Australia and Papua New Guinea. Austral. J. Bot., suppl. vol. 7.
Webb, L. J., and J. G. Tracey. 1981. Australian rain forests: Patterns and change. In Biogeography and Ecology in Australia, eds. A. Keast et al., 1:607–694. The Hague.
Williams, W. D., ed. 1974. Biogeography and Ecology of Tasmania. The Hague. W. Junk.
Wood, J. G. 1959. The phytogeography of Australia. In Biogeography and Ecology in Australia, eds. A. Keast et al., 291–302. The Hague.

VI. HOLANTARCTIC KINGDOM

Akhmetiev, M. A. 1972. The mysterious island of Lord Howe (in Russian). Nature (Priroda) 3:33–41.
Alboff, N. 1896. Observations sur la végétation du canal de Beagle. Revista d. Museo de la Plata. 7.
———. 1902. Essai de la flore raisonnée de la Terre de Feu. Ann. del Museo de la Plata, Sect. Bot. 1:I-XXIII.
———. 1904 (1903). A Contribution to the Comparative Study of the Flora of Tierra del Fuego (in Russian). Supplement to the journal "Geography" 10 (4):1–126.
Aleksandrova, V. D. 1977. Geobotanical Subdivision of the Arctic and Antarctica (in Russian). Leningrad.
Beetle, A. A. 1943. The phytogeography of Patagonia. Bot. Rev. 9:667–679.
Bray, W. L. 1898. On the relation of the flora of the Lower Sonoran Zone in North America to the flora of the arid zones of Chile and Argentina. Bot. Gaz. 26:121–147.
———. 1900. The relations of the North American flora to that of South America. Science 12:709–716.
Cabrera, A. L. 1971. Fitogeografía de la República Argentina. Bol. Soc. Arg. Bot. 14, 1–2:1–42.
Cabrera, A. L., and A. Willink. 1973. Biogeographia de America Latina. Washington.
Carlquist, S. 1974. Island Biology. New York.
Christophersen, E. 1934. Plants of Gough Island (Diego Alvarez). Oslo.
———. 1939. Problems of plant geography in Tristan da Cunha. Saertrykk av Norsh Geografisk Tidskr. 7, 5–8:106–112.
———. 1968. Flowering Plants from Tristan da Cunha. Oslo.
Cockayne, L. 1921. The vegetation of New Zealand. In Die Vegetation der Erde, A. Engler and O. Drude. Leipzig.

———. 1928. The vegetation of New Zealand. In Die Vegetation der Erde, A. Engler and O. Drude, 14. 2 Aufl. Leipzig.
Cookson, I. C. 1947. Plant microfossils from the lignites of Kerguelen Archipelago. British-Australian-New Zealand Antarctic Res. Exp. Rep. 2, 8:129–142.
Corner, R. W. M. 1971. Studies in *Colobanthus quitensis* (Kunth) Bartl. and *Deschampsia antarctica* Desv. IV. Distribution and reproductive performance in the Argentine Islands. Brit. Antarct. Surv. Bull. 26:41–50.
Cour, P. 1959. Flora et vegetation de l'archipel de Kerguelen. Terres Australes et Antarctiques Franç. 8–9:3–40.
Cranwell, L. M. 1959. Fossil pollen from Seymour Island, Antarctica. Nature 184:1728–1785.
Cranwell, L. M., H. J. Harrington, and I. G. Speden. 1960. Lower Tertiary microfossils from McMurdo Sound, Antarctica. Nature 186:700–702.
Donat, A. 1934. Zur Begrenzung der Magellanischen Florengebietes. Ber. Deutsch., bot. Ges. 53:131–142.
Dusén, P. 1901–1903, 1908. Über die Tertiäre Flora der Seymour-Insel. Wiss. Ergeb. Schwed. Südpolar Exped. 3, 3:1–27.
Engler, A. 1914. Pflanzengeographie. In Kultur der Gegenwart, III, ed. P. Hinneberg, Abt. IV, 4:187–263.
Fleming, C. A. 1962. New Zealand biogeography: A palaeontologist's approach. Tuatara 10:53–108.
Florin, R. 1940. The tertiary fossil conifers of Southern Chile and their phytogeographical significance, with a review of the fossil conifers of southern lands. Kungl. Svensk. Vet. Akad. Handl. 19, 2:1–107.
Godley, E. J. 1960. The botany of Southern Chile in relation to New Zealand and the Subantarctic. Proc. Roy. Soc. London, ser. B, 152:457–475.
———. 1975. Flora and vegetation [of New Zealand]. In Biogeography and Ecology in New Zealand, ed. I. G. Kuschel, 177–229. The Hague.
Good, R. A. 1933. A geographical survey of the flora of temperate South America. Ann. Bot. 47:691–725.
Goodspeed, T. H. 1945. Notes on the vegetation and plant resources of Chile. In Plants and Plant Science in Latin America, ed. F. Verdoorn, 145–149. Waltham, Mass.
Gordon, H. D. 1949. The problems of Subantarctic plant distribution. Rep. Austral. and New Zealand Assoc. Adv. Sci. 27:142–149.
Gray, A., and J. Hooker. 1880. The vegetation of the Rocky Mountain Region and a comparison with that of other parts of the world. Bull. U.S. Geol. and Geogr. Surv. Territories 6:1–77.
Green, P. S. 1970. Notes relating to the floras of Norfolk and Lord Howe Islands. Part I. J. Arnold Arbor. 51:204–220.
———. 1979. Observations on the phytogeography of the New Hebrides, Lord Howe Island and Norfolk Island. In Plants and Islands, ed. D. Bramwell, 41–53. London.
Greene, D. M., and A. Holtum. 1971. Studies in *Colobanthus quitensis* (Kunth) Bartl. and *Deschampsia antarctica* Desw. III. Distribution, habitats and performance in the Antarctic botanical zone. Brit. Antarct. Surv. Bull. 26:1–29.

Greene, S. W. 1964. The vascular flora of South Georgia. Brit. Antarct. Surv. Sci. Rep. 45:1–58.

———. 1967. Vascular plant distribution. In Terrestrial Life in Antarctica, ed. V. Buschnell, Antarctic Map Folio Ser., Folio 5:15–16.

———. 1969. The records for South Georgian vascular plants. Brit. Antarct. Surv. Bull. 22:49–59.

Haden-Guest, S., J. K. Wright, and E. M. Teclaff. 1956. A World Geography of Forest Resources. New York.

Holdgate, M. W. 1960. The Royal Society expedition to southern Chile. Proc. Roy. Soc. B152:434–441.

Hooker, J. D. 1844–1847. Flora Antarctica. London.

———. 1853. Introductory Essay to the Flora of New Zealand. London.

———. 1853–1855. Flora Novae-Zelandiae. London.

Hueck, K. 1966. Die Wälder Südamerikana. Stuttgart.

Hueck, K., and P. Seibert. 1972. Vegetationskarte von Südamerika. Stuttgart.

Johnson, P. N. 1975. Vegetation and flora of the Solander Islands. New Zealand J. Bot. 13:189–213.

Johnson, P. N., and D. J. Campbell. 1975. Vascular plants of the Auckland Islands. New Zealand J. Bot. 13:665–720.

Johnston, I. M. 1929. Papers on the flora of Northern Chile. Contrib. Gray Herb. Harvard. Univ. 85, 1:1–172.

———. 1940. The floristic significance of shrubs common to North and South American areas. J. Arnold Arbor. 21:356–363.

Lindsay, D. C. 1971. Vegetation of the South Shetland Islands. Brit. Antarct. Surv. Bull. 25:59–83.

Moore, D. M. 1968. The vascular plants of the Falkland Islands. Brit. Antarct. Surv. Sci. Rep. 60:1–202.

———. 1972. Connections between cool temperate floras, with particular reference to Southern South America. In Taxonomy, Phytogeography and Evolution ed. D. H. Valentine, 115–138. London, New York.

———. 1974. Catálogo de las plantas vasculares nativas de Tierra del Fuego. Anal. Inst. Patagonia, Punta Arenas (Chile) 5:105–121.

———. 1975. The alpine flora of Tierra de Fuego. Anales del Inst. Bot. Madrid 2, 32:419–440.

———. 1979. Origins of temperate island floras. In Plants and Islands, ed. D. Bramwell 69–85. London.

Oberdorfer, E. 1960. Pflanzensoziologische Studien in Chile. Weinheim, 1–208.

Oliver, W. R. B. 1910. The vegetation of the Kermades Islands. Trans. and Proceed. New Zealand Inst. 42:118–175.

———. 1911. The vegetation and flora of Lord Howe Islands. Trans. and Proceed. New Zealand Inst. 49:94–161.

———. 1925. Biogeographical relations of the New Zealand region. J. Linn. Soc. Lond. 47:99–140.

———. 1953. Origin of the New Zealand flora. Proc. Seventh Pacific Sci. Congr. Auckland, 1949, 5:131–146.

———. 1955. History of the flora of New Zealand. Svensk. Bot. Tidskr. 49:9–18.

Oye, P. van, and J. Miegham. 1965. Biogeography and Ecology in Antarctica. The Hague.
Paramonov, S. J. 1963. Lord Howe Island, a riddle of the Pacific. Part III. Pacific Sci. 17:361–373.
Parodi, L. R. 1964. Las regiones fitogeográficas argentinas. In Enciclopedia Argentina de Agriculture y Jardineria ed. L. R. Parodi, 1:1–14. Buenos Aires.
Raven, P. H. 1973. Evolution of subalpine and alpine plant groups in New Zealand. New Zealand J. Bot. 11:177–200.
Reiche, K. 1907. Grundzüge der Pflanzenverbreitung in Chile. In Die Vegetation der Erde, A. Engler and O. Drude, 8:1–371. Leipzig.
Schmithüsen, J. 1953. Die Grenzen der chilenischen Vegetationsgebiete. "Dt. Geographentag Essen 1953." Wiesbaden.
———. 1956. Die räumliche Ordnung der chilenischen Vegetation. Bonner Geogr. Abh. 17.
———. 1960. Die Nadelhölzer in den Waldgesellschaften der südlichen Anden. Vegetatio 9, 4–5:313–327.
Skottsberg, C. 1913. Botanische Ergebnisse der schwedischen Expedition nach Patagonien und dem Feurlande 1907–1909. III. A botanical survey of the Falkland Islands. Kungl. Svenska Vetensk. Akad. Handl. 50, 3:1–129.
———. 1916. Die Vegetationsverhältnisse längs der Cordillera de Los Andes S. von 41° S. Br. Svensk. Vetensk. Akad. Handl. 56, 5:1–82.
———. 1921. The vegetation in South Georgia. Wiss. Ergebn. Schwed. Südpolar-Exped. IV. Abt. 2. Stockholm.
———. 1922. The Natural History of Juan Fernandez and Easter Island. II. Uppsala.
———. 1931. Zur Pflanzengeographie Patagoniens. Ber. Deutsch. bot. Ges. 49:481–493.
———. 1937. Die Flora der Desventuradas-Inseln. Göteborg.
———. 1945a. The Falkland Islands. In Plants and Plant Science in Latin America, ed. F. Verdoorn, 315–318. Waltham, Mass.
———. 1945b. The Juan Fernandez and Desventuradas Islands. In Plants and Plant Science in Latin America, ed. F. Verdoorn, 150–153. Waltham, Mass.
———. 1954. Antarctic flowering plants. Bot. Tidsskr. 51, 4:330–338.
———. 1956. Derivation of the flora and fauna of Juan Fernandez and Easter Island. In The Natural History of Juan Fernandez and Easter Island, ed. C. Skottsberg, 1, 3, 5:193–438. Stockholm.
———. 1960. Remarks on the plant geography of the southern cold temperature zone. Proc. Roy. Soc. Lond. B152:447–457.
Stein, A. H. 1956. Natural forests of Chile. Unasylva 10, 4:155–160.
Sykes, W. R. 1977. Kermadec Islands Flora. Wellington.
Taylor, B. W. 1955. The flora, vegetation and soils of Macquaris Island. Austral. Nat. Antarctic Res. Exped. Rep., ser. B, 2:1–192.
Thomson, M. R. A., and R. W. Burn. 1977. Angiosperm fossils from latitude 70° S. Nature, Lond. 269:139–141.
Tryon, A. F. 1966. Origin of the fern flora of Tristan da Cunha. Brit. Fern. Gaz. 9:269–276.
Turner, J. S., C. N. Smithers, and R. D. Hoogland. 1968. The conservation of

Nordfolk Island. Carlton, Victoria: Melbourne Univ. Press.
Urban, I. 1934. Botánica de las plantas endémicas de Chile. Cancepción.
Wace, N. M. 1961. The vegetation of Gough Island. Ecol. Monogr. 31, 4:337–367.
———. 1965. Vascular plants. In Biogeography and Ecology in Antarctica, eds. J. Mieghem van and P. van Oye, 201–266. The Hague.
Wace, N. M., and J. H. Dickson. 1965. The terrestrial botany of the Tristan da Cunha Islands. Phil. Trans. Roy. Soc. Lond. B249:273–360.
Wace, N. M., and M. W. Holdgate. 1958. The vegetation of Tristan da Cunha. J. Ecol. 46:593–620.
Wallace, A. R. 1880. Island Life. London.
Walton, D. W. H., and S. W. Greene. 1971. The South Georgian species of *Acaena* and their probably hybrid. Brit. Antarct. Surv. Bull. 25:29–44.

INDEX

Aaronsohnia, 122, 131
Abelia, 55, 60, 67
  biflora, 62
  chinensis, 64
  coreana, 51
  triflora, 155
  uniflora, 64
Abeliophyllum, 45, 53
Abies, 12, 65, 69, 101
  alba, 17, 18
  balsamea, 35, 36, 87, 107
  chensiensis, 63
  concolor, 104
  densa, 71
  firma, 54
  forgesii, 63
  fraseri, 85, 87
  grandis, 105
  holophylla, 50, 51
  homolepsis, 54
  kawakamii, 60
  koreana, 54
  lasiocarpa, 36, 106, 107, 167
  magnifica, 104, 105
  mariesii, 54
  nebrodensis, 128
  nephrolepis, 33, 51
  nordmanniana, 23, 25
  numidica, 125
  pindrow, 154
  pinsapo, 125
  religiosa, 185
  sachalinensis, 52
  semenovii, 153
  sibirica, 29, 30, 31, 36, 153
  spectabilis, 154
  veitchii, 54
Abies fraseri-Picea rubens community, 88
Abkhazia, 23
Abolboda, 355
Abolbodaceae, 254, 355
Abrophyllum, 338
Abrotanella, 221, 278, 282, 296
  filiformis, 301
  muscosa, 301
  rosulata, 301
  spathulata, 301
Abutilon, 211
Acacia, 181, 182, 185, 188, 202, 204, 206, 212, 232, 233, 268, 275
  abyssinica, 207
  catechu, 211, 232
  caven, 287
  flava, 211
  gummifera, 124
  heterophylla, 225
  jacquemontii, 211
  leucophloea, 211

nilotica, 211
nubica, 205
planifrons, 233
senegal, 205, 211
seyal, 205
tortilis, 210, 213
Acacia-Commiphora bushland, 208
Acacia community, 271
Acacia-Maerua bushland, 208
Acaena, 267, 279, 281, 282, 286, 296
 adscendens, 277
 californica, 174
 decumbens, 290
 insularis, 292
 magellanica, 290
 minor, 301
 sanguisorbae, 292
 sarmentosa, 292
 stangii, 292
 tenera, 290
 trifida, 174
Acalypha, 61
 reticulata, 215
 rubra, 215
Acamptopappus, 163
Acanthaceae, 344
Acanthocalyx, 341
Acanthocardamum, 136, 149
Acanthocarpus, 351
Acanthocephalus, 140
Acanthochiton, 160
Acantholepis, 140
Acantholimon, 136, 142, 145, 149
 araxanum, 146
 diapensioides, 158
 hohenackeri, 146
 karelinii, 146
 khorassanicum, 147
 raddeanum, 147
Acanthomintha, 163, 173
Acanthopanax, 55
 aculeatus, 76
 henryi, 64
 senticosus, 50
 sessiliflorus, 50
 simonii, 64
 wilsonii, 66
Acanthophoenix, 220, 224
Acanthophyllum, 135, 142, 145
 bungei, 146
 macronatum, 148
 versicolor, 146

Acanthosicyos horridus, 213
Acanthothamnus, 162
Acanthus carduaceus, 72
Acareosperma, 235
Acer, 12, 61, 66, 335
 amplum, 64
 barbatum, 86
 barbinerve, 50
 caesium, 151
 campbellii, 69
 campestre, 28
 chionophyllum, 70
 chloranthus, 70
 circinatum, 102
 davidii, 64
 flabellatum, 64
 franchetii, 64
 fulvescens, 64
 ginnala, 51
 glabrum, 168
 grandidentatum, 168
 grosseri, 64
 heldreichii, 21
 henryi, 64
 hookeri, 69, 72
 insigne, 151
 itoanum, 59
 laevigatum, 69, 75
 leucoderme, 86
 macrophyllum, 102
 mandshuricum, 50, 51
 maximowiczii, 64
 mono, 51
 monspessulanum, 149
 negundo, 168
 nigrum, 85
 oblongum, 75
 oliverianum, 64
 palmatum, 55
 pensylvanicum, 85, 88
 pinnatinervum, 70
 platanoides, 27, 28
 pseudosieboldianum, 50
 robustum, 64
 saccharum, 85, 86
 semenovii, 153
 sikkimense, 69
 sinense, 64
 skutchii, 187
 stachyophyllum, 72
 tataricum, 28
 tegmentosum, 50, 51

INDEX

tetramerum, 69
trautvetteri, 26
triflorum, 50
turkestanicum, 153
ukurunduense, 51
velutinum, 150, 151
wilsonii, 64
Aceraceae, 335
Acerales, 334
Aceratorchis, 47
Achaenipodium, 163
Achariaceae, 326
Acharitea, 219
Achatocarpaceae, 318
Achatocarpus, 318
Achillea, 142, 144, 186
  alpina, 34
  atrata, 20
  cuneatiloba, 147
  eriophora, 210
  glaberrima, 27
Achlyphila, 355
Achlys japonica, 54
  triphylla, 102
Achradotypus, 249
Achyraechaenia, 163
Achyranthes arborescens, 298
Achyronichia, 160
Acicalyptus, 250
Acidonia, 273
Acinos corsicus, 129
Aciphylla, 296
  traillii, 301
  traversii, 301
Ackama rosifolia, 300
Acleisanthes, 159
Acmopyle, 250
Aconitum, 54, 66, 69, 71
  alboviolaceum, 50
  altaicum, 30
  arcuatum, 50
  bartletii, 60
  columbianum, 98
  crassifolium, 50
  desoulavyi, 50
  firmum, 18
  jaluense, 50
  kirinense, 50
  krylovii, 30
  kurilense, 52
  monanthum, 50
  montibaicalensis, 32

neosachalinense, 52
noveboracense, 85
ochotense, 34
paishanense, 50
paniculatum, 18
pekinense, 62
raddeanum, 50
reclinatum, 85
sachalinense, 52
sichotense, 50
uncinatum, 85
variegatum, 18
yezoense, 52
Acoraceae, 356
Acradenia, 272
Acranthera grandiflora, 230
Acrocomia totai, 261
Acronema, 44, 67
Acropogon, 249
Acrymia, 241
Actaea alba, 85
  erythrocarpa, 29
Actinanthus, 138
Actinidia, 61, 66, 69, 321
  hypoleuca, 54
  kolomikta, 50
  melanandra, 63
  strigosa, 72
Actinidiaceae, 321
Actinidiales, 321
Actiniopteridaceae, 307
Actinocarya bhutanica, 72
  tibetica, 67
Actinodaphne lanata, 230, 231
  reticulata, 74
  sikkimensis, 74
Actinokentia, 249
Actinolema, 138
Actinostachys boninensis, 57
Actinostemma lobatum, 54
  paniculatum, 62
Actinostrobus, 273
Adansonia fony, 223
  gregorii, 270
Adelosa, 219
Adenanthos, 273
Adenia, 212, 217, 223
Adenium socotranum, 209
Adenocarpus, 117, 126
  mannii, 199
  viscosus, 116
Adenocaulon, 100

bicolor, 37, 105
Adenolobus, 212
Adenoon indicum, 230
Adenopappus, 163
Adenopeltis, 284
Adenophora, 64, 67
  crispata, 51
  gmelinii, 51
  grandiflora, 51
  jakutica, 32
  morrisonensis, 61
  pinifolia, 63
  tashiroi, 56
Adenosciadium, 133
Adenostoma, 161
  fasciculatum, 177
  sparsifolium, 173
Adenostyles leucophylla, 20
Adenothamnus, 163
Adesmia, 286, 288
  canescens, 288
Adiantaceae, 307
Adiantum, 65, 307
  capillus-veneris, 170
  ogasawarense, 57
Adina cordifolia, 232
Adinandra, 61
  ryukyuensis, 59
Adlumia asiatica, 50
  fungosa, 85
Adolphia, 162
  californica, 173
Adoxa, 341
  moschatellina, 99
Adoxaceae, 9, 44, 341
Adriatic Province, 130
Adzharia, 23, 25
Aegean Islands, 122, 131
Aegiceras, 325
Aegicerataceae, 325
Aegle marmelos, 232
Aeoniopsis, 136
Aeonium, 113, 117
  glandulosum, 115
  glutinosum, 115
  gorgoneum, 118
Aeranthes, 219, 224
Aeschynanthus bracteatus, 76
  superba, 78
  wardii, 70
Aesculus, 335
  assamica, 69

  californica, 176
  glabra, 85
  indica, 155
  neglecta, 85
  turbinata, 55
  wilsonii, 64
Aëtheocephalus, 236
Aëtheolirion, 236, 238
Aetheorhiza bulbosa, 128
Aethionema, 142, 144
  diastrophis, 146
Aextoxicaceae, 283, 329
Aextoxicon, 284, 329
Afgekia, 235
Afghanistan, 135
Aframmi, 203
Africa, northern, 133
  northwestern, 120
  southern, 263
African-European affinities, 199
African-South America relationships, 258
African Subkingdom, 197
African-Tethyan relationship, 111
African-Tristan relationships, 292
Afroalpine floras, 199, 206, 207, 208
Afrocrania, 340
Afroguatteria, 198
Afromontane flora, 206, 207
Afrostyrax, 324, 328
Afrovivella, 207
Agallis, 284
Aganonerion, 235
Agapanthaceae, 349
Agapetes, 66, 69, 70, 271
  bhutanica, 72
Agastachys, 272
Agasyllis, 11, 23, 25
Agathelpis, 264
Agathis, 251, 311
  australis, 295, 300
Agathosma, 264
Agauria salicifolia, 207
Agavaceae, 350
Agavales, 349
Agave, 180, 181, 184
Agdestidaceae, 317
Ageratinastrum, 203
Agiabampoa, 163
Aglaia anamallayana, 230
Agonandra, 337
Agoseris aurantiaca, 37, 107
  gaspensis, 37

Agriophyllum, 136
  latifolium, 150
Agropyron, 167
  coxii, 301
  kingianum, 298
  sajanense, 31
  spicatum, 108
  thoroldianum, 158
  yukonense, 35
Agropyropsis lolium, 126
Agrostis, 73, 203, 208, 240
  antarctica, 278
  canariensis, 118
  magellanica, 278
  nipponensis, 56
Agrostophyllum, 222
Aguaria salicifolia, 199
Aichryson, 112, 117
  divaricatum, 115
  dumosum, 115
  villosum, 115
Ailanthus kurzii, 237
  vilmoriniana, 66
Ainsliaea, 56, 67
  angustifolia, 78
  glabra, 64
  ramosa, 64
Aira, 203
  caryophyllea, 208
Aisandra, 234
Aitchisonia, 139
Aitoniaceae, 334
Aizoaceae, 212, 264, 318
Aizoanthemum, 203
Ajania, 157
Ajuga, 56, 203
  acaulis, 129
  boninsimae, 58
  dictyocarpa, 61
  lobata, 70, 76
  macrosperma, 76
Akania, 335
Akaniaceae, 271, 335
Akebia, 40
  longeracemosa, 60
Alai Range, 139
Aland Islands, 20
Alangiaceae, 340
Alangium, 340
Alaska, 37, 100
Alaska Panhandle, 102
Alaska-Yukon, 12

Alazan Valley, 136
Albania, 21, 131
Albizia julibrissin, 151
  schimperana, 207
Alboff, N., 289
Alcea, 142
  abchasica, 23
  freyniana, 148
  hyrcana, 151
  karsiana, 145
  sophiae, 145
Alchemilla, 186, 207
  cryptantha, 208
  fulgens, 17
  minima, 17
  nemoralis, 27
Alchemilla scrub, 200, 208
Alcimandra, 40
  cathcartii, 69, 74
Alectorurus, 47, 53
Alectris gracilis, 73
Alectryon, 296
Aleisanthia, 241
Aletes humilis, 109
Aletris, 64, 68
Aleutian Islands, 34
Alexander Selkirk Island, 281, 282
Alexandra, 136, 152
Alexeya, 135
Algeria, 121, 125, 133
Alismataceae, 346
Alismatales, 346
Alismatidae, 346
Alkanna, 22
  leptophylla, 149
Allagopappus, 113, 117
Allantospermum, 335
Alleizettella, 235
Alliaceae, 10, 349
Allium, 20, 56, 63, 68, 70, 126, 142, 173, 175, 177
  aaseae, 170
  acidoides, 70
  akaka, 147
  bodeanum, 148
  canadense, 99
  candolleanum, 25
  cristophii, 148
  derderianum, 147
  dictyoprasum, 147
  giganteum, 148
  graciliensis, 25

grande, 26
grossii, 128
helicophyllum, 148
horvatii, 131
kopetdagense, 148
lalesarensis, 149
lenkoranicum, 151
leonidii, 147
mariae, 147
melanantherum, 22
mongolicum, 157
palentinum, 17
parciflorum, 130
phariense, 73
pleianthum, 108
polyrhizum, 157
pumilum, 31
pyrenaicum, 17
rhabdotum, 73
senescens, 52
sosnovskyanum, 145
stenopetalum, 145
tricoccum, 86
vavilovii, 148
vineale, 18
woronowii, 147
Allmaniopsis, 206
Allocalyx, 219, 224
Allochrusa, 135, 146
Allolepis, 164
Alloschmidia, 249
Allospondias, 235
Allotropa, 100, 102
Allowoodsonia, 243
Alluaudia, 217
Alluaudiopsis, 217
Alniphyllum, 41
eberhardtii, 66
Alnus, 12, 60, 187, 321
cremastogyne, 66
crispa, 34, 36
fruticosa, 33
glutinosa, 28
hirsuta, 34
incana, 36
maritima, 94
nepalensis, 75
nitida, 155
rubra, 102
sinuata, 102
subcordata, 151, 152
viridis, 30

Alococarpum, 138
Aloe, 206, 212, 223
bainesii, 201
dichotoma, 213
perryi, 209
Alona, 284, 343
Alopecurus antarcticus, 290
howellii, 176
Aloysia gratissima, 182
Alpandia, 250
Alpinia bilamellata, 58
boninsimensis, 58
Alps, endemism in, 12, 18
Alrawia, 142, 148
Alseuosmia, 294, 296, 339
atriplicifolia, 300
ligustrifolia, 300
linariifolia, 300
palaeiformis, 300
Alseuosmiaceae, 339
Alsinidendron, 247
Alsophila, 309
capensis, 265
Alstroemeria, 286
patagonica, 288
Alstroemeriaceae, 351
Alstroemeriales, 351
Altai, 11, 30, 137
Altai-Sayan Province, 30
Althaea, 142
Althenia, 347
Altingiaceae, 319
Alvordia, 163
Alyssopsis, 136
Alyssum, 142, 144
alpestre, 19
doerfleri, 21
leucadeum, 131
macrocalyx, 126
nebrodense, 129
robertianum, 129
wulgenianum, 19
Alzatea, 332
Alzateaceae, 254, 332
Amaracus, 122
Amaranthaceae, 318
Amaryllidaceae, 350
Amaryllidales, 349
Amaryllis, 264
Amauria, 163
Amazonian Province, 259
Amazonian Region, 258

INDEX

Ambavia, 217
Amblyocarpum, 140
　inuloides, 151
Amblyopappus pusillus, 174
Ambongia, 219
Amborella, 248, 250, 313
Amborellaceae, 248, 313
Ambrella, 219
Ambrosia dumosa, 180, 183
Ambrosina, 123
Amelanchier, 85
　asiatica, 55
　fernaldii, 35, 37
　florida, 102
　gaspensis, 37
Amentotaxus, 311
　formosana, 60
　yunnanensis, 65
Amitostigma, 56
Ammanthus, 122
Ammi crinitum, 129
Ammiopsis, 121
　aristidis, 126
Ammobroma, 162, 343
Ammodaucus, 133
Ammodendron, 138
Ammoides, 121
　atlantica, 126
Ammopiptanthus, 138, 157
　nanus, 156
Ammosperma, 133
Amoora lawii, 230
　manii, 237
Amoreuxia, 328
Amorpha nana, 96
Amorphophallus kiusianus, 56
Ampalis, 218
Ampelocissus elephantina, 223
　nervosa, 72
　sikkimensis, 76
Ampelodesmos, 123
　tenax, 119
Ampelopsis, 64, 338
Ampelosicyos, 217
Amphianthus, 90
Amphibolis, 347
Amphimas, 198
Amphipappus, 163
Amphisiphon, 264
Amphoricarpos elegans, 24
　neumayeri, 22
Amphorocalyx, 218

Amphorogyne, 249
Amsinckia douglasiana, 173
Amsterdam Island, 292, 293
Amur Basin, 34, 42
Amygdalus, 146, 211
　wendelboi, 149
Anabasis brevifolia, 157
　eugeniae, 146
Anacampseros, 212
Anacardiaceae, 334
Anacyclus, 134
　linearilobus, 126
Anagallis serpens, 207
　tenella, 188
Anagyris foetida, 119, 124
　latifolia, 117
Anaphalis, 67
Anaphyllum, 229
Anarrhinum, 122
　corsicum, 129
Anarthria, 266, 273
Anarthriaceae, 355
Anarthrophyllum rigidum, 288
Anastrabe, 201
Anatolia, 11, 23, 121, 131, 136, 143
Anatolian Diagonal, 143, 144
Anatolian Plateau, 143
Anchonium, 136
Anchusa crispa, 129
Ancient Mediterranean Subkingdom, 110
Ancistrocactus, 160
Ancistrocarya, 45, 53
　japonica, 55
Ancistrocladaceae, 197, 323
Ancistrocladales, 323
Ancistrocladus, 323
Ancrumia, 285
Ancylostemon, 45, 67
Andaman Islands, 235, 237
Andamanese Province, 237
Andean Region, 261
Andersonia, 273
Andes, 262, 277, 279, 284, 287
Andrachne chinensis, 62
　colchica, 23
　maroccana, 124
Andresia, 241
Androcorys japonensis, 56
Androcymbium, 111
　psammophilum, 118
Andropogon, 99, 204, 205, 261
　amethystinus, 208

gerardii, 97
scoparius, 97
Androsace, 19, 66, 69, 72
  andersonii, 15
  ciliata, 17
  gorodkovii, 32, 33
  koso-poljanskii, 27
  lehmanniana, 26
  pyrenaica, 17
  tapete, 158
  triflora, 15
Androsiphon, 264
Androya, 219
Andryala, 117, 122, 126
  crithmifolia, 115
  levitomentosa, 20
Aneilema spp., 68
Anelsonia, 100
Anemarrhena, 47, 50, 351
  asphodeloides, 51
Anemiaceae, 307
Anemone, 54, 66, 71, 240
  baicalensis, 30
  canadensis, 36
  edwardsensis, 99
  elongata, 74
  kuznetzowii, 26
  rivularis, 74, 231
  thomsonii, 207
  uralensis, 27
  vitifolia, 69
Anemonella, 79
Anemonopsis, 40, 53
  macrophylla, 54
Anemopsis, 159
Anetium, 307
Aneulophus, 335
Angelica, 55, 66
  ampla, 109
  anomala, 50
  gigas, 50
  grayi, 109
  heterocarpa, 17
  hirsutiflora, 61
  morii, 61
  morrisonicola, 61
  pachycarpa, 17
  razulii, 17
  triquinata, 85
Angianthus, 269
Angiopteris, 65
  boninensis, 57

Angkalanthus, 209
Angola, 203
Angraecum sesquipedale, 222
Anguillaria, 348
Angylocalyx, 198
Anigozanthos, 273
Aningueria adolfi-friedericii, 207
Anisachne, 47
Anisadenia, 43
  pubescens, 75
  saxatilis, 75
Anisocoma, 163
Anisodus luridus, 67
Anisomallon, 249
Anisomeria, 283, 286, 317
Anisophyllea, 332
Anisophylleaceae, 227, 332
Anisopoda, 219
Anisotome, 296
  acutifolia, 301
  antipoda, 301
  flabellata, 301
  latifolia, 301
Ankyropetalum, 136
Annam, 234
Annamese Province, 238
Annamocarya, 234
Annonaceae, 313
Annonales, 313
Annulodiscus, 235
Anodopetalum, 272
  biglandulosum, 272
Anogeissus, 204, 205
  latifolia, 232, 233
  pendula, 211, 232
Antarctic Islands, 276
Antarctic Peninsula, 291
Antennaria, 34
  affinis, 15
  canescens, 15
  glabrata, 15
  hansii, 15
  intermedia, 15
  suffrutescens, 103
  virginica, 89
Anthaenantia, 82
  lanata, 261
Antheliacanthus, 236, 238
Anthemis, 22, 129, 142
  chrysantha, 126
  grossheimii, 147
  saguramica, 26

tigreensis, 207
zyghia, 24
Anthericaceae, 351
Anthericum yedoense, 56
Antherolophus, 236
Antheroporum, 235
Anthocarapa, 249
Anthochlamys, 136
Anthocleista, 342
Anthodiscus, 322
Anthostema madagascariensis, 222
Anthoxanthum, 208
　hookeri, 73
　sikkimense, 73
Anthyllis fulgurans, 128
　hermanniae, 119
　kuzeneviae, 29
　lemanniana, 115
Antiaris, 269
Antidaphne, 338
Antinisa, 217
Antiotrema, 45, 65
　dunnianum, 67
Antipodes Islands, 277, 294, 302
Antirrhinum siculum, 129
Antistrophe serratifolia, 230
Antongilia, 220
Antonia, 259
Antoniaceae, 341
Antrophyum, 307
Anulocaulis, 159
　eriosolenus, 180
　gypsogenus, 180
　leisolenus, 180
　reflexus, 180
Anura, 140
Anurosperma, 315
Anvilleina, 122
Apacheria, 161, 331
Apalachicola center, 94
Apaloxylon, 218
Apama barberi, 230
　siliquosa, 230
Apennines, 18
Apetahia, 246
Aphanisma, 160
　blitoides, 173
Aphanopetalum, 330
Aphanopleura, 138
　trachysperma, 146
Aphanostelma, 45
Aphelexis, 219
Aphelia, 356
Aphloia, 200
　theiformis, 226
Aphragmus involucratus, 30
Aphyllanthaceae, 120, 351
Aphyllanthes, 123, 351
Apiaceae, 10, 26, 159, 340
Apiaceae, arborescent, 281
Apiales, 340
Apiastrum, 162
Apiopetalum, 249
Apium australe, 293
　crassipes, 129
　goughense, 293
　ikenoi, 55
Apocynaceae, 342
Apocynum basikurumon, 55
Apodanthaceae, 315
Apodanthes, 315
Apodicarpum, 44
Apodocephala, 219
Apodytes beddomei, 230
　dimidiata, 207, 265
Apollonias arnottii, 113, 230
　barbusana, 113, 116
Aponogeton, 346
Aponogetonaceae, 346
Aponogetonales, 346
Aporosa lindleyana, 230
Aporostylis, 295
Apostasia, 353
　nipponica, 56
Apostasiaceae, 353
Appalachian-Eastern Asian relationship, 83
Appalachian Mountains, southern, 88
Appalachian-Ozark relationship, 84
Appalachian Plateau Geological Province, 90
Appalachian Province, 84
Aptandraceae, 337
Aquifoliaceae, 337
Aquilegia, 21, 69
　alpina, 18
　aurea, 21
　bernardii, 128
　borodinii, 30
　colchica, 23
　einseleana, 18
　litardierei, 128
　saximontana, 109
　thalictrifolia, 18
　yabeana, 62
Arabian Peninsula, 132, 133, 134, 141, 208

Arabian subregion, 205
Arabidopsis thaliana, 207
  tschuktschorum, 15
Arabis, 19, 54, 61, 168
  alpina, 207
  brownii, 17
  colchica, 23
  ferdinandi-corburgii, 21
  mcdonaldiana, 177
  macloviana, 290
  nordmanniana, 23
  sakhokiana, 23
  turczaninovii, 33
Araceae, 356
Arachnioides, 53
Arachnitis, 285, 348
Araeoandra, 284
Arakan Yoma, 78
Arales, 356
Aralia, 66
  continentalis, 50
  elata, 51
  glabra, 55
  malabarica, 230
  nudicaulis, 99
  spinosa, 86
  taiwaniana, 61
Araliaceae, 10, 340
Araliales, 340
Aralidiaceae, 227, 339
Aralidiales, 339
Aralidium, 339
Aralo-Caspian Province, 152
Araucaria, 311
  angustifolia, 260
  araucana, 286, 287
  columnaris, 251
  cookii, 252
  heterophylla, 251, 295, 298
Araucariaceae, 311
Araucariales, 311
Arbutus, 159
  andrachne, 119, 131, 132
  canariensis, 116
  menziesii, 102, 105, 176
  unedo, 119, 124, 131
Archangiopteris, 39, 65
  itoi, 60
Archeria, 273
Archirhodomyrtus, 249
Arctic, 11
Arctic Province, 12

Arctomecon, 159
Arctostaphylos, 16, 172, 177
  pungens, 182
  uva-ursi, 36
Arcto-Tertiary geoflora concept, 83
Ardisia, 64, 66
  andamanica, 237
  blatteri, 230
  cornudentata, 61
  khasiana, 77
  macrocarpa, 75
  polycephala, 77
  quinquangularis, 77
  rhynchophylla, 77
  virens, 77
Areca, 271
Arecaceae, 356
Arecales, 356
Arecastrum romanzoffianum, 261
Arecidae, 356
Arenaria, 71, 168, 186
  balearica, 127
  brevifolia, 90
  cinerea, 128
  cumberlandensis, 90
  howellii, 177
  huteri, 18
  juncea, 52
  lateriflora, 36
  musciformis, 158
  provincialis, 129
  rossii, 106
  yunnanensis, 66
Arenga wightii, 230
Arfeuillea, 235
Argania, 121, 125
  spinosa, 111, 124, 125
Argentina, 283
Argylia, 285, 286
Argyranthemum, 113, 117
Argyrocytisus, 121
Argyrolobium, 111
  kotschyi, 210
Argyronerium, 235
Argyroxiphium, 247
Arillastrum, 249
Ariocarpus, 160
Arisaema, 56, 65, 70, 73, 77
  manshuricum, 51
Arisarum vulgare, 119
Aristea, 264
Aristida, 135, 174, 211, 288

adscensionis, 216
boninensis, 58
oligantha, 175
pallens, 261
paradoxa, 119
pogonoptila, 210
pungens, 125
roemeriana, 182
stricta, 92
venusta, 261
Aristolochia, 60, 286
 baetica, 124
 bianorii, 128
 chilensis, 286
 croatica, 131
 griffithii, 69, 71
 heterophylla, 63
 macrophylla, 85
 manshuriensis, 50
 nakaoi, 71
 onoei, 54
 pontica, 23
 sempervirens, 119
 sicula, 128
Aristolochiaceae, 314
Aristolochiales, 314
Arizona, 159
Armeniaca vulgaris, 156
Armenian Highlands, 145
Armenian Subprovince, 144
Armeno-Iranian Province, 144
Armeria, 129
 maderensis, 115
Arnica, 101, 163
 alpina, 29
 cernua, 103
 cordifolia, 37, 107
 latifolia, 107
 lonchophylla, 37
 mallotopus, 56
 sachalinensis, 34
 spathulata, 103
Arnicastrum, 163
Arnottia, 219, 224
Aromadendron, 239
Arophyton, 220
Artedia, 138
Artemisia, 34, 51, 56, 63, 67, 142, 149, 155, 157
 arborescens, 124
 argentea, 115
 canariensis, 117

comata, 15
flava, 15
glacialis, 20
gorgonea, 119
herba-alba, 125
hyperborea, 15
limosa, 53
oelandica, 20
pattersonii, 110
porteri, 109
richardsonii, 15
salsoloides, 158
senjavinensis, 15
sibirica, 52
tridentata, 108, 167
umbelliformis, 20
wellbyi, 158
Artemisia community, 149, 167
Arthraerua, 212
 horrida, 213
Arthrocarpum, 207
Arthroclianthus, 249
Arthromeris himalayensis, 74
 wardii, 74
Arthrophytum, 136
Arthropteris, 281, 309
 altescandens, 282
Arum pictum, 127
Aruncus dioicus, 85
 parvulus, 50
Arundinaria, 56, 65
 alpina, 199, 208
 usawai, 62
Arundinella, 68
 hirta, 52
Asarca, 285
Asarum, 54, 60, 63
 canadense, 85
 europaeum, 27, 31
 himalaicum, 71
Ascarina, 222, 251, 314
 lanceolata, 299
Ascarinopsis, 217, 222, 314
Ascension Island, 214
Aschisma kansanum, 96
Asclepiadaceae, 342
Asclepias, 93
 ovalifolia, 99
 perennis, 84
 texana, 85
Ascopholis, 229
Ashtonia, 241

Asia, central, 136, 155
  eastern, 39
  western, 138, 142
Asian-Balkan relationships, 21
Asimina, 79, 313
  triloba, 85
Asir, 208
Aspalathus, 264
Asparagaceae, 349
Asparagales, 349
Asparagus, 349
  acutifolius, 124
  albus, 124
  aphyllus, 124
  arborescens, 118
  capensis, 265
  kiusianus, 56
  longiflorus, 63
  stipularis, 124
Aspen-grass woodland, 95
Asperula, 142
  capitata, 21
  crassifolia, 129
  garganica, 131
  odorata, 31
  staliana, 131
  taurica, 132
  trifida, 55
Asphodelaceae, 10, 350
Asphodelales, 350
Asphodeline lutea, 132
  szovitsii, 147
  tenuiflora, 145
  tenuior, 26
Aspidiaceae, 10, 309
Aspidiales, 309
Aspidistra, 47
Aspidistraceae, 349
Aspidium canariense, 117
Aspidosperma polyneuron, 261
Aspleniaceae, 10, 309
Asplenium, 60, 281, 309
  alvarezence, 292
  ascensionis, 216
  friesiorum, 292
  hermannii-christii, 23
  ikenoi, 57
  micantifrons, 57
  petrarchae, 119
  platybasis, 292
  seelosii, 18
  woronowii, 23

Assam, 40, 71, 236
Astelia, 221, 251, 286, 296
  banksii, 300
  chathamica, 301
  pumila, 290
  trinervia, 300
Asteliaceae, 349
Aster, 56, 61, 64, 67, 83, 86, 93, 97, 240
  acuminatus, 88
  alpinus, 106
  avitus, 90
  brickellioides, 103
  chapmanii, 95
  crenifolius, 35
  curtisii, 89
  drummondii, 85
  eryngiifolius, 95
  gypsophilus, 180
  helenae, 70
  jessicae, 109
  lowrieanus, 88
  macrophyllus, 88
  miyagii, 59
  paludicola, 103
  peirsonii, 104
  porteri, 110
  puniceus, 88
  pygmaeus, 15
  reticulatus, 95
  sagittifolius, 85
  sibiricus, 99
  siskiyouensis, 103
  spinulosus, 95
  surculosus, 88
  wattii, 78
  yukonensis, 35
Asteraceae, 10, 15, 26, 30, 93, 94, 96, 159, 172, 175, 180, 264, 268, 274, 281, 288, 346
Asterales, 346
Asteranthaceae, 323
Asteranthera, 285
Asteridae, 345
Asteriscus spp., 117
Asteropeia, 217, 223, 322
Asteropeiaceae, 222, 322
Asteropyrum, 40
Astilbe, 54, 69, 240
  biternata, 85
  koreana, 50
  macroflora, 61
  rivularis, 75

rubra, 75
Astilboides, 42, 50
  tabularis, 50
Astiria, 218, 224
Astoma, 121, 131
Astragalus, 21, 26, 55, 62, 66, 72, 101, 126, 142, 145, 158, 168, 172
  alpinus, 99
  anisus, 109
  arenarius, 27
  arnoldii, 158
  arrectus, 108
  arthurii, 109
  aszharicus, 23
  bachmarensis, 23
  baionensis, 17
  balearicus, 128
  barrii, 96
  breweri, 177
  clerceanus, 27
  cusickii, 109
  dahuricus, 52
  drabelliformis, 109
  gorodkovii, 15
  gracilis, 96
  gypsodes, 180
  henningii, 27
  heutii, 129
  hyalinus, 96
  japonicus, 53
  karakuschensis, 146
  karelinianus, 27
  latifolius, 146
  leibergii, 108
  leontinus, 19
  malcolmii, 158
  maritimus, 129
  mesites, 146
  molybdenum, 109
  nelsonianus, 109
  olchonensis, 30
  ovalis, 143
  pallescens, 27
  paradoxus, 146
  parryi, 109
  pectinatus, 96
  physocalyx, 21
  plattensis, 96
  proimanthus, 109
  pubiflorus, 27
  richardsonii, 15
  roemeri, 19
  sachalinensis, 53
  satoi, 50
  setsureianus, 50
  simplicifolius, 109
  sirinicus, 129
  speirocarpus, 108
  strictifolius, 146
  szovitsii, 146
  tanaiticus, 27
  tennesseensis, 91
  tibeticus, 158
  trigonocarpus, 32
  vallaris, 109
  verrucosus, 129
  williamsii, 35
  yamamotoi, 53
Astrantia, 11
  bavarica, 19
  carniolica, 19
  major, 18
  pontica, 24
Astrebla, 274
Astrocodon, 11
Astrotrichilia, 218
Astydamia latifolia, 117
Atamisquea emarginata, 289
Atelanthera, 136
Athamanta cortiana, 19
Atherolepis, 235
Atherospermataceae, 314
Athertonia, 271
Athrotaxis, 270, 272, 296
  cupressoides, 272
  selaginoides, 272
Athyriaceae, 309
Athyrium, 53, 60
Atlantic and Gulf Coastal Plain Province, 91
Atlantic-European Province, 16
Atlantic Province, 260
Atractocarpus, 249
Atractylodes, 46
Atragene, 34
Atraphaxis, 142
Atrichodendron, 235
Atrichoseris, 163
Atriplex, 158, 167, 181, 275
  californica, 173
  halimus, 125
  reptans, 180
Atropa komarovii, 148
Atropatenia rostrata, 146
Atropatenian Subprovince, 145, 147

Auckland Islands, 277, 294, 295, 302
Aucuba, 44, 59, 340
  himalaica, 75
Aucubaceae, 39, 340
Augea, 212
Augouardia, 198
Aureolaria, 82
  grandiflora, 86
  laevigata, 86
  patula, 86
Australia, 268, 277
Australia-New Caledonia relationships, 250
Australia-New Zealand relationships, 296
Australian Kingdom, 268
  notes, 275
Austria, 19
Austrobaileya, 313
Austrobaileyaceae, 271, 313
Austrobaileyales, 313
Austrocactus, 283
Austrocedrus, 283
  chilensis, 287
Austromuellera, 271
Austrotaxaceae, 248, 310
Austrotaxus, 248, 310
Avellanita, 284
Avena, 126
  barbata, 175
  canariensis, 118
  fatua, 175
  occidentalis, 118
  saxatilis, 130
Averrhoaceae, 336
Avetra, 219, 352
Avicennia, 296
  officinalis, 233
  resinifera, 300
Avicenniaceae, 344
Axinandra, 332
Aylacophora, 285
Azara, 281
  fernandeziana, 282
Azima, 337
Azolla, 310
Azollaceae, 310
Azorean Province, 114
Azorella, 261, 277, 279, 286
  caespitosa, 288
  filamentosa, 290
  lycopodioides, 290
  selago, 277, 293
Azores, 112, 114

Azorina, 113, 114

Babbagia, 274
Babiana, 212, 264
Baccaurea coutrallensis, 230
Baccharis, 286, 287
  marginalis, 288
Bachmannia, 201
Badula, 217, 220, 224
Baeckea, 251
Baeolepis, 228
  nervosa, 230
Baeometra, 264
Baeriopsis, 163
Bafutia, 198
Bahia, 286
Baikal, 30
Baikal Range, 32
Baileya, 163
Baileyoxylon, 271
Baja California, 159, 181
Balaka, 244
Balanitaceae, 334
Balanites, 333, 334
  aegyptiaca, 211
  roxburghii, 232
Balanopaceae, 320
Balanopales, 320
Balanophora, 64, 271
  spicata, 61
Balanophoraceae, 315
Balanophorales, 315
Balanops, 320
Balansaea, 121
Balbisia, 286, 336
Bald cypress, 92
Baldellia alpestris, 17
Balds, 89
Balduina, 82
  angustifolia, 95
Balearic Islands, 121, 127
Balearic Province, 127
Balfourodendron riedelianum, 261
Balgooy, M. M. J. van, 253
Baliospermum nepalense, 72
Balkan Peninsula, 11, 21, 25
Balkan Province, 21
Ballochia, 209
Ballota hirsuta, 124
Balsaminaceae, 336
Balsaminales, 336
Balsamorhiza sagittata, 98

Baluchistan, 138, 153
Bambusa, 62
Bandar Abbas, 210
Banksia, 268, 273
Barbarea rupicola, 129
 taiwaniana, 61
Barberetta, 353
Barbeuia, 217, 318
Barbeuiaceae, 216, 318
Barbeya, 329
Barbeyaceae, 202, 206, 329
Barbeyales, 329
Barclayaceae, 227, 316
Barguzin Range, 32
Barleria aucheriana, 210
 lichtensteiniana, 213
Barlia, 123
Barneoudia, 283
Barringtonia racemosa, 59
Barringtoniaceae, 323
Barroetia, 163
Bartelettia, 163
Barthea, 43
 formosana, 61
Bartonia, 82
Bartschella, 160
Bartsia, 207
 spicata, 17
Basellaceae, 318
Basselinia, 249
Bassia latifolia, 232
Bataceae, 327
Batales, 327
Bathiaea, 218
Bathiorhamnus, 219
Batis, 327
Battandiera, 134
Baudouinia, 218
Bauera, 330
Baueraceae, 270, 330
Bauhinia, 66
 hupeana, 64
 racemosa, 211
 variegata, 233
Baxteria, 273
Bdallophytum, 315
Bear River Range, 169
Beaufortia, 273
Beaumontia jerdoniana, 230
Beauprea, 249
Beaupreopsis, 249
Bebbia, 163

Beccariophoenix, 220
Becheria, 241
Bedfordia, 272
Beesia, 40
Begonia, 61, 66, 326
 aliciae, 230
 cathcartii, 72
 gemmipara, 72
 griffithii, 72
 henryi, 63
 hymenophylloides, 70
Begoniaceae, 326
Begoniales, 326
Begoniella, 326
Beguea, 218
Behnia, 352
Behria, 164
Beilschmiedia, 187
 tarairi, 300
 yunnanensis, 66
Belamcanda, 47
Bellendena, 270, 273
Bellevalia coelestis, 145
 makuensis, 147
 pomelii, 126
Belliolum, 250
Bellis bernardii, 129
Bellium, 122
 bellidioides, 127
 crassifolium, 129
Belorussia, 29
Bembicia, 217
Bembiciopsis, 217
Bencomia, 112, 117
Benedictella, 121, 125
Bengal Province, 233
Benitoa, 163
Benthamia, 219, 224
Benthamidia, 340
Benthamiella, 284
Bentinkia coddapanna, 230
Berardia, 11, 18
 subacaulis, 20
Berberidaceae, 9, 317
Berberidopsidaceae, 325
Berberidopsis, 271, 283, 325
Berberis, 54, 60, 63, 66, 69, 71, 158, 240,
 261, 279, 281, 286
 aquifolium, 102, 105
 burmanica, 70
 canadensis, 85
 coxii, 70

feddei, 77
griffithiana, 74
heterophylla, 288
hypokerina, 70
khasiana, 77
maderensis, 115
manipurana, 77
nilghirensis, 231
rufescens, 70
sublevis, 74
tinctoria, 231
venusta, 70
wallichiana, 74
wardii, 77
Berchemia, 55
  flavescens, 72
  yunnanensis, 67
Berenice, 219, 224, 345
Bergenia pacifica, 50
  purpurascens, 69
Bergerocactus, 160
Bergia, 323
Berginia, 162
Beringian Arctic, 13
Berkheya chamaepeuce, 213
Berlandiera subacaulis, 95
Berlinianche, 315
Berneuxia, 41, 65
  thibetica, 66, 69
  yunnanensis, 66
Berroa, 288
Bersama, 201, 335
Berteroella, 42
Beschorneria, 164
Beta patula, 115
  webbiana, 117
Betonica abchasica, 24
Betula, 12, 16, 54, 69, 85
  alnoides, 75
  costata, 50
  dahurica, 52
  delavayi, 66
  ermannii, 34
  fruticosa, 33
  glandulosa, 36
  kellerana, 30
  lutea, 88
  medwedewii, 23
  megrelica, 23
  middendorfii, 33, 34
  nigra, 86
  papyrifera, 36, 87, 98, 106
  pendula, 29, 30
  platyphylla, 52
  pubescens, 30, 36
  raddeana, 26
  rotundifolia, 33, 36
  tatewakiana, 53
  tianschanica, 156
Betulaceae, 10, 321
Bhidea, 229
Bhutan, 40, 71
Biarum, 123
Bidens, 215
Biebersteinia, 336
  heterostemon, 62
  odora, 155
Biebersteiniaceae, 9, 336
Bienertia, 136
Bigelowia, 82
Bignoniaceae, 344
Bignoniales, 343
Billia, 335
Biondia, 45
  yunnanensis, 67
Biscutella rotgesii, 129
Bismarck Archipelago, 243
Bismarckia, 220
Bismarckian Province, 243
Bistella, 331
Biswarea, 42, 71
Bivinia, 217
Bivonaea, 121
Bixa, 328
Bixaceae, 328
Bixales, 328
Blabeia, 249
Blachia andamanica, 237
Black Hills, 98
Black Sea shores, 24
Blaeria spicata, 207
Blandfordia, 272
Blandfordiaceae, 270, 350
Blechnaceae, 310
Blechnum, 54, 281, 310
  cycadifolium, 282
  norfolkianum, 298
  penna-marina, 276, 290, 295
  schottii, 282
  spicant, 188
Blennosperma nanum, 176
Blepharistemma, 227
  membranifolia, 230
Blepharizonia, 163

Blepharocarya, 270, 334
Blephilia, 82
 ciliata, 86
Bletilla, 47
Bloomeria, 164
 clevelandii, 173
 crocea, 173
Blue Ridge Geological Province, 88
Blumea, 67
Blyxa leiosperma, 56
Bobea, 247
Bocquillonia, 249
Boemeria, 54, 281
 australis, 298
 boninensis, 57
 dealbata, 299
 excelsa, 282
 hamiltoniana, 77
 macrophylla, 77
 polystachya, 77
 sidaefolia, 77
 taiwaniana, 60
Boenninghausenia albiflora, 69
Boissier, E., 120
Bolandra, 100
Bolanosa, 163
Bolanthus, 121, 131
Bolbitidaceae, 309
Bolbitis boninensis, 57
Bolbostemma, 42
Boleum, 121, 126
Bolivia, 285
Bombacaceae, 329
Bombax ceiba, 233
 mosambicense, 203
Bonannia, 122
Bonatia, 249
Bongardia, 135
Bonin Islands, 42, 56
Boninia, 43, 57
 glabra, 57
 grisea, 57
Boniodendron, 235
Bonnayodes, 228
Bonnetiaceae, 322
Bonniera, 220, 224
Bonnierella, 245
Bontia, 344
Boquila, 283, 316
 trifoliata, 286
Bor, N. L., 77
Boraginaceae, 10, 159, 343

Boraginales, 343
Borago pygmaea, 129
Borassus, 271
 flabellifer, 232
Boreal Subkingdom, 10
Borneo, 239, 241
Borodinia, 11, 32
Boronella, 249
Boronia, 269
Borsczowia, 136
Boscia, 206
Bossiaea, 269
Bostrychanthera, 46
Boswellia, 206
 serrata, 211, 232
Bothriochloa, 233, 288
Bothriospermum chinense, 63
 decumbens, 51
Botrychium, 306
 nipponicum, 53
Botryomeryta, 249
Botschantzevia, 136
Botswana, 203
Bottegoa, 207
Bottionea, 285, 351
Bougeria, 344
Bounty Islands, 303
Boussigonia, 235
Bouteloua, 99
 gracilis, 97
 hirsuta, 182
Boutonia, 219
Bouzetia, 249
Bowenia, 270, 271
Bowlesia, 113
Boykinia aconitifolia, 85
 lycoctonifolia, 55
Brachanthemum, 140
 baranovii, 31
 gobicum, 157
Brachiaria villosa, 78
Brachybotrys, 45, 50
 paridiformis, 51
Brachycereus, 257
Brachycome wardii, 70
Brachyelytrum japonicum, 56
Brachyglottis, 295
Brachypodium arbuscula, 118
Brachystachyum, 47
Brachystegia, 202, 204
Bracteanthus, 314
Brandegea, 160

Brandisia, 344
Brandzeia, 218, 220
Brasenia, 316
Brassaiopsis, 76
 alpina, 72
 trilobata, 70
Brassica, 62, 126, 129
 balearica, 128
 cretica, 132
 insularis, 129
Brassicaceae, 10, 15, 16, 26, 159, 264, 274, 327
Braya alpina, 19
 humilis, 110
 intermedia, 15
 linearis, 15
 purpurascens, 15
 thorildwulfii, 15
Brazil, 260, 277, 284
Brazilian Region, 259
Bredia, 61
Bremontiera, 218, 224
Brenierea, 218
Breonia, 219, 220
Bretschneidera, 43, 335
Bretschneideraceae, 39, 335
Brexia, 338
Brexiella, 218
Breynia, 61
Brickellia greenei, 103
Bridgesia, 284, 286
Briggsia, 45, 67
 kurzii, 70
Brighamia, 247
Brintonia, 82
Briza, 288
Brochoneura, 217
Brodiaea, 172, 173, 175
Bromeliaceae, 255, 353
Bromeliad epiphytes, 187
Bromeliales, 353
Bromus, 203, 288
 cognatus, 208
 himalaicus, 73
 hordeaceus, 175
 interruptus, 17
 moesiacus, 22
 rubens, 175
Bronchoneura, 222
Brongniartikentia, 249
Brossardia, 136, 148
Brousmichea, 236

Broussaisia, 247
Broussonetia kaempferi, 54
Bruguiera, 233
 gymnorrhiza, 59
Brunellia, 330
Brunelliaceae, 254, 330
Bruniaceae, 264, 330
Bruniales, 330
Brunnera sibirica, 31
Brunnichia, 79
Brunonia, 346
Brunoniaceae, 268, 346
Brylkinia, 47
Bryocarpum, 41, 71
 himalaicum, 72
Bryodes, 219, 224
Bryonia verrucosa, 117
Bubbia, 250
 howeana, 297
Bucephalophora, 121
Buchanania lanzan, 232, 233
Buchingera, 136
Buchloe, 99
 dactyloides, 97
Buckinghamia, 271
Buckleya distichophylla, 85
 lanceolata, 54, 55
Buddleja, 67
 albiflora, 64
 bhutanica, 72
 colvilei, 72
 formosana, 61
 japonica, 55
 lindleyana, 64, 158
 macrostachya, 76
 myriantha, 70
 racemosa, 99
 tibetica, 72, 158
Buddlejaceae, 343
Buffonia chevalieri, 125
 teneriffae, 117
Buglossoides calabra, 129
 gastonii, 17
 minima, 129
Bulbine, 351
Bulbinella, 266, 350
 rossii, 301
Bunium, 126
 brevifolium, 115
 mauritianum, 124
Bulbinopsis, 351
Bulbophyllum, 56, 73

INDEX

tuberculatum, 297
Bulgaria, 11, 18, 21, 23
Bulleyia, 47
Bumelia lycioides, 86
  tenax, 93
Bunch-grass community, 186
Bungea, 140
Buphthalmum inuloides, 130
Bupleurum barceloi, 128
  dumosum, 124
  elatum, 129
  euphorbioides, 50
  kaoi, 61
  karglii, 21
  lanceolatum, 155
  martjanovii, 31
  mucronatum, 231
  nipponicum, 55
  petraeum, 19
  plantagineum, 126
  rischavii, 24
  salicifolium, 117
Buraeavia, 250
Burasaia, 217
Burkilliodendron, 241
Burma, 40, 68, 234
Burragea, 161
Burretiokentia, 250
Burroughsia, 163
Bursera, 181, 182, 188
Burseraceae, 334
Bustillosia, 284
Butea frondosa, 232
Butomaceae, 9, 346
Butomopsis, 346
Butomus, 346
Butumia, 198
Butyrospermum, 204
Buxaceae, 320
Buxales, 320
Buxus balearica, 127
  colchica, 23
  hyrcana, 151
  microphylla, 54
  sempervirens, 22
  wallichiana, 155
Byblidaceae, 340
Byblis, 271, 340
Byronia, 337
Byrsophyllum, 228
Bystropogon, 114, 117
  maderensis, 115

piperitus, 115
Byttneriaceae, 328
Bzyb, 24

Caatinga Province, 259
Cabomba, 316
Cabombaceae, 316
Cabralea cangarena, 261
  oblongifolia, 261
Cabucala, 219
Cacalia, 51, 56, 67
  floridana, 95
  mortonii, 72
  nokoensis, 61
  rugelia, 88, 89, 187
  sulcata, 95
Caccinia, 139
  kotschyi, 149
Cactaceae, 159, 318
Cacti, 179, 181
Cactus parklands, 260
Cadaba, 206
Caesarea, 284
Caesalpinia bonduc, 58
  szechuensis, 64
Caesalpiniaceae, 333
Caesia, 267, 269, 351
Calabria, 122, 128
Calacanthus grandiflorus, 230
Caladenia, 240
Calamagrostis, 36, 56
  caucasica, 26
  chordorrhiza, 15
  distantiflora, 51
  hsinganensis, 51
  hyperborea, 15
  kalarica, 32
  mongolicola, 51
  poluninii, 15
  scotica, 17
  tolucensis, 186
Calamintha, 126
Calamus, 230
Calandrinia, 286
  feltonii, 290
Calandriniopsis, 283
Calanthe, 56, 73
  hattorii, 58
Calathodes, 40
Calceolaria, 286
Caldcluvia, 284
Calectasiaceae, 351

Calendula maderensis, 115
  monardii, 126
Calephyton, 220
Calicorema, 212
Calicotome, 121
California, 100, 159
California coast ranges, 105
Californian Province, 171
Calispepla, 138
Callianthemum anemonoides, 18
  kerneranum, 18
  mitabeanum, 52
  sachalinense, 52
  sajanense, 30
Callicarpa, 61
  bodinieri, 64
  glabra, 58
  lobata, 72
  mollis, 56
  nishimurae, 58
  psilocalyx, 78
  rubella, 70
  subpubescens, 58
Callicephalus, 140
Calligonum, 142
  polygonoides, 211
Callistemon, 251
Callistephus, 46
Callitrichaceae, 344
Callitriche, 344
  antarctica, 277, 290
  aucklandica, 301, 302
  stagnalis, 207
Callitris, 111, 275
Calocedrus formosana, 60
Calochlaena, 308
Calochortaceae, 349
Calochortus, 175, 349
  catalinae, 173
  clavatus, 173
  nuttallii, 99
  pulchellus, 173
  umbellatus, 173, 177
Calodecaryia, 218
Calopappus, 285
Calophaca wolgarica, 27
Calophyllum apetalum, 230
  inophyllum, 58, 226
  tacamahaca, 225
Calopyxis, 218
Calorhabdos, 45, 64
  sutchuensis, 67

Calospatha, 241
Calotropis procera, 211, 232
Caltha, 69, 277, 279, 286
  sagittata, 261
Calvaria, 217
Calycadenia, 163
Calycanthaceae, 9, 314
Calycanthus, 314
Calycera, 285
Calyceraceae, 346
Calycerales, 346
Calycocarpum, 79
Calycoseris, 163
Calycotome infesta, 119
Calymmatium, 136
Calypso bulbosa, 36
Calypsogyne, 219, 224
Calyptrosciadium, 138
Calyptrotheca, 206
Calytrix, 269, 273
Camassia cusickii, 109
Camchaya, 236
Camelina yunnanensis, 66
Camelinopsis, 136
Camellia, 61, 66
  caduca, 75
  irrawadiensis, 70
  kissi, 69, 75
  rusticana, 54
  wardii, 70
Cameroon, 198, 199
Camoensia, 198
Campanula, 19, 22, 67, 126
  alpestris, 19
  andina, 26
  ardonensis, 26
  aurita, 35
  bornmuelleri, 147
  bravensis, 119
  calcarea, 24
  carpatica, 19
  chinganensis, 51
  dolomitica, 26
  dzaaku, 24
  dzyschrica, 24
  forsythii, 129
  fulgens, 231
  hondoensis, 56
  jacobaea, 119
  jadvigae, 24
  karakushensis, 147
  kluchorica, 24

INDEX

kolakovskyi, 24
kryophila, 26
ledebouriana, 145
longistyla, 24
makaschwilii, 24
massalskyi, 147
mirabilis, 24, 25
ossetica, 26
panjutinii, 24, 25
paradoxa, 24, 25
persicifolia, 27
radula, 147
reverchonii, 99
schistosa, 24
sphaerocarpa, 24
symphytifolia, 24
trachelium, 31
tridentata, 145
tschuktschorum, 15
uniflora, 38, 106
wightii, 230, 231
zoysii, 20
Campanulaceae, 10, 264, 345
Campanulales, 345
Campbell Island, 277, 294, 302
Campecarpus, 250
Campimia, 241
Campnosperma seychellarum, 226
Campsidium, 285
Camptocarpus, 219, 220
Camptotheca, 44, 339
Campylandra, 47
Campylanthus benthamii, 118
  glaber, 118
  salsoloides, 117
Campylopetalum, 235, 238, 334
Campylotropsis, 66
Campynema, 272, 348, 353
Campynemanthe, 249, 272, 348, 353
Campynemataceae, 348
Canaca, 248
Canacomyrica, 248, 321
Canada, 12
Canadian Province, 34
Cananga, 271
Canarian Province, 115
Canarina abyssinica, 114
  canariensis, 114, 117
  eminii, 114
Canarium denticulatum, 237
  manii, 237
Canary Islands, 112, 115, 117

Canbya, 159
Cancrinia, 140
Cancriniella, 140
Canellaceae, 313
Canephora, 219
Cankiri, 144
Canna, 354
Cannabaceae, 9, 329
Cannaceae, 354
Canotia, 162, 327
Canotiaceae, 337
Cape Kingdom, 263, 267
  notes, 267
Cape Province, 267
Cape Region, 263
Cape Verde Islands, 112, 117, 118
Cape Verde Province, 118
Cape York Peninsula, 271
Capensis, 263
Capitanya, 207
Capparaceae, 327
Capparales, 327
Capparis aphylla, 233
  decidua, 211
  elliptica, 210
  formosana, 61
  nobilis, 298
  spinosa, 132
  yunnanensis, 66
Capraia, 129
Caprifoliaceae, 341
Capsicum boninense, 58
Captaincookia, 249
Capurodendron, 217
Capuronia, 218
Caragana altaica, 30
  arborescens, 31
  franchetiana, 66
  fruticosa, 50
  litwinowii, 62
  ussuriensis, 50
  zahlbruckneri, 62
Carallia integerrima, 233
Cardamine, 54, 61, 72, 240, 281
  conferta, 33
  digitata, 15
  hirsuta, 199, 207, 231
  hyperborea, 15
  obliqua, 207
  pedata, 34
  schinziana, 53
  sphenophylla, 15

yezoensis, 53
yunnanensis, 66
Cardaminopsis neglecta, 19
Cardiandra, 43
　alternifolia, 54
　amamioshimensis, 59
　formosana, 61
Cardiocrinum, 47
　cathayanum, 64
　giganteum, 76
Cardiomanes reniforme, 295
Cardionema ramosissima, 174
Cardiopteridaceae, 337
Cardioteucris, 46, 65
Carduncellus, 122, 126
Carduus, 20, 117
　balansae, 126
　fasciculiflorus, 130
　nervosus, 147
　squarrosus, 115
Cardwellia, 271
Carex, 16, 20, 36, 51, 56, 61, 68, 70, 73 76, 86, 101, 156, 168, 200, 208, 272, 281
　augustinii, 58
　binervis, 188
　bipartita, 110
　canariensis, 118
　capitata, 106
　chathamica, 301
　distans, 18
　durieui, 17
　flavocuspis, 34, 53
　gayana, 288
　hattoriana, 58
　hookerana, 96
　kermadecensis, 299
　latifrons, 25
　longiculmis, 301
　lowei, 115
　malato-belizii, 115
　malyschevii, 32
　manipurensis, 78
　mingrelica, 25
　nardina, 168
　neesiana, 298
　oreocharis, 110
　paupercula, 110
　perraudieriana, 118
　phaeocephala, 107
　pontica, 25

pyrophila, 34, 53
ramenskii, 34, 53
rorulenta, 128
rupestris, 110
sabulosa, 158
scandinavica, 29
tatjanae, 31
toyoshimae, 58
trifida, 278
trinervis, 17
ursina, 15
ventosa, 299, 301
Caribbean Region, 255
Carica, 286
Caricaceae, 326
Caricales, 326
Cariniana estrellensis, 261
Carlemanniaceae, 341
Carlesia, 44
Carlina atlantica, 126
　canariensis, 117
　macrocephala, 130
　xeranthemoides, 117
Carlowrightia, 162
Carmichaelia, 294, 296
　exsul, 297
Carnarvonia, 271
Carnegiea, 160
　gigantea, 179
Carnegieodoxa, 248
Carpacoce, 264
Carpathians, 11, 12, 18
Carpenteria, 162
Carpesium minus, 64
　rosulatum, 56
Carphalea, 219
Carphephorus, 82, 93
　carnosus, 95
　corymbosus, 95
Carphochaete, 163
Carpinaceae, 321
Carpinus, 12, 54, 60
　austro-yunnanensis, 66
　betulus, 152
　caroliniana, 86, 187
　chowii, 62
　cordata, 51
　fargesii, 63
　oblongifolia, 63
　viminea, 69, 75
Carpobrotus chilensis, 174

Carpodetus, 338
Carpolyza, 264
Carpoxylon, 244
Carrierea, 41
Carrisoa, 203
Carterothamnus, 163
Carthamus strictus, 126
Cartonemataceae, 355
Carum, 126
  khasianum, 78
Carvia, 228
Carya, 85, 86
  aquatica, 93
  floridana, 93
  glabra, 84, 92
  illinoensis, 187
  mexicana, 187
  ovata, 187
  texana, 84
  tomentosa, 92
  tonkinensis, 66
Caryocar, 322
Caryocaraceae, 254, 322
Caryophyllaceae, 15, 16, 26, 30, 318
Caryophyllales, 317
Caryophyllidae, 317
Caryopteris, 46
  forrestii, 67
Caryota, 271
Cascade Mountains, 101, 103
Cassandra calyculata, 36
Cassia, 286
  angustifolia, 233
  aphylla, 289
  auriculata, 211
  fistula, 233
  ovata, 233
Cassidispermum, 243
Cassinia amoena, 300
Cassiope, 66
  tetragona, 106
Cassythaceae, 314
Castanea alnifolia, 93
  crenata, 54
  dentata, 85
  ozarkensis, 84, 85
  pumila, 84, 86
  sativa, 17
Castanopsis, 60, 66, 69, 236
  chrysophylla, 105
  cuspidata, 54

Castela texana, 182
Castilleja, 101, 168, 169, 173
  annua, 35
  arctica, 15
  neglecta, 177
  nivea, 106
  puberula, 110
  schrenkii, 29
  septentrionalis, 15
  vorkutensis, 15
  xanthotricha, 108
  yukonis, 35
Casuarina, 268, 271, 275, 296, 320
Casuarinaceae, 320
Casuarinales, 320
Catalpa fargesii, 67
  tibetica, 67
Catananche, 122
Catenularia, 136
Catha dryandri, 115
  edulis, 207
Cathaya, 40
Cathayanthe, 236, 237
Caucalis, 203
  melanantha, 207
  stocksiana, 210
Caucasian Province, 25
Caucasus, 11, 12, 136
Caulanthus, 172
Caulocarpus, 203
Caulophyllum thalictroides, 85
Cavea, 46
Cayratia japonica, 76
  thomsonii, 76
Ceanothus, 172, 177
  arboreus, 173
  greggii, 182
Cecropiaceae, 329
Cedar glades community, 90
Cedrela fissilis, 261
  glaziovii, 261
  toona, 232
Cedrelopsis, 218, 334
Cedronella, 113
Cedros Island, 160
Cedrus atlantica, 125
  deodara, 154
Celastraceae, 337
Celastrales, 337
Celastrus, 67
  hookeri, 70, 76

hypoleuca, 64
loeseneri, 64
stephanotiifolius, 55
Celebes, 240, 243
Celebesian Province, 243
Celmisia, 296
  glabrescens, 301
  rigida, 301
  vernicosa, 301
Celsia cystolithica, 118
  insularis, 118
Celtidaceae, 329
Celtis, 12, 66, 267
  biondii, 63
  boninensis, 57
  jessoensis, 54
  julianae, 63
  kraussiana, 265
  labilis, 63
  nervosa, 60
  occidentalis, 85
  pallida, 182
Cenarrhenes, 272
Cenchrus, 211
Cenocentrum, 234
Centaurea, 20, 22, 27, 117, 126, 130, 142, 144, 280
  abchasica, 24
  albovii, 24
  amblyolepis, 26
  appendicigera, 24
  balearica, 128
  barbeyi, 24
  carduchorum, 147
  declinata, 132
  erivanensis, 147
  fenzlii, 145
  gracillima, 147
  helenioides, 24
  kolakowskyi, 24
  kurdica, 145
  massoniana, 115
  pseudoleucolepis, 28
  pseudoscabiosa, 147
  schischkinii, 147
  sessilis, 145
  talievii, 28
  tomentella, 145
  vanensis, 147
  xanthocephala, 147
Centaurium chloodes, 17

scilloides, 17
Centaurodendron, 280
  dracaenoides, 282
Centauropsis, 219
Central American Province, 256
Central Anatolian Province, 143
Central Andean Province, 262
Central Asia, 137
Central Asiatic Subregion, 155
Central Australian Region, 274
Central Chinese Province, 63
Central European Province, 17
Central Iranian Subprovince, 149
Central Neozeylandic Province, 300
Central Tien Shan Province, 156
Central Valley, California, 174
Centranthera grandiflora, 76
Centrantheropsis, 45
Centranthus trinervis, 129
Centrolepidaceae, 356
Centrolepis, 356
Centrolobium robustum, 260
Centropogon, 215
Cephalantheropsis, 236
Cephalaria, 111
  balearica, 128
  calcarea, 24
  litvinovii, 27
  squamiflora, 127
Cephalopentandra, 207
Cephalopodum, 138
Cephalorrhizum, 136
Cephalorrhynchus kossynskyi, 148
  takhtadzhianii, 147
Cephalosphaera, 201
Cephalostachyum, 221
Cephalostigma, 345
Cephalotaceae, 273, 330
Cephalotaxaceae, 39, 310
Cephalotaxales, 310
Cephalotaxus, 40, 310
  griffithii, 74
  mannii, 77
  wilsoniana, 60
Cephalotus, 330
Ceraria, 212
Cerastium, 19, 60, 186, 203
  afromontanum, 207
  aleuticum, 34
  argenteum, 26
  beeringianum, 106

fontanum, 290
gorodkovianum, 29
kasbek, 26
multiflorum, 26
octandrum, 199, 207
regelii, 15
schizopetalum, 54
soleirolii, 129
sventenii, 117
uralense, 27
vagans, 115
Ceratiola, 79, 324
Ceratocapnos, 120
Ceratocnemum, 121, 125
Ceratoides lenensis, 32
Ceratonia siliqua, 124
Ceratophyllaceae, 316
Ceratophyllales, 316
Ceratophyllum, 316
  manshuricum, 50
Ceratopteris, 307
Ceratostigma griffithii, 71, 158
Cerberiopsis, 249
Cercidiphyllaceae, 39, 319
Cercidiphyllales, 319
Cercidiphyllum, 41, 319
  magnificum, 54
Cercidium, 179, 181, 182
Cercis, 159
  canadensis, 187
  chinensis, 64
  racemosa, 64
  siliquastrum, 119, 148
  yunnanensis, 66
Cercocarpus ledifolius, 170
Cereus, 184
Cerinthe glabra, 131
Ceropegia, 117, 206
  bhutanica, 72
Ceterachopsis, 39
Cevallia, 162
Ceylon Province, 229
Chacoan Province, 260
Chadsia, 218
Chaenactis alpigena, 104
  evermannii, 106
  nevadensis, 104
  nevii, 108
Chaenomeles, 42
  japonica, 55
Chaerophyllopsis, 44, 65

huai, 67
Chaerophyllum elegans, 19
  meyeri, 151
  schmalhausenii, 24
  villarsii, 19
Chaetadelpha, 163
Chaetolimon, 136
Chaetopappa, 163
Chaetopogon, 123
Chaetosciadium, 122, 131
Chaetotropis, 285
Chalcanthus, 136
Chamaebatia, 161
  australis, 173
Chamaeacanthus, 207
Chamaebatiaria, 161
Chamaechaenactis, 163
Chamaecyparis, 101
  formosensis, 60
  obtusa, 54
  pisifera, 54
Chamaecytisus blockianus, 27
  paczoskii, 27
  podolicus, 27
  proliferus, 116
  skrobiszewskii, 27
Chamaedaphne, 36
Chamaegeron, 140
Chamaelaucium, 273
Chamaele, 44
Chamaemeles, 112, 115
  coriacea, 115
Chamaelirium, 82
Chamaepericlimenum, 340
Chamaerhodos sabulosa, 158
Chamaerops, 123
  humilis, 111, 119, 124
Chamaesium, 44
Chamaesphacos, 140
  illicifolius, 150
Chamaexeros, 351
Chambeyronia, 250
Chamelum, 285
Chamorchis, 11
Championia, 228
Chaney, R. W., 83
Chang Tang, 158
Changium, 44, 63
  smyrnioides, 64
Changnienia, 47
Chaparral, 5, 165, 167, 174, 176, 182

Chaptalia, 93, 288
Charadrophila, 264
Chartoloma, 136
Chatham Islands, 277, 294, 301
Chathamian Province, 301
Chaydaia, 235, 237
Cheilanthes californica, 173
  krameri, 53
Cheilanthopsis, 39
Cheiranthus aurantiacus, 62
  scoparius, 116
Cheirolaena, 218, 224
Cheiropleuria, 308
Cheiropleuriaceae, 226, 308
Cheirostylis okabeana, 56
Cheju Island, 54
Chelone, 82
Chelonopsis, 46, 67
  longipes, 56
Chenopodiaceae, 9, 143, 152, 274, 318
Chenopodium, arborescent, 281
  koraiense, 54
Chesneya, 138, 142
Chesniella, 138
Chevreulia lycopodioides, 290
  sarmentosa, 293
Chihuahuan Desert, 160, 181
Chihuahuan subprovince, 180, 182
Chikusichloa, 47
  brachyanthera, 62
Chile, 277, 283
Chile-California relationships, 173, 174
Chile-Patagonian Region, 283
Chiliotrichum, 285
  diffusum, 290
Chilocarpus malabaricus, 230
Chilopsis, 162
Chiloschista pusilla, 230
  usneoides, 73
Chimaphila maculata, 85
  umbellata, 36
Chimonanthus, 40, 314
  yunnanensis, 66
Chin Hills, 75
China, 39, 52, 62, 63, 65, 137, 234
Chionanthus virginiana, 86
Chionocharis, 45
Chionochloa antarctica, 302
  conspicua, 297
Chionographis, 46
  japonica, 56

  koidzumiana, 56
Chionophila, 100
  jamesii, 110
  tweedyi, 106
Chionothrix, 206
Chirita, 67
  fauriei, 64
  primulacea, 72
  pumila, 76
Chloanthaceae, 344
Chloranthaceae, 314
Chloranthales, 314
Chloris, 261
Chlorocarpa, 227
Chlorogalum, 164
  pomeridianum, 173
Chlorophytum, 68
  chinense, 64
Chloroxylon, 228
Choisya, 161
Cholla, 179
Chonopetalum, 198
Chordospartium, 294, 300
Choria, system of, 6
Choriantha, 139, 148
Chorion, defined, 7
Choristylis, 339
Chorizanthe, 173
  brewerii, 177
  chilensis, 174
  coriacea, 174
  uniaristata, 177
Chosenia, 326
  arbutifolia, 33, 34
Chota Nagpur plateau, 232
Chouxia, 218
Christisonia bicolor, 230
  calcarata, 230
  cytinoides, 230
  lawii, 230
Chroësthes, 236
Chromolepis, 163
Chronanthus, 121
Chrozophora gracilis, 150
Chrysactinia, 163
Chrysalidocarpus, 220
Chrysanthemum, 51, 56, 61, 117
  barretii, 115
  dissectum, 115
  haematomma, 115
  jucundum, 63

INDEX

mandonianum, 115
namikawanum, 63
pinnatifidum, 115
Chrysithrix, 269
Chrysobalanaceae, 331
Chrysobraya, 42
Chrysocoma, 212
Chrysogonum, 82
Chrysoma, 82
Chrysopappus, 141
Chrysophae, 122
Chrysophthalmum, 140
Chrysopsis camporum, 98
  flexuosa, 95
  latisquamea, 95
  oligantha, 95
  scabrella, 95
  subulata, 95
Chrysosplenium, 55, 61
  alberti, 30
  baicalense, 30
  filipes, 30
  forrestii, 69
  lanuginosum, 69, 75
  nepalense, 69, 75
Chrysothamnus, 169
Chukot Autonomous Okrug, 15
Chunechites, 45
Chunia, 234, 237
Chusquea, 261, 281
Chymsidia, 11, 23
  agasylloides, 24
Cicerbita macrantha, 72
  pancicii, 22
  uralensis, 27
Cimicifuga, 54
  americana, 85
  heracleifolia, 50
  racemosa, 85
  rubifolia, 85
Cineraria grandiflora, 207
Cinnamomum, 60
  porosum, 260
  pseudopedunculatum, 57
  wightii, 231
Cinnamosma, 217, 222
Cionidium, 248
Circaea alpina, 75, 99
  lutetiana, 31
Circaeaster, 40
Circaeasteraceae, 39, 317

Circumboreal flora, 123, 150
Circumboreal Region, 10
Circumboreal species, 35, 36, 38, 106, 110, 168
Cirrhopetalum boninense, 58
Cirsium, 51, 56, 61, 67, 240
  adjaricum, 24
  boninense, 58
  brevifolium, 109
  congestum, 147
  eriophoroides, 73
  kamtschaticum, 34
  kirbense, 126
  latifolium, 115
  nivale, 186
  parryi, 110
  rydbergii, 170
  sommieri, 145
  spinosissimum, 20
  toyoshimae, 58
  tweedyi, 106
Cissampelos ochiaiana, 60
Cissarobryon, 284
Cissus, 286
  quadrangularis, 208
Cistaceae, 328
Cistales, 328
Cistanche mauritanica, 126
Cistus, 116
  creticus, 131
  incanus, 132
  salviifolius, 131
  tauricus, 132
Cithareloma, 136
Citrus latipes, 75
Cladanthus, 122
Cladium scirpoideum, 281
Cladochaeta, 11, 26
Cladopus, 55
Cladostigma, 207
Cladothamnus, 100
Cladrastis delavayi, 66
  kentuckea, 85
  sikokiana, 55
  wilsonii, 64
Claoxylon centenarium, 57, 58
Clappia, 163
Clarkia, 172, 175
  davyi, 174
  delicata, 173
  rubicunda, 173

tenella, 174
Clastopus, 136
Clausena yunnanensis, 66
Claytonia bostockii, 35
  caroliniana, 85
  virginica, 85
Claytoniella, 11
Cleistanthus collinus, 232
Cleistes, 82
Cleistogenes kitagawae, 52
  thoroldii, 158
Clematis, 54, 60, 66, 71, 74, 85, 296
  albicoma, 89
  alpina, 29
  apiifolia, 63
  boninensis, 57
  buchaniana, 69
  cirrhosa, 124
  coactilis, 89
  cocculifolia, 298
  fascicularis, 69
  hexapetala, 52
  koreana, 50
  mandschurica, 50
  nepalensis, 69
  nutans, 232
  ochotensis, 34
  serratifolia, 50
  simensis, 199
  viticaulis, 89
  wightiana, 231
Clematoclethra, 41, 321
Cleomaceae, 327
Cleome oxypetala, 210
  yunnanensis, 66
Cleonia, 122
Clermontia, 215, 247
Clerodendrum spp., 67
Clethra, 323
  acuminata, 85
  arborea, 113
  barbinervis, 54
  delavayi, 66, 69
  monostachya, 63
Clethraceae, 323
Clevelandia, 162
Cleyera japonica, 75
Cliffortia, 264, 267
Cliftonia, 79
Clinopodium laxiflorum, 61
  macranthum, 56

  micranthum, 56
Clinosperma, 250
Clinostigma savoryana, 58
Clintonia borealis, 88
  umbellata, 86
Cloëzia, 249
Closia, 285
Clusiaceae, 322
Clutia brassii, 203
Clymenia, 243
Cnemidaria, 309
Cnemidiscus, 235
Cneoraceae, 333
Cneoridium, 161
  dumosum, 173
Cneorum pulverulentum, 117
Coastal Plain Province, 91
Cobaeaceae, 343
Cocconerion, 249
Cocculus leaba, 233
Cochemiea, 160
Cochlearia formosana, 61
  groenlandica, 15
  tatrae, 19
Cochlianthus, 43
Cochlospermaceae, 328
Cochlospermum, 328
Cocos Island, 256
Codia, 249
Codonacanthus, 46
Codonoboea, 241
Codonocarpus cotonifolius, 275
Codonocephalum, 141
Codonopsis, 67
  kawakamii, 61
  subsimplex, 72
  thalictrifolia, 72
  viridis, 76
Coelachyropsis, 229
Coelogyne, 76
  ecarinata, 70
  longipes, 73
  occultata, 73
Coelonema, 42, 65
Cohnia, 222, 349
Cola greenwayi, 207
Colania, 236
Colchicaceae, 348
Colchicum, 111
  callicymbium, 22
  corsicum, 130

lactum, 26
liparochiadys, 24
Colchis, 23
Colea, 219, 220
  seychellarum, 226
Coleactina, 198
Coleogyne, 161
  ramosissima, 170
Colletia, 281
Collinsia, 172
  verna, 86
Collinsonia, 82
Collomia biflora, 174
  cavanilesii, 174
  grandiflora, 174
  linearis, 174
Collospermum microspermum, 300
Colobanthium, 123
Colobanthus, 261, 277, 279, 296
  hookeri, 290, 301
  quitensis, 290, 291
  subulatus, 290
Colobogyne, 236
Colophospermum, 203
  mopane, 203
Colorado, 110
Colorado Plateau, 165
Colorado River, 169
Colpogyne, 219
Colquhounia, 67
  vestita, 76
Columbia River Gorge, 103
Columellia, 339
Columelliaceae, 261, 339
Colutea atabaevii, 148
  komarovii, 146
Comandra elegans, 22
Combera, 284
Combretaceae, 332
Combretum, 202, 204, 207, 208
Cometes surratensis, 149, 210
Commander Islands, 34
Commelina sikkimensis, 76
Commelinaceae, 355
Commelinales, 355
Commicarpus, 206
Commidendrum, 215
  rugosum, 216
Commiphora, 204, 206, 207, 213, 223
  wightii, 211
Comoranthus, 219, 220

Comoro Province, 224
Comoros Islands, 220, 224
Compositae, 346
Comptonella, 249
Comptonia, 79, 321
  peregrina, 85
Conanthera, 285, 348
Conchopetalum, 218
Congolian Province, 199
Conimitella, 100
  williamsii, 106
Conioselinum victoris, 34
Connaraceae, 333
Connarales, 333
Conospermum, 269
Conostylidaceae, 353
Conostylis, 273
Conradina, 82, 93
  verticillata, 86
Consolida stenocarpa, 143
Continental drift, 114, 266
Convallaria majalis, 27
Convallariaceae, 349
Convolvulaceae, 342
Convolvulales, 342
Convolvulus, 111, 113, 142
  acanthocladus, 211
  argyracanthus, 149
  canariensis, 117
  durandoi, 126
  gracillimus, 146
  mascatensis, 210
  massonii, 115
  spinosus, 211
  ulicinus, 210
  zargarianus, 149
Conyza angustifolia, 73
  variegata, 207
Cook Islands, 246
Copianthus, 227
Copiapoa, 283, 286
Coprosma, 281, 296, 297
  acutifolia, 299
  arborea, 300
  baueri, 298, 299
  chathamica, 301
  cuneata, 301
  dodonaeifolia, 300
  hookeri, 282
  macrocarpa, 300
  petiolata, 299

pilosa, 298
pumila, 277
pyrifolia, 281
repens, 299
spathulata, 300
Coptidipteris, 309
Coptis, 54, 69
  trifolia, 36
Corallodiscus, 45, 67
  bhutanicus, 72
  lanuginosus, 76
Corallorhiza trifida, 36
Corallospartium, 294
Corbassona, 249
Corchoropsis, 42
Cordeauxia, 207
Cordemoya, 218, 224
Cordia, 188, 286
  gharaf, 211
  myxa, 232
Cordiaceae, 343
Cordilleran Forest Province, 99
Cordylanthus nidularius, 177
Cordyline, 251
  australis, 299
  kaspar, 300
  obtecta, 298, 299
  pumilio, 300
Cordylocarpus, 121
Corema, 324
Coreocarpus, 163
Coreopsis grandiflora, 90
  latifolia, 89
  pulchra, 90
Corethrogyne, 163
Coriandropsis, 138
Coriaria, 334
  japonica, 55
  nepalensis, 75
Coriariaceae, 334
Coriariales, 334
Coridaceae, 325
Coris monspeliensis, 119
Corispermum algidum, 29
  stenolepis, 62
Cornaceae, 10, 339
Cornales, 339
Cornatic Province, 232
Cornus, 12, 64, 66, 339
  alba, 36
  canadensis, 36

capitata, 70, 75
controversa, 75
florida, 86, 187
macrophylla, 75
nuttallii, 102
oblonga, 75
sericea, 36
stricta, 86
suecica, 36
Corokia, 339
  buddleoides, 300
  cotoneaster, 300
  macrocarpa, 301
Coromandel Subprovince, 232
Coronanthera, 294
Coronilla atlantica, 126
Corsia, 348
Corsiaceae, 348
Corsica, 121, 128
Cortaderia pilosa, 290
Cortiella, 44
Corydalis, 54, 60, 62, 63, 66, 71, 317
  buschii, 50
  chaerophylla, 74
  chionophila, 147
  claviculata, 17
  curvicalcarata, 53
  emanuelii, 26
  flavula, 85
  gorodkovii, 33, 34
  halleri, 32
  himalayana, 74
  leptocarpa, 69
  longipes, 74
  paczoskii, 27
  persica, 146
  smirnowii, 26
  tashiroi, 59
  vittae, 23
  watanabei, 50
Corylaceae, 321
Corylopsis, 41, 54
  himalayana, 71, 74
  manipurensis, 69, 77
  platypetala, 63
  stenopetala, 60
  veitchiana, 63
  yunnanensis, 66
Corylus avellana, 27
  colchica, 23
  ferox, 69

heterophylla, 49, 52
  jacquemontii, 155
  pontica, 23
  yunnanensis, 66
Corymborchis subdensa, 58
Corymbostachys, 219
Corynanthera, 273
Corynocarpaceae, 337
Corynocarpus, 251, 337
Cosmelia, 273
Cossonia, 121
Costaceae, 354
Costantina, 235
Cotoneaster, 61, 64, 66
  acutifolia, 158
  alaunicus, 27
  cinnabarinus, 29
  discolor, 148
  distichus, 69
  horizontalis, 69
  lucidus, 30
  persica, 149
  rubens, 69
  simonsii, 78
  taylorii, 72
  turcomanicus, 148
  tytthocarpus, 148
Cottonia, 229
Cotula, 278, 296
  abyssinica, 207
  featherstonii, 301
  lanata, 301
  potentillina, 301
  plumosa, 278, 293, 301
  renwickii, 301
Cotylanthera yunnanensis, 67
Cotyledon, 212
Coulterella, 163
Coulterophytum, 162
Cousinia, 141, 142, 147, 148
  bicolor, 145
  brachyptera, 145
  euphratica, 145
  fragilis, 149
  hablizii, 151
  woronowii, 145
Cousiniopsis, 141
Cowania, 161
Coxella, 294, 301
Craibiodendron henryi, 66
  yunnanense, 66

Craigia, 42, 65
Crambe, 113
  arborea, 117
  armena, 146
  aspera, 27
  croatica, 131
  fruticosa, 115
  gigantea, 117
  koktebelica, 132
  kralikii, 125
  maritima, 131
Crambella, 121, 125
Craniospermum, 139
Craniotome furcata, 76
Craspedolobium, 43
Craspedospermum, 219
Crassula, 208, 212, 264
  alba, 199
  granvikii, 207
  schimperi, 207
Crassulaceae, 330
Crataegus, 12, 64, 85, 158
  kansuensis, 62
  komarovii, 50
  ucrainica, 27
Crateranthus, 198
Cremanthodium, 46, 67
Cremastra, 47
Crematosciadium, 138
Cremocarpon, 219, 224
Crenosciadium, 138, 143
Creosote bush, 179
Crepis, 64
  albanica, 22
  bodinieri, 67
  canariensis, 117
  divaricata, 115
  khorassanica, 148
  lampsanoides, 17
  macedonica, 22
  nana, 106
  noronhaea, 115
  rhaetica, 20
  schachtii, 22
  triasii, 128
  turcomanica, 148
Crete, 111, 122, 131
Crimean-Novorossiysk Province, 131
Crinodendron, 284
  patagua, 287
Crispiloba, 339

Cristaria, 284, 286
Crithmum maritimum, 132
Crocidium, 100
Crocus, 22
  adamii, 26
  autranii, 25
  cambessedesii, 128
  caspius, 151
  corsicus, 130
  etruscus, 130
  hyrcanus, 151
  karsianus, 145
  longiflorus, 130
  michelsonii, 148
  minimus, 130
  scharojanii, 25
  tauricus, 132
  vallicola, 25
Croomia, 56
  pauciflora, 93
Croomiaceae, 352
Crossonephelis, 218, 222
Crossosoma, 161, 331
  californicum, 173
Crossosomataceae, 159, 331
Crossosomatales, 331
Crotalaria, 66, 206, 231
  burhia, 211
  cunninghamii, 275
  furfuracea, 210
  iranica, 210
  vialattei, 126
Croton alabamensis, 85
Crotonopsis, 82
Crotonopsis elliptica, 85, 90
Crozet Islands, 278, 291
Crucianella sintenisii, 148
Cruciferae, 327
Cruckshanksia, 284, 286
Crypsinus, 60
Cryptadenia, 264
Cryptantha, 169, 172, 173
  mariposae, 177
  shacklettiana, 35
  virgata, 109
  weberi, 109
Crypteronia, 332
Crypteroniaceae, 227, 332
Cryptocapnos, 135
Cryptocarya andamanica, 237
  rubra, 287
  yunnanensis, 66
Cryptocodon, 140
Cryptodiscus, 138
Cryptogramma crispa, 105
Cryptogrammaceae, 307
Cryptomeria, 40
Cryptopus, 220
Cryptospora, 136
Cryptostegia, 219
Cryptotaenia elegans, 117
Cryptothladia, 341
Ctenitis, 60
  microlepigera, 57
Ctenolophon, 335
Ctenolophonaceae, 335
Cuba, 256
Cucurbitaceae, 326
Cucurbitales, 326
Culcita, 308
  macrocarpa, 114, 125
Culcitaceae, 308
Cullenia, 227
Cumberland Mountains, 86
Cumberland Plateau Geological Province, 90
Cuminia, 280, 281
Cunninghamia, 40
  konishii, 60
Cunonia, 266
Cunoniaceae, 330
Cunoniales, 330
Cupheanthus, 249
Cuphocarpus, 219
Cupressaceae, 10, 311
Cupressales, 311
Cupressus, 101, 159, 173
  abramsiana, 178
  bakeri, 178
  corneyana, 71
  duprezziana, 134
  forbesii, 178
  goveniana, 178
  macnabiana, 178
  macrocarpa, 173, 178
  nevadensis, 178
  pygmaea, 178
  sargentii, 178
  sempervirens, 151
  stephensonii, 178
  torulosa, 154
Curroria decidua, 213
Curtisia, 340

INDEX

dentata, 265
Curtisiaceae, 340
Cuscuta, 342
Cuscutaceae, 342
Cyananthus, 46, 67
  pedunculatus, 72
Cyanastraceae, 349
Cyanastrum, 349
Cyanea, 247
Cyanella, 348
Cyanopsis, 122
Cyanotis barbata, 208
Cyathea, 187, 295, 309
  dregei, 265
  kermadecensis, 299
  mertensiana, 57
  milnei, 299
  ogurae, 57
Cyatheaceae, 309
Cyathobasis, 136, 143
Cyathocline, 228
Cyathodes, 251
  robusta, 301
Cyathopus, 47, 249
Cybistetes, 264
Cycadaceae, 311
Cycadales, 311
Cycas, 221, 311
Cycladenia, 162
Cyclamen colchicum, 23
  elegans, 151
  parviflorum, 23
Cyclanthaceae, 254, 356
Cyclanthales, 356
Cyclocarya, 41
Cyclophyllum, 250
Cycnogeton, 347
Cylicomorpha, 326
Cylindrocarpa, 140
Cylindrocline, 219, 224
Cymatocarpus, 136
  grossheimii, 146
Cymbalaria aequitriloba, 127
Cymbaria borysthenica, 27
  hepaticifolia, 129
  muelleri, 129
Cymbidiella, 220, 222
Cymbidium, 56
  hookerianum, 73
Cymbocarpum, 138
Cymbochasma borysthenicum, 27

Cymbolaena, 141
Cymbopogon, 210, 211
  khasianus, 78
  schoenanthus, 211
  travancorensis, 230
Cymodocea nodosa, 347
Cymodoceaceae, 347
Cymodoceales, 347
Cymophora, 163
Cymophyllus, 82
Cymopterus, 169
Cynanchum, 55, 67
  bungei, 62
  chinense, 62
  linearifolium, 64
  stenophyllum, 64
Cynomoriaceae, 9, 345
Cynomoriales, 345
Cynomorium, 111, 345
  coccineum, 119
Cynosciadium, 82, 207
Cynosurus, 126
Cyperaceae, 10, 15, 16, 30, 264, 268, 355
Cyperales, 355
Cyperus, 56, 68, 70, 93
Cyphiaceae, 345
Cyphocardamum, 136
Cyphocarpus, 285, 345
Cyphochlaena, 220
Cyphokentia, 250
Cyphomeris, 159
Cyphostigma, 229
Cyphotheca, 43
  betacea, 67
Cypress, 178
Cypripediaceae, 353
Cypripedium, 64
  acaule, 37
  debile, 56
Cyrenaica, 120
Cyrillaceae, 324
Cyrillopsis, 335
Cyrtandroidea, 245
Cyrtocarpa, 182
Cystacanthus yunnanensis, 67
Cystodium, 308
Cystopteris fragilis, 290
Cytinaceae, 315
Cytinales, 315
Cytinus, 315
Cytisopsis, 121, 131

433

Cytisus, 117
　aeolicus, 129
　albidus, 124
　cantabricus, 17
　commutatus, 17
　emeriflorus, 19
　hypocistis, 111
　maderensis, 115
　mollis, 124
　paivae, 115
　tener, 115
Czechoslovakia, 19

Daboecia, 17
　azorica, 114
　cantabrica, 17
Dacrydium, 251, 295
　franklinii, 272
　kirkii, 300
Dactylanthus, 294
Dactylicapnos grandifolia, 70
Dactylis smithii, 118
Dactylocladus, 332
Dactylorhiza foliosa, 115
Dactylostalix, 47
　ringens, 56
Daenikera, 249
Dagestan, 26, 136
Dahuria, 52
Daiswa, 64, 68, 352
Dalbergia, 66, 223
　nigra, 260
　sissoo, 232
Dalea filiciformis, 180
　gattingeri, 91
Dalmatia, 131
Dalzellia, 227
Damnacanthus, 44
　macrophyllus, 55
Dampiera, 269
Danaë, 349
Danaea, 306
Danais, 219, 220
Dankia, 234
Danthonia pungens, 301
Danthonidium, 229
Daphne, 54, 66, 69
　arisanensis, 61
　koreana, 50
　papyracea, 75
　pontica, 25

　rechingeri, 151
　rodriguezii, 128
　shillong, 77
Daphniphyllaceae, 320
Daphniphyllales, 320
Daphniphyllum, 320
　himalayense, 75
　neilgherrense, 230, 231
Darlingia, 271
Darlington, P. J., 6
Darlingtonia, 94, 100, 323
Darwinia, 269
Darwiniothamnus, 257
Dasistoma, 82
　macrophylla, 86
Dasydesmus, 236, 237
Dasylirion, 164
　heteracanthium, 99
Dasynotus, 100
Dasyphora fruticosa, 158
Dasyphyllum, 286
Dasypogon, 273, 351
Dasypogonaceae, 351
Dasysphaera, 206
Datisca, 326
Datiscaceae, 326
Datiscales, 326
Daucosma, 79, 82
　laciniatum, 96
Davallia canariensis, 116, 125, 126
　tasmanii, 300
Davalliaceae, 310
Daveaua, 122
Davidia, 44, 339
Davidiaceae, 39, 339
Davidsonia, 330
Davidsoniaceae, 270, 330
Daviesia, 269, 273
Davis, P. H., 143
Decaisnea, 40
　fargesii, 69
Decalepis, 228
Decaptera, 284
Decarydendron, 217
Decaryia, 217
Decaryochloa, 220
Deccan Province, 232
Deckenia, 220, 225
　nobilis, 226
Decodon, 82
Dedeckera, 160

Degeneria, 244, 312
Degeneriaceae, 244, 312
Degenia, 121, 130
  velebitica, 131
Deinanthe, 43
  bifida, 54
Deinostema, 45
  adenocaulum, 55
Delavaya, 43, 65
  yunnanensis, 66
Delissea, 247
Delmarva peninsula, 94
Delonix regia, 223
Delphinium, 66, 69, 71, 101, 142
  altissimum, 74
  carduchorum, 146
  dubium, 18
  exaltatum, 85
  inconspicuum, 30
  maackianum, 50
  mirabile, 30
  oxysepalum, 18
  penicillatum, 209
  pictum, 127
  ramosum, 109
  sajanense, 30
  stapeliosmum, 74
  tricorne, 85
  uralense, 27
  venulosum, 143
Deltaria, 249
Dendriopoterium, 112
Dendrobenthamia, 340
Dendrocacalia, 46, 57
  crepidifolia, 58
Dendrocalamus affinis, 65
  hamiltonii, 70
  strictus, 233
Dendromecon, 159
Dendropanax burmanicus
  pellucidopunctatus, 61
Dendrophyllanthus, 249
Dendrosenecio, 207, 208
Dendroseris, 280, 282
Dendrosicyos, 209
  socotranus, 209
Dennettia, 198
Dennstaedtiaceae, 309
Depanthus, 249
Deschampsia, 208, 240
  antarctica, 278, 291

  argentea, 115
  danthonoides, 175
  gracillima, 302
  penicellata, 302
Descurainia, 113, 117
  bourgaeana, 116
Desfontainia, 339, 341
Desfontainiaceae, 261, 339
Desideria, 136
Desmodium, 66
Desventuradas Islands, 280, 282
Dethawia, 11, 17
Deutzia, 54, 66
  amanoi, 59
  amurensis, 50, 51
  compacta, 69
  discolor, 64
  glabrata, 50
  globosa, 64
  glomeriflora, 69
  hypoglauca, 64
  mollis, 64
  purpurascens, 69
  taiwanensis, 61
  wardiana, 69
Deutzianthus, 234
Deyeuxia, 64
Dialyceras, 218, 328
Dialypetalanthaceae, 258, 342
Dialypetalanthus, 342
Dialypetalum, 219
Diamorpha, 82, 90
Dianella, 350
Dianellaceae, 350
Dianthus, 21, 54, 111
  cancescens, 146
  carbonatus, 27
  chimanimaniens, 203
  ciliatus, 131
  crossopetalus, 146
  eugeniae, 27
  fragrans, 26
  gallicus, 17
  grossheimii, 146
  humilis, 27
  imereticus, 23
  krylovianus, 27
  marschallii, 27
  morii, 50
  nitidus, 19
  nodiflorus, 145

pygmaeus, 60
robustus, 145
sessiliflorus, 145
uralensis, 27
Diapensia, 66
  himalaica, 69
  lapponica, 38
Diapensiaceae, 9, 324
Diapensiales, 324
Diaphanoptera, 136, 147
Dibrachionostylus, 207
Dicarpellum, 249
Dicellostyles, 227
Dicentra canadensis, 85
  cucullaria, 85
  eximia, 85
  paucinervia, 69, 74
  roylei, 74
  scandens, 74
Diceratella canescens, 210
  floccosa, 210
Dichaetaria, 229
Dichaetophora, 163
Dichapetalaceae, 329
Dichapetalum andamanicum, 237
Dicheranthus, 112
  plocamoides, 117
Dichondraceae, 342
Dichotomanthes, 42
  tristaniicarpa, 66
Dichroa febrifuga, 75
Dichrocephala alpina, 207
Dickinsia, 44, 65
  hydrocotyloides, 67
Dicksonia, 295, 308
  arborescens, 215
  berteriana, 280, 282
  sellowiana, 260
Dicksoniaceae, 308
Dicksoniales, 308
Diclidanthera, 259
Diclidantheraceae, 336
Dicliptera longiflora, 61
Dicoelospermum, 227
Dicoria, 163
Dicoryphe, 217, 220, 222
Dicraeopetalum, 207
Dicranocarpus, 163
  parviflorus, 180
Dicranopteris, 307
Dicrastylidaceae, 344

Dicraurus, 160
Dictyolimon, 136
Dictyosperma, 220, 224
  album, 225
Dicyclophora, 138
Didelta, 212
Didesmus, 131
Didiciea, 47
  japonica, 56
Didierea, 217, 223
Didiereaceae, 223, 318
Didissandra, 64
Didymelaceae, 222, 320
Didymelales, 320
Didymeles, 217, 320
Didymocarpus, 72
  ovalifolia, 230
  yunnanensis, 67
Didymophysa, 137
Didyplosandra, 228
Diegodendraceae, 222, 323
Diegodendron, 217, 323
Diellia, 247
Diels, L., 12
Dielsiocharis, 137
Diervilla, 82
  rivularis, 86
  sessilifolia, 86
Dietes robinsoniana, 297
Digitalis atlantica, 126
  dubia, 128
Digitaria platycarpha, 58
Digoniopterys, 218
Dilatris, 264, 353
Dillenia, 221
  andamanica, 237
  ferruginea, 226
  pentagyna, 232
Dilleniaceae, 321
Dilleniales, 321
Dilleniidae, 321
Dilobeia, 219, 222
Dilophia, 137
Dimeresia, 163
Dimetra, 236, 238
Dinemagonum, 284
Dinkladeodoxa, 198, 199
Dionaea, 82
  muscipula, 93
Dioncophyllaceae, 198, 325
Dionycha, 218

Dionysia, 138, 142, 149
  kossinskyi, 147
  mira, 210
  revoluta, 149
  teucrioides, 146
Dioscorea, 56, 68, 70, 286, 352
  balcanica, 22
  caucasica, 25
  kumaonensis, 76
  zingiberensis, 64
Dioscoreaceae, 352
Dioscoreales, 352
Diospyros, 66, 201, 233, 324
  armata, 64
  comorensis, 224
  diversifolia, 224
  lotus, 22
  melanida, 225
  seychellarum, 226
  tesselaria, 225
Dipcadi, 111
  serotinum, 124
Dipelta, 44
  floribunda, 64
  yunnanensis, 67
Dipentodon, 42, 325
  sinicus, 70
Dipentodontaceae, 39, 325
Diphylleia cymosa, 85
Dipidax, 264
Diplarche, 41
Diplazium, 60
  bonincola, 57
  caudatum, 125
  longicarpum, 57
  subtripinnatum, 57
Diplazoptilon, 46, 65
Diplobryum, 235
Diplocentrum, 229
  congestum, 230
  recurvum, 230
Diplolabellum, 47
Diplolaena, 273
Diplomeris chinensis, 64
  hirsuta, 73
Diplopanax, 44
Diplopogon, 274
Diplospora fruticosa, 64
Diplotaenia, 138
Diplycosia alboglauca, 70
  pauciseta, 70

Dipoma, 42
Dipsacaceae, 341
Dipsacales, 341
Dipsacus asper, 78
  atratus, 72
  pinnatifidus, 207
Dipteranthemum, 273
Dipteridaceae, 197, 308
Dipteris, 308
Dipterocarpaceae, 257, 328
Dipterocarpus indicus, 230
  kerrii, 237
Dipterocome, 141
Dipteronia, 43, 335
  sinensis, 64
Dipterygium glaucum, 211
Diptychocarpus, 137
Dirachma, 209, 236
  socotrana, 209
Dirachmaceae, 202, 206, 209, 336
Dirca occidentalis, 173
Disanthus, 41
Discanthelium californicum, 173
Discocapnos, 264
Diselma, 270, 273
  archeri, 272
Disjunctions
  Alnus maritima, 94
  Appalachian—Ozarkian, 84
  Arctic—Southern Rocky Mountain, 110
  Atlantic North America—Eastern Asia, 83, 94
  California—Chile, 173, 174
  Canadian—Eurasian, 36
  Cape Region—Other Gondwanaland, 266
  Hawaii—Southwest Pacific, 246
  Holarctic, 276, 278
  Hyrcanian—Circumboreal, 150, 151
  Lord Howe—New Zealand, 297
  Macaronesia—Other areas, 113
  Madagascar—Other areas, 221
  Mexican—Appalachian, 187
  New Caledonia—New Hebrides—Fiji, 250
  Old World Tropics—New World Tropics, 255
  Rocky Mountains—Gaspé Peninsula, 37
  St. Helena—Africa, 215
  Sarraceniaceae, 94
  Sonoran—South American, 179
Disporopsis, 47, 68

Disporum, 56, 240
  lanuginosum, 86
  maculatum, 86
  ovale, 51
  uniforum, 64
Dissochondrus, 247
District, floristic, 3
Distylium, 60
  gracile, 60
  indicum, 77
  lepidotum, 57
  pingpiensis, 66
Ditepalanthus, 217
Dithyrea, 161
Diuranthera, 47, 351
  minor, 68
Dizygotheca, 249, 250
Djaloniella, 198
Dobera, 337
Dobinea, 43, 334
  vulgaris, 69
Dobineaceae, 334
Docynia indica, 69, 75
Dodonaea viscosa, 210, 226
Dolichlasium, 285
Dolicokentia, 250
Dombeya brachystemma, 203
  calantha, 203
  leachii, 203
Domeykoa, 284
Donatia, 269, 277, 345
Donatiaceae, 345
Donets Range, 28
Doniophyton, 285
Doratoxylon, 218, 224
Dorema, 139
  glabrum, 146
Doronicum, 20
  balansae, 24
  corsicum, 129
  macrolepis, 24
  tobeyi, 24
Dorstenia gigas, 209
Doryalis spinosa, 203
Doryanthaceae, 270, 350
Doryanthales, 350
Doryanthes, 350
Dorycnium, 117
Dorystoechas, 122, 131, 140
Dossifluga, 236, 238
Douepia, 137

Douglas firs, 103
Douglasia gormanii, 35
Downingia, 172
  bella, 175
  concolor, 175
  cuspidata, 173, 175
  elegans, 176
  humilis, 174
  insignis, 176
  ornatissima, 176
  pulchella, 176
  pusilla, 174, 176
  yina, 176
Draba, 19, 54, 66, 101, 168, 279
  argyraea, 106
  baicalensis, 32
  bellii, 15
  bhutanica, 72
  bryoides, 26
  chamissonis, 15
  crassifolia, 106
  elisabethae, 26
  exunguiculata, 109
  glacialis, 15
  grayana, 109
  gredinii, 15
  groenlandica, 15
  kjellmanii, 15
  macrocarpa, 15
  maguirei, 169
  mollissima, 26
  oblongata, 15
  ossetica, 26
  pohlei, 15
  sekiyana, 61
  sphaerocarpa, 106
  subcapitata, 15
  subsecunda, 23
  supranivalis, 26
  taimyrensis, 15
Drabopsis, 137
Dracaena cinnabari, 114, 209
  draco, 114, 117, 118
Dracaenaceae, 349
Dracocephalum, 67, 70
  faberi, 64
  fragile, 31, 32
  henryi, 64
Dracophyllum, 251, 296, 297
  arboreum, 301
  lessonianum, 300

matthewsii, 300
paludosum, 301
patens, 300
pyramidale, 300
scoparium, 301
viride, 300
Dracopis, 82
Dracunculus, 123
  canariensis, 118
Drakebrockmania, 207
Drakenberg system, 203
Draperia, 100
Drapetes, 269
Drepanocaryum, 140
Drimys, 250, 281
  brasiliensis, 260
  winteri, 281
Droseraceae, 331
Droserales, 331
Drosophyllum, 121, 125, 331
Drusa, 113
  glandulosa, 117, 193
Dryandra, 269, 273
Dryas, 16
  integrifolia, 106
  octopetala, 106
Drymaria elata, 180
  lyropetala, 180
Drymophila, 270, 352
Drymotaenium, 39
Dryobalanops, 239
Dryopetalon, 161
Dryopteridaceae, 309
Dryopteris, 53, 60, 280
  aquilina, 292
  filix-mas, 31
  insularis, 57
  liliana, 23
  spectabilis, 292
Drypetes andamanica, 237
  integerrima, 57
  karapinensis, 61
Duabangaceae, 239, 332
Dubautia, 247
Dubyaea, 46, 67
Ducampopinus, 234, 238
Duckeodendraceae, 258, 343
Duckeodendron, 343
Dudleya, 161, 172
Dulongiaceae, 254, 339
Dumasia truncata, 55

Dunnia, 44, 65
Dupontia, 11
  fischeri, 15
Dutaillyea, 249
Dypsis, 220
Dyscritothamnus, 163
Dysophylla yatabeana, 56
Dysoxylum, 296
  andamanicum, 237
  malabaricum, 230
  patersonianum, 298
Dysphania, 274
Dystaenia ibukiensis, 55
Dzungarian Alatau, 137
Dzungarian Altai, 135, 138
Dzungaro—Tien Shan Province, 156

East Mediterranean Province, 131
East Usambara Mountains, 201
Eastern America—Eastern Asia relationships, 93, 94
Eastern Asiatic Region, 39
Eastern European Province, 27
Eastern Himalayan Province, 70
Eastern Madagascan Province, 222
Eastwoodia, 163
Eatonella, 163
Ebenaceae, 324
Ebenales, 324
Ebenus stellata, 210
Eberhardtia, 234, 237
Eccoilopus, 47
Ecdeiocolea, 273
Ecdeiocoleaceae, 355
Echinacea, 82
  laevigata, 85, 88
  pallida, 91
  purpurea, 85
  tennesseensis, 91
Echinocactus, 160
Echinofossulocactus, 160
Echinophora orientalis, 146
Echinops, 111, 142
  dissectus, 51
  humilis, 31
  kotschyi, 210
  manshuricus, 51
  melitenensis, 145
Echinopterys, 162
Echiochilon, 133
  persicum, 210

Echiochilopsis, 122
Echiostachys, 264
Echium, 113, 116, 117
  glabrescens, 118
  lindbergii, 118
  nervosum, 115
  stenosiphon, 118
  suffruticosum, 126
Ectadium, 212
Ecuador, 277
Edgaria, 42, 71
Edgeworthia, 42
  gardneri, 69, 75
  tomentosa, 75
Edraianthus, 22
  owerinianus, 26
Edwards Plateau, 99
Egyptian—Arabian Province, 134
Ehrendorfer, F., 341
Ehretia, 67
  wallichiana, 72
Ehretiaceae, 343
Elaeagnaceae, 338
Elaeagnales, 338
Elaeagnus, 55, 61, 67, 338
  caudata, 72
  pungens, 158
  rotundata, 57
Elaeocarpaceae, 328
Elaeocarpus, 296
  pachycarpus, 57
  photiniifolius, 57
  tuberculatus, 230
Elaeodendron orientale, 224
Elaeoselinum, 122
Elaeosticta, 139
  glaucescens, 146
Elaphoglossaceae, 309
Elaphoglossum randii, 293
Elatinaceae, 323
Elatinales, 323
Elatine, 323
Elatostema, 54
Elburz, 139
Elburzia, 137, 146
Eleiotis, 228
Eleocharis, 56, 68, 70
  globularis, 34
Eleorchis, 47
Eleusine, 211
Eliaea, 217

Elignocarpus, 218
Elingamita, 294, 300
Elionurus, 261
Elizaldia, 122
Ellertonia rheedii, 230
Elliottia, 79
Ellisiophyllaceae, 343
Elmera, 100
Elmerrillia, 239
Elsholtzia, 67
  ciliata, 76
  integrifolia, 63
  serotina, 51
  stauntonii, 63
Elymus, 174
  junceus, 158
  lanuginosus, 158
  uralensis, 28
  virescens, 15
Elytranthe, 296
  adamsii, 300
Elytropus, 284
Embadium, 274
Embelia lenticellata, 61
Emblica fischeri, 230
  officinalis, 232, 233
Emblingia, 327, 335
Emblingiaceae, 273, 335
Embothrium, 286
Eminium, 142
Emmenanthe, 162
Emorya, 162
Empetraceae, 324
Empetrum, 324
  rubrum, 290
Enarthrocarpus, 134
Enceliopsis, 163
Encephalartos, 201
Enchylaena, 274
Endocaulos, 218
Endomallus, 235
Endressia, 11, 17
Endymion non-scriptus, 17
Engelhardtia spicata, 69
Englerodendron, 200, 201
Enicosanthellum, 234
Enkianthus, 41, 54, 69
  pauciflorus, 66
Enochoria, 249
Ensete, 354
Entelea, 294

Enterolobium contortisiliquum, 260
Eomecon, 40, 63
Epacridaceae, 268, 324
Ephedra, 312
  altissima, 124, 125
  cossoniana, 124
  distachya, 158
  gerardiana, 158
  przewalskii, 157
Ephedraceae, 312
Ephedrales, 312
Ephemeral vegetation, 153, 177
Ephippiandra, 217
Ephippianthus, 47
  sawadanus, 56
Ephippiocarpa, 201
Epifagus virginiana, 187
Epigaea asiatica, 54
  gaultherioides, 23
Epilasia, 141
Epilobium, 272, 277, 296
  antipodum, 301
  brevifolium, 75
  confertifolium, 301
  fleischeri, 19
  kermodei, 70
  latifolium, 38
  montanum, 31
  numidicum, 126
  royleanum, 75
  sikkimense, 69
  stereophyllum, 207
  wallichianum, 75
Epimedium, 54
  macrosepalum, 50
  perralderianum, 125
  pinnatum, 23, 151
  pubescens, 62
  pubigerum, 23
Epipactis dunensis, 17
Epithelantha, 160
Equisetaceae, 306
Equisetales, 306
Equisetum, 306
Eragrostis, 288
Eranthis pinnatifida, 54
  sibirica, 30
  stellata, 50
Ercilla, 317
Erechtites kermadecensis, 299
Eremaean Region, 274

Eremobium, 135
Eremoblastus, 137
Eremocrinum, 164, 351
Eremodaucus, 139
Eremolepidaceae, 254, 338
Eremolepis, 338
Eremopanax, 249
Eremophila, 269, 275, 344
Eremophyton, 133
Eremopogon foveolatus, 211
Eremosparton, 138
Eremostachys, 140, 142
Eremosynaceae, 273, 331
Eremosyne, 331
Eremurus, 142
  kopetdaghensis, 148
  subalbiflorus, 148
Ergocarpon, 139
Eriachaenium, 285, 289
Eriaxis, 249
Erica, 111, 204, 264
  arborea, 116, 131, 207, 208
  cinerea, 115, 188
  mackaiana, 17
  multiflora, 119
  scoparia, 114
  vagans, 17
Ericaceae, 10, 16, 264, 323
Ericales, 323
Erigenia, 82
  bulbosa, 85
Erigeron, 56, 101, 168, 169, 282
  alpinus, 207
  basalticus, 108
  cabrerae, 117
  cervinus, 103
  cronquistii, 169
  delicatus, 103
  disparipilus, 109
  evermannii, 106
  flabellifolius, 106
  flexuosus, 103
  humilis, 106
  incertus, 290
  mancus, 166
  melanocephalus, 110
  mimegletes, 99
  miser, 104
  muirii, 15
  nanus, 20
  petiolaris, 104

pinnatisectus, 110, 166
piperianus, 108
rydbergii, 106
sanctarum, 173
vetensis, 110
Erinna, 285
Erinocarpus, 227
   nimmonii, 230
Erinus, 11
Eriobotrya, 60
   angustissima, 75
   hookeriana, 72
   japonica, 64
   platyphylla, 70
   wardii, 70
Eriocaulaceae, 355
Eriocaulon, 56, 73, 77, 355
   chinorossicum, 51
   faberi, 64
   schimperi, 208
   ussuriense, 51
Eriocephalus, 212
Eriocycla albescens, 62
Eriodictyon, 162
Eriogonum, 101, 168, 169, 172
   allenii, 89
   argillosum, 177
   correllii, 96
   covilleanum, 177
   vischeri, 96
Eriophorum, 16, 36
   polystachion, 110
   scheuchzeri, 110
   triste, 15
Eriophyllum nubigenum, 104
Eriosorus cheilanthoides, 292
Eriospermaceae, 351
Eriospermum, 264, 351
Eriosyce, 283
Eriothrix, 219, 224
Erismadelphus, 336
Eritreo-Arabian Subregion, 205
Eritrichium borealisinense, 63
   czekanowskii, 32
   mandshuricum, 51
   uralense, 27
Erodium, 126
   beketowii, 27
   cicutarium, 175
   corsicum, 129
   gussonii, 129

manescavi, 17
   reichardii, 128
   rodiei, 129
Erucastrum canariense, 117
   virgatum, 129
Eryngiophyllum, 163
Eryngium aristulatum, 175
   bupleuroides, 282
   leavenworthii, 96
   palmatum, 21
   pinnatisectum, 175
   serbicum, 21
   spinalba, 19
   vaseyi, 175
   viviparum, 17
Erysimum, 19, 142
   arbuscula, 115
   buschii, 146
   contractum, 23
   feodorovii, 146
   inense, 30
   nachyczevanicum, 146
   nanum, 146
   stigmatosum, 62
   tenuifolium, 115
   wagifii, 146
   yunnanense, 66
Erythea, 182
Erythrina arborescens, 66
   boninensis, 57
   suberosa, 233
Erythrochlamys, 207
Erythronium japonicum, 56
Erythropalaceae, 337
Erythropalum populifolium, 230
Erythroselinum, 207
Erythrospermum, 221
Erythroxylaceae, 335
Erythroxylum, 335
Escallonia, 281, 286
Escalloniaceae, 338
Eschscholzia, 159, 175
   californica, 175
   lemmonii, 173
Ethiopian Province, 206
Eubotrys recurva, 85
Eubrachion, 338
Eucalyptus, 251, 268, 270, 271, 275
   coccifera, 272
   diversicolor, 273
   ovata, 272

INDEX 443

urnigera, 272
Eucarya, 275
Euchresta japonica, 55
Euclea, 201
Eucommia, 41, 63, 319
  ulmoides, 63
Eucommiaceae 39, 63, 319
Eucommiales, 319
Eucryphia, 277, 286, 330
Eucryphiaceae, 330
Eucrypta, 162
Eugenia, 187
  bracteata, 233
Eulophia toyoshimae, 58
Eulychnia, 283, 286
Euonymus, 55, 61, 64, 67, 70, 76
  attenuatus, 78
  boninensis, 57
  crenulatus, 231
  griffithii, 70
  kachinensis, 70
  kiautschovicus, 62
  maackii, 52
  macrocarpus, 72
  obovatus, 85
  pauciflora, 50
  tibeticus, 72
  verrucosus, 27
Eupatorium, 56, 86, 93, 184
  luciae-brauniae, 90
Euphorbia, 54, 64, 85, 117, 142, 206, 207, 212, 213, 232
  aequoris, 214
  alpina, 30
  altaica, 30
  arbuscula, 209
  beaumierana, 124
  caducifolia, 211
  canariensis, 116
  ceratocarpa, 129
  coniosperma, 146
  croizatii, 62
  dendroides, 119
  echinus, 124
  formosana, 61
  glauca, 298
  gregersenii, 21
  grossheimii, 146
  hakutosanensis, 50
  hieroglyphica, 126
  himalayensis, 72
  komaroviana, 50, 52
  larica, 210, 211
  lucorum, 50
  mandshurica, 50
  maresii, 128
  marschalliana, 148
  mauritanica, 266
  neriifolia, 232
  nivulia, 211
  norfolkiana, 298
  officinarum, 124
  origanoides, 214, 216
  piscatoria, 115
  pontica, 23
  reboudiana, 126
  regis-jubae, 116
  resinifera, 124
  rigida, 132
  savaryi, 50
  sikkimensis, 72
  spiralis, 209
  tashiroi, 61
  terracina, 124
  triflora, 131
  trinervia, 216
  tshuiensis, 30
  tuckeyana, 118
  valliniana, 19
  velenovskyi, 21
Euphorbiaceae, 10, 268, 329
Euphorbia-Didierea thorn scrub, 223
Euphorbiales, 329
Euphrasia, 17, 19, 55, 240, 272, 279, 296
  bhutanica, 72
  simplex, 72
Euphronia, 336
Eupomatia, 312
Eupomatiaceae, 268, 312
Eupomatiales, 312
Euptelea, 41, 319
  polyandra, 54
Eupteleaceae, 39, 319
Eupteleales, 319
Eurasiaticum, 12
Europe, 17, 27, 29
Eurotia ceratoides, 157, 158
  compacta, 157, 158
Eurya, 61, 66, 69, 113
  boninensis, 57
  cerasifolia, 75
  urophylla, 70

wardii, 70
Eurycarpus, 137
Eurycorymbus, 43
Euryops pinifolius, 210
Eurypetalum, 198
Eurytaenia texana, 96
Euscaphis, 43
Eustigma, 41
Euterpe edulis, 261
Eutetras, 163
Euthamia remota, 91
  tenuifolia, 91
Eutrema bracteata, 54
  himalaicum, 72
  parviflorum, 30
  yunnanensis, 66
Euxine Province, 22
Euzomodendron, 121, 125
Evacidium, 122
Evandra, 273
Evax caulescens, 176
  rotundata, 130
Eversmannia, 138
Evodia, 66
  fraxinifolia, 75
  hupehensis, 64
  inermis, 57
  kawagaiana, 57
  littoralis, 298
  merrillii, 61
  mishimurae, 57
Evodiopanax, 44
  innovans, 55
Evonymopsis, 218
Excavatia hexandra, 57, 58
Excremis, 350
Exocarpos, 275
Exochorda giraldii, 50
  serratifolia, 50
Exospermum, 248, 250

Fabaceae, 10, 26, 30, 97, 159, 264, 268, 289, 333
Fabales, 333
Faberia, 46, 65
  ceterach, 67
  lancifolia, 67
  sinensis, 64
Fabiana, 286
Fagaceae, 10, 321
Fagales, 321

Fagara, 281
  boninsimae, 57
  externa, 282
  mayu, 281
Fagaropsis angolensis, 207
Fagonia, 135
  acerosa, 210
  arabica, 211
  subinermis, 210
Fagraea, 342
Faguetia, 218
Fagus, 12
  crenata, 54
  engleriana, 63
  grandifolia, 86, 92, 187
  hayatae, 60
  japonica, 54
  mexicana, 187
  orientalis, 25, 26, 152
  sylvatica, 18
Falkland Islands, 277, 285, 289
Fallugia, 161
Fargesia, 47
Farmeria, 227
Faroe Islands, 17, 188
Fars, 149
Farsetia, 206
Fars-Kermanian Subprovince, 148
Fascicularia, 285
Fatsia, 44
  oligocarpella, 57
  polycarpa, 61
Faucherea, 217
Faxonia, 163
Fedia, 122
  sulcata, 126
Fedorov, An. A., 12
Feeria, 122
Fendlera, 162
Fendlerella, 162
Feneriva, 217
Fergania, 139
Fergusonia, 228
Fernald, M. L., 37
Fernándezian Region, 280
Fernelia, 219, 224
Ferns, 265, 282
Ferocactus, 160, 181
"Fertile Crescent," 143
Ferula, 142, 153, 155
  lancerottensis, 117

linkii, 117
oopoda, 148
persica, 146
turkomanica, 148
Festuca, 20, 73, 158, 167, 174, 203, 204
  abyssinica, 208
  albida, 115
  algeriensis, 126
  bornmuelleri, 118
  donax, 115
  erecta, 278, 290
  gigantea, 31
  idahoensis, 108
  morisiana, 130
  sardoa, 130
  takedana, 56
  tolucensis, 186
Fezia, 121
Ficinia, 264
Ficus, 60, 260
  boninsimae, 57
  iidaiana, 57
  kopetdagensis, 147
  mishimurae, 57
  salicifolia, 208, 210
  sycomorus, 208
Fiji Islands, 244
Fijian Province, 244
Fijian Region, 243
Filago, 111, 126
Filifolium sibiricum, 52
Filipendula, 55
Fimbristylis ochotensis, 34, 53
Fire trees, 35, 87, 103, 104, 106, 176, 178
Fires, role of, 92, 97, 103, 176
Fitchia, 246
Fitzalania, 271
Fitzroya, 283
  cupressoides, 287
Flacourtia, 233
Flacourtiaceae, 325
Flagellaria, 355
Flagellariaceae, 355
Flatrocks, 90
Flaveria anomala, 180
  oppositifolia, 180
Fleurydora, 198
Flindersiaceae, 333
Floerkea, 336
Florida, 94
Floristic system, principles of, 1

Flueggea, 211
Foetidia mauritiana, 225
  rodriguesiana, 224
Fokienia, 65
Folded Appalachians Geological Province, 89
Foleyola, 133
Fontanesia phillyreoides, 148
Fontquera, 122, 125
Forest, arid tropical, 260
  beech, 152
  beech-maple, 86
  beech-oak, 26
  boreal, 38
  conifer-broadleaved, 28
  coniferous, 29, 35, 101
  dark conifer, 31
  deciduous, 17, 86, 223
  Douglas fir, 103
  Eucalyptus, 270, 272, 274
  fir, 31
  hemlock-hardwood, 87
  larch, 31, 32, 33, 52
  laurel, 114, 116
  maple-basswood, 86
  mixed mesophytic, 86
  monsoon, 232
  oak, 28, 51, 148, 152, 185
  oak-chestnut, 87
  oak-hickory, 86
  pine, 29, 32, 87, 91, 103, 104, 116, 185
  rain, 102, 198, 222, 235, 258, 260, 295
  savanna, 238
  sclerophyll, 5, 124, 174, 176, 223
  shola, 231, 232
  spruce, 28
  spruce-cedar-hemlock, 102
  spruce-fir, 35, 38, 87, 88, 89, 107, 108
  subtropical, 261
  summergreen, 62
  thorn, 232
  walnut, 153
  Yucca, 182
Forest-savanna, 198
Forest steppes, 26, 62
Forestiera acuminata, 86
  reticulata, 99
Forgesia, 219, 224
  borbonica, 225
Formania, 48, 65
  mekongensis, 67
Formosia, 45, 60

benthamiana, 61
Forsellesia, 161, 331, 337
Forsskaolea angustifolia, 117
　tenacissima, 211
Forsythia europaea, 21
　japonica, 55
　koreana, 55
　mandschurica, 51
　ovata, 55
Fortunearia, 41, 63
　sinensis, 63
Fortuynia, 137
Fothergilla, 79
　major, 85
Fouquieria, 160, 181, 327
　columnaris, 180
　shrevei, 180
　splendens, 180
Fouquieriaceae, 159, 327
Fouquieriales, 327
Fragaria concolor, 50
　nipponica, 55
　yezoensis, 53
France, 11, 17, 19, 121, 128
Francoa, 284, 331
Francoaceae, 283, 331
Frangula azorica, 114
Frankenia gypsophila, 180
　jamesii, 180
　johnstonii, 180
　portulacifolia, 215
Frankeniaceae, 327
Franklandia, 273
Franklinia, 79
　alatamaha, 93
Fraxinus, 12, 64, 67
　bungeana, 62
　caroliniana, 92, 93
　excelsior, 28
　floribunda, 70, 76
　latifolia, 102
　mandshurica, 50, 51
　micrantha, 155
　quadrangulata, 86
　sieboldiana, 55
Fredolia, 133
Fremontodendron, 161
Frerea, 228
Freycinetia, 296
　baueriana, 298, 299
　boninensis, 58

Fritillaria amabilis, 56
　biflora, 173
　burnatii, 20
　delavayi, 68
　drenovskii, 22
　falcata, 177
　glauca, 177
　grandiflora, 151
　japonica, 56
　kotschyana, 151
　maximowiczii, 51
　purdyi, 177
　raddeana, 148
　tubiformis, 20
　ussuriensis, 51
Fuchsia, 286
Fumaria occidentalis, 17
　purpurea, 17
Fumariaceae, 317
Fumariola, 135
Funkiaceae, 39, 350

Gabon, 198
Gagea helenae, 26
　hiensis, 51
　improvisa, 147
　mauritanica, 126
　nakaiana, 51
　pauciflora, 51
　samojedorum, 29
　vaginata, 53
Gagnebina, 218, 220
Gagnepainia, 236
Gaillardia, 96
　gypsophila, 180
　henricksonii, 180
　multiceps, 180
　powellii, 180
Gaimardia, 356
Galactites mutabilis, 126
Galanthus caucasicus, 26
　krasnovii, 25
　lagodechianus, 26
　latifolius, 26
Galapageian Province, 256
Galapagos Islands, 256, 284
Galax, 79
　aphylla, 85
Galbulimima, 312
Galiniera coffeoides, 207
Galitzkya, 137

Galium, 55, 67, 85, 126, 129, 142, 240, 282
  antarcticum, 290
  asperifolium, 231
  balearicum, 128
  bullatum, 146
  bungei, 63
  ceratopodum, 210
  correllii, 99
  crespianum, 128
  echinocarpum, 61
  elegans, 76
  formosense, 61
  glaciale, 207
  hyrcanicum, 146
  manshuricum, 51
  pauciflorum, 63
  platygalium, 51
  productum, 115
  saxatile, 188
  stojanovii, 21
Gallesia guararema, 260
Galloway, R. W., 269
Galopina, 114
Galpinia, 201
Galvezia speciosa, 173
Gamanthus, 136
Gamblea, 44
  longipes, 70
Ganges, 233
Gantelbua, 228
Garaventia, 285, 349
Garberia, 82
Garcinia, 233, 271
  cambogia, 230, 231
Gardenia boninensis, 58
Gardner, C. A., 274
Gardneria insularis, 55
  shimadai, 61
Garhadiolus, 141
Garhwal, 46
Garnieria, 249
Garo Hills, 75
Garrya, 340
Garryaceae, 340
Garuga pinnata, 232
Gaspé Peninsula, 37, 38
Gasteria, 212, 264
Gastrochilus affinis, 73
  dasypogon, 73
  distichus, 73
Gastrodia, 56

  boninensis, 58
Gastrolepis, 249
Gaultheria, 66, 69, 286, 296
  antarctica, 290
  griffithiana, 75
  japonica, 54
  nummularoides, 75
  shallon, 102, 105
Gaylussacia brachycera, 85
Geisorhiza, 264
Geissoloma, 264, 330
Geissolomataceae, 264, 330
Geissolomatales, 330
Geitonoplesiaceae, 352
Geitonoplesium cymosum, 252
Geniostoma, 221, 251
  glabrum, 57
Genista, 126, 129
  acanthoclada, 128
  berberidea, 17
  cinerea, 119
  demnutensis, 124
  dorycnifolia, 128
  ferox, 124
  hystrix, 17
  lucida, 128
  webbii, 124
Genistidium, 161
Gentiana, 19, 55, 64, 67, 70, 72, 86, 240, 279, 296
  algida, 110
  antarctica, 301
  antipoda, 301
  arctica, 15
  arisanensis, 61
  campanulacea, 78
  cerina, 301, 302
  chathamica, 301
  concinna, 301, 302
  detonsa, 15
  gibbsii, 301
  glauca, 106
  gradata, 70
  grossheimii, 26
  kolakovskyi, 24
  komarovii, 51
  lagodechiana, 26
  manshurica, 51
  marcowiczii, 26
  paradoxa, 24, 25
  speciosa, 76

sugawarae, 34, 53
verna, 145
yuparensis, 53
Gentianaceae, 10, 342
Gentianales, 341
Gentianella, 19
anglica, 17
Gentianopsis, 67
Gentianothamnus, 219
Gentrya, 162
Geoffraya, 235
Geopanax, 219, 225
Georgia, 90, 94
Geosiridaceae, 222, 348
Geosiris, 219, 348
Geptner, V. G., 5
Geraea, 163
Geraniaceae, 336
Geraniales, 336
Geranium, 66, 186, 240
arabicum, 207
endressii, 17
hattai, 50
koreanum, 50
maximowiczii, 50
montanum, 151
phaeum, 18
renardii, 26
richardsonii, 107
traversii, 301
tripartitum, 55
Gerbera, 67
piloselloides, 76, 208
Gesneriaceae, 344
Gesnouinia, 112
arborea, 117
Gethyum, 285
Geum, 85, 203, 277, 279
bulgaricum, 21
parviflorum, 277, 301, 302
sikkimense, 72
Ghats, 227, 230, 232
Gibraltar, 125
Gilbertiella, 198
Gilbertodendron, 198
Gileae, 172
Gilia, 169, 173
Gillenia, 82
stipulata, 85
trifoliata, 85
Gillespiea, 244

Gilletiella, 344
Gilliesia, 285
Gilliesiaceae, 349
Gilmania, 160
Ginkgo, 40, 310
biloba, 63, 191
Ginkgoaceae, 39, 63, 310
Ginkgoales, 310
Girardinia cuspidata, 50
Girgensohnia, 136
Gissar Mountains, 138, 153
Githopsis, 173
Givotia madagascarensis, 223
Gladiolus, 111
Glaucidiaceae, 39, 53, 317
Glaucidiales, 317
Glaucidium, 40, 317
palmatum, 54
Glaucococarpum, 161
Glaziocharis abei, 56
Gleadovia yunnanensis, 67
Gleditsia aquatica, 93
caspia, 151, 152
delavayi, 66
japonica, 151
macracantha, 64
sinensis, 64
Gleichenia, 307
Gleicheniaceae, 307
Gleicheniales, 307
Glenniea, 228
Glia, 264
Globularia, 117, 344
alypum, 119
combessedesii, 128
neapolitana, 129
Globulariaceae, 344
Globulostylis, 198
Glochidion andamanicum, 237
Glossocalyx, 314
Glossocardia, 228
Glossopappus, 122
Gluta travancorica, 230
Glycyrrhiza yunnanensis, 66
Glyphaea tomentosa, 203
Glyphosperma, 351
Glyphostylus, 234
Glyptopetalum lawsonii, 230
Glyptopleura, 163
Glyptostrobus, 65
Gmelina arborea, 232

Gmelina delavayana, 67
Gnaphalium affine, 290
  sarmentosum, 186
  schulzii, 207
  vulcanicum, 186
  webbii, 117
Gnetaceae, 312
Gnetales, 312
Gnetum, 312
Gnidia glauca, 207
Gobi, 157
Goethalsia, 328
Goetzeaceae, 254, 343
Golaea, 207
Gomortega, 283, 314
Gomortegaceae, 283, 314
Gomphostemma, 67
Gonatanthus ornatus, 77
  pumilus, 77
Gondwanaland, 94, 240, 266, 279
Gongrodiscus, 249
Goniocladus, 244
Gonioma, 203
  camassi, 265
Gonocytisus, 121, 131
Gonospermum, 113, 117
Gontscharovia, 140
Goodeniaceae, 268, 345
Goodeniales, 345
Goodmania, 160
Goodyera, 56
  augustini, 58
  boninensis, 58
  hemsleyana, 73
  henryi, 64
  macrophylla, 115
  vittata, 73
Gordonia axillaris, 69, 70
  lasianthus, 93
  obtusa, 231
Gorodkovia, 11, 33
Gosela, 264
Gouania andamanica, 237
Gough Island, 292
Goupiaceae, 337
Gourliea, 284, 286
Gozo, 122, 130
Graellsia, 137
Gramineae, 356
Grammitidaceae, 308
Grammitis billardieri, 295

  kerguelensis, 289, 290
  sakaguchiana, 53
Grammosciadium, 139
Grandidiera, 201
Grangeria, 218, 220
  borbonica, 225
Grantia, 141
  aucheri, 210
Graphandra, 236, 238
Graphistemma, 235, 237
Graptopetalum, 161
Grassland Province, 95
Grasslands, 97, 153, 200, 202, 203, 223, 238, 261, 275, 282, 288, 290
Gravesia, 218
Gray, A., 83
Great Basin, 159, 160
Great Basin Province, 165
Great Plains, 95
Great Smoky Mountains, 39, 86, 89
Greater Antilles, 256
Greater Caucasus, 26
Greece, 11, 22, 121, 131
Green, P. S., 298
Greenland, 15
Greenmaniella, 163
Greenovia, 112, 117
Greslania, 249
Grevea, 339
Grevillea, 250
Grewia, 66, 207
  makranica, 210
  rhombifolia, 61
  tenax, 211
Greyia, 264, 331
Greyiaceae, 264, 331
Grielum, 212, 331
Griffithella, 227
  hookeriana, 230
Griselinia, 277, 286, 339
Griseliniaceae, 276, 339
Grisollea, 218, 220
Grisseea, 242
Groenlandia, 347
Grossularia, 330
Grossulariaceae, 330
Grubbia, 264, 324
Grubbiaceae, 263, 324
Grubov, V. I., 157
Guadelupe Island, 162
Guadua, 261

Guamia, 245, 246
Guardiola, 163
Guatemala, 255
Guayana Highlands, 257
Guayana Province, 258
Guillainia, 250
Guinea, 199
Guineo-Congolian Region, 198
Guiraoa, 121, 126
Guldenstaedtia maritima, 62
Gulf Coastal Plain Province, 91
Gundelia, 141
Gunnera, 240, 281, 286, 340
 peltata, 282
Gunneraceae, 340
Gunnerales, 340
Guttiferae, 322
Gymnemopsis, 235
Gymnocarpos decander, 211
 przewalskii, 157
Gymnochilus, 220, 224
Gymnocladus chinensis, 64
 dioicus, 85
Gymnocranthera canarica, 230
Gymnogrammitidaceae, 310
Gymnophyton, 284
Gymnopteris borealisinensis, 62
Gymnospermium altaicum, 30
 microrrhinchum, 50
 smirnovii, 26
Gymnosporia berberoides, 67
 buxifolia, 265
Gymnostemon, 198
Gymnotheca, 40
 chinensis, 63
 involucrata, 66
Gynandriris, 111
Gynophorea, 137
Gypothamnium, 285, 286
Gypsophila, 142
 acutifolia, 26
 aretioides, 148
 belorossica, 27
 briquetiana, 145
 steupii, 23
 tuberculosa, 145
 uralensis, 27
Gyrocarpaceae, 314
Gyrocarpus, 314
 americanus, 223
Gyroptera, 207

Gyrostemonaceae, 268, 327
Gyrothyraceae, 99

Haastia, 295
Habenaria, 56, 64
 juncea, 73
 obtusata, 37
 tridactylites, 118
Haberlea, 11, 21
 rhodopensis, 22
Habropetalum, 198, 199
Hachettea, 248, 315
Hackelia cronquistii, 170
 hispida, 109
Hacquetia, 11, 18
Haematodendron, 217
Haematostaphis, 204
Haemodoraceae, 353
Haemodorales, 353
Hagenbachia, 353
Hagenia abyssinica, 207
Hagghier Massif, 209
Hainan, 40, 234
Hainania, 234, 237
Hakea, 269, 272
Hakonechloa, 47, 53
 macra, 56
Halacsya, 11, 21
 sendtneri, 22
Halanthium, 136
Halarchon, 136
Halesia diptera, 93
 parviflora, 93
 tetraptera, 85
Halimium, 121
Halimocnemis, 136
Halleria lucida, 207, 265
Halocharis, 136
Halophytaceae, 283, 318
Halophytum, 283, 318
Haloragaceae, 333
Haloragales, 333
Haloragis, 240, 281
 walkeri, 55
Halosciastrum, 44, 50
 melanotilingia, 50
Halostachys, 136
Halotis, 136
Haloxylon ammodendron, 157
Hamadera Hills, 209
Hamadryas, 283, 289

argentea, 290
Hamamelidaceae, 10, 319
Hamamelidales, 319
Hamamelididae, 319
Hamamelis japonica, 54
 mollis, 63
 vernalis, 84, 85
 virginiana, 84, 87
Hammada, 134
Hammatolobium, 121
Hanabusaya, 46, 50
 asiatica, 51
Hanceola, 46
 sinensis, 64, 67
Hancockia, 47
Handelia, 141
Handeliodendron, 43
Hanghomia, 235
Hanging gardens, 169
Hanguana, 351
Hanguanaceae, 351
Hannonia, 123
Haplocalymma, 163
Haplocarpha, 207
Haploesthes, 163
 greggii, 180
Haplopappus, 168, 173
 contractus, 109
 engelmannii, 96
 eximius, 104
 johnstonii, 180
 liatriformis, 109
 macleanii, 35
 ophitidis, 177
 peirsonii, 104
 pygmaeus, 110
 radiatus, 109
 whitneyi, 104
Haplophyllum, 142
 kowalenskyi, 146
 laeviusculum, 146
 obtusifolium, 148
 schelkovnikovii, 146
 tenue, 146
Haplosciadium abyssinicum, 207
Haploseseli, 44
Haplosphaera, 44, 65
 phaea, 67
Haplostachys, 247
Haplostichanthus, 271
Haplothismia, 229

 exannulata, 230
Harbouria, 100
 trachyphylla, 109
Hardwickia binata, 232
Harfordia, 160
Harmandiella, 235
Harpanema, 219
Harpephyllum, 201
Harrimanella hypnoides, 38
Harrysmithia, 44, 65
 dissecta, 67
Hartogiopsis, 218
Hartwrightia, 82
 floridana, 95
Hawaiian Islands, 246
Hawaiian Province, 248
Hawaiian Region, 246
Haworthia, 212, 264
Haya, 209
Hazunta, 219, 220
Heard Island, 278, 291
Hebe, 269, 278, 296
 barkeri, 301
 benthamii, 301
 bollonsii, 300
 breviracemosa, 299
 chathamica, 301
 dieffenbachii, 301
 elliptica, 277, 290
 insularis, 300
 ligustrifolia, 300
 obtusata, 300
 pubescens, 300
Hebenstretia dentata, 207
Heberdenia excelsa, 113
 penduliflora, 113
Hecastocleis, 163
Hecistopteris, 307
Hectorella, 294, 318
Hectorellaceae, 276, 318
Hedera canariensis, 116
 colchica, 24, 25
 formosana, 61
 helix, 17, 18, 25
 nepalensis, 70, 76
Hederopsis, 241
Hedinia, 137
Hedstromia, 244
Hedycaria, 251
Hedycaryopsis, 217
Hedychium spp., 68, 78

Hedyotis adscensionis, 214, 216
  arborea, 216
  grayi, 58
  leptopetala, 58
  mexicana, 58
  pachyphylla, 58
Hedysarum, 126, 142
  atropatanum, 146
  boutignyanum, 19
  candidum, 132
  cretaceum, 27
  grandiflorum, 27
  ucrainicum, 27
  ussuriense, 50
  zundukii, 30
Hedyscepe, 295, 297
Heeria, 264, 268
Helenium brevifolium, 85
  campestre, 85
  virginicum, 86
Heliamphora, 94, 323
Helianthemum, 117
  arcticum, 29
  canariense, 124
  caput-felis, 127
  gorgoneum, 118
  origanifolium, 127
  scoparium, 174
  sessiliflorum, 211
  spartioides, 174
Helianthus, 83, 96
  angustifolius, 85
  atrorubens, 85
  glaucophyllus, 89
  laevigatus, 88, 89
  pumilus, 110
  salicifolius, 85
  silphioides, 85
Helicanthes, 228
Helichrysum, 111, 142, 146, 204, 207, 264, 296
  abyssinicum, 208
  ambiguum, 128
  devium, 115
  frigidum, 130
  gossypium, 117
  kopetdagense, 148
  melanopthalmum, 115
  monizii, 115
  monogynum, 117
  montelinasum, 130
  obconicum, 115
  saxatile, 130
  whyteanum, 203
Helichrysum scrub, 200, 208
Helicia, 67, 250
  formosana, 61
  nilagirica, 230
  travancorica, 230
Heliciopsis henryi, 67
Helicodiceros, 127
Heliconia, 354
  indica, 354
Heliconiaceae, 254, 354
Helicteropsis, 218
Helictotrichon, 20, 158
  elongatum, 208
  hideoi, 56
  sulcatum, 115
  trisetoides, 51
Heliocarya, 139, 149
Heliotropiaceae, 343
Heliotropium, 142, 286
  borasdjunense, 149
  gaubae, 149
  gracillimum, 146
  gypsaceum, 146
  lasiocarpum, 210
  litwinowii, 148
  mesinanum, 148
  persicum, 211
  schahpurense, 146
  szovitsii, 146
Helipterum, 269
Helleborus abchasicus, 23
  trifolius, 127, 128
Hellenocarum, 122
Helminthostachys, 306
Helonias, 82
Heloniopsis, 46, 59, 82
  arisanensis, 61
Helwingia, 44, 340
  himalaica, 70, 75
  lanceolata, 76
Helwingiaceae, 39, 340
Helwingiales, 340
Hemerocallidaceae, 9, 350
Hemerocallis, 56, 350
  coreana, 51
  forrestii, 68
  minor, 33, 52
Hemiandra, 273

Hemiboea henryi, 64
Hemicrambe, 121, 125
Hemilophia, 42
Hemionitidaceae, 307
Hemiphragma heterophyllum, 76
Hemiptelea, 42
Hemitomes, 100, 102
Hemizonia, 163
Hemsleya, 66
Henrardia, 141
Henriqueziaceae, 342
Henryettana, 45, 65
Hepatica acutiloba, 85
  americana, 85
  transsilvanica, 18
Heptacodium, 44, 63
  jasminoides, 64
  miconioides, 64
Heracleum, 66
  abyssinicum, 207
  aconitifolium, 24
  austriacum, 19
  calcareum, 24
  canescens, 155
  carpaticum, 19
  hookerianum, 231
  lanatum, 50
  mantegazzianum, 24
  minimum, 19
  moellendorffii, 55
  obtusifolium, 76
  paphlagonicum, 24
  rigens, 231
  scabrum, 24
  sphondylium, 19
  wallichii, 72
Hermannia, 212
Hermas, 264
Hermidium, 160
  souliei, 64
Hermodactylus, 123, 193
Hernandia, 314
  peltata, 58
Hernandiaceae, 314
Herniaria canariensis, 117
  mascatensis, 209
  mauritania, 125
Herpetospermum, 42
Herpolirion, 351
Herreria, 222, 349
Herreriaceae, 349

Herreriopsis, 219, 222, 349
Herya, 218, 224
Herzegovina, 22
Hesperaloë, 164
Hesperantha, 264
  petitiana, 208
Hesperelaea, 162
Hesperis nivea, 19
  oblongifolia, 19
  steveniana, 132
Hesperocallidaceae, 9, 350
Hesperocallis, 164, 350
Hesperogreigia, 281
Hesperolaburnum, 121
Hesperomannia, 247
Hesperoseris, 280
Hesperothamnus, 161
Hetaeria spp., 56
Heteracia, 141
Heteranthelium, 141
Heteranthemis, 122
Heterocaryum, 139
Heterochaenia, 219, 224
Heterocodon brevipes, 67
Heterodraba, 161
Heterogaura, 161
  heterandra, 173
Heterolamium, 46
Heteromeles, 161
Heteromma, 203
Heterophragma roxburghii, 232
Heteroplexis, 46
Heteropogon contortus, 233
Heteropyxidaceae, 332
Heterosmilax, 68, 352
  seisuiensis, 61
Heterostemma brownii, 61
Heterozostera, 347
Heuchera, 85
  bracteata, 109
  hallii, 109
Hewardia, 272
Hexaneurocarpon, 235
Hexapora, 241
Hexaptera, 284
Hexaspora, 271
Hexastylis, 79
  shuttleworthii, 85
Heywoodiella, 113
Hibbertia, 221
Hibiscadelphus, 247

Hibiscus glaber, 57
  insularis, 298
  ponticus, 23
  taiwanensis, 61
  tiliaceus, 58
Hickelia, 220
Hieracium, 15
  bolanderi, 103
  coreanum, 51
  glaciale, 20
  greenei, 103
  japonicum, 56
  krameri, 56
  megacephalum, 95
  robinsonii, 35
  trailii, 89
Hierochloë brunonis, 301
  clarkei, 78
Hieronymiella, 285
Hijaz, 208
Hildegardia erythrosiphon, 223
Hillebrandia, 247, 326
Himalayas, 39, 70, 135, 137, 142, 154
Himantandraceae, 312
Himantoglossum caprinum, 132
Himantostemma, 162
Hippeastrum, 286
Hippocastanaceae, 335
Hippocratea andamanica, 237
Hippocrateaceae, 337
Hippocrepis balearica, 128
Hippophaë, 338
  rhamnoides, 158
Hippuridaceae, 9, 344
Hippuridales, 344
Hippuris, 344
Hispidella, 122, 126
Hitchcockella, 220
Hladnikia, 11, 18
  pastinacifolia, 19
Hofmeisteria, 163
Hohenackeria polyodon, 126
Hoheria, 294
  populnea, 300
Hokkaido, 52
Holacantha, 162
Holantarctic flora, 261, 269, 279, 287
Holantarctic Kingdom, 276
  notes, 302
Holantarctis, 276
Holarctic flora, 78, 203, 237, 261, 279

Holarctic Kingdom, 9
  notes, 186
Holarctis, 9
Holboellia, 40
  fargesii, 63
  grandiflora, 66
  latifolia, 74
  marmorata, 66
Holcus notarisii, 130
Hollandaea, 271
Hollisteria, 160
  lanata, 173
Holmbergia, 283
Holocalyx balansae, 260
  glaziovii, 260
Holocarpha, 164
Holocheila, 46, 65
Holodiscus discolor, 102, 105
Holographis, 162
Holostemma sinense, 64
Holostyla, 249
Holozonia, 164
Homalachne, 125
Homalanthus, 296
  polyandrus, 299
Homalium chasei, 203
  napaulense, 72
Homalocarpus, 113
Homalodiscus, 138, 210
Homollea, 219
Homolliella, 219
Hong Kong, 41, 234
Hooker, J. D., 83, 215, 232, 279
Hopea andamanica, 237
  utilis, 230
Hoplestigma, 343
Hoplestigmataceae, 198, 343
Hoppea, 228
Horaninovia, 136
  ulicina, 150
Hordelymus europaeus, 18
Hornea, 218, 224
  mauritiana, 225
Hortonia, 227, 229
Hosiea, 43
Hosta, 47, 56, 350
  ensata, 51
Hovenia, 43
Howeania, 298
Howeia, 295, 297
Howellia, 100

INDEX

Hoya bhutanica, 72
  polyneura, 72
  serpens, 72
  yunnanensis, 67
Hua, 328
Huaceae, 198, 328
Huanaca, 284
Huarpea, 285
Hubbardia, 229
  heptaneuron, 230
Hudsonia, 79
Hugoniaceae, 335
Hulsea brevifolia, 104
Hulthemia, 138
Humbertia, 219
Humbertiaceae, 342
Humbertianthus, 218
Humbertiella, 218
Humbertiodendron, 218, 222, 336
Humbertioturraea, 218
Humblotiella, 309
Humblotiodendron, 218, 220
Humboldtia, 228
  oligocantha, 230
Humiriaceae, 335
Hungary, 19
Huodendron, 41, 66
Hutera, 121, 126
Hyacinthaceae, 10, 349
Hyacinthella litwinowii, 148
  transcaspica, 148
Hyalea, 141
Hyalisma, 229
Hyalolaena, 139
Hybanthus concolor, 85
Hydatella, 354
  inconspicua, 300
Hydatellaceae, 354
Hydatellales, 354
Hydnocarpus alpina, 231
Hydnora, 315
Hydnoraceae, 315
Hydnorales, 315
Hydrangea, 54, 66, 69, 261, 279, 286
  heteromalla, 75
  longifolia, 61
  macrophylla, 75
  sargentiana, 64
  scandens, 59
  strigosa, 64
Hydrangeaceae, 339

Hydrangeales, 338
Hydrastidaceae, 9, 317
Hydrastis, 317
  canadensis, 85
Hydrobryopsis, 227
Hydrobryum, 55
  griffithii, 75
Hydrocharitaceae, 346
Hydrocharitales, 346
Hydrochloa, 82
Hydrocleys, 346
Hydrocotyle capitata, 292
  marchantioides, 292
  monticola, 207
  ramiflora, 55
  yabei, 55
Hydrophylax maritima, 233
Hydrophyllaceae, 159, 343
Hydrophyllum appendiculatum, 86
Hydrostachyaceae, 345
Hydrostachyales, 345
Hydrostachys, 345
Hydrostemma, 316
Hydrotriche, 219
Hygea, 285
Hygrochloa, 270
Hylaea, 258
Hylomecon, 40
  japonicum, 54
Hymenaea stilbocarpa, 260
Hymenanthera chathamica, 301
Hymenoclea, 164
Hymenocrater, 140
Hymenolyma, 139
Hymenonema, 122, 131
Hymenophyllaceae, 308
Hymenophyllales, 308
Hymenophyllopsidaceae, 257, 308
Hymenophyllopsidales, 308
Hymenophyllopsis, 308
Hymenophyllum, 280, 295
  aeruginosum, 292
  alishanense, 60
  falklandicum, 290
  ferrugineum, 292
  wilsonii, 188
Hymenopogon oligocarpus, 67
Hymenostemma, 122
Hymenothrix, 164
Hymenoxys acaulis, 98
Hyophorbe, 220, 224

verschaffeltii, 225
Hyoseris, 122
 taurina, 130
Hyparrhenia, 202, 204, 205
Hypecoaceae, 9, 317
Hypecoum, 317
 chinense, 62
 erectum, 62
Hyperaspis, 207
Hypericaceae, 322
Hypericopsis, 138, 149
Hypericum, 54, 85, 203, 240
 aegyptiacum, 124
 afrum, 125
 balearicum, 128
 bithynicum, 23
 canariense, 117
 coadunatum, 117
 elodeoides, 69, 75
 foliosum, 114
 formosanum, 61
 formosissimum, 146
 hookerianum, 69, 75
 lanceolatum, 199
 mysorense, 231
 oligandrum, 203
 peplidifolium, 207
 pseudomaculatum, 84
 pulchrum, 188
 punctatum, 84
 reflexum, 117
 revolutum, 207
 sampsonii, 75
 scabroides, 145
 splendens, 90
 tenuicaule, 72
 uniglandulosum, 145
 xylosteifolium, 23
 yunnanensis, 66
Hypertelis, 212
Hypocalymma, 273
Hypochoeris, 126
 oligocephala, 117
 robertia, 130
Hypogomphia, 140
Hypolepidiaceae, 309
Hypopytis monotropa, 102
Hypoxidaceae, 353
Hypsela, 277
Hypselandra, 234, 236
Hypseloderma, 207

Hypseocharis, 336
Hypseocharitaceae, 261, 336
Hypseochloa, 198
Hypsithermal period, 97
Hyrcania, 136, 140, 142, 150
Hyrcanian Province, 150
Hyssopus cretaceus, 27
Hystrix coreana, 51
 japonica, 56

Iberian Peninsula, 121, 126
Iberian Province, 126
Iberis peyerimhoffii, 126
 semperflorens, 129
 stricta, 129
Icacinaceae, 337
Ichnocarpus himalaicus, 72
Idaho, 106, 170
Idesia, 41
Idiospermaceae, 271, 314
Idiospermum, 275, 314
Idria, 160, 180
Ifloga, 111
Ikonnikovia, 136
Ileostylus, 294
 micranthus, 298
Ilex, 55, 61, 76, 267, 337
 amelanchier, 93
 aquifolium, 16, 17, 22, 151
 beecheyi, 57
 canariensis, 113, 116
 colchica, 24
 corallina, 70
 coriacea, 93
 cornuta, 64
 cyrtura, 70
 denticulata, 231
 dimorphophylla, 59
 hyrcana, 151
 matanoana, 57
 mertensii, 57
 mitis, 207
 montana, 85
 myrtifolia, 93
 paraguayensis, 261
 perado, 114
 percoriacea, 57
 platyphylla, 117
 poneantha, 59
 szechwanensis, 64
 wightiana, 231

INDEX

yunnanensis, 67
Iljinia, 136
Illiciaceae, 313
Illiciales, 313
Illicium, 66, 313
  arborescens, 60
  burmanicum, 70
  floridanum, 93
  griffithii, 74
  henryi, 63
  mexicanum, 187
  parviflorum, 93
  simonsii, 69, 74
  wardii, 70
Illigera, 314
Illyrian Province, 21
Imbricaria seychellarum, 226
Impatiens, 66, 69, 72, 75, 78, 230, 231
Impatientella, 218
Incarvillea, 67
India, 136, 227
Indian Region, 227
Indigofera, 66, 206, 233
  pulchella, 231
Indobanalia, 227
Indochina, 40, 234
Indochinese Province, North, 238
Indochinese Province, South, 238
Indo—Chinese Region, 234
Indochloa, 229
  oligocantha, 230
Indokingia, 219, 225
Indomalesian Subkingdom, 226
Indomalesian—New Caledonian relationships, 250
Indoneesiella, 228
Indopoa, 229
Indorouchera, 335
Indosinia, 234
Indostan Peninsula, 221
Indotristicha, 227
  ramosissima, 230
Inga cynometroides, 230
  edulis, 260
Interior drainage basins, 166
Intsia bijuga, 226
Inula, 67
  discoidea, 145
  macrocephala, 145
  magnifica, 24
  nervosa, 76

Ionopsidium, 121
Iostephane, 164
Iphigenia, 348
Iphiona horrida, 210
Ipomoea, 206
  pes-caprae, 58, 225, 226, 233
  tree, 185
  yunnanensis, 67
Ipsea, 229
Iran, 133, 135, 136, 144, 149, 210
Irania, 137
Iranian Plateau, 135, 142
Irano—Turanian Region, 135
Iraq, 133, 136
Iridaceae, 10, 264, 348
Iriomote Island, 62
Iris, 22, 68, 70, 86, 111, 142
  bakeri, 78
  barnumae, 147
  bracteata, 103
  clarkei, 73
  dichotoma, 52
  fosterana, 148
  gracilipes, 56
  henryi, 64
  hexagona, 93
  innominata, 103
  kobayashii, 63
  koreana, 51
  latifolia, 17
  lazica, 25
  mandshurica, 51
  minutiaurea, 51
  nepalensis, 76
  oxypetala, 158
  pseudocaucasica, 147
  thompsonii, 103
  tripetala, 93
  typhifolia, 51
  wattii, 76, 78
Irvingbaileya, 270, 271
Irvingiaceae, 333
Isachne debilis, 61
Isatis, 142, 144
  erzurumica, 145
  ornithorhynchus, 146
  pachycarpa, 149
Ischaemum ischaemoides, 58
Ischnocarpus, 294
Ischnogyne, 47
Iskandera, 137

"Islands of Linden," 31
Isoberlinia, 203, 204
  doka, 205
Isodendrion, 247
Isoëtaceae, 305
Isoëtes, 305
  howellii, 175
  japonica, 53
  lacustris, 110
  malinverniana, 18
  melanospora, 90
  taiwanensis, 60
  tegetiformans, 90
Isomeris, 160
Isometrum, 45
Isonandra lanceolata, 230
Isophysis, 270, 272
Isoplexis, 113
  sceptrum, 115
Isopogon, 269, 273
Isopyrum adiantifolium, 69
  arisanense, 60
  biternatum, 99
  thalictroides, 18
Isotria, 82
Istranca Range, 24
Italy, 18, 19, 121, 129, 131
Itea, 339
  chinensis, 75
  ilicifolia, 64
  japonica, 54
  macrophylla, 75
  parviflora, 61
Iteaceae, 339
Iturup Island, 53
Ivesia rhypara, 169
Ivodea, 218
Ixanthus, 112
  viscosus, 117
Ixerba, 294, 300, 339
Ixeris, 56
  ameristophylla, 58
  grandicolla, 58
  linguaefolia, 58
  longirostrata, 58
  stebbinsiana, 67
Ixia, 264
Ixianthes, 264
Ixioliriaceae, 9, 351
Ixiolirion, 142, 351
Ixonanthaceae, 335

Ixonanthes, 335
Ixoplexis canariensis, 117
Ixora henryi, 67
  kingdon-wardii, 70
  yunnanensis, 67

Jaintia Hills, 74, 78
Jaliscoa, 164
Jamesianthus, 82
Jankaea, 11, 21
  heldreichii, 22
Japan, 39, 53
Japanese-Korean Province, 53
Japonolirion, 46, 53
Jasione bulgarica, 22
Jasminocereus, 257
Jasminum, 67
  amplexicaule, 76
  azoricum, 115
  dispermum, 76
  dumicola, 78
  farreri, 70
  fruticans, 124
  giraldii, 64
  hemsleyi, 61
  lanceolarium, 76
  nepalense, 76
  odoratissimum, 115
Jatropha, 181, 206
Jaubertia aucheri, 210, 211
  calycoptera, 211
  hymenostephana, 210
  szovitsii, 146
Java, 239
Jeffersonia diphylla, 85
  dubia, 50
Jeffrey, C., 226
Jepsonia, 161
  parryi, 173
Jerdonia, 228, 343
  indica, 230
Jodocephalus, 236
Johnsonia, 273
Johrenia, 139
Joinvillea, 355
Joinvilleaceae, 355
Jovellana, 277
Juan Fernandez Islands, 277, 280
Juania, 280, 281
  australis, 282
Jubaea, 286

chilensis, 287
Jubaeopsis, 201
Juglandaceae, 10, 321
Juglandales, 321
Juglans, 12
  ailanthifolia, 54
  californica, 176
  cathayensis, 63
  cinerea, 85
  kamaonia, 154
  regia, 22, 75, 153, 154
  mandshurica, 50, 51
Julianiaceae, 334
Julbernardia, 203, 204
Julostylis, 227
Jumelleanthus, 218
Juncaceae, 10, 354
Juncaginaceae, 346
Juncaginales, 346
Juncales, 354
Juncus, 36, 56, 64, 68, 73, 76, 279, 354
  biglumis, 110
  capitatus, 208
  georgianus, 90
  inconspicuus, 290
  mertensianus, 104
  requienii, 130
  scheuchzerioides, 278, 290
  uncialis, 176
John Day Valley, 108
Juniper-sagebrush community, 167
Juniperus, 101
  brevifolia, 114
  cedrus, 116
  changii, 65
  communis, 33, 36
  coxii, 70
  excelsa, 149
  macropoda, 210
  osteosperma, 167
  oxycedrus, 124
  phoenicea, 124
  procera, 207, 208
  pseudosabina, 158
  scopulorum, 98, 108
  silicicola, 93
  squamata, 158
  utahensis, 167
  virginiana, 90
Jurinea, 142
  antoninae, 148

  antonowii, 148
  armeniaca, 147
  pulchella, 147
  sintenisii, 148
  tzar-ferdinandii, 22
Justicia, 67
  khasiana, 78
  latiflora, 64

Kabulia, 136
Kadsura, 66, 313
  interior, 69
  longipedunculata, 63
  roxburghiana, 74
Kajewskiella, 243
Kalahari, 213
Kalakia, 139, 146
Kalanchoë gracilis, 61
  robusta, 209
  rosea, 77
Kalidiopsis, 136, 143
Kalidium, 136
  gracile, 157
Kalimantan Province, 241
Kaliphora, 219, 339
Kalmia polifolia, 36
Kalmiopsis, 100
  leachiana, 103
Kalopanax, 44
  septemlobus, 51
Kamchatka, 33, 34
Kanahia, 207
Kandelia candel, 59
Kaniaceae, 332
Kanjarum, 228
Kaokochloa, 212
Kara Kum, 136
Karakorum, 136
Karanataka-Kerala Subprovince, 231
Karelian Isthmus, 28
Karelinia, 141
Karoo-Namib Region, 211
Karoo Province, 214
Karoo vegetation, 266
Karvandarina, 141
Karwinskia humboldtiana, 182
Kaschgaria, 141
Kashgaria, 135, 137
Kashmir, 44, 135, 154
Kaufmannia, 138
Kazakhstan, 136

Keiskea, 46
japonica, 56
Kelleronia, 207
Kelseya, 100
uniflora, 106
Kemp, E. M., 269
Kendrickia, 228
Keniochloa, 207, 208
Kentiopsis, 250
Keracia, 122
Keraudrenia, 221, 269
Kerguelen Islands, 277, 291, 293
Kerguelenian Province, 293
Kermadec Islands, 295, 299
Kermadecia, 250
Kerman Mountains, 148
Kermedecian Province, 299
Kerria, 42
Kerriochloa, 236
Keteleeria, 65
fortunei, 63
Keyserlingia koreensis, 50
Khabarovsk Krai, 32, 47
Kharkov, 28
Khasi Hills, 40, 41, 74, 77
Khasi-Manipur Province, 73
Khaya comorensis, 224
senegalensis, 205
Khidaka, 55
Khorasan, 136
Khorassan Subprovince, 147
Kickxia, 117
Kielmeyeroideae, 322
Kigelianthe, 219
Kimberley district, 270
Kingdoms, floristic, 2
Kingdonia, 40
Kingdoniaceae, 317
Kingia, 273
Kingiaceae, 351
Kinugasa, 47, 53, 352
japonica, 56
Kirengeshoma, 44, 339
koreana, 54
Kirengeshomaceae, 39, 339
Kirgisian Alatau, 139
Kirilowia, 136
Kirkia, 334
Kirkiaceae, 202, 334
Kirkianella, 295
Kirkophytum, 294

Kitaibela, 11
Klamath region, 103
Kleinia scottii, 209
Klossia, 241
Kmeria, 234
Knautia, 19
tatarica, 27
Knema attenuata, 230
yunnanensis, 66
Knightia, 251, 296
Kniphofia foliosa, 208
Knowltonia capensis, 265
Kobresia, 73
capilliformis, 156
hyperborea, 15
myosuroides, 110, 158
simpliciuscula, 106
stenocarpa, 158
Kobresia meadows, 156
Kochia prostrata, 157
Kodar Range, 32
Kodiak Island, 102
Koeberlinia, 160, 327
spinosa, 288
Koeberliniaceae, 327
Koeleria, 20, 174, 203
atroviolacea, 31
balansae, 126
capensis, 208
geniculata, 31
Koelreuteria bipinnata, 66
henryi, 61
Koenigia islandica, 110
Kokia, 247
Kolkwitzia, 44, 63
amabilis, 64
Komarovia, 139
Konkan Subprovince, 231
Konya Ovasi Plain, 143
Kopet Dagh, 136, 138, 144, 147
Korea, 39, 50, 53
Korovinia, 139
Korshinskia, 139
Korthalsella disticha, 298
Kosopoljanskia, 139
Kostermansia, 241
Krameria, 337
Krameriaceae, 337
Krasnodar Krai, 23
Krasnovia, 139
Kremeria grandis, 126

INDEX

multicaulis, 126
Kremeriella, 121, 125
  cordylocarpus, 126
Kreysigia, 271
Krigia, 82
  montana, 89
Kughitangia, 136
Kumaun, 43, 45
Kunamannia, 122
Kunashir Island, 41
Kunkeliella, 112
  canariensis, 117
  psilotoclada, 117
Kurdistan, 139
Kurdo-Zagrosian Subprovince, 148, 149
Kurile Islands, 34, 41, 52
Kurram, 154
Kurtzamra, 285
Kutaisi, 23
Kuznetsk Alatau, 31
Kvarner Archipelago, 131
Kyushu Island, 55
Kyzyl Kum, 136

Labiatae, 344
Labordia, 247
Labramia, 217
Laburnum platycarpum, 124
Lacaitaea khasiana, 78
Laccodiscus, 198
Lachanodes arborea, 215
Lachenalia, 264
Lachnanthes, 353
Lachniphyllum, 141
Lachnocapsa, 209
Lachnocaulon, 93
Lachnoloma, 137
  lehmannii, 150
Lachnostylis, 264
Lacistema, 325
Lacistemataceae, 254, 325
Lactoridaceae, 280, 314
Lactoridales, 314
Lactoris, 280, 314
  fernandeziana, 281, 282
Lactuca, 67, 117
  auriculata, 210
  azerbaijanica, 147
  elata, 64
  gracilipetiolata, 70
  longidentata, 130

  patersonii, 115
  triflora, 64
  yemensis, 208
Lactucosonchus, 113
  webbii, 117
Lafuentia, 122, 125
Lagenandra, 229
Lagenantha, 207
Lagenophora lanata, 300
Lagerstroemia fauriei, 55
  hypoleuca, 237
  parviflora, 232
Lagochilus, 140
Lagophylla, 164
Lagotis clarkei, 72
  hultenii, 15
  yunnanensis, 67
Lagousia, 345
Lake Baikal, 11
Lake Superior, 38, 100
Lake Van, 145
Lamarckia, 123
Lamellisepalum, 207
Lamiaceae, 10, 26, 30, 344
Lamiales, 344
Lamiidae, 341
Lamium ambiguum, 56
  corsicum, 129
  humile, 56
Lamprachaenium, 228
  microcephalum, 230
Lamyropappus, 141
Lamyropsis microcephala, 130
Lanaria, 264, 348, 353
Lancea tibetica, 67
Laos, 40, 235
Lapageria, 285, 352
Lapiedra, 123
Laportea canadensis, 85, 187
  macrostachya, 54
Lardizabala, 283, 316
  biternata, 286
Lardizabalaceae, 316
Laretia, 284
Larix, 12, 101
  dahurica, 32, 52
  gmelinii, 32, 36, 51
  griffithiana, 71
  laricina, 35, 36, 107
  leptolepis, 54
  olgensis, 50

potaninii, 69
 sibirica, 29, 30, 31, 32
Larrea, 179, 181
 divaricata, 179, 288
 tridentata, 179, 183
Larrea-Ambrosia community, 170
Larsenia, 236, 238
La Sal Mountains, 166, 168
Laserpitium nitidum, 19
 peucedanoides, 19
Lasianthus, 61, 67
 andamanicus, 237
 satsumensis, 55
 wardii, 70
Lasiocaryum, 140
Lasiochlamys, 249
Lasiurus, 211
Lasthenia burkei, 176
 chrysantha, 176
 fremontii, 176
 glaberrima, 174, 176
 kunthii, 174
 platycarpha, 176
Latace, 285
Latania, 220, 224
 lontaroides, 225
 verschaffeltii, 225
Lathraea japonica, 55
 rhodopea, 22
Lathyrus frolovii, 30
 karsianus, 145
 krylovii, 30
 litvinovii, 27
 odoratus, 129
 pancicii, 21
 sylvestris, 27
 venosus, 99
 vernus, 32
Latouchea, 45, 65
Launaea cervicornis, 128
 melanostigma, 119
 picridioides, 119
 spinosa, 118, 124
Lauraceae, 9, 314
Laurales, 313
Laurasia, 94, 240
Laurelia, 277, 314
 aromatica, 286
Laurentia bicolor, 126
 canariensis, 117
Laurisilvae, 114

Laurocerasus officinalis, 25
Laurophyllus, 264
Laurus azorica, 114, 116
 nobilis, 131
Lautembergia, 218, 220
Lavandula canariensis, 117
 dentata, 124
 maroccana, 124
 minutolii, 117
 multifida, 124
 rotundifolia, 119
Lavatera assurgentiflora, 173
 phoenicea, 117
 stenopetala, 126
 triloba, 127
Lavioxia, 250
Layia, 173
 chrysanthemoides, 176
 discoidea, 177
Lazistan, 23, 25
Leavenworthia, 79
 stylosa, 91
Lebanon, 123, 137
Lebetanthus, 283
Lebronnecia, 245
Lecanorchis, 56
Lecardia, 249
Lecocarpus, 257
Lecomptedoxa, 198
Lecomptella, 220
Lecythidaceae, 198, 323
Lecythidales, 323
Ledebouriella divaricata, 52
Ledocarpaceae, 261, 336
Leea, 338
Leeaceae, 227, 338
Legenere limosa, 174, 176
 valdiviana, 174
Leguminosae, 333
Leiophyllum, 79
 buxifolium, 85
Leiospora, 137
Leitneria, 82, 334
 floridana, 93, 94
Leitneriaceae, 79, 92, 334
Leitneriales, 334
Lemaireocereus, 181
Lemmonia, 162
Lemnaceae, 356
Lemurella, 220
Lemuriosicyos, 217

Lemuropisum, 218
Lennoa, 343
Lennoaceae, 343
Lenophyllum, 161
Lentibulariaceae, 344
Lenzia, 283
Leonticaceae, 317
Leontochir, 285
Leontodon djurdjurae, 126
  schischkinii, 20
  siculus, 130
Leontopodium, 56, 67
  kamtchaticum, 34
  kurilense, 53
  ochroleucum, 31
  palibinianum, 51
  sinense, 64
Lepechinia, 173
Lepechiniella, 139
Lepidium lyratum, 146
  nitidum, 174
  rigidum, 126
Lepidobotryaceae, 336
Lepidobotrys, 198
Lepidolopha, 141
Lepidolopsis, 141
Lepidophorum, 122
Lepidorrhachis, 295, 297
Lepidospartum, 164
Lepidostemon, 42
Lepidozamia, 270
Lepilaena, 347
Lepironia, 221
Leptacanthus, 228
Leptadenia pyrotechnica, 211
Leptagrostis, 207
Leptarrhena pyrolifolia, 105
Leptocodon, 46
Leptodactylon californicum, 173
Leptodermis, 67
  oblonga, 63
  pulchella, 55
Leptolaena, 218
Leptolepia, 309
Leptomischus, 235
Leptonychiopsis, 241
Leptopteris superba, 295
Leptorhabdos, 140
Leptospermum, 268, 273
Leptostylis, 249
Leptunis, 139

Lepuropetalaceae, 331
Lepuropetalon, 331
Lereschia thomasii, 129
Leslie Gulch, 169
Lesotho, 204
Lespedeza, 66
  bicolor, 52
  davurica, 52
  homoloba, 55
  juncea, 52
  leptostachya, 96
  maximowiczii, 55
  pubescens, 61
Lesquerella angustifolia, 96
  arenosa, 96
  auriculata, 96
  calcicola, 96
  engelmannii, 96
  fremontii, 109
  macrocarpa, 109
  ovalifolia, 96
  perforata, 91
  recurvata, 96
  sessilis, 96
  stonensis, 91
Lesser Antilles, 256
Lessingia, 164
Letestua, 198
Letestudoxa, 198
Letestuella, 198
Leucadendron, 264
  argenteum, 264
Leucanthemum corsicum, 130
  discoideum, 20
Leucheria, 285
  suaveolens, 290
Leucocodon, 228
Leucocoryne, 285, 286, 349
Leucocrinum, 350
Leucogenes, 295
Leucojum longifolium, 130
  roseum, 130
Leucophrys, 212
Leucophyllum, 162
Leucosceptrum, 46, 56
  plectranthoideum, 67
  sinense, 64
Leucospermum, 265
Leucosphaera, 212
  bainesii, 213
Leucospora, 82

multifida, 86
Leucostegane, 241
Leucothoe fontanesianum, 85
  tonkinensis, 66
Leunisia, 285
Leuzea, 122
Leycesteria, 44, 70
  formosa, 76
  insignis, 70
Leymus chinensis, 32
Li, H. L., 60
Liatris, 82, 93
  chapmanii, 95
  ohlingerae, 95
  provincialis, 95
  turgida, 88
Libanotis sibirica, 27
Liberia, 199
Libocedrus, 101, 251, 288, 295
  plumosa, 300
Libya, 133
Libyella, 123
Lifago, 134
Lignariella, 42
Lignocarpa, 294
Ligularia, 51, 56, 67
  kojimae, 61
  pachycarpa, 73
  sinica, 63
  splendens, 51
Liguria, 129
Liguro—Tyrrhenian Province, 128
Ligusticum, 67
  caucasicum, 24
  corsicum, 129
  physospermifolium, 24
  tsusimense, 55
Ligustrum, 61, 64, 67
  confusum, 76
  micranthum, 57
  myrsinites, 78
  salicinum, 55
  suave, 62
  travancoricum, 230
  walkeri, 230
Lilaeaceae, 347
Lilaeopsis macloviana, 290
Liliaceae, 9, 349
Liliales, 348
Liliidae, 348
Liliopsida, 346

Lilium, 56, 64, 68
  bakerianum, 70
  cernuum, 51
  formosanum, 61
  giganteum, 70
  grayi, 86
  jankae, 22
  ledebourii, 151
  mackliniae, 78
  monadelphum, 26
  pyrenaicum, 17
  rhodopaeum, 22
  sherriffiae, 73
  szovitsianum, 25
Limnanthaceae, 9, 172, 336
Limnanthales, 336
Limnanthes, 336
  douglasii, 176
Limnocharis, 346
Limnocharitaceae, 346
Limnopoa, 229
Limnosciadium, 82
Limoniastrum, 121
Limonium, 111, 125, 129, 142
  anatolicum, 143
  anfractum, 131
  arabicum, 210
  arborescens, 117
  asterotrichum, 21
  biflorum, 128
  braunii, 118
  brunneri, 118
  californicum, 174
  caprariense, 128
  franchetii, 62
  globuliferum, 143
  guaicuru, 174
  japygicum, 131
  majoricum, 128
  recurvum, 17
Limosella africana, 207
Linaceae, 335
Linales, 335
Linanthus pusillus, 174
  pygmaeus, 174
Linaria, 126
  arenaria, 17
  biebersteinii, 27
  brunneri, 118
  capraria, 129
  faucicola, 17

loeselii, 19
macroura, 27
microsepala, 131
thymifolia, 17
ventricosa, 124
yunnanensis, 67
Lindauea, 207
Lindelofia kandavanensis, 151
Lindenbergia bhutanica, 72
Lindera, 63
  akoensis, 60
  cercidifolia, 69
  heterophylla, 71
  latifolia, 74
  nacusua, 74
  praecox, 54
  pulcherrima, 74
  sericea, 54
  tonkinensis, 66
  venosa, 71
  vernayana, 69
  wardii, 70
Lindleya, 161
Lindsaea repanda, 57
Lindsaeaceae, 309
Linnaea borealis, 36
Linociera, 219
  henryi, 67
  malabarica, 230, 233
  parkinsonii, 237
Linostoma andamanica, 237
Linum, 111
  adenophyllum, 177
  amurense, 50
  bicarpellatum, 177
  californicum, 177
  clevelandii, 177
  violascens, 30
  virginianum, 85
Lipari Islands, 128
Liparis, 56
  hostaefolia, 58
  perpusilla, 73
  pygmaea, 73
Lipochaeta, 247
Lipskya, 139
Lipskyella, 141
Liquidambar styraciflua, 187
Liriodendron, 86
  chinense, 63, 83
  tulipifera, 83, 86

Liriope, 47
Lisaea, 139
Lissocarpa, 324
Lissocarpaceae, 254, 324
Listera borealis, 37
  cordata, 36
  makinoana, 56
Lithocarpus, 60, 66, 236
  densiflora, 105, 176
Lithophragma bulbifera, 98
Lithophytum, 342, 343
Lithospermum minimum, 129
Lithostegia, 39
Lithraea caustica, 287
Litsea, 60, 63, 155
  brachypoda, 70
  calicaris, 300
  cubeba, 74
  cuttingiana, 70
  elongata, 74
  kingii, 74
  nigrescens, 230
  oblonga, 74
  sericea, 74
Littledalea, 142
Littorella, 344
"Living fossils," center of, 48
Livistona chinensis, 58
  mariae, 275
Llanos Province, 259
Loasaceae, 342
Loasales, 342
Lobelia, 67, 215, 286
  boninensis, 58
  nubigena, 72
  pyramidalis, 70, 76
  rhynchopetalum, 207
Lobeliaceae, 345
Lobostemon, 267
Lochia, 209
Lodoiceae, 220, 225
  maldivica, 226
Loewia, 207
Loganiaceae, 341
Loiseleuria procumbens, 38
Lolium lowei, 115
Lomandra, 351
  banksii, 351
  multiflora, 351
Lomandraceae, 351
Lomariopsidaceae, 309

Lomariopsis boninensis, 57
Lomatia, 286
Lomatium, 101, 169
  howellii, 177
  rollinsii, 109
  serpentinum, 109
  tuberosum, 108
Lomatophyllum, 219, 220
Lonas, 122
Lonchitis pubescens, 265
Lonchocarpus, 260
Lonchophora, 133
Londesia, 136
Lonicera, 12, 64, 67, 70, 156, 240
  aucheri, 210
  biflora, 124
  caerulea, 36
  chrysantha, 52
  etrusca, 132
  glabrata, 76
  kabylica, 126
  kawakamii, 61
  kungeana, 62
  ligustrina, 76
  macrantha, 76
  maximowiczii, 51
  myrtilloides, 72
  oiwakensis, 61
  purpurascens, 155
  pyrenaica, 127, 128
  spinosa, 158
  vidalii, 55
  villosa, 36
Lophiocarpus, 318
Lophiola, 82, 353
Lophira, 323
Lophiraceae, 323
Lopholepis, 229
Lophomyrtus, 294
Lophopappus, 285
Lophophora, 160
Lophopyxidaceae, 239, 337
Lophopyxis, 239, 337
Lophoschoenus, 222
Lophosoria, 309
Lophosoriaceae, 309
Loranthaceae, 338
Loranthus odoratus, 76
  tanakae, 55
Lord Howe Island, 252, 294, 297
Lordhowea, 295, 297

Lord Howean Province, 297
Loropetalum, 41
  chinense, 74
Lotus, 117, 126, 172
  helleri, 90
  jacobaeus, 118
  loweanus, 115
  macranthus, 115
  purpureus, 118
  subpinnatus, 174
  tetraphyllus, 128
Louvelia, 220
Lowiaceae, 227, 354
Loxocalyx, 46
  urticifolius, 64, 67
Loxococcus, 229
Loxodiscus, 249
Loxogrammaceae, 308
Loxogramme biformis, 60
  boninensis, 57
  toyoshimae, 57
Loxostemon, 42
  delavayi, 66
Loxostigma, 45
  griffithii, 76
Loxostylis, 201
Loxsoma, 294, 300, 308
Loxsomaceae, 308
Loxsomales, 308
Loxsomopsis, 308
Lozania, 325
Luculia, 44, 67
  grandifolia, 72
  intermedia, 70
  pineana, 76
Ludia, 200
Ludwigia, 55
Luehea divaricata, 260
Luetkea, 100
Luina, 100
Luisia boninensis, 58
  brachycarpa, 58
  teres, 56
Lumnitzera racemosa, 233
Lupinus, 101, 173, 175
  spectabilis, 177
Luronium, 11
Luxemburgiaceae, 323
Luzon, 43
Luzula, 36, 70, 76, 279, 282, 296, 354
  abyssinica, 208

INDEX

beringensis, 15
canariensis, 118
chinensis, 64
johnstonii, 208
tundricola, 15
Luzuriaga, 278, 352
parviflora, 352
Luzuriagaceae, 352
Lyallia, 291, 318
kerguelensis, 293
Lyauteya, 121, 125
ahmedi, 126
Lycapsus, 280, 282
Lychnis, 54
flos-jovis, 19
nivalis, 19
yunnanensis, 66
Lycianthes boninensis, 58
Lycium barbarum, 211
griseolum, 58
intricatum, 124
Lycocarpus, 121, 126
Lycochloa, 123, 131
Lycopodiaceae, 305
Lycopodium, 305
cernuum, 298
contiguum, 292
diaphanum, 292
magellanicum, 290, 293
saururus, 293
subuliferum, 71
Lycoris, 47
squamigera, 56
Lygeum, 121, 123
spartum, 125
Lyginia, 273
Lygodesmia aphylla, 95
Lygodiaceae, 307
Lyonia, 66
ferruginea, 93
ovalifolia, 69
Lyonothamnus, 161
floribundus, 173
Lyperia, 117
Lyrocarpa, 161
Lyrolepis, 122, 131
Lysidice, 43
Lysimachia, 54, 64, 66, 69, 75, 85
acroadenia, 54
fragrans, 61
leucantha, 54

minoricensis, 128
nemorum, 18
rubida, 57
Lysionotus, 67
ophiorrhizoides, 65
serrata, 70, 76
Lythanthus amygdalifolius, 119
Lythraceae, 332
Lythrum ovatum, 99

Maackia, 43
amurensis, 51
chinensis, 64
Macadamia, 221, 250
Macaranga andamanica, 237
indica, 231
kilimandscharica, 207
Macarisia, 218
Macaronesia, 111
Macaronesian Region, 112
Macbridea, 82
Macdonell Ranges, 275
Macedonia, 21, 22
Macgregoria, 337
Machaeranthera glabriuscula, 109
gypsophila, 180
restiformis, 180
Machaerium, 260
Machaerocarpus californicus, 176
Machilus, 63, 155
boninensis, 57
edulis, 71
gammieana, 71
kobu, 57
odoratissima, 69
pseudokobu, 57
yunnanensis, 66
Mackaya, 201
Mackeea, 250
Mackenziea, 228
Macleaya, 40
microcarpa, 62
Maclura, 79
pomifera, 85
Macphersonia, 200
Macquarie Island, 277, 294, 302
Macrachaenium, 285, 289
Macraea, 257
Macrochlaena, 44
Macropanax undulatus, 76
Macropiper, 296

Macrostelia, 218
Macrosyringion, 122
Macrozamia macdonellii, 275
  riedlei, 273
Madagascan Region, 216
Madagascan Subkingdom, 216
Madagascar, 220, 222
Maddenia, 42
  himalaica, 72
  yunnanensis, 66
Madeira, 112, 115, 117
Madeiran Province, 115
Madia hallii, 177
  sativa, 174
  yosemitana, 104
Madiinae, 172
Madrean Region, 159
Madrean Subkingdom, 158
Madrone, 102, 176
Madro-Tertiary flora, 172
Maerua, 206
  welwitschii, 213
Maesa, 75
  andamanica, 237
  lanceolata, 207
  marianae, 70
Magallana, 284, 336
Magellanian Province, 283, 284, 289
Magnolia acuminata, 85
  campbellii, 69, 74
  cylindrica, 63
  dawsoniana, 65
  dealbata, 187
  delavayi, 65
  fraseri, 85
  globosa, 69
  grandiflora, 93
  griffithii, 69
  henryi, 65
  kachirachirai, 60
  kobus, 54
  nitida, 70
  pyramidata, 93
  rostrata, 69
  salicifolia, 54
  schiedeana, 187
  shangpaensis, 65
  sieboldii, 54
  sinensis, 65
  sprengeri, 63
  stellata, 54
  virginiana, 92
  wilsonii, 65
  zenii, 63
Magnoliaceae, 9, 312
Magnoliales, 312
Magnoliidae, 312
Magnoliopsida, 312
Maguire, B., 257
Magydaris, 122
Maharashtra, 228
Mahonia, 66
  aristata, 70
  griffithii, 71
  hicksii, 71
  leschenaultii, 231
  lomariifolia, 69
  magnifica, 77
  manipurensis, 77
  monyulensis, 71
  oiwakensis, 60
  pycnophylla, 77
  sikkimensis, 71, 74
  simonsii, 77
Mahya, 219, 224
Maianthemum canadense, 36, 88, 99
  dilatatum, 105
Maihuenia, 283
Maillardia, 218, 220
Maingaya, 241
Maireana, 269, 275
Malabaila lasiocarpa, 145
Malabar Province, 230
Malacca, 239
Malacocarpus, 138, 334
Malawi, 203
Malaxis boninensis, 58
Malay Province, 241
Malcolmia arenaria, 126
  illyrica, 21
Malesherbia, 283, 326
Malesherbiaceae, 283, 326
Malesian Region, 238
Malesian Subregion, 241
Malleastrum, 218
Mallotus andamanicus, 237
Malococera, 274
Malperia, 164
Malpighiaceae, 336
Malta, 120, 129
Malus, 12, 55
Malva corsica, 129

INDEX   469

Malvaceae, 10, 329
Malvales, 328
Malvastrum, 286
Mamillopsis, 160
Mammillaria, 181
Manchurian Province, 48
Mandragora turkomanica, 148
Manekia, 314
Mangifera andamanica, 237
Manglietia caveana, 74
  chingii, 65
  forrestii, 65
  grandis, 65
  insignis, 69, 74
  megaphylla, 65
  rufibarbata, 65
  tenuipes, 65
  wangii, 65
  yunnanensis, 65
Mangrove, 59, 225, 233
Manipur, 44, 73, 74, 77
Mantisalca, 122
  delestrei, 126
Maps
  Map 1. Floristic Regions of the World Frontispiece
  Map 2. Floristic Provinces of Europe, Asia Minor, and the Caucasus, 13
  Map 3. Floristic Provinces of the Eastern Asiatic Region, 49
  Map 4. Floristic Regions and Provinces of North America, 80
Maquis, 5, 125, 159, 252, 262
Marantaceae, 354
Marattia boninensis, 57
  fraxinea, 265
  tuyamae, 57
Marattiaceae, 306
Marattiales, 306
Marcetella, 112
Marcgraviaceae, 254, 322
Mardin, 138
Maresia, 135
Margaranthus, 162
Margyricarpus, 288
Mariana Islands, 245
Mariupol-Berdyansk, 28
Marojejya, 220
Marquesas Islands, 245
Marquesia, 258, 328
Marrubium alyssoides, 126

  friwaldskyanum, 22
  parviflorum, 147
  persicum, 147
Marshallia, 82
  caespitosa, 85
  grandiflora, 86, 88
  ramosa, 85
Marshalljohnstonia, 164
Marsileaceae, 307
Marsileales, 307
Marsippospermum, 278
Marticorenia, 285
Martyniaceae, 344
Mascarena, 220
Mascarene Islands, 220, 224
Mascarene Province, 224
Mascarenhasia, 200
Maschalocephalus, 199, 255, 355
Masoala, 220
Mastigiosciadium, 139
Mastixia, 340
Mastixiaceae, 227, 340
Matelea edwardsensis, 99
Mathurina, 217, 224, 326
  penduliflora, 224
Matonia, 239, 308
Matoniaceae, 239, 308
Matoniales, 308
Matthiola bolleana, 117
  capoverdeana, 118
  flavida, 149
  maderensis, 115
Mattiastrum, 140
  gorganicum, 148
  gracile, 148
  turcomanicum, 148
Mauloutchia, 217, 222
Maundia, 347
Maundiaceae, 346
Mauritius, 220, 224
Maurocenia, 264
Maxwellia, 249
Mayaca, 355
Mayacaceae, 355
Maytenus canariensis, 117
  dryandri, 115
  emarginatus, 211
Mazettia salweenensis, 67
Mazus, 64, 67
  miquelii, 55
Mecodium dilatatum, 295

Mecomischus, 122
  pedunculatus, 126
Meconella californica, 173
Meconopsis, 66, 69, 71
  cambrica, 17
  racemosa, 62
  violacea, 70
Medeola, 82
Mediasia, 139
Medicago pironae, 19
  rupestris, 132
Mediterranean, 110, 124, 131, 199, 265, 272
Mediterranean climate, 171
Mediterranean Islands, 123
Mediterranean Region, 119
Medusagynaceae, 225, 322
Medusagynales, 322
Medusagyne, 217, 225, 322
  oppositifolia, 226
Medusandra, 338
Medusandraceae, 198, 338
Meehania cordata, 86
  montis-koyae, 56
Megacaryon, 11, 23
Megacodon, 45
Megadenia bardunovii, 30
  speluncarum, 50
Megalachne, 280
Megaleranthis, 40
Megastylis, 249
Megistostegium, 218
Meiomeria, 160
Melaleuca, 268
  leucadendron, 252
Melampyrum alboffianum, 24
  subalpinum, 19
Melanodendron, 215
Melanophylla, 219, 222, 339
Melanophyllaceae, 339
Melanosciadum, 44
Melanoselinum, 112
  edulis, 115
  insulare, 118
Melanoxylon brauna, 260
Melanthiaceae, 348
Melastoma tetramerum, 57, 58
Melastomataceae, 333
Meliaceae, 334
Melianthaceae, 335
Melianthus, 335
Melica, 174, 288

  canariensis, 118
  teneriffae, 118
Melicope, 297
Melicytus, 247
  latifolius, 298
  novae-zeylandiae, 297
  ramiflorus, 299
Meliosma, 55, 66, 187
  beaniana, 64
  callicarpaefolia, 61
  flexuosa, 64
  microcarpa, 231
  veitchiorum, 64
  wrightii, 231
Meliosmaceae, 335
Melitella, 128
Mellichampia, 162
Melliodendron, 41
  xylocarpum, 66
Mellissia, 215
Melodinus baueri, 298
Melosperma, 285
Melothria japonica, 54
Melville, R., 215
Memecylaceae, 333
Memecylanthus, 249
Memecylon andamanicum, 237
  eleagni, 226
  umbellatum, 233
Mendoncia, 344
Mendonciaceae, 344
Mendoravia, 218
Mengrelia, 23
Meniscogyne, 234
Menispermaceae, 316
Menispermum, 316
  canadense, 85
Menorca Island, 127
Mentha, 272
  japonica, 56
  requienii, 129
Mentzelia, 173
  packardiae, 169
Menyanthaceae, 342
Menyanthes trifoliata, 105
Menziesia, 54
  pilosa, 85
Meopsis, 122
Mercurialis corsica, 129
Merendera rhodopaea, 22
Merismostigma, 249

Merostachys, 261
Merriam, C. Hart, 166
Merrill, E. D., 239, 245
Merrillanthus, 235
Merrilliopanax, 44
Mertensia, 168
  drummondii, 15
  serrulata, 32
Meryta, 251
  angustifolia, 298
  sinclairi, 300
Merxmuellera, 204
Mesembryanthemaceae, 318
Mesembryanthemum, 212
Mesopotamian Province, 142
Mesoptera, 241
Mesquite-grassland, 182
Messerschmidia argentea, 58
  fruticosa, 117
Metalasia, 264
Metaplexis, 45
Metasequoia, 40, 63
Metastachydium, 140
Metatrophis, 245
Metaxya, 309
Metaxyaceae, 309
Meteoromyrtus, 228
  wynaadensis, 230
Metharme, 284
Metrosideros, 251, 266, 267, 296
  albiflora, 300
  boninensis, 57
  kermadecensis, 299
Mexacanthus, 162
Mexican Altiplano, 160, 184, 185
Mexican Highlands, 159, 180
Mexican Highlands Province, 183
Mexican-Appalachian affinities, 187
Mexico, 159, 255
Meyenia, 228
Mezonevron andamanicum, 237
Michauxia, 140
  laevigata, 148
Michelia chapaensis, 65
  doltsopa, 69, 74
  floribunda, 69
  manipurensis, 74
  nilgirica, 231
  punduana, 74
  velutina, 74
  wilsonii, 65

  yunnanensis, 66
Micraeschynanthus, 241
Micrantha, 137
Microbiota, 40, 50
  decussata, 50
Microcachrys, 270, 272
  tetragona, 272
Microcala quadrangularis, 174
Microcephala, 141
Microcycas, 256
Microdactylon, 162
Microdon, 264
Microgynoecium, 136
Microlaena, 240
Microlema, 212
Microlepia, 60
  yakusimensis, 53
Micromeria, 117
  elliptica, 145
  filiformis, 127, 129
  frivaldszkyana, 22
Micronesian Province, 246
Micronychia, 218
Micropeplis, 136
Microphyes, 283
Microphysa, 139
Micropus, 111
Microsciadium, 122, 131
Microseris howellii, 103
Microsisymbrium axillare, 72
Microsorium masaskei, 57
  subnormale, 57
Microsteira, 218
Microsteris gracilis, 98
Microstigma, 11, 30, 137
  deflexum, 30
Microstrobos, 270, 272
  hookerianus, 272
Microstylis, 58
Microtoena, 67
Microtropis ramiflora, 231
Microula, 67, 140
  bhutanica, 72
  tibetica, 157
Mida, 277
  salicifolia, 300
Middle Chilean Province, 286
Middle Siberian Province, 31
Miersia, 285
Miliusa tectona, 237
Millettia, 66

taiwaniana, 61
Milligania, 272, 349
Milula, 47, 71, 349
Milulaceae, 349
Mimosa hamata, 211
Mimosaceae, 333
Mimulus, 172, 173, 175
　brachiatus, 177
　eastwoodiae, 169
　stolonifer, 51
　szechuanensis, 67
　tricolor, 175
Mimusops, 233
　hexandra, 233
　petiolaris, 225
Minuartia, 19, 117
　abchasica, 23
　helmii, 27
　hondoensis, 54
　inamoena, 26
　krascheninnikovii, 27
　rhodocalyx, 23
Miombo, 204
Mirbelia, 269
Miricacalia, 46, 56
Miscanthus, 56
　boninensis, 58
Mischodon, 227
Mischogyne, 203
Mishmi, 47
Mishmi Hills, 41, 42
Misodendraceae, 283, 338
Misodendrum, 284, 338
Mississippi Embayment, 84, 91
Mitchella repens, 187
　undulata, 55
Mitella, 55
　breweri, 105
　caulescens, 105
　formosana, 61
Mitolepis, 209
Mitraria, 285
Mitrasacme, 271
Mitrastemon, 315
　yamamotoi, 74
Mitrastemonaceae, 315
Mitrephora, 271
Mitrogyna rubrostipulata, 207
Mixed-grass prairie, 97
Miyakea, 11, 34
　integrifolia, 34

Mobilabium, 271
Modestia, 141
Moehringia dielsiana, 18
　glaucovirens, 19
　markgrafii, 19
Mogollon Rim, 166
Mogoltau, 139
Mogoltavia, 139
Mohave Desert, 180, 182
Mohavea, 162
Mohavean district, 159, 160
Mohria, 307
Mohriaceae, 307
Moldenkea, 198
Moldenkeanthus, 220
Molinaea, 218, 220
Molluginaceae, 318
Moltkia doerfleri, 21
　suffruticosa, 19
Moluccan Province, 243
Monachyron, 113, 118
Monanthes, 117, 193
Monardella, 173
Monelytrum, 212
Moneses uniflora, 36
Mongolia, 42, 47, 52, 137
Mongolian Province, 156
Monimia, 217, 220
Monimiaceae, 313
Monimopetalum, 43
Monnina angustifolia, 288
Monocelastrus monosperma, 67
Monochasma, 45
　japonicum, 55
Monococcus, 317
Monocosmia, 283
Monodiella, 133, 134
Monogramma, 307
Monolopia, 164
Monoporus, 217
Monoptilon, 164
Monotes, 258, 328
　discolor, 203
Monotropa uniflora, 102
Monotropaceae, 323
Monotropastrum, 41
　humile, 69, 75
Monotropsis, 79
　odorata, 85
Monsonia commixta, 210
Monsoon climate, 154

Montagueia, 249
Montana, 106
Montecristo, 129
Montia fontana, 207, 290
Montinia, 339
Montiniaceae, 339
Montrouziera, 249
Monttea, 285
  aphylla, 289
Mooria, 250
Moraceae, 329
Moraea, 264
Moratia, 250
Moriera, 137
Morierina, 249
Morina, 341
  delavayi, 67
Morinaceae, 9, 341
Morinda boninensis, 58
Moringa, 206, 328
  concanensis, 211
Moringaceae, 328
Moringales, 328
Morisia, 121, 128
  monanthos, 129
Morkilla, 162
Morroco, 111, 120, 124, 125, 133
Mortonia, 162
Morus boninensis, 57
  tiliifolia, 54
  yunnanensis, 66
Moscharia, 285
Moschopsis, 285
Mosdenia, 203
Mosla japonica, 56
Motherwellia, 271
Mount Eddy, 103
Mount Gardez, 154
Mount Kuli-e Genou, 149
Mount Mitchell, 88
Mount Ruwenzori, 199, 200
Mount Shasta, 103
Mount Victoria, 78
Mouretia, 235
Mozambique, 203
Muantum, 235
Muehlenbeckia, 251
  australis, 298
  bryopoda, 181
  quadridentata, 186
Muilla, 164

  maritima, 173
Mukdenia, 42, 50
  acanthifolia, 50
  rossii, 50
Mulinum, 284
  spinosum, 289
Munroidendron, 247
Muntingia, 328
Munzothamnus, 164
Muraltia, 264, 267
Murbeckiella zanonii, 19
Muricaria, 133
Murraya euchrestifolia, 61
Musa, 354
Musaceae, 354
Musales, 354
Musanga, 329
Muscari alpanicum, 24
  colchicum, 24
  dolichanthum, 24
  gussonei, 130
  longipes, 147
  pendulum, 24
Musgravea, 271
Musineon lineare, 169
Mussaenda, 67
  shikokiana, 55
  taiwaniana, 61
Musschia, 113, 115
  aurea, 115
  wollastonii, 115
Myodocarpus, 249
Myonima, 219, 224
Myoporaceae, 268, 344
Myopordon, 141
Myoporum, 344
  boninense, 58
  obscurum, 298
Myoschilos, 284
Myosotidium, 294, 301
Myosotis, 70, 203, 207, 240, 296
  alpina, 17
  antarctica, 301
  capitata, 301
  corsicana, 129
  matthewsii, 300
  rehsteineri, 19
  ruscinonensis, 129
  soleirolii, 129
Myrceugenia, 281, 286
  fernandeziana, 281

schulzei, 282
Myriactis, 67
  japonensis, 56
Myrianthus, 329
Myrica, 321
  californica, 105
  farquhariana, 75
  faya, 114, 116, 126
  gale, 36
  inodora, 93
  pringlei, 187
  salicifolia, 207
  sapida, 75
Myricaceae, 321
Myricales, 321
Myricaria prostrata, 158
Myrioneuron faberi, 64
Myriophyllaceae, 333
Myriophyllum oguraense, 55
Myripnois, 46
  dioica, 63
Myristica andamanica, 237
  malabarica, 230
Myristicaceae, 313
Myrmechis chinensis, 64
Myrmecosicyos, 207
Myrocarpus frondosus, 260
Myrothamnaceae, 320
Myrothamnales, 320
Myrothamnus, 320
Myroxylon peruiferum, 260
Myrsinaceae, 325
Myrsine, 64, 296
  africana, 114
  maximowiczii, 57
  okabeana, 57
  semiserrata, 69, 75
Myrtaceae, 268, 332
Myrtales, 332
Myrtastrum, 249
Myrtopsis, 249
Myrtus communis, 131, 151, 154
  nivellei, 134
Mystropetalon, 264, 315
Mytilaria, 234
Myxopyrum serrulatum, 230

Nablonium, 272
Naga Hills, 40, 43, 74, 78
Najadaceae, 347
Najadales, 347

Najas, 56, 347
Nama canescens, 180
  carnosum, 180
  purpusii, 180
  stevensii, 180
  stewartii, 180
Namaland Province, 213
Namib Province, 213
Namibia, 212
Nananthea, 122
  perpusilla, 130
Nandina, 40, 317
Nandinaceae, 39, 317
Nannoglottis, 46
  yunnanensis, 67
Nannorrhops, 111
  ritchieana, 211
Nannoseris schimperi, 207
Nanocnide, 42
Nanodea, 284, 289
Nanophyton, 136
Nanothamnus, 228
  sericeus, 230, 231
Napaea, 79
  dioica, 85
Narcissus bicolor, 17
  cyclamineus, 17
Narduroides, 123
Nargedia, 228
Nartheciaceae, 348
Narthecium balansae, 24
  ossifragum, 17, 188
  reverchonii, 130
  scardicum, 22
Nassauvia, 289
  gaudichaudii, 290
  serpens, 290
Nastanthus falklandicus, 290
Nasturtiicarpa, 137
Nasturtiopsis, 133
Nastus, 220, 224
Nathaliella, 140
Naufraga, 122, 127
  balearica, 128
Navarretia leucocephala, 175
Nectandra membranacea, 260
Nectaropetalaceae, 335
Nectaropetalum, 335
Nectaroscordum tripedale, 147
Negria, 294, 297
Negripteridaceae, 307

Neillia, 66
  sinensis, 64
  thyrsiflora, 69, 75
  uekii, 50
Nelsoniaceae, 344
Nelumbo, 316
Nelumbonaceae, 316
Nelumbonales, 316
Nemacaulis, 160
Nemacladus, 163, 173, 345
Nemopanthus, 82, 337
Nemophila kirtleyi, 109
Nemuaron, 248, 314
Neobeguea, 218
Neocaledonian Province, 253
Neocaledonian Region, 248
Neocaledonian Subkingdom, 248
Neocallitropsis, 248
Neocentema, 206
Neochamaelea, 112, 333
  pulverulenta, 117
Neochevalierodendron, 198
Neodielsia, 43
Neodregea, 264
Neodypsis, 220
Neofinetia, 47
Neofranciella, 249
Neogoesia, 162
Neoguillauminia, 249
Neohenrya, 45
Neolitsea, 60, 155
  boninensis, 57
  lanuginosa, 74
  umbrosa, 74
Neolloydia, 160
Neoluffa, 42, 71
Neomartinella, 42
Neomolinia japonica, 56
  mandshurica, 51
Neomyrtus, 294
Neopanax kermadecense, 299
Neoparrya, 100
Neophloga, 220
Neopilea, 218
Neoporteria, 286
Neorapinia, 249
Neorites, 271
Neoroepera, 271
Neoschimpera, 219, 225
Neostapfia, 164
Neostapfiella, 220

Neostrearia, 271
Neothorelia, 234
Neotina, 218
Neotropical Kingdom, 254
  notes, 262
Neotropis, 254
Neoveitchia, 244
Neowawraea, 247
Neowormia, 217, 222, 225
Neozeylandic Region, 293
Nepal, 40, 43, 71, 136, 154
Nepenthaceae, 197, 315
Nepenthales, 315
Nepenthes, 221, 271, 315
  khasiana, 78
  pervillei, 226, 315
Nepeta, 142
  agrestis, 129
  algeriensis, 126
  assurgens, 149
  erivanensis, 147
  foliosa, 129
  kochii, 147
  koreana, 51
  manchuriensis, 51
  meyeri, 147
  pseudokoreana, 63
  subsessilis, 56
  teydea, 117
  trautvetteri, 147
Nephelea, 309
Nephelotrophe polystichoides, 66
Nephrodesmus, 249
Nephrolepidaceae, 310
Nephrolepis, 310
Nephropetalum, 161
Nephrosperma, 220, 225
Neraudia, 247
Nerisyrenia, 161
  castillonii, 180
  gracilis, 180
  incana, 180
  linearifolia, 180
Nerium mascatense, 210
  oleander, 124
Nertera, 240, 278
  assurgens, 293
  depressa, 293
  holmboei, 293
  sinensis, 64
Nervilia macroglossa, 73

nipponica, 56
scottii, 73
Nesiota, 215
Nesocaryum, 280, 282
Nesodoxa, 249
Nestegis apetala, 299
Nestlera humilis, 214
Nestronia, 82
Neuracanthus, 206
Neurachne, 274
Neurada, 331
Neuradaceae, 331
Neuradopsis, 331
Neurotropis armena, 146
szovitsiana, 146
Neuwiedia, 353
Nevada, 160
Neviusia, 82
alabamensis, 85
New Guinea, 239
New Guinea—New Caledonia relationship, 251
New Hebridean Province, 244
New Hebrides, 244
New Mexico, 162
New Zealand, 276, 294
New Zealand—New Caledonian relationship, 251
New Zealand—Norfolk Island relationship, 298
Nicolletia, 164, 281
Nidorella varia, 119
Niederleinia, 284
Nigeria, 199
Nigerian-Cameroonian Province, 199
Nikitinia, 141
Nilgiri Subprovince, 231
Nilgirianthus, 228
Nirarathamnos, 209
Nitraria, 333, 334
sphaerocarpa, 157
Nitrariaceae, 334
Nolana, 284, 343
Nolanaceae, 254, 343
Nolina, 164
interrata, 173
Nolinaceae, 349
Nomocharis, 68, 70
Nonea anchusoides, 146
hypoleia, 149
karsensis, 145

taurica, 132
Norfolk Island, 251, 295, 298
Norfolkian Province, 298
Normanbya, 271
Normandia, 249
Noronhia, 219, 220
North American Atlantic Region, 78
North American Prairies Province, 95
North Australian Province, 270
North Indochinese Province, 238
North Island, 294, 299
Northea, 217
seychellana, 226
Northeast Australian Region, 270
Northeastern Siberian Province, 33
Northern Andean Province, 262
Northern Baluchistanian Province, 153
Northern Burmese Province, 68
Northern Chilean Province, 286
Northern Chinese Province, 62
Northern European Province, 29
Northern Neozeylandic Province, 299
Nossi-Bé Island, 222
Notelaea, 114
Notes
  Australian Kingdom, 275
  Cape Kingdom, 267
  Holantarctic Kingdom, 302
  Holarctic Kingdom, 188
  Neotropical Kingdom, 262
  Paleotropical Kingdom, 253
Nothocestrum, 247
Nothochelone, 100
Nothofagus, 250, 251, 269, 286, 296
  alessandri, 287
  antarctica, 287
  cunninghamii, 272
  dombeyi, 287
  glauca, 287
  gunnii, 272
  leonii, 287
  moorei, 271
  obliqua, 287
  procera, 287
  pumilio, 287
Notholaena bryopoda, 180
Notholirion macrophyllum, 73
Nothomyrcia, 280
Nothopegia, 228
  travancorica, 230
Nothophoebe konishii, 60

Nothopodytes foetida, 230
Nothoscordum inutile, 56
Nothosmyrnium, 44
Notochaete hamosa, 70
Notodontia, 235, 238
Notopterygium, 44
  forrestii, 67
Notospartium, 294
Notothlaspi, 294
Nototrichium, 247
Nouelia, 46, 65
  insignis, 67
Nucularia, 133
Numaeacampa, 236
Nunatak theory, 37
Nuphar, 54
  shimadai, 60
  variegatum, 36
Nuristan, 154
Nuxia congesta, 207
Nuytsia, 273
Nyctaginaceae, 318
Nyctanthaceae, 342
Nymania, 212
Nymphaeaceae, 316
Nymphaeales, 316
Nyssa, 339
  aquatica, 92
  ogeche, 93
  sinensis, 64
  sylvatica, 86, 93, 187
Nyssaceae, 339

Oberonia japonica, 56
Obolaria, 82
Oceanopapaver, 249, 327
Ochagavia, 281
  elegans, 282
Ochlandra, 221
  travancorica, 230
Ochnaceae, 322
Ochnales, 322
Ochotonophila, 136
Ochradenus aucheri, 210
Ochrocarpos, 217
Ochropteris, 217, 220
Ochrosia nakaiana, 57, 58
Ochrothallus, 249
Ochthodium, 121, 131
Ochtocosmus, 335
Ocotea bullata, 207, 265

  comoriensis, 224
  foetens, 116
  obtusata, 225
Octoceras, 137
Octoknema, 198
Octoknemaceae, 337
Octomeles, 326
Octotheca, 249
Octotropis travancorica, 230
Odontelytrum, 207
Odontites, 126
  bocconei, 129
  corsica, 129
  holliana, 115
Odontochilus hatusimanus, 56
Odontospermum daltonii, 119
  vogelii, 119
Odontostomum, 164, 348
Oedibasis, 139
Oemlera, 100
Oenanthe lisae, 129
  pteridifolia, 115
Oenothera argillicola, 89
  caespitosa, 170
  fremontii, 96
  psammophila, 170
  silesiaca, 19
Oenotheridium, 284
Oenotrichia, 309
Oeonia, 220, 224
Oeoniella, 220, 224
Ofaiston, 136
Oftia, 343, 344
Ogasawara region, 56
Oianthus, 228
Oistanthera, 219, 224
Okha, 52
Okhotia, 34, 44
Okhotsk Sea, 11, 33
Okhotsk—Kamchatka Province, 33
Okinawa, 41, 58
Oklahoma, 94
Olacaceae, 337
Olax wightiana, 230
Oldenlandia, 67
  retrorsa, 210
Old World-New World tropics, relationship, 254
Olea, 67, 267
  africana, 265
  apetala, 300

aucheri, 149
capensis, 265
europaea, 124, 208, 210
ferruginea, 155
laperrinii, 134
laurifolia, 265
Oleaceae, 342
Oleales, 342
Oleandra, 309
Oleandraceae, 309
Olearia, 296, 297
  albida, 300
  chathamica, 301
  semidentata, 301
  traversii, 301
Olgaea, 141
Oliganthes, 219
Oligoceras, 235
Oligochaeta divaricata, 147
Olinia, 332
  cymosa, 265
Oliniaceae, 332
Olivaea, 164
Oliviera, 139
Olneya, 161
Olympic Mountains, 102, 103
Olympic Peninsula, 103
Olymposciadium, 122
Oman, 209
Oman Mountains, 138
Omania, 134
Omano-Sindian flora, 149
Omano-Sindian Subregion, 209
Ombrocharis, 46
Omphalodes, 55
  cappadocica, 24
  caucasica, 24
  cordata, 64
  kuznetzovii, 24
  littoralis, 17
  nitida, 17
Omphalogramma, 41, 66
  elwesiana, 72
Omphalothrix, 50
  longipes, 51
Onagraceae, 159, 332
Oncinema, 264
Oncostemum, 217, 220
Oncotheca, 248, 322
Oncothecaceae, 248, 322
Ondinea, 270

Onobrychis, 142
  buhseana, 146
  cornuta, 145
  heterophylla, 146
  subacaulis, 146
Onocleaceae, 309
Ononis, 126
  costae, 115
Onopordum algeriense, 126
  armenum, 147
  nogalasii, 117
Onosma, 19, 67, 142
  arcuatum, 145
  bhutanica, 72
  bubanii, 17
  gaubae, 146
  guberlinensis, 27
  lucana, 129
  polyphyllum, 132
  rigidum, 132
  thracica, 21
Onuris, 284
Onychosepalum, 273
Operculicarya, 218
Ophelia tscherskyi, 51
Ophiocephalus, 162
Ophiocolea, 219, 220
Ophioglossaceae, 306
Ophioglossales, 306
Ophioglossum, 280, 306
  californicum, 173
  crotalophoroides, 290
  kawamurae, 53
Ophiopogon, 56, 68, 70
  clavatus, 64
  intermedius, 76
  leptophyllus, 76
Ophiopogonaceae, 349
Ophiorrhiza fasciculata, 76
  lignosa, 70
  prostrata, 72
  steintonii, 72
  treutleri, 76
Opiliaceae, 337
Opisthiolepis, 271
Opithandra, 45
  primuloides, 55
Oplonia, 221
Oplopanax elatus, 50
  elatus, 51
  horridum, 102

japonicus, 55
Opuntia, 179, 181, 286
Opuntia, tree, 185
Orange Island refugium, 95
Orchadocarpa, 241
Orchidaceae, 10, 61, 68, 78, 217, 264, 268, 353
Orchidales, 353
Orchidantha, 354
Orchids, 223, 231, 296, 298
Orchis, 56
　canariensis, 118
　scopulorum, 115
　viridifusca, 25
Orcuttia, 164, 175
　californica, 173, 176
　pilosa, 176
　tenuis, 176
Ordos, 137
Orectanthe, 355
Oregon, 103, 159, 161, 169, 170
Oregonian Province, 101
Oreobliton, 121, 125
　thesioides, 125
Oreobolus, 278
Oreocalamus, 47
Oreocharis, 45, 67
　auricola, 64
Oresitrophe, 42, 62
　rupifraga, 62
Orias, 43
Oriciopsis, 198
Origanum, 126
　hyrcanum, 151
Orinus, 142
Orites, 273
Orixa, 43
Orkney Islands, 291
Ormenis, 122
Ormopterum, 139
Ormosciadium, 139
Ormosia formosana, 61
　travancorica, 230
　yunnanensis, 66
Ornithogalum, 111
　arcuatum, 26
　bungei, 151
　magnum, 26
　schelkownikowii, 147
　tempskyanum, 145
　visianicum, 131

Oreochorte, 44
Oreodendron, 271
Oreogenia, 140
Oreoloma, 137
Oreomyrrhis, 278
　involucrata, 61
Oreophysa, 138
Oreophyton falcatum, 207
Oreosolen, 45
Oreosphacus, 285
Oreostylidium, 294
Ornithostaphylos, 160
Orobanchaceae, 343
Orobanche, 126
　boninsimae, 58
　chironii, 129
　lucorum, 19
　yunnanensis, 67
Orochaenactis, 100
　thysanocarpha, 104
Orontium, 82
　aquaticum, 86
Orophea, 253
　hexandra, 237
Orphium, 264
Ortachne, 285
Ortegia, 121
Orthocarpus, 175
　attenuatus, 174
　campestris, 176
Orthodon, 67
Orthogynium, 217
Orthosiphon debilis, 64
　incurvus, 70, 76
Orthosphenia, 162
Orychophragmus violaceus, 62
Oryctes, 162
Oryzopsis obtusa, 64
Osbeckia leschenaultiana, 231
Osmanthus, 67
　decorus, 24
　fragrans, 70
　lanceolatus, 61
　suavis, 70, 76
Osmorhiza aristata, 31
Osmunda, 306
Osmundaceae, 306
Osmundales, 306
Ostenia, 346
Osteomeles boninensis, 57
　lanata, 57

schwerinae, 66
subrotunda, 64
Ostrearia, 271
Ostrowskia, 140
Ostrya, 12
  carpinifolia, 22
  knowltonii, 170
  liana, 62
  virginiana, 98, 187
Osyris quadrialata, 124
  wightiana, 231
Otanthera scaberrima, 61
Otocarpus, 121, 125
  virgatus, 126
Otonephelium, 228
Otospermum, 122
Otostegia, 206
  aucheri, 210
Ottelia yunnanensis, 67
Oudneya, 133
Ougeinia dalbergioides, 232
Ourisia, 261, 277, 279, 296
Ovidia, 284
Oxalidaceae, 336
Oxalis, 212, 288
  acetosella, 36, 75, 88
  gigantea, 286
  montana, 36
  oregona, 105
Oxanthera, 249
Oxera, 249
Oxydendrum, 79
Oxypappus, 164
Oxyphyllum, 285, 286
Oxyria digyna, 16, 105, 106
Oxyspora yunnanensis, 66
Oxystylidaceae, 327
Oxystylis, 161
Oxytenanthera monostigma, 230
Oxytropis aciphylla, 158
  adenophylla, 32
  ajanensis, 34
  arctica, 15
  arctobia, 15
  arnertii, 50
  bellii, 15
  erecta, 34
  gmelinii, 27
  heterotricha, 32
  hidakamontana, 53
  hippolyti, 27

hyperborea, 15
japonica, 55
jurtzevii, 30
kamtschatica, 34
kodarensis, 32
komarovii, 32
kudoana, 53
kusnetzovii, 30
megalantha, 53
owerinii, 26
oxyphylloides, 32
prenja, 21
rishiriensis, 53
sajanensis, 30
schokanbetsuensis, 53
terrae-novae, 15
tilingii, 34
trautvetteri, 34
yezoensis, 53
yunnanensis, 66
Ozark-Appalachian relationship, 84
Ozark refugium, 94

Pachycentria formosana, 61
Pachycereus, 181
Pachycladon, 294
Pachycormus, 162
Pachyctenium, 122
Pachylaena, 285
Pachynema, 270
Pachyphragma, 11, 23
  macrophyllum, 23
Pachyplectron, 249
Pachypodium, 223
Pachypterygium, 137
Pachysandra axillaris, 61
Pachystegia, 295, 300
Pachystima canbyi, 85
Pachytrophe, 218
Paederota bonarota, 19
  lutea, 19
Paedicalyx, 235, 237
Paeonia, 317
  anomala, 29
  californica, 173
  cambessedesii, 128
  czechuanica, 66
  delavayi, 66
  lactiflora, 50, 52
  lutea, 66
  macrophylla, 23

mairei, 66
mlokosewitschii, 26
potaninii, 66
wittmanniana, 23
yui, 66
yunnanensis, 66
Paeoniaceae, 9, 317
Paeoniales, 317
Pag Island, 131
Pakaraima Plateau, 257
Pakaraimaea, 328
dipterocarpaceae, 257
Pakaraimoideae, 257
Pakistan, 136
Palaeocyanus, 122, 128
crassifolia, 130
Palaeodicraeia, 218
Palagruza Island, 131
Palaquium, 271
ellipticum, 230
Paleotropical Kingdom, 197
notes, 253
Paleotropis, 197
Palestine, 121, 133, 136
Paliurus hemsleyanus, 64
Palmettos, 92
Palm Valley, 275
Palmae, 356
Palms, 222
Palo verde, 179
Palouse, 108
Pamburus, 228
Pamiro-Alai, 135, 136, 137, 140
Pamirs, 136
Pampas, 288
Pampean Province, 288
Panax ginseng, 50
japonicus, 55
pseudoginseng, 70, 76
quinquefolium, 85
trifolium, 85
Pancheria, 249
Pancratium canariensis, 117
illyricum, 130
Pandaceae, 197, 329
Pandanaceae, 356
Pandanales, 356
Pandanus boninensis, 58
farusei, 58
heterocarpus, 225
hornei, 226

Panderia, 136
Panicum spp., 62, 93, 211, 288
arbusculum, 213
Pantathera, 280
Pantelleria Island, 130
Pantropical distribution of families, 254
Paolia, 207
Papaver angustifolium, 15
anjuicum, 15
bipinnatum, 146
bracteatum, 26
burseri, 18
californicum, 173
corona-sancti-stephani, 18
dahlianum, 15
fauriei, 53
gorgoneum, 118
gorodkovii, 15
keelei, 15
maeoticum, 27
malviflorum, 125
oreophilum, 26
pulvinatum, 15
rivale, 33
sendtneri, 18
ushakovii, 15
walpolei, 15
Papaveraceae, 317
Papaverales, 317
Paphiopedilum wardii, 70
Papilionanthe, 229
Paponolirion, 53
Papualand, 239
Papuan Province, 243
Papuan Subregion, 242
Parabarium, 235
Paraberlinia, 198
Paracephaelis, 219
Paracheila, 65
Paracryphia, 248, 322
Paracryphiaceae, 248, 322
Paracryphiales, 322
Paracyclea, 60
Paradisea, 11, 351
Parahebe, 294, 296
Parajaeschkea, 45
Parajusticia, 236
Paralamium, 46, 65
gracile, 67
Paramammea, 217
Paraná Province, 260

Paranneslea, 234
Parantennaria, 271
Paraphlomis hispida, 67
  robusta, 67
Parapiptadenia rigida, 260
Parasitaxus, 248
Paratrophis smithii, 300
Paravitex, 236, 238
Parietaria debilis, 207
  filamentosa, 117
Paris, 70, 352
  incompleta, 25
  manshurica, 51
Parishella, 163, 345
Parkeriaceae, 307
Parlatoria, 137
Parnassia, 69
  alpicola, 55
  fimbriata, 107
  grandifolia, 85
  kotzebuei, 110
  mysorensis, 231
Parnassiaceae, 9, 331
Parochaetus communis, 207
Parolinia, 112, 117
Paronychia canariensis, 117
  franciscana, 174
  illecebroides, 118
Paropsiaceae, 325
Paropyrum, 135
Paroxygraphis, 40, 71
  sikkimensis, 71
Parrotia, 136
  persica, 150, 152
Parrotiopsis, 136, 150, 154
Parrya arctica, 15
  grandiflora, 30
  rydbergii, 110
Parryella, 161
Parryodes, 42
Parryopsis, 137
Parsonsia, 296
Parthenice, 164
Parthenium alpinum, 109
  auriculatum, 85, 88
  hispidum, 85
Parthenocissus, 338
  henryana, 64
  laetivirens, 64
  neilgheriensis, 231
  semicordata, 76

Parvisedum, 161
Parvotrisetum, 123
Pasithea, 351
Paspalidium tuyamae, 58
Paspalum, 93, 288
  stellatum, 261
Passaea, 121
Passiflora henryi, 66
  pinnatistipula, 287
Passifloraceae, 325
Pastinaca aurantiaca, 24
  latifolia, 129
  lucida, 128
Pastinacopsis, 139, 156
Patagonia, 277, 283
Patagonian Province, 288
Patagonula americana, 261
Patkoi Hills, 47
Patosia, 285
Patrinia angustifolia, 64
  heterophylla, 63
  triloba, 55
Paulia, 139
Paulownia, 45, 344
  duclouxii, 67
  fortunei, 67
Pauridia, 264
Pectocarya dimorpha, 174
  ferocula, 174
  peninsularis, 174
  pusilla, 174
Pedaliaceae, 344
Pedicularis, 16, 19, 55, 64, 67, 70, 72, 261, 279
  amoena, 15
  brachystachys, 31
  capitata, 106
  dasyantha, 15, 86
  furbishiae, 35
  groenlandica, 36, 104
  hirsuta, 15
  lanata, 106
  mandshurica, 51
  numidica, 126
  olympica, 24
  perottetii, 230
  pulchella, 106
  tatarinowii, 63
Pedilanthus, 181
Pedinogyne, 45
Pediomelum subacaule, 91

Peganaceae, 9, 334
Peganum, 333, 334
Pelagodoxa, 246
Pelargonium, 111, 212, 264
  acugnaticum, 292
  cotyledonis, 215
  inodorum, 297
Pelecyphora, 160
Peliosanthaceae, 349
Peliosanthes longibracteata, 70
Pellaea, 281
Pelliceria, 322
Pelliceriaceae, 254, 322
Pellionia, 54
Peltandra, 82
Peltanthera, 343
Peltaria turkmena, 147
Peltariopsis, 137
  drabicarpa, 146
  grossheimii, 146
  planisiliqua, 146
Peltiphyllum, 100
Peltoboykinia, 42, 53
  tellimoides, 54
  watanabei, 55
Peltophorum dubium, 260
Pelucha, 164
Pemba Island, 221
Penaeaceae, 264, 332
Pennantia corymbosa, 299
  endlicheri, 299
Penstemon, 86, 101, 168, 169
  acaulis, 109
  alpinus, 109
  compactus, 169
  elegantulus, 109
  hallii, 109
  harbourii, 109
  montanus, 106
  procerus, 107
  secundiflorus, 110
  triphyllus, 109
  uintahensis, 110
  virens, 110
Pentactina, 42
Pentadiplandraceae, 327
Pentamerista, 258, 322
Pentanopsis, 207
Pentapanax castanopsisicola, 61
  leschenaultii, 231
  racemosus, 76

  sinense, 67
  yunnanensis, 66
Pentaphragma, 345
Pentaphragmataceae, 227, 345
Pentaphylacaceae, 227, 322
Pentaphylax, 322
Pentaphylloides, 50
Pentapleura, 140
Pentaptilon, 273
Pentaschistis, 204
  mannii, 208
Pentastemenodiscus, 136
Pentathymelaea, 42
Penthoraceae, 330
Penthorum, 330
Pentopetia, 219
Pentzia, 212
Penza, 28
Peperomia, 281
  berteroana, 292
  boninsimensis, 57
  heyneana, 74
  makaharai, 60
  retusa, 265
Peperomiaceae, 314
Peponidium, 219, 220
Peracarpa carnosa, 70, 76
Perakanthus, 241
Peranemataceae, 309
Pereskia, 184
Perezia microsephala, 173
Pericampylus trinervatus, 60
Perichlaena, 219
Pericome, 164
Peridiscaceae, 325
Peridiscus, 325
Periestes, 219, 220
Perilimnastes, 241
Periomphale, 249
Peripentadenia, 271
Periploca aphylla, 211
  calophylla, 76
Periplocaceae, 342
Peripterygia, 249
Peripterygium, 271, 337
Perispermum, 247
Peristrophe andamanica, 237
Pernettya pumila, 290
Pernettyopsis, 241
Perovskia, 140
Perplexia, 141

microcephala, 148
Perralderia, 122, 134
Perriera, 218
Perrierastrum, 219
Perrierbambus, 220
Perrierella, 220
Perrierophytum, 218
Perrottetia arisanensis, 61
Persea, 60
  borbonia, 93
  indica, 113, 114, 116
  palustris, 93
Persoonia, 296
  toru, 300
Pertya, 56
Peru, 278, 283
Petagnia, 122, 128
  saniculifolia, 129
Petalonyx, 162
  crenatus, 180
Petasites formosanus, 61
Petchia, 228
Petelotiella, 234, 238
Peteria, 161
Petermannia, 270
Petermanniaceae, 352
Petitmenginia, 235
Petiveriaceae, 317
Petrobium, 215
Petrocodon, 45
Petrocoptis, 11, 17
Petrocosmea, 67
Petrogenia, 162
Petromarula, 122, 131
Petrosaviaceae, 348
Petteria, 21
  ramentacea, 21
Peucedanum boninense, 57
  elegans, 50
  formosanum, 61
  gallicum, 17
  lowei, 115
  paishanense, 50
  paniculatum, 129
  pauciradiatum, 146
  ramosissimum, 76
Peucephyllum, 164
Peumus, 283
  boldus, 287
Phacelanthus, 45
Phacelia, 169, 173

  dubia, 90
  fimbriata, 86
  greenei, 177
  gypsogenia, 180
  lyallii, 106
  mollis, 35
Phaenosperma, 47
Phaeonychium, 137
Phaeoptilum, 212, 318
Phagnalon arabicum, 210
  calycinum, 124
  metlesicsii, 130
  purpurascens, 117
  umbelliforme, 117
Phalacrocarpum, 122
Phalacroseris bolanderi, 104
Phalaris canariensis, 118
  maderensis, 115
Phanerodiscus, 219
Phanerogonocarpus, 217
Phanerophlebiopsis, 39
Phanerosorus, 308
Phaulothamnus, 159, 318
Phebalium nudum, 300
Phellinaceae, 248, 337
Phelline, 249, 337
Phellodendron, 43, 60
  amurense, 50
  chinense, 64
Phellolophium, 219
Phenakospermum, 222, 354
Phenax, 221
Philadelphaceae, 339
Philadelphus, 66, 85
  lewisii, 105
  satsumi, 54
  tenuifolius, 51
Philip Island, 294, 295, 298
Philesia, 285, 352
Philesiaceae, 352
Philgamia, 218
Philippia, 207, 223
  comorensis, 224
  keniensis, 208
  mannii, 199
  montana, 225
Philippiella, 283
Philippinean Province, 241
Philippines, 239, 241
Phillyrea angustifolia, 119
  latifolia, 119

media, 124
Philydraceae, 353
Philydrales, 353
Phippsia algida, 110
Phlebocarya, 273
Phlebophyllum, 228
Phleum alpinum, 105, 290
Phloga, 220
Phlomis, 67, 142
  albiflora, 64
  bovei, 126
  breviflora, 70
  cabbaleroi, 126
  ferruginea, 129
  gracilis, 64
  italica, 128
Phlox, 168
  aculeata, 170
  andicola, 96
  buckleyi, 89
  colubrina, 109
  oklahomensis, 96
  viscida, 109
Phoebanthus, 82
  grandiflorus, 95
  tenuifolius, 95
Phoebe, 187, 260
  formosana, 60
  sheareri, 62
Phoenicanthus, 253
Phoenicophorium, 220, 225
Phoenicoseris, 280
Phoenix canariensis, 111, 117
  robusta, 232
  sylvestris, 232
  theophrastii, 111, 131
Pholidota recurva, 73
Pholisma, 343
Pholistoma, 162
Phormiaceae, 350
Phormium, 295, 350
  tenax, 298
Photinia, 66
  amphidoxa, 64
  chingshuiensis, 61
  glabra, 55
  integrifolia, 69
  lasiopetala, 61
  lucida, 61
  myriantha, 70
  parvifolia, 64

Phrodus, 284
Phrymaceae, 344
Phryna, 136
Phrynellia, 136
Phtheirospermum, 45
Phylica, 215, 264, 267
  arborea, 292
Phyllachne, 277
Phyllanthus boninsimae, 57
  griffithii, 77
  takaoensis, 61
Phyllarthron, 219, 220
Phyllis, 112, 114
  viscoa, 117
Phyllitis hybrida, 131
Phylloboea henryi, 67
  sinensis, 64
Phyllocara, 140
Phyllocladaceae, 310
Phyllocladus, 273, 295, 310
  asplenifolius, 272
Phyllocosmus, 335
Phylloctenium, 219
Phyllodoce nipponica, 54
Phylloglossum, 305
Phyllonoma, 339
Phyllonomaceae, 339
Phyllospadix, 347
Phyllostachys, 48, 62, 65
Phyllotrichum, 235
Phylloxylon, 218, 220
Phymaspermum, 212
Physaliastrum, 45
  savatieri, 55
Physalis, 215
  chamaesarachoides, 55
Physandra, 136
Physaria bella, 109
  condensata, 109
Physena, 218, 325, 327, 335
Physenaceae, 216, 335
Physocardamum, 137, 138
Physocarpus amurensis, 50
  ribesifolius, 50
Physocaulis, 122
Physoleucas, 134
Physoplexis, 11
Physoptychis, 137
Physospermopsis forrestii, 67
Physostegia correllii, 99
Phyteuma, 20

serratum, 129
Phytochorionomy, 3, 4, 7
Phytocrenaceae, 337
Phytolaccaceae, 317
Picconia, 112, 114
  azorica, 114
Picea, 12, 65, 101
  abies, 29, 30, 36
  bicolor, 54
  breweri, 103
  engelmannii, 36, 106, 107, 167
  glauca, 35, 36, 87, 98, 107
  jezoensis, 34, 51
  koyamae, 54
  manshurica, 50
  mariana, 35, 107
  mastersii, 62
  maximowiczii, 54
  meyeri, 62
  morinda, 153
  morrisonicola, 60
  obovata, 50, 51
  omorika, 21
  orientalis, 23
  polita, 54
  pungens, 108, 167
  rubens, 87
  schrenkiana, 153
  sitchensis, 102
  smithiana, 154
Pichonia, 249
Pickeringia, 161
  montana, 173
Picris echinoides, 289
Picrodendron, 329
Piedmont Geological Province, 90
Pieris floribunda, 85
  formosa, 69, 75
  forrestii, 66
  taiwanensis, 61
Pilea, 54
Pileostegia, 43
Pilgerochloa, 142
Pilgerodendron, 283
  uviferum, 287
Piliocalyx, 250
Pilopleura, 139
Pilostyles, 315
  hamiltonii, 273
  haussknechtii, 111
Pimelea, 296, 297
Pimia, 244
Pimpinella, 117
  battandieri, 126
  bicknellii, 128
  brachystyla, 62
  calycina, 55
  diversifolia, 76
  flaccida, 78
  idae, 24
  nikoensis, 55
  oreophila, 207
  serbica, 21
  sikkimensis, 76
Pinacantha, 139
Pinaceae, 10, 311
Pinacopodium, 335
Pinales, 311
Pinanga dicksonii, 230
Pinckneya, 82
  pubens, 93
Pinellia, 48
  cordata, 65
  integrifolia, 65
Pinguicula corsica, 129
Pinopsida, 310
Pintoa, 284
Pinus, 12, 65, 101, 185
  albicaulis, 104, 168
  aristata, 168
  arizonica, 185
  armandii, 69
  attenuata, 178
  ayacahuite, 185
  banksiana, 35, 107
  bhutanica, 71
  brutia, 26, 119, 132, 148
  canariensis, 113, 116
  cembra, 18
  clausa, 93
  contorta, 93, 101, 104, 106, 107, 110, 178
  coulteri, 173, 176
  densiflora, 54
  echinata, 91
  edulis, 167
  elliottii, 91, 93
  engelmannii, 185
  flexilis, 98, 104, 168
  gerardiana, 154
  glabra, 93
  griffithii, 21
  halepensis, 119, 126

hartwegii, 185, 186
jeffreyi, 104, 105
koraiensis, 49
kwangtungensis, 65
lambertiana, 104
longaeva, 104, 168
merkusii, 311
michoacana, 185
monophylla, 167
montezumae, 185
monticola, 104, 105
morrisonicola, 60
muricata, 173, 178
nigra, 126
oocarpa, 185
palustris, 91, 92, 93
parviflora, 54
patula, 185
peuce, 21
pinaster, 126
pinea, 119, 126, 131
pityusa, 194
ponderosa, 98, 103, 104, 107, 167
pseudostrobus, 185
pumila, 33
pungens, 85
radiata, 173, 178
remorata, 178
resinosa, 87
rigida, 85, 91
roxburghii, 113, 116
sabiniana, 173, 176
serotina, 93
sibirica, 30, 31
stankewiczii, 194
strobus, 87, 187
sylvestris, 27, 29, 31, 126, 145
taeda, 91
taiwanensis, 60
tenuifolia, 185
teocote, 185
thunbergii, 54
torreyana, 173, 178
virginiana, 85
wallichiana, 154
Pinyon-juniper community, 167
Pionocarpus, 164
Piper, 60
  nepalense, 74
  postelsianum, 57
Piperaceae, 314

Piperales, 314
Piperanthera, 314
Piptadenia, 261
Piptanthus, 43
  tomentosus, 66
Piptocalyx, 271
Piptochaetium, 288
Piptoptera, 136, 152
Pistacia, 159
  khinjuk, 149
  lentiscus, 124, 131
Pistiaceae, 356
Pitardia, 122
Pitavia, 284
Pitcairnia, 353
  feliciana, 199, 255
Pithecellobium, 181
  hassleri, 261
  quaranticum, 261
Pittosporaceae, 268, 340
Pittosporales, 340
Pittosporopsis, 235
Pittosporum, 251, 267, 296, 297
  abyssinicum, 208
  beecheyi, 57
  boninense, 57
  bracteolatum, 298
  chichijimense, 57
  dalycaulon, 230
  daphniphylloides, 61
  ellipticum, 300
  fairchildii, 300
  illicioides, 54
  napaulense, 75
  parvifolium, 57
  pimeleoides, 300
  senacia, 225
Pityopus, 100, 102
Placea, 285, 286
Placopoda, 209
Placospermum, 271
Pladaroxylon leucadendron, 215
Plaesianthera, 228
Plagiobasis, 141
Plagiobothrys acanthocarpus, 176
  austiniae, 175
  distantiflorus, 176
  gracilis, 174
  greenei, 174
  humistratus, 175
  hystriculus, 175

myosotoides, 174
scouleri, 174
trachycarpus, 175
undulatus, 175
Plagiogyria, 307
Plagiogyriaceae, 307
Plagiogyriales, 307
Plagiopetalum, 43
Plagiopteraceae, 234, 328
Plagiopteron, 234, 237, 328
Plagioscyphus, 218, 222
Plagiospermum, 42
Plagius, 122
 flosculosus, 130
Planchonella, 297
 costata, 298, 299
 novo-zelandica, 299, 300
 vitiensis, 300
Planchonia andamanica, 237
Planera, 79
 aquatica, 93
Plantaginaceae, 344
Plantago, 279, 281, 286, 344
 arborescens, 117
 aucklandica, 301, 302
 coronopus, 128
 fernandezia, 281, 282
 hakusanensis, 56
 leiopetala, 115
 malato-belizii, 115
 pentasperma, 293
 robusta, 215
 stauntonii, 293
 subspathulata, 115
 triantha, 301
 tunetana, 126
 webbii, 117
Platanaceae, 9, 320
Platanthera, 56
 boninensis, 58
 exelliana, 73
 juncea, 73
 leptocaulon, 73
 sikkimensis, 73
Platanus, 320
 kerrii, 238
 occidentalis, 187
Plathymenia foliosa, 261
Platycarya, 41
 tonkinensis, 66
Platyceriaceae, 308

Platychaeta, 210
Platycodon, 46
Platycraspedum, 42
Platycrater, 43
 arguta, 54
Platydesma, 247
Platylophus, 203, 264
 trifoliatus, 265
Platyspermation, 249
Platystemon, 159
Platyzoma, 307
Platyzomataceae, 271, 307
Plazia, 285
Plectaneia, 219
Plectrachne, 274
Plectranthus, 64, 67, 70
 excisus, 51
 japonicus, 56
 lasiocarpus, 61
 serra, 51
Pleea, 82
Plegmatolemma, 236, 238
Pleiokirkia, 218
Pleiomeris, 112
 canariensis, 116
Pleocarphus, 285
Pleocaulus, 228
Pleopeltis boniniensis, 57
 onoei, 53
Pleuricospora, 100, 102
Pleurocalyptus, 249
Pleurophyllum criniferum, 301
 hookeri, 301
 speciosum, 301
Pleuropterantha, 207
Pleurosoriopsis, 39
Pleurospermum spp., 67
Plinthus, 212
Plocama, 112
 pendula, 117
Plocosperma, 342
Plocospermataceae, 255, 342
Ploiarium, 322
Pluchea longifolia, 95
Plumbaginaceae, 10, 143, 319
Plumbaginales, 319
Plummera, 164
Poa, 56, 167, 168, 203, 208, 240, 279
 abbreviata, 15
 alpina, 168
 altaica, 31

arctica, 106, 168
aucklandica, 302
breviglumis, 302
commungii, 174
douglasii, 173, 174
flabellata, 290
granitica, 20
guthriesmithiana, 301
hamiltonii, 302
hartzii, 15
incrassata, 302
ircutica, 31
napensis, 173
novareae, 293
pirinica, 22
prolixior, 64
ramosissima, 302
sajanensis, 31
sandbergii, 108
shumushuensis, 34
sublanata, 15
tenerrima, 173
trautvetteri, 15
vrangelica, 15
Poaceae, 10, 15, 16, 26, 30, 264, 268, 356
Poacynum, 139
Poales, 356
Podistera yukonensis, 35
Podoaceae, 334
Podocarpaceae, 310
Podocarpales, 310
Podocarpus, 251, 267, 273, 286, 295
 andinus, 288
 falcatus, 207, 265
 lambertii, 260
 latifolius, 265
 macrophyllus, 69
 nagi, 65
 salignus, 288
 sellowii, 260
Podochrosia, 249
Podonephelium, 249
Podophorus, 280
Podophyllaceae, 317
Podostemaceae, 331
Podostemales, 331
Poeciloneuron, 227
 indicum, 230
 pauciflorum, 230
Poecilostachys, 220
Pogogyne, 163

 abramsii, 173
 nudiuscula, 173
 zizyphoroides, 176
Pogonachne, 229
Pogonatherum rufobarbatum, 78
Pogostemon amaranthoides, 76
 brachstachys, 70
 tuberculosus, 72
Poikilospermum, 329
Poilanedora, 234, 238
Poilaniella, 235
Poland, 11
Polemoniaceae, 159, 343
Polemoniales, 343
Polemonium, 55
 pectinatum, 109
 pulchellum, 31
 racemosum, 51
Poliomintha, 163
Poliothyrsis, 41, 63
 sinensis, 63
Polyachyrus, 285
Polyalthia parkinsonii, 237
Polycardia, 218
Polycarpaea, 117
 gayi, 118
Polychrysum, 141
Polygala, 66, 231
 apiculata, 129
 arillata, 231
 balansae, 124
 hohenackeriana, 146
 mariesii, 64
 mascatensis, 210
 preslii, 129
 reinii, 55
 sardoa, 129
 serpyllifolia, 188
 wattersii, 64
Polygalaceae, 336
Polygalales, 336
Polygonaceae, 10, 319
Polygonales, 319
Polygonanthaceae, 332
Polygonanthus, 332
Polygonataceae, 349
Polygonatum, 56, 64, 68, 70
 brevistylum, 73
 cathcartii, 76
 cirrifolium, 76
 kansuense, 63

obtusifolium, 25
platyphyllum, 63
pubescens, 86
punctatum, 76
stenophyllum, 51
Polygonella, 79
Polygonum, 54, 66, 75, 85
  alopecuroides, 52
  amgense, 32
  paleaceum, 77
  rude, 77
  scoparium, 129
  sphaerostachyum, 158
  viviparum, 158
Polylophium, 139
Polynesian Province, 246
Polynesian Region, 245
Polynesian Subkingdom, 245
Polyosma, 339
Polypleurella, 235, 238
Polypodiaceae, 10, 308
Polypodiales, 308
Polypodium, 60, 65, 308
  lachnopus, 74
  someyae, 53
Polypogon, 281
Polypremum, 341, 343
Polypteris feayi, 95
  integrifolia, 95
Polystichum, 53, 60, 65, 281
  drepanum, 115
  falcinellum, 115
  maderense, 115
  mohrioides, 290, 295
  munitum, 102, 105
  scopulinum, 37
  webbianum, 115
Polytaenia, 82
Polyzygus, 228
  tuberosus, 230
Pomaderris kumeraho, 300
Pomatosace, 41
Pommereschea, 236, 237
Pommereulla, 229
Poncirus, 43
Pontederiaceae, 353
Pontederiales, 353
Popoviocodonia, 46
Populus, 12, 156
  angustifolia, 168
  balsamifera, 36, 98

bonatii, 66
caspica, 151, 152
ciliata, 69
deltoides, 98
fremontii, 168
grandidentata, 36, 85
hopeinensis, 62
suaveolens, 33, 34
szechuanica, 66
tremula, 29, 30, 31, 34, 36
tremuloides, 36, 87, 95, 98, 101, 106, 108, 168
trichocarpa, 102, 106
yunnanensis, 66
Portenschlagiella, 122, 130
  ramosissima, 131
Porto Santo Island, 115
Portugal, 11, 121
Portulaca boninensis, 57
  smallii, 90
Portulacaceae, 318
Posidonia, 347
Posidoniaceae, 347
Posidoniales, 347
Poskea, 206, 344
Postia, 141
Potalia, 342
Potaliaceae, 342
Potamogeton, 36, 347
  fryeri, 56
  mandschuriensis, 51
Potamogetonaceae, 347
Potamogetonales, 347
Potaninia, 138
  mongolica, 157
Potentilla, 16, 55, 66, 75, 85, 101, 168, 240
  adenotricha, 32
  anachoretica, 15
  anadyrensis, 33
  ancistrifolia, 62
  beringensis, 15
  bhutanica, 72
  bryoides, 72
  crassinervis, 129
  diversifolia, 107
  eversmanniana, 27
  fruticosa, 36, 105
  mandshurica, 50
  manipurensis, 77
  miyabei, 53
  montana, 17

INDEX

nivea, 106
palustris, 105
peduncularis, 69
porphyrantha, 146
pulchella, 15
reptans, 208
rubella, 15
rubricaulis, 15
tollii, 32, 33
vahliana, 15
Poterium spinosum, 119, 131
Pothos, 221
Pottingeria, 43, 337, 339
Pouslowia, 229
Pouteria boninensis, 57
Prairie, short-grass, 97
Prairie, tall-grass, 97
Prairie peninsula, 98
Prairies Province, North America, 95
Prangos pabularia, 155
 uloptera, 146
Prasium, 122
 majus, 119, 124
Pratia physaloides, 300
Pre-Balkhash deserts, 137
Prehankian Valley, 52
Premna, 67
 ligustroides, 64
Prenanthes acerifolia, 56
 aspera, 85
 barbata, 85
 blinii, 51
 faberi, 64
 formosana, 61
 pendula, 117
 roanensis, 85, 88, 89
 serpentaria, 85
 tanakae, 56
 volubilis, 70
Preslia, 122
Primorski Krai, 34, 39
Primula, 54, 66, 69, 72, 208, 240
 abchasica, 23
 angustifolia, 109
 aucheri, 210
 bayernii, 26
 beringensis, 15
 bumanica, 70
 darialica, 26
 densa, 70
 deorum, 21

 distyophylla, 70
 egaliksensis, 110
 faberi, 64
 frondosa, 21
 heterochroma, 151
 hidakana, 53
 juliae, 26
 listeri, 75
 longipes, 23
 maguirei, 169
 megaseifolia, 23
 palinuri, 129
 renifolia, 26
 saxatilis, 50
 scotica, 17
 spectabilis, 19
 specuicola, 169
 vulgaris, 128
Primulaceae, 10, 325
Primulales, 324
Prince Edward Islands, 278, 291, 293
Pringlea, 291
 antiscorbutica, 293
Prinsepia, 42
 scandens, 61
 sinensis, 50
 utilis, 69, 75
Prionium, 354
Prionotes, 270, 272
Prionotrichon, 137
Pritchardiopsis, 250
Prockiopsis, 217
Procopiania, 122, 131
Procris boninensis, 57, 58
Prolongoa, 122
Prosopanche, 315
Prosopis, 179, 188, 286
 chilensis, 179
 cineraria, 211
 glandulosa, 99, 179, 182
 juliflora, 179
 reptans, 179, 182
 spicigera, 211
 strombulifera, 289
Prostanthera, 269
Protarum, 220, 225
Protea, 264
Proteaceae, 250, 264, 266, 268, 296, 338
Proteales, 338
Province of the Mexican Highlands, 183
Province of New Zealand Subantarctic Is-

lands, 301
Province of Oman, 209
Province of Uplands of Central Brazil, 260
Prunella prunelliformis, 56
Prunus, 12, 55, 61, 64, 66, 69, 75, 85, 142
  africana, 207
  arabica, 211
  armeniaca, 156
  bigantina, 19
  caroliniana, 93
  davidiana, 62
  glandulosa, 52
  lusitanica, 114
  mandschurica, 50
  nairica, 146
  nakaii, 50
  padus, 156
  scoparia, 149
  serotina, 187
  sibirica, 52
  tenella, 31
  urumiensis, 146
  wendelboi, 149
Psammiosorus, 217, 309
Psammochloa, 142
Psammogeton, 139
Psathyrotes, 164
Pseudannona, 217, 224
Pseudartabotrys, 198
Pseuderucaria, 133
Pseudoanastatica, 137
Pseudobahia, 164
Pseudobetckea, 11, 26
Pseudocamelina, 137
  aphragmodes, 149
  szowitsii, 146
Pseudocedrela, 204
Pseudoclappia, 164
Pseudoclausia, 137
Pseudocoix, 220
Pseudocorchorus, 218
Pseudodichanthium, 229
Pseudodigera, 207
Pseudodistichum, 290
Pseudodracontium, 236
Pseudofortuynia, 137
Pseudoglochidion anamalayanum, 230
Pseudohandelia, 141
Pseudolarix, 40, 63
  amabilis, 63
Pseudolithos, 207

Pseudolotus, 138
Pseudomarrubium, 140
Pseudonemacladus, 163
Pseudopanax, 277
  chathamicus, 301
  discolor, 300
  laetevirens, 286
Pseudoparis, 220
Pseudopteris, 218
Pseudopyxis, 45, 53
  depressa, 55
  heterophylla, 55
Pseudosalacia, 201
Pseudosciadium, 249
Pseudosedum, 138
Pseudoselinum, 203
Pseudostellaria europaea, 19
  palibiniana, 54
Pseudostenosiphonium, 228
Pseudostreptogyne, 220, 224
Pseudostriga, 235
Pseudotaxus, 40, 63, 311
  chienii, 63
Pseudotsuga, 101
  japonica, 54
  macrocarpa, 173
  menziesii, 101, 103, 105, 106, 107, 167, 185
  wilsoniana, 60
Pseudovesicaria, 11, 26
Pseudowintera, 294
Pseudozoysia, 207
Psiadia rotundifolia, 215
Psilocarphus berteri, 174
  brevissimus, 174, 176
  tenellus, 174
Psiloësthes, 236
Psilopeganum, 43, 63
  sinense, 64
Psilotaceae, 306
Psilotrichum, 206
Psilotum, 306
  nudum, 125, 298
Psiloxylaceae, 224, 332
Psiloxylon, 218, 224, 332
Psoralea cuspidata, 96
  hypogaea, 96
  linearifolia, 96
Psorothamnus, 161
Psychine, 121
Psychotria, 67

boninensis, 58
homalosperma, 58
Ptaeroxylaceae, 334
Ptaeroxylon, 334
Ptelidium, 218
Pteracanthus lachenensis, 72
Pterachaenia, 141
Pteralyxia, 247
Pteridaceae, 10, 307
Pteridales, 307
Pteridium, 309
Pteridoblechnum, 271
Pteridophyllum, 40, 53
  racemosum, 54
Pteris, 60, 281, 307
  boninensis, 57
Pternopetalum, 44, 67
  tanakae, 55
Pterocactus, 283
Pterocarpus marsupium, 232
Pterocarya hupehensis, 63
  paliurus, 63
  pterocarpa, 152
  rhoifolia, 54
Pteroceltis, 42
  tatarinowii, 62
Pterocephalus, 117
Pterocyclus rivularum, 67
Pteronia, 212
  glauca, 214
Pteropyrum, 136
  aucheri, 150
  scoparium, 210
Pterospermum, 233
  acerifolium, 232
Pterospora andromedea, 102
Pterostegia, 160
Pterostemon, 162, 339
Pterostemonaceae, 159, 339
Pterostyrax, 41, 66
  hispida, 54
Pteroxygonum, 40
Pterygiella, 45, 67
Pterygopappus, 272
Pterygopleurum, 44, 53
  neurophyllum, 55
Pterygostemon, 137
Ptilagrostis concinna, 158
Ptilotrichum halimifolium, 129
  macrocarpum, 129
Pubistylus, 235, 237

andamanensis, 237
Puccinellia andersonii, 15
  angustata, 15
  antipoda, 302
  brugermanii, 15
  byrrangensis, 15
  gorodkovii, 15
  groenlandica, 15
  lenensis, 16
  macquarensis, 302
  nipponica, 56
  poacea, 16
  porsildii, 16
  rosenkrantzii, 16
  svalbardensis, 16
  tenella, 16
  vaginata, 16
  vahliana, 16
Puccionia, 207
Pueraria, 66
Puget Trough, 103
Pugionium, 137
Pulicaria, 126
  burchardii, 117
  canariensis, 117
Pulmonaria, 11, 19
  affinis, 17
Pulsatilla halleri, 21
  magadanensis, 34
  nipponica, 54
  sachalinensis, 53
Pultenaea, 269
Punica granatum, 332
  protopunica, 209, 332
Punicaceae, 332
Punjab, 140
Puria, 228
Purpureostemon, 249
Purpusia, 161
Putoria, 122
  calabrica, 119
Puya, 287
Pycnandra, 249
Pycnanthemum californicum, 173
  montanum, 86
Pycnocycla aucheriana, 210
  nodiflora, 210
Pycnoplinthus, 137
Pycnorhachis, 241
Pycnospatha, 236
Pycreus colchicus, 25

Pygeum cordatum, 70
Pyracantha angustifolia, 66
  koidzumii, 61
Pyramidium, 137
Pyramidoptera, 139
Pyrenaria shinkoensis, 61
  yunnanensis, 66
Pyrénées, 11, 12, 20
Pyriluma, 249
Pyrola, 36, 69
  decorata, 75
  sikkimensis, 72
Pyrolaceae, 323
Pyrostria, 219, 220
Pyrrorhiza, 353
Pyrrosia boothii, 71
Pyrularia pubera, 85
Pyrus, 12, 64, 66
  balansae, 23
  betulaefolia, 62
  boissierana, 151
  bretschneideri, 62
  communis, 28
  corymbifera, 50, 62
  grossheimii, 151
  hyrcana, 151
  kawakamii, 61
  pashia, 75
  raddeana, 146
  rossica, 27
  turcomanica, 148
  ussuriensis, 50, 51
  zangezura, 26
Pyxidanthera, 79
  barbulata, 85

Queensland Province, 271
Quercus, 12, 22, 54, 60, 63, 66, 69, 85, 86, 185, 187, 236, 261
  acutissima, 65
  afares, 125, 151
  agrifolia, 173, 176
  alba, 92
  arkansana, 93
  baloot, 154
  brantii, 145, 148
  canariensis, 127
  castaneifolia, 150, 151, 152
  chapmanii, 93
  chrysolepis, 105, 176
  coccifera, 119, 127, 131
  congesta, 128
  dealbata, 74
  dilatata, 154
  douglasii, 173, 176
  dumosa, 177
  durata, 177
  engelmannii, 176
  faginea, 127
  falcata, 92
  fenestrata, 74
  gambelii, 108, 167
  garryana, 102
  georgiana, 90
  griffithii, 74
  hartwissiana, 23
  ilex, 119, 124, 127, 131
  incana, 154
  infectoria, 145, 148
  kelloggii, 102, 105
  laevis, 93
  laurifolia, 92, 93
  libani, 145, 148
  lineata, 74
  lobata, 173, 176
  macranthera, 26, 145, 152
  macrocarpa, 98
  mongolica, 50, 51, 52
  nigra, 92
  petraea, 18, 26, 29, 145
  pontica, 23
  pyrenaica, 127
  robur, 27, 28, 145
  rotundifolia, 127
  semecarpifolia, 154
  sicula, 128
  suber, 127
  texana, 99
  tomentella, 173
  turbinella, 182
  variabilis, 65
  virginiana, 92, 93
  wislizenii, 176, 177
Quetta Basin, 154
Quezelia, 133
Quiinaceae, 254, 323
Quillaja, 284
  saponaria, 287
Quinchamalium majus, 288
Quintinia, 251

Racemobambos, 241
Radermachera yunnanensis, 67
Radiola linoides, 199

Raffenaldia, 121
Rafflesia, 239
Rafflesiaceae, 197, 315
Rafflesiales, 315
Rafinesquia, 164
Raillardella, 100, 247
Rain forest
  temperate, 102, 295
  tropical, 198, 222, 238, 258, 260
Rajania, 352
Ramelia, 249
Rameya, 217, 220
Ramonda, 11
  nathaliae, 22
  serbica, 22
Randia, 67, 233
Randonia, 134
Ranevea, 220, 224
Ranunculaceae, 9, 30, 316
Ranunculales, 316
Ranunculus, 54, 60, 66, 69, 71, 85, 101, 168, 186, 240, 272, 279, 296
  acaulis, 277
  amurensis, 50
  biternatus, 277, 290, 292
  caprarum, 281
  carolii, 292
  carpaticus, 18
  chamissonis, 15
  cortusifolius, 117
  diffusus, 74
  dolosus, 151
  eschscholtzii, 104
  hainganensis, 50
  kirkii, 301
  marschlinsii, 128
  moscleyi, 293
  pinguis, 301
  pulsatillifolius, 147
  punctatus, 15
  reniformis, 231
  revelierei, 128
  sabinii, 15
  sajanensis, 30
  sintenisii, 145
  stagnalis, 207
  subscaposus, 301
  tripartitus, 17
  weyleri, 128
  yesoensis, 53
Ranzania, 40, 53
  japonica, 54

Raoulia, 296
  goyenii, 301
Rapa Island, 246
Rapanea, 187
  chathamica, 301
  crassifolia, 299
  daphnoides, 230
  dentata, 300
  kermadecensis, 298, 299
  melanophloes, 207
  wightiana, 230
Rapateaceae, 255, 355
Raphidophyton, 136
Raphiocarpus, 236, 237
Rastrophyllum, 203
Ratibida, 97
Ratzeburgia, 236, 237
Ravenala, 220, 222, 354
Ravenea, 220, 221
Ravensara, 217, 222
Rea, 280
Readea, 244
Reaumuria, 135
  kaschgarica, 157, 158
  songarica, 157
  trigyna, 157
Reboudia, 133, 135
Rechinger, K. H., 144
Redowskia, 11, 32
  sophiifolia, 32
Reederochloa, 164
Reedia, 273
Reevesia, 42
  formosana, 61
Region of the Guayana Highlands, 257
Region of the South Subantarctic Islands, 291
Regions, floristic, 2
Regnellidium, 307
Rehderodendron, 41
  fengii, 66
  tsiangii, 66
Rehmannia, 45
  glutinosa, 63
  rupestris, 64
Reichardia, 117
Reicheëlla, 283, 286
Reineckea, 47
Reinwardtia indica, 75
Relict areas, 28
Relict species, 22
Relicts, glacial, 17

Remya, 247
Reptonia, 138
  mascatensis, 210
Reseda armena, 145
  crystallina, 117
  scoparia, 117
Resedaceae, 327
Restio, 264, 269
Restionaceae, 264, 265, 266, 355
Restionales, 355
Retama, 135
Retzia, 264, 344
Retziaceae, 264, 344
Réunion, 220, 224
Revillagigedo Islands, 256
Reynoldsia, 245
Rhabdodendraceae, 258, 333
Rhabdodendron, 333
Rhabdosciadium, 139
Rhabdothamnopsis, 46
Rhabdothamnus, 294
Rhammatophyllum, 137
Rhamnaceae, 10, 338
Rhamnales, 338
Rhamnella forrestii, 67
  franguloides, 64
  longifolia, 67
Rhamnoluma, 249
Rhamnoneuron, 235
Rhamnus, 12, 55, 61, 67
  alaternus, 124
  caroliniana, 187
  crenulata, 117
  diamantiaca, 51
  glandulosa, 116
  imeretinus, 24
  integrifolia, 117
  kanagusuki, 59
  ludovici-salvatoris, 127
  lycioides, 124
  oleoides, 124
  persicifolia, 129
  ussuriensis, 51, 52
  utilis, 64
Rhamphicarpa medwedewii, 24
Rhaphidophyton, 152
Rhaphiolepis, 42
  integerrima, 57
Rhaphithamnus, 281
  venustus, 282
Rhaponticum canariensis, 117

Rhaptonema, 217
Rheedia, 221
Rheopteris, 307
Rhetinodendron, 280
  berteroi, 282
Rheum, 69
  altaicum, 30
  coreanum, 50
  nobile, 71
  rhaponticum, 21
  spiciforme, 158
Rhexia, 83
Rhigosum, 213
Rhinanthus, 19
Rhipogonaceae, 352
Rhipogonum, 352
Rhipsalis, 318
  fasciculata, 224
  horrida, 223
  madagascariensis, 222
Rhizantella, 273
Rhizobotrya, 11, 18
  alpina, 19
Rhizocephalus, 142
Rhizophora, 233
Rhizophoraceae, 332
Rhizophorales, 332
Rhodiola, 72
  algida, 30
Rhodocodon, 219
Rhodocolea, 219
Rhododendron, 12, 54, 61, 63, 66, 72
  adamsii, 33
  afghanicum, 154
  anthopogon, 154
  arborescens, 85
  arboreum, 75, 154
  barbatum, 154
  boninense, 57
  calendulaceum, 85
  campanulatum, 154
  catawbiense, 85
  caucasicum, 25
  colletianum, 155
  cumberlandense, 85
  dahuricum, 31
  delavayi, 69
  dendricola, 70
  elliottii, 77
  hirsutum, 19
  imperator, 70

INDEX

insculptrum, 70
johnstoneanum, 77
lapponicum, 38
lindleyi, 75
luteum, 29
macropyllum, 102, 105
magnificum, 70
manipurense, 77
maximum, 85
myrtilloides, 70
nilagiricum, 231
parvifolium, 33
periclymenoides, 85
ponticum, 25
prinophyllum, 85
schlippenbachii, 50
simsii, 69
smirnowii, 23
sulfureum, 69
taggianum, 70
tschonoskii, 54
ungernii, 23
vaseyi, 88
wattii, 77
weyrichii, 54
yunnanense, 69
Rhododon, 82
Rhodohypoxis, 203
Rhodolaena, 218
Rhodoleiaceae, 319
Rhodomyrtus tomentosa, 231
Rhodope Mountains, endemism in, 21
Rhodosepala, 218
Rhodothamnus chaemaecistus, 19
Rhodotypos, 42
Rhoicissus, 201
  capensis, 265
Rhoiptelea, 41, 65, 321
  chiliantha, 66
Rhoipteleaceae, 39, 65, 321
Rhopalocarpus, 218, 328
Rhopalostylis, 295
  baueri, 298, 299
  cheesemannii, 299
  sapida, 299
Rhuacophila, 350
Rhus, 111
  albida, 124
  aucheri, 210
  chinensis, 75
  delavayi, 66

monticola, 203
oxyacantha, 124
pentaphylla, 124
typhina, 85
Rhynchocalycaceae, 201, 332
Rhynchocalyx, 201, 332
Rhynchophora, 218
Rhynchosia acuminatifolia, 55
Rhynchosinapis monensis, 17
  wrightii, 17
Rhynchospora, 93
  boninensis, 58
  caucasica, 25
  saxicola, 90
Rhynchotheca, 336
Rhynchothecaceae, 261, 336
Rhysolepis, 164
Rhyticarpus, 264
Rhytidocaulon, 207
Ribes, 54, 85, 101, 261, 286, 330
  diacantha, 52
  formosanum, 61
  glaciale, 69, 75
  glandulosum, 36, 88
  graveolens, 30
  henryi, 64
  howellii, 105
  kolymense, 33
  komarovii, 50
  lacustre, 36
  laxiflorum, 105
  maximowiczianum, 50
  missouriense, 84
  rotundifolium, 84
  sardoum, 129
  triste, 36
  ussuriense, 50
Richardsiella, 203
Richea, 273
Rif, 125
Rindera gymnandra, 126
  media, 146
Riseleya griffithii, 226
Rishiri Island, 53
Ritchiea albersii, 207
Robeschia, 137
Robinia hispida, 85
  pseudoacacia, 85
  viscosa, 85
Robinson Crusoe Island, 280, 281, 282
Robinsonia, 280

gayana, 282
Roborowskia, 135
"Rock-houses," 90
Rocky Mountain Province, 38, 100, 105
Rocky Mountain Region, 99
Rocky Mountains, southern, 109
Rodgersia, 42, 69
　nepalensis, 72
　podophylla, 54
Rodriguez, 220, 224
Rohdea, 47
Rollandia, 247
Romanzoffia, 100
Romnalda, 351
Romneya, 159
　coulteri, 173
Romulea, 126, 264
　columnae, 208
　fischeri, 208
　ligustica, 130
　melitensis, 130
　requienii, 130
　revelierei, 130
Roridula, 264, 267
Roridulaceae, 263, 339
Rorippa nikoensis, 54
Rosa, 52, 55, 61, 64, 66, 69, 75, 85, 208
　abyssinica, 208
　sericea, 158
Rosaceae, 10, 26, 30, 331
Rosales, 331
Roscheria, 220, 225
　melanochaetes, 226
Roscoea, 68
Rosidae, 330
Rosmarinus, 122
　officinalis, 119
　tournefortii, 126
Rostkovia magellanica, 278, 290
　tristanensis, 278, 293
Rosularia modesta, 149
Rotala elatinomorpha, 55
Rothmaleria, 122
Rothrockia, 162
Rouchera, 335
Roupellina, 219
Roussea, 218, 224, 331
Rousseaceae, 216, 331
Rousseauxia, 218
Rouya, 122
Rubia, 61, 67

hexaphylla, 55
　manjith, 76
　peregrina, 124
　rigidifolia, 146
　sikkimensis, 76
　sylvatica, 51
　wallichiana, 76
Rubiaceae, 217, 342
Rubus, 55, 61, 64, 66, 69, 72, 75, 115, 240, 286
　assamensis, 77
　chaetocalyx, 70
　chamaemorus, 36
　hochstetterorum, 114
　hyrcanus, 151
　moluccanus, 231
　nakaii, 57
　opulifolium, 77
　persicus, 151
　petitianus, 208
　spectabilis, 102
　ursinus, 105
　wardii, 70
Ruby Mountains, 168
Rudbeckia, 83, 96
　graminifolia, 95
Ruizia, 218, 224
Rulingia, 221, 269
Rumania, 18
Rumex, 279
　balcanicus, 21
　bequetii, 207
　limoniastrum, 210
　lunaria, 117
　madaio, 54
　regelii, 53
　rupestris, 17
Rumicarpus, 207
Rupicapnos, 121, 125
　muricaria, 125
　numidicus, 125
Ruppia, 347
Ruppiaceae, 347
Ruscaceae, 9, 349
Ruscus, 349
　aculeatus, 17
　colchicus, 25
　hypoglossum, 132
　hyrcanus, 151
　streptophyllus, 115
Russia, 138

Russian Plain, 28
Russowia, 141
Ruta corsica, 129
  oreojasme, 117
  pinnata, 117
Rutaceae, 264, 268, 333
Rutales, 333
Ruthea herbanica, 117
Rutidosis, 274
Rytidocarpus, 121, 125
Ryukyu Islands, 40, 41
Ryukyu Province, 58
Rzedowskia, 162

Sabal, 93
  minor, 92
  palmetto, 92
Sabaudiella, 207
Sabia transarisanensis, 61
Sabiaceae, 335
Saccifoliaceae, 257, 342
Saccifolium, 342
Saccocalyx, 122
  satureoides, 126
Saccoglotis, 335
Sadleria, 247
Sagebrush, 109, 167
Sageretia, 67
  randaiensis, 61
Sagina abyssinica, 207
  afroalpina, 207
  pilifera, 129
Saguaro, 179
Sahara, 132, 133, 134
Saharan Province, 134
Saharo-Arabian Region, 132
Sahelian Province, 204
Sahelo-Sudanian Subregion, 204
St. Helena and Ascension Region, 214
Saint Paul Island, 293
Sakhalin, 11, 34, 41, 52
Sakhalin-Hokkaido Province, 52
Sakishima Islands, 60
Sal tree, 232
Salaciopsis, 249
Salazaria, 163
Salicaceae, 10, 326
Salicales, 326
Salix, 12, 16, 19, 33, 34, 54, 61, 66, 85, 101
  arctophila, 15

  berberifolia, 32, 34
  bhutanensis, 72
  eriophylla, 75
  floridana, 93
  glauca, 36
  heterochroma, 63
  hibernica, 17
  jurtsevii, 33
  kangensis, 50
  khokhrjakovii, 33
  lindleyana, 69
  maximowiczii, 50
  nasarovii, 30
  paludicola, 53
  pauciflora, 53
  pedicellata, 116
  phylicifolia, 36
  planifolia, 36
  plectilis, 72
  polaris, 16
  sajanensis, 30
  sitchensis, 105
  uva-ursi, 15
  yezoalpina, 53
Salpiglossidaceae, 343
Salsola, 156, 158
  cana, 146
  laricifolia, 157
  passerina, 157
Saltia, 133
Salvadora, 211, 337
  oleoides, 211
  persica, 213, 233
Salvadoraceae, 197, 337
Salvadoropsis, 218
Salvage Islands, 113, 115
Salvia, 56, 67, 111, 126, 142, 144, 159, 173, 207
  brachyantha, 147
  broussonetii, 117
  canariensis, 117
  dracocephaloides, 147
  fominii, 147
  forskaohlei, 24
  glutinosa, 18
  halophila, 143
  interrupta, 124
  khorassanica, 148
  limbata, 147
  macilenta, 210
  maximowicziana, 64

mizrayanii, 210
sikkimensis, 72
triloba, 131
umbratica, 63
vermifolia, 144
Salvinia, 310
Salviniaceae, 310
Salviniales, 310
Salweenia, 43
Sambirania, 309
Sambirano Subprovince, 223
Sambucaceae, 341
Sambucus, 12, 341
  adnata, 76
  formosana, 61
  palmensis, 117
  peninsularis, 63
  pubens, 36
  racemosa, 36
  wightiana, 67
Sameraria, 137
  odontophora, 146
San Ambrosia Island, 282
San Bernardino Mountains, 100
San Clemente Island, 161
San Felix Island, 282
San Francisco Mountain, 166
San Jacinto Mountains, 100
San Pietro Island, 127
Sanguinaria, 82
  canadensis, 99
Sanguisorba, 55, 112
  maderensis, 115
  magnifica, 50
Sanicula azorica, 114
  crassicaulis, 174
  europaea, 31
  graveolens, 174
  kaiensis, 55
  petagnioides, 61
  tuberculata, 55
Saniella, 203
Sansevieriaceae, 349
Santa Catalina Island, 161
Santa Cruz Island, 161
Santa Rosa Island, 161
Santalaceae, 338
Santalales, 337
Santalum, 248
  album, 232
  boninense, 57

  fernandezianum, 281
Santaupaua, 228
Santolina, 122
Sapindaceae, 335
Sapindales, 334
Sapindopsis, 235, 237
Sapindus, 182, 233
  boninensis, 57
Saponaria brodeana, 151
  pumilio, 19
Sapotaceae, 324
Sapotales, 324
Sapria himalayana, 315
Sararanga, 356
Sarcanthidion, 249
Sarcanthus scolopendrifolius, 56
Sarcocapnos, 121
Sarcocaulon, 212
  mossamedense, 213
Sarcochilus japonicus, 56
Sarcococca hookeriana, 75
  saligna, 69, 155
Sarcodes, 100, 102
Sarcolaena, 218
Sarcolaenaceae, 222, 328
Sarcomelicope, 249
Sarcopoterium, 121
Sarcopygme, 244
Sarcospermataceae, 324
Sarcostemma daltonii, 118
Sardinia, 121, 128
Sargentia, 161
Sargentodoxa, 40, 316
Sargentodoxaceae, 9, 316
Sarmienta, 285
Sarracenia, 82, 93, 323
  alata, 93
  flava, 93
  leucophylla, 93
  minor, 93
  oreophila, 94
  psittacina, 93
  purpurea, 93
  rubra, 93
Sarraceniaceae, 323
Sarraceniales, 323
Sartwellia, 164
  flaveriae, 181
  mexicana, 181
  puberula, 181
Saruma, 40

Sasa, 48, 56
  rivularis, 53
Sassafras randaiense, 60
  tzumu, 63
Satakentia, 48, 60
  riukiuensis, 62
Satureja, 207
  barceloi, 127
  bzybica, 24
  chandleri, 173
  rumelica, 22
Saurauia, 321
  fasciculata, 72
  napaulensis, 69, 75
  subspinosa, 70
Saurauiaceae, 321
Saururaceae, 314
Saussurea, 34, 51, 56, 61, 63, 67
  chionophylla, 53
  conica, 73
  fauriei, 53
  frolovii, 31
  lamprocarpa, 64
  microcephala, 64
  poljakovii, 32
  porcii, 20
  pygmaea, 20
  sajanensis, 31
  squarrosa, 31
  tridactyla, 158
  tschuktschorum, 15
  wellbyi, 158
Sauvagesiaceae, 323
Savanna, 202, 232, 260, 272, 275
Saxegothaea, 283
  conspicua, 288
Saxifraga, 16, 54, 69, 72, 85, 101, 168, 208, 279
  abchasica, 23
  adscendens, 168
  aizoides, 106
  algissii, 32
  anadyrensis, 33
  anisophylla, 70
  brachypoda, 75
  bronchialis, 106, 168
  caespitosa, 106, 168
  calopetala, 70
  canaliculata, 17
  cernua, 99, 106, 168
  conifera, 17

  ferdinandi-coburgii, 21
  flagellaris, 106
  foliolosa, 110
  hariotii, 17
  hartii, 17
  heteroclada, 70
  hirculus, 110
  maderensis, 115
  manshuriensis, 50
  multiflora, 33
  nathorstii, 15
  nivalis, 106
  oppositifolia, 38, 106
  portosanctana, 115
  praetermissa, 17
  pubescens, 17
  redowskii, 33
  rivularis, 110
  sichotensis, 50
  spathularis, 17
  stribrnyi, 21
  subverticillata, 26
  tricuspidata, 106
  virgularis, 70
Saxifragaceae, 330
Saxifragales, 330
Saxifragella, 284, 289
Saxifragodes, 284, 289
Saxiglossum, 39
Sayan, 30
Scabiosa, 111, 126
  columbaria, 207
  japonica, 55
  lachnophylla, 51
  olgae, 24
  parviflora, 129
  rhodopensis, 21
  vestina, 19
Scaevola, 225
  frutescens, 58
  gracilis, 299
Scalesia, 257
Scaligeria lazica, 24
Scandia, 294
Scassellatia, 207
Schefflera, 251
  abyssinica, 207
  capitata, 230
  myriantha, 207
  racemosa, 231
  taiwaniana, 61

Schefflerodendron, 200
Schellenbergia, 235, 237
Scheuchzeria, 346
Scheuchzeriaceae, 9, 346
Scheuchzeriales, 346
Schickendantzia, 285
Schiedea, 247
Schiekia, 353
Schima argentea, 69
  khasiana, 69, 77
  mertensiana, 57
  wallichii, 59, 75
Schimpera, 133
Schimperella verrucosa, 207
Schinus dependens, 288
Schisandra spp., 66, 155, 313
  arisanensis, 60
  chinensis, 49
  coccinea, 93
  neglecta, 69, 74
  propinqua, 74
  repanda, 54
  rubriflora, 69
  sphenanthera, 63
Schisandraceae, 313
Schischkinella, 136
Schischkinia, 141
Schismatoclada, 219
Schistocarpaea, 271
Schistocaryum, 45, 65
Schivereckia, 11
  berteroides, 27
  kusnezovii, 27
  monticola, 27
  podolica, 27
Schizachyrium, 288
Schizaea, 57
Schizaeaceae, 307
Schizaeales, 307
Schizanthus, 284, 286
Schizeilema, 296
  reniforme, 301
Schizenterospermum, 219
Schizogyne, 113, 117
Schizolaena, 218
Schizolobium excelsum, 261
Schizomeryta, 249
Schizonepeta annua, 31
  tenuifolia, 63
Schizopepon, 42
Schizopetalon, 284

Schizophragma, 43, 66, 69
  hydrangeoides, 54
Schizostigma, 228
Schlagintweitiella, 40
Schmalhausenia, 141
Schmithüsen, J., 4, 5
Schnabelia, 46
Schoenomorphus, 236
Schoepfiaceae, 337
Scholtzia, 273
Schouw, J. F., 1
Schouwia, 133, 135
Schrameckia, 217
Schrebera swietenioides, 232
Schrenkia, 139
Schtschurowskia, 139
Schumacheria, 227, 229
Schwalbea, 82
Schweinfurthia papilionacea, 210
Sciadopanax, 219
Sciadopityaceae, 9, 39, 53, 311
Sciadopityales, 311
Sciadopitys, 40, 311
Sciaphila, 56
  boninensis, 58
  okabeana, 58
Scilla, 111, 286
  atropatana, 147
  haemorrhoidalis, 118
  madeirensis, 115
  monanthos, 24
  verna, 17, 188
  winogradowii, 24
Scirpus, 36, 208
Sclerachne, 242
Scleranthopsis, 136
Scleria, 93
Sclerocactus, 160
Sclerolepis, 82
Sclerophylacaceae, 343
Sclerophylax, 343
Sclerorhachis, 141
Sclerotheca, 246, 247
Sclerotiaria, 139, 156
Scoliopus, 100
Scopolia carniolica, 18
  japonica, 55
  sinensis, 64
Scopulophila, 160
Scorpiothyrsus, 235
Scorzonera, 142, 155

INDEX 503

armeniaca, 147
candavanica, 147
undulata, 124
Scribneria, 100
Scrofella, 45
Scrophularia spp., 67, 117, 142
   aestivalis, 22
   altaica, 31
   amgunensis, 51
   atropatana, 146
   cretacea, 27
   henryi, 64
   hirta, 115
   hyrcana, 151
   kakudensis, 55
   mandshurica, 51
   megalantha, 151
   nachitschevanica, 147
   ningpoensis, 64
   pallescens, 115
   racemosa, 115
   rostrata, 151
   sikkimensis, 72
   sosnovskyi, 24
   tenuipes, 126
   versicolor, 145
Scrophulariaceae, 10, 30, 159, 343
Scrophulariales, 343
Scurula spp., 61
Scutellaria spp., 56, 67, 86
   baicalensis, 52
   balearica, 128
   churchilleana, 35
   discolor, 76
   helenae, 24
   khasiana, 78
   longituba, 58
   moniliorrhiza, 51
   obtusifolia, 64
   pontica, 24
   sessilifolia, 64
   tournefortii, 151
Scutia, 233
Scyphanthus, 284
Scyphocephalium, 198
Scyphochlamys, 219, 224
Scyphostachys, 228
Scyphostegia, 239, 325
Scyphostegiaceae, 239, 325
Scytopetalaceae, 198, 323
Sebaea brachyphylla, 207

Sebertia, 249
Secamone alpini, 265
Securinega leucopyrus, 211
   suffruticosa, 52
Sedopsis, 203
Sedum, 21, 54, 184
   boninense, 57
   brissemoretii, 115
   crassularia, 207
   dumulosum, 62
   epidendrum, 207
   farinosum, 115
   fusiforme, 115
   lanceolatum, 107
   multicaule, 69
   multiceps, 126
   nudum, 115
   populifolium, 30
   purpurascens, 34
   pusillum, 85, 90
   stellariifolium, 62
   stephanii, 34
   stevenianum, 26
   tatarinowii, 62
   viviparum, 50
Seemannaralia, 203
Seidlitzia florida, 146
Selaginaceae, 343
Selaginella, 305
Selaginellaceae, 305
Selinocarpus, 160
   purpusianus, 180
Seikirkia, 280, 281
   berteroi, 282
Selliera, 277
Selloa, 164
Semeiandra, 161
Semele, 113, 349
   androgyna, 116
   maderensis, 115
   menezesii, 115
   pterygophora, 115
   tristonis, 115
Semenovia, 139
Semiaquilegia, 40
   manshurica, 50
Semiarundinaria, 56
Semibegoniella, 326
Sempervivella, 138
Sempervivum, 21
Senecio, 34, 51, 56, 61, 63, 64, 96, 113, 117,

168, 184, 207, 264, 296, 297
anonymus, 86
antennariifolius, 89
antipodus, 301
calcarius, 186
cambrensis, 17
chrysanthemoides, 76
clarkianus, 104
clevelandii, 177
cordatus, 20
ertterae, 169
fendleri, 110
filifolius, 78
gallerandianus, 126
gerberaefolius, 186
greenei, 177
hadiensis, 208
harazianus, 208
huntii, 301
igoschinae, 28
incanus, 20
kleinia, 116
lazicus, 24
littoralis, 290
lyonii, 173
maderensis, 115
millefolium, 89
nagensium, 78
pattersonensis, 104
pauciflorus, 104
pentanthus, 7
persoonii, 20
prenanthiflorus, 215
procumbens, 186
radiolatus, 301
resedifolius, 106
rhabdos, 78
rodriguezii, 128
roseus, 186
schweinitzianus, 86
taraxacoides, 110
tetranthus, 73
trapezunticus, 24
triangularis, 107
vaginatus, 290
Senecio woodlands, 200
Sennia, 207
Septamericanum, 12
Septogarcinia, 242
Sequoia, 100, 101
sempervirens, 105

Sequoiadendron, 100, 101
giganteum, 104
Serapias nurrica, 130
Serbia, 21
Serenoa, 93
repens, 93
Sergia, 140
Sericodes, 162
Sericomopsis, 207
Sericostoma pauciflorum, 211
Serpentine habitat, 103, 177, 252
Serratula coriacea, 147
cupiliformis, 63
forrestii, 67
komarovii, 51
ortholepis, 63
Serresia, 250
Sesamoides, 121
Sesamum prostratum, 233
Seseli, 21
cantabricum, 17
foliosum, 24
leptocladum, 146
leucospermum, 19
nanum, 17
rupicola, 24
yunnanense, 67
webbii, 117
Seselopsis, 139, 156
Seshagiria, 228
Sesleria insularis, 127
ovata, 20
Seychellaria, 219, 220
Seychelles, 220
Seychelles Province, 225
Seymour Island, 291
Seyrigia, 217, 223
Shale-barrens, 89
Sheareria, 46
Shepherdia, 338
Shibataea, 48
Shikotan Island, 43, 53
Shillong Plateau, 77
Shimba Hills, 201
Shiraz, 149
Shiuyinghua, 45
Shola, 231, 232
Shorea robusta, 232, 233
Short-grass prairie, 97
Shortia, 89
galacifolia, 83, 85

sinensis, 66, 83
thibetica, 66
uniflora, 54, 83
Sibbaldia procumbens, 106
Siberia, 11, 29, 31, 33
Sibiraea tomentosa, 66
Sibthorpia africana, 128
  peregrina, 115
Sicily, 121, 128
Sicrea, 234
Sicyosperma, 160
Sideritis, 113, 117
  balansae, 147
  maura, 126
  scardica, 22
Sideroxylon galeatum, 224
  marmulano, 111, 118
Siebera, 141
Sierra Madre del Sur, 188
Sierra Madre Occidental, 184, 185
Sierra Madre Oriental, 180, 184, 186, 187
Sierra Nevada, 103
Sikang-Yünnan Province, 65
Sikhote-Alin, 40, 50
Sikkim, 42, 56, 71
Silene, 21, 54, 61, 101, 111, 125, 129, 142, 168, 203
  acaulis, 38, 106, 110, 168
  akinfievii, 26
  ajanensis, 34
  baschkirorum, 27
  berthelotiana, 117
  boissieri, 23
  capitata, 50
  cretacea, 27
  dinarica, 19
  eremetica, 146
  hellmannii, 27
  koreana, 50
  lacera, 26
  nivea, 99
  nocteolens, 117
  olgae, 50
  otites, 32
  physocalyx, 23
  prilipkoana, 146
  reichenbachii, 131
  rotundifolia, 90
  salsuginea, 143
  soczaviana, 15
  tatarinowii, 62

triflora, 15
turgida, 30
williamsii, 35
yunnanensis, 66
zawadzkii, 19
Siliquamomum, 236
Silphium, 82, 83, 96
  brachiatum, 90
  mohrii, 90
Silvianthus, 44
Simaroubaceae, 333
Simenia, 207
Simla, 43
Simmondsia, 160, 320
Simmondsiaceae, 159, 320
Sinadoxa, 44, 341
Sinai Peninsula, 137
Sinapidendron, 112
  angustifolium, 115
  frutescens, 115
  glaucum, 118
  hirtum, 118
  rupestre, 115
Sindechites henryi, 64
Sinderopsis, 198
Sindian Province, 211
Sindroa, 220
Sinia, 41
Sinobambusa, 48
  elegans, 77
Sinocalycanthus, 314
Sinocarum, 44, 67
Sinochasea, 48, 142
Sinodielsia, 44, 65
  yunnanensis, 67
Sinofranchetia, 40, 63
  chinensis, 63
Sinojackia, 41
  rehderiana, 63
  xylocarpa, 63
Sinojohnstonia, 45
Sinolimprichtia, 44, 65
  alpina, 67
Sinomenium, 40
Sinomerrillia, 43, 65
  bracteata, 67
Sinopanax, 44, 60
  formosanus, 61
Sinopteridaceae, 307
Sinopteris, 39
Sinosideroxylon, 234, 237

Sinowilsonia, 41, 63
  henryi, 63
Siparuna, 314
Siparunaceae, 314
Siphokentia, 243
Siphonodontaceae, 337
Siphonosmanthus delavayi, 67
Siphonostylis, 123
Sirhookera, 229
Sisymbrella, 121
Sisymbriopsis, 137
Sisymbrium volgense, 27
  yunnanensis, 66
Sisyrinchium, 286
  filifolium, 290
  groenlandicum, 15
Sisyndite, 212
Sisyocaulis, 257
Sitkan Province, 101
Sivas, 144
Skimmia, 43
  laureola, 69, 75
  melanocarpa, 69
Skottsberg, C., 281
Skottsbergia, 290
Skythanthus, 286
Sladeniaceae, 322
Sleumerodendron, 249
Sloanea, 250
Slovenia, western, 11
Smelophyllum, 264
Smelowskia jurtzevii, 15
  media, 15
  porsildii, 15
  pyriformis, 35
  tilingii, 34
Smilacaceae, 352
Smilacales, 352
Smilacina, 56, 64, 68
  formosana, 61
  fusca, 70, 76
  oleracea, 70
  trifolia, 36
Smilax, 61, 64, 70
  aspera, 124
  californica, 173
  excelsa, 25
  myrtillus, 78
  pekingensis, 63
Smirnowia, 138, 152
  turkestana, 150

Smithia gracilis, 231
Smithiella, 42, 71
Smithorchis, 47
Smitinandia, 236
Snake River Canyon, 109
Snake River Plains, 165, 170
Sobennikoffia, 220
Sobolewskia caucasica, 26
Society Islands, 245
Socotra, 206, 207, 209
Socotran Province, 209
Socotranthus, 209
Socratina, 219
Solanaceae, 343
Solanales, 343
Solanum, 286
  kierseritzkii, 151
  nava, 117
  pittosporifolium, 64
  trisectum, 115
  vespertilio, 117
Solaria, 285
Soldanella, 11
  austriaca, 19
  carpatica, 19
  villosa, 17
Soleirolia, 121, 127
Solenanthus formosus, 146
  scardicus, 22
  tubiflorus, 126
Solenomelus, 285
Solidago, 83, 86, 93, 97
  albopilosa, 90
  arguta, 89
  calcicola, 35
  curtisii, 88
  glomerata, 89
  lancifolia, 88
  multiradiata, 36
  spithamea, 89
Sollya, 273
Solmsia, 249
Solms-Laubachia, 42
Solomon Islands, 239, 243
Somalia, 206
Somalo-Ethiopian Province, 206
Sonchus, 113, 117
  arboreus, 117
  canariensis, 117
  grandifolius, 301
  pinnatus, 115

squarrosus, 115
tenerrimus, 124
ustulatus, 115
Sonneratia, 332
Sonneratiaceae, 332
Sonora, 160
Sonoran Desert, 181
Sonoran Province, 178
Sophiopsis, 137
Sophora, 278
  benthamii, 75
  bhutanica, 72
  fernandeziana, 281
  howinsula, 297
  masafuerana, 281
  praseri, 69
  tetraptera, 297
  viciifolia, 158
Sorbus, 12, 55, 66
  americana, 88
  apicidem, 70
  baldaccii, 21
  cashmiriana, 155
  colchica, 23
  detergibilis, 70
  dunnii, 64
  folgneri, 64
  griffithii, 72
  hedlundii, 72
  khasiana, 78
  koehneana, 64
  kurzii, 72
  lanata, 155
  maderensis, 115
  manshurensis, 50
  paucinervia, 70
  randaiensis, 61
  schneideriana, 50
  wardii, 69
  xanthoneura, 64
  zahlbruckneri, 64
Sorghastrum nutans, 97
Sorindeia rhodesica, 203
Soroseris, 67
Soulamea terminalioides, 226
Souliea, 40
  vaginata, 69
South Africa, 278
South America, 276, 277
South America-Tristan relationships, 292
South Arabian Province, 208

South Burmese Province, 236
South Chinese Province, 237
South Dakota, 98
South Georgia Island, 276, 290
South Indochinese Province, 238
South Iranian Province, 210
South Island, 294, 301
South Malesian Province, 242
South Mediterranean Province, 125
South Orkney Islands, 278
South Sandwich Islands, 291
South Shetland Islands, 278, 290
South Subantarctic Islands Region, 291
Southeast Australian Province, 271
Southeastern Chinese Province, 65
Southern and Southwestern Madagascan Province, 223
Southern Appalachian Mountains, 88
Southern Moroccan Province, 124
Southern Neozeylandic Province, 301
Southern Rocky Mountains, 109
Southwest Australian Province, 274
Southwest Australian Region, 273
Southwestern Mediterranean Province, 124
Soviet Central Asia, 136
Sowerbaea, 351
Soymida febrifuga, 232
Spain, 11, 120, 125
Sparattanthelium, 314
Sparattosyce, 248
Sparganiaceae, 356
Spartium junceum, 131
Spartocytisus, 112
  supranubius, 116
Spathionema, 207
Spathodeopsis, 235
Spathoglottis ixioides, 73
Spatholirion, 236
Specularia julianii, 126
Speea, 285
Speirantha, 47
Spenceria, 42
Speranskia, 42
Spergularia, 125
  rupicola, 17
Sphaenolobium, 139
Sphaerocionium ferrugineum, 295
Sphaerocoma aucheri, 211
Sphaerophysa, 138
Sphaeropteris, 309
Sphaerosepalaceae, 222, 328

Sphagnum, 35
Sphenoclea, 345
Sphenocleaceae, 345
Sphenostemon, 337
Sphenostemonaceae, 337
Sphinotospermum, 161
Sphyranthera, 235
Spigeliaceae, 341
Spinifex, 275
  squarrosus, 233
Spiraea, 12, 52, 55, 60, 61, 64, 66
  arcuata, 69
  canescens, 69
  dasyantha, 62
  micrantha, 75
  nishimurae, 62
  pubescens, 50
  ussuriensis, 50
  virginiana, 85
Spiraeanthus, 138
Spiranthes romanzoffianum, 104
Spirella, 235
Spiroceratium, 194
  bicknellii, 128
Spirolobium, 235
Spirorrhynchus, 137
Spirospermum, 217
Spirostegia, 140
Spodiopogon sibiricus, 52
Spongiosyndesmus, 139
Sporadanthus, 295
Sporobolus arabicus, 210
  durus, 214, 216
  minutiflorus, 210
  nealleyi, 181
  robustus, 118
  virginicus, 58
Sporoxeia, 235
Spruce-cedar-hemlock forest, 102
Spruce-fir community, 35, 87, 88
Spryginia, 137
Spuriodaucus, 203
Spyridium, 269, 272
Squamellaria, 244
Sredinskya, 11, 23, 25
Sri Lanka, 221, 227
Sri Lanka Province, 229
Stachyopsis, 140
Stachys, 86, 111, 126, 142, 173
  adulterina, 64
  burchelliana, 213

  corsica, 129
  glutinosa, 129
  iva, 22
  macrophylla, 24
  milanii, 22
  persica, 151
  talyschensis, 151
  trapezuntica, 24
Stachyuraceae, 39, 322
Stachyurus, 41, 60, 322
  cordatula, 70
  himalaicus, 69, 75
  macrocarpus, 57
  retusus, 63
  szechuanensis, 63
Stackhousia, 337
Stackhousiaceae, 268, 337
Stadmannia oppositifolia, 225
Staehelina, 122
Staintoniella, 42
Stangeria, 200, 201, 311
Stangeriaceae, 201, 311
Stapelianthus, 219, 223
Stapelieae, 206
Stapfiophyton, 235
Staphylea, 12
  colchica, 24
  emodi, 155
  holocarpa, 64
  pinnata, 22
  trifolia, 85
Staphyleaceae, 334
Stauntonia, 40, 60
  brunoniana, 74
Stauracanthus, 121, 125
Steenis, C. G. G. J. van, 241, 242
Stegia, 121
Stegnosperma, 318
Stegnospermataceae, 318
Steinmannia, 349
Stellaria, 54, 61, 71, 240, 279
  alaskana, 35
  crassipes, 15
  davidii, 62
  decipiens, 301
  decumbens, 158
  delavayi, 66
  imbricata, 30
  irrigua, 30
  longipes, 36
  mannii, 207

INDEX

martjanovii, 30
neo-palustris, 50
yunnanensis, 66
Stellera caucasica, 26
  chamaejasme, 158
  formosana, 61
Stelleropsis, 138
Stemmatium, 285
Stemmatodaphne, 241
Stemona erecta, 64
  vagula, 68
  wardii, 70
Stemonaceae, 352
Stemonoporus, 227
Stenandriopsis, 219
Stenanthella sachalinense, 53
Stenanthium, 82
Stenisiphonium, 228
Stenocarpha, 164
Stenocarpus, 250
Stenochlaenaceae, 310
Stenogyne, 247
Stenomeridaceae, 352
Stenomeris, 239, 352
Stenosiphon, 82
Stenotaenia, 139
Stenothyrsus, 241
Stephanandra, 42
  incisa, 55
  tanakae, 55
Stephania elegans, 74
  glandulifera, 74
Stephanocaryum, 140
Stephanodaphne, 218, 220
Stephanodoria, 164
Stephanolirion, 285
Stephanotis, 219
Steppes, 12, 28, 31, 51, 145
  desert, 155
  forest, 29, 147
  meadow, 158
  mountain, 155
  North American, 95
  Stipa, 145
  tragacanthic, 145, 147, 148
Sterculiaceae, 328
Stereocaryum, 249
Sterigmostemum acanthocarpum, 146
  laevicaule, 149
Stevia, 184
Stewart Island, 277, 294, 301

Stewartia, 54, 66
  malacodendron, 93
  monadelpha, 54
  ovata, 85
  sinensis, 63
Stewartiella, 139
Stigmatella, 133
Stigmatodactylus sikokianus, 56
Stilbaceae, 264, 344
Stilbanthus scandens, 75
Stilbocarpa, 294
  polaris, 301
Stillingia, 221
Stilpnolepis, 141
Stimpsonia, 41
Stipa, 167, 288
  cernua, 174
  glareosa, 157
  gobica, 157
  henryi, 64
  mayeri, 131
  pulchra, 174
  purpurea, 158
  tenacissima, 124, 125
Stipagrostis, 212, 213
Stipulicida, 79
Stirlingia, 273
Stizolophus, 141
Stocksia, 138, 149
Stoebe, 264, 267
  passerinoides, 225
Stokesia, 82
Stone Mountain (Georgia), 90
Stonesia, 198
Storckiella, 250
Storthocalyx, 249
Strangweia, 123
Stranvaesia niitakayamensis, 61
  nussia, 75
  scandens, 66
Strasburgeria, 248, 323
Strasburgeriaceae, 248, 323
Stratiotes, 11
Strausiella, 137
Streblorrhiza, 294
Strelitzia, 354
Strelitziaceae, 354
Strephonemataceae, 198, 332
Streptanthus, 46, 172
  barbiger, 177
  batrachopus, 177

bracteatus, 99
breweri, 177
howellii, 177
insignis, 177
platanifolia, 99
polygaloides, 177
Streptolirion, 47
cordifolium, 76
volubile, 70
Streptoloma, 137
Streptopus simplex, 70
Streptothamnus, 271, 325
Streptotrachelus, 162
Strobilanthes, 64, 67
acrocephalus, 78
formosanus, 61
glandulosus, 237
laevigatus, 72
maculatus, 78
oligantha, 56
stramineus, 70
thomsonii, 72
Strobilopanax, 249, 250
Strobilopsis, 203
Stroganowia litwinowii, 148
Stromatopteridaceae, 248, 307
Stromatopteris, 248, 307
Strophostyles, 82
Strotheria, 164
gypsophila, 181
Strychnaceae, 341
Strychnopsis, 217
Strychnos nux-vomica, 233
Stubendorffia, 137
Stuhlmannia, 201
Stylidiaceae, 345
Stylisma, 82
Stylites, 305
Stylobasiaceae, 273, 335
Stylobasium, 275, 335
Styloceras, 320
Stylocerataceae, 261, 320
Stylocline, 164
Stylodon, 82
Stylomecon, 159
Stylophorum diphyllum, 85
Stypandra, 350
Styracaceae, 324
Styrax, 66, 324
dasyantha, 63
hookeri, 75

matsumuraei, 61
officinalis, 119, 131
serrulatus, 69
shiraiana, 54
tenax, 99
veitchiorum, 63
Styrophyton, 43
Suaeda, 54, 158
vermiculata, 118
Subantarctic Province, 301
Subularia monticola, 207
Succisa trichocephala, 199
Succowia, 121
Suchtelenia, 140
Sudanian Province, 205
Sudano-Zambezian Region, 202
Suksdorfia, 100
Sukunia, 244
Sulawesian Province, 243
Sumatra, 239
Sumatran Province, 242
Sundarban Subprovince, 234
Sunipia paleacea, 73
Supushpa, 228
Suregada procera, 207
Suriana, 333
Surianaceae, 333
Sutera canariensis, 117
Sventenia, 113
bupleuroides, 117
Swallenia, 164
Swat, 154
Swertia, 55, 64, 67, 70, 72, 207, 231
chirayita, 76
macrosperma, 76
nervosa, 76
Swida, 339
Syagrus romanzoffianus, 261
Sycopsis dunnii, 66
formosana, 60
griffithiana, 74
Symbegonia, 326
Sympegma, 136
regelii, 157
Symphochaeta, 280
Symphoricarpos sinensis, 64
Symphyandra lazica, 24
Symphyllarion, 235
Symphyllocarpus, 46
exilis, 51
Symphyoloma, 11, 26

INDEX

Symphyosepalum, 47
Symphytum cordatum, 19
  gussonei, 129
  sylvaticum, 24
Symplocaceae, 322
Symplocos, 54, 61, 66, 75, 322
  araioura, 70
  boninense, 57
  coreana, 54
  kawakamii, 57
  pergracilis, 57
  pyrifolia, 71
  tanakana, 54
Synandra, 82
  hispidula, 86
Synaphea, 273
Syncephalum, 219
Syneilesis, 46
Synelcosciadium, 122, 131, 139
Synstemon, 42
Synthlipsis, 161
Synthyris borealis, 35
  canbyi, 106
Syntrichopappus, 164
Synurus spp., 56
Syrdarinian Karatau, 139
Syreitschikovia, 141
Syrenia taljevii, 27
Syria, 121, 133, 136
Syrian Desert, 138
Syringa, 70
  emodi, 155
  josikaea, 19
  microphylla, 62
  patula, 51
  pekinensis, 62
  potaninii, 67
  pubescens, 62
  reflexa, 64
  reticulata, 51
  tomentella, 67
  vulgaris, 22
  wolfii, 51
  yunnanensis, 67
Syzygium, 61, 75, 233
  cleyeraefolium, 57
  cumini, 232
  stenurum, 70
Szechuan, 40
Szovitsia, 139, 146
  callicarpa, 146

Tabebuia ipe, 261
  pulcherrima, 261
Tacca, 352
Taccaceae, 352
Tachiadenus, 219
Taeniandra, 228
Taeniantherum caput-medusae, 175
Taenidia, 82
  integerrima, 85
  montana, 89
Taeniophyllum aphyllum, 56
Taenitidaceae, 307
Tahitia, 245
Taiga, 30, 32, 157
Taiwan, 39, 45, 59
Taiwania, 40
  cryptomerioides, 60
  flousiana, 69
Taiwanian Province, 59
Tajikistan, 136
Takeikadzuchia, 46
Takhtajania, 217, 313
  perrieri, 221
Takhtajanoideae, 222, 313
Talassian Alatau, 139
Talguenea, 284
Talinella, 217
Talinopsis, 160
Talinum calcaricum, 91
  teretifolium, 90
Talish, 136
Tamaricaceae, 327
Tamaricales, 327
Tamarix, 211
  gallica, 118
  juniperina, 62
Tamaulipan district, 160
Tamaulipan subprovince, 180, 182
Tamaulipan Thorn-scrub, 181, 182
Tambourissa, 217, 220
  gracilis, 222
Tamus, 352
Tanacetopsis, 141
Tanacetum, 67, 142, 155
  akinfievii, 26
  canescens, 147
  gracilis, 158
  lanuginosum, 31
  mucroniferum, 145
  simplex, 109
  tabrisianum, 147

tibeticum, 158
turcomanicum, 148
uniflorum, 147
Tanakea, 42
 radicans, 54
Tangtsinia, 47
Tanulepis, 219, 220
Tapeinia, 285, 289
Taphrospermum, 137
Tapiscia, 43, 334
Taraxacum, 34, 51, 56
 amphiphron, 15
 hyparcticum, 15
 magellanicum, 278
 pumilum, 15
 tundricola, 15
Tarchonanthus camphoratus, 208
Tarenna subsessilis, 58
Tasmania, 276
Tasmanian Province, 272
Tatra Mountains, 19
Tauscheria, 137
Taverniera glabra, 210
Taxaceae, 311
Taxales, 310
Taxillus kaempferi, 55
 matsudai, 61
Taxodiaceae, 311
Taxodium distichum, 92, 93
Taxus, 101, 311
 baccata, 18
 celebica, 311
 floridana, 93, 94
 globosa, 187
 wallichiana, 69, 74
Tecomanthe speciosa, 300
Tecophilaea, 285, 286, 348
Tecophilaeaceae, 348
Tectiphiala, 220, 224
Tectona grandis, 232
Telekia, 11
 speciosissima, 20
Telephium oligospermum, 146
Tellima, 100
 grandiflorum, 105
Telopea, 272
Temnopteryx, 198
Tenagocharis, 346
Tenerife Island, 116
Tengia, 46, 65
Tenimber, island of, 253

Tennessee, 90
Tephrosia falciformis, 211
 hausknechtii, 210
 persica, 210
Tepualia, 284
Tepuianthaceae, 257, 334
Tepuianthus, 334
Terauchia, 47, 50, 351
 anemarrhenefolia, 51
Terminalia, 202, 204, 205, 206, 208
 alata, 232
 bellirica, 233
 bentzoe, 224
 catappa, 58
Terminalia chebula, 232
Ternstroemia, 187, 271
 gymnanthera, 231
Ternstroemiaceae, 322
Tessmannia, 198
Tethyan-African relationship, 111
Tethyan Subkingdom, 110
Tetilla, 284, 331
Tetracarpaea, 272, 338
Tetracarpaeaceae, 272, 338
Tetracentraceae, 39, 319
Tetracentron, 41, 319
 sinense, 69
Tetracera akara, 230
Tetrachondra, 277
Tetrachondraceae, 344
Tetraclinis, 120
 articulata, 111, 124, 125, 126
Tetracme, 137
Tetracoccus, 161
Tetradiclidaceae, 9, 333
Tetradoxa, 44, 341
Tetraena, 138, 157
Tetragonia, 212, 318
Tetragoniaceae, 318
Tetrameles, 326
Tetramerista, 239, 258, 322
Tetrameristaceae, 322
Tetrapanax, 44
Tetrapathaea, 294
Tetrapterocarpon, 218
 geayi, 223
Tetraria, 267
Tetrastigma, 67
 rumicispermum, 76
 umbellatum, 61
Tetrataxis, 218, 224

Tetrathyrium, 41, 65
Tetroncium, 285, 289, 347
Teucridium, 294
Teucrium, 64, 67, 113, 126
 abutiloides, 115
 asiaticum, 128
 betonicum, 115
 cossonii, 128
 fruticans, 124
 japonicum, 56
 mascatense, 210
 multinodum, 24
 quadrifarium, 70, 76
 subspinosum, 128
 trapezunticum, 24
Texas, 99, 159
Texiera, 137
Teyleria, 242
Thailand, 43, 234
Thailandian Province, 237
Thalictrum, 54, 60, 66, 69, 71, 85, 240
 calabricum, 128
 dioicum, 99
 foliolosum, 74
 integrilobum, 53
 mirabile, 90
 prczewalskii, 62
 punduanum, 74
 rhynchocarpum, 199
 uncinatum, 27
Thamnoseris, 280, 282
Thapsia, 122
 decussata, 124
Thaspium, 82
Theaceae, 10, 322
Theales, 322
Thecocarpus, 139
Thelepaepale, 228
Thelesperma ramosius, 181
 scabridulum, 181
Thelethylax, 218
Theligonaceae, 9, 342
Theligonum, 342
 japonicum, 55
Thelocactus, 160
Thelymitra, 240
Thelypodium flavescens, 177
Thelypteridaceae, 309
Thelypteris, 53
 boninensis, 57
 gongylodes, 300

 ogasawarensis, 57
Themeda, 233
 triandra, 204
Theophrastaceae, 324
Theriophonum, 229
Thermopsis alpina, 158
 mollis, 85
Theropogon, 47
 pallidus, 76
Thesidium, 264
Thesium emodi, 72
 hystrix, 214
 italicum, 129
 kerneranum, 19
 kilimandscharicum, 207
 rostratum, 19
Thevenotia, 141
Thismiaceae, 348
Thladiantha, 66
 nudiflora, 63
 punctata, 61
Thlaspi, 19
 bellidifolium, 21
 brevistylum, 129
 macranthum, 132
 stenocarpum, 147
 yunnanensis, 66
Thomasia, 269
Thoracostachyum, 222
Thoreldora, 235
Thorella, 11, 17
Thorn-scrub vegetation, 182
Thorne, R. F., 243, 251
Thottea, 239
Three Kings Islands, 294, 300
Threlkeldia, 274
Thryothamnus, 285
Thuarea involuta, 58
Thuja, 12, 101
 koraiensis, 50
 mertensiana, 102
 plicata, 102, 103, 105
 standishii, 54
Thujopsis, 40, 53
 dolabrata, 54
Thunbergiaceae, 344
Thunbergiella, 264
Thurnia, 355
Thurniaceae, 254, 355
Thurya, 121, 131
Thuspeinanta, 140

Thylacospermum, 136
  caespitosum, 158
Thymelaea antatlantica, 124
  myrtifolia, 128
  velutina, 128
Thymelaeaceae, 10, 329
Thymelaeales, 329
Thymus, 27, 126, 142, 207
  capitatus, 119
  herba-barona, 129
  origanoides, 117
  przewalskii, 51
  richardii subsp. *ebusitanus*, 128
  satureoides, 124
Thyrocarpus, 45
  sampsonii, 67
Thyrsanthella, 82
Thyrsanthera, 235
Thyrsopteridaceae, 280, 308
Thyrsopteris, 280, 308
  elegans, 281, 282
Thyrsostachys, 236
Thysanospermum, 235, 237
Thysanostigma, 236, 238
Tianschaniella, 140, 156
Tiarella cordifolia, 85
Tibet, 40, 71, 137, 157
Tibetan Province, 157
Tibesti, 133, 134
Tibestina, 134
Tien Shan, 135, 142, 153, 156
Tienmuia, 45
Tierra del Fuego, 277, 283, 289
Tilia, 12, 54
  americana, 85, 86
  amurensis, 50
  cordata, 28, 30
  henryana, 64
  heterophylla, 85
  longipes, 187
  mandshurica, 50
  oliveri, 64
  semicostata, 50
  sibirica, 31
  tuan, 64
Tiliaceae, 328
Tillaea moschata, 277
Tillandsia usneoides, 92, 93
Timonius, 221
  seychellensis, 226

Tina, 218
Tinguarra, 112
  cervariifolia, 117
  montana, 117
Tinopsis, 218
Tipularia japonica, 56, 73
Tirania, 234
Tirpitzia, 43, 65
  sinensis, 66
Tisonia, 217
Titanotrichum, 46
Tmesipteridaceae, 306
Tmesipteris, 306
  forsteri, 298
Tobolsk, 30
Todaroa, 112
  aurea, 117
Todea, 306
  barbara, 267
Tofieldia, 56
  himalaica, 73
  yunnanensis, 68
Tokara-Okinawa Province, 58
Tolmatchev, A. I., 5
Tolmiea, 100
Tolpis, 117
  macrorhiza, 115
Tomanthea, 141
  daralaghezica, 147
  phaeopappa, 147
Tomanthera, 82
  densiflora, 96
Tonella, 100
Tongoland-Pondoland Province, 201
Tonkin, 39, 40, 234
Toricellia, 44, 340
  angulata, 66
Toricelliaceae, 39, 340
Toricelliales, 340
Tornabenea, 112, 118
Torreya, 311
  grandis, 63
  nucifera, 54
  taxifolia, 93, 94
  yunnanensis, 65
Touchardia, 247
Tourneuxia, 123
Toussaintia, 198
Tovaria, 327
Tovariaceae, 254, 327

INDEX

Townsendia, 168
  eximia, 110
  fendleri, 110
  glabella, 110
  rothrockii, 110
Toxicodendron diversilobum, 177
  vernix, 86
Trachelanthus, 140
Trachelospermum asiaticum, 55
  foetidum, 57
Trachycarpus, 48
  fortunei, 56
  martianus, 77
Trachydium spp., 67
Trachypogon, 261
Trachypteris, 221
Trachyspermum scaberulum, 67
Trachystemon, 11
  orientale, 22
  orientalis, 24
Trachystoma, 121, 125
Tractopevodia, 235, 237
Traganopsis, 121, 133
Traganum, 134
Tragopogon, 142
  kopetdagensis, 148
  marginatus, 147
  sosnovskyi, 147
  tomentosulus, 148
Trailliaedoxa, 45
  gracilis, 67
Transbaikalia, 51
Transbaikalian Province, 32
Transcaucasia, 136, 138, 144
Trans-Mexican Volcanic Belt, 163, 185, 186
Transvaal, 203
Transylvania, 19
Trapa, 333
  colchica, 24
  incisa, 55
Trapaceae, 333
Trapales, 333
Trapella, 45, 344
  sinensis, 64
Trapellaceae, 39, 344
Trautvetteria carolinensis, 85
Trechonaetes, 284
Tree cacti, 184
Tree euphorbias, 201
Tree ferns, 222, 282

Tree *Ipomoea*, 185
Tree *Opuntia*, 185
Tremacron, 46
  forrestii, 67
  rubrum, 67
Tremandraceae, 268, 337
Trematocarpus, 247
Trematolobelia, 247
Trepocarpus, 82
Treutlera, 45
Triadenum virginicum, 85
Triaenophora, 45
Trianthema heteroensis, 213
Tribelaceae, 283, 339
Tribeles, 284, 339
Tribonanthes, 273
Tricardia, 162
Triceratella, 203
Trichachne hitchcockii, 182
Trichanthemis, 141
Trichipteris, 309
Trichocalyx, 209
Trichocereus, 286
Trichochiton, 137
Trichocladus crinitus, 285
  elliptica, 207
Trichocoronis, 164
Tricholaser, 139
Trichomanes, 280, 295, 308
  auto-obtusum, 57
  bonincola, 57
  boninense, 57
  palmifolium, 60
  titibuensis, 53
Trichopodaceae, 227, 352
Trichoptilium, 164
Trichopus, 221, 352
Trichosandra, 219, 224
Trichosanthes boninensis, 57
  multiloba, 54
Trichotaenia, 235
Tricomaria, 284
Tricyclandra, 217
Tricyrtis, 46, 56
  formosana, 61
  puberula, 63
Tridactylina, 11, 30
  kirilowii, 31
Tridens buckleyanus, 99
  texanus, 182

Trifolium, 66, 186, 203, 207
　barbigerum, 176
　bivonae, 129
　brutium, 129
　cyathiferum, 176
　depauperatum, 174, 176
　euxinum, 24
　fucatum, 176
　grandiflorum, 132
　macraei, 174
　microdon, 174
　polyphyllum, 24
　reflexum, 85
　saxatile, 19
　sintenisii, 24
　speciosum, 132
　virginicum, 85, 89
Triglochin, 347
Triglochinaceae, 346
Trigonella, 142
　uncata, 210
Trigonia, 336
Trigoniaceae, 336
Trigoniastrum, 336
Trigonocaryum, 11, 26
Trigonopyren, 219, 220
Trigonosciadium, 139
　intermedium, 145
Trigonotis, 55, 67
　amblyosepala, 63
　mollis, 64
Trilliaceae, 9, 352
Trillium, 86, 352
　ovatum, 102
Trilobachne, 229
Trimeniaceae, 313
Trimeris, 215
Triodia, 274
Triosteum sinuatum, 51
Tripetaleia, 41
Triplachne, 123
Triplasis, 82
Triplopogon, 229
Triplostegia, 341
Triplostegiaceae, 341
Tripterococcus, 337
Tripterygiaceae, 337
Tripterygium, 43
　doianum, 55
　wilfordii, 70
Triptilion, 285

Trisema, 248
Trisepalum, 236, 237
Trisetaria nitida, 126
Trisetum fuscum, 20
　gracile, 130
Tristagma, 285
Tristan da Cunha, 278, 292
Tristan-Goughian Province, 292
Tristichaceae, 331
Triteleia, 172
　clementina, 173
　versicolor, 173
Triteleiopsis, 164
Trithuria, 354
Triticum araraticum, 147
Triuridaceae, 348
Triuridales, 348
Trivalvaria dubia, 237
Trochetia, 215, 218
Trochetiopsis, 215
Trochocodon, 131
Trochodendraceae, 39, 319
Trochodendrales, 319
Trochodendron, 40, 319
Trochomeriopsis, 217
Trollius europaeus, 18
　hondoensis, 54
　laxus, 85
　macropetalus, 50
　micranthus, 69
　pulcher, 52
　pumilus, 69, 74
　yunnanensis, 66, 69
Tropaeolaceae, 336
Tropaeolales, 336
Tropaeolum, 286
Trophaeastrum, 336
Tropical America-tropical Africa relationships, 255
Tropidia nipponica, 56
Tropidocarpus, 161
Trouettea, 249
Tsaiorchis, 47
Tschihatschewia, 137
Tsebona, 217
Tsingya, 218
Tsoongia, 236
Tsoongiodendron, 40, 65
Tsu Island, 55
Tsuga, 12, 101
　canadensis, 85, 87

caroliniana, 85, 89
diversifolia, 54
heterophylla, 102, 103, 105
longibracteata, 65
mertensiana, 104
sieboldii, 54
yunnanensis, 65, 69
Tsushima island, 55
Tsusiophyllum, 41, 53
Tuapse, 24
Tugarinovia, 141
Tula, 28
Tulestea, 198
Tulipa, 142
 florenskyi, 147
 hoogiana, 148
 kaghyzmanica, 145
 latifolia, 56
 mucronata, 145
 pavlovii, 22
 urumoffii, 22
 wilsoniana, 148
Tulotis iinumae, 56
Tumamoca, 160
Tundra, 12, 16, 38
 alpine, 33, 101, 171
Tunisia, 121, 125
Tupeia, 294
Tupidanthus calyptratus, 76
Turan, 138
Turanian Province, 152
Turanian Region, 135
Turgeniopsis, 139
Turkestanian Province, 152
Turkestan Range, 139
Turkey, 11, 23, 137
Turkmenistan, 136, 138
Turneraceae, 326
Turpinia, 187
 formosana, 61
 nepalensis, 69, 75
Turraea holstii, 207
 robusta, 207
Turricula, 162
Tuscan Archipelago, 121, 129
Tutcheria, 65
Tyan'mushan, 191
Tylophora, 55
 yunnanensis, 67
Typhaceae, 356
Typhales, 356

Tyrrhenian Province, 128
Tytthostemma, 136

Uapaca bojeri, 223
Uebilinia spathulaefolia, 207
Uechtritzia, 141
 armena, 145
Ugamia, 141
Ugni, 281
Uinta Basin, 165
Uinta Mountains, 110
Ukraine, 18, 19, 29
Uldinia, 274
Ulex gallii, 17
 micranthus, 17
Ulmaceae, 329
Ulmus, 12
 davidiana, 62
 lanceaefolia, 69
 macrocarpa, 50
 serotina, 85
 thomasii, 85
 uyematsui, 60
Umbelliferae, 340
Umbellularia, 159
 californica, 102, 105, 176
Umbilicus botryoides, 199, 207
 heylandianus, 117
 schmidtii, 118
Umtiza, 201
Uncaria rhynchophylla, 55
Uncarina, 219
Uncinia, 251, 281, 296, 297
 dikei, 293
 hookeri, 301
 smithii, 290
Ungeria floribunda, 298
Ungernia, 142
Upper Assam, 74
Upper Gangetic Plain Province, 233
Upper Guinea Province, 199
Urals, 28, 29
Urbania, 285
Urbinella, 164
Urginea, 111
Urmenetea, 285
Uromyrtus, 249
Urophysa, 40
Urtica aucklandica, 301, 302
 rupestris, 128
Urticaceae, 329

Urticales, 329
Uruguay, 283
Usnea, 102
Utah, 110, 159, 168, 169
Utleria, 228
Utricularia, 271
  novaezelandiae, 300
  pilosa, 56
  protrusa, 300
Uvularia, 82
Uvulariaceae, 348
Uzambara—Zululand Region, 200

Vacciniaceae, 323
Vaccinium, 12, 16, 54, 61, 63, 66, 69, 75
  amamianum, 59
  arboreum, 86
  arctostaphylos, 25, 114
  boninense, 57
  cylindraceum, 114
  erythrocarpum, 85, 88
  japonicum, 54
  leschenaultii, 230
  maderense, 115
  neilgherense, 231
  ovatum, 105
  oxycoccos, 36
  padifolium, 25
  pallidum, 85
  parvifolium, 102
  uliginosum, 36
  vitis-idaea, 36
Vahliaceae, 331
Valdivia, 284
Valenzuelia, 284
Valeriana, 67, 70, 203, 279
  ajanensis, 34
  celtica, 19
  elongata, 19
  flaccidissima, 55
  petrophila, 31
  pyrenaica, 17
  sitchensis, 107
  subpennatifolia, 51
  texana, 96
Valerianaceae, 341
Valerianella, 126, 240
  amblyotis, 146
Valvanthera, 271
Vanasushava, 228
Vancouveria, 100

  parviflora, 105
Vancouverian Province, 100, 101
Vaseyanthus, 160
Vaseyochloa, 164
Vasiliev, V. N., 12
Vateria, 227
  indica, 230
Vateriopsis, 218, 221, 225
  seychellarum, 226, 253
Vatovaea, 207
Vauquelinia, 161
Vavilov, N. I., 7
Vavilovia, 138
Veillonia, 250
Velebit, 131
Velloziaceae, 354
Velloziales, 354
Venegasia, 164
Venezuela, 257
Veprecella, 218
Veratrilla, 45
  baillonii, 67
Veratrum, 56
  dahuricum, 51
  dolichopetalum, 51
  maackii, 51
  mandshuricum, 63
  parviflorum, 86
  tenuipetalum, 110
  woodii, 86
  yunnanensis, 68
Verbascum, 22, 142, 147
  akdarensis, 210
  hajastanicum, 145
  helianthemoides, 143
  siculum, 129
Verbena, 286
  mendocina, 288
Verbenaceae, 344
Verbesina, 184
  chapmanii, 95
  lindheimeri, 99
Verhuellia, 314
Vernal pools, 175
Vernonia, 67, 93
  adenophylla, 70
  cylindriceps, 78
  glauca, 86
Veronica, 67, 148, 207, 240, 272, 279
  allionii, 19
  caucasica, 26

INDEX

dahurica, 52
deltigera, 72
kiusiana, 55
lanuginosa, 72
linariifolia, 52
microcarpa, 147
monticola, 24
morrisonicola, 61
rhodopaea, 22
robusta, 72
sajanensis, 31
saturejoides, 22
turrilliana, 24
Verschaffeltia, 220, 225
splendida, 226
Verticordia, 269
Veselskya, 137
Vesey-FitzGerald, L. D. E. F., 226
Vestia, 284
Vexillabium, 47, 53, 56
Viburnaceae, 9, 341
Viburnum, 12, 61, 64, 67, 159, 261, 279, 341
alnifolium, 86, 88
boninsimense, 57
carlesii, 55
coriaceum, 70
cuttingianum, 70
cylindricum, 76
edule, 36
foetidum, 76
obovatum, 93
rigidum, 116
tashiroi, 59
Vicariant species, 83, 187, 299
Vicatia conifolia, 67
Vicia, 66, 117, 288
argentea, 17
atlantica, 115
bifoliolata, 128
capreolata, 115
cassubica, 27
freyniana, 24
lilacina, 30
pectinata, 115
Vieraea, 113
laevigata, 117
Vietnam, 40, 234
Viguethia, 164
Viguiera, 184
porteri, 90

Viguierella, 220
Vinca major, 24
Vincetoxicum mukdenense, 62
versicolor, 62
volubile, 51
Viola, 19, 21, 54, 66, 72, 231, 240
abyssinica, 199
aethnensis, 129
amamiana, 59
amurica, 50
biflora, 110
cheiranthifolia, 116
cinerea, 210
diamantica, 50
formosana, 61
hispida, 17
incisa, 30
interposita, 50
jaubertiana, 128
kusnezowiana, 50
lactea, 17
maculicola, 59
nagasawai, 61
nebrodensis, 129
orthoceras, 23
palmensis, 117
paradoxa, 115
rhodosepala, 62
savatieri, 50
serpens, 231
sikkimensis, 75
thomsonii, 75
utchinensis, 59
yamatsutai, 50
Violaceae, 325
Violales, 325
Viridivia, 203
Virunga Volcanoes, 199, 200
Viscaceae, 338
Viscainoa, 162
Viscum alniformosanae, 61
multinerve, 61
Visnea, 112, 113
maconera, 116
Vitaceae, 338
Vitales, 338
Vitex agnuscastus, 132
rotundifolia, 58
yunnanensis, 67
Vitis, 55, 67, 338
amurensis, 50, 51

vinifera, 22
wilsonae, 64
Vittaria, 307
  bonincola, 57
  isoëtifolia, 265
  mediosora, 60
  ogasawarensis, 57
  vittarioides, 292
Vittariaceae, 307
Viviania, 284, 336
Vivianiaceae, 283, 336
Vladimiria, 46, 65
  salwinensis, 67
Voatamalo, 218
Vochysia, 261
Vochysiaceae, 336
Volcano-Bonin Province, 56
Volcano Islands, 57
Volga River Basin, 28
Volkensinia, 207
Volkiella, 203
Vonitra, 220
Voronezh, 28
Vulpia, 175
Vulpiella, 123
Vvedenskiella, 138
Vvedenskya, 139

Wachendorfia, 264, 353
Wagatea, 228
  spicata, 230
Wagenitzia, 123
Wahlenbergia, 215, 345
  bernardii, 126
  linifolia, 216
  pusilla, 207
Waldsteinia fragarioides, 85
Walidda, 228
Wallace's line, 240
Walleria, 348
Wangenheimia, 123
Warburgia, 313
Warburgina, 122, 131, 139
Ward, F. Kingdon, 69
Wardaster lanuginosus, 67
Wardenia, 241
Warea, 79
Warionia, 134
Warner Mountains, 101
Washington, 103
Washingtonia, 164

Waziristan, 154
Weigela, 44, 55
Weinmannia, 221, 222, 251, 330
  silvicola, 300
  trichosperma, 286
Wellstedia, 343
Wellstediaceae, 343
Welwitschia, 212
  mirabilis, 213
Welwitschiaceae, 212, 312
Welwitschiales, 312
Wenatchee Mountains, 103
Wenchengia, 236, 237
Wendelbo, P., 149
Wendelboa, 141
Wendtia, 284, 336
Werdermannia, 284
Werger, J. J. A., 212
West Indian Province, 256
Western Asiatic Subregion, 142
Western Cape Province, 213
Western Himalayan Province, 154
Western Madagascan Province, 223
Western Siberian Province, 29
Whipplea, 100
  modesta, 105
White, F., 200
White Mountains (of New Hampshire), 87
Whitmore, T. C., 240
Whitmorea, 243
Whitneya, 164
Whittaker, R. N., 183
Whittonia, 325
Whytockia, 46
  chiritaeflora, 67
Wickstroemia australis, 298
  floribunda, 70
Widdringtonia, 111
  cedarbergensis, 265
  cupressoides, 265
  schwarzii, 265
Wielandia, 218, 225
Wightia, 344
  speciosissima, 70, 76
Wikstroemia, 54, 64, 66
  mononectaria, 61
  pseudoretusa, 57
  trichotoma, 54
Wilkesia, 247
Willisia, 227
  selaginoides, 230

Winklera, 138
Winteraceae, 313
Winterales, 313
Wisconsin glaciation, 38
Wissmannia, 206
Wisteria floribunda, 55
  japonica, 55
Withania frutescens, 124
Wittsteinia, 271, 323, 324, 339
Wizlizenia, 161
Woodburnia, 44, 70
  penduliflora, 70
Woodfordia fruticosa, 233
Woodland, 205, 275
  oak, 174, 176, 182
  open, 202
  Reptonia-Olea, 210
  sclerophyllous, 165, 272
  xerophilic, 260
Woodrowia, 229
Woodsia rosthorniana, 62
Woodsiaceae, 309
Wormia, 221
Wulfenia baldaccii, 22
Wurmbea, 269, 348
Wyoming, 109

Xanthoceras, 43
Xanthonneopsis, 235
Xanthopappus, 46
Xanthophyllaceae, 336
Xanthophyllum, 271
Xanthophytopsis, 235, 237
Xanthorhiza, 79
Xanthorrhoea, 351
Xanthorrhoeaceae, 268, 351
Xanthostemon, 251
Xenacanthus, 228
Xerochlamys, 218
Xerocladia, 212
Xeronema, 251, 350
  callistemon, 300
Xerosicyos, 217, 223
Xerothermic period, 97
Xerotia, 133
Xiphidium, 353
Xylanthemum, 141
Xylinabariopsis, 235
Xylocalyx, 207
Xylococcus, 160
Xylonagra, 161

Xyridaceae, 355
Xyris, 93, 355

Yakutia, 11, 32
Yeatesia, 82
Yoania, 47
  australis, 300
Young, S. B., 13
Youngia, 67
  yoshinoi, 56
Ypsilandra, 46
  yunnanensis, 68
Yucca, 180, 181
  brevifolia, 170, 182
  gloriosa, 93
  rupicola, 99
Yuccaceae, 350
Yugoslavia, 11, 19, 21, 121, 131
Yukon, 37, 100
Yünnan, 39, 65, 235
Yunquea, 280

Zagros, 138, 139, 142, 148
Zalacella, 236
Zambezian Province, 204
Zambezian Region, 202
Zambia, 203
Zambitsia, 217
Zamiaceae, 311
Zannichellia, 347
Zannichelliaceae, 347
Zanthoxylum, 61, 75
  amamiense, 59
  fauriei, 55
  seyrigii, 223
  yunnanense, 66
Zanzibar-Inhambane Province, 201
Zataria, 140
Zehneria baueriana, 298
Zelkova carpinifolia, 148, 152
  schneideriana, 63
Zenkeria, 229
Zenobia, 79
Zephyra, 285, 348
Zeravschania, 139
Zerdana, 138, 149
Zeugandra, 140, 148
Zeuxine boninensis, 58
Zhiguli Hills, 28
Zhumeria, 210
Zieridium, 249

Zigadenus vaginatus, 170
Zilla, 133, 135
Zimbabwe, 203
Zingeria biebersteiniana, 28
Zingiber mioga, 56
Zingiberaceae, 354
Zingiberales, 354
Ziziphora rigida, 147
Ziziphus, 67
   lotus, 124
   mauritiana, 232
   nummularia, 211
   obtusifolia, 182
   spina-christi, 208, 210
Zollinger's floristic line, 240
Zonation, 31, 32, 104, 107, 171, 174, 208
Zosima, 139
Zostera caulescens, 56

Zosteraceae, 347
Zosterales, 347
Zuccagnia, 284
Zuckia, 160
Zygochloa, 274
   paradoxa, 275
Zygogynum, 248, 250
Zygophyllaceae, 333
Zygophyllum, 135, 212
   cornutum, 126
   dregeanum, 213
   fontanesii, 117, 118
   gilfillani, 214
   kaschgaricum, 156, 157
   macrocarpum, 213
   xanthoxylon, 157
Zygosicyos, 217
Zygostelma, 235, 238

Designer: U.C. Press Staff
Compositor: Prestige Typography
Printer: Thomson-Shore, Inc.
Binder: John H. Dekker & Sons
Text: 10/12 Galliard
Display: Quorum

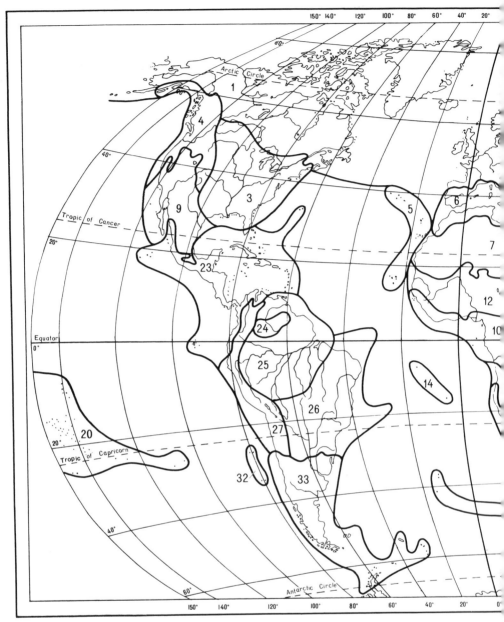

Map 1. FLORISTIC REGIONS OF THE WORLD

1, Circumboreal Region. 2, Eastern Asiatic Region. 3, North American Atlantic Region. 4, Rocky Mountain Region. 5, Macaronesian Region. 6, Mediterranean Region. 7, Saharo-Arabian Region. 8, Irano-Turanian Region. 9, Madrean Region. 10, Guineo-Congolian Region. 11, Uzambara-Zululand Region. 12, Sudano-Zambezian Region. 13, Karoo-Namib Region. 14, St. Helena and Ascension Region. 15, Madagascan Region. 16, Indian Region. 17, Indochinese Region. 18, Malesian Region. 19, Fijian Region. 20, Polynesian Region. 21, Hawaiian Region. 22, Neocaledonian Region. 23, Caribbean Region. 24, Region of the Guayana Highlands. 25, Amazonian Region. 26, Brazilian Region. 27, Andean Region. 28, Cape Region. 29, Northeast Australian Region. 30, Southwest Australian Region. 31, Central Australian or Eremaean Region. 32, Fernándezian Region. 33, Chile-Patagonian Region. 34, Region of the South Subantarctic Islands. 35, Neozeylandic Region.